Salts of Amino Acids

Michel Fleck • Aram M. Petrosyan

Salts of Amino Acids

Crystallization, Structure and Properties

Springer

Michel Fleck
Institute of Mineralogy
and Crystallography
University of Vienna
Vienna, Austria

Aram M. Petrosyan
National Academy of Sciences
of Armenia
Yerevan, Armenia

ISBN 978-3-319-06298-3 ISBN 978-3-319-06299-0 (eBook)
DOI 10.1007/978-3-319-06299-0
Springer Cham Heidelberg New York Dordrecht London

Library of Congress Control Number: 2014942328

Printed on acid-free paper

Springer is part of Springer Science+Business Media (www.springer.com)

Preface

On asking somebody what they are made of, the reply frequently heard is 'mostly water'. Although this statement is definitely correct, it is only the boring background material, so to say, on a par with the answer 'air' when asked what you have in your home. The really interesting stuff is there in smaller percentages, as for example carbohydrates, lipids, and proteins (in the case of our body). The first two of these three groups certainly fulfill important functions, but mostly structural ones. The actual machinery of every living cell consists almost completely of proteins. Therefore, it is no surprise to learn that the only building instructions written in our genetic make-up, i.e. the DNA, are those for proteins. But where are the instructions for the assembly of carbohydrates, lipids, and other material of our body, when there is DNA – and DNA only – to contain information? The answer is obvious: DNA codes for proteins, and these do everything else, such as all disassembling, rearranging, combining, constructing and every other conceivable chemical procedure to manufacture or alter lipids, carbohydrates, steroids, etc.

This brief illustration is supposed not only to show the importance of proteins, but also the vast diversity of functions of these molecules. And as form follows function, the variety of forms is similarly vast. This results from the way proteins are built: As encoded in the DNA, a chain of amino acids is constructed, and as a consequence of the chemical properties of these buildings units, the originally linear protein folds into a highly complex, three dimensional object. Thus, the multitude of protein structures arises simply from the structural variety of amino acids.

Consequently, it is no wonder that proteins and amino acids have been intensively studied by scientists ever since the necessary methods have become available. Besides this avalanche of material produced on biological, biochemical and chemical aspects of proteins and amino acids, there has been a much smaller community of researchers developing and studying compounds of amino acids. These materials, which exist in the solid state at ambient conditions, crystallize in a wide variety of forms and symmetries, display an equally wide range of interesting chemical and physical properties and have therefore become a topic very actively researched recently. As there is no comprehensive work on the entirety of these species so far, we feel that this book should prove an overview and insight for all students, teachers and scientists involved in the study of amino acid compounds.

Vienna, Austria Michel Fleck
Yerevan, Armenia Aram M. Petrosyan

Contents

Chapter 1
Introduction

Abstract Within this chapter, the fundamental properties of amino acids are presented. The basic chemical features of amino acids are the ubiquitous amino and acid groups; the residue of the molecule is usually referred to as side chain. Of an infinite number of conceivable amino acids, twenty (plus a few more or less frequent) members are found in proteins of living beings and thus play a crucial role in the chemistry of life. From a chemical point of view, the chirality of most amino acids is an important feature, which is discussed in regard with the nomenclature systems conventionally employed. Chirality is related with symmetry, both of the molecule and the crystal structure of amino acids (or their salts). Moreover, the chemical flexibility of amino acids, both in terms of symmetry and in terms of their amphoteric nature, is reviewed, thus forming the frame of reference for the following chapters which deal with the actual amino acid salts.

Keywords Amino acids • Standard amino acids • Non-standard amino acids • Nomenclature • Configuration • Chirality • Optical activity • Enantiomer • Racemate • Conformation • Torsion angle • Cations • Anions • Zwitterions

1.1 What Are Amino Acids?

Chemically speaking, amino acids compose just one among many families of organic compounds. Any organic molecule possessing at least one amino and one acid group belongs to this family (Fig. 1.1). Generally, this means any hydrocarbon chain, possibly branched, with or without other functional groups, aromatic rings, or any other imaginable organic structure. Consequently, it is clear the number of amino acids conceivable is infinite.

Nevertheless, the number "20" is found in many lists of amino acids. This of course results from the fact that amino acids are probably the most important molecules for biological functions, and every living cell on the planet has been believed to use the same set of these 20 standard amino acids (Fig. 1.2) to

M. Fleck and A.M. Petrosyan, *Salts of Amino Acids: Crystallization, Structure and Properties*, DOI 10.1007/978-3-319-06299-0_1,

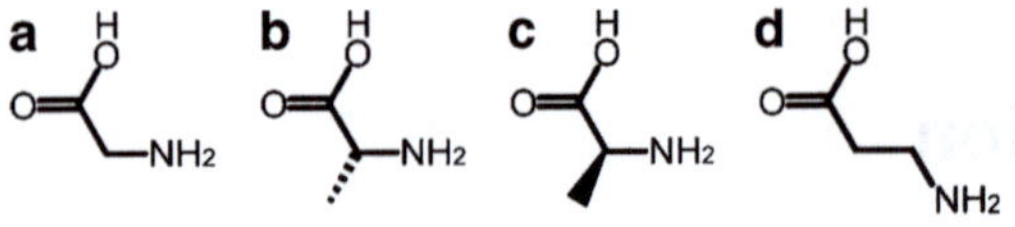

Fig. 1.1 *Simple amino acids*. Glycine, the simplest amino acid, has the minimal requirements of amino acids: one acid and one amino group (**a**). With an additional methyl group, the resulting amino acid is alanine, which exists in three forms: D-α-alanine (**b**) and L-α-alanine (**c**) are enantiomers, whereas in third isomer, β-alanine (**d**) has the amino group bonded to the C3 atom

Fig. 1.2 *The twenty standard amino acids. Yellow* represents unpolar, *blue* polar amino acids, although there is a smooth transition (as, e.g., from Gly to Ala to Val, etc.). Note that apart from the ubiquitous amino and acid groups, there are two (Asp and Glu) with a further acid group (so-called acidic amino acids) as well as three (His, Lys, Arg) with a further amino group (so-called basic amino acids). All amino acids are shown in their neutral state, i.e., with no acidic deprotonation or basic protonation. This does not represent the actual state of any amino acid (neither in solution nor solid state), as protonation or deprotonation (or both) occurs – depending on the pH (see Sect. 1.4)

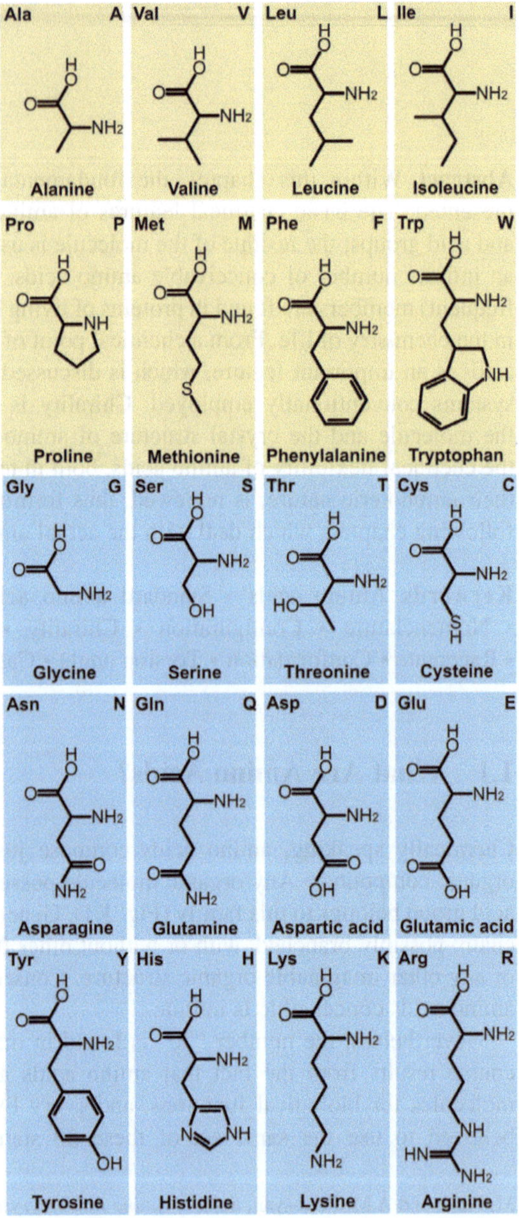

manufacture their proteins. There has been much active discussion on a possible extension of this list since the 1980s, when it has been found that there is another amino acid used in protein production, namely, selenocysteine, which retains a special status (see box below). Posttranslational alterations are frequent, however, so more than 120 amino acids are found in proteins and more than three times as many have been known to occur in other, nonprotein functions in living cells (Wagner and Musso 1983). Presently, it is assumed that there are more than 500 amino acids in nature (Gutiérrez-Preciado et al. 2010).

Although the standard amino acids are listed in most textbooks on organic chemistry and molecular biology, one rarely finds an explanation why exactly these 20 out of an infinite number of amino acids constitute the biological machinery of every cell. In 1981, Weber and Miller presented a detailed analysis of possible reasons, including aspects such as stability, steric reasons, frequency in the primordial soup, etc. Most importantly, the ability to form a strong but flexible chain, with the possibility of reversible lateral bonding, is a main prerequisite for the construction of proteins. Weber and Miller reckon that if life were to develop on another planet, about three fourths of the standard amino acids would be the same as on earth.

Selenocysteine: The 21st Amino Acid?

In the 1980s, it was discovered that the geneticist's model bacterium *Escherichia coli* uses selenocysteine in its protein synthesis. This amino acid is coded by one of the STOP codons, namely, UGA. Additional research found that selenocysteine is incorporated in proteins in the same way as all the other standard amino acids and that this amino acid does not appear in *E. coli* only, but in many animals and even humans (Günzler et al. 1984). Culminating with a paper in Nature (Söil 1988), many scientists happily heralded selenocysteine as the "21st amino acid."

This excitement has deflated since, as it has been found that selenocysteine has several differences in regard with the standard amino acids (apart from the fact that it contains selenium). First, it does not possess its own code in the DNA, but is rather coded by a STOP codon, followed by a so-called selenocysteine insertion sequence (SECIS). Second, it is not ubiquitous but synthesized on demand, as it is highly reactive and thus not easily stored in cells. In fact, selenocysteine attaches to its tRNA as a serine molecule, which then is processed into selenocysteine before being used in protein assembly.

The main reason why this amino acid is not regarded as part of the standard set is the fact that it is not universal in nature, as are the other 20 amino acids. Although ubiquitous in animals (including humans), it does not occur in every domain of life. Only about one sixth of the bacteria and archaea sampled genetically produce selenocysteine, and in higher plants and fungi, this amino acid has not been found at all. Therefore, this 21st amino acid is not considered as part of the standard set (Longtin 2004).

(continued)

Among the many other amino acids that have been found in nature, another one has been considered as an addition to the standard set (Krzycki 2005): pyrrolysine, an amino acid that methanogenic archaea utilize in the methane metabolism and encode via the codon UAG (usually another STOP codon). As pyrrolysine is even less ubiquitous than selenocysteine, this amino acid is also not generally accepted as an addition to the standard 20.

Whatever the reason for their success, there is one feature all standard amino acids have in common: They all are α-amino acids (the nomenclature of amino acid is described in Sect. 1.2). This appears to have been a major advantage in the chemical evolution of proteins, as it allows the formation of the well-known peptide chain, with the peptide bonds along the backbone and the side chains facing outward, thus facilitating the establishment of different types of intrachain bonds, which in turn make the diversity and flexibility of protein structures possible (Weber and Miller 1981).

Nevertheless, other amino acids exist in nature (presumably they were even present in the primordial soup) and play important roles in our biochemistry. Many of these species are not α-amino acids. Maybe the best known example is γ-aminobutyric acid (GABA, Fig. 1.3), which is the main inhibitory neurotransmitter in the central nervous system of all mammals, including us humans. Another prominent non-α-amino acid is β-alanine (Fig. 1.1d), which is a constituent of vitamin B5 (pantothenic acid) as well as the antioxidants carnosine and anserine that are ubiquitous in mammal muscle tissue (Zapp and Wilson 1938). Some more examples are presented in Sect. 1.2.

Amino Acids and Proteins

As said above, amino acids are the monomers that constitute proteins – the macromolecules that make up the majority of biological machinery in every living cell, from bacteria to human beings. This immense diversity in functionality (proteins do not only act as enzymes, there are also structural proteins, motile proteins, protective proteins, transport proteins, membrane proteins, etc.; see Gutteridge and Thornton 2005) has two main reasons: Firstly, amino acids themselves differ structurally (and thus chemically) from one another via the differences of their side chains. Secondly, amino acids bond to each other by a condensation reaction, forming an amide bond, which is more specifically called peptide bond. Units of two amino acids are thus referred to as dipeptides; molecules made from three amino acids are called tripeptides, etc. Such short units are generally referred to as oligopeptides, larger chains as polypeptides, although there is no strict boundary. Larger units still are proteins, although again there is no sharp boundary. In fact, it is not the length of the chain that defines the protein, but

(continued)

the structure that arises from the folding of the chain (consequently, the term polypeptide is often used to indicate a linear chain of amino acids, whereas a protein is a polypeptide that is folded).

This folding results in the actual protein structure and follows from the sequence of amino acids as defined in the genes of living beings. In fact, the genetic code defines so-called codons of three "letters" of the DNA, i.e., three of the bases forming the genetic information within the double helix. A complicated biological process, protein synthesis, is responsible for the formation of the peptide chain in the correct sequence: DNA is copied onto messenger RNA (or mRNA for short), which is moved to the ribosomes (in higher organisms, some additional editing of the RNA occurs before). In the ribosomes, mRNA is brought in contact with amino acids mounted on short RNA strands, so-called transfer RNA (or tRNA). Sterically constrained within the ribosome, two amino acids are positioned close enough to each other so that the condensation reaction occurs. As the mRNA moves through the ribosome, one amino acid after another is added, resulting in a peptide chain. As stated above, the chain has to fold correctly so that it becomes an actual protein. For small proteins, this folding happens by itself; larger chains need help from so-called chaperones, proteins themselves, which assist in correct folding.

The structure of a protein is hierarchical: The peptide chain (i.e., the primary structure) folds locally to helices or sheets simply by hydrogen bonds (α-helices or β-sheets, the secondary structure). These parts again fold up to create the actual protein structure (or tertiary structure), stabilized by nonlocal interactions. These are commonly hydrophobic interactions, hydrogen bonds, ionic bonds, covalent bonds (disulfide bridges between two cysteine side chains), or additional modifications of the amino acids that occur after the synthesis of the chains. Frequently, a functional biological unit does not comprise one lone protein, but an aggregation of more than one protein, interlocked in the so-called quaternary structure. For instance, collagen, the main structural protein of connective tissue in animals, is an arrangement of three protein strands or immunoglobulins (antibodies) consisting of four protein parts.

As the folding of proteins is governed by many parameters, the prediction of the actual structure from the primary sequence is not trivial. As protein molecules are enormous, the structure with the minimal energy cannot be simply calculated, only estimated. As Bohannon (2009) states, there are "more ways to fold a protein than there are atoms in the universe." Funnily, a group of innovative scientists devised an online computer game called *FoldIt*, where players all over the world can engage in protein structure prediction, as they compete and collaborate to optimize energy of a given sequence (Cooper et al. 2010). As it turned out, the community of online players was able to solve problems much more efficiently than expert scientist or computer algorithms could.

(continued)

Apart from prediction or molecular modeling, experimental determination of protein structures is a challenging task for crystallographers. Not only are proteins huge molecules, the crystallization of suitable samples is not as straightforward as for smaller molecules (especially membrane proteins turned out to be hard to crystallize). Still, research in this area has blossomed over the last years and is still growing (for an overview of Protein Crystallography, see Drenth 2007, and references therein).

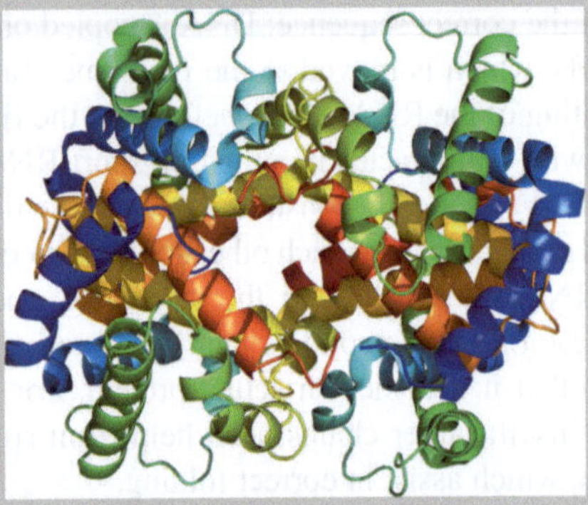

(Image from Wikimedia Commons, http://creativecommons.org/licenses/by-sa/3.0/deed.en)

1.2 Naming Amino Acids

Since there is a large – theoretically an infinite – number of amino acids, several rules have been devised to deal with amino acid nomenclature (besides the standard IUPAC regulations). Apart from the fact that the most common amino acids have trivial names[1] (see Fig. 1.2), there are principles for the formation of semisystematic names as well as atom labeling systems for designating locants on amino acids (IUPAC 1979, 1983).

The employment of semisystematic names applies to substituted standard amino acids and works on the same general principles used for all organic molecules, i.e., the name of the substituent group is attached to the trivial name of the amino acid, whereas the position of the substituent is designated as specified below (e.g., 5-nitrohistidine).

Nevertheless, there are several trivial names in use for other common amino acids, mostly derivatives of standard amino acids, but also some other simple

[1] The etymological background of the trivial names is explained in the respective parts of Chap. 2.

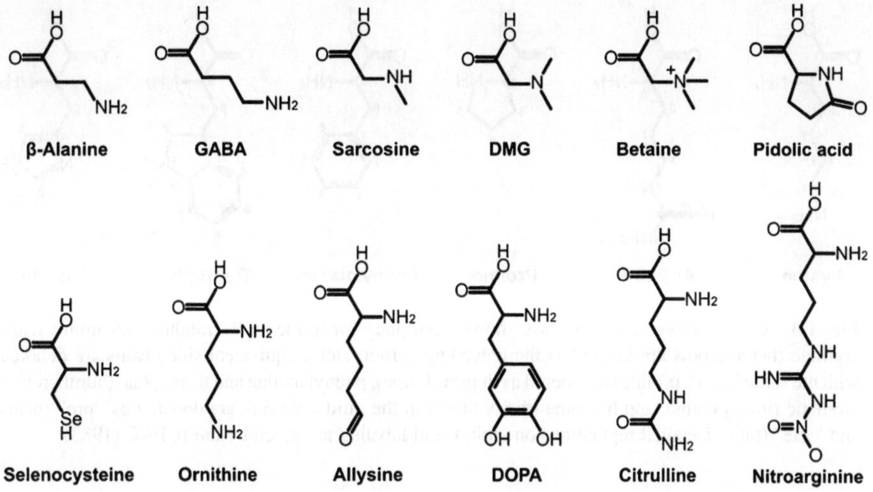

Fig. 1.3 *Some common amino acids beyond the standard twenty*. Only the first two examples are β- and γ-amino acids, namely, β-alanine and γ-aminobutyric acid (GABA). The rest are α-amino acids, as the standard twenty. Together with N-methylglycine (sarcosine), N-dimethylglycine, and N-trimethylglycine (betaine), β-alanine and GABA are examples of non-chiral amino acids. Note also that the structure for betaine is actually the cationic form, betainium (this issue is explained in Sect. 1.4). All other examples shown here are optically active α-amino acids. Pidolic acid (pyroglutamic acid or 5-oxoproline), a derivate of proline, is found in some proteins, as, e.g., bacteriorhodopsin. The aforementioned selenocysteine is the Se analog of cysteine, ornithine appears in the metabolism of urea, as a product of arginine. Allysine is a lysine derivate that plays a vital role as part of structural proteins (e.g., collagen or elastin) and therefore of our connective tissue. DOPA (3,4-dihydroxyphenylalanine) is produced from tyrosine, and acts as precursor for many important hormones and neurotransmitters, such as dopamine, adrenaline, or noradrenalin. Citrulline, as ornithine, is another part of the urea cycle and was first isolated in water melons (Citrullus lanatus, hence the name). Nitroarginine is another derivate of arginine (as is citrulline). In biological systems, all the optically active amino acids referred to here are L-forms (see Sect. 1.3)

examples. Some of these names, along with the semisystematic ones and the formulae, are shown in Fig. 1.3. Moreover, there are several instances which can be regarded as combinations of two standard amino acids. One prominent example is cystine, which is a disulfide of two cysteine moieties (as found in the keratin of our hair). Likewise, lanthionine is a thioester of two alanine residues (and also is found in hair, or wool, where it was first discovered).

As far as atom labeling is concerned, two alternative systems are employed: The atoms are either numbered according to the normal chemical system for designating locants, i.e., starting with the carboxylate carbon atom as C1, or by the utilization of Greek letters, starting with the C2 atom as α (Fig. 1.4). The latter system is of importance not only for defining possible additional substituent positions but to characterize the position of the eponymous amino group: As said above, the standard twenty amino acids have the amino group adjacent to the carboxylic

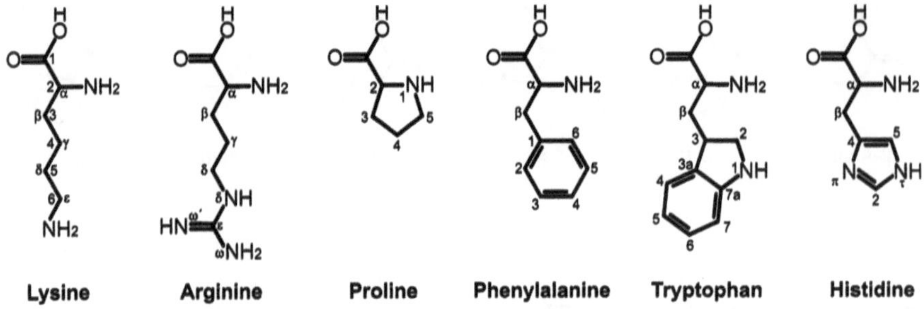

Fig. 1.4 *Atom labeling in amino acids*, shown exemplary for lysine (as a straight-chain amino acid), arginine (heteroatoms are labeled as the preceding carbon atom, equivalent side chains are denoted with the same letter), proline (numbered as in pyrrolidine), phenylalanine and tryptophan (numbered as aromatic ring systems), and histidine (the N-atoms in the imidazole ring are denoted by "pro" (near) and "tele" (far)). Detailed regulations on amino acid labeling are given by the IUPAC (1983)

acid group, i.e., in α position. Therefore, they are all referred to as α-amino acids. Likewise, there are β-amino acids, γ-amino acids, etc.

This system used for designating positions is simple for aliphatic, straight-chain amino acids. When heteroatoms, branching, aliphatic or aromatic cycles, etc. are present, the rules are somewhat more complicated. Some examples for designating locants are shown in Fig. 1.4. For the complete rules, the reader is referred to the details given by the IUPAC (1983).[2]

For the standard amino acids, two specific abbreviations are commonly used, namely, a three- and a one-letter symbol for each amino acid (Fig. 1.2). For general purposes, only the three-letter-symbol is used, the employment of the one-letter-symbol should be restricted to denote long amino acid sequences, i.e., when discussing primary structures of proteins. In this book, we will use the three-letter-symbol only.

---

[2] The numbering of atoms within an amino acid molecule should be done following the rules set by the IUPAC (1983). In acyclic amino acids, the carbon atoms are numbered as follows:

The C atom of the carboxyl group next to the C atom which carries the amino group is C-1, the rest follows consequently. As an alternative, Greek letters can be used, where the C-2 atom is designated α. Although the IUPAC recommend the numbers, not the Greek letters, the terms "α-amino acids" or "α-carbon atom" are frequently used in the literature.

Heteroatoms are given the same number as the carbon atom to which they are attached. The nitrogen atom of the "α-amino group" would thus be N-2. When identical side chains occur, such as in valine or leucine, they are given the same number, one with an apostrophe. In said examples, these methyl atoms are thus labeled C-4 and C-4' (valine) and C-5 and C-5' (leucine). In arginine, where the terminal amino groups are formally different (see scheme in Fig. 1.4) but in fact identical due to resonance, this system is also employed.

Amino acids with rings are numbered along the rules of the systematic nomenclature of rings; for rings with heteroatoms, the numbering scheme of the mother compound is used (see scheme for Pro, Phe, Trp, His in Fig. 1.4).

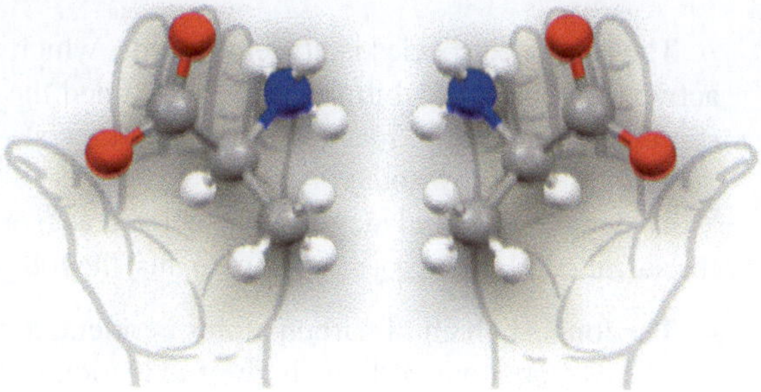

Fig. 1.5 *Chirality in amino acids*. Most amino acids are chiral, i.e., they exist in "left-handed" L- and "right-handed" D-form. These two molecules of opposing chirality are enantiomers; a mixture of both is called a racemate

1.3 Configuration: Chirality and Optical Activity

There is one puzzling fact in the question of the standard amino acids that scientist have pondered for decades: The issue of chirality. Apart from glycine, all of the standard twenty are chiral, i.e., they possess (at least) one carbon atom with four different substituents. Thus they exist in two enantiomers, commonly referred to as left- and right-handed (or L- and D-forms, respectively, Fig. 1.5; for nomenclature of chiral molecules see box below). This alone would not make a puzzle, but all biological systems employ L-amino acids, and L-amino acids nearly exclusively. So far, no reason for this has been found, as chemically both enantiomers are equivalent (only reactions with other chiral systems can distinguish between L- and D-form). Ideas have been proposed, such as simply the random development of L-forms as the first system capable of self-reproduction – a feature conserved by evolution. Another theory discusses the idea that circular polarized light from a quickly rotating star destroyed D-amino acids and thus promoted the other enantiomers (Flores et al. 1977; Bailey et al. 1998). One more possibility is the influence of meteorites, which is based on the fact that analysis of the Murchison meteorite (which contains amino acids) found an excess of L-enantiomers (Cronin and Pizzarello 1997).

D-/L- and *R*-/S-Nomenclature

Organic chemists need to distinguish between left- and right-handed forms when dealing with chiral molecules. However, the simple terms "left handed" and "right handed" cannot be easily employed when speaking of molecules. Today, two systems are in use, the D/L- and *R*/*S*-nomenclature.

The D/L-system was devised by Emil Fischer in 1891 and employed for carbohydrates (in fact monosaccharides, i.e., sugars) and amino acids. Although possible to use for other classes of molecules, these two families are the only ones for which the D-/L-nomenclature is used today.

(continued)

The reference species is glyceraldehyde, which was known to be optically active (see below). Arbitrarily, Fischer named the (+) dextrorotatory form D (for *dexter*) and the (−) levorotatory L (for *laevis*) – a connection which is not general, as we know today!

To decide if a chiral atom is to be defined as D or L, a molecule is viewed in the so-called Fischer projection. To do this, the following rules have to be used:

- The longest chain of carbon atoms is oriented vertically.
- The carbon atom with the highest oxidation state is located on top.
- All bonds are depicted as vertical or horizontal lines.
- Horizontal bonds project toward the viewer; vertical bonds project away from the viewer.
- If the group with the highest priority (i.e., highest oxidation state) is facing right, this form is termed D, and if facing left, it is termed L.

As an example, L-alanine is shown in Fischer projection below.

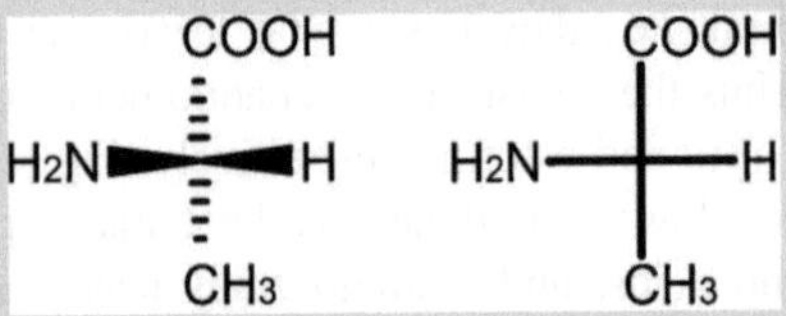

Although this system works well for sugar and amino acid molecules (as well as several other organic species), it cannot be employed generally. For instance, a molecule without a carbon chain cannot be oriented according to Fischer projection.

Thus, another nomenclature system was introduced by Cahn, Ingold, and Prelog: The so-called *R*/*S*-system (for *rectus* and *sinister*) is a method usable for any chiral molecule; it does not need a reference molecule as glyceraldehyde. To decide if a carbon atom is to be termed *R* or *S*, the following rules need to be regarded:

- The substituents of the chiral atoms are assigned priority according to their atomic number (when identical, the atomic numbers of the substituents' substituents are considered).
- The molecule is viewed so that the substituent with the lowest atomic number is facing away from the viewer.
- The remaining substituents are thus forming a circle; if their priority is ordered clockwise, the chiral atom is termed *R*, if counterclockwise, *S*.

Again, this projection is shown for L-alanine, which, in terms of this nomenclature, is called *S*-alanine.

(continued)

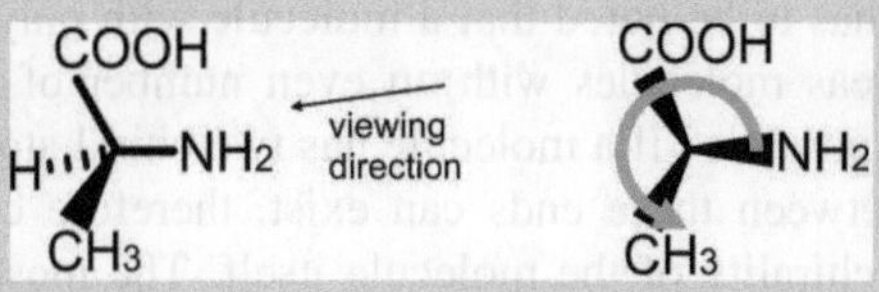

Although the *R*/*S*-nomenclature is the system which can be used for all chiral species, the D/L-system still persists for amino acids (and sugars) out of tradition and comparability with the amount of work published. We therefore stick to this nomenclature system within this book.

Whatever the reason, L-amino acids prevailed and proliferated – once established biochemical pathways of life on earth stuck to this enantiomer.[3] As said above, almost all amino acids in living beings exist in their L-form, as do – consequently – all proteins, and thus all cellular machinery. Nevertheless, over the last decades, some remarkable exceptions have been discovered: D-amino acids were found in the toxin of the platypus (Torres et al. 2002), D-aspartic acid was found to be a neurotransmitter, not only in mammals but also in mollusks (Snyder and Kim 2000; D'Aniello et al. 2011 and references therein). In bacteria, D-alanine and D-glutamine are found in the peptidoglycan cell wall, D-methionine, and D-leucine and other amino acids are used for regulatory signals (Cava et al. 2011). Perhaps most curious is the effect of the D-amino acids found in the toxin of the South American tree frog *Phyllomedusa bicolor*: A certain indigenous people in Peru, the Matsés, employ the frogs' poison for a specific ritual, where they burn their own chests in order to apply an extract gained from the frogs' skin. Severe diarrhea results, followed by a short blackout. When the hunters come to, they experience sharpened senses and the feeling of invulnerability, a hallucination based on the D-amino acid within the drug (Jilek et al. 2005). A brief but concise overview of the role of D-amino acids in animal peptides is given by Jilek and Kreil (2008). Recently, Kantrowitz et al. (2010) reported that D-serine might be utilized for medical treatment of schizophrenia, as this D-amino acid plays a crucial role in the human brain and was found to be deficient in brains of people suffering from schizophrenia.

This spatial difference between molecules of opposite chirality is called *configuration*, in contrast to the term conformation (see below). A very important feature of chiral materials is the fact that they are optically active, i.e., they alter the plane

[3] One might wonder why a small excess of one form resulted in chiral biochemistry, as the other enantiomer would still be present, only in lower quantities. In other words, how come the chemical machinery of proteins is chiral at all? What prevented the development of achiral proteins, i.e., proteins made of amino acids of both chiralities? It was found that homochirality is necessary for the origin of life. Peptides of both L- and D- amino acids cannot fold into bioactive configurations, as, for instance, the α-helix (Bada 1996). Proteins from D- or L-amino acids work equally well, but proteins from racemates do not form effective enzymes.

of polarization of light as it passes through a solution or crystal of this compound (see box below). It has to be noted that a molecule with only one chiral center is always chiral, whereas molecules with an even number of chiral atoms can be achiral. How is this possible? If a molecule has two chiral atoms at opposite ends, mirror symmetry between these ends can exist, therefore canceling out optical activity, as well as chirality of the molecule itself. The most famous example is mesotartaric acid, which in contrast to L- or D-tartaric acid is not optically active.

When synthesized from achiral precursors, the resulting amino acid can assume both configurations, as they are equal to all energetic and chemical purposes. Consequently, such a reaction produces a mixture of D- and L-enantiomers in a 1:1 ratio, a so-called *racemate*. Clearly, a solution containing such a racemic mixture is not optically active, as both enantiomers are present and the effect is canceled out. When growing crystals of such a mixture, it is possible to produce enantiopure crystals (e.g., distinct D- and L-crystals) by adding the respective chiral seeds (in fact, this represents one method of racemic resolution, i.e., the separation of a racemate into its optically active components). A small fraction of racemates (approximately between 5 and 10 %, Jacques et al. 1981) spontaneously crystallize as conglomerates of enantiopure crystals, whereas the majority forms racemic or DL-crystals.

Optical Activity

When plane-polarized light passes through a chiral medium – be it a solution or a crystal – the plane of polarization is altered by a certain value. The degree of alteration is specific for the material (in liquids the concentration must be known, of course) and is usually called the specific rotation α, given as the angle of rotation observed when passing through a solution of a sample with the concentration of 1 g/ml and a path length of 10 cm. If the polarization plane is rotated clockwise (when looking into the light), the compound is labeled (+) or dextrorotatory, if counterclockwise, it is labeled (−) or levorotatory. This is not to be confused with the chirality of the molecule itself, as there is no correlation between configuration and rotational behavior. For instance, out of the 19 optically active standard l-amino acids (glycine is not chiral), 9 are dextrorotatory.

Physically speaking, it is of course not true to say that the plane of polarization is rotated – just as it is not true to assume that photons bounce back off a surface when reflected or are slowed down when passing through matter. In fact, photons do not pass through a medium at all; rather, they are absorbed and reemitted all the time as they interact with matter. Usually, this reemission does not alter the plane of polarization, as the incoming photon's E-vector produces oscillations in the atom in the same direction. Consequently, the reemitted photon's E-vector direction is unchanged, only the propagation of light is delayed by this process. This delay is, of course, described by the refractive index of the material.

(continued)

In the case of optically active materials, we need to take into account the fact that any plane-polarized light can be described as an interference of left-handed and right-handed circularly polarized light. In optically active material, the delay of reemission is different for left-handed and right-handed circularly polarized light. Still, the interference of both does produce plane-polarized light, although the E-vector of the emerging photon has a different direction than the original one. In other words, optical activity is a kind of birefringence: The material has two different refractive indices – for both kinds of circularly polarized light.

The configuration of the molecules has an important impact on molecular and crystal symmetry: As said above, chirality rules out mirror or inversion symmetry (as these symmetry operations invert the chirality of a system and would therefore cancel it). Likewise, if a crystal comprises only one enantiomer of a given species, this crystal cannot possess inversion or mirror symmetry and is therefore necessarily non-centrosymmetric. Racemates or DL-crystals, on the other hand, can very well possess these inversion symmetries, as the respective L- and D-molecules in the crystal are frequently related to each other via mirror planes or inversion centers. This feature will be reflected strongly in the following chapters.

This might appear to be a trivial point. In fact, the symmetry of a crystal is of supreme importance as it correlates with the symmetry of all physical effects. More precisely, the symmetry elements of any physical property of a given crystal must include all symmetry elements of the point group of this crystal (Neumann's principle). This implies that certain effects are possible only in special point groups. Probably the most prominent examples are effects like piezoelectricity (see pp. 28, 32, 38, 72,144, 234, 238, 252, 304, 332, 343f, 351f, 387), which cannot occur in centrosymmetric crystals. Thus, the question of chirality and crystal symmetry is not only an issue of crystallography but also of possible physical applications. Therefore, as most amino acids are chiral, they are possible candidates for all effects that require non-centrosymmetric crystals.

1.4 Conformation: Molecular Shape and Stability

Regrettably, the term configuration is often confused with the term *conformation*, which basically describes the shape of the molecule. In other words, the various conformations of a molecule represent all conceivable shapes of this molecule that are possible due to free rotation around covalent single bonds. Thus, the conformation of a molecule may alter constantly (as it does in liquid or gas phase simply because of thermal motion) and can be "frozen" when incorporated in a crystal lattice. In contrast, the configuration cannot be changed by simple rotation around a bond. In order to change the configuration (from L- to D- or vice versa) a chemical

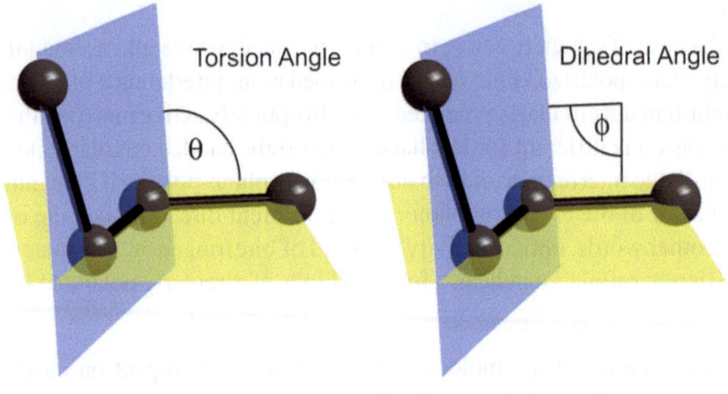

Fig. 1.6 *Torsion and dihedral angle*. When describing conformations of molecules, torsion angles are usually given – representing the angle between the planes through atoms. Alternatively, the angles between the orthogonal vectors of the planes can be given. In this case the term dihedral angle should be used to avoid confusion

reaction has to take place, i.e., bonds must be broken and formed again (McNaught and Wilkinson 1997).

The fact that a molecule may assume any conformation just by rotation about its bonds suggests that all conformations are energetically equal and therefore equally probable. This is not the case. Steric reasons are responsible for different energy levels of different conformations (Mo et al. 2004), although other reasons have been proposed earlier (Weinhold 2001, 2003).

As the rotation around single bonds is continuous, any conformation is possible. Consequently, a precise description of the conformation can be presented by the value of the torsion angle or the dihedral angle (these two terms are sometimes confused, see Fig. 1.6 for definition).

Nevertheless, some conformations are more likely (i.e., energetically favorable) than others, depending on shapes and dimensions of all parts of the molecule. For hydrocarbon chains, some conformations occur frequently and have therefore been given specific names (Fig. 1.7). In the solid state, only these favorable conformations are found.

1.5 Cations, Anion, Zwitterions

Apart from the structural flexibility of amino acids due to their conformational freedom, there is another way in which amino acids can adapt to a given chemical environment: As amphoteric species, amino acids can act as both acids and bases. Thus, they can donate or accept protons (i.e., form anions or cations), or even both at the same time. In this case, the acid proton is formally transferred to the amino group, thus turning the truly neutral molecule into a zwitterion (actually an

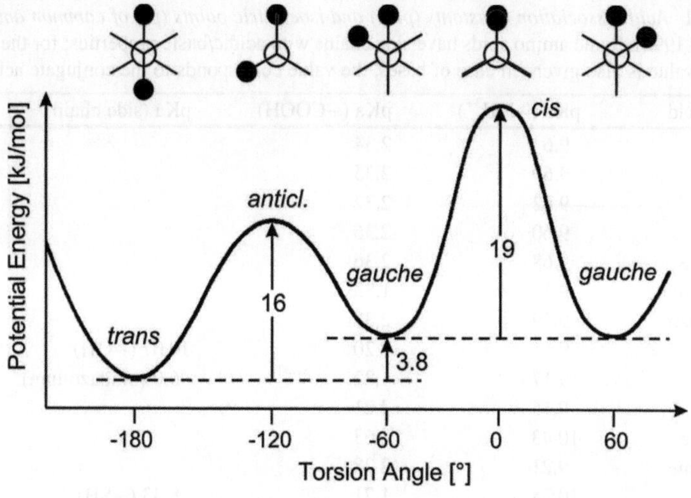

Fig. 1.7 *Conformations of n-butane*, shown as a correlation of torsion angle and potential energy. The molecule is shown on *top*, viewed along the C2–C3 bond (Newman projection), with the terminal methyl groups depicted as black circles. Depending on the relative position of the terminal methyl groups and the hydrogen atoms, staggered conformations are energetically favored in contrast to eclipsed conformations. When the methyl groups are located on opposite ends ("trans"), the potential energy is at a minimum. Another local minimum is found for the "gauche" conformation, at torsion angles of ±60° (McMurry 2007)

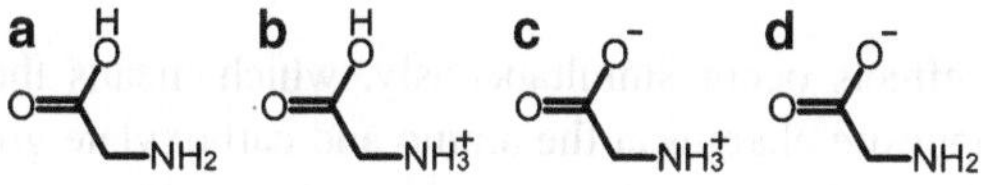

Fig. 1.8 *Protonation and deprotonation of amino acids*. As shown exemplary for glycine, the truly neutral form (**a**) is a very rare one. In solution, the true form is determined by the pH, in solid state by the other ions or molecules present in the crystal. At low, medium, or high values of the pH, the amino acid is a cation (**b**), zwitterion (**c**), or anion (**d**). As usual for cations or anions, the amino acids in these states are named with the conventional endings (-ium or -ate, respectively). Thus, the protonated glycine moiety (**b**) is referred to as a glycinium cation; the deprotonated form (**d**) is called glycinate anion

intramolecular acid–base-reaction, Fig. 1.8). In total, a zwitterion is still neutral, although it distinctly carries a positive and negative charge each (therefore, zwitterions are sometimes referred to as "inner salts"). This zwitterionic state is in fact a very common one: In solution, only a very minor fraction of all amino acid molecules exists in the truly neutral form. Only from an environment of noble gases, amino acids in the non-ionized form can be obtained.

Of course, the pH value of the solution plays a crucial role in determining the state of the amino acid: When below the acid dissociation constant pKa of the amino group (e.g., 9.6 for glycine), this group is protonated; thus, the amino acid carries a positive charge. At values over the pKa of the carboxylic acid group (e.g., 2.34 for glycine), this group is deprotonated, resulting in a negative charge. If the pH is between these

Table 1.1 *Acid dissociation constants (pKa) and isoelectric points (pI) of common amino acids* (Pogliani 1992). Some amino acids have side chains with acidic/basic properties; for these groups the pKa value is also given (in case of bases, the value corresponds to the conjugate acid)

Amino acid	pKa ($-NH^{3+}$)	pKa ($-COOH$)	pKa (side chain)	pI
Glycine	9.6	2.34		5.97
Alanine	9.69	2.35		6.02
Valine	9.62	2.32		5.97
Leucine	9.60	2.36		5.98
Isoleucine	9.68	2.36		6.02
Phenylalanine	9.13	1.83		5.48
Tryptophan	9.39	2.38		5.88
Tyrosine	9.11	2.20	10.07 ($-OH$)	5.65
Histidine	9.17	1.82	6.0 (imidazolium)	7.58
Serine	9.15	2.21		5.68
Threonine	10.43	2.63		6.53
Methionine	9.21	2.28		5.75
Cysteine	10.78	1.71	8.33 ($-SH$)	5.02
Aspartic acid	9.82	2.09	3.86 ($-COOH$)	2.87
Glutamic acid	9.67	2.19	4.25 ($-COOH$)	3.22
Asparagine	8.08	2.02		5.41
Glutamine	9.13	2.17		5.65
Lysine	8.95	2.18	10.53 ($-NH_3^+$)	9.74
Arginine	9.04	2.17	12.48 ($-NH_3^+$)	10.76
Proline	10.60	1.99		6.30

two values, both effects occur simultaneously, which means the amino acid is a zwitterion, with opposite charges at the amino and carboxylate groups.

In a solution of a medium pH, there is a dynamic equilibrium of protonation and deprotonation of the amino acid, with the majority of the molecules in zwitterionic state, and only a small fraction in cationic or anionic state. For every amino acid, there is a pH where the amount of cations and anions exactly balances out. This equilibrium value is called the *isoelectric point* (pI). For amino acids with only the ubiquitous amino and acid group, the pI is exactly between the two pKa values, for those with additional acidic and basic side chains, all pKa values must be taken into account. The pKa as well as the pI values for the amino acids discussed in this book are given in Table 1.1.

In the structural figures shown in the previous sections, the amino acids are always depicted in the truly neutral state, although this does not represent the actual situation. In fact, the diagram as given in Fig. 1.8c, d does also not give a realistic representation, as far as the carboxylate group is concerned. More accurately, the π-electrons of the C=O bond as well as the electron of the deprotonated hydrogen atom are delocalized about the whole carboxylate group, a state well known from aromatic systems and commonly referred to as *mesomeric effect* or *resonance effect.* As the electrons are delocalized, the formal difference between single C–O and double C=O bonds disappear in the molecule. This is clearly expressed in the bond lengths found in crystalline state, as shown for several instances in the following chapters.

1.6 Concluding Remarks

As demonstrated in this chapter, amino acids show – despite the common structural features – a very high degree of diversity. This diversity will become even more apparent in the following chapters, where the pure amino acids as well as their salts are discussed in some details. Several interesting, structural, physical, and chemical aspects will be presented, along with some indications on possible applications of amino acid salts. Although most of the properties that allow exploitation as materials with physical, technical, chemical, or pharmaceutical value occur in salts, mostly in solid state, there has been an interesting finding on amino acids in a curious state: Recently, Fukumoto et al. (2005) reported the synthesis of ionic liquids from 20 natural amino acids (for the utilization of ionic liquids, see box below). Within this paper, the authors showed that amino acid ionic liquids were able to dissolve native amino acids under anhydrous conditions and are themselves soluble in organic solvents (e.g., chloroform). This finding is but one which shows the versatility of amino acids – another unexpected discovery of amino acid properties made nearly 200 years after the first report of a member of this family.

Ionic Liquids

Although ionic liquids have been known for over a century, their real impact on science has been made when Wilkes and Zaworotko (1992) presented ionic liquids with weakly coordinating anions (as PF_6^- and BF_4^-), which have made a wide range of applications possible. Ionic liquids are generally salts with a low melting point, i.e., liquid at more or less ambient conditions. Thus, they are powerful solvents that have made the leap from laboratory to industry (Plechkova and Seddon 2008), both in the chemical and pharmaceutical community. Maybe even more important seems the utilization of ionic liquids as promising solvents for cellulose. As the most abundant chemical on earth, cellulose is part of every kind of plant tissue harvested and yet hard to process chemically. Dissolving cellulose at ambient conditions might be a chance to develop biofuel on a basis that does not compete with food production (Ohno and Fukaya 2009). Another very promising application of ionic liquids is their use as short-time energy storage medium. This is of importance in connection with the increasing development of photovoltaics: As it is necessary to absorb excess energy (e.g., in sunshine) which has to be released later (e.g., at night), there is heavy demand for materials that can store energy reversibly. Ionic liquids with a high liquid-phase temperature range are possible candidates for this application. An overview of physical and chemical properties of ionic liquids is given by Zhang et al. (2009).

References

Bada JL (1996) Amino acid homochirality on earth and mars. Orig Life Evol Biosph 26:518–519
Bailey J, Chrysostomou A, Hough JH, Gledhill TM, McCall A, Clark S, Menard F, Tamura M (1998) Circular polarization in star-formation regions: implications for biomolecular homochirality. Science 281:672–674
Bohannon J (2009) Gamers Unravel the Secret Life of Protein. http://www.wired.com/medtech/genetics/magazine/17-05/ff_protein. Accessed 20 Jan 2014
Cava F, Lam H, de Pedro MA, Waldor MK (2011) Emerging knowledge of regulatory roles of D-amino acids in bacteria. Cell Mol Life Sci 68:817–831
Cooper S, Khatib F, Treuille A, Barbero J, Lee J, Beenen M, Leaver-Fay A, Baker D, Popović Z, Players F (2010) Predicting protein structures with a multiplayer online game. Nature 466:756–760
Cronin JR, Pizzarello S (1997) Enantiomeric excesses in meteoritic amino acids. Science 275:951–955
D'Aniello S, Somorjai I, Garcia-Fernandez J, Topo E, D'Aniello A (2011) D-Aspartic acid is a novel endogenous neurotransmitter. FASEB J 25:1014–1027
Drenth J (2007) Principles of protein X-ray crystallography, 3rd edn. Springer, Heidelberg
Flores JJ, Bonner WA, Massey GA (1977) Asymmetric photolysis of (RS)-leucine with circularly polarized ultraviolet light. J Am Chem Soc 99:3622–3624
Fukumoto K, Yoshizawa M, Ohno H (2005) Room temperature ionic liquids from 20 natural amino acids. J Am Chem Soc 127:2398–2399
Günzler WA, Steffens GJ, Grossmann A, Kim SMA, Ötting F, Wendel A, Flohé L (1984) The amino-acid sequence of bovine glutathione peroxidase. Hoppe Seyler's Z Physiol Chem 365:195–212
Gutiérrez-Preciado A, Romero H, Peimbert M (2010) An evolutionary perspective on amino acids. Nat Educ 3:29
Gutteridge A, Thornton JM (2005) Understanding nature's catalytic toolkit. Trends Biochem Sci 30:622–629
IUPAC (1979) Nomenclature of organic chemistry, sections A, B, C, D, E, F and H. Pergamon Press, Oxford
IUPAC (1983) Nomenclature and symbolism for amino acids and peptides. http://www.chem.qmul.ac.uk/iupac/AminoAcid. Accessed 12 Feb 2012
Jacques J, Collet A, Wilen SH (1981) Enantiomers, racemates and resolutions. Wiley, New York
Jilek A, Kreil G (2008) D-amino acids in animal peptides. Monatsh Chem 139:1–5
Jilek A, Mollay C, Tippelt C, Grassi J, Mignogna G, Müllegger J, Sander V, Fehrer C, Barra D, Kreil G (2005) Biosynthesis of a D-amino acid in peptide linkage by an enzyme from frog skin secretions'. Proc Natl Acad Sci U S A 102:4235–4239
Kantrowitz JT, Malhotra AK, Cornblatt B, Silipo G, Balla A, Suckom RF, D'Souza C, Saksa J, Woods SW, Javitt DC (2010) High dose D-serine in the treatment of schizophrenia. Schizophr Res 121:125–130
Krzycki JA (2005) The direct genetic encoding of pyrrolysine. Curr Opinion Microbiol 8:706–712
Longtin R (2004) A forgotten debate: is selenocysteine the 21st amino acid? J Natl Cancer Inst 96:504–505
McMurry JC (2007) Organic chemistry, 7th edn. Thomson Brooks/Cole, Belmont
McNaught AD, Wilkinson A (1997) Compendium of chemical terminology, 2nd edn. Blackwell Scientific Publications, Oxford
Mo Y, Wu W, Song L, Lin M, Zhang Q, Gao J (2004) The magnitude of hyperconjugation in ethane: a perspective from Ab initio valence bond theory. Angew Chem 116:2020–2024
Ohno H, Fukaya Y (2009) Task specific ionic liquids for cellulose technology. Chem Lett 38:2–7
Plechkova NV, Seddon KR (2008) Applications of ionic liquids in the chemical industry. Chem Soc Rev 37:123–150

Pogliani L (1992) Molecular connectivity model for determination of isoelectric point of amino acids. J Pharm Sci 81:334–336

Snyder SH, Kim PM (2000) D-amino acids as putative neurotransmitters: focus on D-serine. Neurochem Res 25:553–560

Söil D (1988) Enter a new amino acid. Nature 331:662–663

Torres AM, Menz I, Alewood PF, Bansal P, Lahnstein J, Gallagher CH, Kuchel PW (2002) D-Amino acid residue in the C-type natriuretic peptide from the venom of the mammal, Ornithorhynchus anatinus the Australian platypus. FEBS Lett 524:172–176

Wagner I, Musso H (1983) New naturally occurring amino acids. Angew Chem 22:816–828

Weber AL, Miller SL (1981) Reasons for the occurrence of the twenty coded protein amino acids. J Mol Evol 17:273–284

Weinhold F (2001) Chemistry: a new twist on molecular shape. Nature 411:539–541

Weinhold F (2003) Rebuttal to the Bickelhaupt-Baerends case for steric repulsion causing the staggered conformation of ethane. Angew Chem Int Ed 42:4188–4194

Wilkes JS, Zaworotko MJ (1992) Air and water Stable 1-ethyl-3-methylimidazolium based ionic liquids. J Chem Soc Chem Commun 13:965–967

Zapp JA, Wilson DW (1938) Quantitative studies of carnosine and anserine in mammalian muscle. J Biol Chem 126:19–27

Zhang S, Lu X, Zhou Q, Li X, Zhang X, Li S (2009) Ionic liquids: physicochemical properties. Elsevier, Amsterdam

[illegible] (1995) [illegible] connectivity model for deformation [illegible]

[illegible] (2000) [illegible]

[illegible] (1988) [illegible] Nature 331:[illegible]

[illegible]

[illegible]

[illegible]

[illegible]

[illegible]

[illegible]

Chapter 2
Amino Acid Structures

Abstract Although this book deals with salts of amino acids, this chapter discusses the structures of pristine amino acids, since these molecular structures are also found in salts, and factors as solubility and conditions of crystal growth are important for the synthesis of amino acid salts as well as the pure form. In this context, the impact of minimal changes in conditions (such as impurities) on the growth of amino acid crystals is noted. The structural variation is very high, as not only amino acids differ from each other, but many amino acids exist in more than one form. This refers to the fact that enantiopure crystals can be grown as well as racemates (so-called DL-amino acids). Moreover, often more than one hydration state is found: Anhydrous forms are common, but many amino acids form hydrated crystals. For some amino acids (e.g., glycine, proline, methionine), more than one polymorph of the same hydration state is found. Some of these polymorphs form at ambient condition, often due to minuscule changes in conditions. For others, variation in temperature and pressure has been found to be the cause of the formation of different polymorphs. High-pressure data are available for some amino acids (e.g., alanine, serine, cysteine). Most amino acids are found to form a so-called head-to-tail motif, where acid and amino groups connect to form infinite chains. In some of the larger, nonpolar amino acids, a bilayer pattern is found, where polar groups of opposing molecules face each other, forming a layer, with the hydrophobic side chain facing outward (this motif is found in phenylalanine, methionine, leucine, or isoleucine). Not all amino acids crystallize readily, as is proved by the crystal structure of L-arginine, which could be determined only very recently, and that of lysine, which could not be solved at all to date. In addition to the standard 20, some nonstandard amino acids are discussed. Finally, notes on remarkable physical effects (such as piezoelectric data) or possible applications (as for instance the interactions of amino acids with carbon nanotubes) are presented.

Keywords Amino acid structures • Solubility • Crystal growth • Hydrophobicity • Polymorphs • Non-ambient conditions • Head-to-tail motif • Polar amino acids • Nonpolar amino acids • Physical properties

M. Fleck and A.M. Petrosyan, *Salts of Amino Acids: Crystallization, Structure and Properties*, DOI 10.1007/978-3-319-06299-0_2,

2.1 Introducing Remarks

As this work is titled *Salts of Amino Acids*, one may be confused to find a whole chapter on amino acids in their pristine forms. Although it is of course true that pure amino acids are not by definition salts of amino acids, the contents of this chapter are definitely necessary for any scientific work dealing with said salts. This includes chemical and structural aspects of amino acids – both of general and specific nature (i.e., regarding all, some, or only individual amino acids). Clearly, these properties must be considered when working with the respective salts. Moreover, it goes without saying that the structural features of any amino acid are best studied when investigating molecular and crystal structures of these compounds in their pure form.

Besides, there is a sad list of inglorious papers in which researchers enthusiastically present "novel compounds of amino acids," although the results (X-ray diffraction data or vibrational spectra, in most cases) clearly show that these "compounds" are actually amino acids used as precursor for the reactions. Many of these mistakes have been commented on (Petrosyan et al. 2013; Fleck and Petrosyan 2009; and Srinivasan and Dhavskar 2014), but similar mistakes continue to appear in scientific journals. Therefore, it as an absolute necessity that each chemical reaction supposed to produce an amino acid salt is considered skeptically, i.e., always bearing in mind the possibility that no reaction has taken place at all and the supposed product is simply the precursor. Consequently, all characterization results of this "product" need to be compared with the respective data of the precursor material, in this case the pristine amino acids. A detailed overview of frequent mistakes along these lines has been published (Fleck and Petrosyan 2010).

2.2 Solubility and Crystal Growth

Generally, amino acids are compounds with rather surprisingly high boiling and melting points (when compared to organic species with similar molecular sizes). In fact, the melting temperatures are usually higher than the temperature of decomposition but have been assessed to be in the range of 200–300 °C (Dunn and Brophy 1932; Olafsson and Bryan 1970). This of course results from the fact that the amino acid molecules assume the zwitterionic form in solid state; therefore, the strongest interaction of neighboring molecules is the ionic attraction (instead of hydrogen bonds or van der Waals interactions in most organic crystals).

As far as the solubility of amino acids is concerned, it can generally be said that amino acids are soluble in water and insoluble in nonpolar organic solvents (such as hydrocarbons). This effect reflects the ionic character (as said in Chap. 1, amino acids usually exist as zwitterions), as the ionic interactions between adjacent amino acid molecules in solid state are replaced by interactions between amino and acid groups toward the water molecules (comparable to the solubility of ionic substances in water). Of course, the side chain influences the extent of solubility

Table 2.1 *Solubility and hydrophobicity of amino acids*. Solubility in g/100 ml water, at 25 °C (Hamend and Helgeson 1997); Hydropathy index (Kyte and Doolittle 1982)

Name	Solubility	Hydropathy	Name	Solubility	Hydropathy
Alanine	16.63	1.8	Leucine	2.19	3.8
Arginine	19.59	−4.5	Lysine	24.66	−3.9
Asparagine	2.51	−3.5	Methionine	5.59	1.9
Aspartic acid	0.51	−3.5	Phenylalanine	2.80	2.8
Cysteine	2.56	2.5	Proline	130.07	1.6
Glutamic acid	0.88	−3.5	Serine	36.57	−0.8
Glutamine	4.25	−3.5	Threonine	9.79	−0.7
Glycine	25.23	−0.4	Tryptophan	1.32	−0.9
Histidine	4.36	−3.2	Tyrosine	0.054	−1.3
Isoleucine	3.17	4.5	Valine	5.87	4.2

in water, with respect to the nature and size of the chain (Boldyreva 2008, p. 174ff). An overview of the solubility of standard amino acids is given in Table 2.1.

It is remarkable that some amino acids possess very low solubilities. Although plausible for nonpolar amino acids like tryptophan or phenylalanine, other values seem counterintuitive, especially for amino acids like aspartic acid, glutamic acid, and tyrosine, which possess polar groups – a feature associated with solubility in water. However, it has to be considered that solubility is a function of lattice energy. Thus, the additional polar groups (phenolic or acid) can participate in hydrogen bonding with the α-acid and amino group, which makes the solid, crystalline state energetically more favorable than the dissolved state.

In addition to the solubility, several scales have been devised to measure the hydrophobicity of amino acids. Depending on the method, the results differ slightly: Wolfenden et al. (1981) performed vapor pressure measurements, while Kyte and Doolittle (1982) devised a computational method, based on experimental observations derived from the literature. The order is similar, however, and helps to assess the hydrophilic/hydrophobic character of an amino acid (the values given by Kyte and Doolittle 1982; the so-called *hydropathy*, are listed in Table 2.1), which is useful for the evaluation of protein structures.

Many works have been published on the crystallization on amino acid crystals. Apart from a multitude of papers reporting growth (and often other data) of individual amino acids (see the respective sections below), several authors have studied amino acid crystallization in general. Black and Davey (1988) have performed such a work, discussing nucleation, crystal growth, morphology, surface kinetics, polymorphism, and effects of impurities. Within their paper, they report that although the growth of relatively large single crystals of amino acids is relatively straightforward, many factors influence the growth process and are often hard to predict. Computational techniques were found to be able to predict morphologies in pure systems, but impurities affect this process, often dramatically and specifically.

An even more extensive review was published by Boldyreva (2008), who simply stated that "the growth of amino acids is puzzling." The author here refers to the

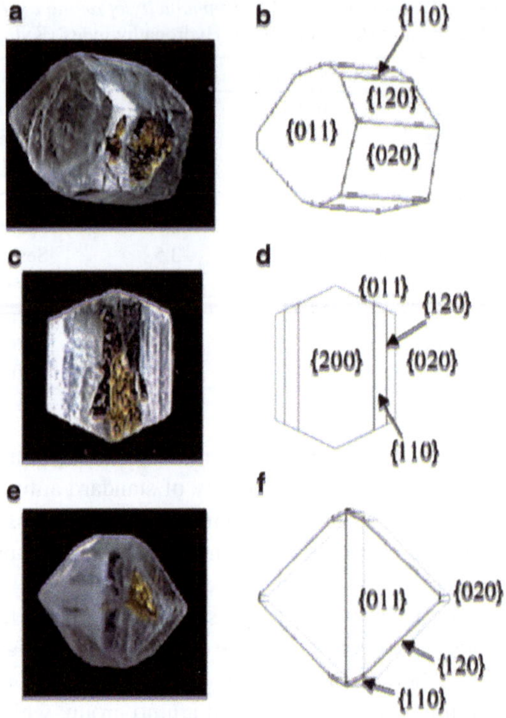

Fig. 2.1 *Crystals of* L-*alanine nucleated on SAMs.* Photographs and morphology of crystals nucleated on 4′-methyl-4-mercaptobipheny (**a**, **b**), 4′-hydroxy-4-mercaptobiphenyl (**c**, **d**), and 4-(4-mercaptophenyl) pyridine (**e**, **f**) (Reprinted with permission from Lee et al. 2002; Copyright (2002) American Chemical Society)

fact that minuscule variations of conditions lead to formation of different crystals, such as different hydrates. For instance, the growth of L-serine by slow evaporation of an aqueous solution yields the monohydrate; addition of L-serine powder to a solution of ethanol and water results in precipitation of anhydrous L-serine. Moreover, when DL-serine powder is dissolved and crystals are grown from this solution, DL-serine crystals are obtained. On the other hand, dissolution of D- and L-serine powder (in equimolar mixture) in water and subsequent crystallization out of this solution yields crystals of both D- and L-serine.

Concluding, these studies have shown that the crystallization of amino acids is highly sensitive to any changes in conditions, impurities, solvent, seeding, aging of the solution, etc. Often, exotic conditions (droplets of particular sizes, nanoparticles, external electric fields, etc.) are necessary to obtain a specific polymorph (for examples and details, see Boldyreva 2008 and references therein).

Recently, additional works have been published on the crystallization on amino acids on so-called self-assembled monolayers or SAMs for short. Possible applications of SAMs in electrochemistry, protein and enantiomer separation, biosensors, etc., have led to increased activity in this field, and amino acids have been considered and investigated (among many other compounds). Many examples have been found: For instance, Lee et al. (2002) published data on the crystallization of L-alanine and DL-valine on SAMs of biphenyl thiols on a gold film. An extensive overview of amino acid crystallization on SAMs is given by Ejgenberg and Mastai (2012) (Fig. 2.1).

2.3 Amino Acid Structures

In this section, the structures of the standard 20 plus several other frequent amino acids are presented. All these structures have been determined by single crystal X-ray diffraction methods, although many other techniques such as vibrational spectroscopy have been employed to verify and support the data. Consequently, the structures presented here represent the molecules in their solid state, usually at ambient conditions. Thus, the following aspects need to be considered:

Clearly, there is one and only one molecular structure of each amino acid (as represented by the formulae shown in Figs. 1.2 and 1.3), conceding that the two enantiomers of optically active amino acids have the same structure, differing only by chirality. Nevertheless, it is possible that one amino acid comes in several different crystalline species, because of the following reasons:

First, for each optically active amino acid, it is conceivable to grow crystals of a solution containing only one enantiomer[1] or both, i.e., from a racemate. Thus, L- (or D-) and DL-crystals can be grown, with the important feature that DL-crystals can possess inversion symmetry (see Sect. 1.3).

Secondly, as amino acid crystals are usually grown from solutions, the incorporation of solvent (i.e., water) molecules in the crystal lattice is possible and occurs frequently. Thus, it is possible that more than one crystal structure of one and the same amino acids exists, differing by hydration state. However, this feature is more frequently found in salts of amino acids than in pure amino acids.

Thirdly, each amino acid can exist in more than one polymorph, depending on temperature, pressure, etc. As said above, the majority of studies dealing with amino acids employ ambient conditions, but there are several examples of phase transitions when temperature or pressure (or other conditions) are altered. The most famous example is the simplest amino acid, glycine, which exists in three polymorphs at ambient conditions. For an overview of amino acids under high-pressure conditions, a comprehensive review was published (Moggach et al. 2008).

All the crystalline species of the amino acids discussed in this book are listed in Tables 2.2–2.7, showing the major crystallographic parameters as space groups and unit cell parameters. A detailed description of these structures is given in the following text.

[1] Usually, the optically active crystals are grown from L-amino acids, as this is the form occurring in nature. The respective D-amino acid crystal has of course the same structure, only with the opposite chirality. For those amino acids where both L- and D-crystals have been investigated, this will be mentioned in the text.

Table 2.2 *Crystallographic properties of standard amino acids, part I.* Space groups and unit cell parameters of glycine, alanine, valine, leucine, isoleucine, proline, and methionine crystals

	s.g.	a	b	c	α	β	γ	Reference
α-Gly	$P2_1/n$	5.105	11.969	5.465	90	111.70	90	Almlöf et al. (1973)
β-Gly	$P2_1$	5.094	6.274	5.385	90	113.19	90	Drebushchak et al. (2002)
γ-Gly	$P3_2$	7.037	7.037	5.478	90	90	120	Shimon et al. (1986)
δ/β'-Gly[a]	$P2_1/a$	11.156	5.864	5.342	90	125.83	120	Dawson et al. (2005)
ε/δ-Gly[a]	*Pn*	4.889	5.754	5.4442	90	116.68	120	Dawson et al. (2005)
"X-phase"Gly	*P2*	6.648	25.867	5.610	90	113.12	120	Surovtsev et al. (2012)
L-Ala	$P2_12_12_1$	5.78	6.03	12.32	90	90	90	Lehmann et al. (1972a)
DL-Ala	$Pna2_1$	12.026	6.032	5.829	90	90	90	Subha Nandhini et al. (2001)
L-Val	$P2_1$	9.71	5.27	12.06	90	90.8	90	Torii and Iitaka (1970)
DL-Val	$P2_1/c$	5.21	22.10	5.41	90	109.2	90	Mallikarjunan and Rao (1969)
DL-Val (120 K)	*P*-1	5.222	5.406	10.838	90.89	92.34	110.02	Dalhus and Görbitz (1996a)
L-Leu	$P2_1$	9.61	5.31	14.72	90	86.2	90	Harding and Howieson (1976)
L-Leu (400 MPa)	*C2*	8.807	5.975	14.193	90	96.17	90	Yamashita et al. (2007)
DL-Leu	*P*-1	14.12	5.39	5.19	111.1	97.0	86.4	DiBlasio et al. (1975)
L-Ile	$P2_1$	9.75	5.32	14.12	90	95.8	90	Torii and Iitaka (1971)
DL-Ile	*P*-1	14.66	5.39	5.27	109.2	114.0	85.2	Benedetti et al. (1973a)
L-Pro	$P2_12_12_1$	11.550	9.020	5.200	90	90	90	Kayushina and Vainshtein (1965)
L-Pro·H_2O	*C2*	20.431	6.192	5.136	90	95.79	90	Janczak and Luger (1997)
DL-Pro	$P2_1/c$	8.991	5.299	11.479	90	97.04	90	Myung et al. (2005)
DL-Pro·H_2O	*Pbca*	5.274	12.087	20.053	90	90	90	Padmanabhan et al. (1995)
L-Met	$P2_1$	9.498	5.189	15.318	90	97.69	90	Torii and Iitaka (1973)
α-DL-Met	$P2_1/a$	9.76	4.79	16.70	90	102	90	Mathieson (1953)
β-DL-Met	*I2/a*	9.94	4.70	33.40	90	106.6	90	Mathieson (1953)
γ-DL-Met	$P2_1/a$	31.76	4.70	9.90	90	90.98	90	Matsuoka et al. (1999)

Note: The Greek letters α, β, etc., in this table denote different polymorphs of the respective amino acids, as all amino acids given here are α-amino acids

[a]The nomenclature of the high-pressure δ- and ε-polymorphs are ambiguous (see text), unit cell dimensions vary due to the applied pressure

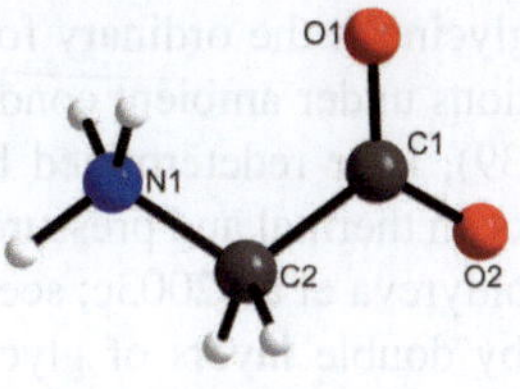

Fig. 2.2 *Glycine*. The molecular structure of glycine in its zwitterionic state, calculated from the data on α-glycine at 288 K (Langan et al. 2002). Note the nearly equivalent C–O-distances, which result from the mesomeric effect (see p. 16), 1.250(1) and 1.254(1) Å in this case. The labeling is not consistent in the literature; for this book we use the labeling given here

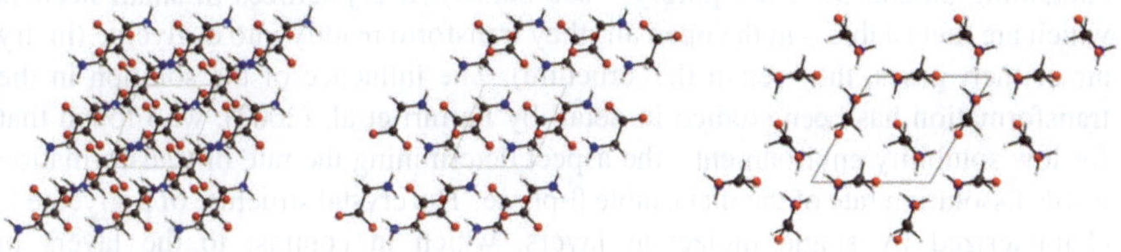

Fig. 2.3 *Polymorphs of glycine*. The crystal structures of α-, β-, and γ-glycine, from left to right (Langan et al. 2002; Drebushchak et al. 2002; Iitaka 1961)

2.3.1 Glycine

As said before, glycine is the simplest amino acid, comprising the minimum of one carboxylic acid group, one amino group, and two hydrogen atoms bonded to the α-carbon atom (Fig. 2.2). Therefore, it is not optically active (the only amino acid of the standard 20) and can assume molecular mirror symmetry. Glycine bears some structural similarity to the smallest sugar molecules (glyceraldehyde) and thus activates the respective receptors on the human tongue, i.e., it tastes sweet – hence its name (*glykeros*, Greek for sweet). The name given originally, however, was "gelatin sugar," as the compound was first prepared from a reaction of gelatin and sulfuric acid by Henri Braconnot (1820), making it one of the first amino acids discovered.

Interestingly, glycine is a very good example for the existence of polymorphs in crystalline form – not one, but three different phases exist at ambient conditions (plus several others at non-ambient conditions, see below), which are referred to as α-, β-, and γ-glycine (Fig. 2.3). Actually, this nomenclature is not really well chosen, as it leads to confusion with the usage of these Greek letters in other amino acids, where they denote the position of carbon atoms (see Sect. 1.2). Thus, one has to bear in mind that α-, β-, and γ-glycine are polymorphs of the same crystal (whereas α- and β-alanine, e.g., are structural isomers – see Figs. 1.2 and 1.3). Moisture, heating, or storage in solution ("solvent-mediated transformations") can result in phase transitions between these forms (Boldyreva et al. 2003a, b).

Of these polymorphs, α-glycine is the ordinary form, i.e., the one most readily obtained from aqueous solutions under ambient conditions. It was first determined by Albrecht and Corey (1939), later redetermined by Marsh (1957) and several other studies, including works on thermal and pressure influence (Legros and Kvick 1980; Langan et al. 2002; Boldyreva et al. 2003c; see also below). This polymorph of glycine is characterized by double layers of glycine molecules parallel (010), connected via hydrogen bonds. The layers themselves are connected to each other via van der Waals interactions only.

The β-form of glycine has been originally described by Fischer (1905), although the structure has been determined much later (Iitaka 1960; redetermined by Drebushchak et al. 2002). β-glycine can be grown from aqueous glycine solutions containing ethanol (but not purely – see below). It crystallizes in small needles which are metastable – in the open air, they transform readily into α-glycine (in dry air or inert gases, they retain the structure). The influence of the solution in the transformation has been studied in detail by Ferrari et al. (2003), who found that for low solubility environments, the aspect determining the rate of transformation is the dissolution rate of the metastable β-phase. The crystal structure of β-glycine is characterized by single molecular layers, which in contrast to the layers in α-glycine are connected via hydrogen bonds, thus representing a three-dimensional framework.

Of the three standard polymorphs, γ-glycine was the last to be discovered (Iitaka 1953). This species turned out to be trigonal and strongly piezoelectric (Iitaka 1958, 1961; Ashok Kumar et al. 2011). In this crystal, the glycine molecules are connected to chains parallel [001], which in turn are connected to a three-dimensional framework via hydrogen bonds. It is interesting to note that γ-glycine is the most stable form under ambient conditions, although it does not crystallize easily from solution.

In many studies of crystalline materials, vibrational spectra are employed for characterization purposes, often in combination with XRD or other data. Besides being a powerful tool for the identification of functional groups, many important structural details can be elucidated from the spectra, as the energies and therefore positions of certain vibrational modes reflect the structural properties. This is shown impressively in the IR spectra of the α- and γ-polymorphs of glycine, which differ slightly but visibly due to the small structural differences (Fig. 2.4). The importance of these details becomes apparent when trying to identify the species by study of the spectrum – something that has led to mistakes in the past when conducted without the necessary care (several instances on misinterpretation of spectral data are given and discussed in Chap. 5). Concerning IR spectra of amino acid, a good overview is given by Matei et al. (2005).

In addition to these three "standard polymorphs," at least two other phases were discovered at higher pressures (which have been named unambiguously[2]): When compressed, β-glycine (space group $P2_1$) undergoes a phase transition to a

[2] Both studies have been published nearly simultaneously (Dawson et al. submitted one month earlier). Without wanting to acknowledge either study as superior, we use both nomenclatures, i.e., we refer to the β′/δ-form and the δ/ε-form, respectively. (The different space groups used for the δ/β′-form represent different settings of the unit cell; the structure is the same.)

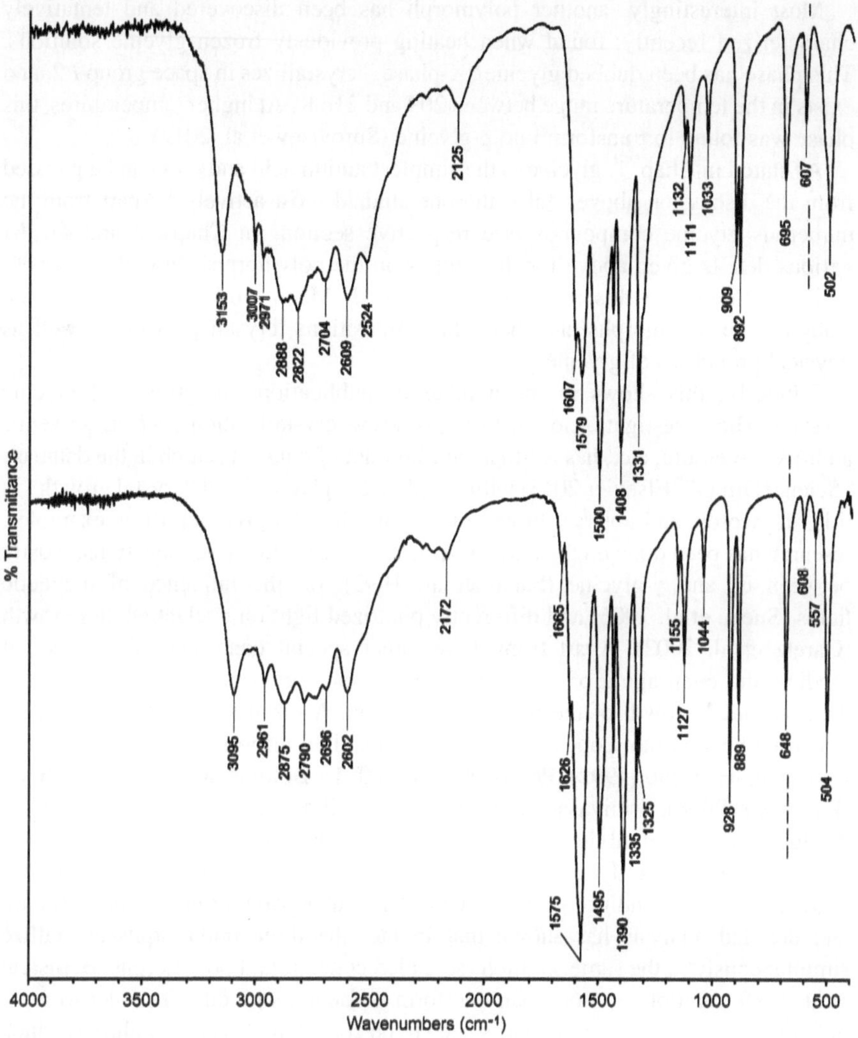

Fig. 2.4 *IR spectra of glycine*. Although similar, the IR spectra of α- and γ-glycine (*top* and *bottom*) differ visibly

centrosymmetric structure, reported as δ-glycine (Dawson et al. 2005, described in space group $P2_1/a$) or β′-glycine (Boldyreva et al. 2003a, b described in space group $P2_1/n$). Similarly, a compression of γ-glycine (space group $P3_1$) results in a phase transition to another monoclinic phase (space group *Pn*), dubbed as ε-glycine (Dawson et al. 2005) or δ-glycine (Boldyreva et al. 2003a, b). For α-glycine, no such phase transition was observed up to at least 23 GPa (Dawson et al. 2005).

Moreover, decompression of the δ/ε-polymorph below approximately 0.62 GPa results in the formation of another phase, which has been detected via Raman spectroscopy and optical microscopy and named ζ-glycine (Goryainov et al. 2006).

Most interestingly, another polymorph has been discovered and tentatively characterized recently, found when heating previously frozen glycine solutions. This phase has been dubbed glycine “X-phase,” crystallizes in space group *P*2, and exists in the temperature range between 209 and 216 K. At higher temperatures, this phase was found to transform into β-glycine (Surovtsev et al. 2012).

As stated in Chap. 1, glycine is the simplest amino acid and – as can be guessed from the discussion above – also the one studied most actively.[3] Apart from the numerous glycine compounds (see respective sections in Chaps. 3 and 4), the various details given above for the studies on the polymorphs show the diversity of (and interest in) this particular amino acid. Much work has been done on studying the polymorphs and their phase transitions, crystal growth, as well as physical properties of glycine.

Primarily, this shows in the number of publications on growth of glycine crystals. The investigation of factors governing crystallization such as solvents, additives, pressure, etc., has led to a huge amount of data – a search in the database “ScienceDirect” (Elsevier 2013) with the phrases “glycine” and “crystal growth” in title, keywords, and abstract turned out 91 articles. To give but a few examples, research has been done on the reaction kinetics of the said phase transitions, mostly between α- and γ-glycine (Sakai et al. 1992), on the influence of magnetic fields (Sueda et al. 2006) and differently polarized light on nucleation and growth (Garetz et al. 2002). Apart from these rather special conditions, the aspect of seeding and even aging of solution (“solution memory”) play important roles in the controlled growth of the desired polymorph. A discussion of the whole data would go beyond the scope of this book; for detailed information see, e.g., Li and Rodríguez-Hornedo (1992), Perlovich et al. (2001), Boldyreva et al. (2003a, b, c), Boldyreva (2008), Srinivasan et al. (2011), as well as the references given above. A short summary of all these results can be given as follows.

As said above, α-glycine crystallizes most easily, i.e., from pristine solutions under almost any conditions. The growth of the other polymorphs is more difficult, and detailed analysis has shown that in fact the three polymorphs crystallize simultaneously in the same solution (so-called concomitant polymorphs, Bernstein et al. 1999), and often, more than one form is found as product. In order to grow only one polymorph, several factors need to be considered, such as additives which promote (or inhibit) the growth of one form. For instance, acids or bases such as acetic acid or ammonia promote the growth of γ-glycine, whereas ethanol stimulates the growth of β-glycine. Nevertheless, presence of seeds appears to be much more important than the presence or absence of additives. Our experiments have shown that with traces of γ-glycine seeds, this polymorph can be crystallized even from pure solutions, whereas pure samples of α-glycine (without any γ-glycine) gave only the α-form, even if the said acids or bases were added. β-Glycine is even harder to grow purely. It was found that using seeds of this form taken from mixtures helps to obtain a higher yield. Boldyreva (2008) reported that spontaneous

[3] This statement applies for physical and chemical research.

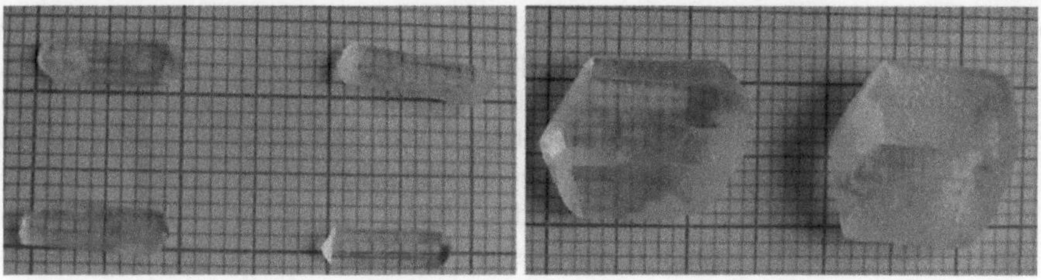

Fig. 2.5 *Crystals of glycine*. Grown crystals of α- and γ-glycine (Reprinted from Sekar and Parimaladevi (2009), with permission from Elsevier)

antisolvent precipitation with acetone from a solution of acetic acid yields the pure β-polymorph, but only if the solution has been left for about 3 days.

We have found that the source of the chemicals makes also a difference (especially for the growth of α- and γ-glycine), as the ratio of these polymorphs varies unpredictably.[4] These subtleties might be responsible for the fact that some experimental preparations following the literature do not always lead to reproducible results. Somehow frustratingly, despite the vast amount of research on crystal growth of these polymorphs, the knowledge is still not complete and the control of the outcome of crystallization is still not perfectly understood (Surovtsev et al. 2012).

Besides the work on growth and polymorphs, a lot of physical data has been published on glycine crystals, such as works on thermal, mechanical, or optical properties.[5] Nevertheless, these reports only convey superficial data, such as thermal stability studies, hardness assessments, or semiquantitative SHG tests, although large single crystals are available (for α- and γ-glycine, Fig. 2.5). Profound measurements of single crystal physical properties have not been reported, although there is one study on the production of polar β-glycine fibers and their nonlinear

[4] It was found that even samples from the same company do not give the same result.

[5] It has to be noted that the number of publications does not match the amount of original work (as – sadly – it frequently happens in science generally). Several papers repeat results of earlier works, in some cases even without citing the older reports. This unfortunate circumstance is commented in a few noteworthy examples only in this book.

Moreover, one has to bear in mind that many papers published on crystalline amino acids and their compounds (not only glycine) are not only repetitive but also exaggerating: A great number of papers reporting "optical," "thermal," or "mechanical" properties of amino acids or their compounds are found, but in most cases these deal with the simplest properties only. For instance, "thermal" properties often refer to investigations of the stability range via TG and DSC, but more subtle properties such as heat capacities, thermal conductivities, thermal expansion, etc., are but rarely found. This problem is even more prominent in the case of "optical" data, especially when referring to "nonlinear optical" crystals (see several instances given in the following chapters). In most cases, results from UV–vis spectra plus a SHG test are employed to herald the discovery of a "novel NLO crystal."

Concluding, one has to bear in mind that the very frequent publications of new papers on optical, thermal, and mechanical studies of novel amino acid crystals are mostly good summaries of superficial data, but rarely more.

optical and piezoelectric properties (Isakov et al. 2011). In this report, the authors present fibers of polyvinyl alcohol approximately 330 nm thick, containing β-glycine crystals. Embedded in this matrix, β-glycine is stable, and the elongated crystals are oriented with the *b*-axis parallel the fiber (this direction also represents the long axis of the needlelike β-glycine crystals). Due to this arrangement, the values of the piezoelectric tensor could not be measured (as would have been possible for samples prepared from large single crystals); nevertheless, the piezoelectric effect along the fibers could be determined. It was found to be exceptionally large ($d_{eff} = 12.5$ pm/V), which corresponds mostly to the tensor coefficient representing the longitudinal effect parallel b (i.e., d_{222}), but other coefficients contribute as well (as the crystals are not perfectly aligned to the fiber axis). This study is an impressive example of possible utilization of amino acid crystals for physical applications, although among amino acid salts there are several examples of even more impressive results (see Chaps. 3 and 4).

Moreover, there are studies of the thermal behavior of α-glycine. Although crystallographic studies are usually conducted at ambient conditions, low-temperature measurements are relatively frequent, as the decreasing thermal motion allows higher precision in the determination of the relevant parameters.[6] For α-glycine, a study at 120 K has been reported by Legros and Kvick (1980), and more recently a paper using neutron diffraction data over the temperature range between 288 and 427 K has been published (Langan et al. 2002). As the determination of the unit cell parameter of a certain range allows the assessment of the anisotropy of thermal expansion, it has been found that this effect is strongest along the *b*-axis (Fig. 2.6). This relates to the fact that structure is layered parallel (010), with only van der Waals interactions between the layers.

2.3.2 *Alanine*

Among the chiral amino acids, alanine is the simplest, with a methyl group as side chain (Fig. 2.7).[7] As stated in the Introduction, the prevalent form in nature is L-alanine (albeit the D-form has been discovered in some instances, as, e.g., in bacterial cell walls). L-Alanine was first synthesized in 1850 and originally isolated in 1875 from silk, ovalbumin, and silk fibroin (Greenstein et al. 1984, pp. 1819–1840), which is a very rich source of this particular amino acid, comprising approximately 30 %.

[6] In addition, the possibility of phase transitions often encourages researchers to conduct crystallographic experiments at lower temperatures.

[7] This relates to α-alanine, as explained in Chap. 1. Except for glycine, where the Greek letters are employed to describe polymorphs, the letter α denotes the fact that the second carbon atom carries the amino group. It has to be emphasized that the β-form of alanine represents a different compound (see p. 64) and must not be confused with α-alanine. β-Alanine does not occur in proteins but is found in two peptides, namely, carnosine and anserine, which are constituents of mammalian muscle. Furthermore, β-alanine is part of the vitamin pantothenic acid.

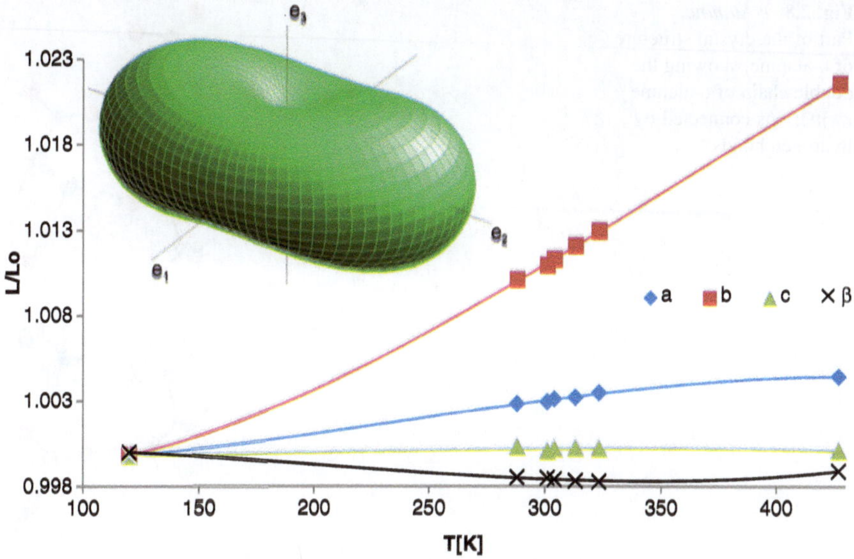

Fig. 2.6 *Thermal expansion of α-glycine*, calculated from the data from Legros and Kvick (1980) and Langan et al. (2002). Regression curves have been fitted by third-order polynomial functions. L is the lattice constant a or b or c or angle ß at T, L_0 is the same lattice constant or angle at 120 K. *Inset*: Representation of the tensor of thermal expansion, calculated from the first derivatives of the polynomial functions (Fleck et al. 2012). Note the strong expansion of b, which is due to the weak interactions between molecular layers parallel (010)

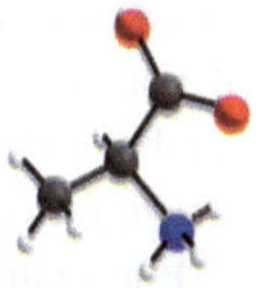

Fig. 2.7 *L-Alanine*. The molecular structure of L-alanine in its zwitterionic state

This high percentage of alanine is in fact responsible for several properties of silk: Alanine occurs in highly repetitive sections of the protein string, which folds to beta-sheets and is thus arranged in compact blocks ("crystalline domains"), with a glycine-rich matrix connecting the blocks. The interaction between the alanine-rich crystalline segments and the elastic, amorphous matrix is responsible for the high tensile strength of silk. Apart from this, the low reactivity of alanine does not allow any stronger interactions between adjacent protein chains, which makes silk flexible and soft (van Beek et al. 2002).

Moreover, L-alanine (as glycine) belongs to the group of nonessential amino acids for human beings (as well as other mammals and birds), i.e., it can be metabolized, in this case from pyruvic acid.

The name *alanine* was given by Strecker (1850) as a derivative from the first letters of the word aldehyde, combined with the word parts "an" (for alkane) and "in" (as conventional ending for amino acids).

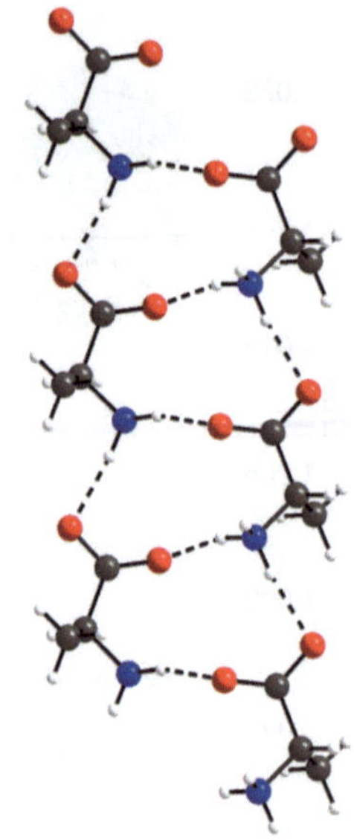

Fig. 2.8 *L-Alanine.* Part of the crystal structure of L-alanine, showing the double chain of L-alanine zwitterions connected by hydrogen bonds

The crystal structure of L-alanine was studied first by Simpson and Marsh (1966) and further refined by Lehmann et al. (1972a). It was discovered that the amino acids connect via N···H–O bonds in typical head-to-tail fashion, to form polar double chains (Fig. 2.8).

As alanine is a chiral amino acid, there are two enantiomers, which occur separately in nature, as they are produced from different biochemical paths from chiral precursors. When synthesized from non-chiral precursors, a racemic mixture of both forms occurs. This mixture can also be crystallized, as the resulting solid can be considered as a cocrystal, as the D- and L-forms occur in a definite stoichiometric ratio, i.e., 1:1.

For many amino acids, DL-crystals have been grown and characterized (see Tables 2.2, 2.3, and 2.4), and many salts of amino acids racemates are known as well (see the following chapters). As alanine is the simplest amino acid which can occur in this form, the crystal structure of the DL-form is discussed here as a representation of DL-amino acid crystals in general.[8]

When investigating the crystal structure of DL-alanine (first reported by Levy and Corey 1941, then refined by Donohue 1950, and Subha Nandhini et al. 2001), it becomes apparent that the D- and L-molecules are arranged symmetrically, connected via glide planes. Consequently, there is only one crystallographically independent molecule in the unit cell. Although this statement is true for many organic crystals, the point here is that the symmetry elements relate molecules of opposite chirality.

[8] The term "DL-alanine" or more generally "DL-amino acid" is not a very good one, as there is no compound "DL-alanine," rather a mixture of D- and L-alanine in solution or solid form. Nevertheless, these terms are used, and bearing in mind that these are not compounds but mixtures, we retain the use of these names in this work.

Table 2.3 *Crystallographic properties of standard amino acids, part II*. Space groups and unit cell parameters of phenylalanine, tryptophan, serine, threonine, and cysteine crystals

	s.g.	a	b	c	α	β	γ	Reference
L-Phe I	*C*2	8.804	6.041	31.564	90	96.6	90	Mahalakshmi et al. (2006)
L-Phe I	*P*2	8.778	6.065	31.597	90	96.78	90	King et al. (2012)
L-Phe II	*P*2	12.063	5.412	13.676	90	99.60	90	Williams et al. (2013)
L-Phe·0.5H_2O	$P2_1$	12.111	5.421	13.770	90	100.01	90	Williams et al. (2013)
L-Phe·H_2O	$P2_1$	13.008	5.430	13.940	90	101.11	90	Williams et al. (2013)
DL-Phe	*P*2[a]	12.93	10.03	6.42	90	99.12	90	Khawas (1971)
L-Trp	*P*1	11.430	11.464	35.606	84.42	87.69	60.10	Görbitz et al. (2012)
D-Trp	$P2_1$	18.387	16.31	6.579	90	95.95	90	Khawas (1985)
DL-Trp (173 K)	$P2_1/c$	18.899	5.745	9.309	90	101.78	90	Hübschle et al. (2004)
L-Ser	$P2_12_12_1$	8.571	9.325	5.615	90	90	90	Benedetti et al. (1973b)
L-Ser·H_2O	$P2_12_12_1$	9.365	12.239	4.835	90	90	90	Frey et al. (1973a, b)
DL-Ser	$P2_1/a$	10.72	9.14	4.825	90	106.45	90	Shoemaker et al. (1953)
L-Cys I	$P2_12_12_1$	8.116	12.185	5.426	90	90	90	Kerr and Ashmore (1973)
L-Cys II	$P2_1$	11.51	5.224	9.52	90	109.8	90	Harding and Long (1968)
L-Cys III	$P2_12_12_1$	7.998	10.288	5.284	90	90	90	Moggach et al. (2006b)
L-Cys IV	$P2_1$	8.105	5.4420	10.916	90	94.90	90	Moggach et al. (2006b)
L-Cys[a]	*P*2[a]	13.26	9.65	7.14	90	96.11	90	Khawas (1971)
DL-Cys I	$P2_1/a$	9.877	4.737	12.877	90	112.04	90	Luger and Weber (1999)
DL-Cys II	$P2_1/a$	9.717	4.987	13.188	90	121.80	90	Minkov et al. (2009)
L-Thr (12 K)	$P2_12_12_1$	13.628	7.618	5.110	90	90	90	Janczak et al. (1997)
L-*allo*-Thr	$P2_12_12_1$	13.69	7.68	5.17	90	90	90	Swaminathan and Srinivasan (1975)

[a]As stated before, this paper does not contain a structural analysis but only assessment of XRD powder patterns; thus, the values and especially the space groups given here are dubious, as no inversion symmetry is present

This has two noteworthy consequences: Firstly, DL-amino acid crystal structures possess symmetry elements which invert the chirality (mirror planes, glide planes, inversion centers) – these elements are impossible in pure D- or L-crystals. Secondly, a representation of the asymmetric unit of the structure is misleading as only one

Table 2.4 *Crystallographic properties of serine crystals*. For non-ambient conditions, pressure or temperature values are noted. Notice that different authors have chosen different cell settings

	s.g.	a	b	c	α	β	γ	Reference
L-Ser	$P2_12_12_1$	8.571	9.325	5.615	90	90	90	Benedetti et al. (1973b)
L-Ser	$P2_12_12_1$	8.599	9.348	5.618	90	90	90	Kistenmacher et al. (1974)
L-Ser (4.0 GPa)	$P2_12_12_1$	5.489	8.339	8.632	90	90	90	Zakharov et al. (2012)
L-Ser (4.2 GPa)	$P2_12_12_1$	5.473	8.309	8.637	90	90	90	Boldyreva et al. (2006a)
L-Ser (4.1 GPa)	$P2_12_12_1$	8.327	8.665	5.485	90	90	90	Moggach et al. (2005a)
L-Ser (4.8 GPa)	$P2_12_12_1$	8.298	8.600	5.466	90	90	90	Moggach et al. (2005a)
L-Ser II (5.2 GPa)	$P2_12_12_1$	6.870	9.640	5.606	90	90	90	Moggach et al. (2006a)
L-Ser II (7.3 GPa)	$P2_12_12_1$	6.874	9.582	5.562	90	90	90	Moggach et al. (2006a)
L-Ser III (8.1 GPa)	$P2_12_12_1$	6.557	9.540	5.609	90	90	90	Moggach et al. (2006a)
L-Ser·H_2O	$P2_12_12_1$	9.365	12.239	4.835	90	90	90	Frey et al. (1973a, b)
L-Ser·H_2O (97 K)	$P2_12_12_1$	4.799	9.306	12.136	90	90	90	King et al. (2010)
DL-Ser	$P2_1/a$	10.72	9.14	4.825	90	106.45	90	Shoemaker et al. (1953)
DL-Ser	$P2_1/a$	10.719	9.136	4.833	90	106.43	90	Frey et al. (1973a, b)
DL-Ser	$P2_1/a$	10.739	9.149	4.830	90	106.42	90	Kistenmacher et al. (1974)
DL-Ser (100 K)	$P2_1/a$	10.756	9.172	4.788	90	106.73	90	Flaig et al. (2002)
DL-Ser	$P2_1/a$	10.736	9.146	4.830	90	106.46	90	Dittrich et al. (2005)
DL-Ser (20 K)	$P2_1/a$	10.776	9.194	4.779	90	106.87	90	Mondal et al. (2012)
DL-Ser (0.1 MPa)	$P2_1/n$	4.832	9.154	10.460	90	99.85	90	Boldyreva et al. (2006b)
DL-Ser (0.2 GPa)	$P2_1/n$	4.843	9.026	10.319	90	100.86	90	Zakharov et al. (2012)
DL-Ser·"$2H_2O$"[a]	$P2_1/n$	4.831	9.150	10.444	90	99.97	90	Chakraborty et al. (2001)

[a]Note the unit cell of this "dihydrate" in comparison with that of anhydrous form (see text for details)

enantiomer is shown – this might lead to the incorrect conclusion that one is viewing a chiral species.

Nevertheless, polar symmetries are possible and often found in DL-amino acid crystals,[9] as in the case of DL-alanine, which crystallizes in the orthorhombic space group $Pna2_1$. A part of the crystal structure of DL-alanine is shown in Fig. 2.9.

[9] A paper by Khawas (1971) reports space group *P*2 for DL-phenylalanine and *P*222 for DL-lysine, but as these results have been derived from visual interpretation of XRD powder patterns, these space groups are dubious.

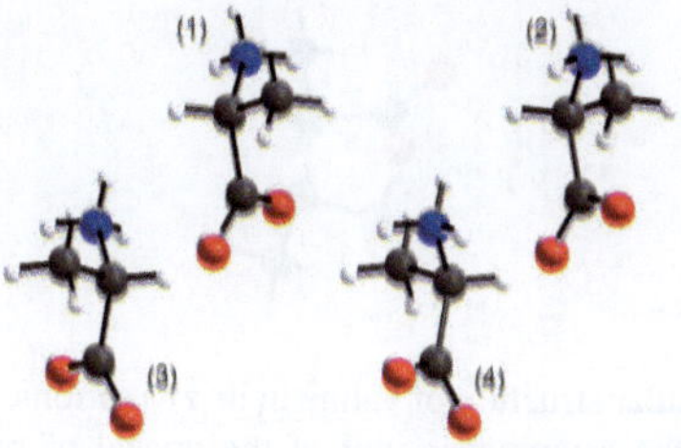

Fig. 2.9 *DL-Alanine*. Part of the crystal structure of LD-alanine, viewed along (010). The a glide plane lies in the plane of projection, with the glide vector oriented horizontally, connecting the molecules *1* and *2* with each other (as well as *3* and *4*). Note the opposing chirality of the molecules *1* and *2* (as well as *3* and *4*). Compare with Fig. 2.8, where all molecules have the same chirality

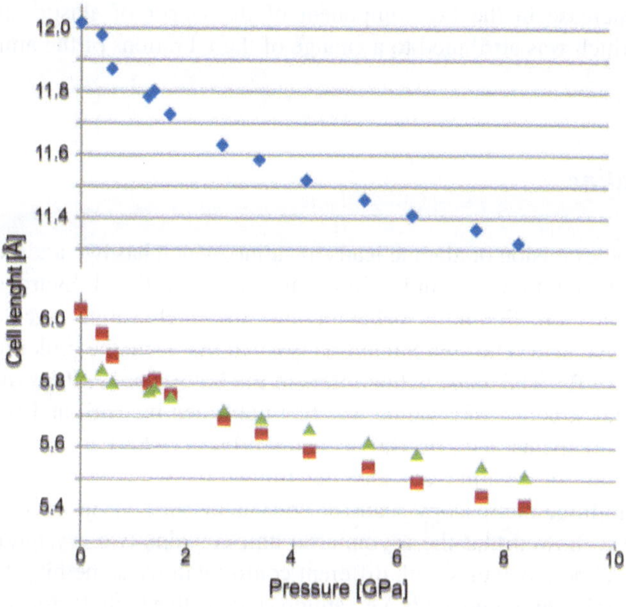

Fig. 2.10 *Unit cell dimensions of DL-alanine at high pressures*. The decrease of the cell lengths a, b, c (shown as diamonds, squares and triangles) as a function of pressure is shown (Data from Tumanov and Boldyreva 2012)

In addition to the studies at ambient conditions, the behavior of DL-alanine at high pressure has been investigated. Although first reports proposed phase transitions at 1.7–2.3 GPa as well as at 6.0–7.3 GPa (Belo et al. 2010), a consecutive study has shown that this is not the case: The unit cell axes' lengths decrease, and cell parameters c and b become equal at about 1.5–2.0 GPa, but the symmetry does not change, as has been shown by powder XRD, single crystal XRD, and Raman spectroscopy (Tumanov and Boldyreva 2012; Fig. 2.10).

Apart from the structural studies, growth of L-alanine crystals was studied. Lechuga-Ballesteros and Rodriguez-Hornedo (1993) reported data on crystal growth, morphology, and the influence of additives. Moreover, the crystal physics

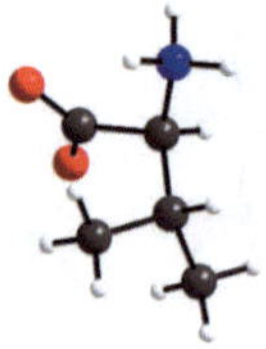

Fig. 2.11 *Valine*. The molecular structure of valine in its zwitterionic state, here in L-configuration. These molecules constitute the asymmetric unit of the crystal of DL-valine; the D-enantiomer is related by glide plane symmetry

of L-alanine were investigated in detail (Tylczyński et al. 2011): The piezoelectric, elastic, and dielectric tensors of L-alanine were measured in the temperature range of 100 to 300 K. L-Alanine was found to be weakly piezoelectric; moreover, a jumpwise increase in the c_{55} component of the tensor of elastic stiffness was observed, which was attributed to a change of the vibrations of the amino group.

2.3.3 *Valine*

Formally, the extension of alanine leads to valine, which has two additional methyl groups compared to the simpler molecule. The additional hydrocarbon parts account for a more lipophilic behavior, due to which valine is classified as a nonpolar amino acid. The compound was isolated first from the milk protein casein by Fischer (1901). The name valine refers to the Latin word *validus* (healthy).

The crystal structure of L-valine was first published by Torii and Iitaka (1970), with later refinements, e.g., by Dalhus and Görbitz (1996b) at 120 K. (Later, the crystal structure of D-valine was also determined by Wang et al. 2002, who also made atomic force microscopy scans of crystals of both enantiomers.) The structural analysis showed that the asymmetric unit contains two crystallographically different L-valine molecules with different conformations, something that is found frequently for larger amino acids, i.e., amino acids with a high degree of freedom in their conformation.

In contrast, the crystal of DL-valine (Mallikarjunan and Rao 1969) contains only one molecule in the asymmetric unit; the other enantiomer is connected via a *c* glide plane. The molecular structure of valine is shown in Fig. 2.11.

In addition to the above papers, Ramachandran and Natarajan (2009) have made studies on solubility and crystal growth of L-valine in silica gel.

Moreover, Dalhus and Görbitz (1996a) found another polymorph of DL-valine when they investigated the crystal at 120 K. They discovered that the crystal which is monoclinic at ambient conditions undergoes a phase transition to a slightly less symmetric, triclinic form (space group *P*-1) upon cooling (thermal data to determine the transition temperature are not available to date). The crystal packing arrangements in both polymorphs is closely related: A transformation of the triclinic *c*-axis into a twofold screw axis (with approximate doubling of the cell length) leads to the monoclinic polymorph.

2.3.4 Leucine

Compared to valine, leucine is slightly more hydrophobic as it contains an additional CH_2 group within the chain. Leucine was discovered along with glycine by Braconnot (1820) and was named for the white color of the powder (from Greek *leukos* for white) – not very originally, as amino acids generally are white when powdered. L-Leucine is an essential amino acid for animals and must be ingested; the biosynthesis in plants is carried out in several stages starting from pyruvic acid, along with L-isoleucine and L-valine (Lehninger et al. 2000).[10]

The crystal structure of L-leucine was first determined by Harding and Howieson (1976), then refined by Coll et al. (1986), and later redetermined at 120 K by Görbitz and Dalhus (1996a). As found for L-valine, the crystal structure of L-leucine comprises two independent molecules in the asymmetric unit. These molecules are oriented toward each other with their hydrophilic ends, while the hydrophobic ends face outward. Multiple hydrogen bonds extending from the amino groups toward the carboxylate oxygen atoms connect the molecules to infinite layers parallel (001), as shown in Fig. 2.12 (top image). Neighboring layers are connected toward each other via van der Waals interactions. This arrangement also reminds of that of L-valine.

The same structural pattern was also found in the crystal structure of DL-leucine by DiBlasio et al. (1975) (Fig. 2.12, bottom). When viewed along the *b*-axis, the structures appear nearly identical. However, in DL-leucine, one-half of the double layers comprise D-leucine molecules, the other half L-leucine molecules, thus accounting for the inversion symmetry, expressed by inversion centers between opposite molecules. As found for DL-valine, there is only independent molecule in the asymmetric unit.

The similarity of the structures is also expressed by the unit cell dimensions of both crystals (compare Table 2.2): The double-layer widths – *c* in L-leucine and *a* in DL-leucine – are about the same, as are the lengths of *b* (the distances between adjacent molecules in viewing direction in Fig. 2.12). The length of *a* in L-leucine is about twice the corresponding length *c* in DL-leucine, since L-leucine comprises two independent molecules in slightly different conformation in the asymmetric unit.

Yamashita et al. (2007) found another polymorph of L-leucine at 400 MPa, in which the general arrangement is the same as in the low-pressure phase, but the molecules assume a slightly different conformation, resulting in denser packing (as is expected) and higher symmetry: The phase is also monoclinic, but crystallizes in space group *C*2, with all molecules identical (i.e., only one independent molecule in the asymmetric unit). Further work on L-Leucine crystals under high pressure and high temperature conditions has been published by Façanha Filho et al. (2008, 2009).

[10] An accumulation of L-leucine, L-isoleucine, and L-valine in human blood and urine has a massive effect on the metabolism and is fatal if untreated. There is a genetic deficiency named maple syrup urine disease (thus termed for the smell of the urine) where these so-called branched-chain amino acids cannot be catabolized.

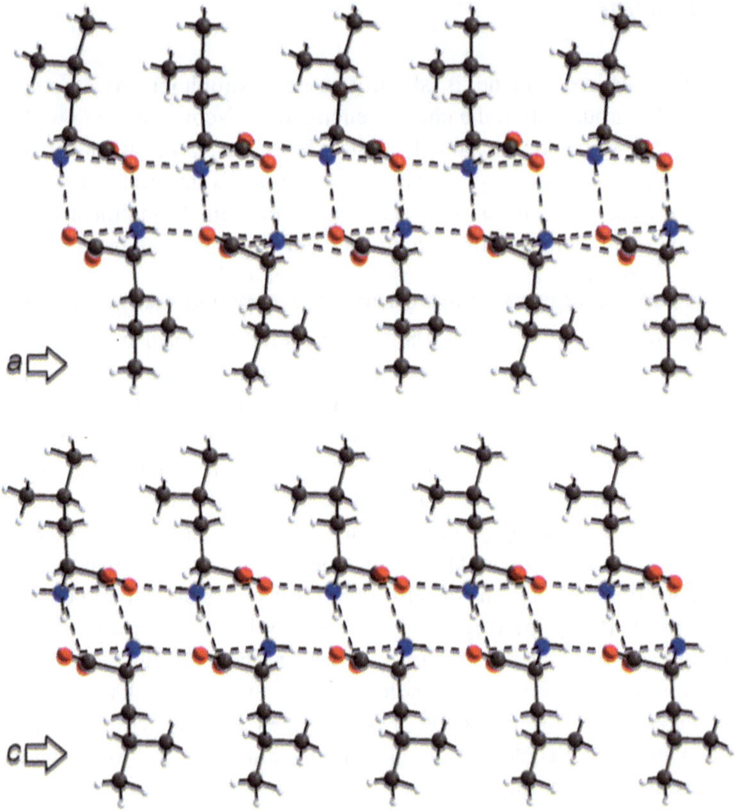

Fig. 2.12 *Leucine*. The crystal structure of L-leucine (*top*) and DL-leucine (*bottom*), viewed along [010]. Note the layered structures, with the hydrophobic chains facing away from the central hydrophilic part, which accounts for the connection of the molecules. The basic topology appears similar in this projection, but the stacking of the molecules along the depth of the viewing direction is different (noticeable in the similarity of two of three unit cell parameters; see Table 2.2). Concerning the molecular structure, the fundamental difference between the two crystals is the configuration of the molecules (in DL-leucine, the top half is the L-form, the bottom half the D-form)

Additional studies by Iwama et al. (2010) have observed the phenomenon of preferential enrichment for leucine (as well as alanine), which refers to a symmetry-breaking enantiomeric resolution phenomenon. Precisely, they observed a solid-to-solid transformation of a metastable polymorphic form into a thermodynamically stable one in the course of crystallization from a solution supersaturated with the racemate. During this polymorphic transition, the crystals are partially disintegrated, releasing the excess enantiomer into the solution. The crystals end up enriched with the opposing enantiomer. This process works without the aid of any other external chiral compound (as is used for chiral resolution).

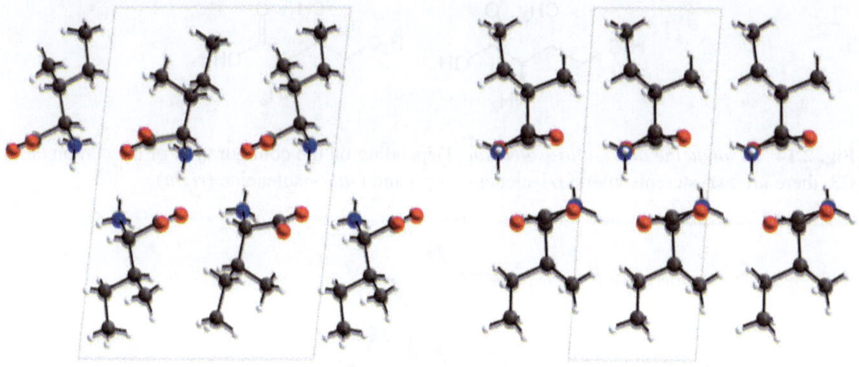

Fig. 2.13 *Isoleucine*. The crystal structure of L-isoleucine (left) and DL-isoleucine (*bottom*), viewed along [010]. Compare with the structures of the leucine crystals in Fig. 2.12, which shows the similarity. The layered arrangement is also visible here, with the hydrophobic chains facing away from the central hydrophilic layer

2.3.5 Isoleucine

As suggested by the name, isoleucine is an isomer of leucine, where one methyl group bonds to C4 instead of C5. L-Isoleucine was discovered in 1904 by the German biochemist Felix Ehrlich in beet sugar molasses. When comparing the unit cell parameters of L-leucine and L-isoleucine (as well as those of DL-leucine and DL-isoleucine; see Table 2.2), one finds that the parameters are similar. Clearly, this reflects the fact that the structures are closely related.

L-Isoleucine crystallizes in space group $P2_1$ as does L-leucine; the structure published by Torii and Iitaka in 1979 (and later redetermined at 120 K by Görbitz and Dalhus 1996b) shows the same arrangement of units: bilayers of isoleucine molecules facing toward each other with the hydrophilic parts (Fig. 2.13). The same phenomenon is found for DL-isoleucine (Benedetti et al. 1973a), as can also be seen in Fig. 2.13 when compared with Fig. 2.12.

It has to be noted that isoleucine has – in contrast to leucine – one additional chiral center at the carbon atom C3 (a feature shared only with threonine; see Sect. 2.3.11). Precisely, only stereoisomer with configuration L/L (or *S*/*S*; see Chap. 1) is referred to as isoleucine (the systematic name is (2*S*, 3*S*)-2-amino-3-methylpentanoic acid). The stereoisomer (2*S*, 3*R*)-2-amino-3-methylpentanoic acid is called L-*allo*-isoleucine (Fig. 2.14). As these molecules are very similar, mixtures containing both isomers can be used to grow crystals containing isoleucine and *allo*-isoleucine. Dalhus and Görbitz (2000) have reported the structure of a crystal grown from a 1:1 mixture of both racemates, i.e., containing four isomers (D- and L-isoleucine as well as D- and L-*allo*-isoleucine) in systematic distribution.

Givand et al. (1998) published a study on the morphology of L-isoleucine crystals, in which they found that the morphology can be predicted by means of molecular mechanics calculations. The calculations were validated against experimentally grown crystals and were found to be of high accuracy.

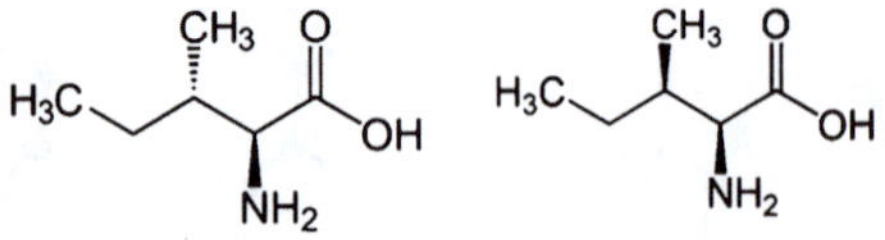

Fig. 2.14 *L-Isoleucine and L-allo-isoleucine*. Depending on the configuration of the carbon atom C3, there are two stereoisomers: L-isoleucine (*left*) and L-*allo*-isoleucine (*right*)

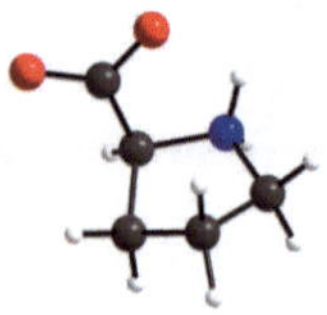

Fig. 2.15 *L-Proline*. Of the standard amino acids, proline is the only example which is a secondary amine (the amine nitrogen is bound to not one but two alkyl groups)

2.3.6 *Proline*

Of all the standard amino acids, proline is unique insofar as the amine nitrogen is part of a five-membered ring, i.e., it is a secondary amine. L-Proline was first isolated by Emil Fischer in 1901 out of casein, who also invented the name (as a derivative of pyrrolidine).

The first crystallographic study of L-proline (Fig. 2.15) was published by Wright and Cole in 1949, who proposed an anhydrous structure, which was confirmed by Kayushina and Vainshtein (1965). Later, a monohydrate was discovered by Janczak and Luger (1997), who studied the structure at 120 K.[11]

The situation is similar for the racemate: DL-Proline also exists as anhydrous crystal and as monohydrate. In contrast to L-proline, the crystal structure of anhydrous form was published later (Myung et al. 2005) than that of the monohydrate (Padmanabhan et al. 1995). In both cases, there is only one molecule in the asymmetric unit, which is related to the other enantiomer via glide planes (space group $P2_1/c$ for DL-proline and *Pbca* for DL-proline monohydrate). Another similarity is the arrangement of proline molecules in both crystals: The five-membered rings are oriented more or less parallel, within the (010) and (100) planes for DL-proline and DL-proline monohydrate, respectively. A detailed comparison of the four polymorphs of proline is given by Seijas et al. (2010).

Although not found in any of the four proline crystals discussed here, there is the possibility of disorder within the ring. In particular, the NH_2^+ group adopts the same geometric form as the CH_2 group, although the typical C–N distance is shorter than

[11] Sakthivel and Anbarasan (2011) published a paper on "structural and optical studies of L-proline hydrate crystals," although the unit cell they give corresponds to the anhydrous form. As no structural study has been performed in this work (despite the title), it seems the authors have mistaken the phase.

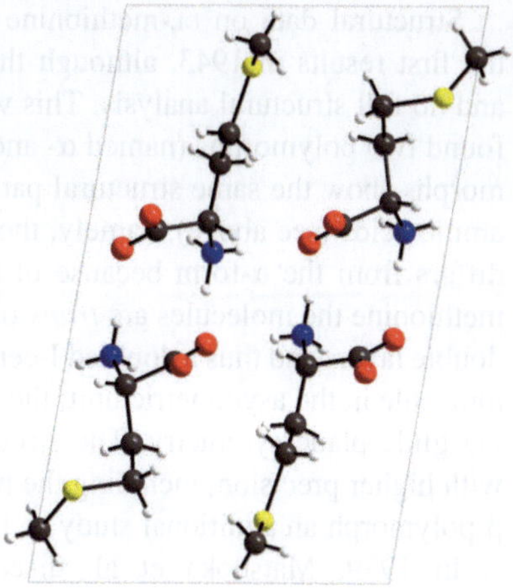

Fig. 2.16 L-*Methionine*. Note the packing of the molecules in double layers, which is also found in crystals of leucine and isoleucine (compare with Figs. 2.10 and 2.11)

the C–C distance (1.47 Å compared to 1.53 Å). Nevertheless, the topology is similar enough that disorder between the two groups is possible, especially if hydrogen bond acceptor exists. This has been found in several salts of proline (see Chaps. 3 and 4), and it is conceivable that it exists in even more, but has been neglected in the refinements.

2.3.7 *Methionine*

Of all standard amino acids, there are only two which contain sulfur (besides the ubiquitous C, H, O, and N), namely, methionine and cysteine. Whereas the thiol group in cysteine is relatively active and therefore prone to reactions (see Sect. 2.3.12), the thioether group is relatively inert. Methionine is another essential amino acid for humans and was first isolated in 1922 from casein (Mueller 1922). The name derives from the groups present in the molecule, i.e., methyl and thio.

The first report on the crystal structure of L-methionine appeared in 1970 (Khawas 1970), but this was merely the analysis of Debye–Scherrer pattern and proposed unit cell dimensions only. The values given here could not be confirmed later and thus appear dubious.

The structure of L-methionine (Torii and Iitaka 1973) is similar to that of L-leucine insofar as there are two independent L-methionine molecules in the asymmetric unit. Moreover, the orientation of the molecules reminds of the arrangement in L-leucine: The hydrophilic ends of the molecules are oriented toward each other, forming a double-layer parallel (001) (Fig. 2.16). As said above, this structural pattern appears in crystals of L-valine, L-leucine, DL-leucine, L-isoleucine, and DL-isoleucine, as well as the α- and β-forms of DL-methionine (see below). This relation is also expressed by the similarity of the unit cell dimensions (Table 2.2).

Structural data on DL-methionine crystals are older: Albrecht et al. published the first results in 1943, although this work only contained unit cell dimensions and no full structural analysis. This was provided 9 years later by Mathieson, who found two polymorphs (named α- and β-DL-methionine; see Table 2.2). Both polymorphs show the same structural pattern as was found for L-methionine and other amino acids (see above), namely, the double-layer arrangement. The β-polymorph differs from the α-form because of the conformation of the molecules: In α-DL-methionine the molecules are *trans*-oriented, resulting in a different stacking of the double layers and thus a doubled I-centered unit cell. In both cases, there is only one molecule in the asymmetric unit; the D- and L-enantiomers are related to each other via glide plane symmetry. The structures of both polymorphs were refined later with higher precision, including the hydrogen atoms (Taniguchi et al. 1980; for the β-polymorph an additional study at 105 K was reported, Alagar et al. 2005).

In 1999, Matsuoka et al. discovered another polymorph with a pseudo-orthorhombic unit cell, depending on the addition of several different acids to the solution and the crystallization temperature, although no explanation of the role of the acid was provided.

2.3.8 Phenylalanine

Phenylalanine was first reported by Schulze and Barbieri in 1879, when they discovered a new compound in seedlings of yellow lupine (Thorpe 1913, pp. 191–193). It turned out to be an amino acid with a phenyl group, hence the name (phenylalanine is the only amino acid of the standard 20 bearing a semi-trivial name).

The crystal structure of the optically active form of phenylalanine was not easily solved. A first study of the D-polymorph was reported by Weissbuch et al. (1990), who reported that the amino acid crystallized in the monoclinic space group *C*2 (Fig. 2.17). This was confirmed by Mahalakshmi et al. (2006) for the other enantiomer, but the data were not unambiguous as there were doubts if the symmetry was correctly established (the growth of suitable crystals was and still is problematic; the R-value given by Weissbuch et al. 1990 exceeds 14 %, and the work of Mahalakshmi et al. 2006 contains no thorough investigation on this, only a brief analysis of a powder XRD pattern). Recently King et al. (2012) calculated and modeled solutions for the other space group that was conceivable, i.e., *P*2, in comparison with a model for *C*2. This study proposed that the primitive space group is the correct one.

In addition (and apparently without knowledge of the work of King et al. 2012), a detailed study working with powder XRD (Williams et al. 2013) yielded additional data by establishing that there are several crystalline phases of optically active phenylalanine, their stability determined by humidity. These phases encompass two polymorphs of anhydrous L-phenylalanine as well as a hemihydrate and a monohydrate (Table 2.3).

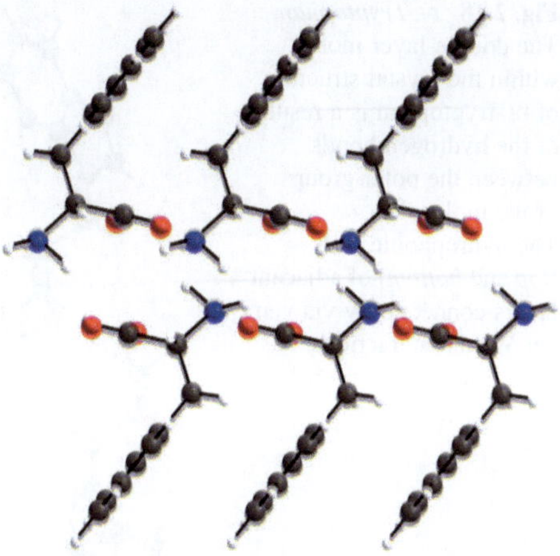

Fig. 2.17 *Phenylalanine*. Part of the crystal structure of D-phenylalanine. The molecules are oriented in bilayers, with the hydrophilic heads oriented toward each other. This structure was reported in space group C2 (as was the structure of the L-enantiomer) but has later been found to actually have a lower symmetry, namely, P2 (see text)

Although the determination of the crystal structure of the racemate is much older than that of the optically active form, the data are also not unambiguous. Khawas (1971) published a paper on the X-ray study of DL-phenylalanine and reported the space group *P*2 for this species. However, as this result comes only from powder XRD data, the space group is highly dubious (in all DL-amino acid crystals whose structures have been determined, inversion symmetry is found).

2.3.9 *Tryptophan*

Tryptophan, the largest of the standard 20, is one more amino acid which was isolated from casein (Hopkins and Cole 1901). It is another essential amino acid and is involved in the biosynthesis of serotonin, which acts as neurotransmitter in the human brain in the regulation of sleep (in the 1980s, tryptophan was used as a sleeping aid before being banned after some health problems). The fact that a rich turkey dinner or a glass of milk before bedtime makes people tired has often been related to the relatively high tryptophan content, but decisive evidence for this is not available.[12] The name tryptophan derives from this property (Greek *tryein*, to wear down; *phanein*, to appear).

In contrast to many other standard amino acids, tryptophan does not crystallize readily; therefore, the growth of crystals and, consequently, the determination of the crystal structure of both the pure enantiomer and the racemate proved difficult. In fact, the crystal structure of L-tryptophan has eluded determination longer than that of most other amino acids (except L-lysine (see Sect. 2.3.19) and nearly simultaneously as L-arginine (see Sect. 2.3.20)); it was determined only a few

[12] There are studies showing that pure L-tryptophan does increase sleepiness (compare Hartmann (1986) and references therein).

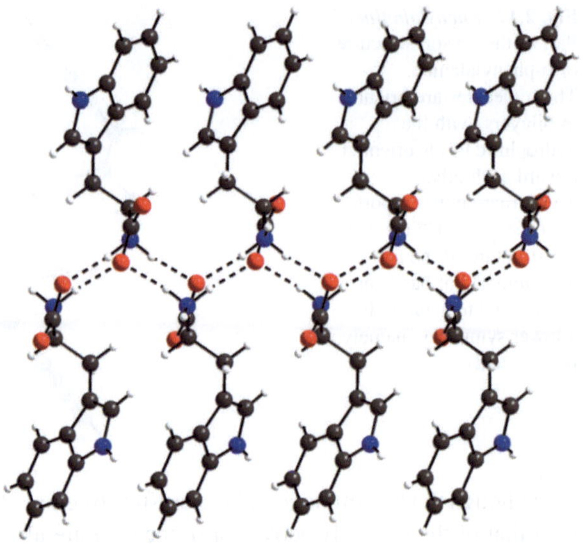

Fig. 2.18 *DL-Tryptophan*. The double-layer motif within the crystal structure of DL-tryptophan is a result of the hydrogen bonds between the polar groups of the molecules. The hydrophobic ends (*top* and *bottom*) of adjacent layers connect only via van der Waals interactions

years ago (Görbitz et al. 2012). The crystal structure is a complicated one, with 16 molecules in the asymmetric unit (i.e., the unit cell as the space group is $P1$). Half of these molecules assume *gauche* conformation, the other half *trans*-conformation, and all molecules of each group have nearly identical geometries. The hydrogen bonding scheme is similar to that found in the racemate, which was structurally characterized earlier (see below). However, the distances are longer, and, therefore, the structure is less densely packed.

The structure of DL-tryptophan (Fig. 2.18) was first determined earlier (Bakke and Mostad 1980), but as tryptophan does not crystallize easily, this experiment did not yield very good data (more than 70 % of the reflections were not observed; thus, no hydrogen atom positions could be determined). Rather recently, Hübschle et al. (2004) performed a study at 173 K where all crystallographic parameters could be determined with satisfactory precision. As found for many of the above DL-amino acid crystals, the asymmetric unit of DL-tryptophan contains only one molecule (assuming an unusual conformation in comparison with other tryptophan compounds), symmetry related to the other enantiomer via an inversion center (thus forming dimers, connected by N–H···O hydrogen bonds). Further N–H···O bonds connect the dimers to a head-to-head double layer. The double layers are connected by weak van der Waals interactions only, thus explaining the reluctance to crystallization.

2.3.10 Serine

Serine is a small polar amino acid (formally alanine with a hydroxyl group, Fig. 2.19) which was first obtained from the silk protein in 1965, which has a high content of this amino acid. The name derives from the Latin word for silk, *sericum*. Although L-serine is the typical natural form as constituent of proteins, D-serine also exists in

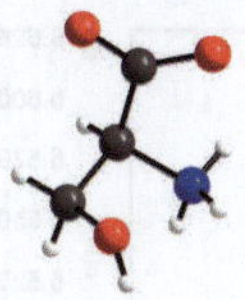

Fig. 2.19 *L-Serine*. Serine is the smallest polar chiral amino acid, formally an alanine with an additional OH group

living beings, particularly in brain tissue. A recent study has found indications that this enantiomer plays an important role in the cellular process involved in learning, referred to as synaptic plasticity (Henneberger et al. 2012).

Along with glycine, serine is the amino acid which has been researched most actively. There are several studies on temperature- and pressure-dependent measurements as well as computational studies for both L- and DL-serine. Moreover, there is a monohydrate form of L-serine (in addition to the anhydrous form). As not all authors use the same setting for the unit cell axes and the values for the unit cell parameters vary with temperature and pressure, an overview of these parameters is given in Table 2.4.

The first crystallographic study of L-serine was performed by Kistenmacher et al. (1974), who found that there is one molecule in the asymmetric unit, which forms an extensive hydrogen bond network to symmetry-related neighbor molecules, thus forming a stable, three-dimensional framework. This pattern also persists at non-ambient conditions, although notable changes have been found by several authors. Moggach et al. (2005a, 2006a) found phase transitions between 4.6 and 5.2 GPa as well as between 7.3 and 8.1 GPa, dubbing the phases L-serine II and III (in addition to the phase found at ambient condition, L-serine I). The variation of the unit cell parameters (and the volume) is shown in Fig. 2.20.

From this pressure-dependent study, it was found that the structure is least compressible along *c*, which corresponds to the direction of molecular chains, connected head to tail by N–H···O hydrogen bonds. In contrast, the highest compressibility was found along the *b*-axis, distorting the structure insofar as to close up voids in the structure. At around 4.8 GPa, a reformation of O–H···O hydrogen bonds occurs (from longer O–H···O–H bonds to shorter O–H···carboxylate bonds) which is accompanied with a conformation change from *gauche* to *trans* in C–C–C–OH backbone, while the general topology persists (Moggach et al. 2005a). For higher-pressure ranges, deuterated samples have been investigated via neutron powder diffraction up to 8.1 GPa (Moggach et al. 2006a), where a mechanism for the second phase transition was proposed: The strain which builds up in the structure around the O–H···O is relieved by a shift as the bond now bifurcates between the original acceptor and another carboxylate oxygen atom of a neighboring molecule. These results were confirmed by ab initio density functional theory calculations (Wood et al. 2008), with only small disagreement between experimental and theoretical data (the maximum root-mean-square deviation between experimental and optimized structures equals 0.121 Å).

As found for L-proline, there exists a monohydrate L-serine crystal besides the anhydrous form. This phase was discovered by Frey et al. (1973a); the structure has

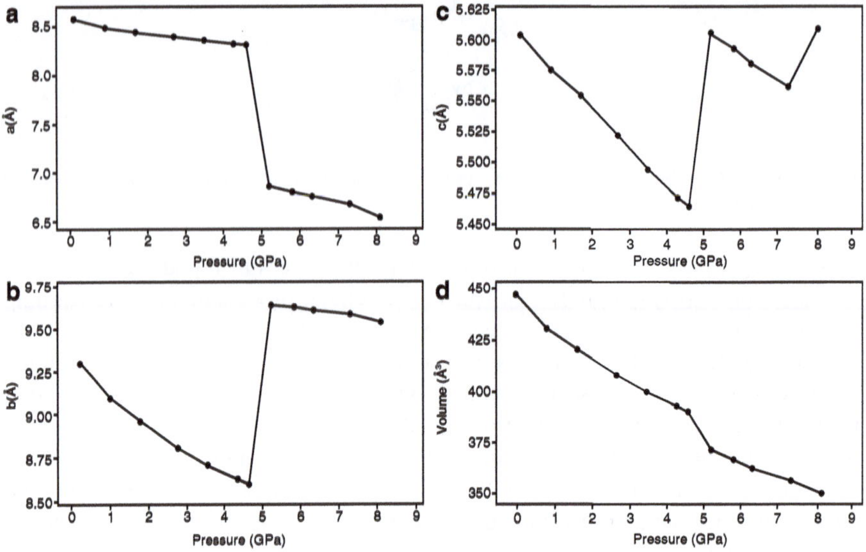

Fig. 2.20 *High-pressure polymorphs of* L*-serine*. The diagrams show the variation of the unit cell parameters (**a–c**) and volume (**d**) of L-serine as function of the pressure. The phase transitions from serine I to II and from II to III at approximately 5 and 8 GPa is clearly visible (Reproduced with permission from the International Union of Crystallography, Moggach et al. 2006a)

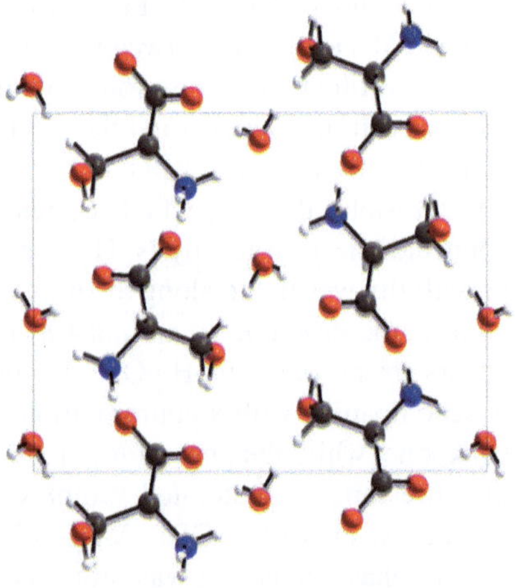

Fig. 2.21 L*-Serine monohydrate*. The orthorhombic crystal structure comprises L-serine zwitterions and water molecules connected by an extensive H-bond network

been refined later via XRD at 97 K and terahertz spectroscopy (King et al. 2010). The asymmetric unit contains one L-serine zwitterion as well as one water molecule, which connect to each other and symmetry-related counterparts in an extensive hydrogen bond network, forming a three-dimensional framework (a projection along (100) is shown in Fig. 2.21).

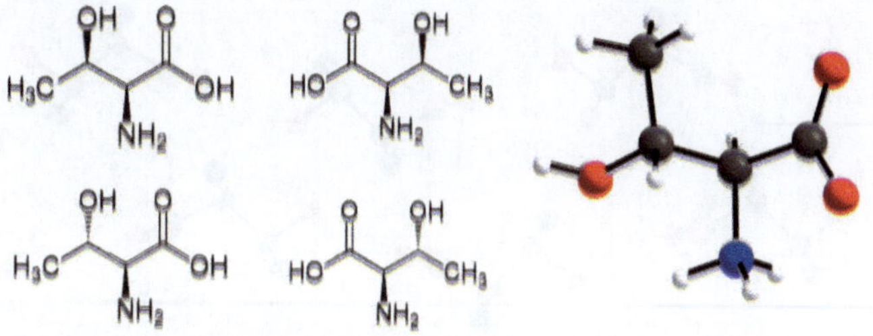

Fig. 2.22 *Isomers of threonine*. Threonine is one of two amino acids with two chiral centers. Thus, in addition to the regular threonine in L- and D-forms (2*S*, 3*R* and 2*R*, 3*S*, *left top*), there are two more isomers, called L- and D-*allo*-threonine (2*S*, 3*S* and 2*R*, 3*R*; *left bottom*). A model of the molecular structure in its zwitterionic state is shown on the *right side*

The crystal structure of DL-serine was determined even before that of the L-form, namely, by Shoemaker et al. in 1953. It also contains one molecule in the asymmetric unit, which is related by an inversion center to its enantiomer counterpart. As discussed for L-serine, there is an extensive hydrogen bond network between the zwitterionic molecules. Later, Frey et al. (1973a) as well as Kistenmacher et al. (1974) refined the structure, as did Flaig et al. (2002) at a temperature of 100 K. A few years later, Dittrich et al. (2005) measured three data sets at 298, 100, and 20 K. They found an improvement in the quality of the molecular geometry (as expected at lower temperatures), but no change in the structure.

As done for L-serine, there have been high-pressure studies of DL-serine crystal (Boldyreva et al. 2006b; Zakharov et al. 2012). Moreover, Mondal et al (2012) performed a comparative study using experimental data with electron density calculations.

2.3.11 Threonine

Apart from serine, threonine is the second of the standard 20 amino acids bearing an OH group. Thus, it is classified as a polar amino acid, and it is essential for humans as well. Of all proteinogenic amino acids, it was discovered last, namely, in the 1930s by W. C. Rose when doing feeding experiments with rats, feeding them all 19 amino acids known up to then and discovering deficiencies (Rose 1931; McCoy et al. 1935). Threonine is one of two standard amino acids possessing two chiral centers (together with isoleucine); therefore, there is another pair of isomers called *allo*-threonine (Fig. 2.22). The name derives from threose, a tetrose sugar of similar structure.

The crystal structure of L-threonine was first studied 20 years after the amino acid was discovered (Shoemaker et al. 1950) and refined later (Ramanadham et al. 1973). A low-temperature study at 120 K is also available (Janczak et al. 1997). Within the crystal structure, there is one molecule in the asymmetric unit (Fig. 2.22 right), which is bonded via several hydrogen bonds to neighboring

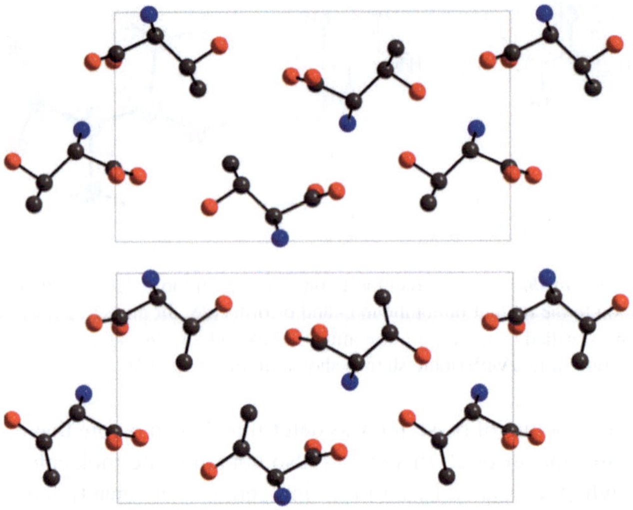

Fig. 2.23 *Crystal structures of* L*-threonine and* L*-allo-threonine*. Both L-threonine (*top*) and L-*allo*-threonine (*bottom*) crystallize in the same packing mode, viewed along [001]. Note the different orientation of the terminal CH_3-group. Hydrogen atoms omitted for clarity

units. The additional OH group extends a hydrogen bond toward a carboxylate group of another molecule and accepts one from an amino group of a third. The resulting three-dimensional network stabilizes the crystal structure.

A very similar structure was found for L-*allo*-threonine (Swaminathan and Srinivasan 1975). The packing mode is virtually identical (Fig. 2.23), since the only difference is the orientation of the methyl group, which does not participate in hydrogen bonding. As the steric impact of this group is not very prominent, the stable packing arrangement does not vary when compared with the structure of L-threonine.

L-Threonine is a crystalline species which proves that amino acid crystals possess noteworthy nonlinear optical properties. Rodriguez et al. (2003) published a study of linear and nonlinear optical properties of this crystal, where they determined refractive indices and their dispersion, along with phase-matching conditions for second harmonic generation at a wavelength of 1064 nm (typical Nd:YAG laser wavelength for SHG experiments), proving the suitability of L-threonine as nonlinear optical material. Recently, Shanti et al. (2013) showed that from a solution of dimethylurea, large single crystals can be grown and confirmed the physical parameters published by Rodriguez et al. (2003).

2.3.12 *Cysteine*

Of the standard 20 amino acids, cysteine is the only one containing a thiol group. This group is responsible for several functions in protein chemistry, most

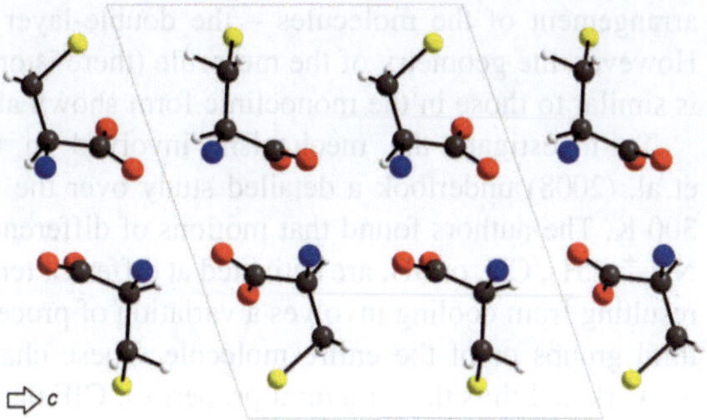

Fig. 2.24 *Crystal structure of L-cysteine (monoclinic form)*, viewed along [010]. Note the bilayer arrangement, as found for other amino acids (e.g., methionine, leucine, isoleucine)

importantly for the formation of disulfide bridges between adjacent protein chains,[13] thus increasing mechanical and thermal stability of proteins (Sevier and Kaiser 2002). Consequently, structure proteins in living beings, such as keratin which constitutes hair, wool, fingernails and toenails, cattle horns, feathers, and reptilian scales, contain a high percentage of cysteine (the sulfur content can even be smelled in burning hair).

Although first discovered in the disulfide form of cystine in 1810 by W. H. Wollaston in bladder stones (who first dubbed it cystic oxide, from Greek *kystis*, bladder pouch), L-cysteine was not immediately identified as protein component. This fact was established in 1899, when L-cysteine was isolated from a cattle horn.

The first determination of the crystal structure of L-cysteine was published by Harding and Long (1968); Görbitz and Dalhus (1996c) later reported a structural study at 120 K. The structure comprises two independent molecules in the unit cell, which were found to have similar interatomic distances and angles but different conformations (N–C–C–S torsion angles of 72.62° and −170.13°, respectively). The molecules are arranged in the frequently found delayer order, where hydrogen bonds between the carboxylate and amino groups connect neighboring units (Fig. 2.24). This structure was later termed as L-cysteine II, as other polymorphs were found (see below).

As commented on before, there is a paper by Khawas (1971), in which XRD powder patterns have been visually analyzed to assess unit cell parameters and space groups of several amino acids (Table 2.3). In this study, Khawas reported a different unit cell than that given by Harding and Long (1968) and Görbitz and Dalhus (1996c), but this result has not been reproduced, although several detailed studies were conducted (see below), and is therefore dubious.

The first new polymorph of L-cysteine has been reported by Kerr and Ashmore (1973), later refined by Kerr et al. (1975). This polymorph represents an orthorhombic form (space group $P2_12_12_1$) and is characterized by a different packing

[13] Cysteine must not be confused with cystine, which is the dimer resulting from a –S–S– bonding of two cysteine molecules.

arrangement of the molecules – the double-layer motif is not represented here. However, the geometry of the molecule (there is only independent L-cysteine unit) is similar to those in the monoclinic form shown above.

To investigate the mechanism involved in the phase transition, Kolesov et al. (2008) undertook a detailed study over the temperature range from 105 to 300 K. The authors found that motions of different parts of the molecule, such as NH_3^+, CH_2, CH, or SH, are activated at different temperatures. The structural strain resulting from cooling involves a variation of processes, e.g., rotations of amino or thiol groups or of the entire molecule. These changes affect the hydrogen bond network and thus the structural properties. CIF data over the temperature range in 25 K steps are available at the Cambridge Structural Database (Allen 2002; Refcodes LCYSTN28 to LCYSTN36). Another set has been measured at 30 K (Moggach et al. 2005b).

The same group later published another series of measurements of L-cysteine, in this case at high pressures (Moggach et al. 2005c, 2006b) up to 4.2 GPa. The authors (who named the orthorhombic phase L-cysteine I; see Table 2.5) discovered two more high-pressure polymorphs (dubbed L-cysteine III and IV). Interestingly, phase III forms upon an increase of pressure over approximately 1.8 GPa and is stable up to 4.2 GPa. Phase IV was only found to form upon decompression down to 1.7 GPa, until at ambient conditions phase I is formed again. The CIFs of these sets are available at the Cambridge Structural Database (Allen 2002; Refcodes LCYSTN23 to LCYSTN27).

The structure of DL-cysteine has been reported by Luger and Weber (1999), who found that there is only one molecule in the asymmetric unit, related to its enantiomer counterpart via inversion symmetry (as commonly found in DL-amino acid crystals). Also, the double-layer motif as realized in the optically active form was found in this crystal as well (Fig. 2.25).

As for L-cysteine, a phase transition was observed in DL-cysteine: Whereas the phase transition found for L-cysteine leaves the crystals intact (Kolesov et al. 2008), the first-order transition observed in DL-cysteine (Minkov et al. 2009) is accompanied by crystal fragmentation. This phase transition has a large hysteresis range (over 100 K); the transition temperature depends strongly on the cooling/heating rate. Interestingly, the low-temperature polymorph (named polymorph II) was also obtained upon a pressure-induced phase transition (at 0.1 GPa, which is the lowest pressure found for a phase transition in an amino acid crystal). More pressure-induced phase transitions were observed at approximately 1.55 and 6.20 GPa (Minkov et al. 2010). Table 2.5 gives an overview of the unit cell dimensions found in the various phases of both L- and DL-cysteine.

2.3.13 Asparagine

Asparagine as the first example of large, hydrophilic amino acids was first isolated from asparagus juice, in which it exists in a relatively high percentage (hence the name), over 200 years ago (Vauquelin and Robiquet 1806). Therefore, asparagine

Table 2.5 *Crystallographic properties of* L*-cysteine crystals.* For non-ambient conditions, pressure or temperature values are noted. Notice that different authors have chosen different cell settings

	s.g.	a	b	c	α	β	γ	Reference
L-Cys I	$P2_12_12_1$	8.116	12.185	5.426	90	90	90	Kerr and Ashmore (1973)
L-Cys I	$P2_12_12_1$	8.116	12.185	5.426	90	90	90	Kerr et al. (1975)
L-Cys I	$P2_12_12_1$	8.111	12.162	5.421	90	90	90	Moggach et al. (2006b)
L-Cys I (30 K)	$P2_12_12_1$	8.116	12.185	5.426	90	90	90	Moggach et al. (2005b)
L-Cys I (105 K)	$P2_12_12_1$	5.420	8.150	12.011	90	90	90	Kolesov et al. (2008)
L-Cys I (125 K)	$P2_12_12_1$	5.421	8.136	12.037	90	90	90	Kolesov et al. (2008)
L-Cys I (150 K)	$P2_12_12_1$	5.418	8.127	12.068	90	90	90	Kolesov et al. (2008)
L-Cys I (175 K)	$P2_12_12_1$	5.419	8.120	12.095	90	90	90	Kolesov et al. (2008)
L-Cys I (200 K)	$P2_12_12_1$	5.420	8.116	12.117	90	90	90	Kolesov et al. (2008)
L-Cys I (225 K)	$P2_12_12_1$	5.422	8.112	12.141	90	90	90	Kolesov et al. (2008)
L-Cys I (250 K)	$P2_12_12_1$	5.423	8.112	12.153	90	90	90	Kolesov et al. (2008)
L-Cys I (275 K)	$P2_12_12_1$	5.425	8.114	12.171	90	90	90	Kolesov et al. (2008)
L-Cys I	$P2_12_12_1$	5.423	8.122	12.180	90	90	90	Kolesov et al. (2008)
L-Cys I (1.8 GPa)	$P2_12_12_1$	7.415	12.000	5.332	90	90	90	Moggach et al. (2006b)
L-Cys II	$P2_1$	11.51	5.224	9.52	90	109.8	90	Harding and Long (1968)
L-Cys II (120 K)	$P2_1$	9.441	5.222	11.337	90	109.00	90	Görbitz and Dalhus (1996c)
L-Cys III (2.6 GPa)	$P2_12_12_1$	8.056	10.488	5.347	90	90	90	Moggach et al. (2006b)
L-Cys III (2.6 GPa)	$P2_12_12_1$	7.998	10.288	5.284	90	90	90	Moggach et al. (2006b)
L-Cys III (4.2 GPa)	$P2_12_12_1$	5.284	7.998	10.298	90	90	90	Moggach et al. (2006b)
L-Cys III (2.6 GPa)	$P2_12_12_1$	7.998	10.288	5.284	90	90	90	Moggach et al. (2006b)
L-Cys III (2.6 GPa)	$P2_12_12_1$	7.998	10.288	5.284	90	90	90	Moggach et al. (2006b)
L-Cys IV (11.7 GPa)	$P2_1$	8.105	5.442	10.916	90	94.90	90	Moggach et al. (2006b)
L-Cys[a]	$P2$[a]	13.26	9.65	7.14	90	96.11	90	Khawas (1971)
DL-Cys I	$P2_1/a$	9.877	4.737	12.877	90	112.04	90	Luger and Weber (1999)
DL-Cys I	$P2_1/a$	9.886	4.735	12.876	90	111.96	90	Minkov et al. (2009)
DL-Cys I (270 K)	$P2_1/a$	9.874	4.728	12.861	90	112.00	90	Minkov et al. (2009)
DL-Cys I (255 K)	$P2_1/a$	9.869	4.723	12.854	90	111.98	90	Minkov et al. (2009)
DL-Cys I (240 K)	$P2_1/a$	9.871	4.722	12.846	90	111.99	90	Minkov et al. (2009)
DL-Cys I (225 K)	$P2_1/a$	9.867	4.719	12.840	90	112.01	90	Minkov et al. (2009)
DL-Cys II	$P2_1/a$	9.717	4.987	13.188	90	121.80	90	Minkov et al. (2009)

(continued)

Table 2.5 (continued)

	s.g.	a	b	c	α	β	γ	Reference
DL-Cys II (263 K)	$P2_1/a$	9.695	4.979	13.155	90	121.88	90	Minkov et al. (2009)
DL-Cys II (200 K)	$P2_1/a$	9.690	4.974	13.124	90	121.75	90	Minkov et al. (2009)
DL-Cys II (150 K)	$P2_1/a$	9.682	4.970	13.090	90	121.72	90	Minkov et al. (2009)
DL-Cys II (100 K)	$P2_1/a$	9.679	4.963	13.066	90	121.70	90	Minkov et al. (2009)
DL-Cys II	$P2_1/a$	9.690	4.974	13.12	90	121.75	90	Minkov et al. (2010)
DL-Cys II (0.25 GPa)	$P2_1/a$	9.712	4.976	13.107	90	121.85	90	Minkov et al. (2010)
DL-Cys II (0.85 GPa)	$P2_1/a$	9.635	4.944	12.885	90	121.81	90	Minkov et al. (2010)

[a]These data are dubious (see footnote of Table 2.3)

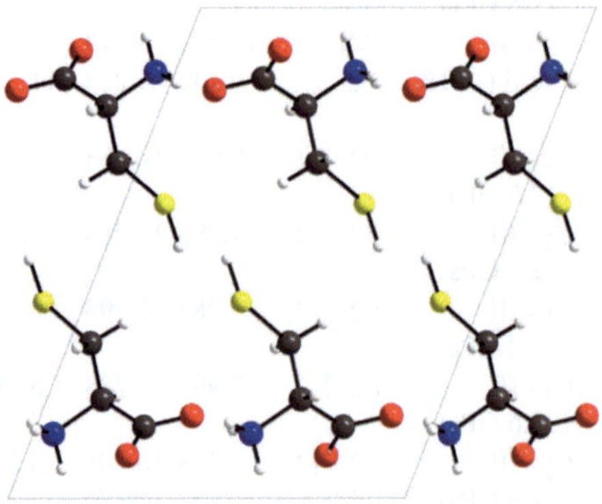

Fig. 2.25 *Crystal structure of DL-cysteine*. The double-layer arrangement as found in L-cysteine is also realized in the racemic form, although there is only one molecule in the asymmetric unit (compare with two independent molecules in Fig. 2.24)

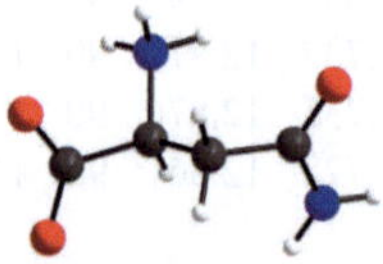

Fig. 2.26 *L-Asparagine*. As glutamine, asparagine possesses an additional amino as well as a keto group, which accounts for the hydrophilic character of the molecule

was actually the first amino acid that was isolated. As all the other long, hydrophilic amino acids (Fig. 2.26), asparagine can form hydrogen bonds with its side chain, which is important in protein structures as it stabilizes the form of the macromolecule (Plimmer and Hopkins 2010). Despite the similarity in structure and name, asparagine (and its anion, asparaginate) must not be confused with aspartic acid (or its anion, aspartate), with the respective abbreviations *asn* and *asp*.

As found for other amino acids, as proline, serine, or phenylalanine, both an anhydrous form and a monohydrate of L-asparagine were found. The crystal structure of the anhydrous form of L-asparagine was described only recently (Yamada et al. 2007), with one molecule in the asymmetric unit. The conformation of the amino acid side chain is *trans*, which is a significant difference in comparison to the monohydrate (see below). The additional amino and carbonyl group provides an extensive hydrogen bond network, which results in a three-dimensional framework.

The structural characterization of L-asparagine monohydrate has been published much earlier, after some discussion on the structure (several papers were published in *Nature* on this; see Kartha and de Vries 1961 and references therein). Later studies have refined this and included hydrogen position in the model (Ramanadham et al. 1972; Verbist et al. 1972). In the following years, more papers have been published on this structure, including studies at 100 K (Wang et al. 1985) and 15 K (Weisinger-Lewin et al. 1989). Arnold et al. (2000) published calculations on dipole moments, electrostatic potential, and electric field gradients in L-asparagine monohydrate. Flaig et al. (2002) additionally calculated topological parameters from data gathered at 100 K using synchrotron radiation. The structure of the D-counterpart has not been published but deposited in the Cambridge Structural Database (Allen 2002; Refcode ILUXOC), whereas no crystal structure of DL-asparagine has been published.

2.3.14 Glutamine

Closely related to asparagine, glutamine can be described as a variation of the former, extended by an additional $-CH_2-$ group in the chain. Glutamine, which is abundant in many plants and also in the human body, was first isolated not as a protein component but in its molecular form from beet juice in 1883. Only in 1932 it was found as protein component. The name derives from gluten (a protein found in wheat and related grains, named for its sticky consistence when mixed with water) and amine. As said above for asparagine, glutamine (or its anion, glutaminate) must not be confused with glutamic acid (or it anion, glutamate), abbreviated *gln* and *glu*, respectively.

The first crystallographic characterization of L-glutamine was published by Cochran and Penfold (1952); a later study reported a refined structure (Koetzle et al. 1973). Furthermore, L-glutamine was investigated under non-ambient conditions, i.e., at low temperatures (130 K, Wagner and Luger 2001) and recently at high pressures (up to 4.9 GPa, Lozano-Casal et al. 2008).

As found for L-asparagine, the crystal structure of L-glutamine is characterized by an extensive hydrogen bond system, with the side chain in extended conformation (Fig. 2.27).

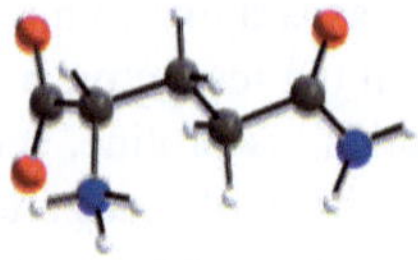

Fig. 2.27 *L-Glutamine*. Similar to L-asparagine, L-glutamine is characterized by additional polar groups and therefore capable of forming a relatively high number of hydrogen bonds. Formally, glutamine can be considered as an asparagine molecule with an additional –CH_2–group

Fig. 2.28 *Crystal structure of DL-glutamine*, viewed along [010]. Note the double-layer motif frequently found in nonpolar amino acids. In contrast to these, the side-chain ends (top and bottom in this view) interconnect by hydrogen bonds to adjacent layers

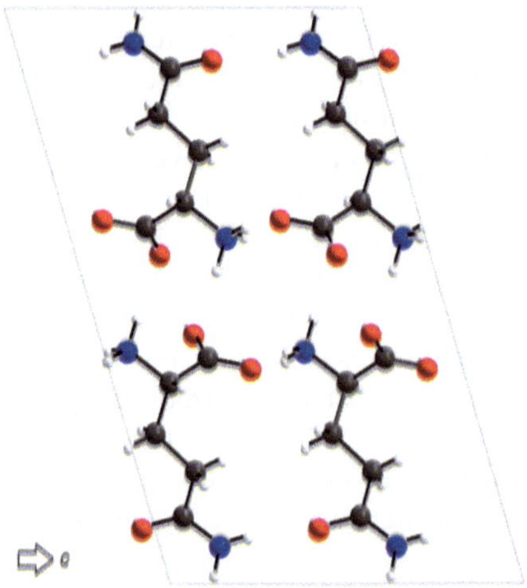

In contrast to asparagine, no hydrated form of glutamine has been found. Instead, the racemate structure was investigated (Suresh et al. 1996). This structure also comprises one molecule in the asymmetric unit, related to its counterpart by inversion centers and glide planes, respectively. In this crystal structure, the molecules are oriented more or less parallel to each other (the molecules long axes parallel [100]), with the polar groups of the α-carbon atom located opposite to each other, thus resembling the double-layer arrangement found in nonpolar amino acids. As the far end of the molecule is also polar, these layers are interconnected to each other by hydrogen bonds at all sides (Fig. 2.28).

2.3.15 *Aspartic Acid*

As said above, aspartic acid must not be confused with asparagine, from which it was first derived in 1827 (Plimmer and Hopkins 2010). Aspartic acid, along with glutamic acid, is a so-called acidic amino acid because of its additional acid group. Its anion is called aspartate.

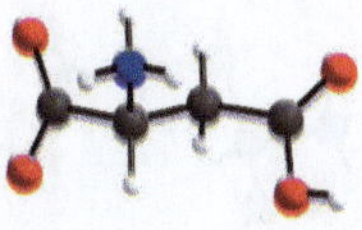

Fig. 2.29 *Aspartic acid.* Note the additional acid group on the right side, which accounts for the acidic behavior of this amino acid (compare also with glutamic acid, Fig. 2.28)

The crystal structure of L-aspartic acid was reported by Derissen et al. (1968) and later refined by Zhang et al. (2006) and by Bendeif and Jelsch (2007); the latter study was performed at 100 K. The structure is monoclinic, with one independent molecule (Fig. 2.29). As in L-asparagine, the high number of hydrogen bonds accounts of the three-dimensional packing of the molecules. In addition to the structure of L-aspartic acid, there is a report on L-aspartic acid monohydrate (Umadevi et al. 2003), which is isotypic to that of L-asparagine monohydrate (Verbist et al. 1972).

In another study, the gyro-optical properties of L-aspartic acid were determined (Asahi et al. 1997). In this work, the components of the gyration tensor and the birefringence as a function of temperature were determined.

In addition to the structure of the enantiomer, the racemate structure was published (Rao 1973). Later, a measurement at 20 K (Flaig et al. 1998) and a refinement of data gathered with neutron diffraction data (Sequeira et al. 1989) were also reported.

Recently, a study proposed another "form of DL-aspartic acid" (Wang et al. 2007). In this paper, the authors stated that the new form "crystallizes in the same space group as the previously reported form but with a slightly different *a* axis." However, when inspecting the structure, it was found that the simple unit cell transformation (transformation matrix 1 0 2/0 1 0/0 0 −1) from the original cell as published by Rao (1973) yields the cell found by Wang et al. (2007). Consequently it is evident that this crystal does not represent a novel phase.

2.3.16 Glutamic Acid

Just as aspartic acid to asparagine, glutamic acid is related to glutamine. Glutamic acid was discovered in 1866 (Plimmer and Hopkins 2010). Its anion, glutamate, is probably the most prominent anion of amino acids, as its sodium salts produce the typical flavor of savory or meaty food. This property of glutamate was found in 1908 and termed umami, a name which stuck and is used nowadays for this taste (Lindemann et al. 2002). Umami is one of only five chemical stimuli which can actually be tasted by the receptors on the human tongue (besides the traditionally known four tastes sweet, sour, bitter, and salty). The role of glutamate in today's food industry is unrivaled by any amino acid, with a world production of more than 2 million tons per year.

The crystal structure of L-glutamic acid has been solved and refined by Bernal (1931). This structure was later termed as α-L-glutamic acid, as another form has

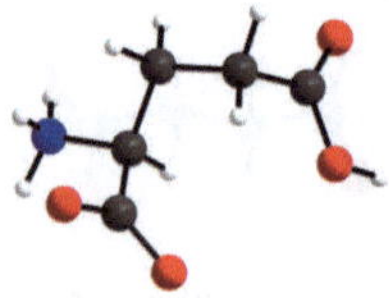

Fig. 2.30 *Glutamic acid*, in zwitterionic form. Compare with aspartic acid (Fig. 2.29), which is smaller by one –CH2– group

been found, namely, β-L-glutamic acid (Hirokawa 1955). An investigation of the transition was published by Hirayama et al. (1980); the relationship of these phases was discussed by Bernstein (1991). Another study employing neutron diffraction, with the focus on the hydrogen bonds, was conducted by Lehmann and Nunes (1980). In addition, there have been several studies on the growth of L-glutamic acid (Sano and Nagashima 1996; Sano et al. 1997; and references therein).

As far as the racemate is concerned, both an anhydrous form and a monohydrate have been reported. Anhydrous DL-glutamic acid (Dunitz and Schweizer 1995) is monoclinic, whereas DL-glutamic acid monohydrate (Ciunik and Glowiak 1983; Flaig et al. 2002) is orthorhombic. In all glutamic acid crystals, there is only one molecule in the asymmetric unit (assuming *gauche* conformation, Fig. 2.30), and in all cases hydrogen bonds connect the units to a three-dimensional framework.

2.3.17 *Tyrosine*

Tyrosine is another amino acid which was discovered in the milk protein casein in 1846 by Justus von Liebig. Since casein is a major constituent of cheese, the name was derived from the Greek word for cheese, *tyros*. Formally, tyrosine is a phenylalanine molecule with an additional hydroxyl group, i.e., 4-hydroxyphenylalanine.

The crystal structure of L-tyrosine was determined relatively late (Mostad et al. 1972), shortly after an initial study by Boggs and Donohue (1971), who determined unit cell dimension and space group. Within the orthorhombic structure, there is one independent tyrosine molecule, with the main chain in *gauche* conformation. Hydrogen bonds extending from the amino and hydroxyl group connect the molecules to a three-dimensional framework. The latter was investigated in detail in a precision neutron diffraction study (Frey et al. 1973b).

Although not published, there is a structure of D-tyrosine, measured at 91 K, deposited in the Cambridge Structural Database (Allen 2002; Refcode FAZHET01), which of course corresponds to that of the L-form in all but absolute configuration.

The racemate DL-tyrosine was investigated shortly after the structure of L-tyrosine was determined (Mostad and Romming 1973). Interestingly, the structures of both the L-form and the racemate are very similar; in fact they appear identical on first glance (Fig. 2.31). The only difference is the alternating configuration of the molecules in DL-tyrosine, which reflects in the higher symmetry (DL-tyrosine crystallizes in the centrosymmetric space group *Pna*2_1).

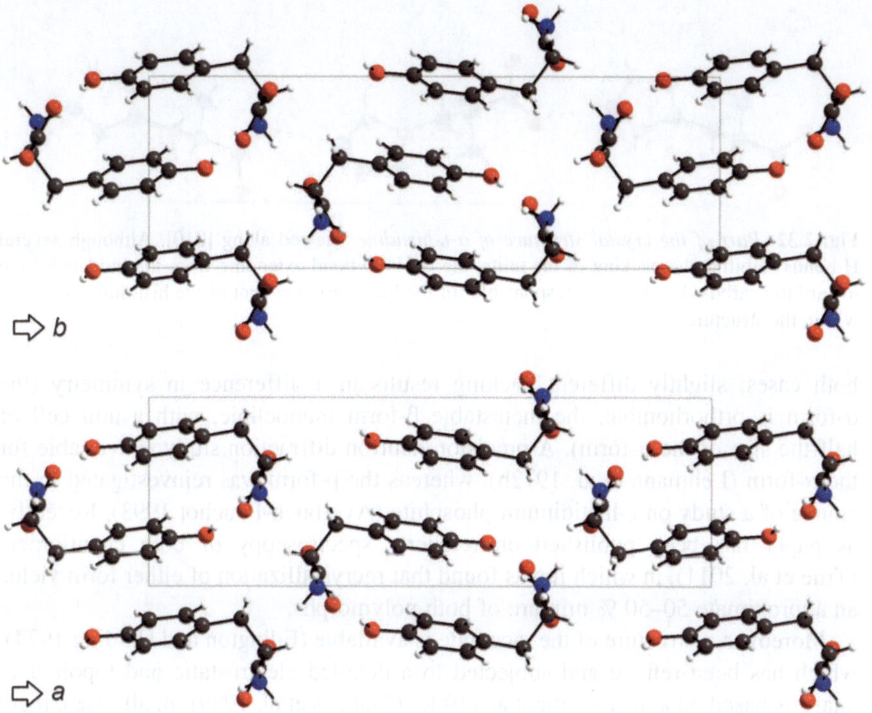

Fig. 2.31 *Crystal structure of tyrosine, viewed along [001].* L-Tyrosine (*top*) and DL-tyrosine (*bottom*) crystallize in nearly identical structures, with the only difference in molecule configuration – in the racemate, D- and L-enantiomers alternate, resulting in the higher orthorhombic symmetry $Pna2_1$. The different setting of the axes a and b results from the arbitrary cell choice in L-tyrosine

2.3.18 Histidine

Along with lysine and arginine, histidine is an amino acid bearing additional amino groups and is thus referred to as a basic amino acid. It was first isolated in 1896 and found in significant quantities in young plant tissues, hence the name (Greek *histos*, tissue). Histidine is an essential amino acid, although this fact was established after some discussion, as it was initially proposed that it is essential only for children (Kopple and Swendseid 1975). Histidine contains an imidazole ring, which can be protonated when in reaction with an acid. Thus, histidine salts can contain a histidinium^{2+} cation, with both the α-amino group and the imidazole group protonated (see Chap. 5). It has to be noted that in protonated imidazole groups, carbon and nitrogen atoms are easily confused, as bond distances in the aromatic system are similar. Usually, hydrogen bonds are reliable indications which of the atoms in the ring are actually the nitrogen atoms (see examples in Sect. 5.1.23).

Crystalline L-histidine was found to exist in two forms, dubbed α- and β-L-histidine (Madden et al. 1972a, b). The molecular structures are similar in

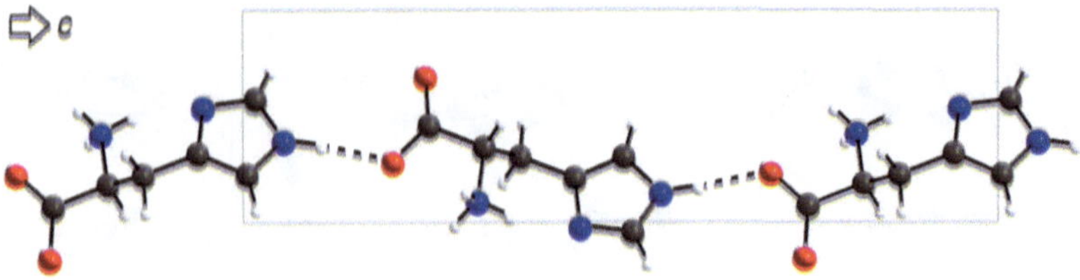

Fig. 2.32 *Part of the crystal structure of α-L-histidine*, viewed along [010]. Although several H-bonds stabilize the packing of the units, the N-H···O bond extending from the imidazole ring toward the carboxylate group is responsible for the linear arrangement of the histidine molecules within the structure

both cases; slightly different packing results in a difference in symmetry (the α-form is orthorhombic, the metastable β-form monoclinic, with a unit cell of half the size of the α-form). A precision neutron diffraction study is available for the α-form (Lehmann et al. 1972b), whereas the β-form was reinvestigated in the course of a study on L-histidinium phosphite (Averbuch-Pouchot 1993). Recently, as paper has been published on terahertz spectroscopy of both enantiomers (True et al. 2011), in which it was found that recrystallization of either form yields an approximate 50–50 % mixture of both polymorphs.

Moreover, a structure of the racemate is available (Edington and Harding 1974), which has been refined and subjected to a detailed electrostatic and topological analysis based on a measurement at 110 K (Coppens et al. 1999). In all cases, there is one independent histidine molecule in the crystal structure, with hydrogen bonds involving amino acid and imidazole groups (Fig. 2.32).

2.3.19 Lysine

Lysine is another basic amino acid, which was first isolated from casein in 1889 by Drechsel (1889). Its name derives from the Greek word *lysis* (for dissolving), as lysine was obtained from protein hydrolysis. From all of the standard amino acids, lysine is the last of which the crystal structure of both the optically active form and the racemate remains undetermined until today. The only work on this issue is over 40 years old (Khawas 1971) and only gives a unit cell as determined from powder XRD data. This data is not very reliable, as the author gives the non-centrosymmetric space group *P*222 for a DL-crystal. This is of course possible but improbable, as all other DL-amino acid crystals possess inversion symmetry. Moreover, this result seems dubious as XRD data do not allow to distinguish between centrosymmetric and non-centrosymmetric symmetry. (The same author also proposes space group *P*2 for DL-phenylalanine, which is similarly dubious; see Sect. 2.3.8.) Although no crystal structure analysis was performed to date, Wang et al. (1996) determined the molecular structure of L-lysine in solution, employing NMR.

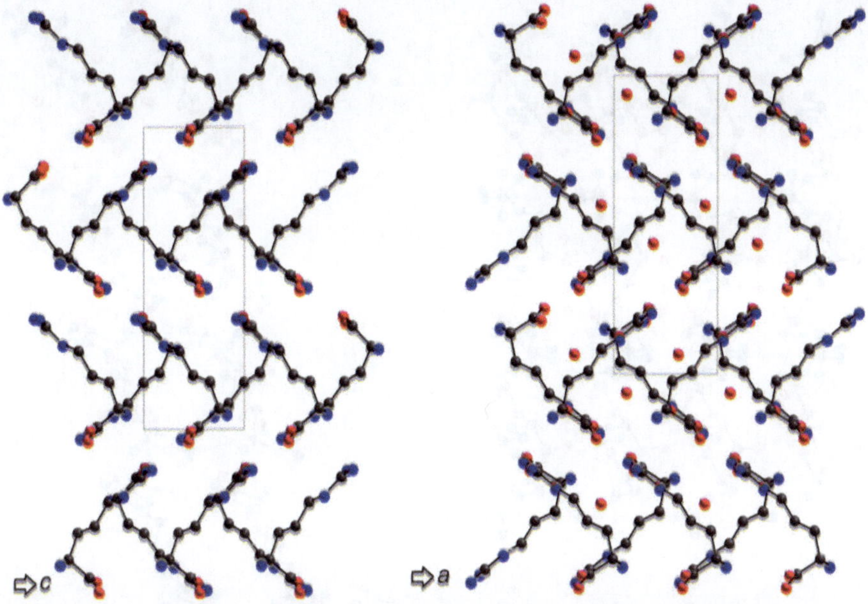

Fig. 2.33 *Crystal structure of* L*-arginine (left) and* L*-arginine dihydrate (right)*, viewed along [100] and [010], respectively. Hydrogen atoms are omitted for clarity. Notice the herring bone pattern formed by the arginine molecules as well as the similarity of the unit cells (the largest difference is the length in the viewing direction, compare values in Table 2.6)

2.3.20 Arginine

L-Arginine is another basic amino acid, which has first been isolated from an extract of lupine seedlings at the end of the nineteenth century (Schulze and Steiger 1886; Schultz and Steiger 1886). Arginine has supposedly been named after the fact that it has first been crystallized not in pure form but as its silver salt (Latin *argentum*, silver). In 1895 L-arginine was identified as component of casein. As in histidine, arginine can be doubly protonated, thus forming an argininium^{2+} cation. Due to the lack in counterbalancing positive charges, this double protonation does not occur in pure arginine (which exists as zwitterion in solid state, as do all amino acids) but is frequently found in arginine salts (see following chapters).

The crystal structure of pure L-arginine has eluded discovery for rather long; only very recently this structure could be determined and refined from XRD powder data, more or less simultaneously by Thomas (2012) and Courvoisier et al. (2012). Apparently, as the large molecule possesses a high conformational flexibility, it is difficult to grow good-quality crystals; therefore, the structure determination from single crystals proved to be not possible.

The monoclinic structure (space group $P2_1$) comprises two independent molecules in the asymmetric unit in slightly different conformations. The molecules are arranged more or less in parallel planes, resulting in a herring bone motif (Fig. 2.33, left).

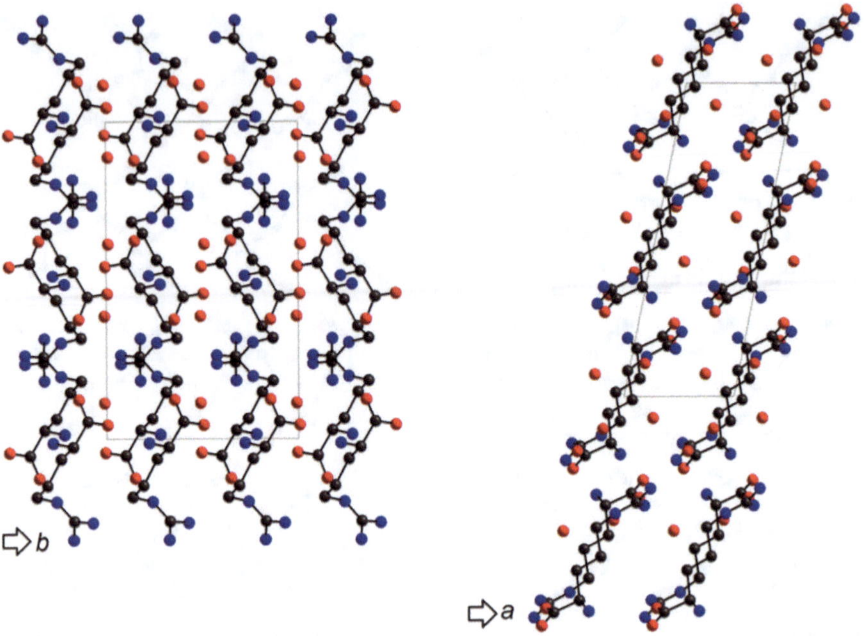

Fig. 2.34 *Crystal structure of* DL*-arginine monohydrate (left) and* DL*-arginine dihydrate (right),* viewed along [100] and [010], respectively. Hydrogen atoms are omitted for clarity. The herring bone pattern found in the structures of L-arginine (compare with Fig. 2.33) can also be seen in the structure of the monohydrate, although less clearly since the molecules assume gauche conformation. In the dihydrate, all molecules are oriented more or less parallel, thus resembling the L-arginine structures (with half of the molecules mirrored)

In contrast to the anhydrous form, there is a dihydrate which crystallizes much more readily. The first description was published by Karle and Karle (1964); a later study by Lehmann et al. (1973) further refined the structure. The structure is orthorhombic (space group $P2_12_12_1$), but the herring bone pattern is found here as well (Fig. 2.33, right). The interstitial water molecules account for the higher stability as well as the increase in the unit cell size (863.77 $Å^3$ in the anhydrous form as compared to 1047.59 $Å^3$ in the dihydrate).

Monaco et al. (1987) conducted a detailed study of several salts of L-arginine (including L-arginine dihydrate) and investigated several optical properties, both linear (refractive indices) and nonlinear (SHG and phase-matching).

As far as the racemate of arginine is concerned, there are two crystal phases, both hydrates: DL-arginine monohydrate (Kingsford-Adaboh et al. 2000, 2002, measured at 100 K) and DL-arginine dihydrate (Suresh et al. 1994). To date, no anhydrous phase of DL-arginine has been reported.

In DL-arginine monohydrate, there is one independent molecule, which is related to its enantiomer via glide planes (the symmetry is orthorhombic, space group *Pbca*). In contrast to the molecules in the optically active crystals, the arginine molecules here assume *gauche* conformation (torsion angle C3–C4–C5–N2 equals 59.01°). Nevertheless, the herring bone pattern is still recognizable (Fig. 2.34 left),

Table 2.6 *Crystallographic properties of standard amino acids, part III.* Space groups and unit cell parameters of asparagine, glutamine, aspartic acid, glutamic acid, tyrosine, histidine, lysine, and arginine crystals

	s.g.	a	b	c	α	β	γ	Reference
L-Tyr	$P2_12_12_1$	6.91	21.12	5.83	90	90	90	Mostad et al. (1972)
DL-Tyr	$Pna2_1$	20.83	6.81	5.90	90	90	90	Mostad and Romming (1973)
α-L-His	$P2_12_12_1$	5.177	7.322	18.87	90	90	90	Madden et al. (1972a)
β-L-His	$P2_1$	5.172	7.384	9.474	90	97.16	90	Madden et al. (1972b)
DL-His	$P2_1/c$	8.983	8.807	9.415	90	97.65	90	Edington and Harding (1974)
L-Asp	$P2_1$	7.617	6.892	5.142	90	99.84	90	Derissen et al. (1968)
L-Asp·H_2O	$P2_12_12_1$	5.587	9.822	11.813	90	90	90	Umadevi et al. (2003)
DL-Asp	$C2/c$	18.947	7.433	9.184	90	123.75	90	Rao (1973)
α-L-Glu	$P2_12_12_1$	7.06	10.30	8.75	90	90	90	Bernal (1931)
β-L-Glu	$P2_12_12_1$	5.17	17.34	6.95	90	90	90	Hirokawa (1955)
DL-Glu	$P2_1/n$	5.743	13.036	8.431	90	103.66	90	Dunitz and Schweizer (1995)
DL-Glu·H_2O	$Pbca$	9.119	15.480	10.627	90	90	90	Ciunik and Glowiak (1983)
L-Asn	$P2_1$	5.062	6.700	8.054	90	91.71	90	Yamada et al. (2007)
L-Asn·H_2O	$P2_12_12_1$	5.593	9.827	11.808	90	90	90	Ramanadham et al. (1972)
L-Gln	$P2_12_12_1$	16.01	7.76	5.10	90	90	90	Cochran and Penfold (1952)
DL-Gln	$P2_1/c$	16.051	4.654	9.937	90	106.48	90	Suresh et al. (1996)
DL-Lys[a]	$P222$[a]	8.89	11.04	7.13	90	90	90	Khawas (1971)
L-Arg	$P2_1$	9.76	16.02	5.58	90	98.1	90	Courvoisier et al. (2012)
L-Arg·$2H_2O$	$P2_12_12_1$	5.68	11.87	15.74	90	90	90	Karle and Karle (1964)
DL-Arg·H_2O (100 K)	$Pbca$	11.464	9.961	16.013	90	90	90	Kingsford-Adaboh et al. (2000)
DL-Arg·$2H_2O$	$P2_1/c$	5.788	11.508	16.347	90	101.51	90	Suresh et al. (1994)

Note: The Greek letters α and β in this table denote different polymorphs of the respective amino acids, as all amino acids given here are α-amino acids

[a]Unit cell and space group have been estimated from powder XRD only and are not reliable (see comments on this paper in the section on phenylalanine and cysteine); no structure determination is available

and as in the hydrated form of L-arginine, the water molecules link the amino acid molecules via hydrogen bonds.

The crystal structure of DL-arginine dihydrate differs from the two structures above insofar as the molecules in this crystal are arranged more or less parallel, i.e., the above-mentioned herring bone motif is not represented here. Nevertheless, the structure resembles that of L-arginine (also visible in the similarity of the unit cell dimensions; see Table 2.6): In L-arginine, the molecules' orientation alternates – thus the herring bone pattern – whereas in DL-arginine dihydrate, the orientation does not alternate (Fig. 2.34 right).

All arginine crystals express a peculiarity in respect with all other amino acids: In the zwitterionic form, the α-amino group is not protonated (in contrast to all other amino acids). Instead, the ω-amino group is protonated, accounting for the zwitterionic character of the molecule. This leads to another peculiarity: Arginine crystals do not include hydrogen bonds from the α-amino to the α-acid groups (the so-called head-to-tail motif), which is also in contrast to all other amino acids, where this hydrogen bonding occurs.

2.3.21 Nonstandard Amino Acids

As discussed in Chap. 1, the number of different amino acids is infinite. Nevertheless, a finite number of different amino acid molecules has been reported of course, and of those, only a smaller number is common enough to be used in chemistry and physics, e.g., as precursor for syntheses of new materials. Still, here we present only an arbitrary selection, always bearing in mind that this represents only a part of those known in the literature. This selection, listed in Table 2.7, represents all of the amino acid structures (from those shown in Fig. 1.4) found in the Cambridge Structural Database (Allen 2002).

Among the nonstandard amino acids, β-alanine is the simplest β-amino acids and as glycine, a non-chiral species. Within human metabolism, β-alanine is part of the dipeptide carnosine (together with histidine), which is abundant in muscles. As β-alanine is the limiting precursor, ingestion of this amino acid has been found to increase carnosine levels and muscular performance and on the other hand decrease muscular fatigue (Harris et al. 2006; Hill et al. 2007).

β-Alanine was the first to be crystallographically characterized (Jose and Pant 1965); a later investigation provided a structure complete with hydrogen positions (Papavinasam et al. 1986). The orthorhombic structure is centrosymmetric, as the molecule is not chiral; therefore, mirror symmetry is possible and in fact realized in the structure (Fig. 2.35). Within the asymmetric unit, there is one molecule, which assumes gauche conformation (torsion angle C1–C2–C3–N1 equals 83.19°).

As β-alanine is the simplest β-amino acid, γ-aminobutyric acid (GABA) is the simplest γ-amino acid and probably the most prominent example, as it has an important biological function: GABA is the major inhibitory neurotransmitter in the brain of vertebrates (including humans; Davidson 1976, pp. 105–108; Watanabe et al. 2002; and references given therein; a search in the online database SCIRUS gives over 200,000 hits from Journal sources).

Structurally, the compound is less interesting – the molecular structure is simple, representing a zwitterionic, linear, non-chiral hydrocarbon chain with amino and acid groups on opposite ends. The crystal structure of GABA has been published almost simultaneously by Steward et al. (1973) and Tomita et al. (1973). The amino acid was found to crystallize in the monoclinic space group $P2_1/c$, with one independent molecule, where the acid group end adopts *gauche* conformation

Table 2.7 *Crystallographic properties of nonstandard amino acids*. Space groups and unit cell parameters of crystals of the nonstandard amino acid discussed in this work

	s.g.	a	b	c	α	β	γ	Reference
β-Ala	*Pbca*	9.865	13.81	6.07	90	90	90	Jose and Pant (1965)
GABA	$P2_1/a$	8.228	10.036	7.210	90	110.63	90	Steward et al. (1973)
GABA	$I4_1cd$	11.963	11.963	15.282	90	90	90	Dobson and Gerkin (1996)
Sarcosine	$P2_12_12_1$	6.687	7.899	8.607	90	90	90	Mostad and Natarajan (1989)
DMG	*Pbca*	11.223	10.010	18.729	90	90	90	Minkov and Boldyreva (2012)
DMG·0.5H_2O	*C2/c*	20.040	10.733	11.112	90	103.78	90	Minkov and Boldyreva (2012)
L-DOPA	$P2_1$	13.619	5.232	6.062	90	97.56	90	Howard et al. (1995)
L-Cystine	$P4_1$	6.710	6.710	21.73	90	90	90	Chaney and Steinrauf (1974)
L-Cystine	$P6_122$	5.420	5.420	55.980	90	90	120	Oughton and Harrison (1959)
L-Citrulline	$P2_1$	9.162	5.143	8.969	90	95.81	90	Toffoli et al. (1987)
L-Citrulline·2H_2O	$P2_12_12_1$	4.671	13.07	16.40	90	90	90	Xianglin et al. (1985)
L-Nitroarginine	*P*1	5.107	7.459	12.221	82.23	89.01	82.52	Apreyan et al. (2008)

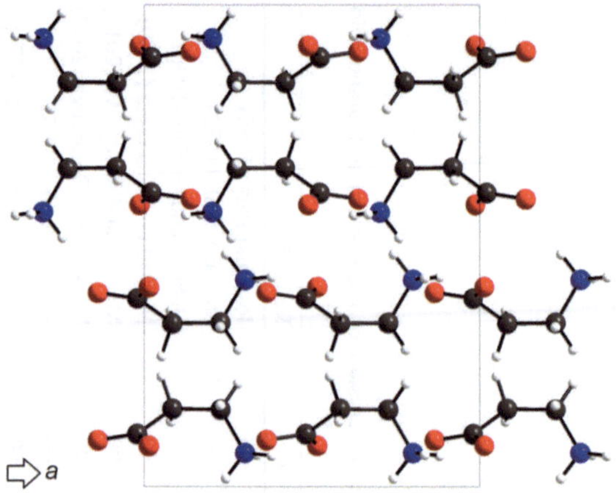

Fig. 2.35 *Crystal structure of β-alanine*, viewed along [001]. Notice that glide planes connect the symmetry-related molecules, as this amino acid is non-chiral

(torsion angle C1–C2–C3–C4 equals −72.6°, Fig. 2.36, left). A later study dealt with a reinvestigation by neutron and X-ray, at 122 K (Craven and Weber 1983). Later still, a tetragonal polymorph was found (Dobson and Gerkin 1996) in which the amino acid's conformation is different: The acid group end is in *trans*-conformation, whereas the amino end exists in gauche conformation (torsion angle C2–C3–C4–N1 equals 62.1°). All molecules are oriented more or less along the tetragonal axis (Fig. 2.36, right), facing in the same direction, i.e., the structure is polar.

Sarcosine (*N*-methylglycine) is an example of a simple, non-chiral amino acid (formally an extension of glycine) which was discovered by Justus von Liebig in 1847 and later found to be a hydrolysis product of creatine, a compound derived from meat extract (hence the name, Greek *sarcos* for meat). Sarcosine has been investigated as possible medication against schizophrenia (in addition to antipsychotic drugs, Lane et al. 2006). Moreover, there has been active discussion on the issue if sarcosine can be used as biomarker for prostate cancer (when found in urine), although the latest data is not conclusive (Pavlou and Diamandis 2009 and references therein).

The crystal structure of sarcosine has been published by Jose and Pant (1965); a later study on deuterated sarcosine provided accurate hydrogen positions (Trzebiatowska-Gusowska and Gagor 2007). It was found that the structure can be described as a herring bone pattern (Fig. 2.37), with a network of hydrogen bonds connecting the molecules to a three-dimensional framework.

The formal extension of sarcosine by another methyl group is realized in the species *N,N*-dimethylglycine (DMG), another non-chiral amino acid. DMG occurs in living beings and was (and still sometimes is) dubbed as a vitamin, namely, vitamin B16, but as DMG deficiency does not result in any health problems, the term vitamin is not justified.

The crystal structure of DMG was published only recently, with two species, an anhydrous form and a hemihydrate (Minkov and Boldyreva 2012). Both forms crystallize in centrosymmetric space groups (*Pbca* and *C*2/*c*, respectively), and in

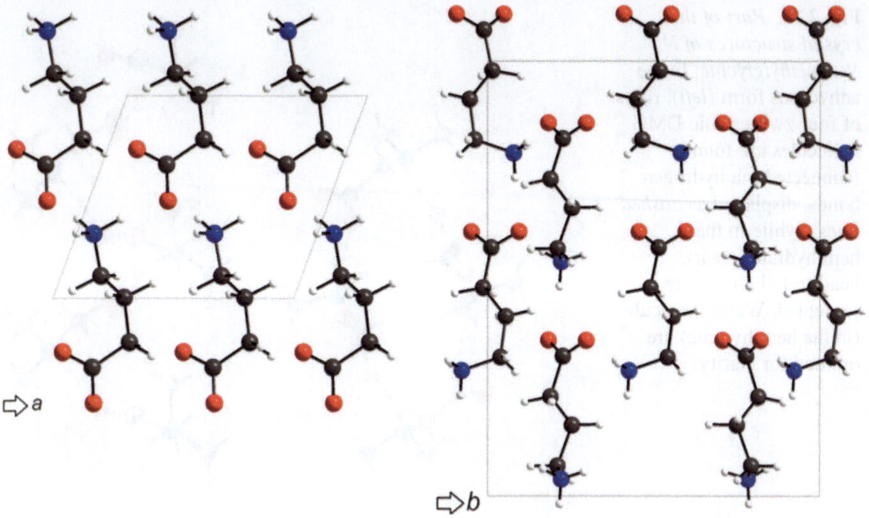

Fig. 2.36 *Part of the crystal structures of the polymorphs of γ-aminobutyric acid (GABA).* Monoclinic form (*left*) and tetragonal form (*right*), viewed along [010] and [100], respectively. Note that for reasons of clarity, only a layer out of the structure is shown. In the monoclinic polymorph, the molecules not displayed are oriented in the other direction (i.e., with the acid groups facing up and amino groups facing down), whereas in the tetragonal polymorph, all molecules are oriented as the ones shown; thus, the structure is clearly polar

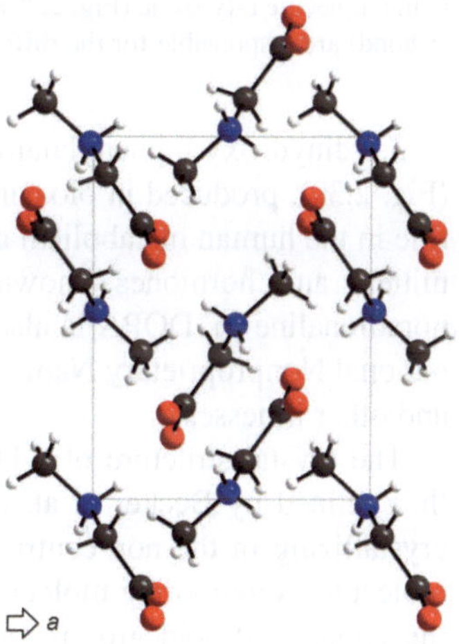

Fig. 2.37 *Crystal structure of sarcosine*, viewed along [010]. The molecules are oriented more or less parallel [101] and [10-1], arranged in a hydrogen pattern, and connected by hydrogen bonds to form a framework

both cases, two zwitterionic DMG molecules are located in the asymmetric unit. In DMG, these molecules form a ring motif (each ring comprises four molecules), whereas in DMG hemihydrate the molecules are connected to head-to-tail chains along *c*, with the water molecules connecting the chains to a three-dimensional framework (Fig. 2.38).

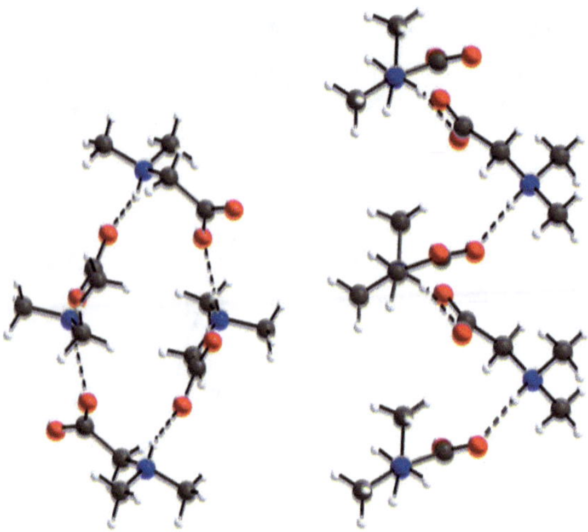

Fig. 2.38 *Part of the crystal structures of N, N-dimethylglycine*. In the anhydrous form (*left*), rings of four zwitterionic DMG molecules are found (connected via hydrogen bonds, displayed as *dashed lines*), while in the hemihydrate (*right*), head-to-tail chains are presented. Water molecules (in the hemihydrate) are omitted for clarity

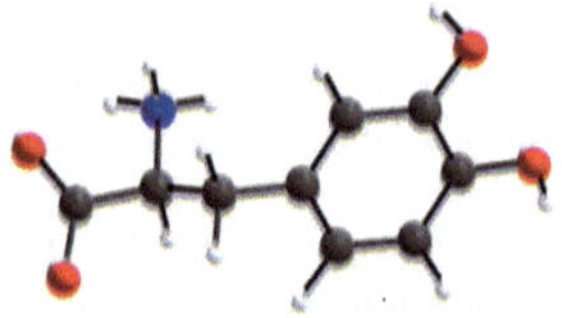

Fig. 2.39 *Molecular structures of 3,4-dihydroxy-*L*-phenylalanine (*L*-DOPA)*. Note the similarity to the molecule L-tyrosine (Fig. 2.29); nevertheless, the addition hydroxyl group (and the resulting H-bond) are responsible for the different packing modes

3,4-dihydroxy-L-phenylalanine (L-DOPA) is a rather large, chiral amino acid (Fig. 2.39), produced in biosynthesis from L-tyrosine. L-DOPA plays an important role in the human metabolism as the amino acid is the precursor of the neurotransmitters and hormones known as catecholamines (dopamine, adrenaline, and noradrenaline). L-DOPA is also employed as a psychoactive drug (with the International Nonproprietary Name "levodopa") in the treatment of Parkinson's disease and other illnesses.

The crystal structure of L-DOPA was discovered by Mostad et al. (1971) and then refined by Becker et al. (1973) and Howard et al. (1995). It is monoclinic, crystallizing in the non-centrosymmetric space group $P2_1$, with one independent molecule. Neighboring molecules are connected via hydrogen bonds, and as far as the amino and acid groups are concerned, a double chain parallel *b* is formed. Additional hydrogen bonds from the hydroxyl groups connect these chains to a three-dimensional framework.

Another example of a nonstandard amino acid is cystine, the dimer of cysteine, formed by oxidation of two of these amino acid molecules, connected by a disulfide bridge. Cystine was discovered even before cysteine in bladder stones, hence the name (see Sect. 2.3.12). The crystal structure of the L-form was determined by Oughton and Harrison (1959) and found to crystallize in the unusually high hexagonal symmetry (space group $P6_122$); later, a second polymorph with tetragonal symmetry (space group $P4_1$) was discovered (Chaney and Steinrauf 1974). In a more recent work, the hexagonal phase was studied in more detail in a charge density study (Dahaoui et al. 1999). Even more recently, Moggach et al. published the results of a high-pressure XRD investigation of the hexagonal polymorph, up to 6.4 GPa. The structure can be described as composed of pairs of glycine-like layers parallel (001). In this structure, the *c*-axis is unusually long (55.980 Å at ambient conditions) and is affected most strongly by the compression (53.549 Å at 3.7 GPa). This anisotropic strain is a result of the closing of voids and a compression of the backbone of the cystine molecules.

As cystine is a component of urinary stones, kidney stones, gallstones, etc., there is medical interest in the growth conditions of this phase. Although the study of cystine crystals in vivo is difficult (due to the high solubility during the fixation), in vitro studies have been undertaken, as nucleation, growth rates, and other parameters can be studied much more easily, as was shown by a work on crystal growth of L-cystine (Girija et al. 1995 and references therein).

The last of the nonstandard amino acids to be discussed in this work are a pair of derivates of arginine. The first of these is L-citrulline, an amino acid first found in water melons (hence the name, Latin *citrullus*, water melon) in 1914 by Koga and Odake, but only 16 years later identified by Wada (1930). Although L-citrulline is not a member of the standard 20 and therefore not utilized for the composition of proteins, there is a cellular process called "citrullination" which refers to the conversation of L-arginine molecules which are already part of proteins to L-citrulline. The biological impact of L-citrulline is actively studied as the amino acid is thought to be responsible for many effects (Curis et al. 2005; Patil et al. 2009; and references therein). Among these, the most popular finding is the effect on L-citrulline on blood vessels, which resulted in headlines like "Watermelon May Have Viagra-Effect" (Texas A&M University 2008).

As far as crystalline L-citrulline is concerned, researchers found an anhydrous form (Toffoli et al. 1987) as well as a dihydrate (Xianglin et al. 1985; Toffoli et al. 1986). Within the anhydrous crystal, there is one L-citrulline molecule in the asymmetric unit, which is extended but for one angle (torsion angle C3–C4–C5–N2 equals 56.78°). In the dihydrate, the conformation of the molecules is *gauche* as well, but counterclockwise (the respective torsion angle equals −72.76°). This difference, along with the additional water molecules, leads to a completely different packing mode (Fig. 2.40).

Apart from L-citrulline, there is another derivate of L-arginine, namely, L-nitroarginine, which has been identified as an inhibitor of the enzyme nitric oxide synthase, by which it is in effect a vasoconstrictor in mammals (Bansinath et al. 1993). Consequently, L-nitroarginine has been employed in the study of biological effects of nitric oxide and its role in vasodilation and constriction (Woodman and Dusting 1991).

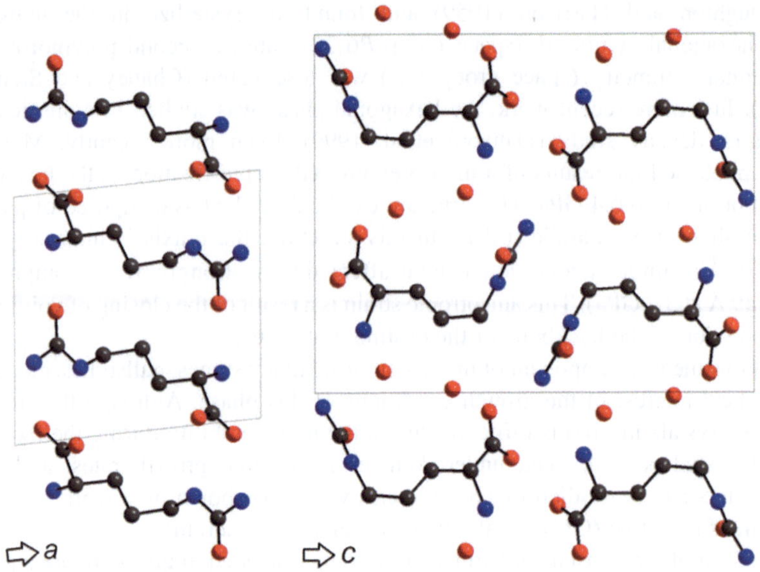

Fig. 2.40 *Crystal structures of* L-*citrulline (left) and* L-*citrulline dihydrate (right).* In both species, the L-citrulline molecules adopt gauche conformation as far as the central bond is concerned. Hydrogen atoms are omitted for clarity

The crystal structure of L-nitroarginine has eluded determination for a long time; only recently a paper on the structure at ambient conditions (Apreyan et al. 2008, Fig. 2.41) was published, followed by a study at 100 K (Borissova et al. 2010). The asymmetric unit (in this case the unit cell as the symmetry is triclinic, space group *P*1) comprises two molecules, which are interconnected by an extensive hydrogen bonding system.

For the other amino acids shown in Fig. 1.4, no crystal structures have been found in the Cambridge Structural Database (Allen 2002), although various salts are known (see following chapters).

2.4 Summary

In this chapter, the molecular and crystal structures of the 20 standard amino acids, plus several other nonstandard amino acids, have been discussed in detail. Clearly, all structures are different, although some motifs occur repeatedly. One of these recurring details is the head-to-tail sequence, established by hydrogen bonds connecting the α-amino and the α-carboxylate groups and its symmetry equivalents. This motif is found in almost all amino acid structures, which is not surprising in amino acid with no polar side chains. However, polar amino acids and amino acid hydrates also show this head-to-tail sequence – the only exception being arginine

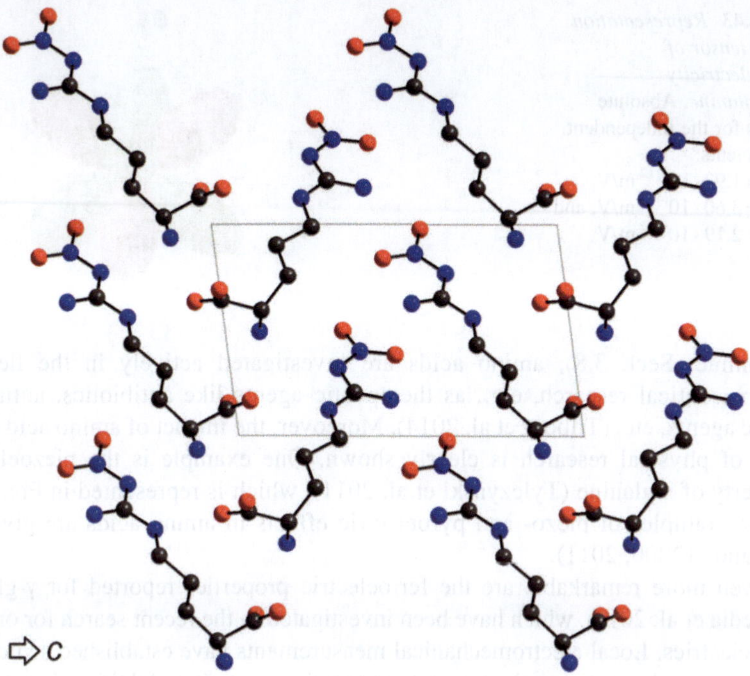

Fig. 2.41 *Crystal structure of* L*-nitroarginine*, viewed along [100]. The crystal is non-centrosymmetric triclinic, i.e., no symmetry elements are present. Hydrogen atoms are omitted for clarity

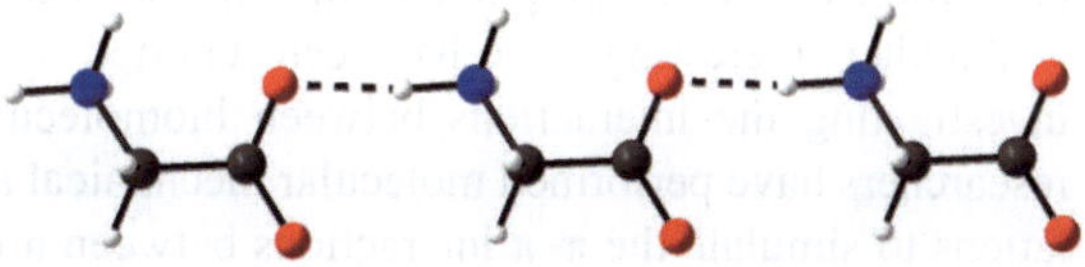

Fig. 2.42 *Head-to-tail sequence in amino acids*, as shown for a straight chain in α-glycine

(in all forms, optically active, racemic, anhydrous, hydrated), where the α-amino group is not protonated as in all other amino acids (see Sect. 2.3.20). These head-to-tail sequences have earlier been proposed to be of relevance for prebiotic polymerization (Suresh and Vijayan 1893, and references therein) and are realized in various forms: More or less straight chains of small amino acids (as, e.g., in glycine, Fig. 2.42) and helices of larger amino acids, where the side chain is facing outward from the helical axis (usually a twofold screw axis), are found. Moreover, amino acid dimers occur, and double layers with the polar groups facing each other and the nonpolar side chains facing outward are frequent motifs.

Apart from the structural discussion presented here (which forms the basis for the analysis of structural data of amino acid salts in the following chapters), amino acids are of interest in life sciences. Apart from the obvious importance as protein components (see Chap. 1) and the nutritional value (see, e.g., monosodium

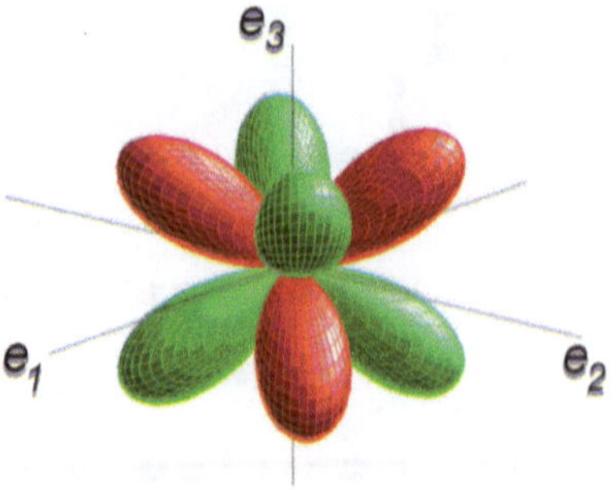

Fig. 2.43 *Representation of the tensor of piezoelectricity for* L*-alanine*. Absolute values for the independent coefficients: $d_{132} = 1.92 \cdot 10^{-12}$ m/V, $d_{231} = 3.60 \cdot 10^{-12}$ m/V, and $d_{312} = 2.19 \cdot 10^{-12}$ m/V

glutamine, Sect. 3.8), amino acids are investigated actively in the field of pharmaceutical research, e.g., as therapeutic agents like antibiotics, antiarrhythmic agents, etc. (Tilborg et al. 2014). Moreover, the impact of amino acid in the field of physical research is clearly shown. One example is the piezoelectric property of L-alanine (Tylczyński et al. 2011), which is represented in Fig. 2.43. Other examples of piezo- and pyroelectric effects in amino acids are given by Lemanov (2000, 2011).

Even more remarkable are the ferroelectric properties reported for γ-glycine (Heredia et al. 2012), which have been investigated in the recent search for organic ferroelectrics. Local electromechanical measurements have established switchable ferroelectric domains in microcrystals grown from solution. Additional molecular simulations show that the polarization vector in this crystals can be switched on the nanoscale level, which might present an opportunity to manufacture a novel class of bioelectronic logic and memory devices. This work follows the concept of amino acid ferroelectric thin films (Balashova and Krichevtsov 2011), thus providing a basis for a very interesting line of material science research.

Another interesting issue has been reported by Yang et al. (2012): In a study investigating the interactions between biomolecules and carbon nanotubes, the researchers have performed molecular mechanical and quantum-mechanical calculations to simulate the π–π interactions between aromatic amino acid analogs and carbon nanotube – proposing a model for likewise connection in biological systems.

One more noteworthy discovery was reported for lysine, which has been used as a curing agent of epoxy in electronic materials. This possible application would represent a novel, eco-friendly way for curing of electronic materials, as the epoxy amino acid ester is thermally degradable and can also be used as a reworkable resin (Li et al. 2006).

References

Alagar M, Krishnakumar RV, Mostad A, Natarajan S (2005) DL-Methionine at 105 K. Acta Cryst E61:o1165–o1167

Albrecht G, Corey RB (1939) The crystal structure of glycine. J Amer Chem Soc 61:1087–1103

Albrecht G, Schnakenberg GW, Dunn MS, McCullough JD (1943) Quantitative investigations of amino acids and peptides. XII. Structural characteristics of some amino acids. J Phys Chem 47:24–30

Allen FH (2002) The Cambridge structural database: a quarter of a million crystal structures and rising. Acta Cryst B58:380–388

Almlöf J, Kvick Å, Thomas JO (1973) Hydrogen bond studies. 77.* Electron density distribution in α-glycine: X-N difference Fourier synthesis vs ab initio calculations. J Chem Phys 59:3901–3906

Apreyan RA, Karapetyan HA, Petrosyan AM (2008) On molecular structure of L-nitroarginine. J Mol Struct 874:187–193

Arnold WD, Sanders LK, McMahon MT, Volkov AV, Wu G, Coppens P, Wilson SR, Godbout N, Oldfield E (2000) Experimental, Hartree–Fock, and density functional theory investigations of the charge density, dipole moment, electrostatic potential, and electric field gradients in L-asparagine monohydrate. J Am Chem Soc 122:4708–4717

Asahi T, Takahashi M, Kobayashi J (1997) The optical activity of crystalline L-aspartic acid. Acta Cryst A53:763–771

Ashok Kumar R, Ezhil Vizhi R, Vijayan N, RajanBabu D (2011) Structural, dielectric and piezoelectric properties of nonlinear optical γ-glycine single crystals. Phys B Condensed Matter 406:2594–2600

Averbuch- Pouchot MT (1993) Crystal structure of L-histidinium phosphite and a structure reinvestigation of the monoclinic form of L-histidine. Z Kristallogr 207:111–120

Balashova EV, Krichevtsov BB (2011) Amino-acid ferroelectric thin films. In: Lallart M (ed) Ferroelectrics – material aspects, InTech. Available from: http://www.intechopen.com/books/ferroelectrics-material-aspects/amino-acid-ferroelectric-thin-films

Bakke O, Mostad M (1980) The structure and conformation of tryptophan in the crystal of the pure racemic compound and the hydrogen oxalate. Acta Chem Scand B 34:559

Bansinath M, Arbabha B, Turndorf H, Garg UC (1993) Chronic administration of a nitric oxide synthase inhibitor, *N* omega-nitro-L-arginine, and drug-induced increase in cerebellar cyclic GMP in vivo. Neurochem Res 18:1063–1066

Becker JW, Thathachari YT, Simpson PG (1973) The crystal and molecular structure of 3, 4-dihydroxy L-phenyl-alanine. Proc Ind Acad Sci A 77:99–129

Belo EA, Lima JA, Freire PTC, Melo FEA, Filho JM, Bordallo HN, Polian A (2010) High-pressure Raman spectra of racemate DL-alanine crystals. Vib Spectrosc 54:107–111

Bendeif E, Jelsch C (2007) The experimental library multipolar atom model refinement of L-aspartic acid. Acta Cryst C63:o361–o364

Benedetti E, Pedone C, Sirigu A (1973a) The crystal structure of DL-isoleucine and structural relations between racemic and optically active pairs in some amino acids. Acta Cryst B29:730–733

Benedetti E, Pedone C, Sirigu A (1973b) Gazzetta Chim Ital 103:555–561

Bernal JD (1931) The crystal structure of the natural amino acids and related compounds. Z Kristallogr Kristallgeom Kristallphys Kristallchem 78(1931):363–369

Bernstein J (1991) Polymorphism of L-glutamic acid: decoding the α-β phase relationship via graph-set analysis. Acta Cryst B47:1004–1010

Bernstein J, Davey RJ, Henck JO (1999) Concomitant polymorphs. Angew Chem Int Ed Engl 38:3440–3461

Black SN, Davey RJ (1988) Crystallization of amino acids. J Cryst Growth 90:136–144

Boggs B, Donohue J (1971) The unit cell and space group of L-tyrosine. Acta Cryst B27:247

Boldyreva E (2008) Crystalline amino acids. In: Boeyens JCA, Ogilvie JF (eds) Models, mysteries and magic of molecules. Springer, Cham

Boldyreva EV, Drebushchak VA, Drebushchak TN, Paukov IE, Kovalevskaya YA, Shutova ES (2003a) Polymorphs of glycine – Part I. Relative stability of the polymorphs. J Therm Anal Calorim 73:409–418

Boldyreva EV, Drebushchak VA, Drebushchak TN, Paukov IE, Kovalevskaya YA, Shutova ES (2003b) Polymorphs of glycine – Part II. Polymorphic transitions. J Therm Anal Calorim 73:419–428

Boldyreva EV, Ahsbahs H, Weber HP (2003c) A comparative study of pressure-induced lattice strain of α- and γ-polymorphs of glycine. Z Kristallogr 218:231–236

Boldyreva EV, Sowa H, Seryotkin YV, Drebushchak TN, Ahsbahs H, Chernyshev V, Dmitriev V (2006a) Pressure-induced phase transitions in crystalline L-serine studied by single-crystal and high-resolution powder X-ray diffraction. Chem Phys Lett 429:474–478

Boldyreva EV, Kolesnik EN, Drebushchak TN, Sowa H, Ahsbahs H, Seryotkin YV (2006b) A comparative study of the anisotropy of lattice strain induced in the crystals of DL-serine by cooling down to 100 K, or by increasing pressure up to 8.6 GPa. A comparison with L-serine. Z Kristallogr 221:150–161

Borissova AO, Antipin MY, Karapetyan HA, Petrosyan AM, Lyssenko KA (2010) Cooperativity effects of H-bonding and charge transfer in an L-nitroarginine crystal with $Z' > 1$. Mendeleev Commun 20:260–262

Braconnot H (1820) Mémoiresur la conversion de matièresanimales en nouvelles substances par le moyen de l'acidesulfurique. Ann Chim Phys 13:113–115

Chakraborty S, Bera AK, Ghosh S, Bhattacharya S, Pal AK, Mukhopadhyay BP, Banerjee A (2001) Hydrogen bonding potentiality of serine in more aquated environment: crystal and molecular structure of dihydrated DL-serine, $C_3H_7NO_3.2H_2O$. Indian J Phys A75:499–501

Chaney MO, Steinrauf LK (1974) The crystal and molecular structure of tetragonal L-cystine. Acta Cryst B30:711–716

Ciunik Z, Glowiak T (1983) DL-Glutamic acid monohydrate, $C_5H_9NO_4.H_2O$. Acta Cryst C39:1271–1273

Cochran W, Penfold BR (1952) The crystal structure of L-glutamine. Acta Cryst 5:644–653

Coll M, Solans X, Font-Altaba M, Subirana JA (1986) Structure of L-leucine: a redetermination. Acta Cryst C42:599–601

Coppens P, Abramov Y, Carducci M, Korjov B, Novozhilova I, Alhambra C, Pressprich MR (1999) Experimental charge densities and intermolecular interactions: electrostatic and topological analysis of DL-histidine. J Am Chem Soc 121:2585–2593

Courvoisier E, Williams PA, Lim GK, Hughes CE, Harris KDM (2012) The crystal structure of L-arginine. Chem Commun 48:2761–2763

Craven BM, Weber H-P (1983) Charge density in the crystal structure of γ-aminobutyric acid at 122 K – an intramolecular methylene H bridge. Acta Cryst B39:743–748

Curis E, Nicolis I, Moinard C, Osowska S, Zerrouk N, Be'nazeth S, Cynober L (2005) Almost all about citrulline in mammals. Amino Acids 29:177–205

Dahaoui S, Pichon-Pesme V, Howard JAK, Lecomte C (1999) CCD charge density study on crystals with large unit cell parameters: the case of hexagonal L-cystine. J Phys Chem A103:6240–6250

Dalhus B, Görbitz CH (1996a) Triclinic form of DL-valine. Acta Cryst C52:1759–1761

Dalhus B, Görbitz CH (1996b) Crystal structures of hydrophobic amino acids I. Redetermination of L-methionine and L-valine at 120K. Acta Chem Scand 50:544–548

Dalhus B, Görbitz CH (2000) Structural relationships in crystals accommodating different stereoisomers of 2-amino-3-methylpentanoic acid. Acta Cryst B56:720–727

Davidson N (1976) Neurotransmitter amino acids. Academic Press, London

Dawson A, Allan RD, Belmonte SA, Clark SJ, David WIF, McGregor PA, Parsons S, Pulham CR, Sawyer L (2005) Effect of high pressure on the crystal structures of polymorphs of glycine. Cryst Growth Des 5:1415–1427

Derissen JL, Endeman HJ, Peerdeman AF (1968) The crystal and molecular structure of L-aspartic acid. Acta Cryst B24:1349–1354

DiBlasio B, Pedone C, Sirigu A (1975) DL-Leucine. Acta Cryst B31:601–602

Dittrich B, Hübschle CB, Messerschmidt M, Kalinowski R, Girnt D, Luger P (2005) The invariom model and its application: refinement of D, L-serine at different temperatures and resolution. Acta Cryst A61:314–320

Dobson AJ, Gerkin RE (1996) γ-Aminobutyric acid: a novel tetragonal phase. Acta Cryst C52:3075–3078

Donohue J (1950) The crystal structure of DL-alanine. II. Revision of parameters by three-dimensional Fourier analysis. J Am Chem Soc 72:949–953

Drebushchak TN, Boldyreva EV, Shutova ES (2002) β-Glycine. Acta Cryst E58:o634–o636
Drechsel E (1889) Zur Kenntniss der Spaltungsprodukte des Caseins. J prakt Chem 39:425–429
Dunitz JD, Schweizer WB (1995) Anhydrous DL-glutamic acid. Acta Cryst C51:1377–1379
Dunn MS, Brophy TW (1932) Decomposition points of the amino acids. J Biol Chem 99:221–229
Edington P, Harding MM (1974) The crystal structure of DL-histidine. Acta Cryst B30:204–206
Ejgenberg M, Mastai Y (2012) Crystallization on self assembled monolayers. In: Mastai Y (ed) Advances in crystallization processes, InTech. Available from: http://www.intechopen.com/books/advances-in-crystallization-processes/crystallization-on-self-assembled-monolayers
Elsevier (2013) ScienceDirect. http://www.info.sciverse.com/. Accessed 3 Apr 2013
Façanha Filho PF, Freire PTC, Lima KCV, Mendes Filho J, Melo FEA (2008) High temperature Raman spectra of L-leucine crystals. Brazilian J Phys 38:131–137
Façanha Filho PF, Freire PTC, Melo FEA, Lemos V, Mendes Filho J, Pizani PS, Rossatto DZ (2009) Pressure-induced phase transitions in L-leucine crystal. J Raman Spectrosc 40:46–51
Ferrari ES, Davey RJ, Cross WI, Gillon AL, Towler CS (2003) Crystallization in polymorphic systems: the solution-mediated transformation of β to α glycine. Cryst Growth Des 3:53–60
Fischer E (1901) Ueber die Hydrolyse des Caseins durch Salzsaeure. Hoppe-Seyler's Z Phys Chem 33:151–176
Fischer E (1905) Synthese von Polypeptiden. XIII. Chloride der Aminosäuren und Polypeptide und ihre Verwendung zur Synthese. Ber Dtsch Chem Ges 38:2914–2925
Flaig R, Koritsanszky T, Zobel D, Luger P (1998) Topological analysis of the experimental electron densities of amino acids. 1. D, L-Aspartic acid at 20 K. J Am Chem Soc 120:2227–2238
Flaig R, Koritsanszky T, Dittrich B, Wagner A, Luger P (2002) Intra and intermolecular topological properties of amino acids: a comparative study of experimental and theoretical results. J Am Chem Soc 124:3407–3417
Fleck M, Petrosyan AM (2009) Comments on papers reporting IR-spectra and other data of alleged L-alanine alaninium nitrate and L-alanine sodium nitrate crystals. Cryst Res Technol 44:769–772
Fleck M, Petrosyan AM (2010) Difficulties in the growth and characterization of non-linear optical materials: a case study of salts of amino acids. J Cryst Growth 312:2284–2290
Fleck MF, Ghazaryan VV, Petrosyan AM (2012) Crystal structure at 296 and 150 K, vibrational spectra and thermal behaviour of sarcosine sarcosinium nitrate. Z Kristallogr 227:819–824
Frey MN, Lehmann MS, Koetzle TF, Hamilton WC (1973a) Precision neutron diffraction structure determination of protein and nucleic acid components. XI. Molecular configuration and hydrogen bonding of serine in the crystalline amino acids L-serine monohydrate and DL-serine. Acta Cryst B29:876–884
Frey MN, Koetzle TF, Lehmann MS, Hamilton WC (1973b) Precision neutron diffraction structure determination of protein and nucleic acid components. X. A comparison between the crystal and molecular structures of L-tyrosine and L-tyrosine hydrochloride. J Chem Phys 58:2547–2556
Garetz BA, Matic J, Myerson AS (2002) Polarization switching of crystal structure in the nonphotochemical light-induced nucleation of supersaturated aqueous glycine solutions. Phys Rev Lett 89:175501–175504
Girija EK, Kalkura SN, Ramasamy P (1995) Crystallization of cystine. J Mat Sci Mat Med 6:617–619
Givand JC, Rousseau RW, Ludovice PJ (1998) Characterization of L-isoleucine crystal morphology from molecular modeling. J Cryst Growth 194:228–238
Görbitz CH, Dalhus B (1996a) Redetermination of L-leucine at 120K. Acta Cryst C52:1754–1756
Görbitz CH, Dalhus B (1996b) L-Isoleucine, redetermination at 120K. Acta Cryst C52:1464–1466
Görbitz CH, Dalhus B (1996c) L-Cysteine, monoclinic form, redetermination at 120K. Acta Cryst C52:1756–1759
Görbitz CH, Törnroos KW, Day GM (2012) Single-crystal investigation of L-tryptophan with $Z' = 16$. Acta Cryst B68:549–557

Goryainov SV, Boldyreva EV, Kolesnik EN (2006) Raman observation of a new (ζ) polymorph of glycine? Chem Phys Lett 419:496–500

Greenstein JP, Winitz M, Robert E (1984) The chemistry of the amino acids, vol 3. Krieger Publishing Company, Malabar

Hamend JP, Helgeson HC (1997) Solubilities of the common L-α-amino acids as a function of temperature and solution pH. Pure Appl Chem 69:935–942

Harding MM, Howieson RM (1976) L-Leucine. Acta Cryst B32:633–634

Harding MM, Long HA (1968) The crystal and molecular structure of L-cysteine. Acta Cryst B24:1096–1102

Harris RC, Tallon MJ, Dunnett M, Boobis L, Coakley J, Kim HJ, Fallowfield JL, Hill CA, Sale C, Wise JA (2006) The absorption of orally supplied β-alanine and its effect on muscle carnosine synthesis in human vastus lateralis. Amino Acids 30:279–289

Hartmann EL (1986) Effect of L-tryptophan and other amino acids on sleep. Nutrition Rev 44:70–73

Henneberger C, Bard L, Rusakov DA (2012) D-Serine: a key to synaptic plasticity? Int J Biochem Cell Biol 44:587–590

Heredia A, Meunier V, Bdikin IK, Gracio J, Balke N, Jesse S, Tselev A, Agarwal PK, Sumpter BG, Kalinin SV, Kholkin AL (2012) Nanoscale ferroelectricity in crystalline γ-glycine. Adv Funct Mater 22:2996–3003

Hill CA, Harris RC, Kim HJ, Harris BD, Sale C, Boobis LH, Kim CK, Wise JA (2007) Influence of beta-alanine supplementation on skeletal muscle carnosine concentrations and high intensity cycling capacity. Amino Acids 32:225–233

Hirayama N, Shirahata K, Ohashi Y, Sasada Y (1980) Structure of α form of L-glutamic acid. α–β transition. Bull Chem Soc Jpn 53:30–35

Hirokawa S (1955) A new modification of L-glutamic acid and its crystal structure. Acta Cryst 8:637–641

Hopkins FG, Cole SW (1901) A contribution to the chemistry of proteids: Part I. A preliminary study of a hitherto undescribed product of tryptic digestion. J Physiol 27:418–428

Howard ST, Hursthouse MB, Lehmann CW, Poyner EA (1995) Experimental and theoretical determination of electronic properties in L-dopa. Acta Cryst B51:328–337

Hübschle CB, Messerschmidt M, Luger P (2004) Crystal structure of DL-tryptophan at 173K. Cryst Res Technol 39:274–278

Iitaka Y (1953) A technique for testing the piezoelectric properties of crystals. Acta Cryst 6:663–664

Iitaka Y (1958) The crystal structure of γ-glycine. Acta Cryst 11:225–226

Iitaka Y (1960) The crystal structure of β-glycine. Acta Cryst 13:35–45

Iitaka Y (1961) The crystal structure of γ-glycine. Acta Cryst 14:1–10

Isakov D, de Matos Gomes E, Bdikin I, Almeida B, Belsley M, Costa M, Rodrigues V, Heredia A (2011) Production of polar β-glycine nanofibers with enhanced nonlinear optical and piezoelectric properties. Cryst Growth Des 11:4288–4291

Iwama S, Horiguchi M, Sato H, Uchida Y, Takahashi H, Tsue H, Tamura R (2010) Observation of the preferential enrichment phenomenon for essential α-amino acids with a racemic crystal structure. Cryst Growth Des 10:2668–2675

Janczak J, Luger P (1997) L-Proline monohydrate at 100 K. Acta Cryst C53:1954–1956

Janczak J, Zobel D, Luger P (1997) L-Threonine at 12 K. Acta Cryst C53:1901–1904

Jose P, Pant LM (1965) The crystal and molecular structure of β-alanine. Acta Cryst 18:806–810

Karle IL, Karle J (1964) An application of the symbolic addition method to the structure of L-arginine dihydrate. Acta Cryst 17:835–841

Kartha G, de Vries A (1961) Structure of asparagine monohydrate. Nature 192:862–863

Kayushina RL, Vainshtein BK (1965) X-ray determination of the structure of L-proline. Kristallografiya 10:833–844

Kerr KA, Ashmore JP (1973) Structure and conformation of orthorhombic L-cysteine. Acta Cryst B29:2124–2127

Kerr KA, Ashmore JP, Koetzle TF (1975) A neutron diffraction study of L-cysteine. Acta Cryst B31:2022–2026

Khawas B (1970) The unit cells and space groups of L-methionine, L-β-phenylalanine and DL-tyrosine. Acta Cryst B26:1919–1922

Khawas B (1971) X-ray study of L-arginine HCl, L-cysteine, DL-lysine and DL-phenylalanine. Acta Cryst B27:1517–1520

Khawas B (1985) D-Tryptophan. Indian J Phys A59:219–226

King MD, Buchanan WD, Korter TM (2010) Investigating the anharmonicity of lattice vibrations in water-containing molecular crystals through the terahertz spectroscopy of L-serine monohydrate. J Phys Chem A 114(35):9570–9578

King MD, Blanton TN, Korter TM (2012) Revealing the true crystal structure of L-phenylalanine using solid-state density functional theory. Phys Chem Chem Phys 14:1113–1116

Kingsford-Adaboh R, Grosche M, Dittrich B, Luger P (2000) Arginine monohydrate at 100 K. Acta Cryst C56:1274–1276

Kingsford-Adaboh R, Dittrich B, Wagner A, Messerschmidt M, Flaig R, Luger P (2002) Topological analysis of DL-arginine monohydrate at 100 K. Z Kristallogr 217:168–173

Kistenmacher TL, Rand GA, Marsh RE (1974) Refinements of the crystal structures of DL-serine and anhydrous L-serine. Acta Cryst B30:2573–2578

Koetzle TF, Frey MN, Lehmann MS, Hamilton WC (1973) Precision neutron diffraction structure determination of protein and nucleic acid components. XIII. Molecular and crystal structure of the amino acid L-glutamine. Acta Cryst B29:2571–2575

Kolesov BA, Minkov VS, Boldyreva EV, Drebushchak TN (2008) Transitions in the crystals of L- and DL-cysteine on cooling: intermolecular hydrogen bonds distortions and the side-chain motions of thiol-groups. 1. L-Cysteine. J Phys Chem B112:12827–12839

Kopple JD, Swendseid ME (1975) Evidence that histidine is an essential amino acid in normal and chronically uremic man. J Clin Invest 55:881–891

Kyte J, Doolittle R (1982) A simple method for displaying the hydropathic character of a protein. J Mol Biol 157:105–132

Lane H, Huang C, Wu P, Liu Y, Chang Y, Lin P, Chen P, Tsai G (2006) Glycine transporter I inhibitor, *N*-methylglycine (sarcosine), added to clozapine for the treatment of schizophrenia. Biol Psychiatry 60:645–649

Langan P, Mason SA, Myles D, Schoenborn BP (2002) Structural characterization of crystals of α-glycine during anomalous electrical behaviour. Acta Cryst B58:728–733

Lechuga-Ballesteros D, Rodriguez-Hornedo N (1993) Growth and morphology of L-alanine crystals: influence of additive adsorption. Pharm Res 10:1008–1014

Lee AY, Ulman A, Myerson AS (2002) Crystallization of amino acids on self-assembled monolayers of rigid thiols on gold. Langmuir 18:5886–5898

Legros JP, Kvick A (1980) Deformation electron density of α-glycine at 120 K. Acta Cryst B36:3052–3059

Lehmann MS, Nunes AC (1980) A short hydrogen bond between near identical carboxyl groups in the α-modification of L-glutamic acid. Acta Cryst B36:1621–1625

Lehmann MS, Koetzle TF, Hamilton WC (1972a) Precision neutron diffraction structure determination of protein and nucleic acid components. I. Crystal and molecular structure of the amino acid L-alanine. J Am Chem Soc 94:2657–2660

Lehmann MS, Koetzle TF, Hamilton WC (1972b) Precision neutron-diffraction structure determination of protein and nucleic-acid components. 4. Crystal and molecular structure of amino acid L-histidine. Int J Pept Protein Res 4:229–239

Lehmann MS, Verbist JJ, Hamilton WC, Koetzle TF (1973) Precision neutron diffraction structure determination of protein and nucleic acid components. Part V. Crystal and molecular structure of the amino-acid L-arginine dihydrate. J Chem Soc Perkin Trans 2:133–137

Lehninger AL, Nelson DL, Cox MM (2000) Lehninger principles of biochemistry, 3rd edn. Worth Publishing, New York

Lemanov VV (2000) Piezoelectric and pyroelectric properties of protein amino acids as basic materials of soft state physics. Ferroelectrics 238:211–218

Lemanov VV (2011) Piezoelectricity in protein amino acids. Phys Solid State 53:1191–1193

Levy HA, Corey RB (1941) The crystal structure of DL-alanine. J Am Chem Soc 63:2095–2108
Li L, Rodríguez-Hornedo N (1992) Growth kinetics and mechanism of glycine crystals. J Cryst Growth 121:33–38
Li Y, Xiao F, Moon KS, Wong CP (2006) Novel curing agent for lead-free electronics: amino acid. J Polymer Sci A 44:1020–1027
Lindemann B, Ogiwara Y, Ninomiya Y (2002) The discovery of umami. Chem Senses 27:843–844
Lozano-Casal P, Allan DR, Parsons S (2008) High-pressure structural study of L-α-glutamine and the use of Hirshfeld surfaces and graph-set notation to investigate the hydrogen bonding present in the structure up to 4.9 GPa. Acta Cryst B64:466–475
Luger P, Weber M (1999) DL-Cysteine at 298K. Acta Cryst C55:1882–1885
Madden JJ, McGandy EL, Seeman NC (1972a) The crystal structure of the orthorhombic form of L-(+)-histidine. Acta Cryst B28:2377–2382
Madden JJ, McGandy EL, Seeman NC, Harding MM, Hoy A (1972b) The crystal structure of the monoclinic form of L-histidine. Acta Cryst B28:2382–2389
Mahalakshmi R, Jesuraja SX, Das DS (2006) Growth and characterization of L-phenylalanine. Cryst Res Tech 41:780–783
Mallikarjunan M, Rao ST (1969) Crystal structure of DL-valine. Acta Cryst B25:296–303
Marsh RE (1957) A refinement of the crystal structure of glycine. Acta Cryst 10:814
Matei A, Drichko N, Gompf B, Dressel M (2005) Far-infrared spectra of amino acids. Chem Phys 316:61–71
Mathieson AML (1953) The crystal structures of the dimorphs of DL-methionine. Acta Cryst 5:332–341
Matsuoka M, Yamanobe M, Tezuka N, Takiyama H, Ishii H (1999) Polymorphism, morphologies and bulk densities of DL-methionine agglomerate crystals. J Cryst Growth 188–189:1299–1306
McCoy RH, Meyer CE, Rose WC (1935) Feeding experiments with mixtures of highly purified amino acids. VIII. Isolation and identification of a new essential amino acid. J Biol Chem 112:283–302
Minkov VS, Boldyreva EV (2012) The effect of partial methylation of the glycine amino group on crystal structure in *N*, *N*-dimethylglycine and its hemihydrate. Acta Cryst C68:o283–o287
Minkov VS, Tumanov NA, Kolesov BA, Boldyreva EV, Bizyaev SN (2009) Phase transitions in the crystals of L- and DL-cysteine on cooling: the role of the hydrogen-bond distortions and the side-chain motions. 2. DL-Cysteine. J Phys Chem B113:5262
Minkov VS, Tumanov NA, Cabrera RQ, Boldyreva EV (2010) Low temperature/high pressure polymorphism in DL-cysteine. CrystEngComm 12:2551–2560
Moggach SA, Allan DR, Morrison CA, Psrsons S, Sawyer L (2005a) Effect of pressure on the crystal structure of L-serine-I and the crystal structure of L-serine-II at 5.4 GPa. Acta Cryst B61:58–68
Moggach SA, Clark SJ, Parsons S (2005b) L-Cysteine-I at 30 K. Acta Cryst E 61:o2739–o2742
Moggach SA, Allan DR, Parsons S, Sawyer L, Warren JE (2005c) The effect of pressure on the crystal structure of hexagonal L-cystine. J Synch Rad 12:598–607
Moggach SA, Marshall WG, Parsons S (2006a) High-pressure neutron diffraction study of L-serine-I and L-serine-II, and the structure of L-serine-III at 8.1 GPa. Acta Cryst B62:815–825
Moggach SA, Allan DR, Clark SJ, Gutmann MJ, Parsons S, Pulham CR, Sawyer L (2006b) High-pressure polymorphism in L-cysteine: the crystal structures of L-cysteine-III and L-cysteine-IV. Acta Cryst B62:296–309
Moggach SA, Parsons S, Wood PA (2008) High-pressure polymorphism in amino acids. Cryst Rev 14:143–184
Monaco SB, Davis LE, Velsko SP, Wang FT, Eimerl D, Zalkin A (1987) Synthesis and characterization of chemical analogs of L-arginine phosphate. J Cryst Growth 85:252–255
Mondal S, Prathapa SJ, van Smaalen S (2012) Experimental dynamic electron densities of multipole models at different temperatures. Acta Cryst A 68:568–581
Mostad A, Natarajan S (1989) Crystal and molecular structure of sarcosine. Acta Chem Scand 43:1004–1006

Mostad A, Romming C (1973) Crystal-structure of DL-tyrosine. Acta Chem Scand 27:401–410
Mostad A, Ottersen T, Romming C (1971) On the structure of L-DOPA (2S-3-(3,4-dihydroxyphenyl)alanine). Acta Chem Scand 25:3549–3560
Mostad A, Nissen HM, Romming C (1972) Crystal structure of L-tyrosine. Acta Chem Scand 26:3819–3833
Mueller JH (1922) A new sulphur containing amino acid isolated from casein. Exp Biol Med 19:161–163
Myung S, Pink M, Baik MH, Clemmer DE (2005) DL-Proline. Acta Cryst C61:o506–o508
Olafsson PG, Bryan AM (1970) Evaluation of thermal decomposition temperatures of amino acids by differential enthalpic analysis. Microchim Acta 58:871–878
Oughton BM, Harrison PM (1959) The crystal structure of hexagonal L-cystine. Acta Cryst 12:396–404
Padmanabhan S, Suresh S, Vijayan M (1995) DL-Proline monohydrate. Acta Cryst C51:2098–2100
Papavinasam E, Natarajan S, Shivaprakash NC (1986) Reinvestigation of the crystal structure of β-alanine. Int J Pept Protein Res 28:525–527
Patil BS, Jayaprakasha GK, Murthy KNC, Vikram A (2009) Bioactive compounds: historical perspectives, opportunities, and challenges. J Agric Food Chem 57:8142–8160
Pavlou M, Diamandis EP (2009) The search for new prostate cancer biomarkers continues. Clin Chem 55:1277–1279
Perlovich GL, Hansen LK, Bauer-Brandl A (2001) The polymorphism of glycine. Thermochemical and structural aspects. J Therm Anal Calorim 66:699–715
Petrosyan AM, Ghazaryan VV, Fleck M (2013) On the existence of "L-threonine formate", "L-alanine lithium chloride" and "bis L-alanine lithium chloride" crystals. Spectrochim Acta 105:623–625
Plimmer RHA, Hopkins FG (eds) (2010) The chemical composition of the proteins. Monographs on biochemistry. Part I. Analysis, 2nd edn. Longmans, Green and Co, London, p 112
Ramachandran E, Natarajan S (2009) Synthesis of L-valine crystals. Cryst Res Technol 44:641–646
Ramanadham M, Sikka SK, Chidambaram R (1972) Structure of L-asparagine monohydrate by neutron diffraction. Acta Cryst B28:3000–3005
Ramanadham M, Sikka SK, Chidambaram R (1973) Structure determination of L_s-threonine by neutron diffraction. Pramana 1:247–259
Rao ST (1973) Refinement of DL-aspartic acid. Acta Cryst B29:1718–1720
Rodriguez JJ Jr, Rodrigues Misoguti L, Nunes FD, Mendonça CR, Zilio SC (2003) Optical properties of L-threonine crystals. Opt Mater 22:235–240
Rose WC (1931) Feeding experiments with mixtures of highly puriefied amino acids: I. The inadequacy of diets containing nineteen amino acids. J Biol Chem 94:155–165
Sakai H, Hosogai H, Kawakita T, Onuma K, Tsukamoto K (1992) Transformation of α-glycine to γ-glycine. J Cryst Growth 116:421–426
Sakthivel N, Anbarasan PM (2011) Growth, structural and optical studies of L-proline hydrate crystals. Int J Curr Res 3:386–389
Sano C, Nagashima N (1996) Effects of L-glutamyl peptide on the growth of L-glutamic acid crystals (α-form). J Cryst Growth 166:129–135
Sano C, Kashiwagi T, Nagashima N, Kawakita T (1997) Effects of additives on the growth of L-glutamic acid crystals (β-form). J Cryst Growth 178:568–574
Schulze E, Steiger E (1886) Ueber einen neuen stickstoffhaltigen Bestandtheil der Keimlinge von Lupinus luteus. Ber Dt Chem Ges 19:1177–1180
Schultz E, Steiger E (1886) Über das Arginin. Z Physiol Chem 11:43–65
Seijas LE, Delgado GE, Moraa AJ, Fitch AN, Brunelli M (2010) On the crystal structures and hydrogen bond patterns in proline pseudopolymorphs. Powder Diff 25:235–240
Sekar C, Parimaladevi R (2009) Effect of KCl addition on crystal growth and spectral properties of glycine single crystals. Spectrochim Acta A74:1160–1164
Sequeira A, Rajagopal H, Ramanadham H (1989) A neutron study on the structure of DL-aspartic acid. Acta Cryst C45:906–908

Sevier CS, Kaiser CA (2002) Formation and transfer of disulphide bonds in living cells. Nat Rev Mol Cell Biol 3:836–847

Shanti A, Krishnan C, Selvarajan P (2013) Optical, mechanical and thermal characterization of L-threonine single crystals grown in dimethyl urea solution. Phys Scr 88:035801

Shimon LJW, Lahav M, Leiserowitz L (1986) Stereoselective etchants for molecular crystals. Resolution of enantiomorphs and assignment of absolute structure of chiral molecules and polar crystals. Nouv J de Chim 10:723–737

Shoemaker DP, Donohue J, Schomaker V, Corey RB (1950) The crystal structure of L-threonine. J Am Chem Soc 72:2328–2349

Shoemaker DP, Barieau RE, Donohue J, Lu CS (1953) The crystal structure of DL-serine. Acta Cryst 6:241–256

Simpson HJ, Marsh RE Jr (1966) The crystal structure of L-alanine. Acta Cryst 20:550–555

Srinivasan BR, Dhavskar KD (2014) Comments on the paper: comparative study of mechanical, dielectric and electrical properties of solution grown semi-organic NLO crystal glycine with additives-ammonium oxalate, potassium and barium nitrate. Indian J Pure Appl Phys 52:60–63

Srinivasan TP, Indirajith R, Gopalakrishnan R (2011) Growth and characterization of α and γ-glycine single crystals. J Cryst Growth 318:762–767

Steward EG, Player RB, Warner D (1973) The crystal and molecular structure of γ-aminobutyric acid determined at low temperature. Acta Cryst B29:2038–2040

Strecker A (1850) Ueber die künstliche Bildung der Milchsäure und einen neuen, dem Glycocoll homologen Körper. Justus Liebigs Ann Chem 75:27–45

Subha Nandhini M, Krishnakumar RV, Natarajan S (2001) DL-Alanine. Acta Cryst C57:614–615

Sueda M, Katsuki A, Fujiwara Y, Tanimoto Y (2006) Influences of high magnetic field on glycine crystal growth. Sci Techn Adv Mater 7:380–384

Suresh CG, Vijayan M (1893) Occurrence and geometrical features of head-to-tail sequences involving amino acids in crystal structures. Int J Pept Protein Res 22:129–143

Suresh S, Padmanabhan S, Vijayan M (1994) X-ray studies on crystalline complexes involving amino acids and peptides. XXVII. Effect of chirality, specific interactions and characteristic aggregation patterns in the structures of arginine and its complexes with formic acid. J Biomol Struct Dyn 11:1425–1435

Suresh S, Padmanabhan S, Vijayan M (1996) DL-Glutamine. Acta Cryst C52:1313–1316

Surovtsev NV, Adichtchev SV, Malinovsky VK, Ogienko AG, Drebushchak VA, Manakov AY, Ancharov AI, Yunoshev AS, Boldyreva EV (2012) Glycine phases formed from frozen aqueous solutions: revisited. J Chem Phys 137:065103-1–065103-10

Swaminathan P, Srinivasan R (1975) Studies in molecular structure, symmetry and conformation. VIII. Crystal and molecular structure of L-allothreonine. Acta Cryst B31:217–221

Taniguchi T, Takaki Y, Sakurai K (1980) The crystal structures of the and forms of DL-methionine. Bull Chem Soc Jpn 53:803–804

Texas A&M University (2008) Watermelon may have viagra-effect. ScienceDaily. Retrieved December 7, 2013, from http://www.sciencedaily.com/releases/2008/06/080630165707.htm

Thomas JM (2012) Determining the structure of L-arginine and other organic molecules. Chem Phys Chem 13:2637–2638

Thorpe TE (1913) A dictionary of applied chemistry. Longmans, Green, & Co., New York

Tilborg A, Norberg B, Wouters J (2014) Pharmaceutical salts and cocrystals involving amino acids: a brief structural overview of the state-of-art. Eur J Med Chem 74:411–426

Toffoli P, Rodier N, Astoin J (1986) Structure cristalline de l'acide L-amino-2-ur_eido-5-pentanoïque dihydrate (L-citrulline dihydrate). Bull Soc Chim Fr 1:116–123

Toffoli P, Khodadad P, Rodier N, Astoin J (1987) Structure de l'acide L-amino-2 uréido-5 pentanoïque (L-citrulline). Acta Cryst 43:945–947

Tomita KI, Higashi H, Fujiwara T (1973) Crystal and molecular structure of ω-amino acids, ω-amino sulfonic acids and their derivatives. IV. The crystal and molecular structure of γ-aminobutyric acid (GABA), a nervous inhibitory transmitter. Bull Chem Soc Jpn 46:2199–2204

Torii K, Iitaka Y (1970) The crystal structure of L-valine. Acta Cryst B26:1317–1326

Torii K, Iitaka Y (1971) The crystal structure of L-isoleucine. Acta Cryst B27:2237–2246

Torii K, Iitaka Y (1973) Crystal structures and molecular conformations of L-methionine and L-norleucine. Acta Cryst B29:2799–2807

True AB, Schroeck K, French TA, Schmuttenmaer CA (2011) Terahertz spectroscopy of histidine enantiomers and polymorphs. J Infrared Milli Terahz Waves 32:691–698

Trzebiatowska-Gusowska M, Gagor A (2007) [$^2H_3$]Sarcosine. Acta Cryst E63:o4694–o4694

Tumanov NA, Boldyreva EV (2012) X-ray diffraction and Raman study of DL-alanine at high pressure: revision of phase transitions. Acta Cryst B68:412–423

Tylczyński Z, Sterczyńska A, Wiesner M (2011) Temperature dependences of piezoelectric, elastic and dielectric constants of L-alanine crystal. J Phys Condens Matter 23:355901

Umadevi K, Anitha K, Sridhar B, Srinivasan N, Rajaram RK (2003) L-Aspartic acid monohydrate. Acta Cryst E59:o1073–o1075

van Beek JD, Hess S, Vollrath F, Meier BH (2002) The molecular structure of spider dragline silk: Folding and orientation of the protein backbone. Proc Natl Acad Sci U S A 99:10266–10271

Vauquelin LN, Robiquet PJ (1806) La découverte d'un nouveau principe végétal dans le suc des asperges. Ann Chim 57:88–93

Verbist JJ, Lehmann MS, Koetzle TF, Hamilton WC (1972) Precision neutron diffraction structure determination of protein and nucleic acid components. VI. The crystal and molecular structure of the amino acid L-asparagine monohydrate. Acta Cryst B28:3006–3013

Wada M (1930) Über Citrullin, eine neue Aminosäure im Presssaft der Wassermelone, Citrullis vulgaris schrad. Biochem Z 224:420–429

Wagner A, Luger P (2001) Charge density and topological analysis of L-glutamine. J Mol Struct 595:39–46

Wang JL, Berkovitch-Yellin Z, Leiserowitz L (1985) Location of 'tailor-made' additives in the crystal and their effect on crystal habit. A study on the host-additive system L-asparagine-L-aspartic acid monohydrate. Acta Cryst B41:341–348

Wang JJ, Wang XC, Ouyang YS, You YJ, Dai AB (1996) The solution structure studies of L-lysine by means of NMR lanthanide shift reagent. Acta Phys Chem Sin 8:647–653

Wang WQ, Gong Y, Liang Z, Sun FL, Shi DX, Gao HJ, Lin X, Jiang P, Wang ZM (2002) Direct observation of surface structure of D-alanine and D/L-valine crystals by atomic force microscopy and comparison with X-ray diffraction analysis. Surf Sci 512:L379–L384

Wang GM, Li ZX, Duan CS, Li H (2007) DL-Aspartic acid. Acta Cryst E63:o4003

Watanabe M, Maemura K, Kanbara K, Tamayama T, Hayasaki H (2002) GABA and GABA receptors in the central nervous system and other organs. Int Rev Cytol 213:1–47

Weisinger-Lewin Y, Frolow F, McMullan RK, Koetzle TF, Lahav M, Leiserowitz L (1989) Reduction in crystal symmetry of a solid solution: a neutron diffraction study at 15 K of the host/guest system asparagine/aspartic acid. J Am Chem Soc 111:1035–1040

Weissbuch I, Frolow F, Addadi L, Lahav M, Leiserowitz L (1990) Oriented crystallization as a tool for detecting ordered aggregates of water-soluble hydrophobic.alpha.-amino acids at the air-solution interface. J Am Chem Soc 112:7718–7724

Williams PA, Hughes CE, Buanz ABM, Gaisford S, Harris KDM (2013) Expanding the solid-state landscape of L-phenylalanine: discovery of polymorphism and new hydrate phases, with rationalization of hydration/dehydration processes. J Phys Chem C117:12136–12145

Wolfenden R, Andersson L, Cullis PM, Southgate CCB (1981) Affinities of amino acid side chains for solvent water. Biochemistry 20:849–855

Wood PA, Francis D, Marshall WG, Moggach SA, Parsons S, Pidcock E, Rohl AL (2008) A study of the high-pressure polymorphs of L-serine using ab initio structures and PIXEL calculations. Cryst Eng Comm 10:1154–1166

Woodman OL, Dusting GJ (1991) *N*-nitro L-arginine causes coronary vasoconstriction and inhibits endothelium-dependent vasodilatation in anaesthetized greyhounds. Br J Pharmacol 103:1407–1410

Wright BA, Cole PA (1949) Preliminary examination of the crystal structure of L-proline. Acta Cryst 2:129–130

Xianglin J, Zuohua P, Ronghan G, Qichen H (1985) Determination of the crystal structure of L-citrulline dihydrate and α-hydroxymethylserine. Acta Chim Sin 43:5–9
Yamada K, Hashizume D, Shimizua T, Yokoyama S (2007) L-Asparagine. Acta Cryst E63:o3802–o3803
Yamashita M, Inomata S, Ishikawa K, Kashiwagi T, Matsuo H, Sawamura S, Kato M (2007) A high-pressure polymorph of L-α-leucine. Acta Cryst E63:o2762–o2764
Yang Z, Wang Z, Tian X, Xiu P, Zhou R (2012) Amino acid analogues bind to carbon nanotube via π-π interactions: comparison of molecular mechanical and quantum mechanical calculations. J Chem Phys 136:0250103
Zakharov BA, Kolesov BA, Boldyreva EV (2012) Effect of pressure on crystalline L- and DL-serine: revisited by a combined single-crystal X-ray diffraction at a laboratory source and polarized Raman spectroscopy study. Acta Cryst B68:275–286
Zhang YM, Cao C, Lin Q, Yang LZ, Wie TB (2006) Research of hydrogen bonding in L-Asp. Chem Res Appl 18:952–954

Chapter 3
Compounds of Amino Acids as Anions

Abstract This chapter deals with salts formed from metal cations and amino acid anions. All amino acids can act as monovalent anion; the acidic members glutamic acid and aspartic acid as well as cysteine and tyrosine can form both monovalent and divalent anions. Due to the large number of available cations, a multitude of combinations is possible and in fact found (for the standard twenty amino acids, over 150 crystal structures are published, and many more species have been characterized by other methods). Different hydration states (as reported for the pristine amino acid crystals) occur as well, and the existence of polymorphs is also documented for some cases. The formation of metal–amino acid complexes has been studied in several works (both for a given amino acid and a given cation), and the flexibility of amino acid molecules as ligands allows coordination of cations with different chemical properties (such as charge or ionic radius). Several coordination modes are found for amino acids – the molecules can act as monodentate, bidentate, tridentate, and bridging ligands. The connectivity of the coordination polyhedra of the metal cations is considered – isolated units are frequent, but chains and layers occur as well. Moreover, these units can connect via amino acid molecules to form higher-dimensional structures. Hydrogen bonds (also involving water molecules, both in coordination or in the interstices as crystal water) further stabilize the units, forming relatively stable phases. The frequency of salts varies among the different amino acids, and most examples are reported for glycinate salts, whereas crystals of cations and basic and weakly soluble amino acids are difficult to obtain (e.g., arginate and lysinate salts have not been reported in crystalline state). Finally, the aspect of symmetry is considered (the chirality of most amino acids playing a crucial role), and, consequently, the properties of these salts have impact on applications in the fields of physics, biology, and medicine.

Keywords Salts of amino acids • Crystal structure • Molecular structure • Polymorphs • Metal cations • Chemical hardness • Coordination • Isolated units • Chains • Layers • Frameworks • Connectivity • Ligands

M. Fleck and A.M. Petrosyan, *Salts of Amino Acids: Crystallization, Structure and Properties*, DOI 10.1007/978-3-319-06299-0_3,

3.1 Introducing Remarks

In the previous chapter, structural aspects of the 20 standard amino acids plus a few nonstandard examples are described. These aspects can be considered as the foundation for the present and the following chapters, where actual salts of amino acids are discussed (as promised by the title of this book). Among these salts, three types of compounds can be distinguished: Firstly, crystals comprising amino acids in anionic form and cations are considered. These species are discussed in the present chapter. Secondly, crystals of amino acid and neutral inorganic salts are reported in Chap. 4. In the latter category, the amino acids exist in overall neutral, zwitterionic form. Finally, there are crystals where amino acids act as cations, to which Chap. 5 is dedicated.

Among species of the first type, monovalent and divalent cations are the most frequent ones, usually in combination with a respective number of monovalent amino acid anions, as, for instance, M^+AA^-or $M^{2+}(AA^-)_2$ (where M is any cation, AA any amino acid anion), although there are also examples of crystals including trivalent cations. Up to date, no crystals of tetravalent cations and amino acids have been reported. The behavior of amino acids in their formation of complexes of cations has been much investigated (Kiss 1990; Laurie 1987, 1995), and it was found that (N,O)-chelation is frequent (see examples in this and the following chapter). The analysis of data showed that amino acids form very weak complexes with alkali metal cations and weak complexes with alkaline earth metal cations. Thus, crystal structural data on alkali metal amino acid salts are rare (there are only a handful of reports on salts of lithium, plus aspartates and glutamates of sodium and potassium, for which crystal structures could be obtained). Alkaline earth metal amino acid salt crystals could be grown more easily, although several species did not form crystals good enough for XRD experiments (as, e.g., the barium salts of arginine, glutamine, proline, serine, and valine, Bush et al. 2008). For a detailed review of alkali and alkaline earth metal amino acid salts, see Daniele et al. (2008). Other cations, especially transition metals, can form strong complexes with amino acids; therefore, crystal structures of these salts have been reported in great numbers.

As a multitude of metals can be combined with a multitude of amino acids (in different ratios) plus different states of hydration, a large number of salts are conceivable (and in fact have been obtained). This diversity is enlarged as both pure enantiomers and racemic mixtures have been used for synthesis.

CHEMICAL HARDNESS

Half a century ago, several theories were published which proposed rules for the prediction of complex formation in inorganic salts, i.e., predicting which kinds of ligands tend to bind to which kind of metals. In the 1960s, Pearson (1963, 1966, 1968) formulated a theory in which he classified Lewis acids and bases according to their binding preferences, defining hard and soft acids

(continued)

and bases (and borderline or intermediate acids and bases in between), an idea which was named HSAB (hard/soft acid/bases) principle. Pearson's categorization was oriented on K_f values, i.e., formation constants for the reaction of a Lewis acid and a Lewis base to a Lewis acid/base complex, and simply stated that hard acids prefer to bind to hard bases and soft acids prefer to bind to soft bases. In addition to earlier works, Pearson extended the traditional inorganic metal/ligand viewpoint to include organic chemistry.

Within this classification, hard acids and bases are generally small, have a high charge, are compact, and thus hold the electrons tightly by the nucleus. Consequently, they are not strongly polarizable. Li^+ and F^- are a typical pair of hard acid and base. Large, highly polarizable ions are, in contrast, considered as soft). Klopman (1968) tried to quantify Pearson's principle, employing frontier molecular orbital theory (FMO) and extended the HSAB model by the following proposition: Hard acids bind to hard bases, giving charge-controlled complexes, i.e., dominated by ionic interactions. Soft acids bind to soft bases, giving FMO-controlled complexes, i.e., dominated by covalent interactions.

Although the HSAB principle works basically well, it is a simplified scheme for describing reaction chemistry. In a book on this topic, Pearson (1997) stated succinctly that

> this principle is simply a restatement of the experimental evidence which led to [the system in the first place]. It is a condensed statement of a very large amount of chemical information. As such it might be called a law. But this label seems pretentious in view of the lack of a quantitative definition of hardness. HSAB is not a theory, since it does not explain variations in the strength of chemical bonds. The word 'prefer' in the HSAB Principle implies a rather modest effect.

Pearson allows that there are many examples of very strong bonds formed between partners of "mismatched" pairs. In fact, there is no physical parameter that correlates with the hardness or softness.

Consequently, there have been attempts to modify or extend the model. Leach proposed a more detailed classification (2014), which tried to take all parameters involved in reaction chemistry into account.

Related to the coordination chemistry of amino acid salts, these models have been brought into discussion into several papers on amino acid salts. Still, the impact of these models on amino acid chemistry is not as important as one might think, as the available ligand groups – carboxylate oxygen atoms and amino nitrogen atoms – are both "hard" bases, so all cations (either "hard" or "soft" acids) are coordinated by these ligands. The only exception are the –SH and –S– groups in cysteine and methionine, and even there the factors of the HSAB model are not paramount – as shown in the example of sodium bis (L-cysteinato)-vanadium dihydrate, which coordinates the cation with all three atoms (N, O, S).

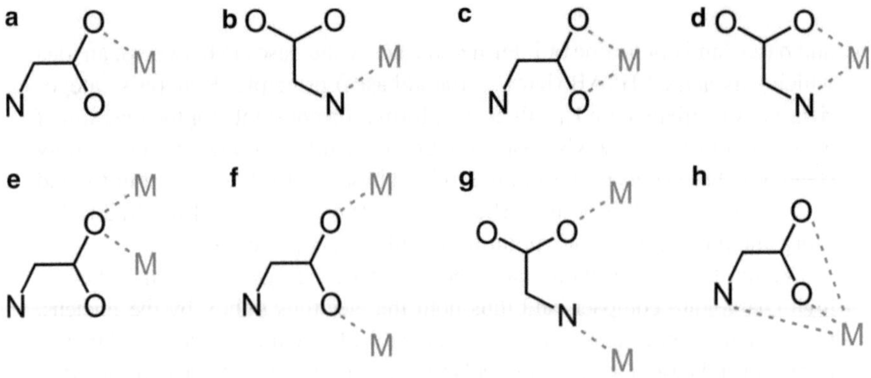

Fig. 3.1 Coordination modes of amino acids around cations, shown exemplary for glycine. The amino acid can act as a monodentate (O) or (N) ligand (**a** and **b**), as a bidentate (O,O′) or (O,N) ligand (**c** and **d**), and as a bridging (O,O) and (O,O′) ligand (**e** and **f**). Moreover, bridging (O,N) and tridentate coordination modes are possible (**g** and **h**); for amino acids with other electronegative atoms, there are other possibilities as well (see text)

For these salts, one has to consider the possibilities of chemical bonding. Whereas the units in amino acid crystals as described in Chap. 2 are connected mainly by hydrogen bonds, the addition of cations introduces the ionic interactions as an important structural feature. Amino acids make very flexible ligands, as each molecule possesses at least three highly electronegative atoms (two oxygen atoms in the carboxylate group plus one nitrogen atom in the amino group) which can act as ligand atoms. Consequently, each amino acid molecule can bind to a cation in several ways, namely, as monodentate, bidentate, or tridentate ligand. Although the tridentate coordination via the three ubiquitous ligand atoms of all amino acids (two carboxyl oxygens plus one amino nitrogen) is sterically unfavorable, some amino acids can act as tridentate ligands via other atoms: Groups like the imidazole ring in histidine, the thiol sulfur in cysteine, the thioether group in methionine, the phenol group in tyrosine, and the additional acid (or carboxylate) groups in glutamic and aspartic acid are potential metal binding sites and therefore often involved in cation coordination. An overview of these mechanisms is given by Kiss (1990) and Laurie (1987, 1995); see also references therein.

Moreover, the amino acid can act as a bridging ligand, both via one bridging atom (denoted as O, O) and two different bridging atoms (denoted as O, N or O, O′). An overview of these coordination modes is shown in Fig. 3.1.

Very often, there is more than one amino acid molecule in coordination around a cation. This arrangement is likely to be symmetric, i.e., the cation is located on a special position (inversion centers if the amino acid is achiral or the structure comprises both enantiomers), with the amino acid molecules symmetry related to each other.

Thus, within amino acid salts containing cations, the coordination polyhedron of the metal cation plus the surrounding molecules represents the dominating building unit. Within the various species, these units can be isolated from each other (i.e., connected only via hydrogen bonds) or form infinite one-dimensional structures as chains or bands, two-dimensional structures (layers), or three-dimensional frameworks. An overview of these motifs is given exemplary for glycine metal halogenide salts (Fleck 2008). Likewise, Hakimi and Aliabadi (2012) published a review, discussing amino acid salts of copper, as did Kremer et al. (2005) for lanthanide salts, focusing on the cation and not the amino acid. A general overview of metal complexes with several amino acids is given by Beck (2009). The compounds discussed in these reviews are listed in the respective parts of Tables 3.1, 3.2, 3.3, 3.4, 3.5, 3.6, 3.7, 3.8, 3.9, and 3.10.

It has to be noted that the amino acid molecules are not the only possible ligands in amino acid salts. Water molecules are often present in coordinating function. Furthermore, in compounds that contain other anions (e.g., halogenides, nitrates, etc.), these units can act as additional ligands (see Chap. 4).

All these aspects clearly show that the possible structural diversity is very high, and this statement has been proved by the experimental determination of a multitude of crystal structures of amino acid salts. Isotypic structures occur (as, e.g., in the pair of bis(glycine)$CoCl_2{\cdot}2H_2O$ and bis(glycine)$NiCl_2{\cdot}2H_2O$, Stenzel and Fleck 2004; Ch'ng et al. 2008), but usually different constituents result in different structures, even if the atoms are similar in radii and charge (as shown by the structure of bis(glycine)$ZnCl_2{\cdot}2H_2O$ and bis(glycine)$MnCl_2{\cdot}2H_2O$, Fleck et al. 2008; Pepinsky et al. 1958, both of which are not isotypic with the aforementioned pair or each other).

To get a grip on the structural diversity, a system has been introduced to designate the connectivity of the building units (Fleck 2008): Two numbers are given, in order to indicate the dimension of connection of the polyhedra with each other as well as that of the polyhedral units via amino acid molecules. Therefore, 0 stands for isolated units (or 0^x, where x stands for the number of polyhedral in the unit), 1 for chains, 2 for layers, and 3 for frameworks. Thus, if, for instance, the polyhedra are isolated from each other and amino acid units connect neighboring polyhedral to a one-dimensional chain, the designation would be 0–1. Chains of polyhedra (connected by common corners, edges, or faces) which are connected by amino acid ligands to a three-dimensional framework would be 1–3. The respective pair of numbers is given in the following tables for each amino acid salt, if a structure has been determined.

As said above, we discuss compounds comprising cations and amino acids in anionic state in this chapter. From the standard amino acids, all are found in such compounds, though not with the same frequency. Lysine and arginine, for instance, as basic amino acids do not form anions as easily as acidic amino acids do. Additionally, the solubility is an important factor, as well as conformational flexibility. Among the nonstandard amino acids discussed in this book, salts of cations and β-alanine, sarcosine, ornithine, and DMG have been reported.

As the reports of compounds of cations and amino acid anions published in the literature are very numerous, the discussion of all these structures would go well beyond the scope of this work, and only exemplary species are presented in the text. An overview of all these crystals as found in the literature is given in Tables 3.1, 3.2, 3.3, 3.4, 3.5, 3.6, 3.7, 3.8, 3.9 (for standard amino acids), and 3.10 (for nonstandard amino acids).

NOMENCLATURE OF AMINO ACID SALTS

Amino acid salts can – as all other salts of organic acids – simply be described by naming the cation and the anion, e.g., silver glycinate. As water molecules are often present in such salts, this has to be reflected in the name as well. Since these water molecules can both be part of the coordination sphere or present in interstices (i.e., exist as crystal water), the addition of the word hydrate (or monohydrate, dihydrate, etc.) is not sufficient to describe this feature. In order to express this difference in the name, the following nomenclature system is used: Ligands are indicated before the cation, crystal water (or other possible compounds), is added at the end. For instance, the name *bis (alaninato)-diaqua-cadmium monohydrate* indicates that two alaninate anions and two water molecules coordinate the central cadmium cation, with one addition water molecule in form of crystal water.

This nomenclature can be extended by including the coordination mode of the amino acid, in this case as *bis(alaninato-N,O)-diaqua-cadmium monohydrate*. Even more informative is the addition of affixes denoting conformational information, like catena- to denote a chain structure poly- for a polymeric structure, etc. Moreover, Greek letters are used to describe the coordination mode more precisely. Thus, κ (kappa) indicates the single ligating atom attachments of a polyatomic ligand (as an amino acid) to a coordination center. Likewise, μ (mu) denotes the fact that a group bridges two or more coordination centers. Other affixes like *cis-* or *trans-* are also used frequently.

Nevertheless, authors do not always utilize the full extent of this system, as the names grow cumbersome, so species like magnesium dinitrate tetrahydrate are properly referred to as *catena-poly[[[tetraaquamagnesium(II)]-μ-glycine-*κ^2*O:O′] dinitrate]* (Fleck and Bohatý 2005). Within this book, we do and/or use employ this nomenclature fully (for reasons of clarity), i.e., we indicate the position of ligands, but omit most affixes (apart from *trans-* and *cis-* when both forms occur) as well as references to coordination modes, since in most cases the structures are described in text or figures.

Table 3.1 *Salts of amino acids and cations, part I*: space groups and unit cell parameters of compounds of cations and glycinate anions

	s.g.	*a*	*b*	*c*	*α*	*β*	*γ*	Conn.	Ref.
Ag Gly	*P*-1	5.24	5.71	6.90	84.21	71.33	84.44	N/A	Acland and Freeman (1971)
Ag Gly·0.5H_2O	$P2_1/n$	8.90	6.43	15.35	90	97.83	90	0–1	Acland and Freeman (1971)
Li Gly	$P2_12_12_1$	4.998	7.864	9.261	90	90	90		Müller et al. (1994)
Ca Gly_2·H_2O	*P*-1	6.436	7.900	8.210	80.21	81.80	73.57	1–1	Fox et al. (2007)
Ca Gly_2·3H_2O	*P*-1	9.657	9.688	5.763	90.588	77.00	97.48	1–1	Le Bail et al. (2013)
Zn Gly·H_2O	*A*1/*A*-1	15.051	5.297	10.006	89.89	87.15	90.39	N/A	Low et al. (1959)
Zn Gly_2·H_2O	*P*-1	9.165	9.571	10.438	105.97	106.19	107.12	0–2	Newman et al. (1990)
Zn Gly_2·H_2O	*C*2/*c*	19.694	10.402	14.915	105.97	92.68	107.12		Konar et al. (2010)
Cd Gly_2·H_2O	*I*2/*a*	14.818	5.297	9.990	90	90.40	90	0–2	Low et al. (1959) and Barrie et al. (1993)
Cu Gly_2·H_2O	$P2_12_12_1$	10.86	5.22	13.50	90	90	90	0–1	Freeman et al. (1964)
Co^{III} Gly_3	$P2_1/c$	5.303	13.616	13.690	90	99.62	90	0–0	Gu et al. (2007)
Co^{III} Gly_3.H_2O	$P2_1/c$	6.248	14.185	12.132	90	101.47	90	0–0	Zhao et al. (2004)
Co^{III} Gly_3·2H_2O	$P2_1/n$	13.860	13.153	13.996	90	112.88	90	0–0	Dewan (1988)
Co^{III} Gly_3·2H_2O	$Pna2_1$	15.331	13.205	11.606	90	90	90	0–0	Miyanaga et al. (1982)
Ni Gly_2·2H_2O	$P2_1/c$	7.625	6.596	9.670	90	116.57	90	0–0	Freeman and Guss (1968)
Cr (OH) Gly_2	$P2_1/n$	5.691	16.920	7.900	90	99.90	90	0^2–0^2	Veal et al. (1973)
Cr Gly_3·H_2O	$P2_1/c$	6.256	14.649	12.267	90	100.39	90	0–0	Bryan et al. (1971)
trans-Pt Gly_2	*P*-1	5.322	4.169	8.332	99.88	98.04	104.44	0–0	Freeman and Golomb (1969)
cis-Pt Gly_2	$P112_1/c$	5.431	10.420	13.133	90	90	107.3	0–0	Pozhidaev et al. (1975)
trans-Pd Gly_2	$P2_1/c$	5.664	9.046	7.796	90	120.33	90	0–0	Baidina et al. (1982)
cis-Pd Gly_2·3H_2O	*I*112/*a*	6.846	17.46	8.029	90	90	91.39	0–0	Baidina et al. (1979)

3.2 Glycine

Among all salts of amino acids and cations, those of glycine (more precisely, glycinate) are most numerous. Several monovalent, divalent, and trivalent cations have been combined with glycine, and in some cases different hydration states have been reported (Table 3.1).

As far as monovalent cations are concerned, only few species have been reported (Table 3.1). The oldest work is by Acland and Freeman (1971), who reported silver glycinate in anhydrous form and as a *hemihydrate*. The hemihydrate was later refined by Nomiya and Yokoyama (2002), who also reported antimicrobial activities of this species (along with the silver salts of aspartic acid and asparagine; see Sect. 3.8). The structure is an example of a linear connection of cations with glycinate ligands in bridging (O,N) mode (Fig. 3.2).

Among the bivalent cations, both earth alkali metals and bivalent transition metals are frequently used for synthesis of amino acid salts, glycine being no exception. A typical example is the crystal of bis(glycinato)-zinc monohydrate, which deserves attention as two different polymorphs have been reported in the literature: Low et al. (1959) published a study on triclinic zinc bis(glycinate) monohydrate in space group *A*1 or *A*-1 (this aspect could not be determined, although this cell choice is rather strange). In this structure, Zn is reported to exist in octahedral coordination. A later study (Newman et al. 1990) found another triclinic polymorph in which the zinc cations are [5]-coordinate. The smaller cell given in this work is derived by a transformation of the reoriented version of the cell given by Low et al. 1959. The authors suggest that this larger cell might be a result of twinning, although this hypothesis remains without proof so far. Konar et al. (2010) opine that "the presence of eight molecules in a triclinic unit cell is very unlikely. It is obvious that these authors used the wrong space group and could only arrive at a tentative solution."

Another interesting salt is diaqua-bis(glycinato)-nickel, which was originally reported by Stosick (1945) and later refined by Freeman and Guss (1968) and Castellano et al. (1982). This last work provided accurate hydrogen positions, as researchers from this group were interested in the structural details, since an antiferromagnetic phase transition was found at 0.88 K, established by the measurement of specific heat and magnetic susceptibility (Calvo et al. 1982a, b, and references therein). The crystal structure is a typical example of isolated cation–ligand units, in this case one nickel cation coordinated by two glycinate anions and two water molecules each (Fig. 3.3).

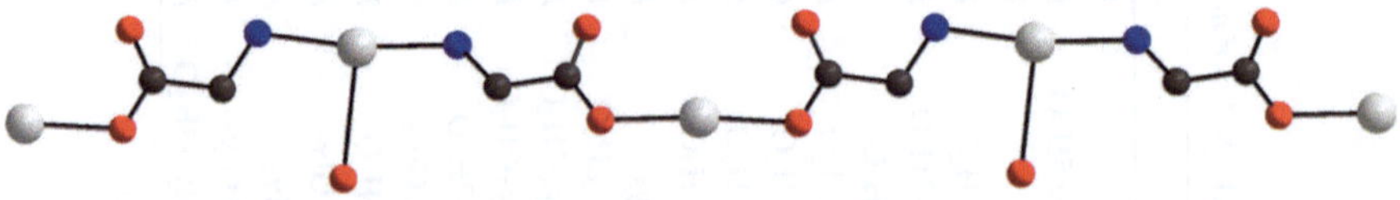

Fig. 3.2 Structure of silver glycinate hemihydrate, an example of a crystal where an amino acid (glycine, in this case) acts as bridging (O,N) ligand to connect the silver cations to infinite chains. Hydrogen atoms are omitted for clarity

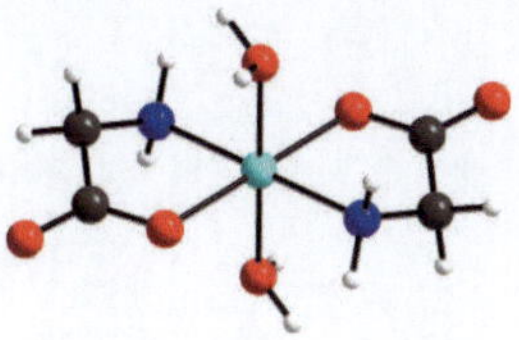

Fig. 3.3 Diaqua-bis(glycinato)-nickel. The Ni^{2+} cation is [6]-coordinated by two bidentate (N,O) glycinate anions as well as two water molecules. These $Ni(C_2H_4NO_2)(H_2O)_2$ units are connected to each other via hydrogen bonds within the crystal

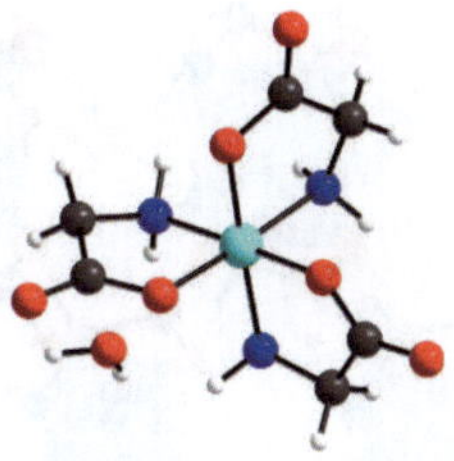

Fig. 3.4 Tris(glycinato)-cobalt monohydrate. The trivalent cobalt cation is [6]-coordinated (as the divalent nickel cation in diaqua-bis(glycinato)-nickel (Fig. 3.3), but in this case all ligands are atoms from glycinate anions. The water molecules are located in the interstices and part of the hydrogen network connecting the units

A different coordination situation is found in the structure of tris(glycinato)-cobalt^{3+} monohydrate (Zhao et al. 2004). In this species, the water molecules are not part of the coordination polyhedron, as three glycinate anions act as ligands. Thus, all corners are occupied by atoms from these molecules, and the water molecules exist as crystal water in the interstices (Fig. 3.4). This difference is due to the trivalent charge of the Co cation, which has similar geometric features as the Ni^{2+} cation (and therefore a similar coordination polyhedron), but an additional glycinate anion is necessary to counterbalance the charge. The same arrangement is found in the structure of tris(glycinato)chromium monohydrate (Bryan et al. 1971).

Another salt of glycine with a trivalent cation is that of aluminum, as reported by Daydé et al. (2002), who investigated aluminum complex formation equilibria of glycine as well as serine, threonine, and histidine, via glass electrode potentiometry and NMR spectroscopy (crystal structures have not been reported for either of these salts).

It is noteworthy that there is more than one cobalt glycinate salt. Apart from the above-mentioned monohydrate, there is an anhydrous phase (Gu et al. 2007) and two dihydrates (an orthorhombic phase, Miyanaga et al. 1982, and a monoclinic phase, Dewan 1988). In all of these phases, the Co^{3+} cation is coordinated by three glycinate ions in bidentate (N,O) mode, although the topological arrangement of the anions toward each other differs.

Veal et al. (1973) reported structural and magnetic properties of the compound hydroxo-di(glycinato) chromium, which is an example of a dimeric structure: In this

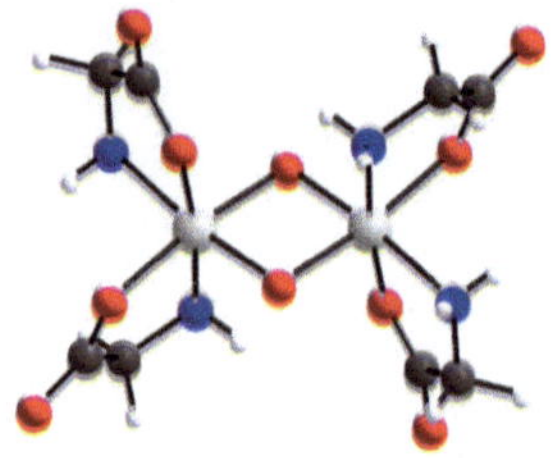

Fig. 3.5 Hydroxo-di (glycinato)-chromium is an example of a dimeric structure. The dimers are isolated units, connected via hydrogen bonds

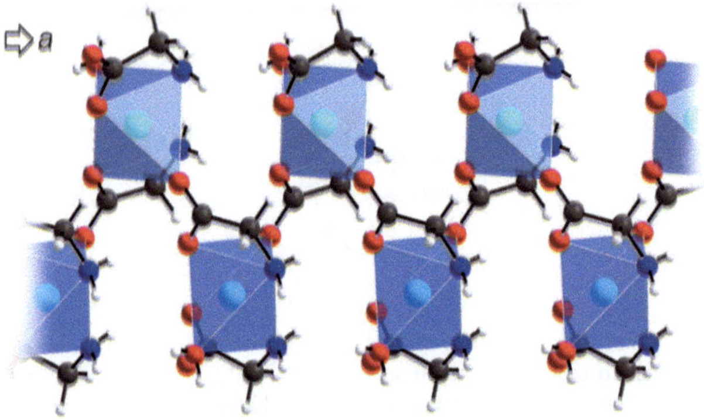

Fig. 3.6 Helical chain in bis(glycinato)-copper monohydrate. For clarity, the CuO_4N_2-octahedra are pictured

species, two $Cr^{3+}(OH)(C_2H_4NO_2)$ units are connected to each other to form a dimer (thus the correct name is di-hydroxo-tetra(glycinato) dichromium, Fig. 3.5). The species is actually not a compound of a metal cation and an amino acid anion (as, e.g., tris(glycinato) chromium monohydrate, Bryan et al. 1971); neither is it a compound of a zwitterionic amino acid and a neutral salt (as the crystals discussed in Chap. 4).

In contrast to the above structures, the crystal structure of *cis*-bis(glycinato)-copper monohydrate (reported first by Tomita and Nitta 1961; then republished by Freeman et al. 1964; later studied by EPR spectroscopy by Martino et al. 1996; and finally redetermined by Casari et al. 2004 and by Konar et al. 2010) contains infinite helical chains (Fig. 3.6), as the glycinate anions act not only as bidentate (N,O) ligands to one Cu^{2+} cation but also as bridging (O,O′) ligands, connecting neighboring cations. It is interesting to note that the helical chains are chiral; therefore the structure is non-centrosymmetric (orthorhombic space group $P2_12_12_1$). This is one (of several) examples for the fact that achiral amino acids (as glycine certainly is) can form non-centrosymmetric crystal structures.

Apart from isolated units and one-dimensional chains, amino acid salts can also form layered structures. An example is bis(glycinato)-cadmium monohydrate (Barrie et al. 1993), where the glycinate ions connect neighboring polyhedral not only to chains but layers parallel (100), as displayed in Fig. 3.7.

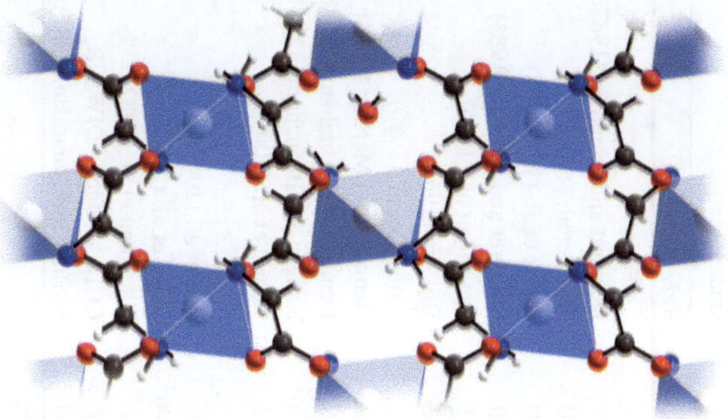

Fig. 3.7 Layers in bis(glycinato)-cadmium monohydrate. For clarity, the CdO_4N_2-octahedra are pictured. Neighboring layers are connected via a hydrogen bond network, including the water molecules in the interstices

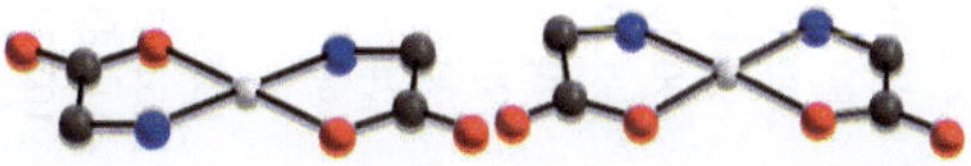

Fig. 3.8 Conformational polymorphs of bis(glycinato)-platinum. *trans*-Bis(glycinato)-platinum (*left*) and *cis*-bis(glycinato)-platinum (*right*). Note that in the *trans*-form, the two glycinate ions are symmetry related (the unit has point group symmetry −1, with the platinum cation located on the inversion center), whereas in the *cis*-form they are independent. Hydrogen atoms are omitted for clarity

Apart from the usually octahedrally coordinate transition metals, the transition metals platinum and palladium are frequently used for syntheses of amino acid salts. Glycinates are no exception. As far as platinum is concerned, there are two similar species which differ only by the relative position of the glycinate ions: *trans*-bis(glycinato)-platinum (Freeman and Golomb 1969; Fig. 3.8 left) and *cis*-bis (glycinato)-platinum (Pozhidaev et al. 1975; Fig. 3.8 right) are both good examples of the planar [4]-coordination of the Pt^{2+} cation, where two glycinate anions act as bidentate (N,O) ligands.

In addition to the salts listed in Table 3.1, Bowman et al. (2010) reported a rubidium salt of glycine (along with serine, threonine, and proline analogs) by mass spectrometry, although no crystals suitable for structure analysis were obtained.

3.3 Alanine

In contrast to glycine, alanine (as well as all other standard amino acids) is chiral; therefore, salts of one enantiomer as well as of the racemate can be synthesized. This has been attempted for many combinations, and many publications on L-alaninates and DL-alaninates are found (Table 3.2).

Table 3.2 *Salts of amino acids and cations, part II*: space groups and unit cell parameters of compounds of cations and alaninate anions

	s.g.	*a*	*b*	*c*	*α*	*β*	*γ*	Conn.	Ref.
L-Alanine									
Ag L-Ala	$P2_1$	5.477	16.941	5.539	90	96.10	90	0–1	Demaret and Abraham (1987a)
Co^{III} L-Ala_3	$P2_12_12_1$	18.55	14.459	5.09	90	90	90	0–0	Drew et al. (1966)
Co^{III} L-$Ala_3 \cdot H_2O$	$P2_12_12_1$	5.247	13.555	19.351	90	90	90	0–0	Herak et al. (1978)
Ni L-$Ala_2 \cdot 4H_2O$[a]	Cc[a]	20.55	6.50	9.70	90	93.15	90	0–0	Demaret and Mercier (1983)
Zn L-Ala_2	$P2_1$	8.637	5.334	9.581	90	90.36	90	0–1	Dalosto et al. (1999)
trans-Cu^{II} L-Ala_2	$P2_1$	9.24	5.05	9.59	90	95.20	90	0–2	Dijkstra (1966)
cis-Cu^{II} L-Ala_2	$P2_12_12_1$	11.645	8.772	8.444	90	90	90	0–1	Moussa et al. (1999)
Cd L-Ala_2	[b]	5.45	9.19	18.08	90	90	90	N/A	Demaret and Mercier (1983)
Cd L-$Ala_2 \cdot 3H_2O$	$P3_121$	6.376	6.376	25.590	90	90	120	0–0	Demaret and Abraham (1987b)
Cr (OH) L-$Ala_2 \cdot 3H_2O$	$P1$	5.369	9.994	11.209	72.65	74.76	77.27	0^2–0^2	Ranger and Beauchamp (1981)
KNi L-$Ala_3 \cdot 3H_2O$	$P3$	10.132	10.132	4.884	90	90	120	0–2	Khatib et al. (2009)
DL-Alanine									
Ca DL-Ala $_2 \cdot 3H_2O$	$C2/c$	14.143	6.433	13.814	90	109.15	90	0–0	Fox et al. (2007)
Co DL-Ala $_2 \cdot 2H_2O$	$P2_1/c$	13.727	4.987	19.476	90	131.13	90	0^2–0^2	Sahin et al. (2008)
Cu^{II} DL-Ala $_2 \cdot H_2O$	$C2/c$	12.087	9.583	8.973	90	110.85	90	1–1	Hitchman et al. (1987)
Ni DL-Ala $_2 \cdot 4H_2O$	P-1	10.788	9.698	6.503	89.96	107.74	93.01	0–0	CCDC-Refcode DIWYIR
Ni DL-Ala $_2 \cdot 4H_2O$	$Pna2_1$	9.750	6.453	20.513	90	90	90	0–0	Mostad and Natarajan (1987)

[a]The space group Cc is not compatible with a structure comprising only one enantiomer; see text for details
[b]The authors only give unit cell dimensions, no space group

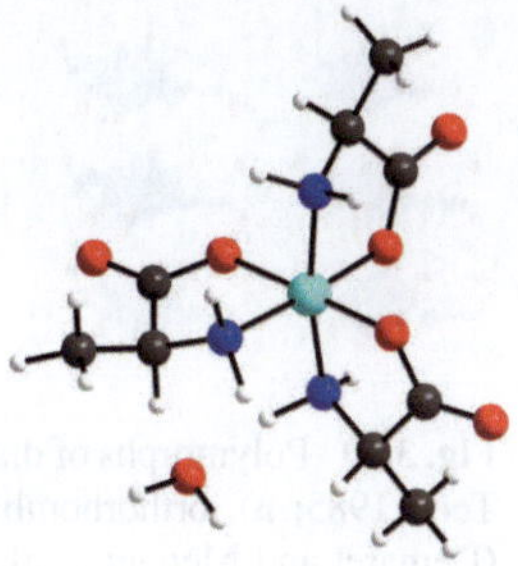

Fig. 3.9 Tris(L-alaninato)-cobalt monohydrate. Note similarity of the coordination topology of the [6]-coordinate trivalent cobalt cation as compared with the respective glycinate salt (Fig. 3.4)

Analog to the silver salts of glycine, the structure of (L-alaninato) silver (Demaret and Abraham 1987a) consists of infinite chains, comprising linear [2]-coordinate silver cations, connected by bridging (N,O) L-alaninate anions. Likewise, there is a species tris(L-alaninato) cobalt monohydrate (Herak et al. 1978; Fig. 3.9), where a trivalent Co cation is coordinated by three L-alaninate anions in bidentate (N,O) mode. Previously, Denning and Piper (1966) had published a work on optical activity and the determination of the absolute configuration in this species (and related salts of L-leucine and L-proline).

Another species which deserves closer attention is that of diaqua-bis(L-alaninato)-nickel dihydrate (first published by Demaret and Mercier 1983, who did not give a space group, refined by Teoh et al. 1987). Although the authors report the salt of L-alanine in their work, the symmetry given by Teoh et al. (1987)is *Cc*. However, as glide plane symmetry is not compatible with the presence of only one enantiomer, this result is not acceptable. Either the authors assigned the symmetry incorrectly or the structure actually comprises both D- and L-alaninate ions (as it was found in the model calculated from the CIF as taken from the Cambridge Structural Database, Allen 2002; Refcode BUNVOV01). In fact, there are two polymorphs of the respective racemate compound, i.e., diaqua-bis(DL-alaninato)-nickel dihydrate reported in the literature, a triclinic form (space group *P*-1, Kamwaya and Teoh 1985) and an orthorhombic form (space group $Pna2_1$, Mostad and Natarajan 1987). A comparison of these two polymorphs with the alleged L-alaninate analog shows very similar, but still different structures (Fig. 3.10). A detailed study comparing crystallization conditions and conducting a redetermination of these species is necessary to decide the question if the alleged L-alaninate is truly a third DL polymorph.

Another motif found for glycinate salts, namely, the dimer of hydroxo-di (glycinato)chromium (Fig. 3.5), was also discovered in two alaninates: Both bis (hydroxo)-bis(is(L-alaninato)-chromium) trihydrate (Ranger and Beauchamp 1981) and (aqua)-bis(DL-alaninato)-cobalt monohydrate (Sahin et al. 2008) comprise dimeric units.

Although no platinum salts of alanine are reported, there is a pair of *trans*- and *cis*-polymorphs of bis(L-alaninato)-copper: The *trans*-form is monoclinic (space group $P2_1$, Dijkstra 1966; Blazic et al. 1992; Hitchman et al. 1987), while the *cis*-form is orthorhombic (space group $P2_12_12_1$, Moussa et al. 1999; of this

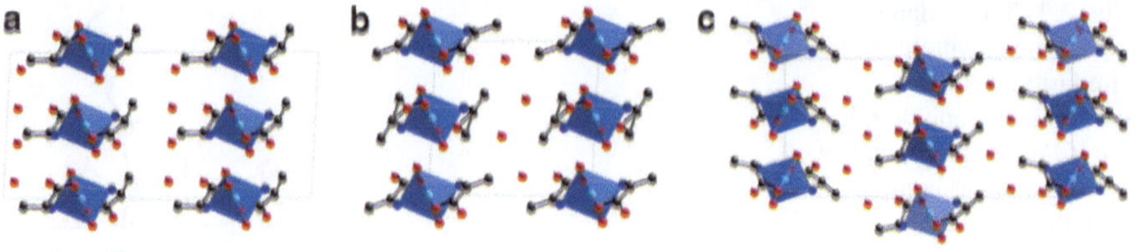

Fig. 3.10 Polymorphs of diaqua-bis(DL-alaninato)-nickel dihydrate. Triclinic form (Kamwaya and Teoh 1985; **a**), orthorhombic form (Mostad and Natarajan 1987; **b**), and alleged "L"-alaninate (Demaret and Mercier et al. 1983; **c**) which appears to be a racemate (see text for explanation). Hydrogen atoms are omitted for clarity

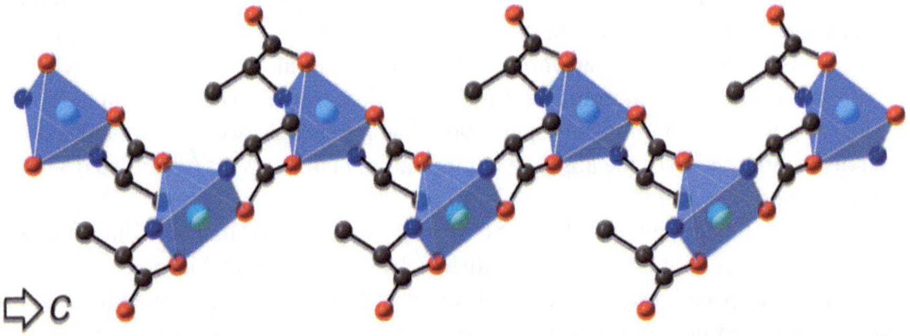

Fig. 3.11 Part of the crystal structure of *cis*-bis(L-alaninato)-copper. The [4 + 1]-coordinate copper cations are part of an infinite chain parallel [010]. The equatorial positions of the coordination sphere are occupied by the *cis*-oriented alaninate anions; the connection to the chain is a result of the occupation of the apical position by neighboring alaninate anions. Hydrogen atoms are omitted for clarity

species an D-alaninate analog is also published, Gillard et al. 1969). Additionally, Presnyak et al. (1972) published a report on tris(L-alaninato)thallium(III) .

In the *trans*-form, the Cu^{2+} cation is [4 + 2]-coordinated, with the atoms form two *trans*-oriented L-alaninate anions in equatorial positions and oxygen atoms from neighboring alaninate ions occupying the apical positions. The resulting structure can be described as a two-dimensional infinite layer. In the *cis*-form, the coordination is different: The copper cations are [4 + 1]-coordinated, and the resulting arrangement is a chain structure (Fig. 3.11).

The crystal structure of aqua-bis(DL-alaninato)-copper (first published by Hitchman et al. 1987; refined by Mirceva et al. 1989 and Zhang et al. 2005; the magnetic interactions were reported by Calvo et al. 1991) is another chain structure, although with different topology: The Cu^{2+} cations are [4 + 2]-coordinated, with the D-and L-alaninate anions in equatorial positions and water molecules in the apical positions. These water molecules act as bridging ligands, thus connecting the neighboring units to an infinite chain along c (Fig. 3.12).

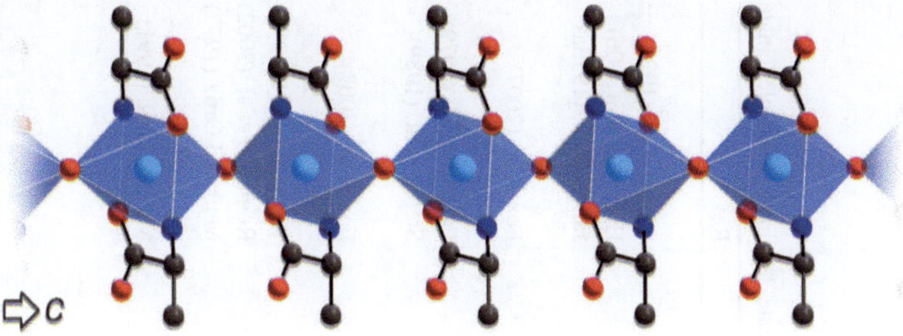

Fig. 3.12 Part of the crystal structure of aqua-bis(DL-alaninato)-copper. The [4 + 2]-coordinate copper cations are part of an infinite chain parallel [010]. The equatorial positions of the coordination sphere are occupied by the alaninate anions; the apical water molecules connect the cations to a chain. Hydrogen atoms are omitted for clarity

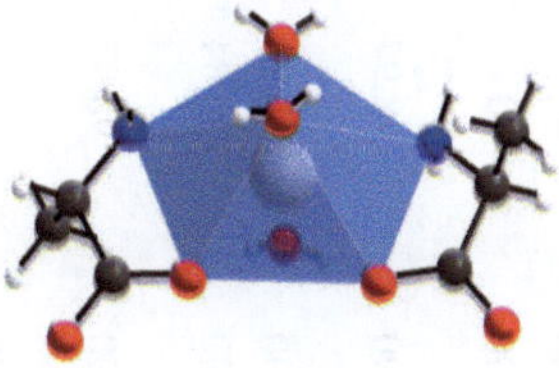

Fig. 3.13 Bis(DL-alaninato)-triaqua-calcium. Two alaninate ligands of opposing chirality in bidentate (N,O) coordination mode as well as three water molecules make up the unit, which is connected to adjacent units by hydrogen bridges only

In a study by Fox et al. (2007), several amino acid salts of calcium have been published, including bis(DL-alaninato)-triaqua-calcium. The authors found a large structural diversity of Ca^{2+} salts, which is due to the variability of the often irregular coordination figures around this cation (in contrast to the relatively rigid coordination figures found around many transition metal cations). The DL-alaninate is characterized by isolated units of a central calcium cation, coordinated by two alaninate ions of opposing chirality as well as three water molecules (Fig. 3.13).

Double salts have been reported as well, an instance is given with the species of tris(L-alaninato)-triaqua-nickel-potassium (Khatib et al. 2009). This structure is remarkable insofar as it crystallizes in the trigonal space group *P*3, i.e., in an unusually high symmetry. In this salt, potassium octahedra (located on threefold axes) are [6]-coordinated by three water molecules and three monodentate L-alaninate anions, which also connect in bidentate (N,O) mode to the nickel cations (also [6]-coordinate). Thus, a layer perpendicular to the threefold axis is formed (Fig. 3.14).

Table 3.3 *Salts of amino acids and cations, part III*: space groups and unit cell parameters of compounds of cations and valinate, leucinate, and isoleucinate

	s.g.	*a*	*b*	*c*	*α*	*β*	*γ*	Conn.	Ref.
L-Valine									
Cu L-Val $_2$·H_2O	*C*2	21.314	9.586	7.417	90	108.89	90	0–0	Steren et al. (1990)
Zn L-Val $_2$·H_2O	$P2_12_12_1$	7.428	9.434	20.586	90	90	90	0–0	Hwang et al. (2006)
Pd L-Val_2·H_2O	$P2_12_12_1$	9.68	9.645	15.401	90	90	90	0–0	Jarzab et al. (1973a)
DL-Valine									
Ca DL-Val_2·H_2O	*C*2/*c*	24.210	6.290	9.841	90	98.97	90	1–1	Fox et al. (2007)
L-Leucine									
Cu L-Leu_2	$P2_1$	9.725	5.127	14.689	90	105.79	90	0–2	Fawcett et al. (1979)
Zn L-Leu_2	$P2_1$	9.584	5.389	14.866	90	106.84	90	0–1	Steren et al. (1989)
DL-Leucine									
Ca DL-Val_2·H_2O	*C*2/*c*	24.210	6.290	9.841	90	98.97	90	1–1	Fox et al. (2007)
L-Isoleucine									
Cu L-Ile_2·H_2O	$P2_12_12_1$	7.635	9.444	21.681	90	90	90	0–0	Weeks et al. (1969)
Zn L-Ile_2·$2H_2O$	*C*2	9.898	6.428	14.227	90	103.71	90	0–0	Rombach et al. (2002)
Ni L-Ile_2·$2H_2O$	*P*1	5.98	13.80	5.86	96.8	112.6	87.5	N/A	Muir and Ortiz (1977)
DL-Isoleucine									
Cu L-Ile_2	*Aba*2	11.165	11.111	25.985	90	90	90	N/A	Martino et al. (1991)

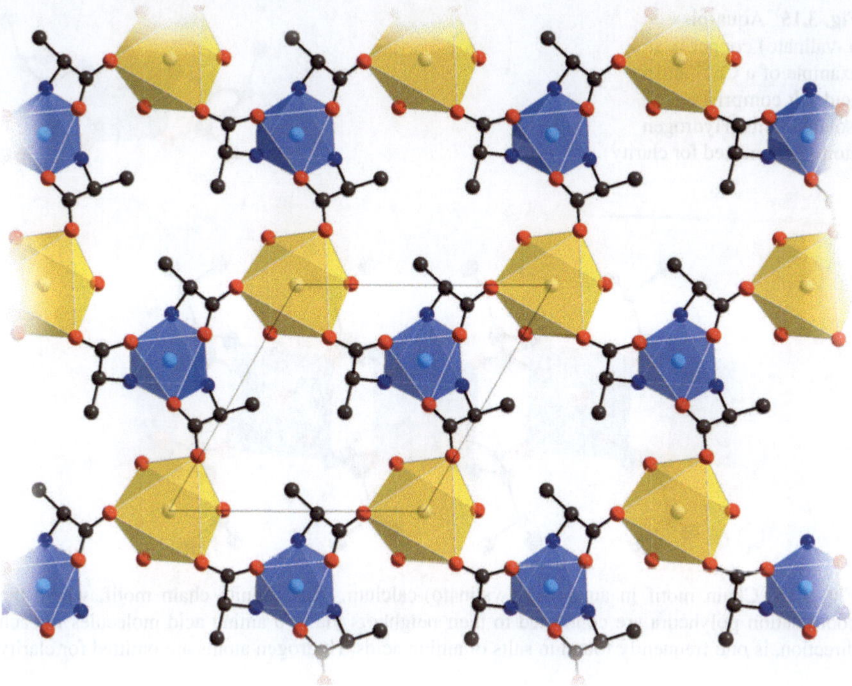

Fig. 3.14 Layers in tris(L-alaninato)-triaqua-nickel-potassium, viewed along (001) (note the larger potassium octahedra (*yellow*, with three water molecules in coordination, located upon threefold axes) and the smaller nickel polyhedra. The trigonal unit cell is indicated by thin lines. Hydrogen atoms are omitted for clarity)

3.4 Valine, Leucine, and Isoleucine

Salts of cations and the nonpolar amino acids valine, leucine, or isoleucine are less common than those of glycine or alanine (and are therefore grouped in this subsection).

Among the salts of valine, aqua-bis(L-valinato)copper has been investigated concerning its structural and magnetic properties (Steren et al. 1990). The structure comprises isolated units of a central Cu^{2+} cation in [4 + 1]-coordination, with two L-valinate anions acting as bidentate (N,O) ligands in equatorial position and one water molecule in apical position (Fig. 3.15).

Isolated units are also the building units in *cis*-bis(L-valinato) palladium monohydrate (Jarzab et al. 1973a; later refined by Pletnev et al. 1992) as well as in aqua-bis(L-valinato)-zinc (Hwang et al. 2006). In the first case, Pd^{2+} is planar [4]-coordinate, and in the latter case Zn^{2+} is pyramidal [5]-coordinate, with a water molecule at the apex. In both structures, the valinate anions are *cis*-oriented, forming the bidentate (N,O) ligands in the equatorial plane as in aqua-bis(L-valinato) copper (as shown in Fig. 3.15). The zinc salt is characterized by disorder of the side

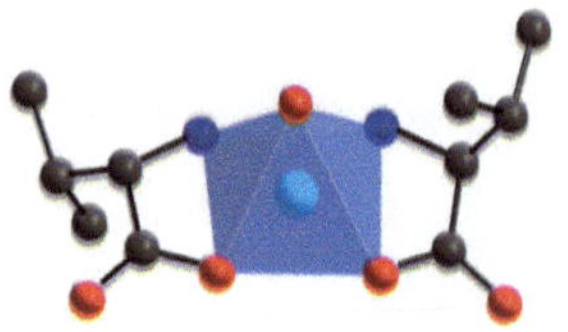

Fig. 3.15 Aqua-bis (L-valinato) copper is an example of a Cu^{2+} amino acid salt comprising isolated units. Hydrogen atoms are omitted for clarity

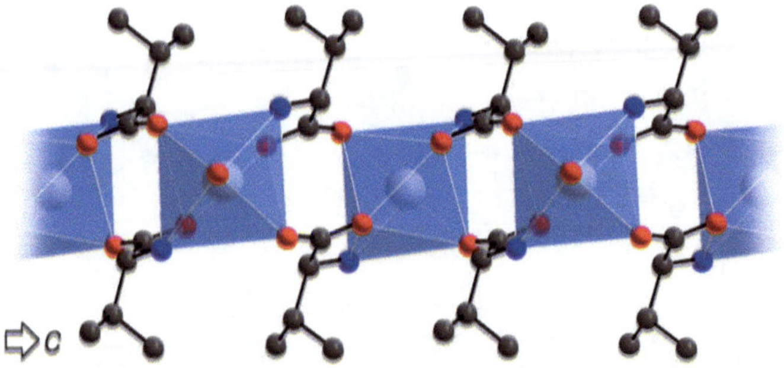

Fig. 3.16 Chain motif in aqua-bis(DL-valinato)-calcium. This infinite chain motif, where the coordination polyhedra are connected to their neighbors via two amino acid molecules in each direction, is one frequently found in salts of amino acids. Hydrogen atoms are omitted for clarity

chain, which can be explained through the lack of interactions of the hydrophobic part of the amino acid.

The only DL-valinate reported in the literature is aqua-bis(DL-valinato)-calcium (Fox et al. 2007). In contrast to the respective DL-alaninate (Fig. 3.11), this compound is a chain structure, where the amino acid anions act both as bidentate (N,O) and bridging (O,O′) ligand. The resulting motif (Fig. 3.16) is one frequently found in salts of cations and amino acids (as well as salts of neutral compounds and amino acids; see Chap. 4): an infinite chain where neighboring polyhedra are connected via two amino acid molecules each, with the amino acid side chain facing away from the chain axis.

Another chain structure is found in bis(L-leucinato)-zinc (Steren et al. 1989): Within this structure, there are two crystallographically different L-leucinate anions. Although both act as bidentate (N,O) ligands, only one is also in bridging (O,O′) function, connecting the [5]-coordinate Zn^{2+} cations to a polar chain along the *b*-axis (Fig. 3.17).

Another L-leucinate, bis(L-Leucinato) copper (Fawcett et al. 1979) represents a layered structure: The L-leucinate anions connect each [4 + 2]-coordinate Cu^{2+} cation with four neighbors, thus forming a two-dimensional connection (Fig. 3.18). It is interesting to note that the structure resembles that of bis(glycinato)-cadmium monohydrate (Fig. 3.7), although both cations and anions are clearly different in their steric properties.

The only DL-leucinate salt is that of calcium (Fox et al. 2007), which is composed from $[Ca_2(leu)_4(H_2O)_4]$ dimers linked to infinite chains by sharing aqua ligands.

Fig. 3.17 Chain in bis(L-leucinato)-zinc, parallel [010]. Of the two coordinating L-leucine molecules, only one is in bridging (O,O′) function (note the polarity of the chain. Hydrogen atoms are omitted for clarity)

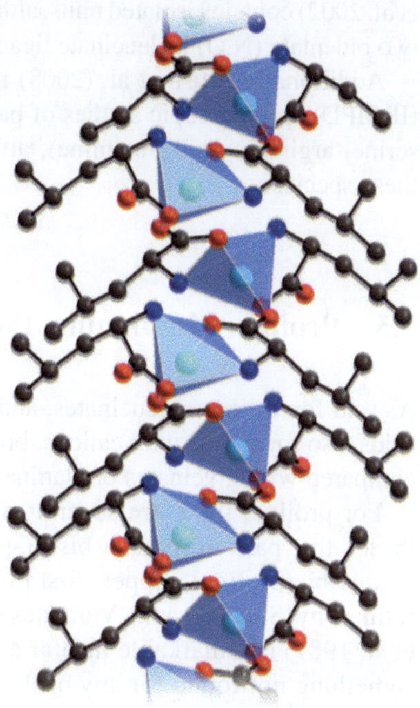

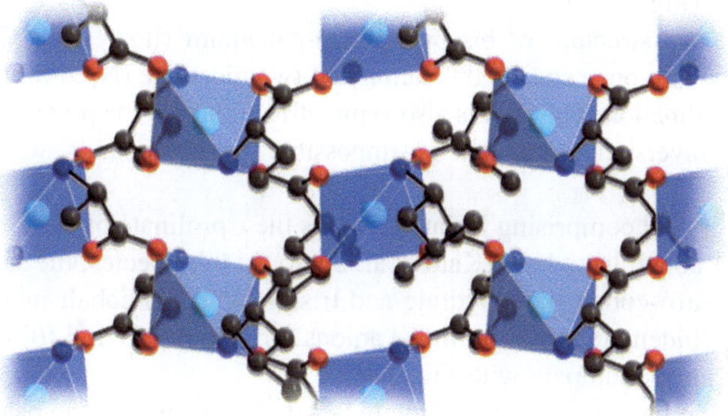

Fig. 3.18 Layers in bis(L-leucinato) copper, parallel (001) (note the topological similarity to the structure of bis(glycinato)-cadmium monohydrate (Fig. 3.7). Hydrogen atoms are omitted for clarity)

Of the salts of isoleucine, the structure of aqua-bis(L-isoleucinato)-copper (Weeks et al. 1969; Liao et al. 2006) comprises isolated [5]-coordinate Cu^{2+} cations, with the isoleucinate anions as *cis*-oriented, bidentate (N,O) ligands and the water molecule in apical position. Likewise, the salt of diaqua-bis(L-isoleucinato)-zinc (Rombach et al. 2002) contains isolated units, although the Zn^{2+} cations are [6]-coordinated, with two bidentate (N,O) isoleucinate ligands and two *cis*-oriented water molecules.

Additionally, Bush et al. (2008) reported infrared multiple photon dissociation (IRMPD) spectroscopic studies of barium complexes of valine (as well as proline, serine, arginine, and glutamine), although no crystal structures are available for these species.

3.5 Proline, Methionine, Phenylalanine, and Tryptophan

As said for valinates, leucinates, and isoleucinates, the remaining nonpolar amino acids also form salts with cations, but the number of species is relatively low when compared with glycinates or alaninates.

For proline, there are several salts of both L-proline and DL-proline. Among these, the pair of diaqua-bis(DL-prolinato)-nickel (Ahmed et al. 1980) and diaqua-bis(proline)-copper (first published by Mathieson and Welsh 1952; later refined by Shamala and Venkatesan 1973; Ahmed et al. 1980; and Hitchman et al. 1987) is remarkable insofar as it represents a pair of isotypic compounds – something not found for any of the salts described above and generally rare for salts of cations and amino acid anions. The structure is characterized by isolated units, which possess point symmetry -1 – the metal cation is located on the inversion center, with symmetry-related prolinate ions and water molecules in coordination (Fig. 3.19).

Similarly, the structure of bis(L-prolinato)-palladium (Ito et al. 1971) contains isolated units with one central Pd^{2+} cation and two bidentate (N,O)-prolinate anions in planar coordination. This unit is also symmetric, although the point symmetry is 2 (of course, inversion symmetry is impossible in compounds with only one enantiomer).

Another crystal comprising isolated units is the L-prolinate of Co^{3+}, namely, tris (L-prolinato)-cobalt trihydrate (Kato et al. 2008). In this species, the same motif as in tris(glycinato)-cobalt monohydrate and tris(L-alaninato)-cobalt monohydrate is found: Three bidentate (N,O)-prolinate anions form an octahedral [6]-coordination around the cation (compare with Fig. 3.4).

There is one more L-prolinate salt with isolated units, which can be described as diaqua-bis(L-prolinato)-copper aqua-bis(L-prolinato)-copper dihydrate. This crystal

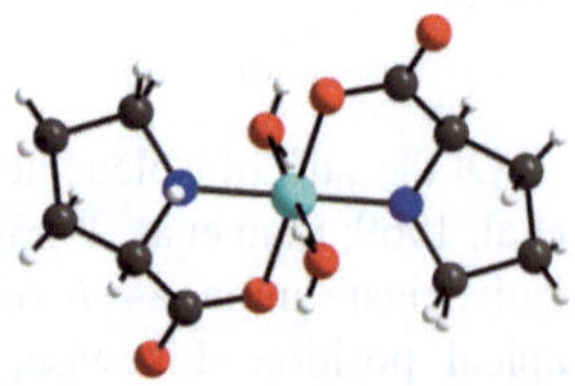

Fig. 3.19 Isolated units in diaqua-bis(DL-prolinato)-nickel, isotypic with the respective copper compound. The unit is symmetric (point symmetry -1)

Table 3.4 *Salts of amino acids and cations, part IV*: space groups and unit cell parameters of compounds of cations and prolinate, methioninate, phenylalaninate, and tryptophanate anions

	s.g.	a	b	c	α	β	γ	Conn.	Ref.
L-Proline									
Zn L-Pro_2	$P2_1$	9.609	5.648	10.491	90	91.63	90	0–1	Ng et al. (1995)
Cu L-$Pro_2 \cdot 5H_2O$	$P2_1$	11.187	12.172	11.661	90	114.96	90	0–0	Sartoris et al. (1999)
Co^{III} L-$Pro_3 \cdot 3H_2O$	$P2_12_12_1$	10.167	10.843	17.216	90	90	90	0–0	Kato et al. (2008)
Pd L-Pro_2	$B22_12$	10.312	12.172	9.707	90	90	90	0–0	Ito et al. (1971)
Pt L-Pro_2	[a]	10.19	11.91	9.74	90	90	90	N/A	Slyudkin et al. (1985)
DL-Proline									
Cu DL-$Pro_2 \cdot 2H_2O$	$P2_1/n$	5.61	17.86	7.12	90	108.0	90	0–0	Mathieson and Welsh (1952)
Ni DL-$Pro_2 \cdot 2H_2O$	$P2_1/n$	5.418	18.334	6.968	90	105.39	90	0–0	Ahmed et al. (1980)
L-Methionine									
Cu L-Met_2	$P2_1$	9.487	5.061	15.563	90	92.46	90	0–2	Ou et al. (1979)
DL-Methionine									
Cu DL-Met_2	$P2_1$/a	9.482	5.018	16.035	90	93.78	90	N/A	Veidis and Palenik (1969)
L-Phenylalanine									
Ag L-Phe	$P2_1$	7.233	6.412	10.164	90	106.70	90	0–1	Wang and Pei (2006)
Cu L-Phe_2	$P2_1$	16.710	5.217	9.509	90	98.40	90	0–2	van der Helm et al. (1971)
Zn L-$Phe_2 \cdot H_2O$	$P2_1$	5.630	31.876	9.518	90	90.07	90	0–1	Rombach et al. (2002)
Mn L-$Phe_2 \cdot H_2O$	$P2_12_12_1$	6.091	10.364	30.432	90	90	90	0–2	Weng et al. (2003)
Pb L-Phe_2	$C2$	9.831	5.888	15.504	90	104.73	90	0–2	Marandi and Shahbakhsh (2007)
DL-Tryptophan									
Fe^{II} DL-Trp_2	$P2_1/c$	19.828	5.423	9.280	90	97.69	90	0–2	Wang et al. (2008a)
Ni DL-Trp_2	$P2_1/c$	19.724	5.614	9.037	90	97.62	90	0–2	Wang et al. (2007)
Co^{II} DL-Trp_2	$P2_1/c$	19.595	5.465	9.055	90	98.02	90	0–2	Wang et al. (2008b)
Mn^{II} DL-Trp_2	$P2_1/c$	19.874	5.360	9.542	90	97.41	90	0–2	Xie et al. (2006)

[a]Only unit cell parameters are given, no structure or space group reported

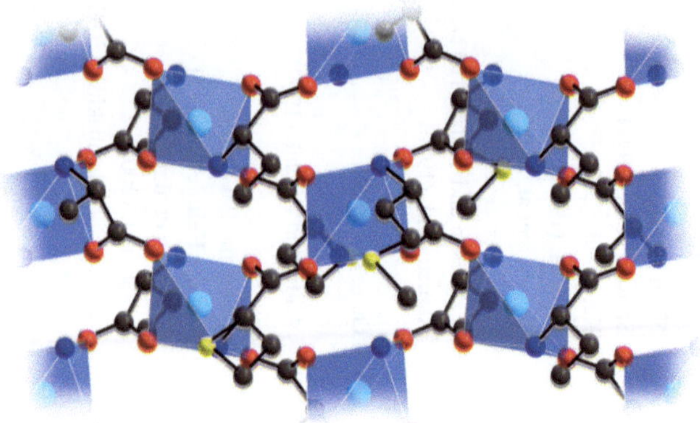

Fig. 3.20 Layers in bis(L-methioninato)-copper, parallel (001) (note the topological similarity to the structures of bis(L-leucinato) copper and bis(glycinato)-cadmium monohydrate (Figs. 3.7 and 3.18). Hydrogen atoms are omitted for clarity)

was reported by Sartoris et al. (1999) and is remarkable as there are two different, separate isolated units, one [4 + 1]- and one [4 + 2]-coordinate (hence the name).

Of all L-prolinates, only the Zn salt bis(L-prolinato)-zinc (Ng et al. 1995) crystallizes in a chain structure.

In comparison to salts of proline, the number of methioninates is even less numerous. As an example, we show the structure of *trans*-bis(L-methioninato)-copper as reported by Ou et al. (1979). This crystal comprises layers parallel (001), analog bis (glycinato)-cadmium monohydrate and bis(L-leucinato) copper . A view upon a detail of the layer is shown in Fig. 3.20. For this salt, a selenium analog was reported as well, i.e., *trans*-bis(L-selenomethioninato)-copper (Baran 2005).

The work of Fox et al. (2007), who reported Ca salts of glycine, DL-alanine, DL-valine, and DL-leucine (see respective sections), yielded crystalline salts of DL-methionine and DL-proline as well. Nevertheless, the authors did not give the unit cell data or structural parameters (they write that "refinement was hampered by disorder and/or twinning problems"). According to their work, the crystallization of these phases is difficult, and up to date, no data on these phases has been reported.

When investigating the salts of cations and phenylalaninates, only structures containing L-phenylalanine are found, but no racemates. Among these is the chain structure of (L-phenylalaninato)-silver (Wang and Pei 2006), which shows the same topology as the respective glycinate and alaninate compounds: a linear chain of [2]-coordinate Ag^+ cations, connected by bridging (O,N) amino acid ligands.

Another chain structure is found in bis(L-phenylalaninato)-zinc (Rombach et al. 2002), which has the same topology as the respective leucinate (Fig. 3.21; compare with Fig. 3.17). This similarity is also expressed in the relation of the unit cell parameters, where the shortest unit cell axis corresponds to the translation distance along the chain axis. The chain (and thus the structure) is polar, as in the respective Zn salt.

Fig. 3.21 Chain in bis (L-phenylalaninato)-zinc. Compare with the respective leucinate salt (Fig. 3.17). Hydrogen atoms are omitted for clarity

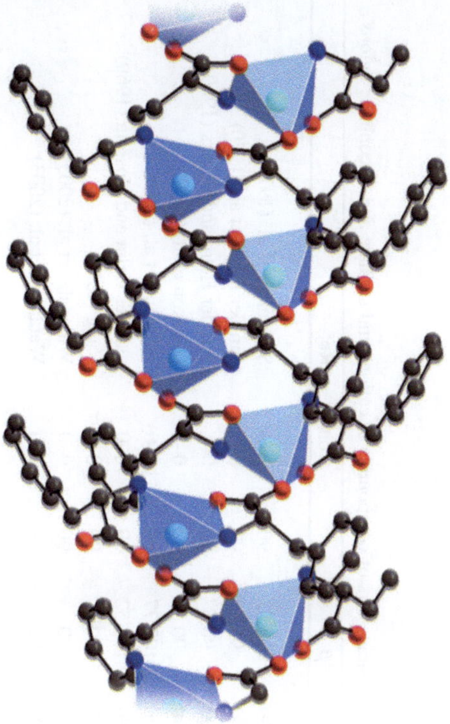

Among L-phenylalaninates, several layered structures have been found as well. While the layers in the Cu^{2+} and Pb^{2+} salts (van der Helm et al. 1971; Marandi and Shahbakhsh 2007) resemble those in the Cu methionate, Cu leucinate, and Cd glycinate salts by their topology, the layers in the Mn^{2+} phenylalaninate (Weng et al. 2003) are of a different topology: These sheets (parallel 001) are slightly corrugated, with the phenylalaninate anions facing away from the layer, leaving holes in the structure, in which the water molecules – connected via hydrogen bonds – are located (Fig. 3.22).

In contrast to the above amino acid salts, the situation for salts of tryptophan is different: Firstly, there are no salts of the pure enantiomer in the literature; only racemates have been characterized. Secondly, all the four structures found in the literature are isotypic, which is unusual (compare with the structures discussed above). In detail, these are salts of the type bis(DL-tryptophan)-Me^{2+}, with the metals nickel (Wang et al. 2007), iron (Wang et al. 2008a), cobalt (Wang et al. 2008b), and manganese (Xie et al. 2006).[1] This seemingly unusual aspect that all structures are isotypic can not only be attributed to the similarity of the metal cations (for other

[1] Despite the fact that the structures are isotypic (when calculated from the CIFs), it is found that the authors of the structure of the manganese salt have published a unit cell which is significantly longer in the *c* direction – an aspect that remains without explanation so far.

Table 3.5 *Salts of amino acids and cations, part V*: space groups and unit cell parameters of compounds of cations and serinate and threoninate anions

	s.g.	*a*	*b*	*c*	*α*	*β*	*γ*	Conn.	Ref.
L-Serine									
Ca L-$Ser_2 \cdot 2H_2O$	$P2_1$	10.076	6.850	8.308	90	94.23	90	N/A	Sicignano et al. (1974)
Zn L-Ser_2	$P2_1$	9.542	8.181	5.666	90	96.72	90	0–1	van der Helm et al. (1970)
Cu L-Ser_2	$P2_1$	9.924	8.413	5.651	90	90.6	90	0–1	van der Helm and Franks (1969)
Cu L-Ser_2	$P2_1$	8.65	9.6	5.08	90	90	92.6	0–2	D'yakon et al. (1991)
Cu L-$Ser_2 \cdot 2H_2O$	$C2$	7.587	8.568	8.826	90	102.77	90	0–0	Versiane et al. (2005)
Ni L-$Ser_2 \cdot 2H_2O$	$C2$	7.776	8.546	8.834	90	101.52	90	0–0	van der Helm and Hossain (1969)
cis-Pd L-Ser_2	$P2_12_12_1$	8.828	9.705	11.315	90	90	90	0–0	Vagg (1979)
cis-Pt L-Ser_2	$P2_12_12_1$	8.710	9.773	11.355	90	90	90	0–0	Watanabe et al. (2003)
trans-Pt L-$Ser_2 \cdot 2H_2O$	$P2_12_12_1$	7.019	7.745	20.946	90	90	90	0–0	Watanabe et al. (2003)
DL-Serine									
Cu DL-Ser_2	$Pna2_1$	9.450	4.970	20.150	90	90	90	0–2	D'yakon et al. (1992)
Pt DL-$Ser_2 \cdot 2H_2O$	$P2_1$	4.627	24.543	5.500	90	111.00	90	0–0	Krylova et al. (2006)
L-Threonine									
Co L-$Thr_2 \cdot 2H_2O$	$C222_1$	5.843	10.120	22.36	90	90	90	0–0	Rizzi et al. (2003)
Zn L-$Thr_2 \cdot H_2O$	$P2_1$[a]	5.875	22.275	5.807	90	119.69	90	0–0	Hamalainen (1977)
Cu L-$Thr_2 \cdot H_2O$	$P2_1$	11.02	4.90	11.16	90	93.5	90	0–1	Amirthalingam and Muralidharan (1975)
DL-Threonine									
Cu DL-Thr_2	$C112_1/b$	22.30	9.56	5.11	90	90	92.6	N/A	Diacon et al. (2000)
Cu DL-Thr_2	$P112_1/b$	11.10	9.56	5.11	90	90	92.6	0–1	Diacon et al. (2000)
Cu DL-Thr_2	C[b]	19.77	11.01	10.64	90	93.5	90	N/A	Amirthalingam and Muralidharan (1973)

[a]Symmetry probably incorrect; see text
[b]No structure given, space group denoted only as Cm or C2/m

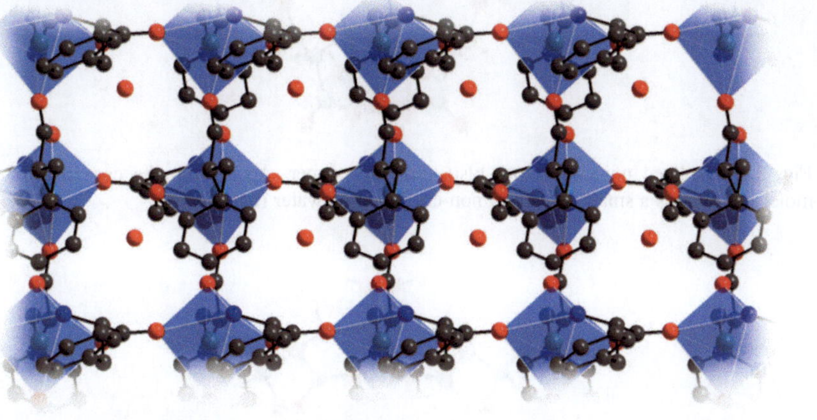

Fig. 3.22 Layers in bis(L-phenylalaninato)-manganese monohydrate (note the interstices in the layers, where the water molecules are located. Hydrogen atoms are omitted for clarity)

amino acids, salts of these cations do not often form isotypic structures). More importantly, as tryptophan is a rather large, sterically rigid molecule, the packing mode is apparently governed by the amino acid, whose steric requirements prevail over that of the metal cations. The structures are layered, of the same topology as bis(L-methioninato)-copper, bis(L-leucinato) copper, and bis(glycinato)-cadmium monohydrate.

In addition to the salts characterized by XRD and thus listed in Table 3.4, several other species have been reported, investigated by alternative methods as no crystals suitable for XRD were available. Among these were the sodium and potassium salts of phenylalanine, tyrosine, and tryptophan (Ruan and Rodgers 2004), whose molecular structures were assessed via mass spectroscopy. Likewise, lithium, sodium, and potassium prolinates (Moision and Armentrout 2006) as well as rubidium prolinate (Bowman et al. 2010) were characterized by the same method.

3.6 Serine and Threonine

The small polar amino acids serine and threonine are more common as anions in salts with metal cations in comparison with their nonpolar relatives (Table 3.5; compare with Table 3.4). For all of these amino acids, salts of both L- and DL-form have been reported. Within both serine and threonine salts, isotypic structures occur (as can be seen from the similarity of unit cell parameters in the case of serinates; for threonine salts, see text below), although most compounds crystallize in a unique arrangement.

Noteworthy are the three Cu^{2+} serinate crystal, two of which are anhydrous and crystallize even in the same space group ($P2_1$). In one polymorph (van der Helm and Franks 1969), the units are infinite chains analog those in *cis*-bis(L-alaninato)-copper, whereas the second polymorph (D'yakon et al. 1991) is a layer structure

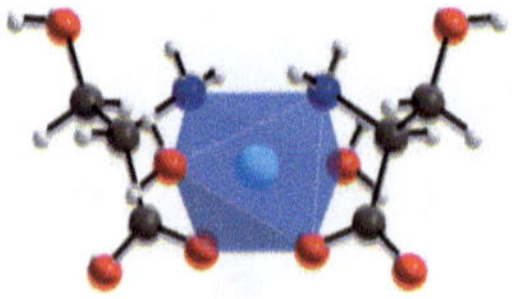

Fig. 3.23 Isolated units in diaqua-bis(L-serinato)-copper. Apart from the coordinating water molecule, there is a small amount of non-coordinating water (not shown)

Fig. 3.24 Isolated units in diaqua-bis(L-threonine)-cobalt. Note that the water molecules are in *cis*-position (in contrast to diaqua-bis(glycinato)-nickel (Fig. 3.3) and respective salts. The respective zinc salt is isotypic (see text)

similar as bis(glycinato)-cadmium monohydrate. The third species was discovered most recently and is a dihydrate (measured at 120 K, Versiane et al. 2005), in which the Cu^{2+} cations are [4 + 2]-coordinated, with two *cis*-oriented serinate anions in bidentate (N,O) mode, plus two water molecules (Fig. 3.23). This species is isostructural with the respective Ni salt, which was discovered much earlier (van der Helm and Hossain 1969).

As said above, there is another pair of isotypic serinate salts (as can be guessed from the unit cell parameters; see Table 3.5), namely, *cis*-bis(L-serinato)-platinum and *cis*-bis(L-serinato)-palladium (Watanabe et al. 2003; Vagg 1979), which correspond structurally to *cis*-bis(glycinato)-platinum.

Furthermore, there is one DL-serinate which was reported to crystallize in a non-centrosymmetric space group (namely, *trans*-(DL-serinato)-platinum dihydrate in space group $P2_1$, Krylova et al. 2006). As racemic amino acid salts usually show inversion symmetry with either planes or centers, this appears dubious. A symmetry check with the program PLATON (Spek 2009) indeed gave the symmetry $P2_1/m$, so it appears the symmetry given is not correct and this crystal is in fact centrosymmetric.

Within the salts of threonine, the structure of diaqua-bis(L-threonine)-cobalt (Rizzi et al. 2003) is noteworthy as this salt does contain isolated units as found for many other likewise salts, with the exception that the coordinating water molecules are *cis*-oriented (Fig. 3.24). The respective zinc salt (Hamalainen 1977) was published with a lower symmetry ($P2_1$ instead of $C222_1$), but similarities in structure and unit cell parameters lead to the assumption that this species was published with a wrong symmetry. A symmetry check by PLATON (Spek 2009) showed that actually the orthorhombic symmetry is correct for both the zinc and cobalt salts. The same arrangement is found in the structure of bis(L-alaninato)-diaqua-cadmium monohydrate (Demaret and Abraham 1987b)

Additionally, salts of alkali metals with serine and threonine were reported, but due to the lack of suitable crystals, no XRD experiments were conducted. Instead,

Bowman et al. (2010) employed mass spectroscopy to investigate the molecular structure of the rubidium salts; Ye et al. (2008) utilized the same method for the lithium, sodium, and potassium salts of serine and threonine.

3.7 Cysteine

The salts of cysteine deserve closer attention, as they differ from all other amino acid salts by the fact the sulfur atom of the thiol group can act as a ligand, which it usually does. Moreover, when deprotonation occurs, both the acid group and the thiol group can be deprotonated; thus, a bivalent cysteinate anion is formed. These properties are responsible for the fact that the cysteine salts published in the literature differ in many respects from all salts discussed above. One unique aspect is the fact that cysteinate anions readily form tridentate (O,N,S) ligands.

The first species listed in Table 3.6 is noteworthy insofar as it is the only amino acid salt with Bi^{3+} from all compounds of cations and amino acids. Tris(L-cysteinato-S)-bismuth monohydrate comprises [3]-coordinate Bi^{3+} cations (Fig. 3.25), where the amino acid is bonded via a Bi–S bond – thus, this is the only salt where neither oxygen nor nitrogen atoms act as ligands. In fact, the molecules could be considered as zwitterionic as far as acid and amino groups are concerned: The protons from the acid groups are formally transferred to the amino groups.

The compounds (L-cysteinato)-zinc and (L-cysteinato)-cadmium (Rebilly et al. 2008) are interesting as both comprise polyhedral chains (distorted tetrahedra and octahedral, in the case of Zn and Cd, respectively), i.e., chains of polyhedral sharing common corners or edges (Fig. 3.26). These one-dimensional structures are further connected via the L-cysteinate^{2-} molecules, which interconnect the chains to a framework. These chains can be described as fragments of bulk structures of CdS (i.e., the rock salt high-pressure phase) and ZnS (modification wurtzite) since the sulfur atom of the L-cysteinate bridges similarly as the sulfur atom in the respective minerals.

Apart from these structural features, Rebilly et al. (2008) studied the band gaps of these species in comparison with the respective minerals and found blue shifts (1.37 eV for Zn, compared to wurtzite, and 2.59 eV for Cd, compared to CdS) which were attributed to the lower dimensionalities of the metal–sulfur arrangements. These assessments were supported by density of states calculations.

The group of L-cysteinates also encompasses double salts, i.e., salts with two different cations, in these cases a combination of alkali and transition metals

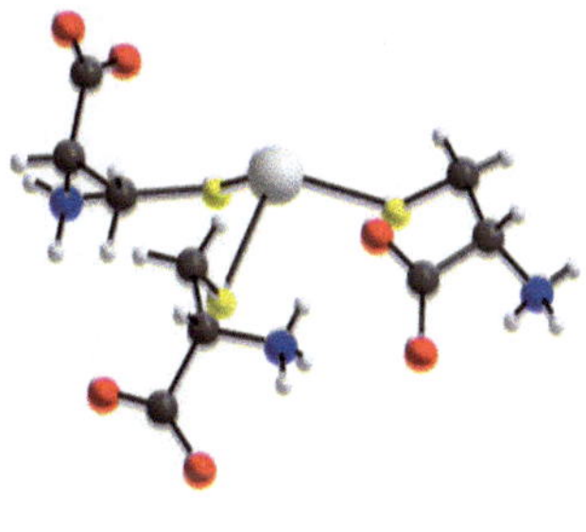

Fig. 3.25 Units in tris (L-cysteinato-S)-bismuth monohydrate. This species is unique insofar as the amino acids coordinate by the sulfur atom only. Note also that the acid groups are deprotonated, while the amino groups are protonated

Table 3.6 *Salts of amino acids and cations, part VI*: space groups and unit cell parameters of compounds of cations and cysteinate anions

	s.g.	*a*	*b*	*c*	*α*	*β*	*γ*	Conn.	Ref.
L-Cysteine									
Bi L-$Cys_3 \cdot H_2O$	$P2_12_12_1$	5.135	11.841	28.120	90	90	90	0–0	Wang and Xu (2008)
Zn L-Cys	$P2_12_12_1$	6.025	8.838	9.402	90	90	90	1–1	Rebilly et al. (2008)
Cd L-Cys	$P2_12_12_1$	5.631	9.483	9.542	90	90	90	1–1	Rebilly et al. (2008)
Na_2Zn L-$Cys_2 \cdot 6H_2O$	$C2$	5.206	23.313	15.052	90	92.53	90	2–2	Bell and Sheldrick (1984)
Na V L-$Cys_2 \cdot 2H_2O$	$P3_1$	18.620	18.620	9.604	90	90	120	1–1	Maeda et al. (1993)
Na Cr L-$Cys_2 \cdot 2H_2O$	$P3_12$	10.644	10.644	9.809	90	90	120	1–1	De Meester et al. (1977)
K Ni L-$Cys_2 \cdot 1.5H_2O$	$P2_1$	17.055	6.424	14.905	90	123.56	90	2–2	Baidya et al. (1991)
DL-Cysteine									
Tl DL-Cys	$P2_1/a$	10.841	7.672	8.391	90	114.58	90	0^2–0^2	Freeman and Moore (1977)

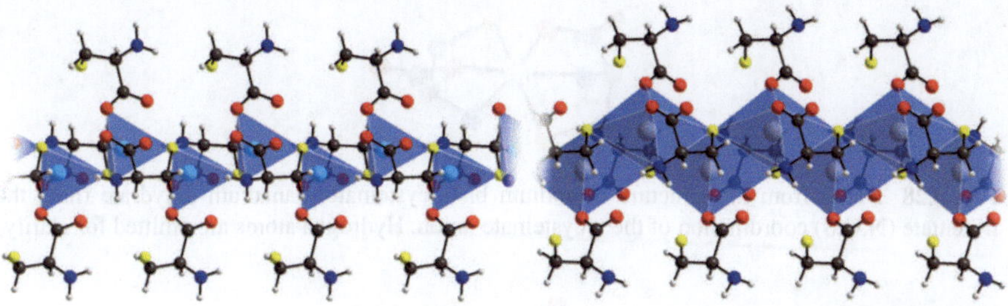

Fig. 3.26 Chains in (L-cysteinato)-zinc (*left*) and (L-cysteinato)-cadmium (*right*). In both compounds, the polyhedra are connected to chains (via corners or edges), with additional connection provided by the bivalent cysteinate anions. Neighboring chains are connected to a three-dimensional framework

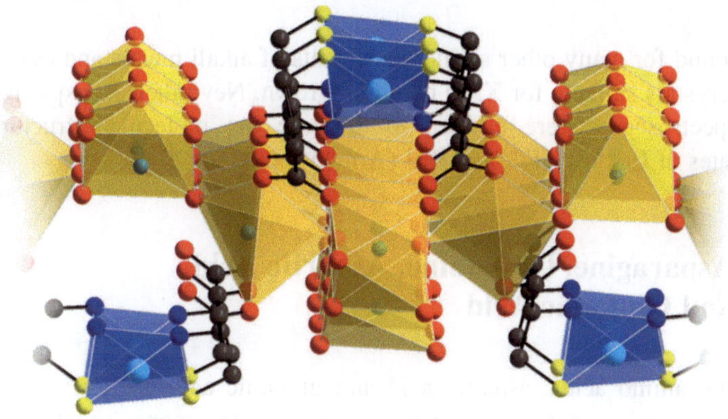

Fig. 3.27 View upon a layer in disodium bis(L-cysteinato)-zinc hexahydrate. The Na polyhedra (*yellow*) form a layer, upon which the zinc polyhedra (*blue*) are connected. Hydrogen atoms are omitted for clarity

(Table 3.6). For all these salts, the polyhedra of the alkali metals are connected via common corners, edges, and faces and form chains (in the case of the isotypic pair of Na–V and Na–Cr salts) or layers (in the cases of the Na–Zn and K–Ni salts). Attached to these chains and layers are the cysteinate anions, which also connect the transition metals. As an example, the layered structure of the Na–Zn salt is shown (Fig. 3.27). In addition, one detail of the Na–V salt is given (Maeda et al. 1993; Fig. 3.28), where the tridentate coordination of cysteinate – here around a vanadium cation – can be seen.

Furthermore, there is on racemic salt, which is also unique insofar as this is the only salt of Tl^{+} among all amino acids. DL-Cysteinato-thallium (Freeman and Moore 1977) comprises dimers of thallium polyhedra, which are located on an inversion center (Fig. 3.29). The lone pair of the monovalent thallium cation influences the coordination polyhedron, which is clearly asymmetric (on this effect see Mudring and Rieger 2005 and references therein).

Fig. 3.28 Detail from the structure of sodium bis(L-cysteinato)-vanadium dihydrate (note the tridentate (N,O,S) coordination of the L-cysteinate anion. Hydrogen atoms are omitted for clarity)

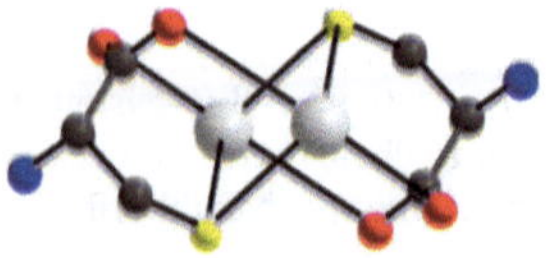

Fig. 3.29 Dimeric unit in DL-cysteinato-thallium. The Tl^{+} cations show an asymmetric coordination, which is due to the lone pair

As found for many other amino acids, salts of alkali metals and cysteine exist, but no crystals suitable for XRD could be grown. Nevertheless, experiments with mass spectroscopy were utilized to describe the molecular structure of the cysteinates of Li^{+}, Na^{+}, K^{+}, and Rb^{+} (Armentrout et al. 2010).

3.8 Asparagine, Glutamine, Aspartic Acid, and Glutamic Acid

As acidic amino acids, aspartic acid and glutamic acid are well suited for the synthesis of amino acid salts. Both molecules can form mono- and divalent cations (even in the first case, both acid groups are deprotonated, but the amino group is protonated), i.e., as hydrogen glutamate and hydrogen aspartate, or glutamate and aspartate. Moreover, both molecules are rather large amino acids with high conformational flexibility, and since the possible ligand atoms are located on opposite ends of the molecules, these amino acids are suitable for connection of cations. This property is clearly established by the structures found in aspartate and glutamate salts: Isolated units are rare, and structures of higher dimensionality are frequently found. Since glutamic acid is the larger molecule, this amino acid is even more suitable for connecting cations; thus, the dimensionality is even higher (for aspartates, a chain structure is the most frequent type, whereas among glutamates layers and framework structures feature most frequently).

Another aspect resulting from the size and flexibility of these molecules is the possibility of tridentate coordination, which was found to be realized via one oxygen atom from the α-acid group, the amino nitrogen atom, as well as one oxygen from the δ- or ε-acid group.

Because of all these features, it is not surprising that the number of aspartate and glutamate salts is rather high – almost all common metal cations have been employed in the syntheses of salts with these amino acids (Tables 3.7 and 3.8). Still, some salts could not be obtained in crystal form (see end of the section),

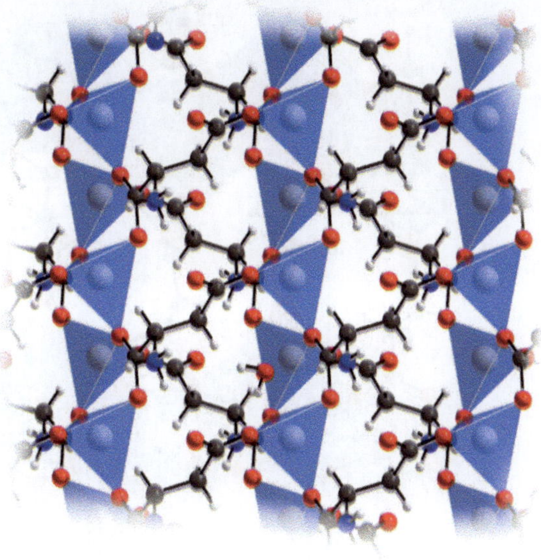

Fig. 3.30 Layers in (hydrogen L-aspartato)-lithium monohydrate (note the corner-linked chains of lithium tetrahedral, connected to layers via the hydrogen aspartate anions)

although much work has been devoted to this task. Especially the group of Hubert Schmidbaur worked very actively with salts of these family, but they found that some crystals took very long (years, in some cases) to grow.

In contrast, salts of asparagine and glutamine, being amides of aspartic acid and glutamic acid, are less frequent, as these amino acids have neutral side chains (as opposed to the acidic side chains of the respective acids).

Among the aspartates, there are several salts of monovalent cations. For instance, (hydrogen L-aspartato)-lithium monohydrate (Schmidbaur et al. 1989a) comprises corner-linked, distorted Li^+ tetrahedral, which are connected by the $HAsp^-$ anions to layers parallel (001). As the nitrogen atom is protonated in the $HAsp^-$ anion, it is not in coordination, instead the oxygen atoms from the acid groups act as ligand atoms (Fig. 3.30).

Among the divalent cations, several structural motifs are found. In the isotypic pair of barium and strontium salts (Schmidbaur et al. 1990a), the same pattern as in the Li salt described above is found: Polyhedral chains are connected via the aspartate anions (in *trans*-conformation, linear between the chains). This is noteworthy insofar as the aspartate anions are divalent, i.e., the amino group is not protonated, and the nitrogen atoms could act as ligand atoms. Still, the steric situation does not allow these atoms as part of the coordination sphere (Fig. 3.31).

In contrast to this motif, the structure of diaqua-(L-aspartate)-magnesium monohydrate (Schmidbaur et al. 1986) is a chain structure, where the isolated Mg^{2+} octahedra (here the aspartate acts as tridentate ligand) are connected to neighboring octahedra also via the other oxygen atom of the δ acid group (Fig. 3.32).

The same motif is found in the calcium salt, tetraaqua-(L-aspartate)-calcium (Schmidbaur et al. 1989b). Within this crystal, there are two independent aspartate

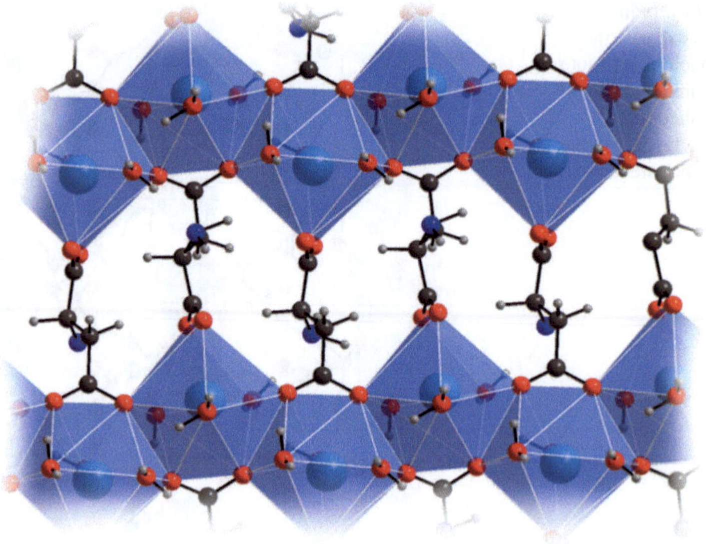

Fig. 3.31 Layers in triaqua-(L-aspartato)-strontium (note the corner-linked chains of lithium tetrahedral, connected to layers via the hydrogen aspartate anions)

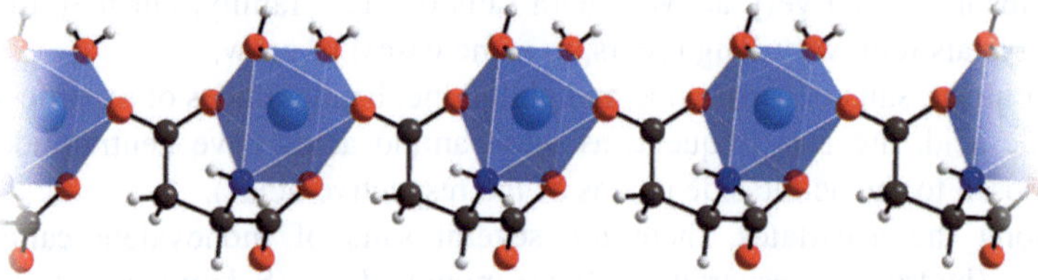

Fig. 3.32 Chains in diaqua-(L-aspartate)-magnesium monohydrate. In this salt, all possible ligand atoms from the divalent aspartate anion act as ligands (note also the tridentate (O,N,O′) coordination mode of the aspartate anion)

molecules, therefore two crystallographically different but topologically identical chains, which show the same arrangement as in the magnesium salt. The same authors reported another phase, namely, a dihydrate, which was synthesized from different precursors (namely, L-aspartic acid and $Ca(OH)_2$, in contrast to $CaCO_3$, which was used for the tetrahydrate).

In addition, there is a group of isotypic salts (in fact the zinc, nickel, cobalt, and manganese analogs of the magnesium salt), which show the same chain motif found in the magnesium compound. Although the topology is identical and the cations are also octahedral, the structure differs slightly from the Mg^{2+} salt (note the difference in the unit cell dimensions, Table 3.7).

The zinc, nickel, and cobalt salts were first reported by Doyne et al. (1957), who only gave the unit cell dimensions. As later works established the structure for the zinc (Kryger and Ramsussen 1973; Zhang et al. 2004) and the manganese analog

Table 3.7 *Salts of amino acids and cations, part VII*: space groups and unit cell parameters of compounds of cations and aspartate and asparaginate

	s.g.	*a*	*b*	*c*	α	β	γ	Conn.	Ref.
L-Aspartic acid									
Li L-Asp·H_2O	$P2_1$	6.417	5.123	10.133	90	98.19	90	1–2	Schmidbaur et al. (1989a)
K L-Asp·$3H_2O$	$P2_12_12$	6.934	24.626	5.008	90	90	90	1–1	Schmidbaur et al. (1989a)
Mg L-Asp·$3H_2O$	$P2_12_12_1$	6.140	9.430	15.015	90	90	90	0–1	Schmidbaur et al. (1986)
Ca L-Asp·$4H_2O$	$C2$	23.298	6.532	14.587	90	118.09	90	0–1	Schmidbaur et al. (1989b)
Ca L-Asp·$2H_2O$	$P2_12_12_1$	6.776	8.695	13.013	90	90	90	0–1	Schmidbaur et al. (1989b)
Sr L-Asp·$3H_2O$	$P2_12_12_1$	7.304	9.914	11.837	90	90	90	1–2	Schmidbaur et al. (1990a)
Ba L-Asp·$3H_2O$	$P2_12_12_1$	7.564	10.227	12.024	90	90	90	1–2	Schmidbaur et al. (1990a)
Zn L-Asp·$3H_2O$	$P2_12_12_1$	9.386	7.92	11.532	90	90	90	0–1	Doyne et al. (1957)
Co L-Asp·$3H_2O$	$P2_12_12_1$	9.39	11.37	7.85	90	90	90	0–1	Doyne et al. (1957)
Mn L-Asp·$3H_2O$	$P2_12_12_1$	9.555	7.906	11.614	90	90	90	0–1	Schmidbaur et al. (1988)
Ni L-Asp·$3H_2O$	$P2_12_12_1$	9.4	11.35	7.83	90	90	90	0–1	Doyne et al. (1957)
Cu L-Asp·$2H_2O$	$C2$	9.504	10.038	7.555	90	94.01	90	0–1	Calvo et al. (1993)
Pd L-Asp_2	$P2_12_12_1$	5.315	9.629	22.233	90	90	90	N/A	Gao et al. (2008)
$CaCo_2$ L-Asp_4·$7.5H_2O$	$P6_3$	15.972	15.972	21.164	90	90	120	0–2	Oonishi et al. (1975)
$CaCo_2$ L-Asp_4·$10H_2O$	$P2_12_12_1$	18.688	10.517	17.309	90	90	90	0–1	Oonishi et al. (1973)
DL-Aspartic acid									
Ag DL-Asp·xH_2O	$P2_1/c$	5.626	22.469	5.611	90	100.77	90	0^2–2	Nomiya and Yokoyama (2002)
Mg DL-Asp_2·$4H_2O$	$P2_1/c$	5.710	14.291	9.117	90	102.57	90	0–0	Ciunik (1987)
Mn DL-Asp_2·$4H_2O$	$P2_1/c$	5.701	14.466	9.253	90	102.36	90	0–0	Ciunik (1987)
KCo^{III} DL-Asp_2·xH_2O	$P2_1/c$	12.485	12.469	9.691	90	92.952	90	0–2	Versiane et al. (2006)
L-Asparagine									
Ag L-Asn_2	$C222_1$	7.425	8.019	21.537	90	90	90	0–1	Nomiya and Yokoyama (2002)
Cu L-Asn_2	$P2_1$	5.028	18.928	6.174	90	96.4	90	0–1	Stephens et al. (1975)
Zn L-Asn_2	$P2_1$	12.322	5.027	9.702	90	99.12	90	0–2	Stephens et al. (1977)
Cd L-Asn_2	$P2_1$	12.42	5.081	9.84	90	101.4	90	0–2	Flook et al. (1973)
Co^{III} L-Asn_3·$3H_2O$	$P2_12_12_1$	39.082	7.523	6.643	90	90	90	0–0	Sekizaki (1979)

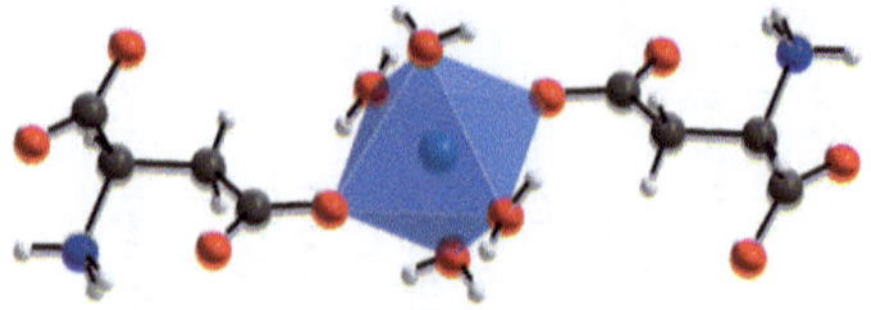

Fig. 3.33 Units in tetra-aqua-bis(DL-hydrogen aspartato)-manganese. In this salt, the amino acids act as monodentate ligands. The unit is symmetric; D- and L-aspartate anions are related via the inversion center upon which the cation is located

(Schmidbaur et al. 1988), the agreement of the unit cell dimensions allows the conclusion that all four salts crystallize in the same structure.[2]

Furthermore, there is an analog Cu^{2+} salt, whose structure shows the same topology, although it is not isotypic (the copper cations are not octahedrally coordinated, but adapt a [4 + 1] pyramidal coordination; Calvo et al. 1993).

Another interesting pair of salts are the two double salts of L-aspartate: Both are Ca^{2+}/Co^{3+} salts of the formula $Ca^{2+}(Co^{3+})_2(Asp^{2-})_4$, although they differ in the degree of hydration (Oonishi et al. 1973, 1975). This results in different crystal packing as well: The decahydrate is orthorhombic and comprises a helical chain structure, whereas the semiheptahydrate is hexagonal (an unusually high symmetry for amino acid salts) with a layered structure.

Furthermore, there are some papers on racemic aspartate salts. Nomiya and Yokoyama (2002), who also tested the antimicrobial activities of silver glycinate, published the structure of (DL-hydrogen aspartato)-silver hydrate. In this crystal, the Ag^+ cations form dimers, which are connected to layers by the hydrogen aspartate anions. In the same study, a salt of asparagine was published, namely, (L-asparaginato)-silver (and its enantiomer counterpart, the respective D-form). This asparaginate salt shows the linear chain motif found for the glycinate salt, where the amino acids act as bridging (N,O) ligands.

Another pair of isotypic compounds is tetra-aqua-bis(DL-hydrogen aspartato)-manganese and magnesium (Ciunik 1987; the latter was refined by Schmidbaur et al. 1989d). These salts are remarkable insofar as they comprise isolated units, in which the Mn^{2+} or Mg^{2+} cations (in octahedral coordination) have four water molecules as ligands, while the aspartate anions act as monodentate (O) ligands only (Fig. 3.33)

As stated above, the salts of asparagine are less numerous. In addition to the silver salt (which has been commented on above), there is an isotypic pair of salts, namely, bis(L-asparaginato)-zinc and bis(L-asparaginato)-cadmium (Stephens et al. 1977; Flook et al. 1973). In these salts, the same structural layer motif as in bis(glycinato)-cadmium monohydrate is found: isolated units, connected via the amino acid anions to layers.

[2] For the nickel salt, a later work established the structure, although the unit cell parameters differ stronger than expected (Antolini et al. 1982). To settle this properly, a study of all four salts under the same conditions would be necessary.

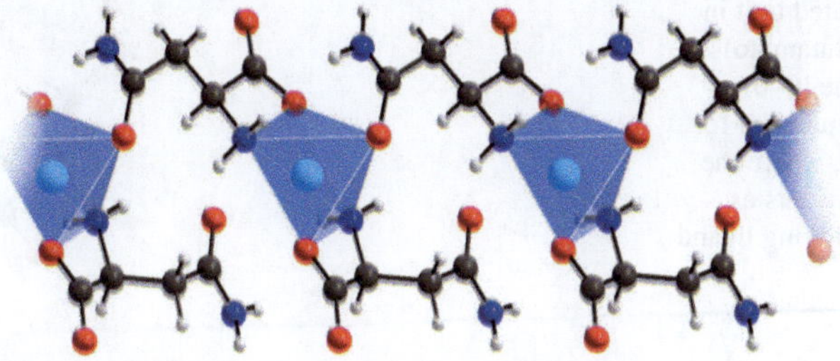

Fig. 3.34 Chains in bis(L-asparaginato)-copper. Note that there are two crystallographically different asparaginate anions, where only one is n bridging function

The respective copper salt (Stephens et al. 1975; redetermined by Venkato et al. 2004) has a chain structure similar to its threoninate analog, where isolated [4 + 1] copper pyramids are connected to a chain. As in the threoninate, there are two different asparaginate anions, one bridging mode between two pyramids, the other one coordinating only one cation (Fig. 3.34). The topology is not identical; however, in the threonine salt, the bridging oxygen atoms both belong to the α-acid group, whereas in the asparaginate the bridging ligand coordinates one cation via an α-acid oxygen atom and the other cation via the δ-amido oxygen atom.

One more asparaginate is tris(L-asparaginato)-cobalt trihydrate (Sekizaki 1979), where three Asp^- anions coordinate the Co^{3+} cation in bidentate (O,N) mode, forming an octahedral coordination, analog to the respective glycine salt.

As said before, the salts of glutamic acid resemble those of aspartic acid, although the larger size of the molecule allows even higher conformational flexibility. Thus, the glutamate salts are higher dimensional than most other amino acid salts.

There are several hydrogen glutamates of monovalent cations (including ammonium, which is isotypic with one potassium polymorph, Nagashima et al. 1992a, b), although one of the sodium salts deviates from all other amino acid salts discussed so far: In (tetrakis-aqua)-sodium (L-glutamate) monohydrate (Kashiwagi et al. 1995), the hydrogen glutamate anion is not in coordination. Instead, this crystal contains a chain of Na^+ octahedra with only water molecules as a ligand; the $HGlu^-$ anions are connected only via hydrogen bonds to this chain.

The proper sodium glutamate (or monosodium glutamate, MSG), where the amino acid acts as a ligand, is in fact the most popular amino acid salt. As addressed in Chap. 2, MSG has become the basis of a worldwide trillion dollar industry and, subsequently, a constituent of many dishes and thus of the diet of most people on earth (Ault 2004). The first crystallographic study was done by Rao and Mallikarjunan (1975), who only reported unit cell data, but no structure. This was determined by Sano et al. (1989), whose group also published a study on the effect of additives on the growth of MSG crystals (Sano et al. 1990).

Of the divalent cations, the structure of tetra-aqua-(L-glutamato)-magnesium is noteworthy as it is the only salt of a cation with octahedral (or higher) coordination that comprises isolated units. The Mg^{2+} cation has four water molecules as ligands, while the glutamate anion acts as bidentate (O,N) ligand only (Fig. 3.35). The only

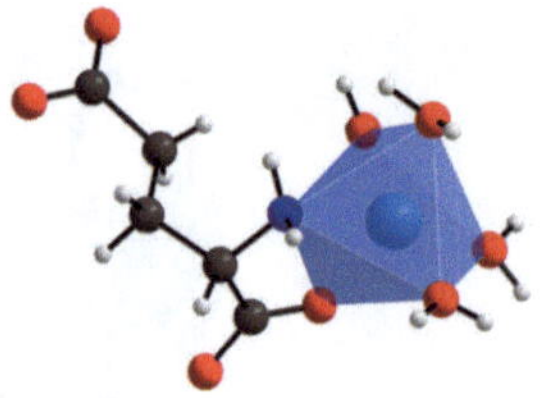

Fig. 3.35 Isolated unit in tetra-aqua-(L-glutamato)-magnesium. The (0–0) connectivity is unusual for glutamate salts, where the anion frequently acts as tridentate or bridging ligand

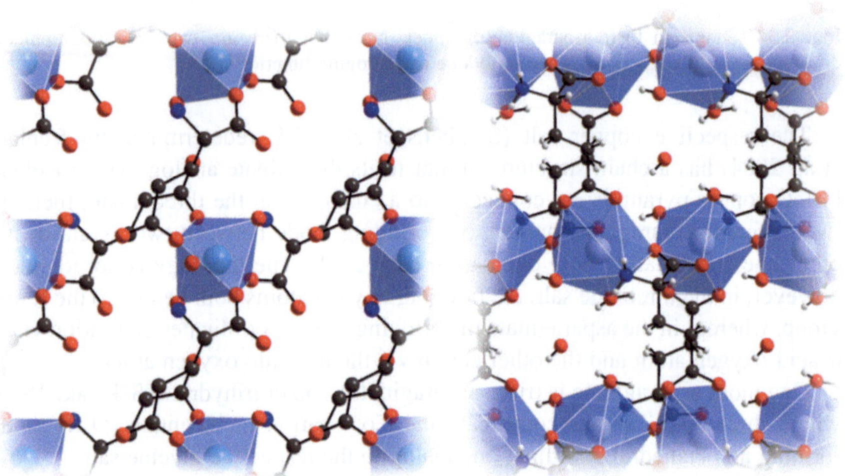

Fig. 3.36 Layers in diaqua-(L-glutamato)-calcium monohydrate (*left*) and dihydrate (*right*). Both salts show layered structures, but while in the monohydrate the polyhedra are isolated, there are polyhedral chains (oriented horizontally in the image) in the dihydrate

other glutamate which contains isolated units only is that of palladium, which assumes the usual planar [4]-coordination.

Einspahr and Bugg reported two different hydrates of calcium glutamate, namely, diaqua-(L-glutamato)-calcium monohydrate (Einspahr and Bugg 1974), which is trigonal, and diaqua-(L-glutamato)-calcium dihydrate (Einspahr and Bugg 1979) which crystallizes in tetragonal symmetry. Both are layered structures, although the coordination of the cation and connectivity of the units are completely different (Fig. 3.36).

Likewise, there are two different hydrates of the strontium salt (Christgau et al. 2005; Schmidbaur et al. 1989c), one of which has an isotypic barium analog (as seen from the unit cell dimensions, Table 3.8). All are characterized by polyhedral chains, connected to layers, as found in the calcium salts.

A different situation is found in the isotypic series of cadmium, zinc, cobalt, and copper salts, as represented by (L-glutamato)-aqua-cadmium monohydrate (determined and refined by Soylu et al. 1974; Flook et al. 1977; Tomlin et al. 1997; and Li et al. 2007b; the zinc salt was published even earlier by Zn Gramaccioli 1966; later refined by Proch et al. 2008 at 173 K; the copper salt was published by

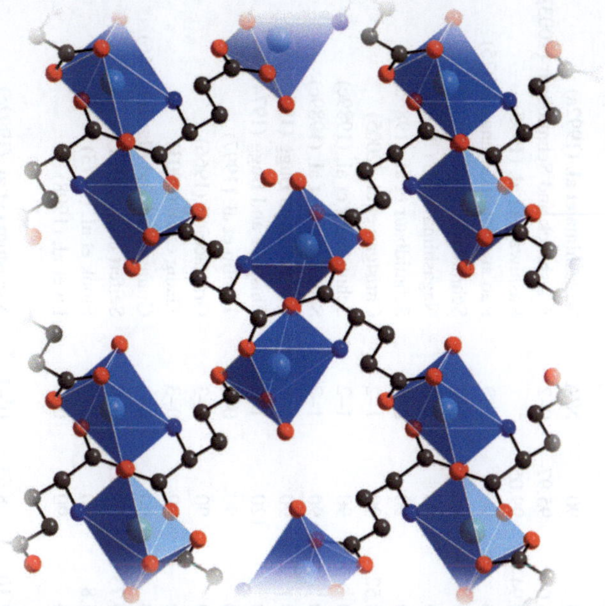

Fig. 3.37 Framework in (L-glutamato)-aqua-cadmium monohydrate and its isotypic counterparts (Zn, Co, Cu), viewed along (001). Although in this projection the polyhedra appear to be sharing common corners, they are truly isolated. The connection is provided by the glutamate anions, thus forming a three-dimensional framework (the connectivity is therefore 0–3)

Gramaccioli and Marsh 1966; Mizutani et al. 1998; and Ray et al. 2004; the cobalt salt was reported by Zhang et al. 2003). This structure is a three-dimensional framework, made up by isolated [6]-coordinate polyhedra, which is a unique motif in amino acid salts (Fig. 3.37).

Moreover, there is one double salt with an established structure, namely, bis (L-glutamato)-penta-aqua-copper-magnesium dihydrate (Lu et al. 1998). This species has an interesting structure, as it comprises chains of alternating copper [4 + 1] pyramids and magnesium [6] octahedra, connected by Glu^{2-} anions. The Cu^{2+} cation is coordinated in bidentate (N,O) mode, the Mg^{2+}, in monodentate (O) mode (Fig. 3.38).

Although there are many different salts of L-glutamate, not much work has been done on racemate salts. Only one DL-salt has been published, sodium DL-glutamate dihydrate (Nagashima et al. 1992c), which comprises dimers of edge-sharing polyhedra, connected to chains.

Finally, there is one glutaminate salt, bis(L-glutamine)-copper (Deschamps et al. 2003a; also measured at 120 K by Schveigkardt et al. 2002), which consists of isolated units which are connected to layers by the glutamate anions – analog to the structure of bis(L-leucinato) copper.

Table 3.8 *Salts of amino acids and cations, part VIII*: space groups and unit cell parameters of compounds of cations and glutamate anions

	s.g.	*a*	*b*	*c*	*α*	*β*	*γ*	Conn.	Ref.
L-Glutamic acid									
(NH_4) L-Glu·H_2O	$P2_12_12$	7.713	21.469	4.987	90	90	90	N/A	Nagashima et al. (1992a)
Li L-Glu	*P*1	5.2961	7.852	7.951	90.03	106.04	95.97	0–2	Wiesbrock and Schmidbaur (2003)
Na L-Glu·$5H_2O$	*P*1	6.116	16.656	6.007	99.29	100.94	99.05	[a]	Kashiwagi et al. (1995)
Na L-Glu·H_2O	$P2_12_12_1$	17.954	15.241	5.559	90	90	90	1–3	Rao and Mallikarjunan (1975)
K L-Glu·H_2O	$P2_12_12_1$	6.289	7.240	18.153	90	90	90	1–2	Schmidbaur et al. (1990b)
K L-Glu·H_2O	$P2_12_12$	7.838	20.725	5.175	90	90	90	1–2	Nagashima et al. (1992b)
Mg L-Glu·$4H_2O$	$P2_12_12_1$	7.838	10.113	13.156	90	90	90	0–0	Schmidbaur et al. (1989c)
Sr L-Glu_2·$5H_2O$	$P2_1$	8.710	7.245	14.585	90	100.52	90	1–2	Christgau et al. (2005)
Sr L-Glu_2·$6H_2O$	$P2_12_12_1$	7.355	8.772	20.283	90	90	90	1–2	Schmidbaur et al. (1989c)
Ba L-Glu_2·$6H_2O$	$P2_12_12_1$	7.54	8.96	20.25	90	90	90	1–2	Schmidbaur et al. (1989c)
Ca L-Glu_2·$4H_2O$	$P4_32_12$	7.562	7.562	29.077	90	90	90	0–2	Einspahr and Bugg (1979)
Ca L-Glu·$3H_2O$	$P3_121$	8.863	8.863	20.863	90	90	120	1–2	Einspahr and Bugg (1974)
Cd L-Glu·$2H_2O$	$P2_12_12_1$	11.575	10.764	7.256	90	90	90	0–3	Tomlin et al. (1997)
Zn L-Glu·$2H_2O$	$P2_12_12_1$	11.190	10.463	7.220	90	90	90	0–3	Gramaccioli (1966)
Co L-Glu·$2H_2O$	$P2_12_12_1$	7.149	10.468	11.295	90	90	90	0–3	Zhang et al. (2003)
Cu L-Glu·$2H_2O$	$P2_12_12_1$	11.084	10.350	7.238	90	90	90	0–3	Gramaccioli and Marsh (1966)
Pd L-Glu_2	*P*1	4.886	5.161	13.365	93.73	99.73	104.25	0–0	Seifert et al. (2011)
Cd L-Glu·$2H_2O$	$P2_1$	12.30	8.45	9.12	90	95.8	90	N/A	Flook et al. (1973)
MgCuL-Glu_2·$7H_2O$	$C222_1$	10.260	16.622	11.609	90	90	90	0–1	Lu et al. (1998)
DL-Glutamic acid									
Na DL-Glu·$2H_2O$	*P*-1	7.129	10.186	6.472	107.04	102.10	98.63	0^2–1	Nagashima et al. (1992c)
L-Glutamine									
Cu L-Glu	*C*2	28.190	5.084	9.420	90	99.01	90	0–2	Schveigkardt et al. (2002)

[a] In this crystal, glutamate is not in coordination around the sodium cation; see text

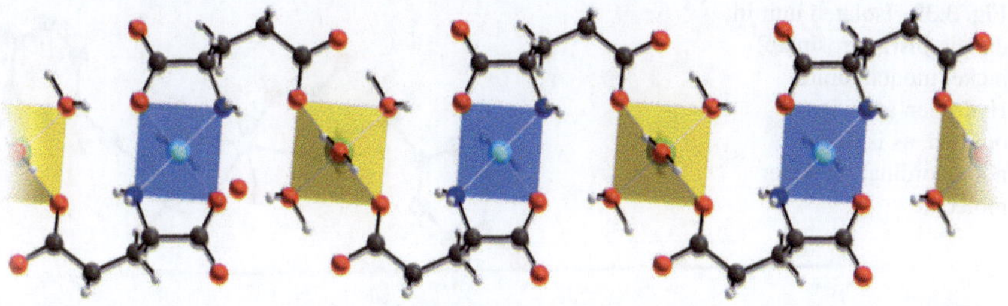

Fig. 3.38 Chain in the Mg-Cu-aspartate double salt bis(L-glutamato)-penta-aqua-copper-magnesium dihydrate. Although the coordination polyhedra of both cations look similar in this projection, they are different – copper is [4 + 1]-coordinated (*blue*), magnesium [6]-coordinated (*yellow*) (note also the different ligands)

It is noteworthy that among all amino acids, aspartic acid and glutamic acid are the only ones which form sodium and potassium salts stable enough that crystals could be obtained which could be used for XRD (in contrast to several other amino acids, where these salts have been reported, but no crystal structures are available). Still, data from mass spectroscopy for both K and Na salts of both acidic amino acids as well as asparagine and glutamine are available (Heaton et al. 2006; Heaton and Armentrout 2008). Schmidbaur et al. (1990c) published a detailed review of aspartic and glutamic acid as a ligand to alkali and alkaline earth metal cations. Apart from structural and chemical considerations, they also discussed toxicity of several aspartates upon ingestion, solution studies, and influence of these amino acids upon transport and distribution of magnesium in the human body.

3.9 Tyrosine, Histidine, Lysine, and Arginine

Although classified as a polar amino acid, tyrosine has a very low solubility (compare Table 2.1); thus, not many salts of tyrosine have been synthesized.

Among the L-tyrosinates, several salts are reported, all of which comprise divalent cations in combination with two monovalent Tyr^- anions. Most species comprise isolated units, as in the nickel (Fig. 3.39) and the palladium salts (Table 3.9), although one salt has a chain structure: In bis-(L-tyrosinato)-copper (first reported by van der Helm and Tatsch 1972; later refined by Weng et al. 2002), chains of the type of the respective asparaginate and threoninate salt are found: Of the two crystallographically different anions, only one acts as a bridging ligand between the Cu^{2+}[4 + 1] pyramids.

Apart from the L-tyrosinates, there is one racemic salt of which the crystal structure has been determined: (DL-tyrosinato)-aqua-zinc (Li et al. 2007b). In this compound, the phenolic group is deprotonated. Thus, another ligand atom is

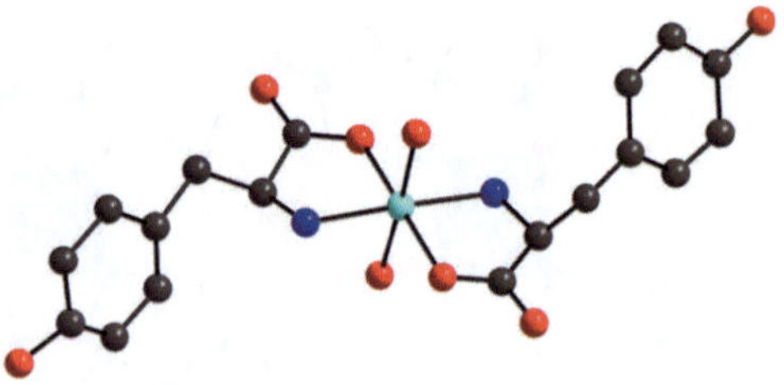

Fig. 3.39 Isolated unit in diaqua-bis(L-tyrosinato)-nickel monohydrate. Hydrogen atoms are omitted, as is the non-coordinating water molecule

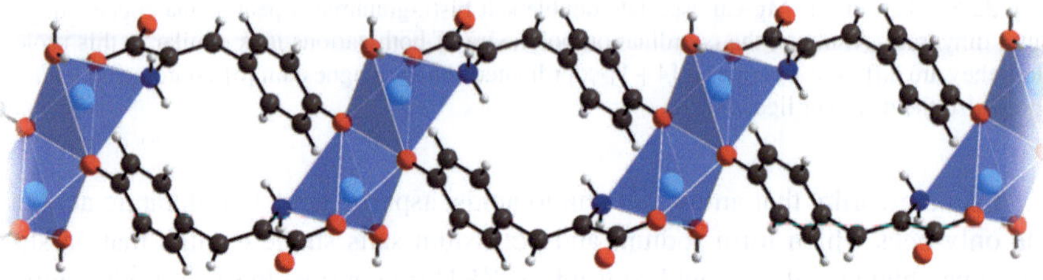

Fig. 3.40 Chains in (DL-tyrosinato)-aqua-zinc. Note the centrosymmetric nature of the chains (inversion centers within the dimers and in the interstices between the Tyr^{2-} anions)

available for the coordination of cations. Consequently, these tyrosinate ions are divalent, with negative charges at opposite ends, which allows connection of units (dimers of Zn^{2+} [5] polyhedra in this case) to chains. The tyrosinate^{2-} anions act as bridging ligands via the functional groups at the α-carbon on one side and the phenolate oxygen on the other side (Fig. 3.40). This motif has not been observed in any other amino acid salt previously.

Other DL-tyrosinates (e.g., of nickel and copper) have been reported, although no crystal structures are available (Hartwell and Bailar 1970).

The number of histidine salts reported in the literature is relatively high, as the nitrogen atoms in the imidazole ring provide an additional ligand atom. Thus, tridentate (O,N,N′) coordination is common and explains the fact that several salts could be crystallized. Moreover, the relatively large histidinate anion can "wrap" itself around the cations; thus, isolated units dominate among these salts (Table 3.9).

This tridentate (O,N,N′) coordination is found in the nickel salts of both L-histidine and DL-histidine. The optical active form was reported first by Sakurai et al. (1978), later refined by Maia et al. (2013) who performed also vibrational spectroscopic measurements and DFT calculations. The racemate salt structure was reported even earlier (Fraser and Harding 1967) and can be described as a structure comprising isolated units: Nickel cations are located upon twofold axes and are octahedrally coordinated by the amino nitrogen, the imidazole nitrogen, and a carboxyl oxygen of each histidinate anion, which possess opposite chirality.

Similar arrangements are found in the cobalt and zinc salts of DL-histidine as well as the cadmium salt of L-histidine (Table 3.9).

Table 3.9 *Salts of amino acids and cations, part IX*: space groups and unit cell parameters of compounds of cations and tyrosinate anions

	s.g.	*a*	*b*	*c*	*α*	*β*	*γ*	Conn.	Ref.
L-Tyrosine									
Ca L-$Tyr_2 \cdot 12H_2O$	$P2_1$	6.121	28.37	9.227	90	108.76	90	N/A	Eriks and Rudnick (1979)
Ni L-Tyr_2	$P2_12_12_1$	6.135	4.844	18.236	89.05	77.69	107.54	0–0	Hamalainen et al. (1978)
Ni L-$Tyr_2 \cdot 2H_2O \cdot CH_3OH$	$P1$	4.975	5.996	21.005	94.23	92.26	106.15	0–0	Hamalainen et al. (1977)
Cu L-$Tyr_2 \cdot 3H_2O$	$P1$	13.049	22.227	6.078	90	90	90	0–1	van der Helm and Tatsch (1972)
Pd L-Tyr_2	$P2_1$	5.530	19.246	8.700	90	105.40	90	0–0	Sabat et al. (1979)
Pd L-$Tyr_2 \cdot 0.5H_2O$	$P2_1$	9.968	17.714	10.549	90	90.144	90	0–0	Jarzab et al. (1973b)
DL-Tyrosine									
Zn DL-$Tyr \cdot H_2O$	P-1	5.863	9.681	9.772	65.34	75.33	78.44	0^2–1	Li et al. (2007a)
L-Histidine									
Ag (L-His)	$P2_1$	5.730	7.586	9.041	90	104.78	90	0–1	Nomiya et al. (2000)
Ni L-$His_2 \cdot H_2O$	$C2$	29.406	8.267	6.314	90	90.01	90	0–0	Sakurai et al. (1978)
Cu (L-His)$_2 \cdot 2H_2O$	$P2_1$	6.96	11.665	9.630	90	94.273	90	0–0	Deschamps et al. (2004)
Cu (L-His)$_2 \cdot 3H_2O$	$P3_121$	10.992	10.992	32.249	90	90	120	0–0	Deschamps et al. (2003b)
Co L-$His_2 \cdot H_2O$	$C2$	29.44	8.324	6.347	90	90.0	90	0–0	Harding and Long (1968)
Cd L-$His_2 \cdot 2H_2O$	$P4_32_12$	7.397	7.397	30.53	90	90	90	0–0	Fuess and Bartunik (1976)
DL-Histidine									
Ag (DL-His)$\cdot 6H_2O$	$C2/c$	12.220	15.201	11.868	90	93.28	90	1–1	Kasuga et al. (2011)
Co (DL-His)$\cdot 2H_2O$	$P2_1/c$	10.04	11.10	15.31	90	111.6	90	0–0	Candlin and Harding (1970)
Cu (DL-His)$_2 \cdot 6H_2O$	P-1	6.894	8.478	9.553	98.61	116.13	95.88	0–0	Camerman et al. (1978)
Zn (DL-His)$_2 \cdot 5H_2O$	$C2/c$	16.41	14.755	10.99	90	129.6	90	0–0	Harding and Cole (1962)
Ni DL-$His \cdot H_2O$	$Aba2$	15.18	13.05	7.72	90	90	90	0–0	Fraser and Harding (1967)

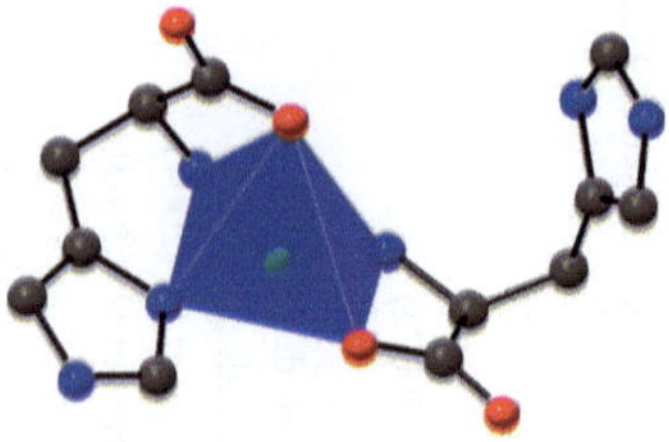

Fig. 3.41 Units in bis(L-histidinato)-copper dihydrate (note the asymmetric coordination polyhedron and the fact that the two L-histidine anions assume different coordination modes, namely, (N,N′,O) and (N,O), respectively. Water molecules in the interstices and hydrogen atoms are omitted for clarity

Furthermore, there are several examples of copper histidinate salts (an overview of this system is given by Deschamps et al. 2005, who even speak of the "saga" of copper histidine). Sigel and McCormick (1971) reported the first salt of Cu^{2+} bis (L-histidinate) in solution, and Deschamps et al. (2003b) published data on a trigonal form of this phase. This crystal is a trihydrate and comprises unusual, asymmetric units: The copper cation is coordinated by two L-histidinate anions, where both are bidentate, but different atoms make the ligands, (N,O) in one anion and (N,N) in the other. Another phase was reported shortly thereafter (Deschamps et al. 2004): In this monoclinic dihydrate salt, the coordination is even more unusual: One ligand is bidentate (N,O), the other tridentate (N,N′,O) (Fig. 3.41).

In addition to the above salts, a racemate salt has been reported as well: bis (DL-histidinato)-diaqua copper tetrahydrate (Camerman et al. 1978), which also comprises isolated units.

The only salt where infinite units were found is the silver salt of L-histidine (Nomiya et al. 2000). In this species, linear [2]-coordinate silver cations are connected to a chain between the amino nitrogen atom and the imidazole nitrogen of each L-histidinate anion, forming a chain along (100). It is noteworthy that the oxygen atoms do not act as ligands in this salt.

Several other salts of histidine have been reported, which could not be grown as crystals. Apart from an indium (III) salt of L-His, which has been characterized by polarography (Jain and Kapoor 1983), the alkali metal salts for Na, K, Rb, and Cs histidinates were recently investigated by mass spectrometry (Armentrout et al. 2012).

Of arginine and lysine salts, no crystal structures were found, as crystallization of such salts proved to be without success. Of arginine, one study was conducted on the barium salt of argininate, characterized by IR spectroscopy (Bush et al. 2008). Lysinate salts were also investigated by spectroscopic, potentiometric, and calorimetric measurements (Conato et al. 2000), who found a copper complex of lysinate in solution (with a ratio of Cu^{2+}/lysinate of 1: 2). As no crystal structure analysis was possible, molecular mechanics calculations were performed to obtain theoretical structures. The calculations showed that double bidentate (O,N) coordination of two L-lysinate anions plus two water molecules in [6]-coordination is most likely. It should be added that copper lysinate (along with the analog glycinate and histidinate) has been discussed as a possible bovine nutritional supplement by Attaelmannan and Reid (1996), who proposed that chelated Cu^{2+} is more easily incorporated than the same cation as part of an inorganic salt.

3.10 Nonstandard Amino Acids

Among the nonstandard amino acids, β-alanine is the smallest, and as glycine another non-chiral amino acid. Likewise, sarcosine is a small, non-chiral amino acid with similar properties as a ligand. This is reflected in the crystal structures of β-alaninate and sarcosinate salts, most of which are centrosymmetric, and often similar to the respective glycinates. Thus, the isolated units in diaqua-bis (β-alanine)-nickel (Jose et al. 1964) and diaqua-bis(β-alanine)-copper tetrahydrate (Tomita 1961) as well as diaqua-bis(sarcosinato)-nickel (Guha 1973) and diaqua-bis(sarcosinato)-copper (Krishnakumar et al. 1994) show the same topology as their glycinate analogs (in some cases, even the unit cell dimensions are comparable; see Table 3.10). The same applies for the Co^{3+} salt (Soling 1978; Ng et al. 2005). Another polymorph of this salt was published by Yoneda et al. (1986).

Nevertheless, not all β-alaninates correspond to glycinate analogs. Bis(β-alaninate)-copper tetrahydrate (Kamwaya and Teoh 1984) is a structure of centrosymmetric chains, with two crystallographically independent Cu^{2+} cations, located upon inversion centers. The coordination of these two cations is different: One is coordinated only via β-alaninate anions; the other one has two water molecules as apical ligands (Fig. 3.42).

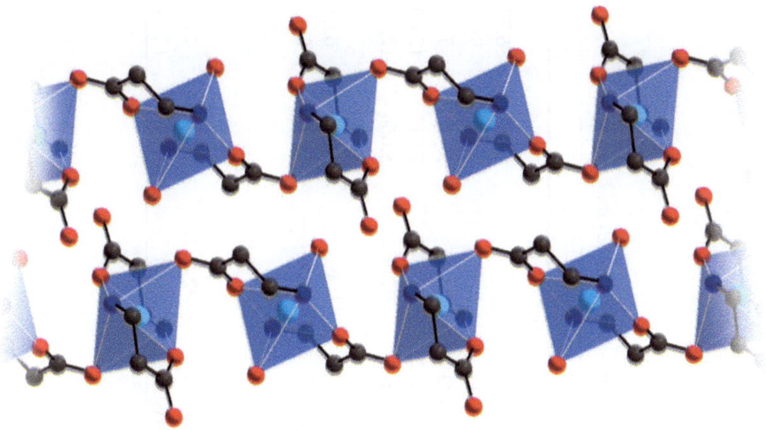

Fig. 3.42 Chains in bis(β-alanine)-copper tetrahydrate. This structure contains two crystallographically independent cations and amino acid molecules. In the chains the copper cations are located upon inversion centers (note the different coordination of the two different copper cations)

Table 3.10 *Salts of amino acids and cations, part X*: space groups and unit cell parameters of compounds of cations and nonstandard amino acid anions

	s.g.	*a*	*b*	*c*	*α*	*β*	*γ*	Conn.	Ref.
β-Alanine									
Ni β-$Ala_2 \cdot 2H_2O$	*P*-1	8.48	6.77	4.93	103.0	95.2	102.3	0–0	Jose et al. (1964)
Cu β-$Ala_2 \cdot 4H_2O$	*P*-1	9.977	9.207	7.068	102.21	79.91	88.65	0–1	Kamwaya and Teoh (1984)
Cu β-$Ala_2 \cdot 6H_2O$	$P2_1/c$	5.46	7.71	18.11	90	92.0	90	0–0	Tomita (1961)
Co^{III} β-$Ala_3 \cdot 4H_2O$	$P2_12_12_1$	22.513	13.001	5.633	90	90	90	0–0	Yoneda et al. (1986)
Co^{III} β-$Ala_3 \cdot 4H_2O$	$P2_1/n$	5.666	12.868	22.215	90	90.66	90	0–0	Soling (1978)
Sarcosine									
Ni $Sar_2 \cdot 2H_2O$	*P*-1	5.35	6.63	8.31	103.6	95.4	99.3	0–0	Guha (1973)
Cu $Sar_2 \cdot 2H_2O$	$P2_1/n$	5.598	8.380	12.346	90	100.25	90	0–0	Krishnakumar et al. (1994)
L-Ornithine									
Pd L-Orn_2	$P2_12_12_1$	10.68	14.207	9.051	90	90	90	N/A	Nakayama et al. (1973)
Dimethyl-glycine									
Cu $DMG_2 \cdot 3H_2O$	$P2_12_12_1$	6.520	7.048	31.800	90	90	90	0–0	Ng et al. (2006)
Cu $DMG_2 \cdot H_2O$	*Pbca*	11.300	12.417	17.531	90	90	90	0–0	Sabolovic and Kaitner (2008)

Apart from β-alaninates, not many salts of these amino acids could be obtained. In an old study, Brubaker and Busch (1966) reported the synthesis of nickel bis (L-ornithinate), although all attempts to grow crystals of this phase remained without success. Similarly, Conato et al. (2000) studied a salt of the composition copper bis(L-ornithinate) by spectroscopic, potentiometric, and calorimetric investigations. As no XRD analysis was possible, molecular mechanics calculations were conducted to obtain theoretical structures. These models gave several possible structures with [6]-coordinate Cu^{2+} cations, ranging from double bidentate (O,N) coordination of two L-ornithinate anions plus two water molecules over mixed coordination of bidentate and tridentate (O,N,N′) plus one water molecule to double tridentate (O,N,N′) coordination.

Aqua-bis(dimethylglycinato)-copper dihydrate (first reported by Ng et al. 2006; later refined by Sabolovic and Kaitner 2008) and aqua-bis(dimethylglycinato)-copper (Sabolovic and Kaitner 2008) show the same units, [4 + 1]-coordinate Cu^{2+} cations, with a water molecule at the apex and two *trans*-oriented DMG ligands in bidentate (O,N) coordination at the equatorial positions. This topology corresponds to the respective alaninate and valinate salts.

3.11 Concluding Remarks

Within this chapter, the diversity of amino acid salts was presented clearly: Several combinations of cations and amino acid anions, different ratios, variation in hydration, as well as instances of polymorphs of compounds prove impressively how rich the structural variety is – a fact which will be underlined by the following chapters even more. In Chap. 4, dealing with salts of amino acids and neutral materials, several of the structural patterns described above will be revisited, but many new structural motifs will be found as well.

Apart from the structural diversity, the variation of properties and – as a result – of applications is remarkable. Several instances of physical properties have been discussed above, e.g., new room-temperature ferroelectric bis(glycine)$MnCl_2 \cdot 2H_2O$ (Pepinsky et al. 1958) or antiferromagnetic diaqua-bis(glycinato)-nickel (Calvo et al. 1982b), and more examples of likewise examples will be presented in the following chapters.

In addition to these interesting physical properties, amino acid salts have been investigated in terms of pharmacological interest. For instance, copper complexes of amino acids have been reported to show an astonishingly wide spectrum of effects, including anti-inflammatory, anticonvulsant, antiulcer, and antitumoral activity (Sorenson 1976, 1982, 1995; Baran 2004; and references therein).

References

Acland CB, Freeman HC (1971) Model compounds for metal–protein interaction: crystal structures of four silver(II) complexes with glycine, glycylglycine, and imidazole. J Chem Soc D, pp 1016–1017

Ahmed KJ, Habib A, Haider SZ, Malik KMA, Hess H (1980) J Bangladesh Acad Sci 4:85–92. (Quoted from CSD, code: AQPRLB)

Allen FH (2002) The Cambridge structural database: a quarter of a million crystal structures and rising. Acta Cryst B 58:380–388

Amirthalingam V, Muralidharan KV (1973) Crystallographic data of copper complexes of threonine. Indian J Chem 11:388

Amirthalingam V, Muralidharan KV (1975) The crystal structure of bis-(L-threonine) copper(II). H_2O. Pramana 4:83–94

Antolini L, Menabue L, Pellacani GC, Marcotrigiano G (1982) Structural, spectroscopic, and magnetic properties of diaqua-(L-aspartato)nickel(II) hydrate. J Chem Soc Dalton Trans, pp 2541–2543

Armentrout PB, Armentrout EI, Clark AA, Cooper TE, Stennett EMS, Carl DR (2010) An experimental and theoretical study of alkali metal cation interactions with cysteine. J Phys Chem B 114:3927–3937

Armentrout PB, Citir M, Chen Y, Rodgers MT (2012) Thermochemistry of alkali metal cation interactions with histidine: influence of the side chain. J Phys Chem A 116:11823–11832

Attaelmannan MA, Reid RS (1996) The speciation of lysine-complexed copper as a bovine nutritional supplement. J Inorg Biochem 64:215–224

Ault A (2004) The monosodium glutamate story: the commercial production of MSG and other amino acids. J Chem Educ 81:347–355

Baidina IA, Podberezskaya NV, Borisov SV, Golubovskaya EV, Shestakova NA, Mal'chikov GD (1979) The crystal structure of palladium(II) *cis*-diglycinate trihydrate. J Struct Chem 20:463–467

Baidina IA, Podberezskaya NV, Borisov SV, Golubovskaya EV (1982) Crystal structure of the β-modification of palladium(II)-trans-diglycinate. J Struct Chem 23:241–244

Baidya N, Ndreu D, Olmstead MM, Mascharak PK (1991) Synthesis, structure, and properties of potassium bis(L-cysteinato-N, S)nickelate(II) sesquihydrate. Inorg Chem 30:2448–2451

Baran EJ (2004) Trace elements supplementation: recent advances and perspectives. Mini Rev Med Chem 4:1–9

Baran EJ (2005) Structural data and vibrational spectra of the copper(II) complex of L-selenomethionine. Z Naturforsch B60:663–666

Barrie PJ, Gyani A, Motevalli M, O'Brien P (1993) Solid-state cadmium-113 NMR studies on cadmium complexes with glycine, L-alanine, and L-cysteine. Inorg Chem 32:3862–3867

Beck W (2009) Metal complexes of biologically important ligands, CLXXII [1]. metal ions and metal complexes as protective groups of amino acids and peptides – reactions at coordinated amino acids. Z Naturforsch B64:1221–1245

Bell P, Sheldrick WS (1984) Preparation and structure of zinc-complexes of cysteine derivatives. Z Naturforsch B39:1732–1737

Blazic B, Bukovec N, Bukovec P, Lazarini F (1992) Vestn Slov Kem Drus [Bull Slovenian Chem Soc] 39:285–290. (Quoted from CSD, code: CUALTE02)

Bowman VN, Heaton AL, Armentrout PB (2010) Metal cation dependence of interactions with amino acids: bond energies of Rb + to Gly, Ser, Thr, and Pro. J Phys Chem B 114:4107–4114

Brubaker GR, Busch DH (1966) Flexidentate chelation of α, ω-diaminocarboxylic acids with copper(II) and nickel(II). Inorg Chem 5:2110–2113

Bryan RF, Greene PT, Stokely PF, Wilson WE Jr (1971) Crystal and molecular structure of tris (glycinato)chromium(III) monohydrate, $Cr(C_2H_4NO_2)_3.H_2O$. Inorg Chem 10:1468–1473

Bush MF, Oomens J, Saykally RJ, Williams ER (2008) Effects of alkaline earth metal Ion complexation on amino acid Zwitterion stability: results from infrared action spectroscopy. J Am Chem Soc 130:6463–6471

Calvo R, Nascimento OR, Torikachvili MS, Maple MB (1982a) Specific heat of nickel diglycine dihydrate between 0.5 and 10 K. J Appl Phys 53:2671–2673

Calvo R, Nascimento OR, Fisk Z, Suassuna J, Oseroff SB, Machin I (1982b) Magnetic susceptibility of antiferromagnetic nickel diglycine dihydrate. J Appl Phys 53:2674–2676

Calvo R, Levstein PR, Castellano EE, Fabiane SM, Piro OE, Oseroff SB (1991) Crystal structure and magnetic interactions in bis(D, L-alaninato)copper(II) hydrate. Inorg Chem 30:216–220

Calvo R, Steren CA, Piro OE, Rojo T, Zuniga FJ, Castellano EE (1993) Crystal structure and magnetic properties of diaqua(L-aspartato)copper(II). Inorg Chem 32:6016–6022

Camerman N, Fawcett JK, Kruck TPA, Sarkar B, Camerman A (1978) L-histidinato-D-histidinatodiaquocopper(II) tetrahydrate. J Am Chem Soc 100:2690–2693

Candlin R, Harding MM (1970) Crystal and molecular structure of D-histidino-L-histidinocobalt (II) dehydrate. J Chem Soc A, pp 384–393

Casari BM, Mahmoudkhani AH, Langer V (2004) A redetermination of *cis*-aquabis(glycinato-κ2N, O)copper(II). Acta Cryst E 60:m1949–m1951

Castellano EE, Nascimento OR, Calvo R (1982) Refinement of diaquabis(glycinato-O, N)nickel (II). Acta Cryst B 38:1303–1305

Ch'ng CD, Teoh SG, Chantrapromma S, Fun HK, Goh SM (2008) catena-Poly[[[diaquanickel (II)]-di-μ-glycine] dichloride]. Acta Cryst E 64:m865–m866

Christgau S, Odderhede J, Stahl K, Andersen JET (2005) Strontium D-glutamate hexahydrate and strontium di(hydrogen L-glutamate) pentahydrate. Acta Cryst C 61:m259–m262

Ciunik Z (1987) Crystal structure of Mg(II) and Mn(II)-DL-aspartate tetrahydrate. Pol J Chem 61:521–527

Conato C, Contino A, Maccarrone G, Magri A, Remelli M, Tabbi G (2000) Copper(II) complexes with l-lysine and l-ornithine: is the side-chain involved in the coordination? A thermodynamic and spectroscopic study. Thermochim Acta 363:13–23

D'yakon IA, Donu SV, Chapurina LF, Avilov AS (1991) Crystal structure of a mixed compound of copper (II) with glycine and L-serine. Kristallografiya (Russ) (Crystallogr Rep) 36:219–221

D'yakon IA, Donu SV, Chapurina LF, Kairyak NL (1992) Crystal structure of copper (II) with racemic serine. Kristallogr (Russ) (Crystallogr Rep) 37:1391–1395

Dalosto SD, Ferreyra MG, Calvo R, Piro OE, Castellano EE (1999) Structure of bis(l-alaninato) zinc(II) and single crystal EPR spectra of Cu(II) impurities. J Inorg Biochem 73:151–155

Daniele PG, Foti C, Gianguzza A, Prenesti E, Sammartano S (2008) Weak alkali and alkaline earth metal complexes of low molecular weight ligands in aqueous solution. Coord Chem Rev 252:1093–1107

Daydé S, Champmartin D, Rubini P, Berthon G (2002) Aluminium speciation studies in biological fluids. Part 8. A quantitative investigation of Al(III)–amino acid complex equilibria and assessment of their potential implications for aluminium metabolism and toxicity. Inorg Chim Acta 339:513–524

de Meester P, Hodgson DJ, Freeman HC, Moore CJ (1977) Tridentate coordination by the L-cysteine dianion. Crystal and molecular structure of sodium bis(L-cysteinato)chromate(III) dihydrate. Inorg Chem 16:1494–1498

Demaret A, Abraham F (1987a) Structure du L-α-alaninate d'argent. Acta Cryst C 43:1519–1521

Demaret A, Abraham F (1987b) Structure du L-α-alaninate de cadmium trihydraté. Acta Cryst C 43:2067–2069

Demaret D, Mercier D (1983) Crystal data for nickel and cadmium α-L-alaninates. J Appl Crystallogr 16:279–281

Denning RG, Piper TS (1966) Optical activity, absolute configuration, and rearrangement reactions of tris amino acid complexes of cobalt(III) with L-alanine, L-leucine, and L-proline. Inorg Chem 5:1056–1065

Deschamps P, Zerrouk N, Nicolis I, Martens T, Curis E, Charlot MF, Girerd JJ, Prange T, Benazeth S, Chaumeil JC, Tomas A (2003a) Copper(II)-L-glutamine complexation study in solid state and in aqueous solution. Inorg Chim Acta 353:22–34

Deschamps P, Kulkarni PP, Sarkar B (2003b) X-ray structure of physiological copper(II) – bis (l-histidinato) complex. Inorg Chem 42:7366

Deschamps P, Kulkarni PP, Sarkar B (2004) X-ray structure of physiological copper(II) – bis (l-histidinato) complex. Inorg Chem 43:3338–3340

Deschamps P, Kulkarni PP, Gautam-Basak M, Sarkar B (2005) The saga of copper(II)–l-histidine. Coord Chem Rev 249:895–909

Dewan JC (1988) Structure of tris(glycinato)cobalt(III) dihydrate. Acta Cryst C 44:2199–2201

Diacon IA, Donu SV, Chapurina LF (2000) Crystal structures of copper(II) compounds with racemic threonine. Crystallogr Rep 45:601–603

Dijkstra A (1966) The crystal structure of the copper complex of L-alanine. Acta Cryst 20:588–590

Doyne TH, Pepinksy R, Watanabe T (1957) Cyclic configuration of the aspartate ion in the crystal structure of zinc, cobaltous and nickelous aspartate trihydrate. Acta Cryst 10:438–439

Drew MGB, Dunlop JH, Gillard RD, Rogers D (1966) The absolute configuration of α(+)-tris-L-alaninatocobalt(III). Chem Commun, pp 42–43

Einspahr H, Bugg CE (1974) Calcium-binding to α-amino acids: crystal structure of calcium L-glutamate trihydrate. Acta Cryst B 30:1037–1043

Einspahr H, Bugg CE (1979) Calcium binding to α-amino acids: the crystal structure of calcium di-L-glutamate tetrahydrate. Acta Cryst B 35:316–321

Eriks K, Rudnick SE (1979) ACA, Ser.2, 6, 54. (Quoted from CSD, code: CALTYR)

Fawcett TG, Ushay M, Rose JP, Lalancette RA, Potenza JA, Schugar HJ (1979) Molecular structures of the copper-amino acid complexes bis(L-leucinato)copper(II) and bis(D, L-2-aminobutyrato)copper(II). Inorg Chem 18:327–332

Fleck M (2008) Compounds of glycine with halogen or metal halogenides: review and comparison. Z Kristallogr 223:222–232

Fleck M, Bohatý L (2005) catena-Poly[[[tetraaquamagnesium(II)]-μ-glycine-κ2O:O′] dinitrate]. Acta Cryst E 61:m1887–m1889

Fleck M, Held P, Schwendtner K, Bohatý L (2008) New compounds of Glycine with metal halogenides. Z Kristallogr 223:212–221

Flook RJ, Freeman HC, Moore CJ, Scudder ML (1973) Model compounds for metal–protein interaction: crystal structures of seven cadmium(II) complexes of amino-acids and peptides. Chem Commun, pp 753–754

Flook RJ, Freeman HC, Scudder ML (1977) An X-ray and neutron diffraction study of aqua (L-glutamato)cadmium(II) hydrate. Acta Cryst B 33:801–809

Fox S, Busching I, Barklage W, Strasdeit H (2007) Coordination of biologically important α-amino acids to calcium(II) at high pH: insights from crystal structures of calcium α-aminocarboxylates. Inorg Chem 46:818–824

Fraser KA, Harding MM (1967) The crystal and molecular structure of bis(histidino)nickel (II) monohydrate. J Chem Soc A, pp 415–420

Freeman HC, Golomb ML (1969) The crystal structure of *trans*-bisglycinatoplatinum(II), $Pt(NH_2CH_2COO)_2$. Acta Cryst B 25:1203–1206

Freeman HC, Guss JM (1968) Refinement of the structure of diaquobisglycinatonickel(II) $Ni(NH_2CH_2COO)_2(OH_2)_2$. Acta Cryst B 24:1133–1135

Freeman HC, Moore CJ (1977) DL-cysteinatothallium(I). Acta Cryst B 33:2690–2692

Freeman HC, Snow MR, Nitta I, Tomita K (1964) A refinement of the structure of bisglycinocopper(II) monohydrate, $Cu(NH_2CH_2COO)_2.H_2O$. Acta Cryst 17:1463–1470

Fuess F, Bartunik H (1976) Neutron diffraction structure determination of bis(L-histidinato) cadmium dihydrate. Acta Cryst B 32:2803–2806

Gao EJ, Yin HX, Zhang WZ, Wang KH, Gu XF, Zhu MC, Wu Q, Liu L, Sun YG (2008) Zigzag-shaped supramolecular networks formed via hydrogen bonding. Crystal structures, DNA binding, and cytotoxic activity of a new palladium(II) complex. Coord Chem 34:516–521

Gillard RD, Mason R, Payne NC, Robertson GB (1969) The isomers of α-amino acids with copper(II). Part II. The crystal and molecular structure of *cis*-bis-[D-alaninato]copper(II). J Chem Soc A, pp 1864–1871

Gramaccioli CM (1966) The crystal structure of zinc glutamate dihydrate. Acta Cryst 21:600–605

Gramaccioli CM, Marsh RE (1966) The crystal structure of copper glutamate dihydrate. Acta Cryst 21:594–600

Gu KQ, Sun YX, Zhang R, Zhang NW, Che HW (2007) Tris(glycinato-κ2N, O)cobalt(III). Acta Cryst E 63:m740–m742

Guha S (1973) The crystal and molecular structure of bis(sarcosinato)nickel(II) dihydrate. Acta Cryst B 29:2167–2170

Hakimi M, Aliabadi TS (2012) Coordination chemistry of copper α- amino acid complexes. World Appl Program 2:431–443

Hamalainen R (1977) Finn Chem Lett, pp 113–117 (Quoted from CSD, code: TREAZN)

Hamalainen R, Lajunen K, Valkonen J (1977) Crystal structure of diaquabis (L-tyrosinato) nickel (II) bismethanol. Finn Chem Lett 4–5:108–112

Hamalainen R, Ahlgren M, Turpeinen U, Raikas T (1978) Diaquabis(L-tyrosinato)nickel(II) monohydrate. Cryst Struct Commun 7:379–384

Harding MM, Cole SJ (1962) The crystal structure of Di(histidino) zinc pentahydrate. Acta Cryst 16:643–650

Harding MM, Long HA (1968) Crystal and molecular structure of bis-(L-histidino)cobalt (II) monohydrate. J Chem Soc A, pp 2554–2559

Hartwell IO, Bailar JCJ (1970) Preparation and characterization of tyrosine and lysine metal chelate polyesters and polyamides. J Am Chem Soc 92:1284–1289

Heaton AL, Armentrout PB (2008) Experimental and theoretical studies of potassium cation interactions with the acidic amino acids and their amide derivatives. J Phys Chem B 112:10056–12065

Heaton AL, Moision RM, Armentrout PB (2006) Experimental and theoretical studies of sodium cation interactions with the acidic amino acids and their amide derivatives. J Phys Chem A 112:3319–3327

Herak R, Prelesnik B, Krstanovic I (1978) The crystal structure and absolute configuration of (−)-mer-tris(L-alaninato)cobalt(III) monohydrate. Acta Cryst B 34:91–95

Hitchman MA, Kwan L, Engelhardt LM, White AH (1987) Electron spin resonance and electronic spectra and crystal and molecular structures of copper(II) amino acid complexes. J Chem Soc Dalton Trans, pp 457–465

Hwang CS, Lee N, Kim YA, Park YB (2006) Synthesis of the water dispersible L-valine capped ZnS:Mn nanocrystal and the crystal structure of the precursor complex: [Zn(Val)2(H_2O)]. Bull Korean Chem Soc 27:1809–1814

Ito T, Marumo F, Saito Y (1971) The crystal structure of bis-(L-prolinato)palladium(II). Acta Cryst B 27:1062–1066

Jain SL, Kapoor RC (1983) Polarographic study of mixed-ligand complexes of In(III) with L-histidine and glutamine/L-proline. Inorg Chim Acta 78:93–96

Jarzab TC, Hare CR, Langs DA (1973a) *Cis*-bis(L-valinato) palladium(II) monohydrate. Cryst Struct Commun 2:395–403

Jarzab TC, Hare CR, Langs DA (1973b) *Cis*-bis(L-tyrosinato)palladium(II) hemihydrate, $C_{36}H_{42}N_4O_{13}Pd_2$. Cryst Struct Commun 2:399–403

Jose P, Pant LM, Biswas AB (1964) The crystal structure of nickel β-alanine dihydrate. Acta Cryst 17:24–28

Kamwaya ME, Teoh SG (1984) J Fiz Malays 5:1 (Quoted from CSD, code: CORZUE)

Kamwaya ME, Teoh SG (1985) J Fiz Malays 6:109 (Quoted from CSD, code: DIWYIR)

Kashiwagi T, Sano C, Kawakita T, Nagashima N (1995) Monosodium L-glutamate pentahydrate. Acta Cryst C 51:1053–1056

Kasuga NC, Tagaki Y, Tsuruta SI, Kuwana W, Yoshikawa R, Nomiya K (2011) Synthesis, structure and antimicrobial activities of meso silver(I) histidinate [Ag 2(D-his)(L-his)] n (Hhis = histidine) showing different self-assembly from those of chiral silver(I) histidinates. Inorg Chim Acta 368:44–48

Kato M, Hayashi M, Fujihara T, Nagasawa A (2008) $(-)_{545}$-*fac*-Δ-Tris(L-prolinato)cobalt(III) trihydrate. Acta Cryst E 64:m684

Khatib A, Aqra F, Deamer D, Oliver A (2009) Crystal structure of [bis(L-alaninato)diaqua]nickel (II) dihydrate. J Chem Res, pp 98–102

Kiss T (1990) Complexes of amino acids. In: Burger K (ed) Biocoordination chemistry: coordination equilibria in biologically active systems. Ellis Horwood, Chichester, pp 56–134.

Klopman G (1968) Chemical reactivity and the concept of charge- and frontier-controlled reactions. J Am Chem Soc 90:223–234

Konar S, Gagnon K, Clearfield A, Thompson C, Hartle J, Ericson C, Nelson C (2010) Structural determination and characterization of copper and zinc bis-glycinates with X-ray crystallography and mass spectrometry. J Coord Chem 63:3335–3347

Kremer C, Torres J, Dominguez S, Mederos A (2005) Structure and thermodynamic stability of lanthanide complexes with amino acids and peptides. Coord Chem Rev 149:567–590

Krishnakumar RV, Natarajan S, Bahadur SA, Cameron TS (1994) Crystal structure of diaquabissarcosinatocopper(II). Z Kristallogr 209:443–444

Kryger L, Ramsussen SE (1973) Walden inversion. III. The crystal structure and absolute configuration of Zn(II) (+)-aspartate trihydrate. Acta Chem Scand 27:2674–2676

Krylova LF, Matveeva LM, Romanenko GV (2006) Pt(II) stereoisomeric complexes with serine. Zh Strukt Khim (Russ) 47:651–662

Laurie SH (1987) In: Wilkinson G, Gillard RD, McCleverty JA (eds) Comprehensive coordination chemistry, vol 2. Pergamon Press, Oxford, pp 739–776

Laurie SH (1995) The Synthesis, Reactions, Properties and Applications of Coordination Compounds. In: Berthon G (ed) Handbook of metal-ligand interactions in biological fluids: bioinorganic chemistry, vol 1. Marcel Dekker, New York, pp 603–619

Le Bail A, Daudon M, Bazin D (2013) A new compound in kidney stones? Powder X-ray diffraction study of calcium glycinate trihydrate. Acta Cryst C 69:734–737

Leach MR (2014) The chemogenesis web book. http://www.meta-synthesis.com/webbook.html. Accessed 17 Jan 2014

Li DQ, Zhou J, Liu X (2007a) The one-dimensional chain polymer of aqua(tyrosinato)zinc(II). Acta Cryst C 63:m371–m373

Li MX, Zhao HJ, Shao M, Miao ZX, Liang SW (2007b) Synthesis, characterization and crystal structures of two binary acid complexes based on L-glutamate and L-aspartate. J Coord Chem 60:2549–2557

Liao SR, Lin QB, Le XY, Feng XL (2006) Synthesis, characterization and superoxide-dismutase-like activity of Cu (II) complex of L-isoleucinate. Huax Yanjiu (Chin) 17:9–12

Low BW, Hirshfeld FL, Richards FM (1959) Glycinate complexes of zinc and cadmium. J Am Chem Soc 81:4412–4416

Lu T, Li X, Mao Z, Qiu W, Ji L, Yu K (1998) Self-assembly of copper and magnesium ions will L-glutamate: spectroscopic and structural properties of [CuMg(L-GLUTAMATO)$_2$ $(H_2O)_3]\cdot 2H_2O$. Polyhedron 17:75–79

Maeda H, Kanamori K, Michibata H, Konno T, Okamoto KI, Hidaka J (1993) Preparation and properties of vanadium (III) complexes with L-cysteinate and D-penicillaminate. Bull Chem Soc Jpn 66:790–796

Maia JR, Lima JA Jr, Freire PTC, Mendes Filho J, Nogueira CES, Teixeira AMR, de Menezes AS, Remédios CMR, Cardoso LP (2013) FT-IR and Raman spectra and DFT calculations on bis (L-histidinato)nickel(II) monohydrate. J Mol Struct 1054–1055:143–149

Marandi F, Shahbakhsh N (2007) Synthesis and crystal structure of $[Pb(phe)_2]_n$: a 2D coordination polymer of lead(II) containing phenylalanine. Z Anorg Allg Chem 633:1137–1139

Martino DM, Steren CA, Calvo R, Piro OE (1991) Magnetic and structural properties of *trans*-bis (d, l-isoleucine) copper(II). J Solid State Chem 90:211–215

Martino DM, Passeggi MCG, Calvo R, Nascimento OR (1996) EPR spectroscopy and exchange interaction parameters in $Cu(glycine)_2\cdot H_2O$. Phys B Condens Matter 225:63–75

Mathieson AML, Welsh HK (1952) The crystal structure of copper proline dihydrate. Acta Cryst 5:599–604

Mirceva A, Thomas JO, Gustafsson T (1989) Structure of *trans*-bis(DL-α-alaninato)copper (II) monohydrate. Acta Cryst C 45:1141–1144

Miyanaga A, Sakaguchi U, Morimoto Y, Kushi Y, Yoneda H (1982) Structural and kinetic trans influence in (amino acidato)cobalt(III) complexes – hydrogen exchange and crystal and molecular structure of mer-tris(glycinato)cobalt(III) dihydrate. Inorg Chem 21:1387–1391

Mizutani M, Maejima N, Jitsukawa K, Masuda H, Einaga H (1998) An infinite chiral single-helical structure formed in Cu(II)-L-/D-glutamic acid system. Inorg Chim Acta 283:105–110

Moision RM, Armentrout PB (2006) The special five-membered ring of proline: an experimental and theoretical investigation of alkali metal cation interactions with proline and its four- and six-membered ring analogues. J Phys Chem A 110:3933–3946

Mostad A, Natarajan S (1987) Crystal and molecular structure of DL-nickel alaninate tetrahydrate. Z Kristallogr 178:263–270

Moussa SM, Fenton RR, Kennedy BJ, Piltz RO (1999) Hydrogen bonding in *cis*-bis(L-alaninato) copper(II): a single crystal neutron diffraction study. Inorg Chim Acta 288:29–34

Mudring AJ, Rieger F (2005) Lone pair effect in thallium(I) macrocyclic compounds. Inorg Chem 44:6240–6243

Muir JA, Ortiz A (1977) Crystal data for bis(L-isoleucinato)nickel(II) dihydrate. J Appl Cryst 10:489

Müller G, Maier GM, Lutz M (1994) Lithium coordination to amino acids and peptides. Synthesis, spectroscopic characterization and structure determination of lithium complexes of neutral and anionic glycine and diglycine. Inorg Chim Acta 218:121–131

Nagashima N, Sano C, Kawakita T, Iitaka Y (1992a) Crystal structure of ammonium L-glutamate monohydrate. Anal Sci 8:119–124

Nagashima N, Sano C, Kawakita T, Iitaka Y (1992b) Crystal structure of potassium L-glutamate monohydrate. Anal Sci 8:115–118

Nagashima N, Sano C, Kawakita T, Iitaka Y (1992c) Crystal structure of sodium DL-glutamate dihydrate. Anal Sci 8:267–274

Nakayama Y, Matsumoto K, Ooi S, Kuroya H (1973) X-ray molecular structure of bis-(L-ornithinato)palladium(II). Conformation of the seven-membered chelate ring. Chem Commun, pp 170–171

Newman JM, Bear CA, Hambley TW, Freeman HC (1990) Structure of bis(glycinato)zinc(II) monohydrate, a five-coordinate zinc(II) complex. Acta Cryst 46:44–48

Ng CH, Fun HK, Teo SB, Teoh SG, Chinnakali K (1995) Bis(L-prolinato-N, O)zinc(II). Acta Cryst C 51:244–245

Ng CH, Ong HKA, Ngai KS, Tan WT, Lim LP, Teoh SG, Chong TS (2005) *mer*-tris(β-alaninato) cobalt(III): crystal structure, solution properties and its DNA cleavage in the presence of ascorbic acid. Polyhedron 24:1503–1509

Ng CH, Ong HKA, Kong CW, Su KK, Ng SW (2006) Synthesis, characterization and nucleolytic property of bis(N, N′-dimethylglycinato)copper(II). J Coord Chem 59:1089–1099

Nomiya K, Yokoyama H (2002) Syntheses, crystal structures and antimicrobial activities of polymeric silver(I) complexes with three amino-acids [aspartic acid (H_2asp), glycine (Hgly) and asparagine (Hasn)]. J Chem Soc Dalton Trans, pp 2483–2490

Nomiya K, Takahashi S, Noguchi R, Nemoto S, Takayama T, Oda M (2000) Synthesis and characterization of water-soluble silver(I) complexes with l-histidine (H_2his) and (*S*)-(−)-2-pyrrolidone-5-carboxylic acid (H_2pyrrld) showing a wide spectrum of effective antibacterial and antifungal activities. Crystal structures of chiral helical polymers $[Ag(Hhis)]_n$ and $\{[Ag(Hpyrrld)]_2\}_n$ in the solid state. Inorg Chem 39:3301–3311

Oonishi I, Shibata M, Marumo F, Saito Y (1973) The crystal structure of calcium *cis*(*N*)-*trans*(O_5)-bis-(L-aspartato)cobaltate(III) *cis*(*N*)-*trans*(O_6)-bis(L-aspartato)cobaltate(III) decahydrate. Acta Cryst B 29:2448–2455

Oonishi I, Sato S, Saito Y (1975) The crystal structure of calcium *cis*(*N*)-*trans*(O_6)-bis-(L-aspartato)cobaltate(III)-water (2/15). Acta Cryst B 31:1318–1324

Ou CC, Powers DA, Thich JA, Felthouse TR, Hendrickson DN, Potenza JA, Schugar HJ (1979) Molecular structure and magnetic properties of *trans*-bis(L-methioninato)copper(II), Cu $(C_5H_{10}NO_2S)_2$. Inorg Chem 17:34–40

Pearson RG (1963) Hard and soft acids and bases. J Am Chem Soc 85:3533–3543

Pearson RG (1966) Acids and bases. Science 151:172–177

Pearson RG (1968) Hard and soft acids and bases, HSAB, Part 1: fundamental principles. J Chem Educ 45:581–587

Pearson RG (1997) Chemical hardness. Wiley, New York

Pepinsky R, Vedam K, Okaya Y (1958) New room-temperature ferroelectric. Phys Rev 110:1309–1311

Pletnev VZ, Zolotarev YA, Galitskii NM, Verenich AI (1992) Crystal and molecular structure of complex *cis*-$[(VAL)_2Pd]\cdot H_2O$. J Struct Chem 33:98–102

Pozhidaev AI, Simonov MA, Kruglik AI, Shestakova NA, Mal'chikov GD (1975) X-ray diffraction study of *cis*-diglycinatoplatinum(II) Pt $(NH_2CH_2COO)_2$. J Struct Chem 16:997–998

Presnyak NT, Grineva AV, Kovalevskaya IP, Kirichenko IN (1972) On interaction of thallium (III) chloride with α-alanine in aqueous and aqueous-alcoholic solutions. Zh Obschej Khim (Russ) (J Gen Chem) 42:729–732

Proch S, Villanueva JMR, Irrgang T, Doring C, Kempe R (2008) Refinement of the crystal structure of aquazinc(II) glutamate hydrate, $Zn(H_2O)(C_5H_7NO_4)\cdot H_2O$. Z Kristallogr NCS 223:55–56

Ranger G, Beauchamp AL (1981) Structure of di-μ-hydroxo-bis{bis[(S)-alaninato]chromium(III)} trihydrate. Acta Cryst B 37:1063–1067

Rao ST, Mallikarjunan M (1975) Model of metal interaction with protein. Acta Cryst A 31:S48

Ray MS, Drew MGB, Ghosh A (2004) Synthesis, crystal structure and thermal studies of catena-(L-glutamato)-aqua copper(II) monohydrate. J Indian Chem Soc 81:543–546

Rebilly JN, Gardner PW, Darling GR, Bacsa J, Rosseinsky MJ (2008) Chiral II – VI semiconductor nanostructure superlattices based on an amino acid ligand. Inorg Chem 47:9390–9399

Rizzi AC, Brondino CD, Calvo R, Baggio R, Garland MT, Rapp RE (2003) Structure and magnetic properties of layered high-spin Co(II)(l-threonine)$_2(H_2O)_2$. Inorg Chem 42:4409–4416

Rombach M, Gelinsky M, Vahrenkamp H (2002) Coordination modes of amino acids to zinc. Inorg Chim Acta 334:25–33

Ruan C, Rodgers MT (2004) Cation-ð interactions: structures and energetics of complexation of Na + and K + with the aromatic amino acids, phenylalanine, tyrosine, and tryptophan. J Am Chem Soc 126:14600–14610

Sabat M, Jezowska M, Kozlowski H (1979) X-ray evidence of the metal ion tyrosine aromatic ring interaction in bis(L-tyrosinato)palladium(II). Inorg Chim Acta 37:L511–L512

Sabolovic J, Kaitner B (2008) The effect of hydrogen bonding in two crystal modifications of aquabis(N, N-dimethylglycinato-κN, κO)copper(II): experimental and theoretical study. Inorg Chim Acta 361:2418–2430

Sahin O, Buyukgungor O, Kose DA, Necefoglu H (2008) Crystal structure of tetrakis(alaninato)di (.MU.)aquadicobalt(II) monohydrate. X-Ray Struct Anal Online 24:x269–x270

Sakurai T, Iwasaki H, Katano T, Nakahashi Y (1978) Bis(L-histidinato)nickel(II) monohydrate. Acta Cryst B 34:660–662

Sano C, Nagashima N, Kawakita T, Iitaka Y (1989) Crystal and molecular structures of monosodium L-glutamate monohydrate. Anal Sci 5:121–122

Sano C, Nagashima N, Kawakita T, Iitak Y (1990) The effects of additives on the crystal habit of monosodium L-glutamate monohydrate. J Cryst Growth 99:1070–1075

Sartoris RP, Ortigoza L, Casado NMC, Calvo R, Castellano EE, Piro OE (1999) Structure, single crystal EPR spectra, and exchange interactions in [Cu(l-proline)2]2·5H_2O and Cu(d, l-proline) 2·2H_2O. Inorg Chem 38:3598–3604

Schmidbaur H, Muller G, Riede J, Manninger G, Helbig J (1986) Elucidation of the structure of pharmacologically active magnesium L-aspartate complexes. Angew Chem Int Ed 25:1013–1014

Schmidbaur H, Bach I, Riede J, Muller G, Helbig J, Hopf G (1988) Synthese und struktur des mangan(II)-L-aspartat-trihydrats, Mn(L-Asp)·3H_2O. Chem Ber 121:795–797

Schmidbaur H, Bach I, Wilkinson DL, Muller G (1989a) Metal ion binding by amino acids. Preparation and crystal structures of lithium hydrogen L-aspartate hydrate and potassium hydrogen L-aspartate dihydrate. Chem Ber 122:1427–1431

Schmidbaur H, Bach I, Wilkinson DL, Muller G (1989b) Metal ion binding by amino acids. Preparation and crystal structures of two calcium L-aspartate hydrates. Chem Ber 122:1439–1444

Schmidbaur H, Bach I, Wilkinson DL, Muller G (1989c) Metal ion binding by amino acids. Preparation and crystal structures of magnesium, strontium, and barium L-glutamate hydrates. Chem Ber 122:1433–1438

Schmidbaur H, Bach I, Wilkinson DL, Muller G (1989d) Metal ion binding by amino acids. The crystal structure of racemic magnesium bis(hydrogen aspartate) tetrahydrate Mg(L-AspH)(D-AspH)·$4H_2O$. Chem Ber 122:1445–1447

Schmidbaur H, Mikulcik P, Muller G (1990a) Metal ion binding by amino acids: strontium and barium l-aspartate trihydrate Sr/Ba-(L)-Asp·$3H_2O$. Chem Ber 123:1599–1602

Schmidbaur H, Mikulcik P, Muller G (1990b) Metal Ion binding by amino acids. Potassium hydrogen L-glutamate monohydrate K(L-GluH)·H_2O. Chem Ber 123:1001–1004

Schmidbaur H, Helbig J, Classen HG (1990c) Aspartic and glutamic acid as ligand to alkali and alkaline-earth metals: structural chemistry as related to magnesium therapy. Angew Chem Int Ed Engl 29:1090–1103

Schveigkardt JM, Rizzi AC, Piro OE, Castellano EE, de Santana RC, Calvo R, Brondino CD (2002) Structural and single crystal EPR studies of the complex copper L-glutamine: a weakly exchange-coupled system with *syn-anti* carboxylate bridges. Eur J Inorg Chem, pp 2913–2919

Seifert A, Wagner C, Merzweiler K (2011) Bis(hydrogen l-glutamato)palladium(II). Acta Cryst E 67:m1362

Sekizaki M (1979) The crystal structure of *fac*(*N*)-Δ-tris(L-asparaginato)cobalt(III) trihydrate. Bull Chem Soc Jpn 52:403–406

Shamala N, Venkatesan K (1973) Copper proline dihydrate, $Cu(C_5H_8NO_2).2H_2O$. Cryst Struct Commun 2:5–8

Sicignano A, Gandhi S, Eriks K (1974) ACS, Abstract papers (Summer) 227. (Quoted from CSD, code: CALSER)

Sigel H, McCormick DB (1971) Structure of the copper(II)-L-histidine 1:2 complex in solution. J Am Chem Soc 93:2041–2044

Slyudkin OP, Minacheva LK, Fedorova GA, Sakharova VG, Khodashova TS, Porai-Koshits MA (1985) Circular dichroism of bis-chelate complexes of platinum (II) with cyclic amino acids. Crystal structure of *trans*-[Pt(Opro)2].$2H_2O$. Koord Khim (Russ)(Coord Chem) 11(7):963–973

Soling H (1978) The crystal and molecular structure of the violet form of tris(beta-alaninato)cobalt (III) tetrahydrate. Acta Chem Scand A 32:361–366

Sorenson JRJ (1976) Copper chelates as possible active forms of the antiarthritic agents. J Med Chem 19(1976):135

Sorenson JRJ (1982) The anti-inflammatory activities of copper complexes. In: Sigel H (ed) Metal ions in biological systems, vol 14. Marcel Dekker, New York, pp 77–124.

Sorenson JRJ (1995) Pharmacological activities of copper compounds. In: Berthon G (ed) Handbook of metal–ligand interactions in biological fluids, vol 2. Marcel Dekker, New York, pp 1128–1139.

Soylu H, Uklu D, Morrow JC (1974) The crystal structure of cadmium glutamate dihydrate $CdC_5H_7NO_4{\cdot}2H_2O$. Z Kristallogr 140:281–288

Spek AL (2009) Structure validation in chemical crystallography. Acta Cryst D 65:148–155

Stenzel K, Fleck M (2004) Poly[[[diaquacobalt(II)]-di-μ-glycine] dichloride]. Acta Cryst E 60: m1470–m1472

Stephens FS, Vagg RS, Williams PA (1975) The crystal and molecular structure of bis (L-asparaginato)copper(II), $[Cu(OOCCHNH_2CH_2CONH_2)_2]n$. Acta Cryst B 31:841–845

Stephens FS, Vagg RS, Williams PA (1977) The crystal and molecular structure of bis (L-asparaginato)zinc(II), $[Zn(OOC.CHNH_2.CH_2.CONH_2)_2]n$. Acta Cryst B 33:433–437

Steren CA, Calvo R, Piro OE, Rivero BE (1989) Molecular structure of bis(L-leucinato)zinc(II) and single-crystal EPR spectra of the substitutionally copper(II)-63-doped complex. Inorg Chem 28:1933–1938

Steren CA, Calvo R, Castellano EE, Fabiane MS, Piro OE (1990) Molecular structure and single crystal EPR spectra of bis(L-valinato)copper(II) monohydrate, $Cu[H_2NCH(CH_3)_2CHCO_2]_2 \cdot H_2O$. Physica B 164:323–330

Stosick AJ (1945) The crystal structure of nickel glycine dihydrate. J Am Chem Soc 67:365–370

Teoh SG, Chan BT, Fun HK, Kamwaya ME (1987) Crystal structure of diaquabis(*L*-alaninato) nickel(II) dihydrate $[Ni^{2+}(NH_2CHCH_3COO^-)_2(H_2O)_2]\cdot 2H_2O$. Z Kristallogr 181:199–204

Tomita K (1961) Über die Kupfersalze der ω-Aminosäuren. III. Über die Kristallstruktur von Kupfersalz des β-Alanins. Bull Chem Soc Jpn 34:297–300

Tomita K, Nitta I (1961) Über die Kupfersalze der ω-Aminosäure. II. Über die Kristallstruktur des Kupfer-Glycin-Monohydrates. Bull Chem Soc Jpn 34:286–291

Tomlin DW, Cooper TM, Cline SM, Hughes JM, Adams WW (1997) Cadmium glutamate dihydrate. Acta Cryst C 53:1815–1816

Vagg RS (1979) The crystal and molecular structure of bis(L-serinato)palladium(II). Acta Cryst B 35:341–344

van der Helm D, Franks WA (1969) The crystal structure of bis-(L-serinato)copper(II). Acta Cryst B 25:451–457

van der Helm D, Hossain MB (1969) The crystal structure of diaquobis-(L-serinato)nickel(II). Acta Cryst B 25:457–463

van der Helm D, Tatsch CE (1972) The crystal structure of bis-(L-tyrosinato)copper(II). Acta Cryst B 28:2307–2312

van der Helm D, Nicholas AF, Fisher CG (1970) The crystal structure of bis-(L-serinato)zinc. Acta Cryst B 26:1172–1178

van der Helm D, Lawson MB, Enwall EL (1971) The crystal structure of bis-(L-phenylalaninato) copper(II). Acta Cryst B 27:2411–2418

Veal JT, Hatfield WE, Jeter DY, Hempel JC, Hodgson DJ (1973) Structural and magnetic properties of di-.mu.-hydroxo-tetraglycinatodichromium(III). Inorg Chem 12:342–346

Veidis MV, Palenik GJ (1969) The structure of a copper complex of an essential sulphur-containing amino-acid; bis(methioninato)copper(II). J Chem Soc D, pp 1277–1278

Venkato I, Lariucci C, Ferreira KD, Santana RC, Carvalho JF (2004) Redetermination of bis (L-asparaginato)copper(II). Acta Cryst E 60:m1428–m1430

Versiane O, Felcman J, de Miranda JL, Howie RA, Skakle JMS (2005) Diaquabis(L-serinato) copper(II) 0.1-hydrate at 120 K. Acta Cryst E 61:m2517–m2519

Versiane O, Felcman J, de Miranda JL, Howie RA, Skakle JMS, Wardell JL (2006) Potassium *cis*-[(*R*)-aspartato(2-)][(*S*)-aspartato(2-)]cobaltate(III) 3.5-hydrate at 120 K. Acta Cryst E 62:m52–m55

Wang L, Pei Y (2006) *catena*-Poly[silver(I)-μ -L-phenylalaninato-$\kappa^2 N{:}O$]. Acta Cryst E 62: m1487–m1488

Wang YJ, Xu L (2008) pH-dependent displacement of $[Bi(citrate)]^-$ with cysteine: synthesis, spectroscopic and X-ray crystallographic characterization of $Bi(cysteine)_3$. J Inorg Biochem 102:988–991

Wang J, Xu X, Ma W, Lu L, Yang X (2007) *rac-catena*-Poly[nickel(II)-di-μ-tryptophanato]. Acta Cryst E 63:m2867–m2868

Wang J, Xu XY, Ma WX, Xu RB, Tong YP, Lu LD, Yang XJ, Wang DQ (2008a) Jiegou Huaxue (Chin) (Chin J Struct Chem) 27:153. (Quoted from CSD, code: SITPUH)

Wang J, Xu XY, Ma WX, Hu XL, Shi PF, Wang MY, Lu LD, Yang XJ, Wang X (2008b) An amino acid coordination polymer[Co(L-trp)(D-trp)]n. Wuji Huaxue Xuebao (Chin) 24:1514–1518

Watanabe M, Kai M, Goto K, Ohmuro H, Furukawa S, Chikaraishi N, Takayama T, Koike Y (2003) Preparation of platinum(II) complexes with L-serine using KI. X-ray crystal structure, HPLC and ^{195}Pt NMR spectra. J Inorg Biochem 97:240–248

Weeks CM, Cooper A, Norton DA (1969) The crystal structure of the copper(II) complex of L-isoleucine. Acta Cryst B 25:443–450

Weng J, Hong M, Shi Q, Cao R, Chan ASC (2002) Chiral supramolecular assemblies derived from the insertion of a segmental ligand into a copper-tyrosine framework. Eur J Inorg Chem, pp 2553–2556

Weng JB, Hong MC, Cao R, Shi Q, Chan ASC (2003) Jiegou Huaxue (Chin) (Chin J Struct Chem) 22:195. (Quoted from CSD, code: AJIZOI)

Wiesbrock F, Schmidbaur H (2003) Lithium L-hydrogen-α-glutamate: a layer structure with asymmetrical tunnels formed by nets with two different macrocycles. Cryst Eng Comm 5:262–264

Xie Y, Wu HH, Yong GP, Wang ZY (2006) *rac*-Poly[bis(μ-tryptophanato)manganese(II)]. Acta Cryst E 62:m2089–m2090

Ye SJ, Clark AA, Armentrout PB (2008) Experimental and theoretical investigation of alkali metal cation interactions with hydroxyl side-chain amino acids. J Phys Chem B 112:10291–10302

Yoneda H, Yukimoto T, Kushi Y, Nakazawa H (1986) Chromatographic separation of isomers of meridional tris(amino-acidato) cobalt(III) homochelates and their absolute configuration. J Liq Chromatogr 9:573–590

Zhang Y, Saha MK, Bernal I (2003) [Cobalt(II)L-glutamate(H_2O)·H_2O]: a new 3D chiral metal–organic interlocking network with channels. Cryst Eng Comm 5:34–37

Zhang Y, Lin Q, Wei T (2004) Microwave-assisted synthesis, crystal structure and toxicity of L-aspartic acid zinc spiral linear supramolecular polymer. Sci China B 47:295–300

Zhang Y, Sun H, Oldfield E (2005) Solid-state NMR fermi contact and dipolar shifts in organometallic complexes and metalloporphyrins. J Am Chem Soc 127:3652–3653

Zhao XJ, Du M, Wang Y, Bu XH (2004) Formation of cobalt(II)–piperazine supramolecular systems under different organic acid mediums: synthesis, characterization and crystal structures. J Mol Struct 692:155–161

Chapter 4
Compounds of Amino Acids and Neutral Salts

Abstract In this chapter we present salts composed of amino acids combined with inorganic cations and anions. These salts generally contain amino acids in neutral, i.e., zwitterionic form, with the charges balanced by the cations and anions. However, some of the phases resented here contain cations, anions, and amino acids, but the amino acids are not in neutral form, as, e.g., glycinate$^-$ $Cu^{2+}(NO_3)^-$ or (betainium$^+$)$_2$ $Cu^{2+}Cl^-{}_4 \cdot H_2O$, but for reasons of similarity, they are included in this chapter. The variation of salts (and thus the number of species) is even greater as for salts of amino acids and cations, as a multitude of combinations of amino acids, cations, and anions is possible. Additionally, different hydration states are often found, as well as different ratios of components. Thus, there are several instances of systems containing more than one phase (e.g., the system glycine–zinc–chloride–water). Despite the greater variation, many coordination patterns found in the salts described in the previous chapter recur in the salts reported here. Likewise, the connectivities of units are similar as those found in salts of amino acids and cations. Instances of materials with noteworthy physical properties are presented.

Keywords Salts of amino acids • Crystal structure • Molecular structure • Polymorphs • Zwitterions • Coordination • Isolated units • Chains • Layers • Frameworks • Connectivity • Frequency of connectivity patterns

4.1 Introducing Remarks

As stated before, this chapter deals with salts of amino acids and neutral, inorganic salts. In these species, the amino acid usually exists in zwitterionic form, but it is still coordinated to a cation. As other anions are present, the amino acids are not the only possible ligands. These anions can act as additional ligands (plus possible water molecules as well). Moreover, in some cases the inorganic anions are not part of the coordination spheres, but located in the interstices between units composed of cations and coordinating amino acids.

M. Fleck and A.M. Petrosyan, *Salts of Amino Acids: Crystallization, Structure and Properties*, DOI 10.1007/978-3-319-06299-0_4,

The structural diversity addressed (and found in Chap. 3) is even greater for the salts discussed in this chapter, as a multitude of cation–anion combinations can be used for synthesis with the different amino acids. Therefore, salts of the type $(AA)_nM^+X^-$, $(AA)_nM^{2+}X^{2-}$, $(AA)_nM^{2+}(X^-)_2$, etc., are possible (where X is any anion), again with various states of hydration and pure enantiomers and racemates. As cations, the most common metals are used, among the anions frequently employed for the syntheses are halogenides, nitrates, sulfates, and perchlorates, but many others have been reported as well.

As for the salts of cations and amino acid anions, the same coordination modes are found (Fig. 3.1), and again, the same system designating the connectivity of the building units is employed here. As many structural motifs of the previous chapter are found in the salts discussed herein as well, we will often refer to compounds (and Figs.) of Chap. 3.

4.2 Glycine

Of all amino acids, glycine has been used most frequently for the synthesis of salts of amino acids and neutral compounds (Table 4.1).The reason for this large amount of data is the fact that glycine is well soluble in water and tends to incorporate readily into crystals – the small molecule makes a good ligand and is sterically flexible. In fact, more than 200 data sets – taken from the Cambridge Structural Database (Allen 2002) – were found, plus several other reports which are not listed in the database. Among these, there are instances of more than one publication on one species, as, for instance, the structure of dichloro-bis(glycine)-zinc dihydrate, which has been reported by Fleck et al. (2008a) and simultaneously by Priya et al. (2008) and then by Tepavitcharova et al. (2009). Additionally, papers on growth and properties are found.

Moreover, there are many instances of more than one phase in one system. Within the system to which the species addressed above belongs, i.e., glycine–Zn–Cl–H_2O, at least four phases exist (Table 4.1), in different ratios of the constituents.

Thus, it is impossible to discuss every species or work in the scope of this book. We have therefore decided to list at least the majority of salts and refer to most of the publications dealing with the said salts. Consequently, we discuss the structures in order of the inorganic anions.

4.2.1 Glycine Metal Halogenides

Of all glycine salts, the majority belongs to the family of halogenides, and within the halogenides, the majority are chlorides, although fluorides, bromides, and iodides have been synthesized as well. As addressed above, within this group

Table 4.1 *Salts of glycine and neutral compounds, part I:* space groups and unit cell parameters of compounds of glycine metal halogenides

	s.g.	*a*	*b*	*c*	*α*	*β*	*γ*	Conn.	Reference
$G^+Sb^{III}F_4$	*P*2$_1$/*b*	8.563	8.916	9.52	90	90	112.85	0–0	Davidovich et al. (1991)
G $Sb^{III}F_3$	*Pb*	6.955	6.462	6.748	90	90	90.33	0–0	Davidovich et al. (1991)
cis-G_2 $Pt^{II}Cl_2.H_2O$	*P*-1	8.698	6.233	11.172	94.36	99.60	113.34	0–0	Baidina et al. (1979)
cis-G_2 $Pt^{II}Cl_2$	*P*2$_1$/*n*	5.119	10.429	18.453	90	91.53	90	0–0	Baidina et al. (1981)
trans-G_2 $Pt^{II}Cl_2.2H_2O$	*P*-1	5.359	7.054	8.441	85.67	75.91	71.71	0–0	Davies et al. (1995)
trans-G_2 $Pt^{IV}Cl_2$	*P*2$_1$/*c*	8.185	10.477	10.899	90	92.03	90	0–0	Davies et al. (1995)
cis-G_2 $Pt^{IV}Cl_2$	*Pbca*	10.163	11.028	16.310	90	90	90	0–0	Steinborn et al. (1997)
trans-G G_2 PdCl	*Pbca*	8.443	10.552	19.356	90	90	90	0–0	Steinborn et al. (1997)
G $LiCl·H_2O$	*P*2$_1$/*c*	10.103	5.064	11.930	90	107.39	90	1–1	Müller et al. (1994)
G $CaCl_2·3H_2O$	*P*2$_1$/*c*	9.964	6.898	13.959	90	104.11	90	1–1	Fleck et al. (2008a)
G_2 $CaCl_2·4H_2O$	*P*-1	14.702	15.483	18.824	88.96	73.62	88.65	1–1	Natarajan et al. (2012b)
G $CaCl_2·4H_2O$	*P*2$_1$/*c*	13.01	6.794	15.375	90	91.4	90	1–1	Jayalakshmi and Vijayan (1967)
G_3 $CaCl_2$	*Pb*2$_1$*a*	14.774	19.573	9.067	90	90	90	0^2–2	Ravikumar et al. (1986)
G_2 $BaCl_2·H_2O$	*Pbcn*	8.31	14.84	9.32	90	90	90	1–1	Narayanan and Venkataraman (1975)
G $SrCl_2·3H_2O$	*Pbcn*	16.42	9.35	8.26	90	90	90	1–1	Narayanan and Venkataraman (1975)
G_2 $SrCl_2·H_2O$	*Pcnb*	8.177	9.056	14.411	90	90	90	1–1	Natarajan et al. (2012a)
G $CdCl_2$	*P*2$_1$/*n*	8.17	8.91	13.62	90	107.0	90	0^2–2	Thulasidhass and Mohana Rao (1980)
G_4 Cd_3Cl_2	*P*-1	8.759	8.931	9.038	68.69	61.85	66.25	1–1	Dan and Rao (2005)
G $MgCl_2·4H_2O$	*Pnma*	7.711	10.584	12.983	90	90	87.0	0–1	Fleck and Bohaty (2005c)
G $MnCl_2$	*P*-1	4.97	7.92	6.98	107.4	115.9	87.0	0–1	Narayanan and Venkataraman (1975)
G $MnCl_2·2H_2O$	*P*2$_1$/*c*	8.400	5.607	16.790	90	90.18	90	0–1	Clegg et al. (1987)
G_2 $MnCl_2·2H_2O$	*P*2$_1$	9.96	8.53	6.860	90	107	90		Pepinsky et al. (1958)
G $MnCl_2·2H_2O$	*P*2$_1$/*n*	6.519	15.981	7.893	90	97.18	90	0–1	Mrozek et al. (1999a)
G $CoCl_2·2H_2O$	*P*2$_1$/*n*	6.376	15.877	7.749	90	97.63	90	0–1	Clegg et al. (1987), Dan and Rao (2005)
G_2 $CoCl_2·2H_2O$	*P*2$_1$/*c*	10.568	5.988	8.844	90	91.55	90	0–2	Stenzel and Fleck (2004)
$G_2NiCl_2·2H_2O$	*P*2$_1$/*c*	10.601	5.858	8.711	90	90.489	90	0–2	Ch'ng et al. (2008)
G $NiCl_2·2H_2O$	*P*2$_1$	8.203	5.475	8.311	90	90.97	90	0–1	Fleck and Bohatý (2004)

(continued)

Table 4.1 (continued)

	s.g.	*a*	*b*	*c*	*α*	*β*	*γ*	Conn.	Reference
G $ZnCl_2 \cdot H_2O$	*P*2$_1$/*a*	7.866	9.124	10.645	90	103.51	90	0–0	Fleck et al. (2008a)
G_2 $ZnCl_2 \cdot 2H_2O$	*C*2/*c*	14.444	6.917	12.968	90	117.90	90	0–0	Fleck et al. (2008a)
G_3 $ZnCl_2$	*Pbn*2$_1$	11.230	15.251	15.564	90	90	90	0–0	Hariharan et al. (1989)
G $ZnCl_2$	*P*2$_1$	4.733	10.865	6.205	90	95.98	90	0–1	Tepavitcharova et al. (2009)
G_3 $CeCl_3 \cdot 3H_2O$	*P*2$_1$2$_1$2$_1$	4.788	12.074	30.949	90	90	90	0–1	Fleck et al. (2008a)
G $NdCl_3 \cdot 7H_2O$	*Pm*	6.030	13.877	8.626	90	100.76	90	0–1	Li et al. (1999)
G $NdCl_3 \cdot 3H_2O$	*P*2$_1$/*n*	10.037	14.630	10.063	90	101.98	90	0^2–0^2	Panyushkin et al. (2003)
G_3 $NdCl_3 \cdot 3H_2O$	*P*2$_1$2$_1$2$_1$	12.0	30.8	4.8	90	90	90	0–1	Balyaeva et al. (1969)
G_3 $LaCl_3 \cdot 3H_2O$	*P*2$_1$2$_1$2$_1$	4.798	12.094	30.974	90	90	90	0–1	Gao et al. (2002)
G_3 $PrCl_3 \cdot 3H_2O$	*P*2$_1$2$_1$2$_1$	4.779	12.051	30.953	90	90	90	0–1	Tianzhu et al. (1989)
G_3 $YbCl_3 \cdot 3H_2O$	*P*2$_1$2$_1$2$_1$	4.749	11.917	30.716	90	90	90	0–1	Ren et al. (2000)
G_3 $GdCl_3 \cdot 3H_2O$	*P*2$_1$2$_1$2$_1$	11.960	30.789	4.765	90	90	90	0–1	Gao et al. (2001)
G_4 $NdCl_6 \cdot 6H_2O$	*P*2$_1$/*n*	10.037	14.630	10.063	90	101.98	90	0^2–0^2	Panyushkin et al. (2003)
G_2 $Mo_2Cl_4 \cdot xH_2O$	*P*-1	11.941	12.022	11.614	92.42	93.04	88.94	0^2–0^2	Bino et al. (1979)
G_2 $Mo_2Cl_4 \cdot xH_2O$	*P*2$_1$/*n*	9.536	14.792	9.339	90	102.99	90	0^2–0^2	Ardon and Bino (1995)
G_2 $Mo_2Cl_4 \cdot xH_2O$	*P*2$_1$/*c*	8.350	9.346	14.807	90	103.42	90	0–0^2	Bino et al. (1979)
G_3 $CaBr_2$	*Pbc*2$_1$	9.15	14.84	20.31	90	90	90	0^2–2	Mohana Rao and Natarajan (1980)
G_2 $LiBr \cdot H_2O$	*P*2$_1$/*c*	7.540	17.417	8.273	90	118.14	90	0–1	Balakrishnan et al. (2012)
G $CdBr_2$	*P*2$_1$/*n*	8.41	9.01	13.98	90	108.0	90	0^2–2	Thulasidhass and Mohana Rao (1980)
G_2 $Mo^{II}Br_2 \cdot xH_2O$	*P*-1	11.941	12.022	11.614	92.42	92.04	88.94	0^2–0^2	Bino et al. (1979)
G_2 $Mo^{II}Br_2 \cdot xH_2O$	*P*2$_1$/*c*	8.350	9.346	14.807	90	103.42	90	0–0^2	Bino et al. (1979)
G_2 $Cr^{II}Br_2 \cdot 2H_2O$	*P*-1	8.455	11.340	6.799	98.74	104.38	77.68	0–1	Ardon et al. (1984)
G_2 $Cr^{II}Br_2 \cdot xH_2O$	*P*2$_1$/*c*	8.325	9.307	14.675	90	103.27	90	0–0^2	Ardon et al. (1984)
cis-G_2 $PtBr_2 \cdot H_2O$	*P*-1	8.735	6.416	11.474	94.54	99.54	115.03	0–0	Baidina et al. (1979)
G_2 $MnBr_2 \cdot 2H_2O$	*Pbca*	21.613	12.649	8.990	90	90	90	0–2	Glowiak and Ciunik (1978a)
G_2 $MnBr_2 \cdot 2H_2O$	*P*2$_1$/*c*	11.943	6.060	8.979	90	111.65	90	0–2	Glowiak and Ciunik (1978a)
G_2 $MnBr_2 \cdot 2H_2O$	*P*2$_1$/*c*	7.961	16.421	9.692	90	136.6	90	0–1	Glowiak and Ciunik (1977)
G_2 $MgBr_2 \cdot 2H_2O$	*P*2$_1$/*c*	11.836	6.070	8.829	90	111.89	90	0–2	Krishnakumar and Natarajan (1995a)

G_2 $CoBr_2 \cdot 2H_2O$	$P2_1/c$	11.771	6.055	8.822	90	111.53	90	0–2	Ravikumar et al. (1985), Dan and Rao (2005)
G_2 $NiBr_2 \cdot 2H_2O$	$P2_1/c$	11.815	6.021	8.750	90	111.80	90	0–2	Fleck and Bohaty (2005c), Dan and Rao (2005)
G $NaI_2 \cdot H_2O$	$P2_1/c$	12.402	6.947	12.193	90	100.4	90	1–1	Verbist et al. (1971)
G_4 KI_3	$C2/c$	22.969	10.683	8.242	90	91.54	90	0–1	Herbstein and Kapon (1980)
G_3 $CaI_2 \cdot H_2O$	$P2_1/c$	12.07	7.37	19.13	90	109.6	90	1–1	Natarajan and Rao (1977)
G_2 $CaI_2 \cdot 3H_2O$	$Pca2_1$	9.862	13.059	22.724	90	90	90	0–1	Fleck et al. (2008a)
G_2 $MgCuCl_2 \cdot 7H_2O$	$P\text{-}1$	6.682	6.749	10.772	72.668	83.441	72.230	0–0	Davies et al. (2003)
G $BaPdCl_2 \cdot 2H_2O$	$P2_1/n$	9.559	14.035	11.868	90	104.09	90	3–3	Baidina et al. (1980)

Fig. 4.1 *Crystals of dichloro-bis(glycine-O)-zinc glycine.* Like many species soluble in water, glycine salts can be grown from aqueous solutions. These crystals were grown by means of the slow evaporation method and were then cut and polished to prepare samples for the measurement of direction-dependent physical properties, e.g., refractive indices or components of the dielectric and piezoelectric tensors

there are several instances of more than one species within one system. The example of the system glycine–Zn–Cl–H_2O has been named above, as four compounds have been reported: Apart from the hydrates aqua-dichloro-glycinato-zinc and dichloro-diglycine-zinc dihydrate (both by Fleck et al. 2008a), there are two anhydrous forms – glycine-dichloro-zinc (Tepavitcharova et al. 2009) and dichloro-bis(glycine)-zinc glycine (Hariharan et al. 1989, later refined by Fleck et al. 2008b; and Tepavitcharova et al. 2009). This latest salt is by far the most interesting member within this family: As the large, optical quality single crystals can be grown (Fig. 4.1), several studies of crystal physical properties such as piezoelectric, pyroelectric, elastic, and dielectric properties were published (Fleck et al. 2008b; Sugandhi et al. 2013; Tylczyński and Wiesner 2014, and references therein). Moreover, optical properties have been investigated, as the species was found to be a many-phonon SRS crystal manifesting different nonlinear laser ($\chi^{(3)}+\chi^{(2)}$) interactions under one-micron picosecond pumping (Kaminskii et al. 2009).

Likewise, in the system glycine–Ca–Cl–H_2O, at least four species exist: The ratio of G–$CaCl_2$ is 1:1 (at least two phases); 1:2 and 1:3 have been found. In glycinato-dichloro-triaqua-calcium (Fleck et al. 2008a; Yusenko et al. 2008), chloride anions are part of the coordination sphere, while in the respective tetrahydrate (Natarajan et al. 2012b) as well as in diglycine calcium dichloride tetrahydrate (Jayalakshmi and Vijayan 1967 reported only unit cell data and space group; Natarajan and Rao 1976 published the structure), they are located in the interstices only. All three compounds comprise chains (Fig. 4.2). In the fourth phase, tris(glycine) calcium dichloride (Ravikumar et al. 1986), isolated units, connected to layers, makes up the structure. The existence of a fifth salt (monoaqua glycine calcium chloride) was proposed, although no crystals suitable for single crystal XRD could be grown; thus, only spectral and powder data are available (Yusenko et al. 2008). A good overview of the phases in this system (as well as likewise bromide and iodide salts) is given by Natarajan et al. (2012b).

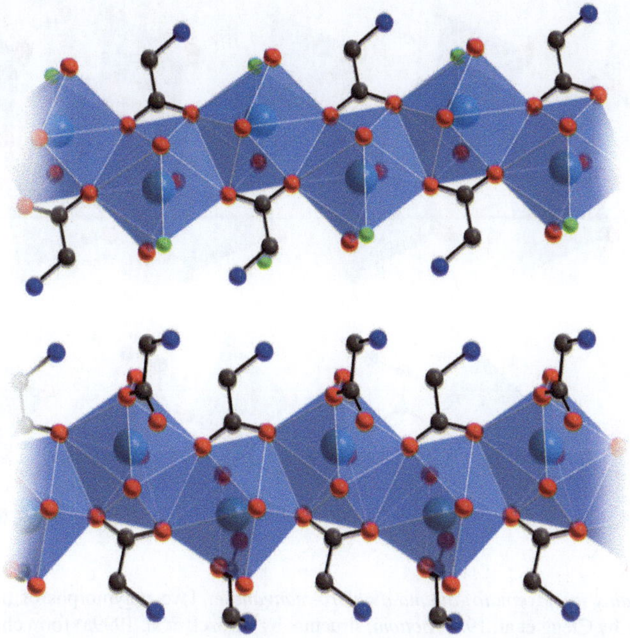

Fig. 4.2 *Chains in glycine calcium chloride salts.* In glycinato-dichloro-triaqua-calcium (*top*), there is one glycine molecule per calcium cation; therefore, chloride anions are part of the coordination sphere. In diglycine calcium dichloride tetrahydrate (*bottom*), there are twice as many glycine zwitterions, which occupy all ligand positions; thus, the chloride anions are located in the interstices (not shown). Hydrogen atoms are omitted for clarity

One more example of a system with more than one phase is that of glycine–Mn^{2+}–Cl–H_2O. Four different species were identified in this system: Three phases consist of glycine and $MnCl_2$ in equimolar ratios and one anhydrous crystal (Narayanan and Venkataraman 1975) and two different dihydrates (one determined by Clegg et al. 1987 and then refined by Ciunik and Glowiak 1980a, the other one published by Mrozek et al. 1999a), all of which comprise chain structures. The chains are topologically identical; still the structure is different (Fig. 4.3). Another dihydrate phase was reported by Pepinsky et al. (1958) and later refined by Lee et al. (1962). Pepinsky et al. found this salt to be ferroelectric at ambient conditions (up to 55 °C).

The above examples are not the only instances of more than one species in one system; others are chloride salts of Mo, Cd, Ni, etc. (see Table 4.1). This very high structural diversity reflects the flexibility of glycine as ligand, as it proves that different packing arrangements with comparable lattice energies are possible. Nevertheless, isotypic structures occur, as, e.g., the 1:1 glycine salts of bivalent manganese and cobalt (Mrozek et al. 1999a; Clegg et al. 1987) or the 2:1 glycine salts of cobalt and nickel (Stenzel and Fleck 2004; Ch'ng et al. 2008). The structure of the latter pair is layered: Because of the higher ratio of glycine, the octahedra are linked via two amino acids to each other (Fig. 4.4). This pattern is found in other glycine metal halogenides, namely, the bromide salts of magnesium, cobalt, and

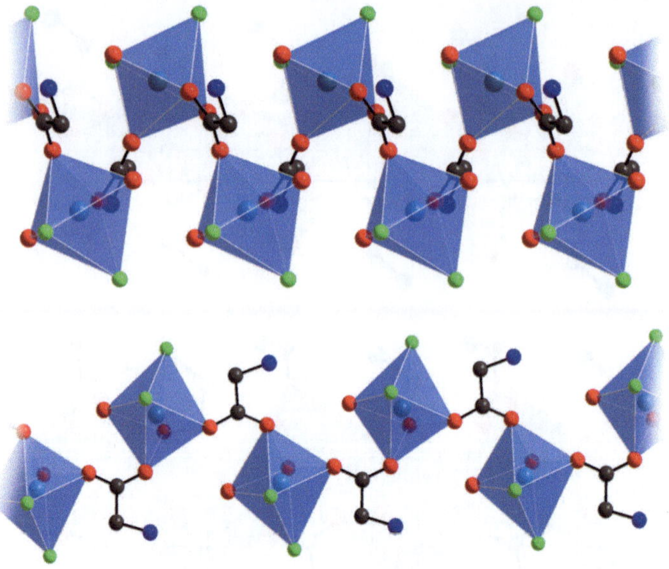

Fig. 4.3 *Chains in (glycinato)-diaqua-dichloro-manganese*. Two polymorphs of this compound (*top*, structure by Clegg et al. 1987; *bottom*, structure by Mrozek et al. 1999a) form chain structures, with identical topology but different packing modes. Hydrogen atoms are omitted for clarity

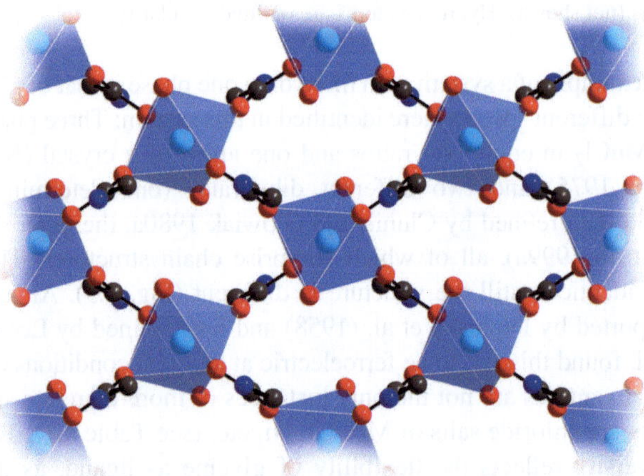

Fig. 4.4 *Layers in bis(glycinato)-diaqua-cobalt dichloride*. Due to the higher ratio of glycine per cation, more ligand positions are occupied by oxygen atoms from the amino acids, accounting for the connection to layers. The chloride anions are located in the interstices (not shown). Hydrogen atoms are omitted for clarity

nickel (compare unit cell parameters as given in Table 4.1). Topologically, it reminds of the layer motif found in metal glycinates (and other cation–amino acid salts; compare Figs. 3.7, 3.15, 3.17), although there is an important difference: In neutral salts, the amino acid exists in zwitterionic state; therefore the amino group is protonated and thus the nitrogen atom is not available as ligand. Other examples for isotypic structures are the chloride salts of rare earth elements, which can also be described as chains.

Apart from the chain and layer structures discussed above (and many others), isolated units are also found in glycine metal halogenides, although not as frequently as in salts of glycinate and cations. Among these are the platinum and palladium salts, in which the coordination resembles that described in Chap. 3 for likewise salts.

Finally, double salts have been reported as well. One instance is the glycine chloride of magnesium and copper, in which both cations have different environments: The Cu^{2+} cations are [4 + 2]-coordinated, with two *trans*-oriented glycine zwitterions in equatorial position, while the Mg^{2+} cations are octahedrally coordinated by six water molecules (Davies et al. 2003).

4.2.2 Glycine Metal Perchlorates

Apart from the large family of glycine metal halogenide salts, there are several examples of perchlorates. Among these salts, isolated units, chains, and layered structures are found (Table 4.2). It is remarkable that trivalent cations feature strongly in this family, such as rare earth elements or scandium. The salt of this latter element, tris(glycinato)-scandium triperchlorate, was first published by Yan et al. (1996) and reported to crystallize in the triclinic space group *P*-1. Later, it was found that this species actually crystallizes in the trigonal space group *R*-3 (Wang et al. 1998). The structure features chains parallel the threefold axes, where adjacent scandium polyhedra are connected to each other via three glycine zwitterions each (Fig. 4.5), with the perchlorate anions located in the interstices between the chains. This topology is not unique – the same pattern is found in tris(glycinato)-diaqua-cerium trichloride monohydrate (Fleck et al. 2008a), although the coordination number of the larger cerium cations is naturally higher, the additional ligands being water molecules. The charge-counterbalancing anions (chloride in that case) are also located in the interstices, as are the perchlorate anions in this structure.

Another chain structure is found for the trivalent salts of the rare earth elements Ho, Dy, Sm, Nd, Pr, and Er (Table 4.2). The structures are isotypic, which is to be expected, but the topology is surprising. In contrast to the above chain, the connection between neighboring cations is provided alternatively via four and two glycine molecules (Fig. 4.6).

A noticeable feature of perchlorate salts is the fact the ClO_4 anions are sometimes disordered, as they are not part of the coordination sphere of the cations, but located in the interstices between the units, where they are bonded by hydrogen

Table 4.2 *Salts of glycine and neutral compounds, part I:* space groups and unit cell parameters of compounds of glycine metal perchlorates

	s.g.	*a*	*b*	*c*	*α*	*β*	*γ*	Conn.	Reference
G $Cu(ClO_4)·xH_2O$	*P*2$_1$/*n*	7.900	10.344	9.566	90	110.88	90	0–2	Estrader et al. (2008)
G_3 $Sc(ClO_4)_3$	*R*-3	19.048	19.048	9.053	90	90	120	0–1	Wang et al. (1998)
$G_3La(ClO_4)_3·2H_2O$	*P*-1	10.730	12.958	8.806	97.10	113.05	71.51	1–1	Ma et al. (1994a)
G_3 $Ho(ClO_4)_3·3H_2O$	*Cc*	20.506	9.245	23.989	90	100.28	90	0–1	Legendziewicz et al. (1989b)
G_3 $Dy(ClO_4)_3·3H_2O$	*Cc*	20.56	9.42	24.16	90	98.7	90	0–1	Legendziewicz et al. (1989b)
$G_3Sm(ClO_4)_3·2.5H_2O$	*P*-1	11.549	14.122	15.654	97.01	102.7	105.43	0–1	Ma et al. (1993a)
G_3 $Nd(ClO_4)_3·4.5H_2O$	*P*-1	11.554	14.108	15.660	97.14	102.82	106.28	0–1	Legendziewicz et al. (1989a)
G_3 $Pr(ClO_4)_3·2.5H_2O$	*P*-1	11.558	14.126	15.667	97.06	102.66	105.32	0–1	Du et al. (1996)
$G_3Er(ClO_4)_3·2.5H_2O$	*P*-1	11.476	14.028	15.443	96.67	102.67	105.51	0–1	Yifan and Kezhen (1987)
$G_2Er_2(ClO_4)_6·10H_2O$	*P*2$_1$/*n*	10.571	10.837	17.728	90	90.09	90	$0–0^2$	Zhao and Jin (1996)
G $ReCl_4·0.5H_2O$	*P*-1	8.213	9.646	12.850	74.16	78.22	70.46	$0^2–0^2$	Shtemenko et al. (2003)

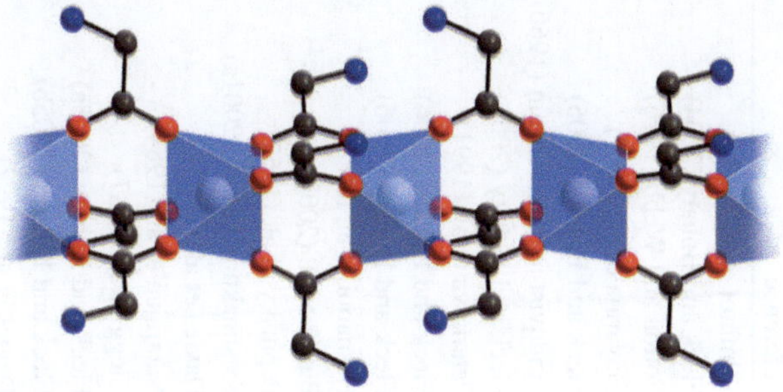

Fig. 4.5 *Chains in tris(glycinato)-scandium triperchlorate.* Each trivalent scandium cation is coordinated by six oxygen atoms from six different glycine molecules; connection between neighboring cations is achieved via three glycine molecules each

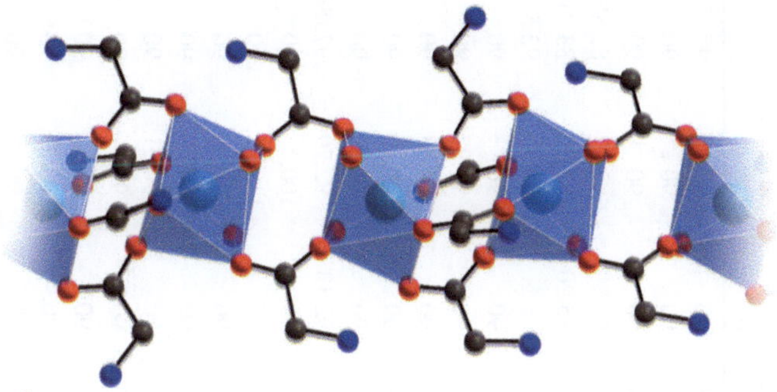

Fig. 4.6 *Chains in tris(glycinato)-diaqua-samarium trichloride monohydrate, as example for the group of isotypic REE salts.* Note the different connection in comparison with the respective scandium compound (Fig. 4.5): Here, the polyhedra are alternatively connected via four and two glycine ligands

bonds only. In some cases, the oxygen positions are clearly distributed over several positions; in others they are located but often show relatively high displacement. As this aspect is of more relevance in the salts of amino acids and anions, it will be discussed in Chap. 5.

4.2.3 Other Glycine Salts

Although halogenide and perchlorate anions are the most common anions in glycine salts, several other anions have been employed in syntheses as well. Nitrates and sulfates have been used frequently, but phosphates, thiosulfates, chromates, and molybdates have been reported as well (Table 4.3).

Among these salts, some very interesting species are found: Probably the most prominent member of this family is glycinato-silver nitrate or simply glycine silver

Table 4.3 *Salts of glycine and neutral compounds, part II:* space groups and unit cell parameters of compounds of glycine metal sulfates, thiosulfates, phosphates, nitrates, chromates, and molybdates

	s.g.	*a*	*b*	*c*	*α*	*β*	*γ*	Conn.	Reference
$G^+ (NH_4)SO_4$	$P2_1/c$	8.262	10.073	8.632	90	92.66	90	0–0	Vilminot et al. (1974)
$G\ Li_2SO_4$	$Pna2_1$	16.423	5.005	7.654	90	90	90	2–2	Fleck and Bohatý (2004)
$G_2\ MoSO_4{\cdot}2H_2O$	I-4	11.290	11.290	9.150	90	90	90	0^2–0^2	Cotton and Webb (1976)
$G\ MgSO_4{\cdot}5H_2O$	P-1	5.988	6.783	13.391	85.39	82.87	82.88	0–0	Elayaraja et al. (2007)
$G\ CoSO_4{\cdot}5H_2O$	P-1	5.570	6.775	13.335	85.20	83.30	83.21	0–0	Fleck and Bohatý (2006)
$G\ FeSO_4{\cdot}5H_2O$	P-1	6.86	13.6	6.07	96.1	96.8	92.5	0–0	Lindqvist and Rosenstein (1960)
$G\ ZnSO_4{\cdot}3H_2O$	$Pca2_1$	8.440	8.278	12.521	90	90	90	0–1	Fleck and Bohatý (2004)
$G\ NiSO_4{\cdot}6H_2O$	$P2_1/c$	5.726	12.302	17.010	90	97.92	90	0–0	Peterkova et al. (1991)
$G\ K_2S_2O_3$	$P2_1/c$	5.630	20.244	7.762	90	94.33	90	3–3	Fleck and Bohatý (2006)
$G\ Na_4(S_2O_3)_2{\cdot}2H_2O$	$C2/c$	17.775	7.311	22.595	90	100.10	90	2–3	Fleck and Bohatý (2006)
$G\ GaPO_4$	$P2_1/c$	9.177	5.154	12.513	90	90.67	90	0–2	Hasnaoui et al. (2005)
$G\ LiNO_3$	P-1	5.590	5.898	8.658	92.13	92.79	97.09	0–2	Baran et al. (2003)
$G\ AgNO_3$	$P2_1/a$	5.451	19.493	5.541	90	100.2	90	2–2	Pepinsky et al. (1957)
$G\ NaNO_3$	Cc	14.321	5.260	9.113	90	119.07	90	2–3	Krishnakumar et al. (2001a)
$G_2\ CuAg_2(NO_3)$	$Pbca$	8.792	8.51	16.317	90	90	90	3–3	Davies et al. (2003)
$G\ Ca(NO_3)_2{\cdot}2H_2O$	$P2_12_12_1$	6.866	11.271	13.248	90	90	90	1–1	Natarajan et al. (1984)
$G\ Co(NO_3)_2{\cdot}4H_2O$	$P2_1/n$	9.243	6.720	19.041	90	96.51	90	0–1	Clegg et al. (1987)
$G\ Ni(NO_3)_2{\cdot}4H_2O$	$P2_1/n$	9.201	6.773	18.977	90	96.7	90	0–1	Fleck and Bohaty (2005a)
$G\ Mg(NO_3)_2{\cdot}4H_2O$	$P2_1/n$	9.230	6.779	19.052	90	96.57	90	0–1	Fleck and Bohaty (2005b)
$Cu\ G^- (NO_3){\cdot}H_2O$	$P2_1/n$	9.278	5.182	13.971	90	103.29	90	1–1	Davies et al. (1992)
$G_2\ Li_2CrO_4{\cdot}H_2O$	$P2_12_12_1$	6.202	9.192	19.294	90	90	90	2–3	Fleck et al. (2006)
$G_2\ Li_2MoO_4$	$P2_1$	5.192	7.734	12.492	90	94.277	90	0^2–3	Fleck et al. (2006)

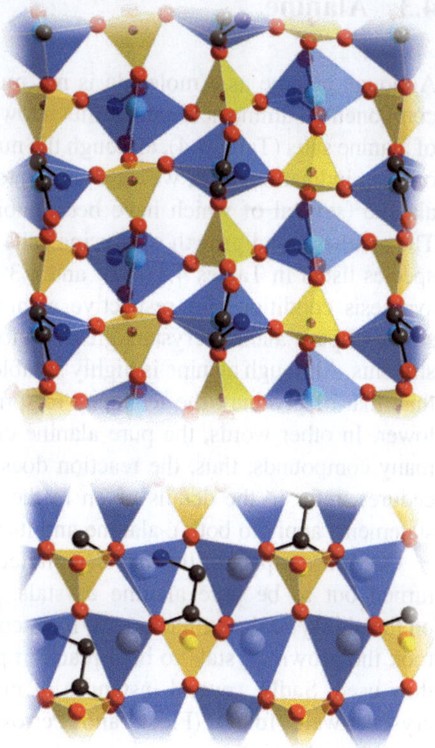

Fig. 4.7 *Layers in phosphato-glycine-gallium and sulfato-glycine-di-lithium (top and bottom).* Although both layers comprise tetrahedral units (both lithium coordination polyhedra, *blue*, and phosphate or sulfate anions, *yellow*), the structures differ noticeably

nitrate (GSN), which was first reported by Pepinsky et al. (1957), who found it to undergo a phase transition at about −55 °C, when the crystal becomes ferroelectric. Pepinsky et al. attributed this spontaneous polarization to disturbance of the planarity of the glycine zwitterion in the crystalline field. Rao and Viswamitra (1972) refined the crystal structure, which is sensitive to light and therefore tended to deteriorate under X-rays. Recently, Choudhury et al. (2013) performed an analysis employing neutron diffraction to circumvent this problem and attain more accurate data on the phase transition.

Apart from this physically interesting example, there are several structurally interesting species within this family. These encompass structures of isolated units, chains, layers, and frameworks, most of which represent their own unique structure type (although a few isotypic structures are found, as can be seen from the unit cell parameters in Table 4.3). For instance, the layers in phosphato-glycine-gallium (Hasnaoui et al. 2005) differ significantly from those found in sulfato-glycine-di-lithium (Fleck and Bohatý 2004; Fig. 4.7).

The latter phase was investigated several times. After the publication of the structure by Fleck and Bohatý (2004), Balakrishnan and Ramamurthi (2006) and later Suresh Kumar et al. (2007) reported the growth of good-quality, large single crystals as well as several physical properties.

4.3 Alanine

Although alanine as a molecule is not much larger than glycine, the utilization as component of amino acid salts is much lower. This can be seen simply by the number of alanine salts (Table 4.4), although the number of publications is actually not proof, only an indication. Still, we have attempted syntheses of many salts of glycine and alanine (several of which have been published; see Tables 4.1, 4.2, 4.3 and 4.4). The synthesis and growth of glycine salts proved to be relatively easy; most of the species listed in Tables 4.1, 4.2, and 4.3 could be synthesized. In contrast, similar synthesis conditions for respective alanine salts often gave no results: From the solution, pure alanine crystals grew without reaction with the other dissolved constituents. Although alanine is highly soluble in water (although the solubility is lower than that of glycine), the tendency to form compounds with cations and anions is lower. In other words, the pure alanine crystals are energetically more stable than many compounds; thus, the reaction does not take place easily. For synthesis procedures, refer to the details given in the references in Table 4.4. (Note that these statements apply to both L-alanine and its racemate.)

One more aspect of this has been noted in publications on "alanine salts" which turned out to be pure alanine crystals. In many cases, researchers have relied on the ability of the amino acid to form compounds with inorganic salts and simply took the grown crystals to be a reaction product of alanine and whatever material they used. Sadly, several instances of publications on such incorrectly identified crystals were found (Fleck and Petrosyan 2009; Petrosyan et al. 2013; and references therein).

Among the salts of L-alanine, isolated units, chains, layers, and framework structures have been reported. Of the chain structures, the $Mn^{2+}Cl_2$ salt is noteworthy: In contrast to the simple structures found in the glycine analogs, this salt has a complicated structure (Mrozek et al. 1999b). The structure can be characterized as an arrangement of two crystallographically different chains, with a total of eight different chloride positions, of which seven are ligands of the cations, the eighth is not but located between the chains (Fig. 4.8).

Other chain structures are simpler, as, for instance, the 1–1 chains in the strontium chloride salt (Fleck et al. 2008c) and the copper sulfate salt (Davies et al. 1993), both shown in Fig. 4.9.

Another interesting structure is the salt nitrato-L-alanine-sodium (Van Hecke et al. 2007), which comprises polyhedral layers, where the nitrate anions fill the interstices between the large sodium polyhedra. These layers are connected via the L-alanine zwitterions, which act as monodentate and bridging ligands to polyhedra of one layer but also to polyhedra of neighboring layers. A view upon a layer is displayed in Fig. 4.10.

A very interesting case is the series of rare earth element perchlorate salts, as represented by the neodymium salt (Glowiak et al. 1996), which was found to crystallize in space group $P1$. The structure is isotypic with its terbium analog as published later by Mohamed et al. (2010), who measured the structure at 100K.

Table 4.4 *Salts of amino acids and neutral compounds, part II:* space groups and unit cell parameters of compounds of neutral compounds with alanine

	s.g.	*a*	*b*	*c*	*α*	*β*	*γ*	Conn.	Reference
L-Alanine									
L-Ala $MnCl_2 \cdot 2H_2O$	*P*1	7.544	10.868	11.207	85.55	87.02	82.84	1–1	Mrozek et al. (1999b)
L-Ala $SrCl_2 \cdot 3H_2O$	$P2_1$	8.540	7.167	8.769	90	95.02	90	1–1	Fleck et al. (2008c)
L-Ala $CdCl_2$	*C*2	16.240	7.272	7.987	90	116.44	90	1–2	Schaffers and Keszler (1993)
L-Ala_2 $HoCl_3 \cdot 4H_2O$	*P*1	10.284	10.970	9.646	100.54	102.11	119.89	0–0^2	Dao et al. (1988), Cong-Ngoan et al. (1988)
L-Ala$DyCl_3 \cdot 3H_2O$	$P2_1$	8.459	15.716	11.518	90	102.11	90	0–1	Glowiak et al. (1991a)
L-Ala_2 $PtCl_2$	$P2_122_1$	7.460	8.544	9.754	90	90	90	0–0	Pavone et al. (1988)
L-Ala $Na(NO_3)$	$P2_12_12_1$	5.348	9.172	13.528	90	90	90	2–3	Van Hecke et al. (2007)
L-$Ala_2Hg_3(NO_3)_2 \cdot 3H_2O$	$P2_1$	10.626	19.729	15.318	90	91.57	90	1–3	Saunders et al. (2008)
L-$Ala_2Cu^{II}{}_2(SO_4) \cdot 5H_2O$	$P2_1$	6.311	23.119	14.064	90	93.42	90	1–1	Davies et al. (1993)
L-$Ala_2Nd(ClO_4)_3 \cdot 4H_2O$	*P*1	11.056	11.151	11.386	78.51	64.92	65.65	0–0^2	Glowiak et al. (1996)
L-Ala_2 $Sm(ClO_4)_3 \cdot 4H_2O$	*P*1	10.822	11.190	10.744	109.88	101.85	112.39	0–0^2	Torres et al. (2001)
L-Ala_2 $Gd(ClO_4)_3 \cdot 4H_2O$	*P*-1[a]	11.057	11.058	11.339	78.420	64.689	65.696	0–0^2	Wang et al. (2003)
L-Ala_2 $La(ClO_4)_3 \cdot 4H_2O$	*P*-1[a]	11.227	11.445	11.014	114.46	114.51	78.62	0–0^2	Ma et al. (1995)
L-Ala_2 $Er(ClO_4)_3 \cdot 4H_2O$[a]	*P*-1[a]	11.028	11.308	10.988	117.10	114.48	77.16	0–0^2	Hu et al. (1995)
L-$Ala_2Tb(ClO_4)_3 \cdot 4H_2O$	*P*1	10.770	10.777	11.352	79.35	65.39	67.66	0–0^2	Mohamed et al. (2010)
DL-Alanine									
DL-Ala $ZnCl_2$	$P2_1/c$	9.996	13.622	8.616	90	90.30	90	0–0	Subha Nandhini et al. (2002a)
DL-Ala_2 $PdCl_4$	*P*-1	6.463	7.420	8.891	77.82	84.54	67.34	0–0	Laligant et al. (1994)
DL-Ala_2 $CdCl_2 \cdot H_2O$	$P2_1/c$	7.119	14.408	8.640	90	107.98	90	1–1	Natarajan (2003)
DL-$Ala_2MnBr_2 \cdot 4H_2O$	$P2_1/c$	4.790	19.596	9.289	90	102.26	90	0–1	Ciunik and Glowiak (1980b)
DL-Ala_2SeBr_6	*C*2/*m*	12.33	7.71	10.26	90	105.5	90	N/A	Hempel et al. (1977a)
DL-Ala_2 $Co(NO_3)_2 \cdot 4H_2O$	$P2_1/n$	5.502	6.640	23.821	90	96.22	90	0–0	Rajagopal et al. (2003)
DL-Ala_2 $Er(ClO_4)_3 \cdot 4H2O$	*C*2/*c*	18.288	14.662	17.508	90	102.67	90	0–0^2	Glowiak et al. (1996)

[a]These space groups are incorrectly assigned; the correct space group is P1 for all these salts. In the case of the erbium salt, the authors reported this salt as racemate, which is most probably wrong (see text)

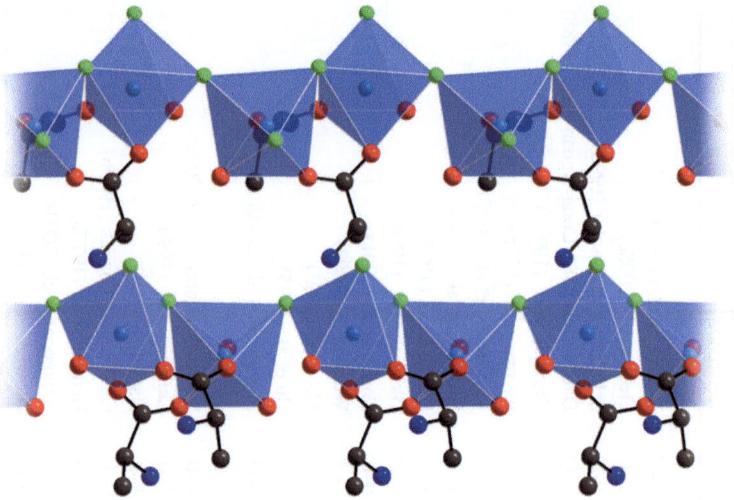

Fig. 4.8 *Chains in the manganese chloride alanine salt.* Note the fact that the chains are topologically and crystallographically different: In the lower chain, only three chloride anions act as ligands; the octahedral corner facing away from the viewer is occupied by a water molecule. The remaining chloride anion is located in the interstices (not shown). Hydrogen atoms are omitted for clarity

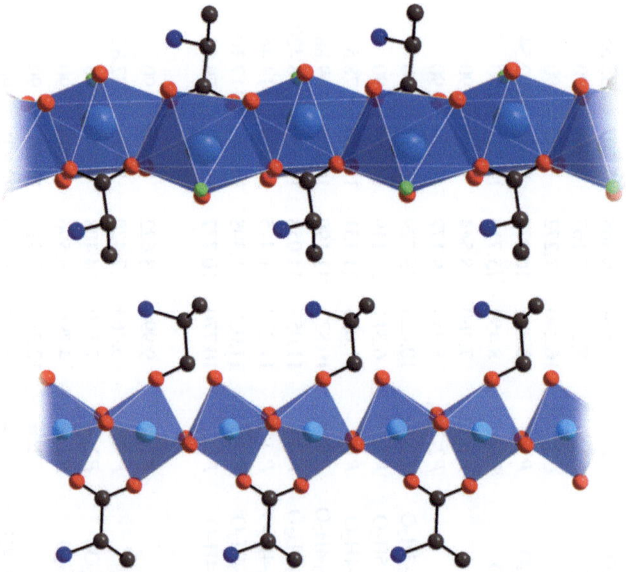

Fig. 4.9 *Chains in the L-alaninates of $SrCl_2$ (top) and $CuSO_4$ (bottom).* The topology is similar, although the Sr^{2+} polyhedra are naturally larger, with more ligand atoms from the L-alanine zwitterions (compare the bidentate coordination with the monodentate in the Cu^{2+} salt) as well as chloride anions as ligands. Other anions are located in the interstices (not shown). Hydrogen atoms are omitted for clarity

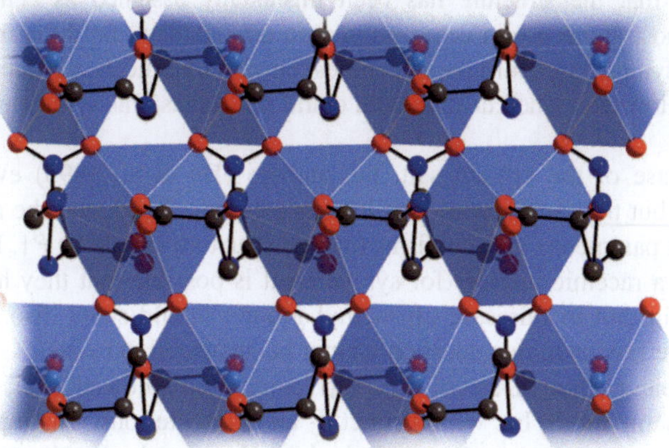

Fig. 4.10 *View upon a layer in nitrato-L-alanine-sodium, parallel (001)*. The Na^+ polyhedra are connected via nitrate anions in the interstices, with L-alanine zwitterions above and below the layers, providing connection to adjacent layers. Hydrogen atoms are omitted for clarity

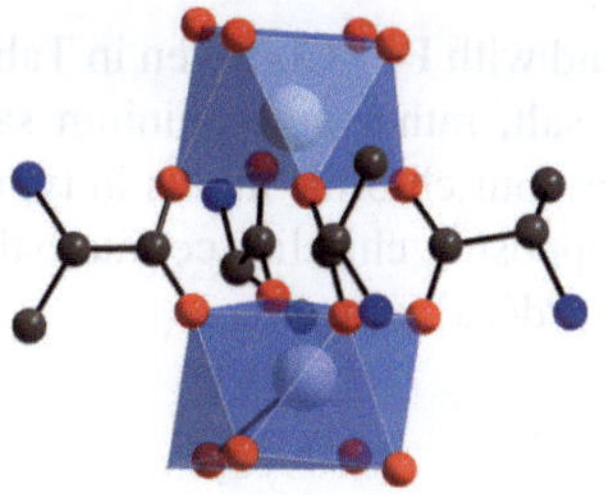

Fig. 4.11 *Dimeric unit in the rare earth element perchlorate salts*. Four L-alaninate zwitterions connect neighboring polyhedra; the ligands on *top* and *bottom* are water molecules. Perchlorate anions are located in the interstices (not shown). Hydrogen atoms omitted for clarity

The species comprises dimeric units of two polyhedra connected by four L-alanine zwitterions and four water molecules as ligand to each cation (Fig. 4.11). The only noteworthy difference between the Nd phase and the Tb phase was the fact that the ClO_4 tetrahedra, located in the interstices between the polyhedral units, which were found to be disordered at ambient conditions, could be refined in ordered positions at 100K.

The remarkable aspect is that other rare earth element perchlorate alanine salts, namely, of Gd, La, and Er, were reported with similar unit cell dimensions, but centrosymmetric space group *P*-1. In the case of the gadolinium salt (Wang et al. 2003), no indication was given whether the amino acid is D or L. The authors only speak of "centrosymmetric units." However, when investigating the CIF as deposited in the Cambridge Structural Database (Allen 2002), it turns out that half of the alanine molecules have been refined as disordered. More interestingly, only the D-enantiomers were refined as disorder, the L-forms are well defined. It therefore seems very likely that the structure has been mistakenly assigned as centrosymmetric (after all, most atoms are in centrosymmetric arrangement; only the methyl groups violate this symmetry), a mistake sometimes found in triclinic, nearly centrosymmetric structures. However, when dealing with chiral molecules, this symmetry is ruled out right away.[1]

In the case of the erbium salt, the authors (Hu et al. 1995) even speak of DL-alanine, but the same disorder as in above case is found, and the agreement of the unit cell parameters also confirms that the structure should be $P1$. If the authors really used a racemic mixture for synthesis, it is possible that they have grown a racemic twin, but in this case all methyl groups should show disorder. Without access to the raw data, this question cannot be answered, however.

Nevertheless, a true DL-salt of $Er(ClO_4)^3$ is reported (Glowiak et al. 1996), which shows the same motif, but with distinct D- and L-zwitterions. The topology of the dimeric units corresponds well to those in the L-alanine salts, but due to the different orientation of the methyl side chains, the packing is different, which shows in the clearly different unit cell (Table 4.4).

The same dimeric unit was also reported by Dao et al. (1988) for thc holmium chloride L-alanine salt. Likewise, the anions are located in the interstices; thus, the unit cell dimensions differ.

The DL-alanine compound with $Pd^{2+}Cl_4$ given in Table 4.4 (Laligant et al. 1994) is not actually an alanine salt, rather an alaninium salt: The bivalent palladium cation is [4]-coordinated by four chloride atoms in typical square geometry, while two alaninium cations of opposing chirality counterbalance the charge, connected via hydrogen bonds to the $[PdCl_4]$ unit.

4.4 Valine, Leucine, and Isoleucine

Of the small, nonpolar amino acids valine, leucine, and isoleucine, salts with neutral compounds have been reported (Table 4.5), although not as many as for alanine. In fact, the number of published structures decreases with the size of the nonpolar amino acid (Gly > Ala > Val > Leu and Ile).

Among the salts of L-valine, the dimeric units in the europium chloride (Zhang et al. 2007) correspond to those found in the rare earth element perchlorates (and chlorides) of alanine (as shown in Fig. 4.11). Chloride anions (and several crystal water molecules) in the interstices connect the units to each other. The same motif is also found in the DL-valine salts of samarium perchlorate (Ma et al. 1993b) and terbium chloride (Zhang et al. 2007).

[1] To be sure, the structures of the Nd and Tb salts which were correctly reported with space groups $P\text{-}1$ were checked with the program PLATON (Spek 2009), which confirmed that no symmetry has been missed.

Table 4.5 *Salts of amino acids and neutral compounds, part III:* space groups and unit cell parameters of compounds of neutral compounds with valine, leucine, and isoleucine

	s.g.	*a*	*b*	*c*	*α*	*β*	*γ*	Conn.	Reference
L-Valine									
L-Val_6 $Cr_3O(NO_3)_7 \cdot 5H_2O$	$P2_1$	13.104	21.460	12.218	90	110.99	90	0^3–0^3	Kato et al. (1991)
L-Val_2 $Pb(NO_3)_2 \cdot 2H_2O$	$P2_12_12_1$	13.609	26.75	5.416	90	90	90	0–1	Burford et al. (2004)
L-Val_5 $Pb_2(ClO_4)_4 \cdot 2H_2O$	$P2_1$	12.075	25.569	15.854	90	90.87	90	1–1	Gasque et al. (2008)
L-Val SbF_3	$P2_1$	5.892	6.239	12.345	90	95.55	90	0–1	Zemnukhova et al. (2005)
"L-Val_2 $NiCl_2$"	[a]	7.594	14.796	29.745	[a]	[a]	[a]	[a]	Sangeetha et al. (2012)
L-Val_2 $EuCl_3 \cdot xH_2O$	*P*1	14.067	18.405	19.234	87.74	82.32	82.51	0–0^2	Zhang et al. (2007)
L-Val_5 $Er_4Cl_8(OH)_4 \cdot 12H_2O$	*P*1	11.631	12.692	13.115	82.47	65.76	87.65	N/A	Wang et al. (2001)
L-Val_2 $AgCo(NO_2)_2$	*C*2	18.851	7.019	6.279	90	102.01	90	2–2	Prelesnik et al. (1990)
L-Val_2 $KCo^{III}CO_3 \cdot 2H_2O$	$P2_12_12_1$	21.41	14.68	5.76	90	90	90	1–2	Price and Russell (1981)
DL-Valine									
DL-Val_2 $CaCl_2 \cdot 2H_2O$	*Pcca*	24.161	10.018	7.742	90	90	90	1–1	Glowiak and Ciunik (1978b)
DL-Val $ZnCl_2$	*C*2/*c*	19.997	6.226	13.503	90	103.02	90	0–0	Subha Nandhini et al. (2001a)
DL-Val_2 $Te^{IV}I_6$	$P2_12_12_1$	18.78	19.88	9.57	90	90	90	N/A	Dauter et al. (1977)
DL-Val_2 Nb^VF_5O	$P2_1/c$	9.802	19.421	10.198	90	116.19	90	0–0	Pushilin et al. (2007)
DL-Val_2 $Sm(ClO_4)_3 \cdot 4H_2O$	*Pnn*2	14.433	20.166	14.489	90	90	90	0–0^2	Ma et al. (1993b)
DL-Val_2 $TbCl_3 \cdot 4H_2O$	*C*2/*c*	26.438	11.356	16.310	90	115.42	90	0–0^2	Zhang et al. (2007)
L-Leucine									
L-Leu $Pb(NO_3)_2$	*C*2	19.864	5.230	19.621	90	98.644	90	1–1	Gasque et al. (2008)
DL-Leucine									
DL-Leu_2 $Te^{IV}I_6$	*P*-1	8.31	12.47	8.25	114.3	110.2	113.2	N/A	Dauter et al. (1977)
L-Isoleucine									
L-Ile_2 $Pb(NO_3)_2 \cdot 2H_2O$	$P2_12_12_1$	5.456	13.656	30.572	90	90	90	1–1	Gasque et al. (2008)
L-Ile_2 $Eu(ClO_4)_3 \cdot 4H_2O$	*C*2	21.704	10.423	15.121	90	120.25	90	0–0^2	Legendziewicz et al. (1999)
DL-Isoleucine									
DL- Ile_2 $Nd(ClO_4)_3 \cdot 4H_2O$	*P*-1	10.483	11.495	13.878	92.16	109.22	110.17	0–0^2	Legendziewicz et al. (1999)

[a]No structure or space group given, probably a different phase. See text for comment

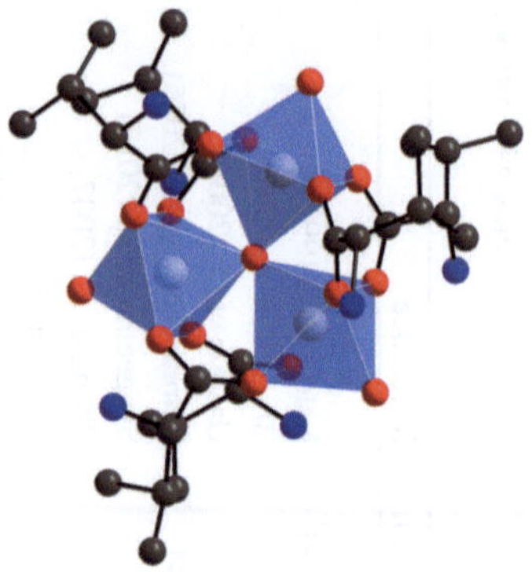

Fig. 4.12 *Trimeric unit in the L-valine Cr_3O nitrate salt*. Each chromium polyhedron is connected to its neighbors via the central oxygen as well as two L-valine molecules. The unit is charge-counterbalanced by nitrate anions (not shown). Hydrogen atoms omitted for clarity

Another noteworthy example of a structure of isolated units is that of oxo-hexakis (L-valine)-tris(aqua-chromium) heptanitrate pentahydrate (Kato et al. 1991). This salt is unique insofar as this topology has not been found in any amino acid salt so far. It consists of a dimeric unit of three Cr^{3+} polyhedra sharing one common corner. Furthermore, each polyhedron is connected to both neighbors via two valine zwitterions; the remaining corner is occupied by a water molecule (Fig. 4.12). Seven charge-counterbalancing nitrate anions as well as additional water molecules in the interstices connect these trimeric units via hydrogen bonds.

One more structure comprising isolated units is that of dichloro-bis(DL-valine)-zinc (Subha Nandhini et al. 2001a). The structure of this unit is not unique. On the contrary, it was found in many amino acid salts, such as the respective DL-alanine salt (Subha Nandhini et al. 2002a) and dichloro-bis(glycine)-zinc dihydrate (Fleck et al. 2008a), both L- and DL-proline zinc chloride (Yukawa et al. 1985; Lutz and Bakker 2003), and the respective betaine and sarcosine salts (Chen and Mak 1991c; Subha Nandhini et al. 2001b).

It is characterized by a tetrahedrally coordinated Zn^{2+} cation, with two chloride anions and two monodentate (O) amino acid zwitterions as ligands (Fig. 4.13).

In addition to the structures of isolated units, several chain structures were reported. Noteworthy are the lead nitrate salts of L-valine, L-leucine, and L-isoleucine (the first was reported by Burford et al. 2004; the structure of the latter pair was published by Gasque et al. 2008). In all three salts, chains were found: In the L-valine and L-isoleucine salts, these chains are topologically identical (and apart from the longer side chain of the amino acid, the species are isostructural; compare unit cell parameters in Table 4.5). In the case of the L-leucine salt, there are double chains, which can be regarded as a connection of two of the single chain as in the L-valine and L-isoleucine salts (Fig. 4.14). It is remarkable that the layers in nitrato-L-alanine-sodium can be considered as a structure consisting of an infinite number of chains connected to each other (compare with Fig. 4.10).

Among the species listed in Table 4.5, there is one salt which actually does not belong to salts of neutral compounds and amino acids, namely, the niobium valine compound reported by Pushilin et al. (2007). This salt comprises $[NbF_5O]^{2-}$ octahedra, which are charge-counterbalanced by two valinium cations, connected by hydrogen bonds only.

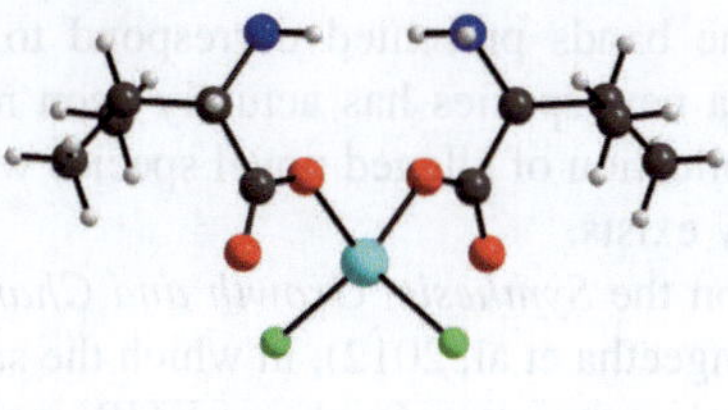

Fig. 4.13 *Isolated unit in the dichloro-bis(DL-valine)-zinc*. This unit appears frequently, e.g., in the respective glycine and alanine salt. In this case, the point symmetry of the unit is 2 (as in the glycine analog)

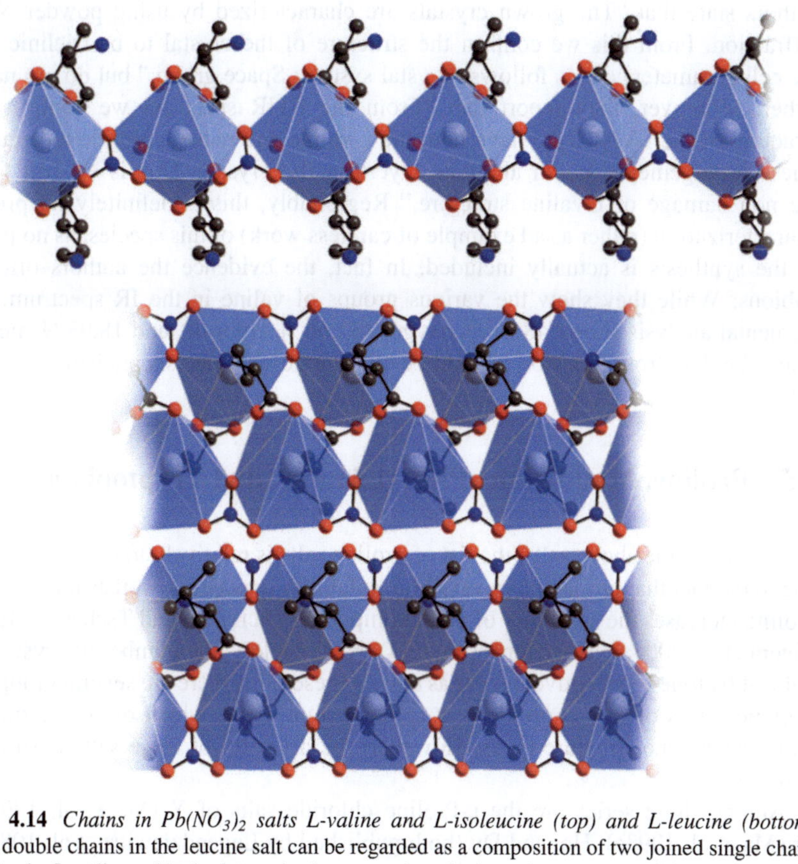

Fig. 4.14 *Chains in $Pb(NO_3)_2$ salts L-valine and L-isoleucine (top) and L-leucine (bottom)*. The double chains in the leucine salt can be regarded as a composition of two joined single chains (as in the L-valine salt), the layers in the respective alanine salt as a composition of an infinite number of chains (see Fig. 4.10). Hydrogen atoms are omitted for clarity

In addition to the crystals listed in Table 4.5, Maadeswaran and Chandrasekaran (2011) reported the synthesis, growth, and characterization of L-valine cadmium chloride monohydrate, although they gave no indication if the synthesis really yielded this material. No crystal structure or even unit cell parameters were given (although large single crystals were grown), and the powder pattern published appears to be of a much simpler phase. The FTIR spectrum shown contains no proof of a reaction (the bands presented correspond to L-valine only), and no evidence is given that a new species has actually been reported. This paper is a sad instance for the publication of alleged novel species without any proof that the supposed phase actually exists.

There is one paper on the *Synthesis, Growth and Characterization of L-Valine Nickel (II)Chloride* (Sangeetha et al. 2012), in which the said phase is reported and characterized by several techniques (such as FTIR and UV–vis spectroscopy, thermal data, etc.). However, the authors did not determine the crystal structure. Even more amazing, they published only the unit cell lengths, but did not give angles, although they state that the crystal is triclinic. In the Conclusion, the authors state that "The grown crystals are characterized by using powder XRD diffraction. From this we confirm the structure of the crystal to be triclinic and the cell parameters are as follows. Crystal system: Space group:" but do not name either. Moreover, they report that "From the FTIR spectrum we confirm the structure of the LVNiCL to have both the L-valine and nickel chloride molecules. These arrangements are in alternate layers in the crystal. This is evident from the non damage of L-valine structure." Regrettably, this is definitely no proper characterization (rather a sad example of careless work) of this species, as no proof of the synthesis is actually included. In fact, the evidence the authors offer is dubious: While they show the various groups of valine in the IR spectrum, the elemental analysis shows 48.41 % oxygen, 32.66 % chloride, and 18.93 % nickel. Thus, the data from this work should be considered with due skepticism.

4.5 Proline, Methionine, Phenylalanine, and Tryptophan

In contrast to the above salts, the list of proline salts is relatively long. This may be due to the fact that proline has a very high solubility in water as well as the fact that proline increases the solubility of other compounds (Schobert and Tschesche 1978; Civera et al. 2005; and references therein). Nevertheless, the number of crystals in Table 4.6 alone is deceptive insofar as among these salts, there are several groups of isotypic series of rare earth element (plus yttrium) salts, which of course do not represent unique structure types. However, the diversity of these salts is surprisingly high.

Among these series are the L-Proline chloride salts of Y (Ma et al. 1994b), Er (Ma et al. 1993c), Ho, and Dy (both published by Legendziewicz et al. 1989c), which crystallize in the monoclinic space group $P2_1$. These salts comprise one-dimensional chains, which are characterized by isolated [MeO_8]-polyhedra, connected via one bridging (O,O′) L-proline zwitterion. Each polyhedron has another monodentate (O) L-proline molecule as ligand; the remaining ligands are water molecules (Fig. 4.15, top). Chloride anions are located in the interstices between the chains.

Another type of rare earth element salts of L-proline can also be described as chain structures, namely, the L-proline perchlorate salts of Sm, Eu, Gd, and Er (Table 4.6). However, this chain is topologically different: Two bridging L-proline zwitterions

Table 4.6 *Salts of amino acids and neutral compounds, part IV:* space groups and unit cell parameters of compounds of neutral compounds with proline

	s.g.	*a*	*b*	*c*	*α*	*β*	*γ*	Conn.	Reference
L-Proline									
L-Pro $LiCl \cdot H_2O$	$P2_1$	7.680	5.074	10.336	90	105.86	90	0–1	Uma Devi et al. (2009)
L-Pro $LiBr \cdot H_2O$	$P2_1$	8.035	5.147	10.622	90	104.25	90		Shkir et al. (2014) SUBM. JCG
L-Pro$_2$ $ZnCl_2$	$P2_12_12_1$	6.599	13.546	16.275	90	90	90	0–0	Yukawa et al. (1985)
L-Pro$_2$ $MnCl_2 \cdot H_2O$	$P2_12_12_1$	7.112	10.196	13.249	90	90	90	1–1	Rzaczynska et al. (1997)
L-Pro$_2$ $CdCl_2 \cdot H_2O$	$P2_12_12_1$	7.277	9.988	13.520	90	90	90	1–1	Yukawa et al. (1983)
L-Pro $CuCl \cdot H_2O$	$P2_1$	9.509	5.206	8.189	90	109.40	90	0–1	Yukawa (1992)
L-Pro Hg_2Cl_4	$P1$	7.274	9.447	10.477	108.62	107.26	97.35	0–1	Kalaiselvi et al. (2008)
L-Pro$_2$ $YCl_3 \cdot 5H_2O$	$P2_1$	8.268	10.980	11.965	90	106.97	90	0–1	Ma et al. (1994b)
L-Pro$_2$ $ErCl_3 \cdot 5H_2O$	$P2_1$	8.294	10.981	11.934	90	107.04	90	0–1	Ma et al. (1993c)
L-Pro$_2$ $HoCl_3 \cdot 5H_2O$	$P2_1$	11.968	11.030	8.309	90	106.97	90	0–1	Legendziewicz et al. (1989c)
L-Pro$_2$ $DyCl_3 \cdot 5H_2O$	$P2_1$	11.968	11.038	8.302	90	107.00	90	0–1	Legendziewicz et al. (1989c)
L-Pro $NdCl_6 \cdot 11H_2O$	$C2$	25.076	8.132	18.361	90	120.68	90	$0–0^2$	Glowiak et al. (1991b)
L-Pro$_2$ $MnBr_2 \cdot MeOH$	$P2_1$	8.286	8.256	9.387	90	92.17	90	0–1	Lamberts and Englert (2012)
L-Pro$_2$ $MnI_2 \cdot 3H_2O$	$P2_12_12_1$	8.126	11.867	13.727	90	90	90	0–1	Lamberts and Englert (2012)
L-Pro$_2$ $Eu(ClO_4)_3 \cdot 6H_2O$	$P42_12$	16.176	16.176	10.227	90	90	90	$0–0^2$	Huskowska et al. (1998)
L-Pro$_3$ $Sm(ClO_4)_3 \cdot 3H_2O$	$P1$	13.052	13.725	9.929	110.34	100.73	109.62	0–1	Torres et al. (2002)
L-Pro$_3$ $Eu(ClO_4)_3 \cdot 3H_2O$	$P1$	9.924	13.033	13.791	62.36	68.08	79.20	0–1	Huskowska et al. (1998)
L-Pro$_3$ $Gd(ClO_4)_3 \cdot 3H_2O$	$P1$	9.906	13.052	13.703	109.63	110.31	100.72	0–1	Pan et al. (1997)
L-Pro$_3$ $Er(ClO_4)_3 \cdot 3H_2O$	$P1$	13.003	13.639	9.861	110.26	100.94	109.35	0–1	Hu et al. (1993)
L-Pro$_3$ $Pr(ClO_4)_3 \cdot 2H_2O$	$P2_1$	9.879	21.883	13.393	90	91.23	90	0–1	Wang et al. (1993)
L-Pro$_3$ $Nd(ClO_4)_3 \cdot 2H_2O$	$P2_1$	13.379	21.912	9.870	90	91.28	90	0–1	Legendziewicz et al. (1988)
L-Pro$_3$ $Y_2(ClO_4)_3(OH)_2 \cdot 7.5H_2O$	$R3$	12.786	12.786	12.786	100.22	100.22	100.22	$0^4–0^4$	Chen et al. (2000)
L-Pro$_3$ $Gd_2(ClO_4)_3(OH)_2 \cdot 7.5H_2O$	$R3$	12.895	12.895	12.895	100.66	100.66	100.66	$0^4–0^4$	Ma et al. (2000)
L-Pro$_3$ $Er_2(ClO_4)_3(OH)_2 \cdot 7.5H_2O$	$R3$	12.781	12.781	12.781	100.07	100.07	100.07	$0^4–0^4$	Hu et al. (2003)
DL-Proline									
DL-Pro$_2$ $ZnCl_2$	$C2/c$	18.671	5.943	13.396	90	104.64	90	0–0	Lutz and Bakker (2003)
DL-Pro$_2$ $MnCl_2 \cdot 2H_2O$	$P2_1/c$	9.375	9.195	10.122	90	106.38	90	0–0	Glowiak and Ciunik (1977)
DL-Pro$_2$ $MnBr_2 \cdot 2H_2O$	$P2_1/c$	9.316	9.132	10.059	90	106.91	90	0–0	Lamberts and Englert (2012)
DL-Pro$_2$ $MnI_2 \cdot 2H_2O$	$P2_1/c$	9.664	9.403	10.023	90	104.45	90	0–0	Lamberts and Englert (2012)
DL-Pro $MnSO_4 \cdot 4H_2O$	Cc	8.165	16.857	9.222	90	92.02	90	N/A	Ciunik and Glowiak (1981)

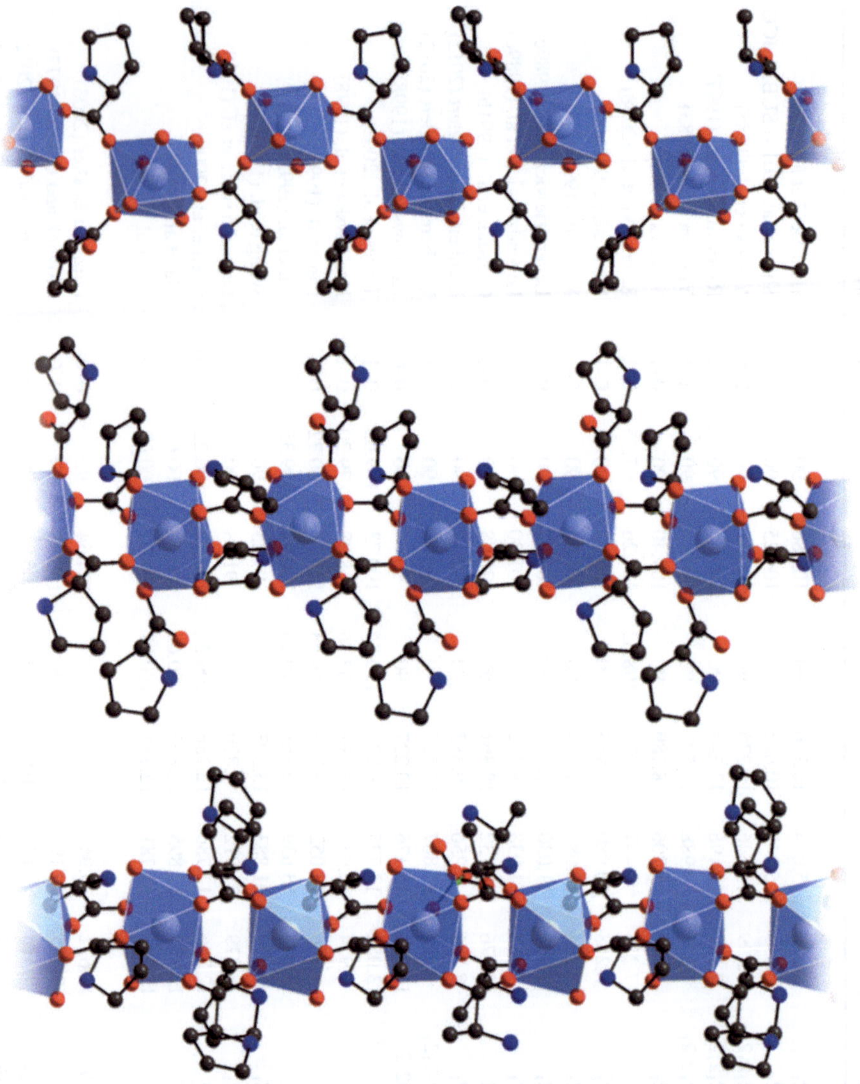

Fig. 4.15 *Chains in L-proline rare earth element salts. Top*: The chains in the chloride salts show a similar topology to that of some chains of compounds of cations with amino acid anions, where one amino acid is a bridging ligand, the other one is not. *Middle*: The chains in the perchlorate salts of Sm, Eu, Gd, and Er show different topology, where polyhedra are connected to each other via two L-proline units each. *Bottom*: Another type of perchlorate salt of Nd and Pr shows a different topology again. Alternatively, two and four proline molecules connect the neighboring cations. In all cases the anions are located in the interstices (not shown). Hydrogen atoms are omitted for clarity

connect neighboring cations, one more L-proline acts as monodentate ligand and only one water molecule is part of the coordination polyhedron (Fig. 4.15, middle). The ratio of L-proline:cation:water is 3:1:3 (in contrast to 2:1:5 in the above described chlorides).

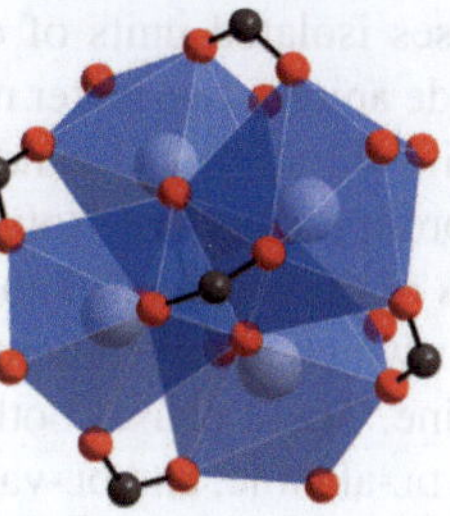

Fig. 4.16 *Tetrameric unit in L-proline rare earth elements perchlorate hydroxy salts.* The polyhedra share common corners (OH groups) and are additionally connected via L-proline zwitterions (of which only the carboxylate group is shown). Hydrogen atoms are omitted for clarity

A third chain type was found in perchlorate salts of praseodymium and neodymium (with a ratio of L-proline:cation:water 3:1:2), where adjacent cations are connected alternatively by two and four L-proline zwitterions each (Fig. 4.15, bottom).

In contrast to the chain structures, there is one L-proline perchlorate salt of Eu (published by Huskowska et al. 1998) which shows the same dimeric unit motif found in the glycine REE perchlorates (and other REE amino acid salts): two [8]-coordinate europium cations connected via four L-proline zwitterions, with the perchlorate anions in the interstices. However, in the case of this salt, the symmetry is higher: The chain (and thus the whole crystal structure) possesses tetragonal symmetry (space group $P42_12$).

Moreover, there is one Nd perchlorate salt of L-proline (Glowiak et al. 1991b) with a ratio of L-proline:neodymium:water of 1:1:11. In this structure, there are also dimeric units as in the above salt, but topologically different: Only two L-prolines connect the neighboring cations in bridging (O,O) mode; the remaining ligands are five water molecules and one chloride anion.

Finally, there is a series of perchlorate hydroxy salts of yttrium (Chen et al. 2000), gadolinium (Ma et al. 2000), and erbium (Hu et al. 2003) which crystallize in another structure type: Again, isomeric units are found, although in this case these units are of a tetrameric nature, with four cations in a tetrahedral arrangement, connected via (OH) groups. The L-proline zwitterions act as further bridging (O,O′) ligands; the remaining corners are occupied by water molecules, resulting in $[\text{L-Pro}_6(H_2O)_8Me_4(OH)_4]^{6+}$ units (Fig. 4.16). These units are connected via perchlorate anions in the interstices, where water molecules provide additional connection via hydrogen bonds.

Apart from the large number of rare earth element salts, other cations have been reported as well. Among these is the isotypic pair of bis-chloro-bis(L-prolinato)-manganese monohydrate (first published by Rzaczynska et al. 1997, later refined by Lamberts and Englert 2012) and its cadmium analog (Yukawa et al. 1983). Another isotypic series encompasses the manganese halogenide DL-proline salts of Cl, Br, and I (Lamberts and Englert 2012; the chloride was first reported by Glowiak and

Ciunik 1977), which comprises isolated units of centrosymmetric [6]-coordinate Mn^{2+} cations, with two bromide anions, two water molecules, and two monodentate proline zwitterions (of opposite chirality) as ligands. Lamberts and Englert (2012) noted that although the DL-proline salts are isotypic, the L-proline salts are not (compare unit cell parameters of these salts in Table 4.6).

Another pair of compounds which deserve attention is the pair of salts of $ZnCl_2$, both of L-proline and DL-proline. These phases both comprise the same unit that is found for the salts of glycine, DL-alanine, and DL-valine, i.e., a tetrahedron with two chloride anions and two monodentate amino acid zwitterions. The L-proline salt was first characterized by Yukawa et al. (1985), later refined by Hoffmuller et al. (1999); Anandha Babu and Ramasamy (2008) published optical and dielectric data on this species.

Uma Devi et al. (2009) reported the structure of L-proline aqua lithium chloride, grew single crystals of a length of several centimeters, and determined several physical parameters of this phase. Among these was a powder SHG test, in which the authors find nonlinear optical activity – a phenomenon encountered in many (non-centrosymmetric) amino acid salts. Nevertheless, the trend of authors heralding the discovery of a new "NLO material," as the authors of the above papers proposed, is actually an exaggeration, as a simple SHG test does not prove the fact that it can be utilized as NLO material. Many more details need to be confirmed before a crystalline species can be pronounced as NLO material (e.g., the refractive indices need to be determined, phase-matching conditions calculated). Thus, authors should be more careful before publishing such a statement.

Of the other large, nonpolar amino acids, not many salts with neutral compounds have been synthesized, as is expected from their chemical behavior (Table 4.7). Among these amino acids, methionine has been used most often and has a unique role: This amino acid is able of coordinating with its sulfur atom, which it does in all structures that have been investigated. Even more interesting, the amino acid exists in truly neutral state (the acid group retains its proton, while the amino group is not protonated), i.e., methionine does not exist as zwitterion. Thus, the methionine compounds listed in Table 4.7 are actually not amino acid salts. In these compounds, the amino acids act as bidentate (N,S) ligands, as shown for dichloro-(L-methionine)-platinum (Freeman and Golomb 1970; the respective D-form was reported by Llorca et al. 2001; Fig. 4.17). The same situation was found in the other platinum and palladium compounds, of both L- and DL-methionine.

The coordination mode in the Co^{3+} salts listed in Table 4.7 is similar, although these phases do actually not represent salts of neutral amino acids: In these cases, the amino acid is deprotonated, so these are actually methioninate salts. In this state, the amino acid acts as tridentate (O,N,S) ligand.[2]

[2] In fact, only in other compounds (i.e., with other organic groups) does methionine act as a "regular" amino acid ligand, for example, in the DL-methionine salt of methyl-mercury (Wong et al. 1977), where bidentate (O, N) and bridging (O, O′) modes are assumed.

Table 4.7 *Salts of amino acids and neutral compounds, part V:* space groups and unit cell parameters of compounds of neutral compounds with methionine, phenylalanine, and tryptophan

	s.g.	*a*	*b*	*c*	*α*	*β*	*γ*	Conn.	Reference
L-Methionine									
L-$Met_2\ Co^{III}Cl$	$P2_12_12$	8.525	9.381	10.225	90	90	90	0–0	Hambley (1988)
L-Met $PtCl_2$	$P2_1$	7.578	10.213	13.563	90	98.77	90	0–0	Wilson et al. (1992)
L-Met $PtCl_2$	$P1$	7.34	8.91	8.39	74.47	78.13	86.40	0–0	Freeman and Golomb (1970)
L-Met $PdCl_2$	$P2_1$	8.272	6.956	8.978	90	105.17	90	0–0	Caubet et al. (1992)
L-$Met_2\ Co^{III}(ClO_4)\cdot H_2O$	$P2_12_12_1$	8.320	10.062	21.495	90	90	90	0–0	Hambley (1988)
L-$Met_2\ NaCo^{III}(ClO_4)_2$	$P22_12_1$	5.495	11.189	16.604	90	90	90	0–0	Hambley (1988)
DL-Methionine									
DL-Met $PtCl_2$	$P2_1/c$	7.52	9.87	15.85	90	118.35	90	0–0	Freeman and Golomb (1970)
DL-Met $PdCl_2$	$P2_1/n$	6.920	13.788	10.531	90	94.69	90	0–0	Warren et al. (1970)
DL-$Met_2\ Hg(ClO_4)_2\cdot 2H_2O$	$P2_1/c$	11.549	16.399	12.964	90	110.21	90	N/A	Carty and Taylor (1976)
L-Phenylalanine									
L-$Phe_2\ Pb(NO_3)_2\cdot 2H_2O$	$P2_12_12_1$	5.385	13.560	34.235	90	90	90	0–1	Bernès and Gasque, (2008)
DL- Phenylalanine									
DL-$Phe_2\ Mo^{II}I_2\cdot 6H_2O$	P-1	11.969	13.746	9.457	104.90	107.49	66.47	0^2–0^2	Apfelbaum-Tibika and Bino (1984)
DL-$Phe_2\ Nd(ClO_4)_3\cdot 7H_2O$	$C2/c$	10.835	22.27	25.85	90	93.10	90	0^2–0^2	Jin et al. (1994b)
L-Tryptophan									
L-$Trp_2\ Hg^{I}Cl_3$	$P2_1$	16.688	5.626	14.022	90	108.17	90	1–1	Book et al. (1981)

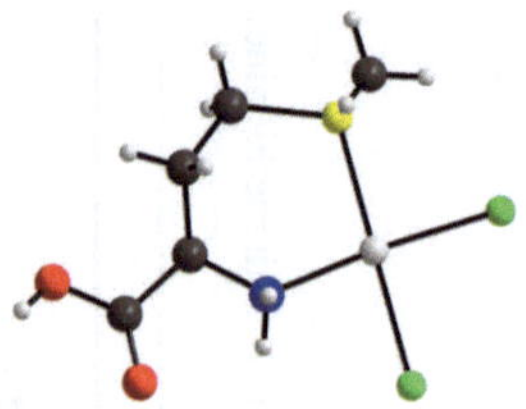

Fig. 4.17 *Units in dichloro-(L-methionine)-platinum.* It is noteworthy that in this salt, the amino acid acts as bidentate (N,S) ligand, as it is not in its zwitterionic state

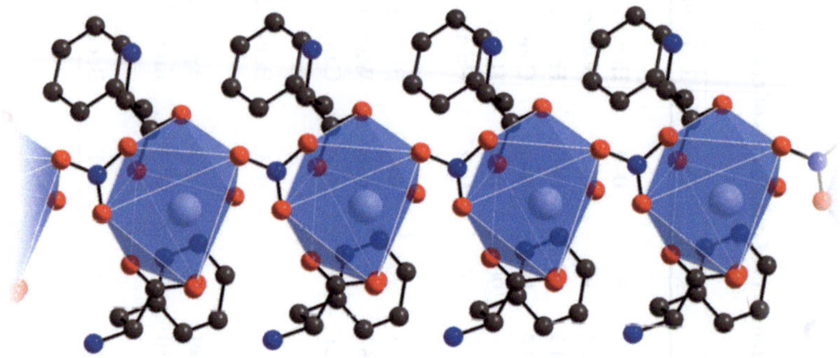

Fig. 4.18 *Chains in diaqua-(nitrato)-bis(L-phenylalaninato)-lead nitrate.* Another, non-coordinating nitrate anion is located between the chains (not shown). Hydrogen atoms omitted for clarity

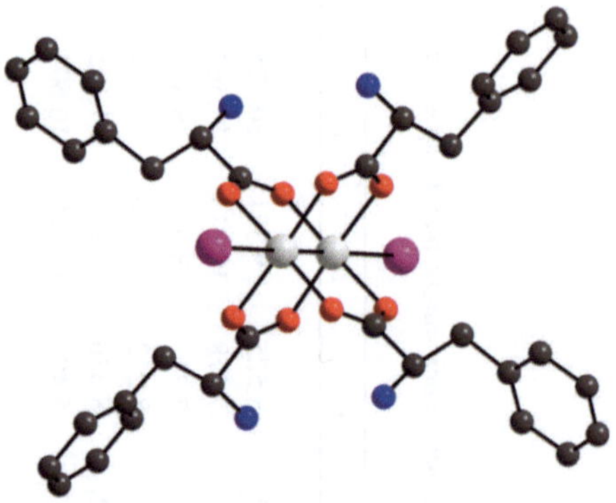

Fig. 4.19 *Dimeric units in the DL-phenylalanine molybdenum iodide salt.* This dimeric unit of two opposing pyramids is a motif found in other amino acid salts as well, as, e.g., β-alanine salts of copper and rhodium or betaine copper chloride. Hydrogen atoms omitted for clarity

Salts of phenylalanine and tryptophan and neutral compounds are rare. One instance is the salt of L-phenylalanine and $Pb(NO_3)_2$, where chains of Pb^{2+} polyhedra are connected via nitrate anions, with the amino acid molecules as bidentate (O,O′) ligands (Fig. 4.18).

Another example is the crystal of di-iodo-tetra(DL-phenylalaninato)-di-molybdenum(ii) di-iodide hexahydrate (Apfelbaum-Tibika and Bino 1984, Fig. 4.19). In this species, dimeric units occur, where four amino acid zwitterions act as bridging (O,O′) ligands between neighboring cations, with the iodide anion as an additional ligand. This motif has been found in other amino acid salts as well, as, for instance, several examples of β-alanine as well as betaine.

The one Trp compound, a salt of Hg^+ and chloride (Book et al. 1981), is another instance of a species where the amino acid is not in coordination. Infinite trichloromercury chains are charge-counterbalanced by non-coordinating tryptophanium cations. Other salts of tryptophanium with additional organic groups have been reported as well.

4.6 Serine, Threonine, and Cysteine

The two small, polar amino acids serine and threonine are soluble in water and possess an additional oxygen atom in the side chain, which can act as a ligand atom. Nevertheless, only few compounds of these amino acids and neutral salts have been reported, as crystals of the pure amino acids crystallize much more readily than the salts (as has been found for alanine). A few instances are listed in Table 4.8.

One of these, (L-serinato)-hexaaqua-holmium trichloride (Glowiak and Dao 1993), is a chain structure with a ratio of cation:amino acid of 1:1. In this species, the coordination of serine via all three of its oxygen atoms can be clearly seen: The carboxylate group represents a bridging (O,O′) ligand, as frequently found in other amino acid salts. Additionally, one of the cations is bidentate (O,O′) coordinated, in this case with one carboxylate oxygen as well as the hydroxyl oxygen atom (Fig. 4.20).

A chain with identical topology is found in (L-threonine)-hexaaqua-praseodymium trichloride (Rzaczynska et al. 1999; note the similarity in the unit cell dimensions which shows that the structure is nearly isostructural, apart from the difference in the amino acid side chain), as well as in the pentahydrate salts of Ho, Eu, and Yb (compare Table 4.8).

Zhang and coworkers have conducted a detailed study of a series of heptanuclear octahedral polyhedra $\{LnNi_6\}$ with glycine, L-alanine, and L-threonine as ligands in aqueous media (Zhang et al. 2004). Within this family, seven phases with complicated compositions were found, most of which could be grown in crystals good enough for XRD experiments. Of the salt given in Table 4.8, L-Thr_{12} $Ni_6La(ClO_4)_3 \cdot 15H_2O$, which is shown as an example, only the unit cell dimensions could be determined, which proved to be cubic, a unique result among amino acid salts. Because of weak intensities, no space group or structure could be elucidated.

In contrast to the relatively rich diversity of cysteinate salts of cations, the number of cysteine salts with neutral compounds is very low. Two mercury salts have been reported (Table 4.8), but no structure was determined. More interesting is the only racemic cysteine salt, tetrakis(DL-cysteine)-octa-aqua-di-terbium dichloride (Xu et al. 2006), which shows the common motif of dimeric units (as found in respective salts of L- and DL-alanine, DL-valine, and L-proline). Although the unit is topologically identical, the packing is different, as expressed by the unique symmetry *Fddd*.

Table 4.8 *Salts of amino acids and neutral compounds, part VI:* space groups and unit cell parameters of compounds of neutral compounds with threonine, serine, methionine, and tyrosine

	s.g.	*a*	*b*	*c*	*α*	*β*	*γ*	Conn.	Reference
L-Serine									
L-Ser $CaPO_4 \cdot 2H_2O$	$P2_1$	5.534	12.759	5.40	90	104.77	90	0–3	Suga and Okabe (1996)
L-Ser $HoCl_3 \cdot 5H_2O$	$P2_1$	8.909	19.847	7.602	90	90.51	90	0–1	Glowiak and Dao (1993)
L-Threonine									
L-Thr $PrCl_3 \cdot 3H_2O$	$P2_1$	9.196	19.877	9.316	90	110.12	90	0–1	Rzaczynska et al. (1999)
L-Thr $HoCl_3 \cdot 5H_2O$	$P2_12_12_1$	10.508	7.395	18.708	90	90	90	0–1	Glowiak and Ngoan (1988)
L-Thr $EuCl_3 \cdot 5H_2O$	$P2_12_12_1$	10.578	7.417	18.815	90	90	90	0–1	Rzaczynska et al. (1999)
L-Thr $YbCl_3 \cdot 5H_2O$	$P2_12_12_1$	10.441	7.356	18.646	90	90	90	0–1	Rzaczynska et al. (1999)
L-Thr_8 $Cu_6Na(ClO_4)_5 \cdot 7H_2O$	$P2_1$	11.682	23.216	13.080	90	94.78	90	0^7–0^7	Xiang et al. (2005)
L-Thr_{12} $Ni_6La(ClO_4)_3 \cdot 15H_2O$	*cubic*	33.696	33.696	33.696	90	90	90	0^7–0^7	Zhang et al. (2004)
L-Cysteine									
L-Cys $HgCl_2$	$P2_12_12_1$	14.699	8.017	7.025	90	90	90	N/A	Taylor and Carty (1977)
L-Cys $HgCl_2 \cdot 0.5H_2O$	*C*2	24.181	5.093	12.006	90	118.83	90	N/A	Taylor and Carty (1977)
L-Cys $Nd(ClO_4)_2 \cdot xH_2O$	$P2_12_12_1$	13.674	18.485	19.334	90	90	90	0–2	Jin et al. (1994b)
DL-Cysteine									
DL-Cys_2 $TbCl \cdot 4H_2O$	*Fddd*	11.462	21.214	35.636	90	90	90	0–0^2	Xu et al. (2006)

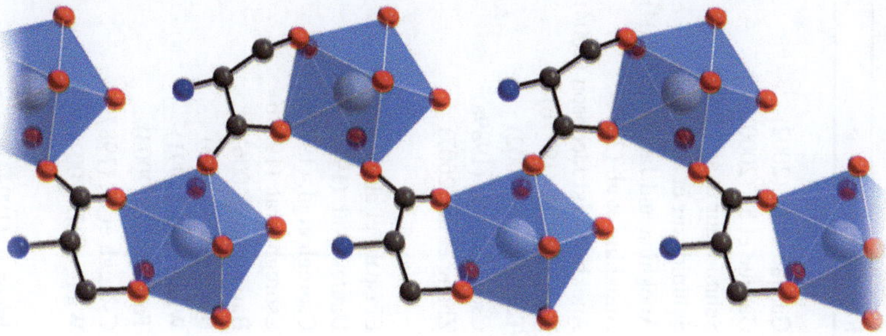

Fig. 4.20 *Chains in (L-serinato)-hexaaqua-holmium trichloride*. This chain reminds of other chains in salts with a cation:amino acid ratio of 1:1 (like glycinato-diaqua-dichloro-manganese). However, in the case of serine, one more oxygen atom can act as a ligand. Non-coordinating chloride anions are located in the interstices (not shown). Hydrogen atoms omitted for clarity

4.7 Asparagine, Glutamine, Aspartic Acid, and Glutamic Acid

The acidic amino acids aspartic acid and glutamic acid have been found to react readily to form compounds with neutral salts, in contrast to their amide counterparts asparagine and glutamine. However, these salts are not truly compounds of amino acids with neutral salts, as the acids are actually monovalent anions (with both acid groups deprotonated and the amino group protonated), i.e., hydrogen aspartates and hydrogen glutamates. Nevertheless, we have included the species here as they compose amino acids as well as cations plus anion. As found for compounds of these acids with cations (see Chap. 3), the molecules with acid groups on either end act as bridging ligands to connect the polyhedra to high-dimensional structures. Thus, layer and framework arrangements are frequent (Table 4.9).

One example is the structure of (L-aspartato)-nitrato-cadmium (Gasque et al. 2002). The Cd^{2+} polyhedra connect to chains via common corners, with nitrate groups further stabilizing the chains. L-Hydrogen aspartate anions in between act as bridging ligands, both between polyhedra within one chain and between neighboring chains (with the α- and δ-carboxylate groups, respectively), thus forming a layered arrangement (Fig. 4.21).

The group of Anokhina and coworkers has published several works on oxo- and hydroxy-nickel salts of L-aspartic acids (Table 4.9). These structures are all characterized by a network of Ni^{2+} octahedra, connected by common corners or edges to three-dimensional frameworks, with holes in the structure. In these holes, the amino acid molecules are located, which also act as ligands of the cations. It has to be noted, however, that the structure given for (L-aspartato)-(oxo)-diaqua-di-nickel dihydrate cannot be correct, as the space group $I4_1/a$ is given (Anokhina and Jacobson 2004). As L-aspartic acid is chiral, centrosymmetry is ruled out; thus, a redetermination of this structure is necessary to decide which symmetry is correct.

Another example of a layered structure is that of (L-aspartato)-tetra-aqua-samarium dichloride (Torres et al. 2002). The connection mode is different compared to that in

Table 4.9 *Salts of amino acids and neutral compounds, part VIII:* space groups and unit cell parameters of compounds of neutral compounds with aspartic acid, glutamic acid, and lysine

	s.g.	*a*	*b*	*c*	*α*	*β*	*γ*	Conn.	Reference
L-Aspartatic acid									
L-Asp $CdNO_3$	$P2_12_12_1$	7.158	7.289	14.951	90	90	90	1–2	Gasque et al. (2002)
L-Asp $PbNO_3$	$P2_12_12_1$	4.589	11.616	15.476	90	90	90	1–3	Gasque et al. (2000)
L-Asp $MgCl{\cdot}3H_2O$	$P2_1$	8.470	8.761	14.163	90	102.76	90	1–2	Schmidbaur et al. (1994)
L-Asp ZnCl	$P2_12_12_1$	8.556	8.566	9.804	90	90	90	0–3	Schmidbaur et al. (1987)
L-Asp $Ni_2O{\cdot}6H_2O$	$P2_12_12_1$	13.932	14.657	12.193	90	90	90	1–1	Anokhina and Jacobson (2004)
L-Asp_4 $Ni_5(OH)_2{\cdot}xH_2O$	$I4_12_12$	18.186	18.186	11.706	90	90	90	3–3	Anokhina et al. (2006)
L-Asp $Ni_2O{\cdot}4H_2O$	$I4_1/a$	19.130	19.130	11.379	90	90	90	3–3	Anokhina and Jacobson (2004)
L-Asp_2 $TeBr_6{\cdot}4H_2O$	$P2_12_12$	8.71	17.62	7.58	90	90	90	N/A	Hempel et al. (1977b)
L-Asp $SmCl_2{\cdot}4H_2O$	$P2_1$	9.137	7.638	9.201	90	91.57	90	0–2	Torres et al. (2002)
L-Asp $HoCl_2{\cdot}6H_2O$	$P3_1$	10.071	10.071	11.845	90	90	120	0–1	Csoregh et al. (1989a)
L-Asp_5 $Eu_4Cl(OH)_4{\cdot}xH_2O$	$P2_12_12_1$	11.559	20.748	23.684	90	90	90	0^4–3	Zhang et al. (2007)
L-Glutamic acid									
L-Glu $CaCl{\cdot}H_2O$	$P2_12_12_1$	5.073	8.217	23.067	90	90	90	1–2	Einspahr et al. (1977)
L-Glu $CaTeI_6$	$P2_12_12$	8.43	22.05	7.87	90	90	90	N/A	Dauter et al. (1977)
L-Glu $Ho(ClO_4)_2{\cdot}2.5H_2O$	$P2_1$	10.863	16.587	19.741	90	102.93	90	0^2–2	Csoregh et al. (1989b)
L-Glu $Dy(ClO_4)_2{\cdot}2.5H_2O$	$P2_1$	11.015	16.560	19.939	90	103.16	90	0^2–2	Csoregh et al. (1989b)
L-Glu $Eu(ClO_4)_2{\cdot}2.5H_2O$	$P2_1$	11.023	16.624	20.032	90	102.87	90	0^2–2	Barja et al. (2006)
L-Glu $Gd(ClO_4)_2{\cdot}2.5H_2O$	$P2_1$	11.020	16.616	19.994	90	102.95	90	0^2–2	Barja et al. (2006)
L-Glu $Er(ClO_4)_2{\cdot}2.5H_2O$	$P2_1$	19.987	16.505	11.040	90	104.69	90	0^2–2	Wang et al. (1991)
L-Glu $Sm(ClO_4)_2{\cdot}2.5H_2O$	$P2_1$	11.069	16.673	20.137	90	102.93	90	0^2–2	Torres et al. (2003)
L-Glu $Pr(ClO_4)_2{\cdot}5.5H_2O$	$P2_12_12_1$	11.065	14.458	22.596	90	90	90	0^2–2	Csoregh et al. (1987)
L-Glu_3 $Er_4(ClO_4)_5(OH)_4{\cdot}14H_2O$	$P2_12_12_1$	14.619	17.211	21.215	90	90	90	0^4–3	Wang et al. (2001)
DL-Glutamic acid									
DL-Glu_2 $Er_2(NO_3)_3{\cdot}9H_2O$	$P1$ (P-1)	7.711	9.643	10.385	81.36	88.33	84.55	0^2–1	Jin et al. (1994a)
L-Asparagine									
L-Asn $Zn(PO_3)$	$P2_12_12_1$	5.035	9.454	18.609	90	90	90	0–1	Gordon and Harrison (2004)

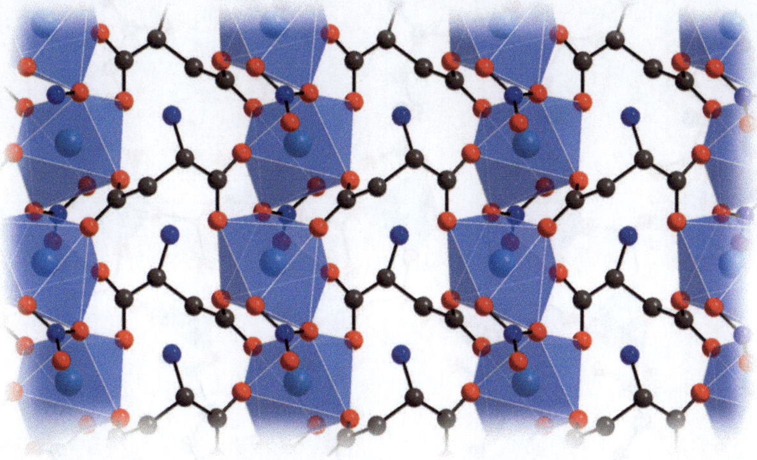

Fig. 4.21 *Layers in (L-aspartato)-nitrato-cadmium, parallel (001).* Typical for salts of aspartic acid, the polyhedra are connected via both acid groups to higher-dimensional structures (layers in this case). Hydrogen atoms omitted for clarity

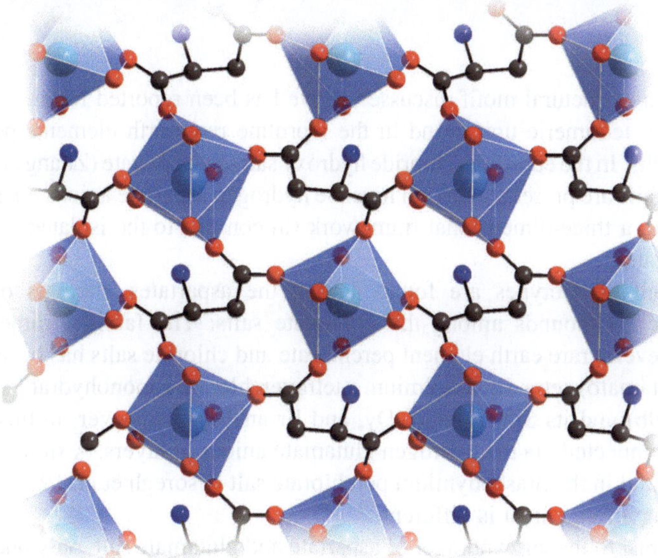

Fig. 4.22 *Layers in (L-aspartato)-tetra-aqua-samarium dichloride, parallel (100).* In contrast to the layers in the lead nitrate salt, here the polyhedra are isolated, connected by amino acid anions to layers. Hydrogen atoms omitted for clarity

the lead nitrate salt: In this species, isolated polyhedra are connected to layers via the L-hydrogen aspartate anions, in an arrangement similar to that of bis(glycinato)-cobalt dichloride and similar salts, although in the present case, each amino acid anion acts as ligand via all four oxygen atoms (Fig. 4.22).

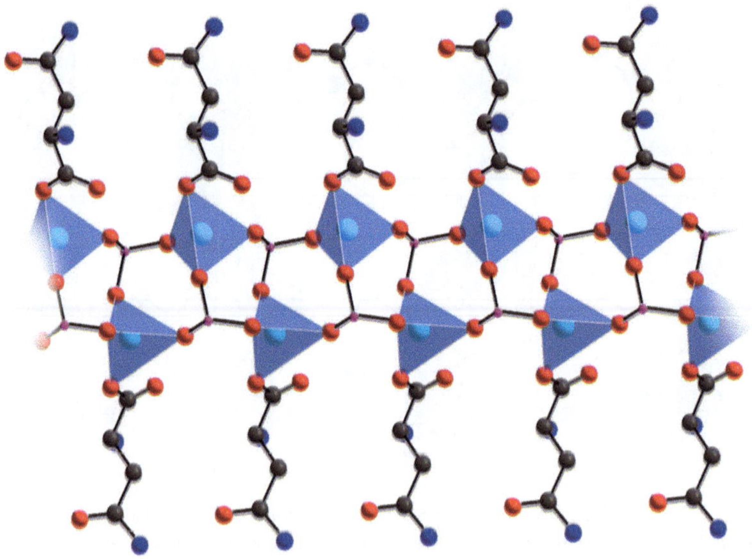

Fig. 4.23 *Bands in phosphito-(L-asparagine)-zinc*. In contrast to the aspartate salts, the asparagine molecule exists in neutral, i.e., zwitterionic state. Hydrogen atoms omitted for clarity

One more structural motif discussed above has been reported for an L-aspartate, namely, the tetrameric unit found in the L-proline rare earth elements perchlorate hydroxy salts: In the europium chloride hydroxy salt of L-aspartate (Zhang et al. 2007), the same units are present, although here the hydrogen aspartate anions connect these tetramers to a three-dimensional framework (in contrast to the isolated units in the L-proline salt).

Although no isotypes are found among the aspartates, there is one series of isotypic compounds among the glutamate salts: The familiar dimeric motif found in several rare earth element perchlorate and chloride salts has been reported in (L-glutamato)-tetra-aqua-holmium tetraperchlorate monohydrate (Csoregh et al. 1989b) and its Sm, Eu, Gd, Dy, and Er analogs. However, in this case the units are connected via the hydrogen glutamate anions to layers. A similar arrangement is found in the praseodymium perchlorate salt (Csoregh et al. 1987), although the packing arrangement is different.

In contrast to the large number of aspartate and glutamate salt, only one instance of an asparagine salt of neutral compounds was found in the literature: Phosphito-(L-asparagine)-zinc (Gordon and Harrison 2004) comprises zinc tetrahedra; phosphite (or phosphonate) anions connect these tetrahedra to a band. L-Asparagine molecules in zwitterionic state are attached as monodentate ligands, facing sideways away to create an infinite planar band (Fig. 4.23).

4.8 Tyrosine, Histidine, Lysine, and Arginine

The last group of the 20 standard amino acids discussed in this chapter is the large amino acids tyrosine, histidine, lysine, and arginine. Of the first member, tyrosine, only one crystalline salt has been reported: iodo-bis(DL-tyrosinato) molybdenum iodide hexahydrate (Apfelbaum-Tibika and Bino 1984). In this salt, Mo^{2+} is charge-counterbalanced by iodide, where one iodide anion is in coordination, the other one is located in the interstices of the dimeric units (Fig. 4.24). This motif has been found in the respective DL-phenylalanine salt. The packing arrangement is comparable as well (as shown by the similarity of the unit cell parameters; compare values in Tables 4.10 and 4.7).

Of the basic amino acids, the majority of salts are those of lysine, for which both salts of L-lysine and the racemate were reported. Still, not for all these salts structural data are available (Table 4.10). Among the salts that were structurally characterized, isolated units dominate: The salts of the divalent cations of nickel, copper, and platinum with chloride have been found to form the usual octahedral [6]-, pyramidal [4 + 1]-, and planar [4]-coordination geometries.

Although Hempel et al. (1977a) reported preliminary data of two arginine salts (along with many other crystalline amino acid salts), no further structural data of these species was published thereafter. However, there are reports of copper salts of L-arginine, which were investigated only recently: Hu et al. (2008) published structural data of *cis*- and *trans*-Cu^{2+}complexes of L-arginine, where the copper cations assume pyramidal [4 + 1]-coordination, and in case of the *cis*-phase are connected to infinite chains. Another salt of arginine and a neutral compound has been structurally characterized (Arularasan et al. 2013): The copper chloride salt of arginine shows tetrameric units where dimers of [4 + 1]- and [4 + 1 + 1]-coordinated polymers are connected via L-arginine zwitterions to a tetrameric unit (Fig. 4.25).

Among the salts of histidine are two more examples of species which actually are not neutral amino acid salts: (DL-histidinato)-mercury-trichloride (Adams et al. 1970). This species comprises monovalent histidinium cations with Hg^{2+} and three chloride anions. In the crystal, mercury cations are coordinated by three chloride anions and one carboxyl oxygen atom of the histidine ion.

The other example is bis(DL-histidinate) $Co^{3+}Br$ (Therup 1977); the cobalt cations show octahedral coordination, where two histidine zwitterions of opposing chirality act as tridentate (N,N',O) ligands.

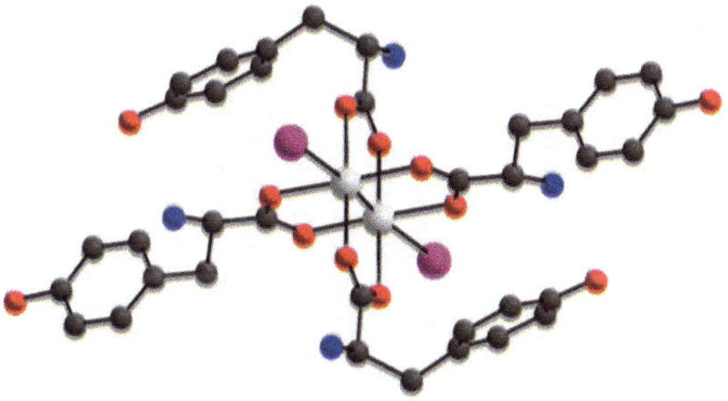

Fig. 4.24 *Units in iodo-bis (DL-tyrosinato) molybdenum iodide hexahydrate.* Compare with the units of the DL-phenylalanine analog (Fig. 4.19). Hydrogen atoms omitted for clarity

Table 4.10 *Salts of amino acids and neutral compounds, part IX:* space groups and unit cell parameters of compounds of neutral compounds with aspartic acid, glutamic acid, and lysine

	s.g.	*a*	*b*	*c*	*α*	*β*	*γ*	Conn.	Reference
DL-Tyrosine									
DL-$Tyr_2Mo^{II}I_2 \cdot 3H_2O$	*P*-1	12.266	12.952	8.918	98.50	92.31	63.58	0^2–0^2	Apfelbaum-Tibika and Bino (1984)
L-Lysine									
L-$Lys_2NiCl_2 \cdot 3H_2O$	*P*1	9.802	7.118	8.533	79.47	112.27	94.48	0–0	Venkatasubramanian and Suresh (1993)
L-$Lys_2CuCl_2 \cdot 2H_2O$	$P2_1$	5.189	16.988	11.482	90	93.57	90	0–0	Duarte et al. (1985)
L-$LysPtCl_2 \cdot H_2O$	$P2_1$	11.262	11.041	9.690	90	102.07	90	0–0	Bino et al. (1988)
L-Lys $SeBr_6$	*C*2	34.00	7.88	12.62	90	93.0	90	N/A	Hempel et al. (1977a)
L-Lys $PtCl_6$	$P2_12_12_1$	10.03	22.45	6.83	90	90	90	N/A	Hempel et al. (1977b)
L-$Lys_2Na_2MoO_4 \cdot 8H_2O$	*P*1	10.783	11.246	10.019	93.21	102.75	112.98	N/A	Inuoe and Yamase (1995)
L-$Lys_2Na_2ZnSO_4 \cdot H_2O$	$P2_1$	10.173	7.276	16.037	90	107.50	90	2–2	Zhu et al. (1990)
DL-Lysine									
DL-$Lys_2CuCl_2 \cdot 2H_2O$	$P2_1/n$	5.152	17.394	11.325	90	96.81	90	0–0	Venkatasubramanian and Saha (1984)
DL-$LysTiNbO_5$	*Pnma*	6.487	3.804	47.32	90	90	90	N/A	Grandin et al. (1985)
DL-$Lys_2Na_2MoO_4 \cdot 8H_2O$	*P*-1	10.785	11.207	9.989	92.88	102.58	112.72	N/A	Inuoe and Yamase (1995)
L-Histidine									
L-$HisCu(NO_3)_2$	*P*1	5.468	7.153	13.844	98.62	87.07	109.83	0–0	Evertsson (1969)
L-$HisCuCl_2$	$P2_12_12_1$	10.01	5.944	17.547	90	90	90	1–1	Colyvas et al. (1982)
DL-Histidine									
(DL-His)$_2$ $Co^{III}Br$	$P2_1/c$	9.834	15.104	10.772	90	107.16	90	0–0	Therup (1977)
DL-$HisHgCl_3$	$P2_12_12_1$	8.116	7.206	19.77	90	90	90	0–0	Adams et al. (1970)
L-Arginine									
L-$Arg_2Cu(NO_3)_2 \cdot 3H_2O$	$P2_1$	10.386	16.885	15.959	90	108.49	90	0–0	Hu et al. (2008)
L-$Arg_2Cu(NO_3)_2 \cdot 3H_2O$	*C*2	26.705	7.319	12.788	90	92.09	90	0–1	Hu et al. (2008)
L-$Arg_2CuCO_3 \cdot 2H_2O$	$P2_1$	8.444	12.096	10.250	90	95.65	90	0–0	Viera et al. (2005)
L-$ArgCuCl_2$	$P2_1$	11.932	12.881	15.395	90	99.27	90	0^2–0^4	Arularasan et al. (2013)
L-Arg $SeBr_6$ ("needles")	*P*1	11.66	7.78	12.43	106.5	114.3	96.7	N/A	Hempel et al. (1977a)
L-Arg $SeBr_6$ ("plates")	$P2_1$	7.43	10.74	11.77	90	102.2	90	N/A	Hempel et al. (1977a)

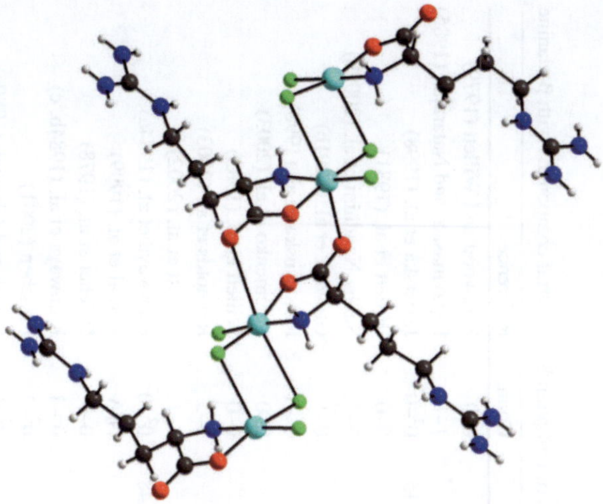

Fig. 4.25 *Tetrameric units in the copper chloride of L-arginine*. Note the different coordination polyhedra of the copper cations

4.9 β-Alanine

Of the nonstandard amino acids, β-alanine is the simplest member. As it has the same size and molecular weight, one would expect it to be as reactive as its "standard cousin," α-alanine, as far as the synthesis of salts with neutral salts is concerned. However, the list of β-alanine salts is much longer than that of α-alanine, which might be due to the fact that the amino group is more distant to the coordination polyhedron (the ligand atoms are the oxygen atoms of the carboxylate group) and does not influence complexation of the cation, but rather connects units via hydrogen bonds. In many cases the β-alanine molecules adopt gauche conformation as some of the hydrogen bonds extending from the amino nitrogen atom reach back toward the closest oxygen of the carboxylate group.

Also, among the list given in Table 4.11, there are several examples of salts which do not represent neutral salts, but actually salts of β-alaninium, such as the $Fe^{III}Cl_4$ compound (Kirkwood and Willett 1979), in which the β-alaninium cation counterbalances the charge of the $[FeCl_4]^-$ cluster, but is not part of the coordination sphere. A similar situation is found in the isotypic pair of bis(β-alaninium) $Cu^{2+}Cl_4$ and the respective bromide (Willett et al. 1981, 1983), also in the Re^{3+} chloride salt, where isolated $[Re_2Cl_8]^{2-}$ clusters are connected to the noncoordinating amino acid cations via hydrogen bonds only (Shtemenko et al. 2003).

Nevertheless, there are several instances of truly neutral β-alanine salts. In the system β-alanine–copper–chloride–water, three phases were reported (Table 4.11), all of which differ by structure as well as the topology of the building units. Among these, the salt bis(β-alanine)-chloro-copper chloride hemihydrate (Jezierska et al. 1998) is noteworthy: In this species, the dimeric motif of two opposing pyramids, connected via amino acid ligands, as reported for the DL-phenylalanine molybdenum iodide salt (and its DL-tyrosine analog), is found. This motif also

Table 4.11 *Salts of amino acids and neutral compounds, part X:* space groups and unit cell parameters of compounds of neutral compounds with β-alanine

	s.g.	*a*	*b*	*c*	*α*	*β*	*γ*	Conn.	Reference
β-AlaFe^{III}Cl$_4$	*P*2$_1$/*c*	6.996	11.994	12.444	90	91.01	90	0–0	Kirkwood and Willett (1979)
β-AlaCuCl$_2$	*P*2$_1$/*c*	7.960	12.249	6.875	90	94.10	90	1–1	Papavinasam and Natarajan (1985)
β-Ala$_2$CuCl$_2$·0.5H$_2$O	*P*-1	9.444	9.498	14.107	104.04	105.90	97.15	0^2–0^2	Jezierska et al. (1998)
β-Ala$_2$ CuCl$_4$	*I*2/*c*	7.402	7.734	24.236	90	91.82	90	0–0	Willett et al. (1981)
β-Ala CdCl$_2$	*Pna*2$_1$	6.939	12.945	7.971	90	90	90	1–1	Subha Nandhini et al. (2002b)
β-Ala NdCl$_3$·3H$_2$O	*P*2$_1$/*c*	6.018	16.652	12.518	90	116.18	90	0–1	Glowiak et al. (1991b)
β-Ala$_3$ NdCl$_3$·2H$_2$O	*P*2$_1$/*c*	10.846	19.552	9.464	90	102.02	90	0^2–1	Rzaczynska et al. (1998)
β-Ala$_2$ Re$_2$Cl$_8$	*C*2/*c*	21.049	7.718	13.201	90	116.40	90	0–0	Shtemenko et al. (2003)
β-Ala$_2$ CuBr$_4$	*I*2/*c*	7.761	8.027	24.295	90	90	90	0–0	Willett et al. (1983)
β-Ala PbBr$_2$·3H$_2$O	*P*2$_1$/*c*	6.007	16.529	8.3057	90	100.56	90	3–3	Reynolds et al. (2003)
β-Ala$_2$SnI$_4$	*P*2$_1$/*c*	20.604	13.146	8.482	90	95.75	90	1–1	Willett et al. (2002)
β -Ala$_2$ Cu(NO$_3$)$_2$·3H$_2$O	*P*-1	12.499	8.717	8.070	77.74	82.44	71.48	0^2–0^2	Kamwaya et al. (1984a)
β -Ala$_2$ Co(NO$_3$)$_2$·H$_2$O	*C*2/*c*	16.882	11.990	15.017	90	115.09	90	N/A	Bernal et al. (1999)
β -Ala$_2$ Zn(NO$_3$)$_2$·2H$_2$O	*P*2$_1$/*n*	23.891	6.576	5.519	90	95.67	90	0–0	Dejehet et al. (1978)
β -Ala AgNO$_3$	*P*2$_1$/*n*	6.656	8.280	12.975	90	94.90	90	0^2–1	Kamwaya et al. (1984b, c)
β -Ala AgNO$_3$	*Cc*	8.455	25.509	6.461	90	96.15	90	0^2–1	Erxleben (2001)
β -Ala$_8$ Mn$_3$(ClO$_4$)$_6$·2H$_2$O	*Pbca*	23.455	21.159	11.187	90	90	90	0^2–1	Ciunik and Glowiak (1980c)
β -Ala$_2$ Rh(ClO$_4$)$_2$·2H$_2$O	*Pccn*	14.435	19.177	13.385	90	90	90	0^2–0^2	Korp et al. (1981)
β -Ala$_3$ La(ClO$_4$)$_3$·2.5H$_2$O	*P*-1	9.46	12.917	21.726	76.79	80.85	83.35	0^2–1	Ma et al. (2002)
β -Ala$_3$ Pr(ClO$_4$)$_3$·2.5H$_2$O	*P*-1	12.898	21.742	9.390	99.33	96.75	76.70	0^2–1	Wang et al. (1996)
β -Ala$_3$ Nd(ClO$_4$)$_3$·2.5H$_2$O	*P*-1	12.913	21.765	9.368	99.07	96.91	76.75	0^2–1	Liang et al. (2004)
β -Ala$_3$ Sm(ClO$_4$)$_3$·2.5H$_2$O	*P*-1	12.926	21.707	9.316	98.89	96.93	76.89	0^2–1	Jianxue et al. (1992)
β -Ala$_3$ Gd(ClO$_4$)$_3$·2.5H$_2$O	*P*-1	12.911	21.679	9.272	98.94	97.00	76.85	0^2–1	Liang et al. (2004)
β -Ala$_3$ Ho(ClO$_4$)$_3$·2.5H$_2$O	*P*-1	9.199	12.870	21.620	76.92	81.14	82.90	0^2–1	Ma et al. (2004)
β -Ala$_3$ Er(ClO$_4$)$_3$·2.5H$_2$O	*P*-1	12.880	21.629	9.210	98.77	97.29	76.92	0^2–1	Liang et al. (2004)

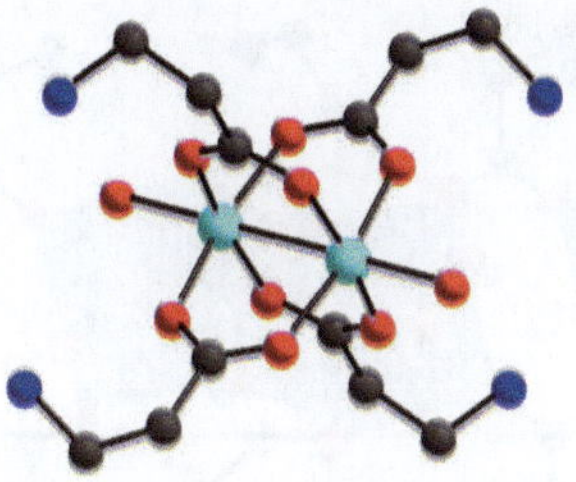

Fig. 4.26 *Dimeric unit in tetrakis(β-alanine)-diaqua-di-copper tetranitrate tetrahydrate.* This motif of opposing pyramids is found also in molybdenum iodide salts of DL-phenylalanine and DL-tyrosine; compare with Fig. 4.19. Hydrogen atoms are omitted for clarity

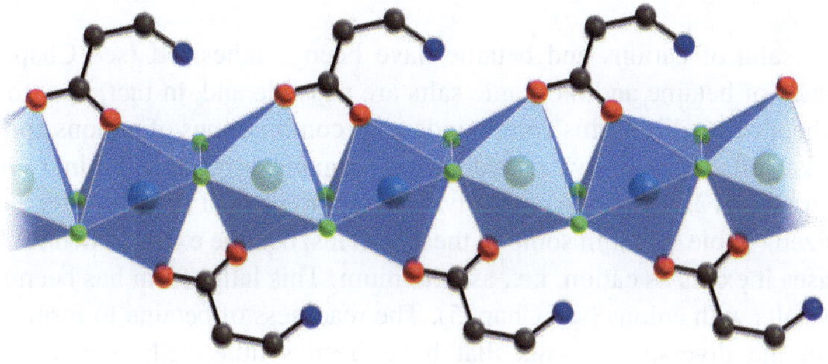

Fig. 4.27 *Chains in β-alaninato-dichloro-cadmium.* Hydrogen atoms are omitted for clarity

occurs in the rhodium perchlorate salt (Korp et al. 1981), as well as in the copper nitrate salt (Kamwaya et al. 1984a; Fig. 4.26).

Furthermore, there are several examples of chain structures among salts of β-alanine. For instance, the cadmium chloride salt (Subha Nandhini et al. 2002b) comprises polyhedral chains, where the chloride anions act as bridging ligands, with the amino acid zwitterions providing additional connection (Fig. 4.27).

A different chain structure is found in the large family of isotypic rare earth perchlorate salts of β-alanine, which were reported for La, Pr, Nd, Sm, Gd, Ho, and Er (Table 4.11). In these species, dimers of cation polyhedra connected by four β-alanine units are found. These dimers are in turn connected to chains via two β-alanine units each (Fig. 4.28), with the perchlorate anions located in the interstices. An analog chain is also found in the neodymium chloride β-alanine salt (Rzaczynska et al. 1998).

4.10 Betaine

As discussed in the Introduction, betaine (*N,N,N*-trimethylglycine) is an amino acid which cannot exist in truly neutral form, since the positive charge at the nitrogen atom cannot be counterbalanced by proton transfer (as in regular amino acids).

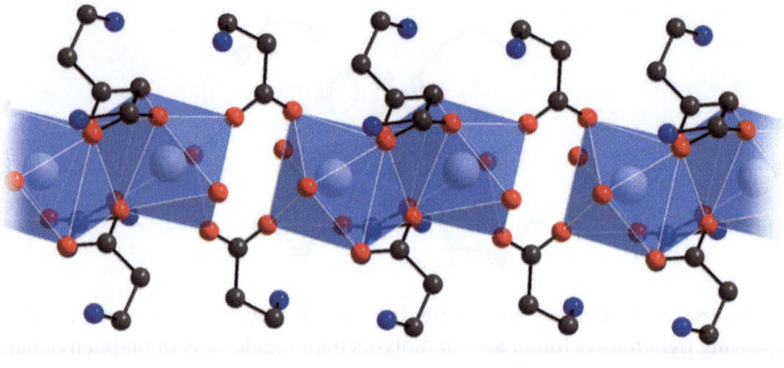

Fig. 4.28 *Chains in the series of rare earth element perchlorate β-alanine salts.* Hydrogen atoms are omitted for clarity

Thus, no salts of cations and betaine have been synthesized (see Chap. 3), but compounds of betaine and inorganic salts are possible and, in fact, easy to synthesize, as betaine readily forms compounds with combinations of cations and anions. Betaine is well soluble in water and various organic solvents and binary mixtures (Wang et al. 2012); therefore, a relatively large number of betaine salts have been synthesized (Table 4.12). In some of these species, betaine exists in neutral form; in other cases it exists as cation, i.e., as betainium. This latter form has been reported in many salts with anions (see Chap. 5). The readiness of betaine to form salts also shows in the diversity of salts that have been synthesized: Nitrates, sulfates, chlorides, bromides, iodides, perchlorates, etc., have been obtained in crystalline form. As betaine is a non-chiral amino acid, centrosymmetry is possible and, in fact, found in almost all cases (see space groups listed in Table 4.12).

As an example for a betaine nitrate salt, the structure of the respective nickel salt (Chen and Mak 1991b) is given here: Nickel cations are in octahedral coordination, with four water molecules as well as two monodentate betaine zwitterions as ligands. These isolated units (Fig. 4.29) are connected to each other via an extensive hydrogen bond network, which involves also the non-coordinating nitrate anions.

Another instance of salts comprising isolated units is the betaine salt of magnesium sulfate (Rodrigues et al. 2005a), where the [6]-coordinate magnesium has only water molecules as ligands, while the betaine zwitterions as well as the sulfate anions are connected only via hydrogen bonds.

In contrast to the few nitrate and sulfate salts, the number of halogenide species is much greater, with several systems containing more than one phase. Such a system is that of betaine–copper–chloride–water, in which several phases were found.

Among these is one salt with a ratio of betaine:$CuCl_2$ of 2:3, i.e., bis-betaine-diaqua-hexachloro-tri-copper (Wiehl et al. 2008b), which comprises building units not found in any other amino acid salt. These units are trimeric, with a central [4 + 2]-coordinated Cu^{2+} cation, sharing common edges with two lateral [4 + 1] polyhedra (Fig. 4.30). Each of the two betaine zwitterions connects the central cation to one of the lateral cations, acting as bridging (O,O′) ligand. At opposite ends of these trimeric units, water molecules are located, from which hydrogen bonds extend toward chloride ions of neighboring units.

Table 4.12 *Salts of amino acids and neutral compounds, part XI:* space groups and unit cell parameters of compounds of neutral compounds with betaine

	s.g.	*a*	*b*	*c*	*α*	*β*	*γ*	Conn.	Reference
Bet $Ag(NO_3)\cdot H_2O$	$P2_1/c$	5.659	15.813	11.549	90	98.10	90	0^2–1	Chen and Mak (1991a)
Bet_2 $Ni(NO_3)\cdot 4H_2O$	P-1	6.434	12.570	13.244	90.68	91.28	91.57	0–0	Chen and Mak (1991b)
Bet_2 $Mg(SO_4)\cdot 9H_2O$	$P2_1/c$	6.373	15.905	24.994	90	97.521	90	0–0	Rodrigues et al. (2005a)
Bet_2 $TaF_6\cdot 2H_2O$	$P2/a$	12.553	6.531	12.587	90	110.25	90	0–0	Silva et al. (1999)
Bet $CaCl_2\cdot 2H_2O$	$Pnma$	10.97	10.15	10.82	90	90	90	0–1	Brill et al. (1985)
Bet $SrCl_2\cdot 2H_2O$	$Pnma$	6.784	13.178	15.060	90	90	90	1–1	Andrade et al. (2002)
Bet $ZnCl_2\cdot H_2O$	$Imm2$	9.269	11.860	8.137	90	90	90	0–0	Chen and Mak (1991c)
Bet $CuCl_2$	$C2/c$	30.1	13.533	11.71	90	103.78	90	0–0	Chen and Mak (1993)
Bet $CuCl_2\cdot H_2O$	$P2_1/n$	6.321	9.8207	17.166	90	99.595	90	0^2–0^2	Wiehl et al. (2008a)
Bet_2 $(CuCl_2)_3\cdot 2H_2O$	$Pbca$	11.181	11.438	18.867	90	90	90	0^3–0^3	Wiehl et al. (2008b)
Bet_2 $CuCl_4\cdot H_2O$	$P2_1/n$	14.053	9.109	16.409	90	105.01	90	0–0	Schreuer et al. (2006a)
Bet_2 $CuCl_2\cdot 2H_2O$	$P2_1/c$	11.051	14.714	11.162	90	107.40	90	0^2–0^2	Silva et al. (2001b)
Bet_2 $CuBr_2\cdot 2H_2O$	$P2_1/c$	11.304	14.766	11.324	90	108.19	90	0^2–0^2	Wiehl et al. (2006)
Bet_2 $(CuBr_2)_3\cdot H_2O$	P-1	11.681	13.285	14.719	64.61	75.23	74.82	0^2–0^2	Schreuer et al. (2006b)
Bet_3 $MnCl_2$	P-3	12.880	12.880	9.1173	90	90	120	0–1	Chen and Mak (1991d)
Bet $CdCl_2\cdot H_2O$	$P2_1/c$	10.225	10.919	10.062	90	107.68	90	0^2–0^2	Chen and Mak (1991e)
Bet $CdCl_2$	$Cmc2_1$	9.652	13.991	7.157	90	90	90	1–1	Chen and Mak (1991e)
Bet_2 $(CdCl_2)_3\cdot 2H_2O$	P-1	6.704	9.338	20.056	101.41	96.65	93.15	1–1	Che et al. (2007)
Bet_2 $EuCl_3\cdot 10H_2O$	$Pbca$	12.566	17.865	22.257	90	90	90	0^2–0^2	Yang et al. (1999)
Bet_2 $RhCl_2\cdot 2H_2O$	$P2_1/c$	11.040	14.881	11.139	90	106.66	90	0–0	Zhou et al. (1991)
Bet_3 $(MnBr_2)_2$	P-1	12.724	12.494	9.112	89.44	93.96	112.76	0–1	Haussühl et al. (2009a)
Bet_3 $MnBr_2\cdot 2H_2O$	P-1	9.382	11.346	13.151	70.62	81.40	70.15	0^2–0^2	Haussühl et al. (2009b)
Bet_3 $MnBr_2$	P-1	9.140	12.700	12.871	66.557	86.063	89.249	0–1	Kocadag et al. (2008)
Bet_2 $NaBr\cdot H_2O$	$C2$	9.108	6.531	13.306	90	105.11	90	0–2	Rodrigues et al. (2005b)
Bet $KBr\cdot 2H_2O$	$Pbmn$	25.218	7.513	5.728	90	90	90	1–2	Haussühl (1989)
Bet $KI\cdot 2H_2O$	P-1	5.653	5.894	14.016	82.23	82.16	73.22	0–1	Andrade et al. (1999)
Bet $RbI\cdot 2H_2O$	P-1	5.687	5.890	14.192	83.14	82.72	73.56	0–1	Andrade et al. (2001)

(continued)

Table 4.12 (continued)

	s.g.	*a*	*b*	*c*	*α*	*β*	*γ*	Conn.	Reference
$Bet_3\ Fe^{III}(CN)_6{\cdot}4H_2O$	*P*-1	9.517	11.553	15.145	93.82	91.00	90.11	0–0	Yan et al. (2001)
$Bet_3\ Zn(ClO_4)_2{\cdot}H_2O$	*P*2$_1$/*c*	6.055	26.373	16.872	90	91.23	90	0–0	Chen et al. (1997)
$Bet_4\ Cu(ClO_4)_2$	*I*-42*m*	11.711	11.711	11.695	90	90	90	0–0	McEwen (1973)
$Bet_2\ Cu(ClO_4)_2$	*C*2/*m*	12.186	12.652	12.875	90	100.0	90	N/A	McEwen (1973)
$Bet_4\ Tb(ClO_4)_6{\cdot}2H_2O$	*P*2$_1$/*c*	11.777	27.725	11.795	90	99.67	90	$0\text{–}0^2$	Yang et al. (1999)
$Bet_4\ Eu(ClO_4)_6{\cdot}2H_2O$	*P*2$_1$/*c*	11.781	27.747	11.798	90	99.50	90	$0\text{–}0^2$	Yang et al. (1999)
$Bet_2\ Cd(ClO_4)(NO_3)$	*Pbcm*	9.283	7.591	26.045	90	90	90	0–1	Tong et al. (1997)
$Bet_4\ Cd_3Cl_4(ClO_4)_2{\cdot}2H_2O$	*P*2$_1$/*c*	13.872	11.656	12.261	90	93.20	90	$0^3\text{–}1$	Tong et al. (1997)
$Bet_2\ ZnCl(ClO_4)$	*Pnma*	26.007	11.510	6.014	90	90	90	0–0	Chen et al. (1997)

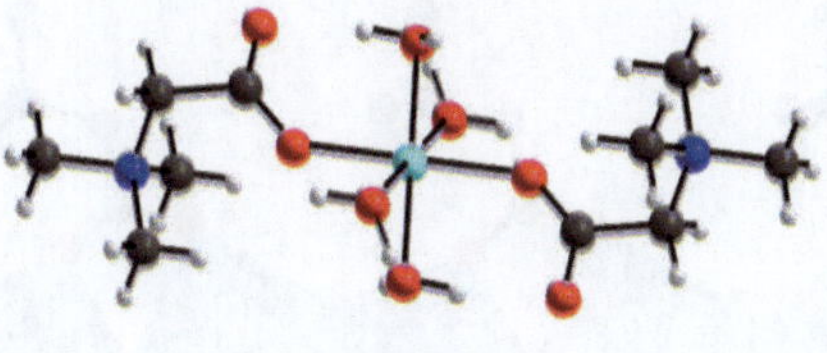

Fig. 4.29 *Isolated unit in bis(betaine) tetraaqua nickel dinitrate.* Notice that in betaine the amino group cannot act as a ligand as it has bonds to four carbon atoms

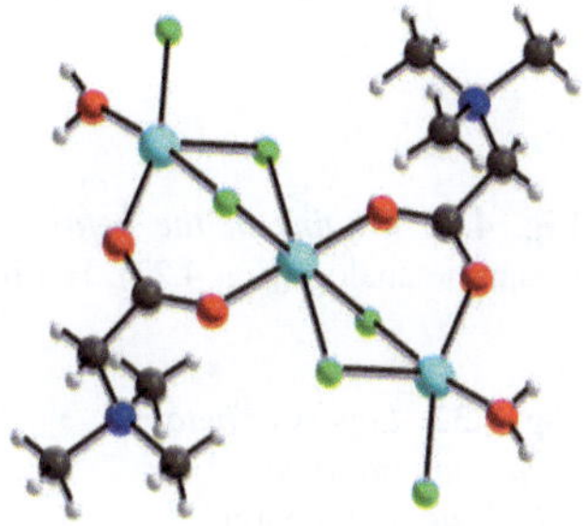

Fig. 4.30 *Units in bis-betaine-diaqua-hexachloro-tri-copper.* The trimeric unit is unique among amino acid salts; note the different coordination modes of the central and the lateral cations

Another salt in the same system exists at a betaine:$CuCl_2$ ratio of 2:1 (along with an isotypic bromide salt, Silva et al. 2001b; Wiehl et al. 2006). In these species, the familiar motif of dimers of opposite pyramids, as found in DL-phenylalanine and DL-tyrosine molybdenum iodide salts as well as several β-alanine copper salts, can be observed.

Moreover, there is a salt at a 1:1 ratio of betaine:$CuCl_2$ (Chen and Mak 1993), which represents another unique structure: This species is an example of two copper cations in different coordination within one crystal, one in tetrahedral [4]-coordination and the other in pyramidal [4 + 1]-coordination (these pyramids again are arranged as dimers as in the above structure).

In addition to these three truly neutral salts, there is a species where betainium cations are combined with $CuCl_4$ anions (Schreuer et al. 2006a).

Another example of a structure with isolated units is the $ZnCl_2$ salt reported by Chen and Mak (1991c), in which the same motif as in other amino acid zinc chloride salts is found: The Zn^{2+} is tetrahedrally [4]-coordinate, with two chloride anions and two monodentate amino acids (betaine in this case) as ligands.

The same authors published a betaine salt of cadmium chloride (Chen and Mak 1991e), in which chains with the same topology as those in the respective β-alanine salt are found (Fig. 4.31). Since the chain corresponds to that of the β-alanine salt (and the chains are similarly packed), the unit cell dimensions of both salts are comparable, but the symmetry differs (compare values in Tables 4.11 and 4.12).

In the same system, other phases exist: One is monohydrate and does not comprise chains, but isolated dimeric units (Chen and Mak 1991e). A third phase has a complicated structure with two topologically and chemically different chains: One type of chain contains cadmium cations, coordinated by both chloride anions and betaine, whereas the second chain is composed of cadmium chloride polyhedra, without other ligands. Both chains compose the triclinic, centrosymmetric structure (Che et al. 2007).

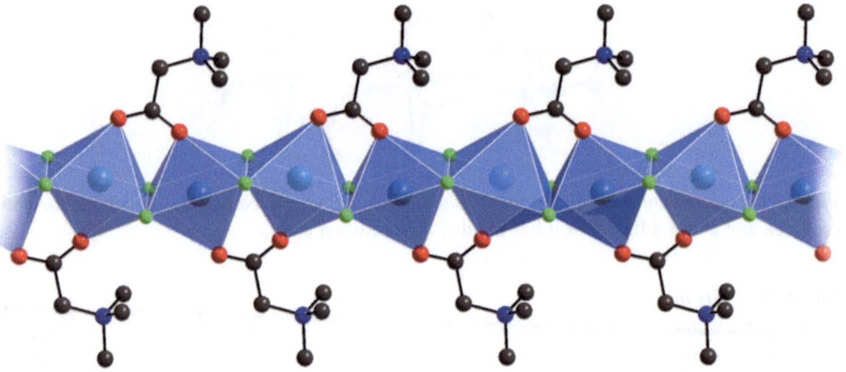

Fig. 4.31 *Chains in the betaine salt of cadmium chloride*. Compare with the chains in the β-alanine analog (Fig. 4.27). Hydrogen atoms are omitted for clarity

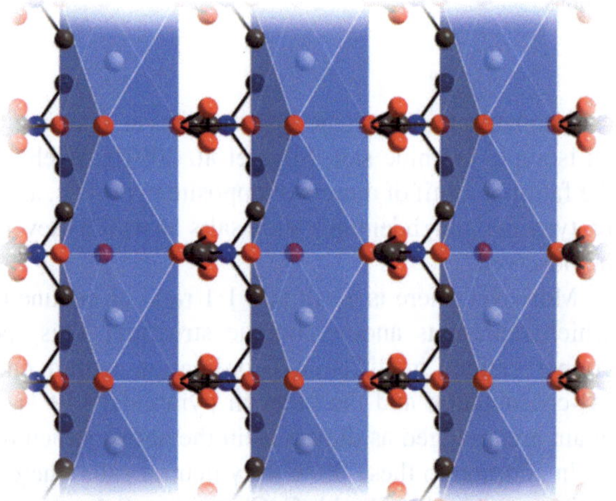

Fig. 4.32 *Layers in betaine potassium bromide dihydrate*. Polyhedral chains are connected to layers parallel (010), with the betaine zwitterions oriented perpendicular to the layers. Hydrogen atoms are omitted for clarity

Apart from the above chloride, there is one crystal phase which has been investigated very actively, namely, the trigonal betaine salt of manganese chloride. The structure was reported by Chen and Mak (1991e) and later refined by Schreuer and Haussühl (1993). Earlier, Siegfried Haussühl grew large single crystals of this phase (along with several other betaine compounds, mostly betainium anion salts, e.g., with chloride, bromide, iodide, nitrate) and prepared samples to measure the components of the elastic and thermoelastic tensors of these salts. Within this work, the only other salt of betaine and a neutral compound was that of betaine potassium bromide (the crystal structure was published later by Andrade et al. 2000), which was found to be a layered structure (Fig. 4.32).

Analog salts of potassium iodide and rubidium iodide were synthesized, which are isotypic but different from the above salt (Andrade et al. 1999; Andrade et al. 2001): These halogenide salts comprise chains instead of layers (Fig. 4.33).

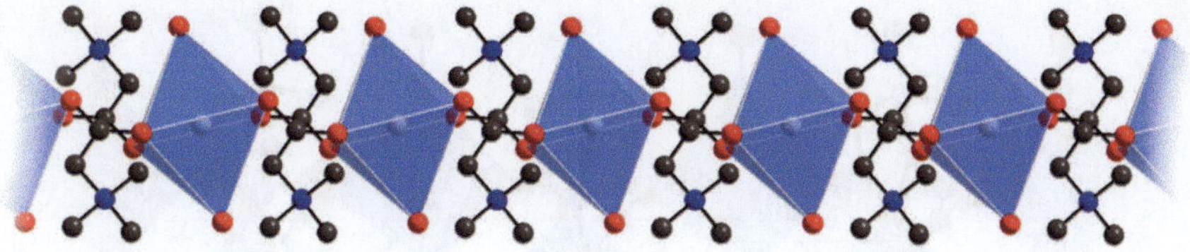

Fig. 4.33 *Chains in the isotypic KI and RbI salts of betaine*. Hydrogen atoms are omitted for clarity

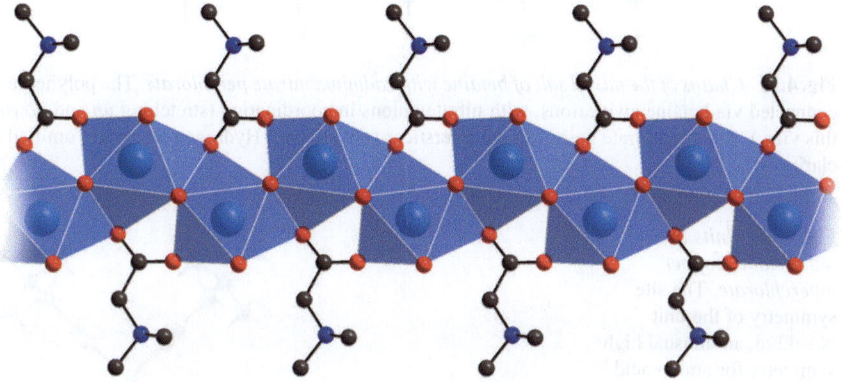

Fig. 4.34 *Chains in the betaine salt of strontium chloride*. Hydrogen atoms are omitted for clarity

Apart from the alkali metal halogenide salts of betaine, there are several salts of alkaline earth metals, such as betaine-diaqua-dichloro-calcium (Brill et al. 1985), which comprises zigzag chains, and the respective strontium salt (Andrade et al. 2002; Fig. 4.34).

Apart from alkali, alkaline earth, and transition metal salts, several rare earth element salts of betaine were obtained. These encompass salts of europium chloride or terbium perchlorate (both reported by Yang et al. 1999), both of which are characterized by dimeric units, similar (but not identical) as in the rare earth element perchlorates of glycine (and other amino acids).

There are several more perchlorate salts of betaine with remarkable structures, as, for instance, the mixed salt of betaine cadmium perchlorate nitrate (Tong et al. 1997), which comprises isolated units connected to chains, where nitrate anions are part of the coordination, while the perchlorate anions are not (Fig. 4.35).

Noteworthy is also the structure of tetrabetaine-copper diperchlorate (first published by McEwen 1973, later refined by Ng et al. 1998), which has a tetragonal symmetry. This symmetric arrangement is found even in the units, where a copper cation with four betaine ligands has site symmetry −42 *m* (Fig. 4.36). The charge-counterbalancing perchlorate anions in the interstices are disordered, a phenomenon frequently found in perchlorate salts where no extensive hydrogen bonding system defines the positions of the atoms in the ClO_4 anion. As the betaine zwitterion (in contrast to other amino acids) cannot provide N–H-type hydrogen bonds and

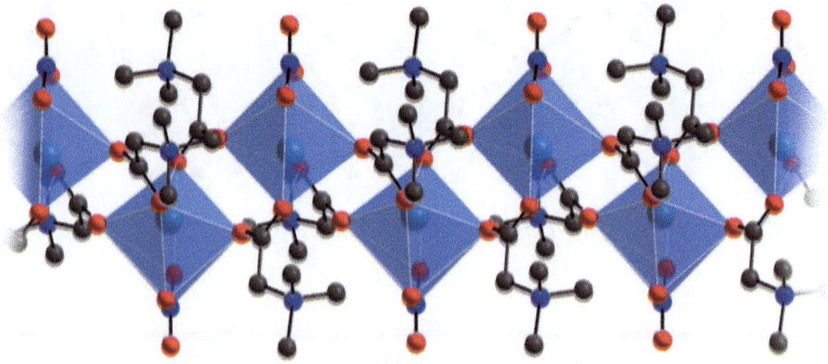

Fig. 4.35 *Chains of the mixed salt of betaine with cadmium nitrate perchlorate*. The polyhedra are connected via betaine zwitterions, with nitrate anions in coordination (stretching *up* and *down* in this view), and perchlorate anions in the interstices (not shown). Hydrogen atoms are omitted for clarity

Fig. 4.36 *Units in tetrabetaine-copper diperchlorate*. The site symmetry of the unit is −42 m, an unusual high symmetry for amino acid compounds. The disordered perchlorate anions are located in the interstices (not shown)

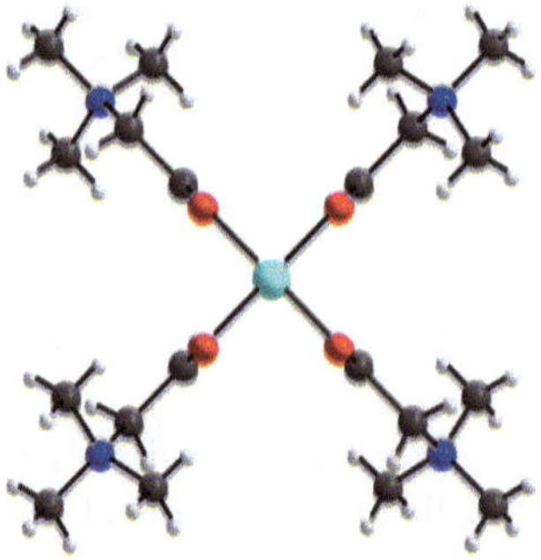

the structure is anhydrous, no hydrogen bonds toward the anions can be formed. Therefore, only van der Waals interactions provide connection between the units.

In addition to the neutral salts of betaine, there are various instances of species in which betainium cations are combined with inorganic cations and anions, as, e.g., the species bis(betainium) tetrachloro-copper monohydrate (Schreuer et al. 2006a) mentioned above. Other examples are tris(betainium) hexacyanoferrate tetrahydrate (Yan et al. 2001), where isolated $(FeCN_6)_3$ groups and three betainium cations comprise the structure. Likewise, betainium hexafluoro-tantalum dihydrate (Silva et al. 1999) is not truly a neutral salt but a betainium salt.

Apart from the multitude of works on structural and physical date on salts of betaine, several phase transitions have been reported within this family. Schaak (1990) reported phase transitions in several betaine compounds; later Yuzyuk et al. (2000) studied the KBr salt of betaine and discovered a phase transition at about 160K, related to the freezing of the reorientation of water molecules. Likewise, a phase transition in betaine potassium iodide dihydrate was found at about 100K (Almeida et al. 1998). More phase transitions were discovered in salts of betaine and anions, e.g., for betaine phosphate and phosphite (see Sect. 5.1.4).

4.11 Other Nonstandard Amino Acids

In addition to the numerous salts of betaine and β-alanine discussed above, there are several other nonstandard amino acids which form salts with neutral compounds. Instances for the non-chiral amino acids sarcosine and dimethylglycine (DMG) as well as the chiral amino acids ornithine and cystine are included in this section (Table 4.13).

Ornithine, as a smaller variant of lysine, could not be used for synthesis of salts with neutral compound easily. There are only a few examples, such as the copper chloride salt of L-ornithine, where the cations show typical [4+2]-coordination, with the amino acids in *trans*-orientation, as bidentate (N,O) ligands. The apical positions are occupied by the chloride anions. The unit is close to centrosymmetric, but for the chiral residues of both ligand molecules. Furthermore, there are some L-ornithine salts with molybdate clusters (the so-called Keggin-type polyoxometalates, Li et al. 2004), in which the clusters dominate the structure, and L-ornithine zwitterions coordinate these clusters. An older work (Brubaker and Busch 1966) identified a salt of the composition bis(L-ornithine) $NiCl_2$ by spectroscopy, but no crystals of this phase could be grown to date.

In contrast to the relatively few salts of L-ornithine, there are more examples of sarcosine salts, which crystallize in structures with isolated units or chains.

The zinc chloride salt of sarcosine (Subha Nandhini et al. 2001b) is characterized by the typical [4]-coordination of zinc found in most amino acid $ZnCl_2$ salts: The cation is coordinated by two chloride anions and two oxygen atoms from different amino acid moieties (Fig. 4.37).

Another phase in the same system has been reported by Krishnakumar et al. (2001b), where one Zn^{2+} cation is coordinated by three chloride and one sarcosinium cation (with a non-coordinating water molecule). Thus, the species of trichloro-(sarcosinio)-zinc monohydrate is not a neutral amino acid salt.

In the system sarcosine–$MnCl_2$–water, two phases were reported. The first salt is monoclinic and has a ratio of 1:1 of sarcosine:$MnCl_2$ (Silva et al. 2001a), while the second phase is triclinic with a ratio of 2:1 (Silva et al. 2001a). Both can be described as chain structures, and the 2:1 phase is topologically identical with the respective glycine salt (Fig. 4.38).

Another remarkable sarcosine salt is that of calcium tris(sarcosine) dichloride. This structure is, which can be regarded as a chain of distorted calcium octahedra, connected via three sarcosine zwitterions in bidentate (O,O′) mode to each neighbor. The chains themselves have a pseudo-trigonal symmetry within the orthorhombic unit cell. The crystal structure was reported by Ashida et al. (1972) at ambient conditions in the centrosymmetric space group *Pnma*. Later, Mishima et al. (1984) have found a phase transition to a non-centrosymmetric, ferroelectric phase at 127K. The ferroelectric structure has symmetry $Pn2_1a$.

Recently, we reported the structure of diaquaiodidotetrasarcosinepotassium (Fleck et al. 2013), which was characterized via single crystal XRD analysis as well as vibrational spectroscopy and was found to be the first example of a sarcosine metal

Table 4.13 *Salts of amino acids and neutral compounds, part XII:* space groups and unit cell parameters of compounds of neutral compounds with ornithine, sarcosine, cystine, and dimethylglycine

	s.g.	*a*	*b*	*c*	*α*	*β*	*γ*	Conn.	Reference
L-Ornithine									
L-$Orn_2CuCl_2 \cdot 2H_2O$	$P2_1$	5.218	15.636	12.058	90	93.37	90	0–0	Stephens et al. (1977)
L-$Orn_2PtCl_2 \cdot H_2O$	$P2_12_12_1$	7.108	10.569	14.554	90	90	90	0–0	Via et al. (2006)
L-$Orn_2(Mo_{12}O_{41}PH) \cdot 8H_2O$	$P2_12_12_1$	14.590	18.654	18.690	90	90	90	[a]	Li et al. (2004)
L-$Orn_2(Mo_{12}O_{41}Ge) \cdot 10H_2O$	$C222_1$	15.604	21.321	16.239	90	90	90	[a]	Li et al. (2004)
L-$Orn_2(Mo_{12}O_{41}Si) \cdot 10H_2O$	$C222_1$	15.519	21.224	16.180	90	90	90	[a]	Li et al. (2004)
DL-Ornithine									
DL-$OrnSeBr_6$	*P*-1	10.89	7.45	11.84	101.8	100.5	103.3	N/A	Hempel et al. (1977a)
Sarcosine									
Sar_3 $CaCl_2$	*Pnma*	9.156	17.460	10.265	90	90	90	0–1	Ashida et al. (1972)
Sar_3 $CaCl_2$ (127K)	$Pn2_1a$	9.122	17.408	10.228	90	90	90	0–1	Mishima et al. (1984)
Sar $MnCl_2 \cdot 2H_2O$	$P2_1/c$	8.314	5.777	18.680	90	95.83	90	0–1	Silva et al. (2001a)
Sar_2 $MnCl_2 \cdot 2H_2O$	*P*-1	4.822	5.379	13.783	83.76	80.55	87.58	0–1	Rzaczynska et al. (2002)
Sar_2 $ZnCl_2$	*Pbca*	14.191	10.655	15.917	90	90	90	0–0	Subha Nandhini et al. (2001b)
Sar $ZnCl_3$	$P2_1$	6.618	7.499	9.900	90	92.62	90	0–0	Krishnakumar et al. (2001b)

Sar $BaCl_2 \cdot 4H_2O$	$P2_12_12_1$	7.235	10.668	15.686	90	90	90	1–1	Krishnakumar and Natarajan (1995b)
Sar $CdCl_2$	$P2_1/n$	7.960	13.844	6.917	90	92.42	90	1–1	Yamada et al. (1994)
Sar_2 $PtBr_2$	$P2_1/a$	10.921	9.562	6.890	90	125.46	90	N/A	Sabo et al. (2005)
Sar_3 $BaBr_2$	$P2_1/c$	18.345	10.668	8.921	90	91.86	90	1–1	Trzebiatowska-Gusowska et al. (2009)
Sar_4 $KI \cdot 2H_2O$	$C2/c$	13.714	22.799	8.079	90	111.94	90	1–1	Fleck et al. (2013)
Sar_2 TeI_6	$P4_2/m$	25.00	25.00	8.26	90	90	90	N/A	Dauter et al. (1977)
Sar_3 $Eu(ClO_4)_3 \cdot 2H_2O$	$P1$	9.146	11.042	14.696	100.42	104.56	109.24	N/A	Gawryszewska et al. (2000)
L-Cystine									
L-Cystin $Nd(ClO_4)_3 \cdot 4H_2O$	$P2_12_12_1$	13.674	18.485	19.334	90	90	90	0^2–2	Qingri et al. (1994)
Dimethylglycine									
DMG $CuCl_2$	$P2_1/c$	9.759	12.861	14.060	90	99.20	90	0^4–0^4	Silva et al. (2005a)
DMG_2 $Cu_2Cl_3 \cdot 2H_2O$	$P2_1/c$	15.678	6.264	22.811	90	124.29	90	0^2–0^2	Silva et al. (2005b)
DMG $MnCl_2 \cdot 2H_2O$	$P2_1/c$	7.580	16.324	16.857	90	106.21	90	1–2	Silva et al. (2006)
DMG $CdCl_2 \cdot 2H_2O$	$P2_1/a$	7.696	21.854	6.253	90	103.69	90	1–1	Yamada et al. (1994)
DMG_2 $PtCl_4$	$P2_1cn$	6.916	11.700	19.723	90	90	90	0–0	Gravenhorst et al. (1999)
DMG_2 $KCo^{III}(CO_3) \cdot H_2O$	$P2_12_12_1$	15.483	15.624	6.092	90	90	90	1–3	Kojima et al. (1994)
DMG_9 $Fe_3O(NO_3) \cdot 2H_2O$	P-1	12.424	16.565	17.627	90.59	101.48	108.97	0^3–0^3	Silva et al. (2003)

[a]Molybdate clusters (Keggin-type polyoxometalates), with L-ornithine zwitterions in coordination

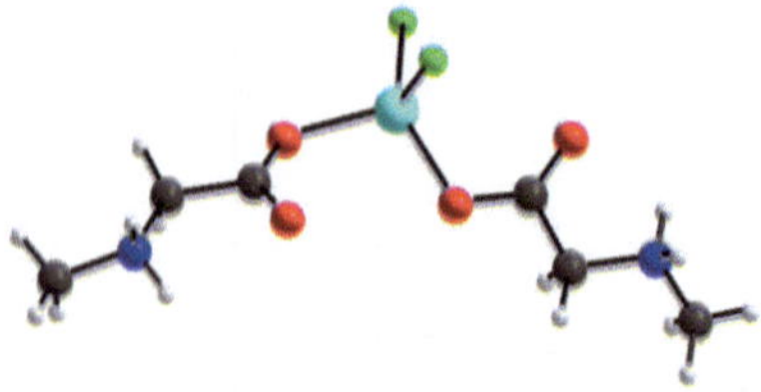

Fig. 4.37 *Units in dichloro-bis(sarcosinato)-zinc.* The unit is characterized by the typical [4]-coordinate zinc cations, found also in the respective salt of glycine, alanine, valine, betaine, etc.

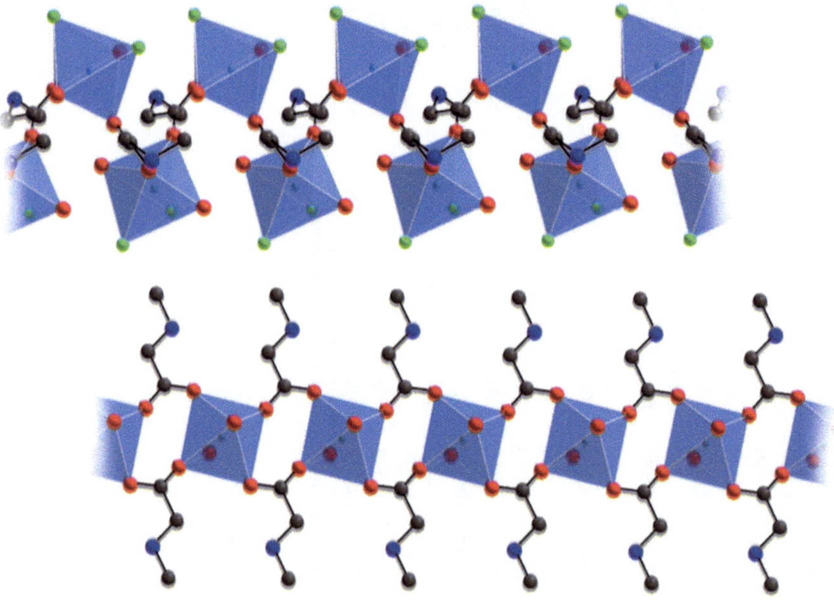

Fig. 4.38 *Chains in salts of sarcosine and $MnCl_2$.* On phase comprises sarcosine and $MnCl_2$ in a ratio of 1:1 (*top*), the other 2:1 (*bottom*). Note the similarity of the 2:1 phase with the respective glycine salt. Hydrogen atoms are omitted for clarity

halogenide salt with a ratio of 4:1. This work also contains a review of sarcosine metal halogenide salts (in comparison with glycine metal halogenide salts), where it was shown that all sarcosine metal halogenide structures have analogs with respective glycine compounds (if the connectivity is considered). The KI salt is an exception, as no likewise glycine metal halogenide salt with a 4:1 ratio is known.

Cystine, the dimer of cysteine, has been used in the synthesis of several larger polypeptides and compounds with organic molecules, but salts of cystine and neutral inorganic salts are rare. One instance is the neodymium perchlorate salt (Qingri et al. 1994), which has dimeric units of neodymium polyhedra (similar as in other rare earth element perchlorate amino acid salts). However, as cystine itself is a large amino acid with acid groups at opposite ends, the molecules provide connection of neighboring polyhedral dimers, thus forming an infinite layer (Fig. 4.39).

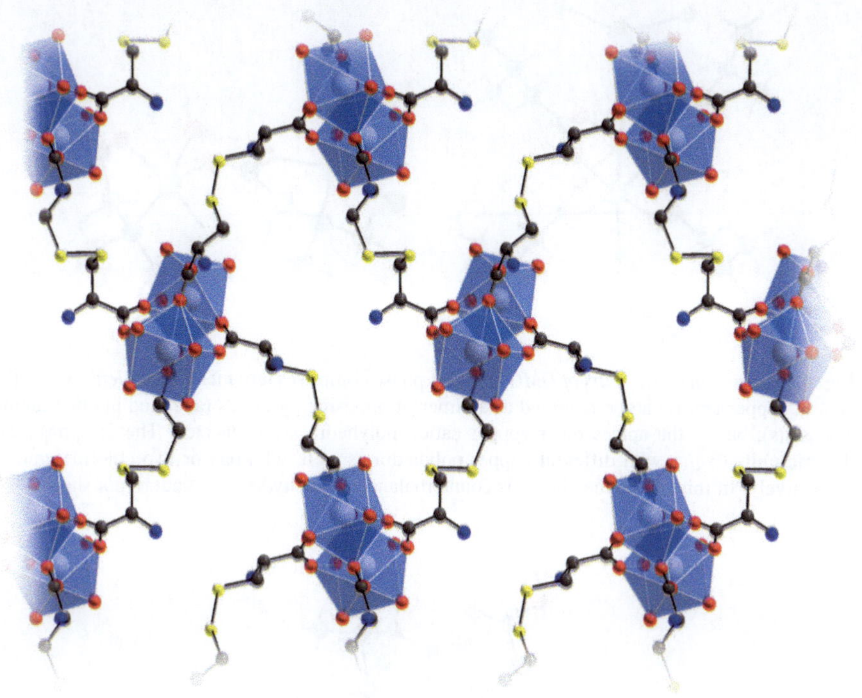

Fig. 4.39 *Layers in the neodymium perchlorate salt of cystine*. The dimeric units are similar as in rare earth element perchlorate salts of smaller amino acids, but as cystine has acid groups at both ends of its long backbone, the units are connected to layers. Hydrogen atoms are omitted for clarity

Of the salts of dimethylglycine, there are two phases of DMG and $CuCl_2$. The first has a ratio of DMG:$CuCl_2$ of 1:1 and is characterized by tetrameric units (Silva et al. 2005a). These units can be regarded as dimers of pyramidal coordinated copper cations of the type found for many amino acids, but in this instance the apices of the pyramids are part of additional copper polyhedra, in this case composed of four chloride anions (Fig. 4.40, left). The second phase has a ratio of 2:1, but only as far as DMG and copper are concerned. In fact, this species is actually a dimethylglycinate salt: The amino groups are deprotonated; the charge for two Cu^{2+} cations, three Cl^- anions, and two DMG anions is counterbalanced by a hydronium cation. The two copper cations are both [5]-coordinate, but have different ligand atoms (Fig. 4.40, right). Silva et al. (2005a) reported also weak antiferromagnetic interactions in this crystalline salt.

The $MnCl_2$ salt of DMG was reported by Silva et al. (2006), who also investigated the magnetic properties of this salt. The structure is characterized by Mn^{2+} polyhedra sharing common corners, which are further connected by DMG zwitterions to stabilize chains along the *a*-axis.

Another chain structure was found by Yamada et al. (1994): In the cadmium chloride salt of DMG, each cadmium cation is coordinated by one carboxylate oxygen atom, two water molecules, and three chloride anions in distorted octahedral geometry. Two chloride anions are bridging ligands, thus forming an infinite chain.

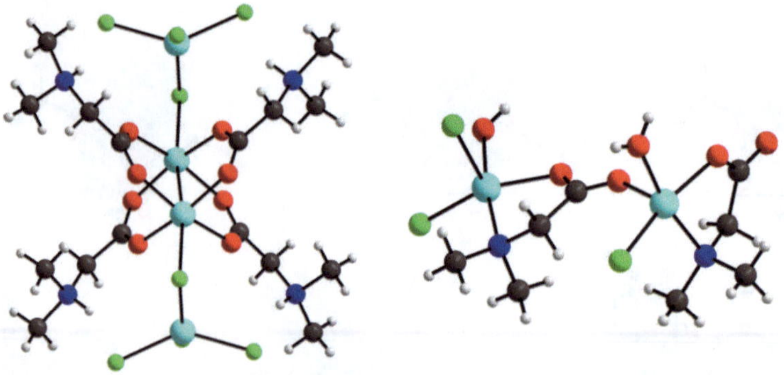

Fig. 4.40 *Units in $CuCl_2$ salts of DMG*. The 1:1 phase comprises tetrameric units (*left*), where the central copper atoms can be regarded as a dimer of opposing pyramids (as found in other amino acid salts), but to the apices other copper cation polyhedra are connected. The 2:1 phase has dimeric units (*right*) with different copper polyhedra as well, with one or two chloride anions, respectively. In this phase, the charge is counterbalanced by a hydronium cation (not shown)

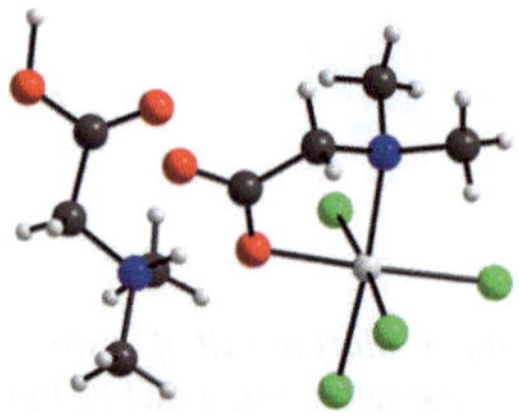

Fig. 4.41 *Dimethylglycinium tetrachloro-dimethylglycinyl-platinum*. This salt comprises units of platinum cations coordinated by four chloride anions as well as one dimethylglycinate anion as well as non-coordinating dimethylglycinium cations. It is thus the only salt where cations and anions of one amino acid occur within one structure

Among the DMG chloride salts in Table 4.13, there is another example of a salt which does not contain neutral amino acid molecules: The Pt^{4+} cation is octahedrally [6]-coordinate, with four chloride anions and one DMG anion in bidentate (O,O′) mode, plus another DMG cation which is not part of the coordination sphere (Fig. 4.41). This species is the only salt where amino acid molecules occur as cations and anions in one structure, which is possible due to the strong complex formed around the platinum cation (Gravenhorst et al. 1999).

4.12 Concluding Remarks

As shown in this chapter, the diversity of salts of amino acids and neutral salts is enormous. Nevertheless, several structural motifs recur, as these represent stable units, formed by the coordination of amino acids around specific cations. Especially, coordination patterns including oxygen and nitrogen atoms of the

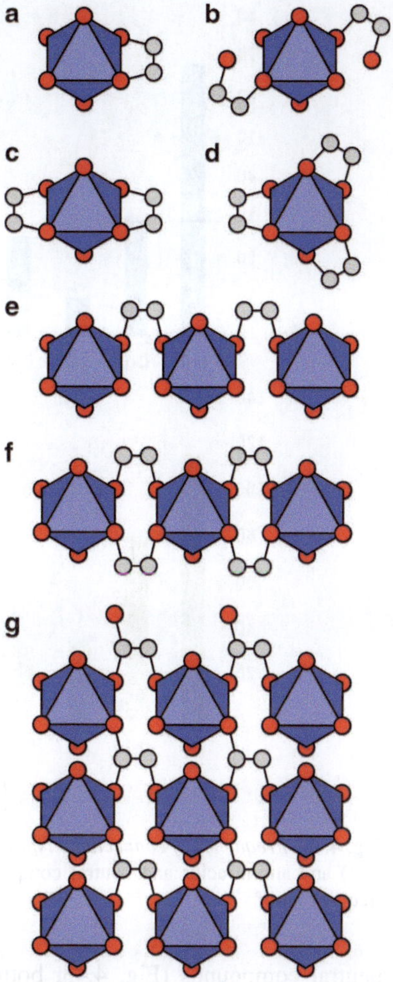

Fig. 4.42 *Schematic representation of common coordination motifs.* Nitrogen and oxygen atoms are not distinguished. Bidentate coordination by one amino acid (**a**) as found in $MetPtCl_2$, double monodentate coordination (**b**) as found in $Bet_2Ni(NO_3)_2.4H_2O$, double bidentate coordination (**c**) as found in L-$Lys_2NiCl_2 \cdot 3H_2O$, and triple bidentate coordination (**d**) as found in Co^{3+} salts of glycinate. Chain structures composed of isolated polyhedra connected by one bridging amino acid moiety (**e**) as found in the glycine manganese chloride salt or connected by two moieties (**f**) as found in the sarcosine manganese chloride salt. Layered structures composed of isolated polyhedra, connected by bridging amino acids (**g**) as found in the L-aspartate manganese chloride salt. Variation of all types occur, as, for instance, with additional monodentate amino acid moieties bonded to the polyhedra

α-carboxylate and amino groups are frequent, as these are relatively independent of the side-chain composition. Of course, side chains with atoms which can act as ligands may form other coordination patterns than side chains comprising only hydrocarbon parts. Thus, the nonpolar amino acids have relatively similar structural motifs, whereas amino acids with unique arrangements (like cysteine) form specific coordination patterns. For some of the most common patterns, schematic representations are shown in Fig. 4.42.

When comparing the frequency of these coordination patterns, it becomes apparent that lower-dimensional structures are most common. An arrangement of structures with regard to the connectivity (as designated by connection of polyhedra and connection of polyhedra via amino acid in the notation used in the tables in this and the previous chapter) yields the following picture: Isolated units are found most frequently, followed by chains, layers, and framework structures (Fig. 4.43). There is a slight discrepancy between salts of amino acids with cations (Fig. 4.43, top) and

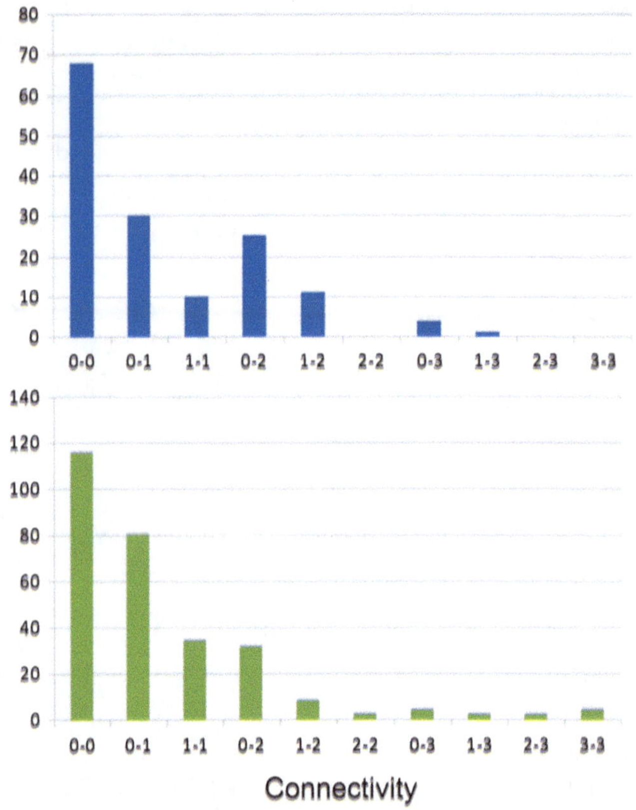

Fig. 4.43 *Frequency of connectivity types.* The connectivities of salts of amino acids and cations (*top*) and amino acids and neutral compounds (*bottom*) show that isolated units are the most frequent motif

neutral compounds (Fig. 4.43, bottom), although the distinction between these groups is not absolute (see examples in this chapter).

At any rate, these numbers cannot be considered as a strict correlation, but rather a trend, as for some amino acids, a large number of isotypic compounds (e.g., rare earth element salts) were synthesized. Still, the trend is obvious: Lower-dimensional connectivities are more likely to form than higher connectivities. (Of course, this consideration does not distinguish between the different amino acids, but apart from glutamic acid and aspartic acid, which are found to form higher-dimensional structures, this trend is general.)

One also has to note that the division of salts of amino acids with cations and such with neutral compounds is difficult (and not in all cases possible). Although some species clearly belong to one or the other category (as, e.g., silver glycinate or glycine silver nitrate), there are other instances where this distinction is not as straightforward. This is of course the case when inorganic cations and anions are both present. Thus, a salt synthesized out of a solution of an amino acid and a neutral compound (like betaine and $CuCl_2$) might give truly neutral salts (i.e., where one or more zwitterionic betaine molecules combine with one bivalent copper cation and two chloride anions) or betainium salts, where additional chloride anions counter the positive charge of the amino acid cations. On the other hand, amino acid anions may also combine with inorganic cations and anions, as, e.g., in copper glycinate nitrate, where one Cu^{2+} cation is balanced by one nitrate and one glycinate anion each. The most dramatic example is the DMG salt of platinum and chloride, where both DMG cations and anions occur within one structure (Gravenhorst et al. 1999). This chemical flexibility sometimes lead to confusion, as several papers (often, but not always older ones) refer to the amino acid always by the name of the neutral form, even when it exists in cationic or anionic form. Researchers should therefore always carefully check charge balances in order to avoid incorrect designation of amino acid cations, anions, or zwitterions.

Sadly, among the great number of publications on amino acid salts, there are some papers which promise more interesting results than they actually contain. Several instances of proposed "nonlinear optical materials" have been indicated, as well as some instances of simply careless works containing insufficient data (e.g., the L-valine salts of cadmium chloride and nickel chloride or the L-proline salt of lithium chloride). Therefore, authors should – despite the enthusiasm they devote to the research of amino acid salts – always be careful and thorough in the characterization of the products they synthesize.

References

Adams MJ, Hodgkin DC, Raeburn UA (1970) Crystal structure of a complexe of mercury(II) chloride and histidine hydrochloride. J Am Chem Soc: 2632–2635

Allen FH (2002) The Cambridge structural database: a quarter of a million crystal structures and rising. Acta Cryst B58:380–388

Almeida A, Chaves MR, Moreira JA, Pinto F, Klöpperpieper A (1998) Betaine potassium iodide dihydrate: a new compound of betaine. J Phys Condens Matter 10:L773–L778

Anandha Babu G, Ramasamy P (2008) Synthesis, crystal growth and characterization of novel semiorganic nonlinear optical crystal: dichlorobis(L-proline)zinc(II). Mater Chem Phys 113:727–733

Andrade LCR, Costa MMR, Paixao JA, Agostinho Moreira J, Almeida A, Chaves MR, Kloppenpieper A (1999) Crystal structure of betaine potassium iodide dihydrate, $(C_5H_{11}NO_2)_2KI.2H_2O$. Z Kristallogr NCS 214:83–84

Andrade LCR, Costa MMR, Pinto F, Paixao JA, Almeida A, Chaves MR, Kloppenpieper A (2000) Crystal structure of betaine potassium bromide dihydrate, $C_5H_{11}NO_2KBr.2H_2O$. Z Kristallogr NCS 215:537–538

Andrade LCR, Costa MMR, Filipa P, Rodrigues VH, Paixao JA, Almeida A, Chaves MR, Kloppenpieper A (2001) Crystal structure of betaine rubidium iodide dihydrate, $(C_5H_{11}NO_2)_2RbI.2H_2O$. Z Kristallogr NCS 216:227–228

Andrade LCR, Costa MMR, Paixao JA, Santos ML, Moreira JA, Chaves MR, Almeida A (2002) Crystal structure of calcium-doped strontium betaine chloride tetrahydrate, $(C_5NO_2H_{11})(Ca_{0.18}Sr_{0.82})Cl_2.4H_2O$. Z Kristallogr NCS 217:79–80

Anokhina EV, Jacobson AJ (2004) [Ni2O(l-Asp)(H2O)2]·4H2O: a homochiral 1D helical chain hybrid compound with extended Ni – O – Ni bonding. J Am Chem Soc 126:3044–3045

Anokhina EV, Go YB, Lee Y, Vogt T, Jacobson AJ (2006) Chiral three-dimensional microporous nickel aspartate with extended Ni – O – Ni bonding. J Am Chem Soc 128:9957–9962

Apfelbaum-Tibika F, Bino A (1984) Complexes of dimolybdenum(II) with DL-amino acids. Inorg Chem 23:2902–2905

Ardon M, Bino A (1995) Structure and reactivity of an intermediate in the two-electron reduction of dimolybdenum(III) by a hydride ligand in the MoHMo system. Inorg Chim Acta 229:417–420

Ardon M, Bino A, Cohen S, Felthouse TR (1984) Two quadruply bonded amino acid complexes of dichromium(II): structures of tetrakis(glycine)tetrabromodichromium(II) tetrahydrate and tetrakis(glycine)tetrachlorodichromium(II) trihydrate. Inorg Chem 23:3450–3455

Arularasan P, Sivakumar B, Chakkaravarthi G, Mohan R (2013) Bis(μ-L-arginine-κ^{3}N^2,O:O')bis (L-arginine-κ^{2}N^2,O)tetra-μ-chlorido-tetrachloridotetracopper(II). Acta Cryst E69:m583–m584

Ashida T, Bando S, Kakudo M (1972) The crystal structure of trissarcosine calcium chloride. Acta Cryst B28:1560–1565

Baidina IA, Podberezskaya NV, Borisov SV, Shestakova NA, Kuklina UF, Mal'chikov GD (1979) The crystal structures of cis-dichlorobis(glycine)platinum(II) and cis-dibromobis(glycine) platinum(II). J Struct Chem 20:467–470

Baidina IA, Podberezskaya NV, Borisov SV (1980) Crystal structure of barium dichloro(glycinato) palladate (II) dihydrate $Ba[Pd(NH_2CH_2COO)Cl_2]_2{\cdot}2H_2O$. J Struct Chem 21:659–662

Baidina IA, Podberezskaya NV, Borisov SV (1981) Crystal structure of cis-dichlorobis (glycine) platinum (II). J Struct Chem 22:465–467

Balakrishnan T, Ramamurthi K (2006) Growth and characterization of glycine lithium sulphate single crystal. Cryst Res Technol 41:1184–1188

Balakrishnan T, Ramamurthi K, Jeyakanthan J, Thamotharan S (2012) catena-Poly[[[aqua(glycine-κO)lithium]-μ-glycine-κ^{2}O:O'] bromide]. Acta Cryst E69:m60–m61

Balyaeva KF, Porai-Koshits MA, Malinovskii TI, Aslanov LA, Sukhanova LS, Martynenko LI (1969) Crystalline structure of neodymium triglycinate acid of the composition $NdCl_3(H_2NCH_2COOH)_3$ 3 H_2O. J Struct Chem 10:470–471

Baran J, Drozd M, Pietraszko A, Trzebiatowska M, Ratajczak H (2003) Crystal structure and vibrational studies of glycine-$LiNO_3$and glycine-$NaNO_3$ crystals. Pol J Chem 77:1561–1577

Barja B, Baggio R, Calvo R, Garland MT, Perec M, Rizzi A (2006) Gadolinium(III) and europium (III) L-glutamates: synthesis and characterization. Inorg Chim Acta 359:3921–3926

Bernal I, Cai J, Somoza F, Massoud SS (1999) The phenomenon of conglomerate crystallization. Part 50. The crystallization behaviour of [*trans*-Co(en)$_2$(NO$_2$)$_2$]ClO$_4$ (I), *meso*-[Co-*trans*-Me-(N-Me-Ethylenediamine)$_2$-trans-(NO$_2$)$_2$]ClO$_4$ (II), *K*[*trans*-Co(β-Alaninato)$_2$(NO$_2$)$_2$] (III) and the isolation and partial structural description of (H$_5$O$_2$)[*trans*-Co(β-Alaninato)$_2$(NO$_2$)$_2$] (IV). J Coord Chem 46:533–550

Bernes S, Gasque L (2008) Diaqua(nitrato-κ2O, O')bis(L-phenylalaninato-κ 2O, O')lead(II) nitrate. Acta Cryst E64:m566–m567

Bino A, Cotton FA, Fanwick PE (1979) Polymorphs of Tetrakis(glycine)tetrachlorodimolybdenum(II). Inorg Chem 18:1719–1722

Bino A, Cohen S, Altman J, Wilchek M (1988) The crystal and molecular structures of dichloro [*N*, *O*–(D, L)-diaminopropionic acid] platinum(II) and Dichloro [*N*, *O*–(L)-lysine] platinum(II) monohydrate. Inorg Chim Acta 147:99–102

Book L, Carty AJ, Chieh C (1981) Five-coordinated halogenomercurate(II) complexes: crystal structures of bis(L-tryptophanium) trichloromercurate(II) and bis(D, L-homocysteine thiolactonium) tetrabromomercurate(II). Can J Chem 59:138–143

Brill W, Schildkamp W, Spilker J (1985) Die Kristallstruktur der Hochtemperaturphase von Betaincalciumchlorid-Dihydrat, $(CH_3)_3NCH_2COO{\cdot}CaCl_2{\cdot}2H_2O$. Z Kristallogr 172:281–289

Brubaker GR, Busch DH (1966) Flexidentate Chelation of α, ω-Diaminocarboxylic acids with Copper(II) and Nickel(II). Inorg Chem 5:2110–2113

Burford N, Eelman MD, LeBlanc WG, Cameron TS, Robertson KN (2004) Definitive identification of lead(II)-amino acid adducts and the solid state structure of a lead–valine complex. Chem Commun: 332–333

Carty AJ, Taylor NJ (1976) Binding of inorganic mercury at biological sites: crystal structures of Hg^{2+} complexes with sulphur amino-acids. Chem Commun: 214–216

Caubet A, Moreno V, Molins E, Miravitlles C (1992) Methionine and histidine Pd(II) and Pt(II) complexes: crystal structures and spectroscopic properties. J Inorg Biochem 48:135–152

Che YX, Luo F, Zheng JM (2007) Supramolecular architectures of metallomacrocyclic and coordination polymers with dicarboxylate and 4,4′-bis(imidazol-1-ylmethyl)biphenyl ligands. J Mol Struct 828:15–18

Chen XM, Mak TCW (1991a) Metal–betaine interactions. Part 3. Crystal structures of polymeric diaquabis(betaine)disilver(I) dinitrate and bis(pyridine betaine)disilver(I) diperchlorate. J Chem Soc Dalton Trans 1219–1222

Chen XM, Mak TCW (1991b) Metal-betaine interactions. IV. Crystal structures of trans-tetraaquabis(betaine)nickel(II) nitrate and trans-diaquatetrakis(pyridine betaine)cobalt(II) bis [trichloro(pyridine betaine)cobaltate(II)]. J Crystallogr Spectrosc Res 21:471–478

Chen XM, Mak TCW (1991c) Metal-betaine interactions VII. Crystal and molecular structures of aquadichloro(pyridine betaine)zinc(II), dichlorobis(pyridine betaine)zinc(II) and dichlorobis (betaine)zinc(II) monohydrate. Inorg Chim Acta 182:139–144

Chen XM, Mak TCW (1991d) A linear polymeric manganese(II) complex bridged by skew-skew bridging carboxylato groups. Crystal structure of *catena*-tris(betaine)manganese(II) tetrachloromanganate (II), $[Mn(Me_3NCH_2COO)_3]_n{\cdot}nMnCl_4$. Inorg Chim Acta 189:3–5

Chen XM, Mak TCW (1991e) Metal-betaine interactions. II. Crystal structures of catena-dichloro (betaine)cadmium(II) and dimeric diaquabis(betaine)tetrachlorodicadmium(II). J Crystallogr Spectrosc Res 21:27–32

Chen XM, Mak TCW (1993) Metal-betaine interactions. Part 17: a study of intradimer Cu···Cu distance variation in copper(II) betaine complexes containing $[Cu_2\,(\text{carboxylato–O, O}')_4L_2]^{n+}$ species. Struct Chem 4:247–259

Chen XM, Huang XC, Tong ML, Tong YX, Ng SW (1997) Synthesis and crystal structures of two Monomeric Zinc(II) complexes containing carboxylate and Aqua Ligands. Aust J Chem 50:865–868

Chen D, Zhang D, Li L, Ma B, Jin T (2000) Changde Shifan Xueyuan Xuebao, Ziran Kexueban (Chin.) (J.Changde Teachers Univ. (Nat.Sci.Ed.)) 12, 14 (Quoted from CSD, code: YITGOX)

Ch'ng CD, Teoh SG, Chantrapromma S, Fun HK, Goh SM (2008) catena-Poly[[[diaquanickel (II)]-di-μ-glycine] dichloride]. Acta Cryst E64:m865–m866

Choudhury RR, Chitra R, Aliouane N, Schefer J (2013) Ferroelectric glycine silver nitrate: a single-crystal neutron diffraction study. Acta Cryst B69:595–602

Ciunik Z, Glowiak T (1980a) Glycinemanganese(II) dichloride dihydrate [*catena*-diaquadichloro-μ-glycinemanganese(II)]. Acta Cryst B36:1212–1213

Ciunik Z, Glowiak T (1980b) Conformational analysis of double carboxylic bridges in bis (DL-α-alanine)manganese(II) dibromide dihydrate and the related compounds. Inorg Chim Acta 44:L249–L250

Ciunik Z, Glowiak T (1980c) The structure of *catena*-octakis-μ-(β-alanine)-trimanganese (II) hexaperchlorate dihydrate. Acta Cryst B36:2029–2033

Ciunik Z, Glowiak T (1981) (DL-Proline)manganese(II) sulphate tetrahydrate [*catena*-diaqua-μ-(DL-proline)manganese(II) sulphate]. Acta Cryst B37:693–695

Civera M, Sironi M, Fornili SL (2005) Unusual properties of aqueous solutions of L-proline: a molecular dynamics study. Chem Phys Lett 415:274–278

Clegg W, Lacy OM, Straughan BP (1987) Structures of three glycine-bridged polymeric complexes: $[Mn(glycine)(H_2O)_2Cl_2]$, $[Co(glycine)(H_2O)_2Cl_2]$, and $[Co(glycine)(H_2O)_4](NO_3)_2$. Acta Cryst C43:794–797

Colyvas K, Tietze HR, Egri SKJ (1982) The structure of dichloro(L-histidine)copper(II). Aust J Chem 35:1581–1586

Cong-Ngoan D, Huskowska E, Glowiak T, Legendziewicz J (1988) Spectroscopy and crystal structure of octaaquatetra-L-α-alaninediholmium(III) hexachloride $[Ho_2(L\text{-}\alpha\text{-}AlaH)_4(H_2O)_8]Cl_6$. J Less Common Met 136:339–347

Cotton FA, Webb TR (1976) Tetrakis(glycine)dimolybdenum disulfate tetrahydrate. Inorg Chem 15:68–71

Csoregh I, Kierkegaard P, Legendziewicz J, Huskowska E (1987) Crystal structure of a praseodymium glutamate perchlorate hydrate, $Pr_2(L\text{-}Glu)_2(ClO_4)_4.11H_2O$. Acta Chem Scand 41:453–460

Csoregh I, Kierkegaard P, Legendziewicz J, Huskowska E (1989a) Crystal structure of holmium aspartate chloride hydrate $Ho(L\text{-}Asp)Cl_2.6H_2O$. Acta Chem Scand 43:636–640

Csoregh I, Czugler M, Kierkegaard P, Legendziewicz J, Huskowska E (1989b) Crystal structure, absorption, and fluorescence spectra of lanthanoid glutamate perchlorate nonahydrates, $Ln_2(C_5H_8NO_4)_2(ClO_4)_4.9H_2O$. Acta Chem Scand 43:735–747

Dan M, Rao CNR (2005) Metal aminocarboxylate coordination polymers with chain and layered structures. Chem Eur J 11:7102–7109

Dao CN, Glowiak T, Huskowska E, Legendziewicz J (1988) Spectroscopy and crystal structure of octaaquatetra-L-α-alaninediholmium(III) hexachloride $[Ho_2(L\text{-}\alpha\text{-}AlaH)_4(H_2O)_8]Cl_6$. J Less Common Met 136:339–348

Dauter Z, Hempel A, Jedrzejczak A (1977) Crystallographic data of hexaiodotellurates(IV) of some amino acids. Z Kristallogr 146:320–321

Davidovich RL, Logvinova VB, Zemnukhova LA, Udovenko AA, Kondratyuk IP (1991) Antimony(III) fluoride and oxyfluoride complexes with glycine.Koord.Khim.(Russ.) (Coord. Chem.) 17:1342–1345

Davies HO, Gillard RD, Hursthouse MB, Mazid MA, Williams PA (1992) Boussingault's mixed copper(II) glycinate nitrate. J Chem Soc Chem Commun 226–228

Davies HO, Gillard RD, Hursthouse MB, Lehmann A (1993) A linear polymer of copper(II) with water and the carboxy group of L-alanine as bridges. Chem Commun: 1137–1139

Davies HO, Brown DA, Yanovsky AI, Nolan KB (1995) Synthesis of trans-bis (glycinehydroxamato) platinum(II) hydrate and trans-dichlorobis (glycylglycine) platinum (II) dihydrate and the crystal and molecular structures of trans-dichlorobis (glycine) platinum(II) dihydrate and trans, trans-dichlorobis (glycinato) platinum (IV), hydrolysis products formed in reactions of the former complexes. Inorg Chim Acta 237:71–77

Davies HO, Park JH, Gillard RD (2003) New biomimetic chemistry in water: the anomalous 'salting-in' of neutral compounds of natural amino acids with transition metals. Inorg Chim Acta 356:69–84

Dejehet F, Debuyst R, Ledieu B, Declercq JP, Germain G, van Meerssche M (1978) Crystal and molecular structure of bis(β-alanine) zinc nitrate tetrahydrate. Electron spin resonance and optical studies of copper(II) diluted in this compound. Inorg Chim Acta 30:197–204

Du BS, Fan YT, Huang XY, Wu JJ (1996) Jiegou Huaxue(Chin.) (Chin J Struct Chem) (1996), 15, 141 (Quoted from CSD, code: ZUFCAE)

Duarte MTLS, deCarrondo MAAF, Goncalves MLSS, Hursthouse MB, Walker NPC, Dawes HM (1985) The preparation and crystal structure analysis of a 2:1 complex between L-lysine and copper(II) chloride. Inorg Chim Acta 108:11–15

Einspahr H, Gartland GL, Bugg CE (1977) Calcium binding to α-amino acids: the crystal structure of calcium *L*-glutamate chloride monohydrate. Acta Cryst B33:3385–3390

Elayaraja K, Parthiban SP, Ramalingom S, Bocelli G, Kalkura SN (2007) Tetraaquadiglycinemagnesium(II) hexaaquamagnesium(II) bis(sulfate). Acta Cryst E63:m2901–m2902

Erxleben A (2001) Synthesis and structure of $\{[Ag(SalGly)]\cdot 0.33H_2O\}n$: an infinite double helical coordination polymer. Inorg Chem 40:2928–2931

Estrader M, Diaz C, Ribas J, Solans X, Font-Bardia M (2008) Synthesis, characterization and magnetic properties of six new copper(II) complexes with amino acids as bridging ligand, exhibiting ferromagnetic coupling. Inorg Chim Acta 361:3963–3969

Evertsson B (1969) The crystal structure of bis-L-histidinecopper(II) dinitrate dihydrate. Acta Cryst B25:30–41

Fleck M, Bohaty L (2004) Three novel non-centrosymmetric compounds of glycine: glycine lithium sulphate, glycine nickel dichloride dihydrate, and glycine zinc sulphate trihydrate. Acta Cryst C60:m291–m295

Fleck M, Bohaty L (2005a) catena-Poly[[[tetraaquanickel(II)]-μ-glycine-κ2O:O'] dinitrate]. Acta Cryst E61:m1890–m1893

Fleck M, Bohaty L (2005b) catena-Poly[[[tetraaquamagnesium(II)]-μ-glycine-κ2O:O'] dinitrate]. Acta Cryst E61:m1887–m1889

Fleck M, Bohaty L (2005c) Two novel glycine metal halogenides: catena-poly[[[diaquanickel(II)]-di-μ-glycine] dibromide] and catena-poly[[[tetraaquamagnesium(II)]-μ-glycine] dichloride]. Acta Cryst C61:m412–m416

Fleck M, Bohaty L (2006) Compounds of glycine with metal sulfates and thiosulfates: glycine cobalt sulphate pentahydrate, glycine sodium thiosulfate dihydrate, and glycine potassium thiosulphate. Acta Cryst C62:m22–m26

Fleck M, Petrosyan A (2009) Comments on papers reporting IR-spectra and other data of alleged L-alanine alaninium nitrate and L-alanine sodium nitrate crystals. Cryst Res Technol 44:769–772

Fleck M, Schwendtner K, Hensler A (2006) Two new non-centrosymmetric lithium salts of glycine: bis (glycine) lithium chromate monohydrate and bis (glycine) lithium molybdate. Acta Cryst C62:m122–m125

Fleck M, Held P, Schwendtner K, Bohatý L (2008a) New compounds of glycine with metal halogenides. Z Kristallogr 223:212–221

Fleck M, Becker P, Bayarjargal L, Ochrombel R, Bohatý L (2008b) Crystal growth, crystal structure and physical properties of polar orthorhombic tris(glycine) zinc chloride. Cryst Res Technol 43:127–134

Fleck M, Lengauer C, Bohatý L, Tillmanns E (2008c) Synthesis, crystal structures and thermal behaviour of novel L-alanine halogenide compounds. Acta Chim Slov 55:880–888

Fleck M, Ghazaryan VV, Petrosyan A (2013) Diaquaiodidotetrasarcosinepotassium: an overview of sarcosine metal halogenide structures. Acta Cryst C69:11–16

Freeman HC, Golomb ML (1970) Model compounds for metal–protein interaction: crystal structure of three platinum(II) complexes of L- and DL-methionine and glycyl-L-methionine. J Chem Soc D: 1523–1524

Gao SL, Ren F, Shi QZ (2001) Yingyong Huaxue(Chin.) (Chin J Appl Chem) (2001), 18, 281 (Quoted from CSD, code: XUCNAK)

Gao SL, Li YZ, Ren F, Shi QZ, Wang LF (2002) *catena*-Poly[[[diaqualanthanum(III)]-tri-μ-glycinato] trichloride monohydrate]. Acta Cryst E58:m234–m236

Gasque L, Bernes S, Ferrari R, de Barbarin CR, Guiterrez MDJ, Mendoza-Diaz G (2000) Complexation of lead(II) by L-aspartate: crystal structure of polymeric Pb(aspH)(NO_3). Polyhedron 19:649–653

Gasque L, Bernes S, Ferrari R, Mendoza-Diaz G (2002) Cadmium complexation by aspartate. NMR studies and crystal structure of polymeric Cd(AspH)NO_3. Polyhedron 21:935–941

Gasque L, Verhoeven M, Bernes S, Barrios F, Haasnoot JG, Reedijk J (2008) The added value of solid-state Pb NMR spectroscopy to understand structures of Pb amino acid complexes. Eur J Inorg Chem: 4395–4403

Gawryszewska PP, Jerzykiewicz L, Sobota P, Legendziewicz J (2000) Spectroscopy and structure of Eu(III) complex with *N*-methylglycine. J Alloys Compd 300:275–282

Glowiak T, Ciunik Z (1977) Bis(DL-proline)manganese(II) dibromide dihydrate. Acta Cryst B33:3237–3238

Glowiak T, Ciunik Z (1978a) Two crystal structures of polymorphic bis(glycine)manganese (II) bromide dihydrate. Acta Cryst B34:1980–1983

Glowiak T, Ciunik Z (1978b) The crystal structure of bis(DL-valine) calcium chloride dihydrate and bis(DL-2- Aminobutyric Acid) calcium chloride dihydrate. Bull Acad Pol Sci Ser Sci Chim 26:43–51

Glowiak T, Dao CN (1993) Structure of pentaaqua(L-serine)holmium(III) trichloride. Acta Cryst C49:1171–1173

Glowiak T, Ngoan DC (1988) Structure of pentaaqua(L-threonine)holmium(III) trichloride. Acta Cryst C44:41–43

Glowiak T, Legendziewicz J, Dao CN, Huskowska E (1991a) Absorption, luminescence and crystal structure studies of dysprosium compound with L-α-Alanine: [Dy(L-α-AlaH)(H_2O)$_6$] Cl_3. J Less Common Met 168:237–248

Glowiak T, Dao CN, Legendziewicz J, Huskowska E (1991b) Structures of neodymium(III) complexes with amino acids: (I) catena-poly{[triaquatrichloroneodymium(III)]-μ-(β-alanine–O, O')}; (II) pentaaquachloro-μ-(L-proline–O, O')-neodymium(III) hexaaqua-μ-(L-proline-O, O')-neodymium(III) pentachloride. Acta Cryst C47:78–81

Glowiak T, Legendziewicz J, Huskowska E, Gawryszewska P (1996) Ligand chirality effect on the structure and its spectroscopic consequences in [Ln_2(Ala)$_4$(H_2O)$_8$] (C104)6 crystals. Polyhedron 15:2939–2947

Gordon LE, Harrison WTA (2004) Amino acid templating of inorganic networks: synthesis and structure of l-asparagine zinc phosphite, C4N2O3H8·ZnHPO3. Inorg Chem 43:1808–1809

Grandin A, Borel MM, Raveau B (1985) Intercalation of primary diamines and amino acid in the layer structure oxide $HTiNbO_5$. J Solid State Chem 60:366–375

Gravenhorst O, Bruhn C, Herzog R, Steinborn D (1999) N,N-Dimethylglycinatokomplexe von Platin(IV). Z Anorg Allg Chem 625:1349–1353

Hambley TW (1988) Structural and strain-energy-minimization study of the geometric and configurational isomers of bis(L-methioninato)cobalt(III). Acta Cryst B44:601–609

Hariharan M, Rajan SS, Srinivasan R, Natarajan S (1989) Crystal structure of a complex of glycine with zinc chloride. Z Kristallogr 188:217–222

Hasnaoui MA, Simon-Masseron A, Gramlich V Patarin J, Bengueddach A (2005) A new organic–inorganic hybrid gallophosphate prepared in the presence of glycine. Eur J Inorg Chem: 536–542

Haussühl S (1989) Elastic and thermoelastic properties of twelve adducts of betaine, $(CH_3)_3NCH_2COO$, with H_2O, HCl, HBr, HI, HNO_3, H_2SO_4, H_3PO_3, H_3PO_4, 1,4-toluene sulfonic acid, $MnCl_2$ and KBr. Z Kristallogr 188:311–320

Haussühl E, Wiehl L, Schreuer J (2009a) Crystal structure of catena-(tris(μ-betaine-O, O') manganese(II) tetrabromomanganate, [Mn($C_5H_{11}NO_2$)$_3$][$MnBr_4$]. Z Kristallogr NCS 224:33–34

Haussühl E, Wiehl L, Schreuer J (2009b) Crystal structure of trisbetaine manganese(II) dibromide dihydrate, ($C_5H_{11}NO_2$)$_3$·$MnBr_2$·$2H_2O$. Z Kristallogr NCS 224:35–36

Hempel A, Dauter Z, Szwabski S (1977a) Preliminary crystal data on hexabromoselenates(IV) of some amino acids. Z Kristallogr 146:318–319

Hempel A, Dauter Z, Pastuszak R (1977b) Crystal data on amino acids salts containing anions of the type MeX_6^{2-}. Z Kristallogr 146:317

Herbstein FH, Kapon M (1980) Crystal structures of the triiodide salt-molecule complexes (Xanthotoxin)$_2$·KI_3 and (Glycine)$_4$·KI_3. Z Kristallogr 153:73–87

Hoffmüller W, Polborn K, Beck W (1999) Metal complexes of biologically important ligands, CXIX. A tetrahedral zinc complex (L-*tert*-Leucine)$_2$$ZnCl_2$ of carboxylate coordinated L-*tert*-leucine in the zwitterionic form. Z Naturforsch B 54:734–736

Hu NH, Wang ZL, Niu CJ, Ni JZ (1993) Structure of an erbium complex with L-proline. Acta Cryst C49:2086–2089

Hu NH, Wang ZL, Niu CJ, Ni JZ (1995) Tetrakis(μ-DL-alanine-*O*:*O*')octaaquadierbium(III) Hexaperchlorate. Acta Cryst C51:1565–1568

Hu SM, Du WX, Xia SQ, Fu RB, Zhang JJ, Wu XT (2003) Jiegou Huaxue(Chin.) (Chin J Struct Chem) (2003), 22, 309 (Quoted from CSD, code: ELIMAN)

Hu R, Yu Q, Liang F, Ma L, Chen X, Zhang M, Liang H, Yu K (2008) Syntheses and crystal structures of *cis* - and *trans* -copper(II) complexes of L-arginine. J Coord Chem 61:1265–1271

Huskowska E, Turowska-Tyrk I, Legendziewicz J, Glowiak T (1998) Two high and low symmetry europium complexes with L-proline: spectroscopy and structure. J Alloys Compd 275:852–858

Inoue M, Yamase T (1995) Synthesis and crystal structures of γ-type octamolybdates coordinated by chiral lysines. Bull Chem Soc Jpn 68:3055–3063

Jayalakshmi K, Vijayan M (1967) Space group and unit cell of diglycine strontium chloride trihydrate and diglycine calcium chloride tetrahydrate. Acta Cryst 23:669

Jezierska J, Glowiak T, Ozarowski A, Yablokov YV, Rzaczynska Z (1998) Crystal structure, EPR and magnetic susceptibility studies of tetrakis [μ-(β-alanine)-O, O′]dichlorodicopper (II) dichloro monohydrate. Inorg Chim Acta 275:28–36

Jianxue L, Ninghai H, Chunji N, Qingbo M (1992) The crystal structure of a samarium complex with β-alanine. J Alloys Compd 184:L1–L3

Jin Q, Wang X, Jin T, Xu G, Zhang S (1994a) Preparation and crystal structure of a new erbium complex with glutamic acid: $[Er_2(Glu)_2(NO_3)_2(H_2O)_4]$ $(NO_3)_2{\cdot}5H_2O$. Polyhedron 13:2957–2961

Jin Q, Wang X, Jin T, Zhang S, Xu G (1994b) Crystal structure of two neodymium complexes with amino acids: $\{[Nd_2(cystine)_2(OH_2)_8](ClO_4)_6{\cdot}3H_2O_n\}$ and $[Nd_2(phenylalanine)_4(OH_2)_8](ClO_4)_6{\cdot}H_2O$. J Alloys Compd 207:400–405

Kalaiselvi D, Kumar RM, Jayavel R (2008) Redetermination of poly[μ-chlorido-heptachlorido-μ_3-L-proline-μ_2-L-proline-tetramercury(II)]. Acta Cryst E64:m1048–m1049

Kaminskii AA, Bohatý L, Becker P, Eichler HJ, Rhee H, Hanuza J (2009) Orthorhombic tris (glycine) zinc chloride $Gly_3{\cdot}ZnCl_2$ – a new semi-organic many-phonon SRS crystal manifesting different nonlinear-laser $(\chi_{(3)} + \chi^{(2)})$ interactions under one micron picosecond pumping. Laser Phys Lett 6:872–885

Kamwaya ME, Papavinasam E, Teoh SG (1984a) Structure of tetrakis[μ-(β-alanine)-O, O']-diaquadicopper(II) tetranitrate tetrahydrate, $[Cu_2(H_2O)_2(C_3H_7NO_2)_4](NO_3)_4.4H_2O$. Acta Cryst C40:2043–2045

Kamwaya ME, Papavinasam E, Teoh SG, Rajaram RK (1984b) Structure of bis[μ-(β-alanine)-O, O']-disilver(I) dinitrate, $[Ag_2(C_3H_7NO_2)_2](NO_3)_2$. Acta Cryst C40:1318–1320

Kamwaya ME, Papavinasam E, Teoh SG, Rajaram RK (1984c) Crystal structure of silver nitrate β -alanine $[Ag^+(NO_3)^-(NH_3^+CH_2CH_2COO^-)]$. Z Kristallogr 169:51–56

Kato H, Nakata K, Nagasawa A, Yamaguchi T, Sasaki Y, Ito T (1991) Preparation and properties of oxo-centered trinuclear chromium(III) complexes with bridging optically active amino acids, $[Cr_3(\mu_3\text{-}O)(\mu\text{-}O, O'\text{-amino-acid})_6(H_2O)_3]^{7+}$. Bull Chem Soc Jpn 64:3463–3465

Kirkwood B, Willett RD (1979) β-Alaninium tetrachloroferrate(III). Acta Cryst B35:1883–1884

Kocadag M, Fleck M, Bohatý L (2008) *catena*-Poly[[manganese(II)-tris(μ-betaine-$\kappa^2 O{:}O'$)] tetrabromidomanganate(IV)]. Acta Cryst E64:m1273–m1274

Kojima M, Saito R, Nakajima K, Ouyang Y, Fujita J (1994) Synthesis of trans(N)-Carbonatobis(N, N-dimethylglycinato)cobaltate(III) by Decarboxylation of N,*N*-Dimethylglycine. Bull Chem Soc Jpn 67:108–112

Korp JD, Bernal I, Bear JL (1981) The crystal and molecular structures of diaquo-tetra-μ-(β-alaninatonium)dirhodium(II) tetraperchlorate tetrahydrate. Inorg Chim Acta 51:1–7

Krishnakumar RV, Natarajan S (1995a) Structure of diaquabisglycinemagnesium (II) bromide. Indian J Phys A69:223–227

Krishnakumar RV, Natarajan S (1995b) Crystal structure of a complex of sarcosine with barium chloride. Cryst Res Technol 30:825–830

Krishnakumar RV, Subha Nandhini M, Natarajan S, Sivakumar K, Varghese B (2001a) Glycine sodium nitrate. Acta Cryst C57:1149–1150

Krishnakumar RV, Subha Nandhini M, Natarajan S (2001b) Trichloro(sarcosinio)zinc (II) monohydrate. Acta Cryst E57:m192–m194

Laligant Y, Laligant A, Emery J, Buzare JY (1994) Complexes of tetrachloro palladium(II) with alanine – synthesis, crystal-structure and solid-state-NMR-study of $(H\text{-}Alanine)_2$ $(PdCl_4)$. Eur J Solid State Inorg Chem 31:325–335

Lamberts K, Englert U (2012) Structures from MnX_2 and proline: isomorphous racemic compounds and a series of chiral nonisomorphous chain polymers. Acta Cryst B68:610–618

Lee M, Okaya Y, Pepinksy R (1962) Crystal structure of ferroelectric (glycine)$_2$·Mn_2·2H_2O. Bull Am Phys Soc 7:177

Legendziewicz J, Glowiak T, Huskowska E, Ngoan DC (1988) Crystal structure and spectroscopic studies of catena-diaqua-tris-μ-(L-proline)neodymium(III) triperchlorate, [Nd(L-pro·H)$_3$(H_2O)$_2$] (ClO_4)$_3$. Polyhedron 7:2495–2502

Legendziewicz J, Huskowska E, Waskowska A, Argay G (1989a) Spectroscopy and crystal structure of neodymium coordination compound with glycine: Nd_2(Gly)$_6$·(ClO_4)$_6$·9H_2O. Inorg Chim Acta 92:151–157

Legendziewicz J, Huskowska E, Argay G, Waskowska A (1989b) Spectroscopy and crystal structure of holmium and dysprosium complex compounds with glycine: Ln(Gly)$_3$(H_2O)$_3$(ClO_4)$_3$. J Less Common Met 146:33–47

Legendziewicz J, Glowiak T, Huskowska E, Ngoan DC (1989c) Crystal structures and spectroscopy studies of lanthanide complexes with L-proline [Ln(L-proH)$_2$(H_2O)$_5$]Cl_3 (Ln = Ho, Dy). Polyhedron 8:2139–2146

Legendziewicz J, Ciunik Z, Gawryszewska P, Sokolnicki J (1999) Structural effect of ligand chirality and its spectroscopic consequences. Polyhedron 18:2701–2709

Li JR, Zhou LP, Jin TZ (1999) Synthesis and crystal structure of neodymium complex with glycine [Nd(Gly)(H_2O)$_7$]Cl_3. Chem Res Chin Univ 15:306–310

Li J, Qi Y, Li J, Wang H, Wu X, Duan L, Wang E (2004) Heteropolymolybdate–amino acid complexes: synthesis, characterization and biological activity. J Coord Chem 57:1309–1319

Liang FP, Ma LF, Huang SW, Huanf YQ (2004) Yingyong Huaxue(Chin.) (Chin J Appl Chem) 21, 58 (Quoted from CSD, code: TEHKOH, TEHKUN, TEHLAU)

Lindqvist I, Rosenstein R (1960) The crystal structure of iron(II) sulphate pentahydrate glycine. Acta Chem Scand 14:1228–1229

Llorca J, Molins E, Espinosa E, Mata I, Miravitlles C, Cervantes G, Caubet A, Moreno V (2001) Dichloro(D-methionine-N, S)platinum(II) at 130 K. Acta Cryst C57:804–806

Lutz M, Bakker R (2003) Dichlorobis(DL-proline-κO)zinc(II). Acta Cryst C59:m18–m20

Ma AZ, Li LM, Lin YH, Xi SQ (1993a) Wuji Huaxue Xuebao(Chin.) (Chin J Inorg Chem) 9, 401 (Quoted from CSD, code: YICSUY)

Ma AZ, Li LM, Lin YH, Juin SC, Xi SQ (1993b) Yingyong Huaxue(Chin.) (Chin J Appl Chem) 10, 110–113 (Quoted from CSD, code: PIZKEO)

Ma AZ, Li LM, Lin YH, Xi SQ (1993c) Structure of an erbium coordination compound with L-proline, {[Er(Pro)$_2$(H_2O)$_5$]Cl_3}*n*. Acta Cryst C 49:865–867

Ma AZ, Li LM, Lin YH, Xi SQ (1994a) Crystal structure and infrared spectra of a lanthanum coordination compound with glycine, {[La(Gly)s·2H_2O]·(ClO_4)$_3$}$_n$. J Coord Chem 33:59–67

Ma AZ, Li LM, Jin SC, Lin YH, Xi SQ (1994b) (Quoted from CSD, code: ZOBCUO)

Ma AZ, Li LM, Lin YH, Xi SQ (1995) Jiegou Huaxue(Chin.) (Chin J Struct Chem) 14, 5 (Quoted from CSD, code: YIVZIM)

Ma BQ, Zhang DS, Gao S, Jin TZ, Yan CH, Xu GX (2000) From cubane to supercubane: the design, synthesis, and structure of a three-dimensional open framework based on a Ln_4O_4 cluster. Angew Chem Int Edit 39:3644–3646

Ma LF, Liang FP, Qin HC, Zhang MN, Hu RX, Yu KB (2002) Wuji Huaxue Xuebao(Chin.) (Chin J Inorg Chem) 18, 205 (Quoted from CSD, code: SASVAK)

Ma LF, Wang LY, Liang FP (2004) Jiegou Huaxue(Chin.) (Chin J Struct Chem) 23, 207 (Quoted from CSD, code: ESOZUH)

Maadeswaran P, Chandrasekaran J (2011) Synthesis, growth and characterization of L-valine cadmium chloride monohydrate – a novel semiorganic nonlinear optical crystal. Optik 122:1128–1131

McEwen RS (1973) Crystal structures of complexes with zwitterionic ligands. Chem Commun: 68–69

Mishima N, Itoh K, Nakamura E (1984) Structure of calcium chloride-sarcosine (1/3), $CaCl_2$.3$C_3H_7NO_2$, in the ferroelectric phase. Acta Cryst C40:1824–1827

Mohamed ME, Chopra D, Venugopal KN, Govender T, Kruger HG, Maguire GEM (2010) Tetrakis-μ-L-alanine-κ^{8}O:O'-bis[tetraaquaterbium(III)] hexaperchlorate. Acta Cryst E66: m193–m194

Mohana Rao JK, Natarajan S (1980) Structure of tris(glycine)calcium(II) dibromide. Acta Cryst B36:1058–1061

Mrozek R, Rzaczynska Z, Sikorska-Iwan M, Glowiak T (1999a) Crystal structure and vibrational spectra of two complexes of manganese(II) with glycine. J Chem Cryst 29:803–808

Mrozek R, Rzaczynska Z, Sikorska-Iwan M, Jaroniec M, Glowiak T (1999b) A new complex of manganese(II) with l-α-alanine: structure, spectroscopy and thermal study. Polyhedron 18:2321–2326

Müller G, Maier GM, Lutz M (1994) Lithium coordination to amino acids and peptides. Synthesis, spectroscopic characterization and structure determination of lithium complexes of neutral and anionic glycine and diglycine. Inorg Chim Acta 218:121–131

Narayanan P, Venkataraman S (1975) Crystal structure analyses of some addition compounds of glycine. Z Kristallogr 142:52–81

Natarajan S (2003) A hydrated complex of DL-alanine with cadmium chloride. Acta Cryst E59: m756–m758

Natarajan S, Rao JKM (1976) The crystal structure of diglycine calcium chloride tetrahydrate. Curr Sci 45:490

Natarajan S, Rao JKM (1977) The crystal structure of triglycine calcium iodide monohydrate. Curr Sci 46:595

Natarajan S, Ravikumar K, Rajan SS (1984) Crystal structure of diaqua nitrato glycine calcium (II) nitrate. Z Kristallogr 168:75–82

Natarajan S, Sundar JK, Athimoolam S, Srinivasan BR (2012a) Synthesis and characterization of anew strontium(II) coordination polymer based on a μ_2-tridentate bridging zwitterionic glycine. J Coord Chem 64:2274–2283

Natarajan S, Srinivasan BR, Sundar JK, Ravikumar K, Krishnakumar RV, Suresh J (2012b) Synthesis and structural characterization of a calcium coordination polymer based on a μ3-bridging tetradentate binding mode of glycine. J Chem Sci 124:781–790

Ng SW, Chen XM, Yang G (1998) Metallo-ligand $[Cu^{II}(carboxylato)_4]^{2+}$ cores: tetrakis (trimethylammonioacetato-*O*)copper diperchlorate and its 1/1 complex with sodium perchlorate. Acta Cryst C54:1389–1393

Pan L, Zheng NW, Jin XL (1997) A comparison of the structures of RE^{3+} with proline. Z Kristallogr 212:601–604

Panyushkin VT, Bukov NN, Abramov DE (2003) Structure and vibration spectra of neodymium biglycinate. Polyhedron 22:271–277

Papavinasam E, Natarajan S (1985) Crystal structure of bis-(*μ*-*β*-alanine-O′O)dicopper tetrachloride $[(C_3H_7NO_2)_2\ Cu_2Cl_4]$. Z Kristallogr 172:251–256

Pavone V, Lombardi A, di Blasio B, Benedetti E, Pedone C (1988) Pt(II) complexes of amino acids and peptides. I. Structural analysis of trans-$[Cl_2$-Pt-(L-HAlaOH$)_2]$. Inorg Chim Acta 153:171–174

Pepinsky R, Okaya Y, Eastman DP, Mitsui T (1957) Ferroelectricity in glycine silver nitrate. Phys Rev 107:1538–1539

Pepinsky R, Vedam K, Okaya Y (1958) New room-temperature ferroelectric. Phys Rev 110:1309–1311

Peterková J, Podlahová J, Loub J, Micka Z (1991) Structure of pentaaqua(glycine)nickel (II) sulfate monohydrate. Acta Cryst C47:2664–2666

Petrosyan AM, Ghazaryan VV, Fleck M (2013) On the existence of "l-threonine formate", "l-alanine lithium chloride" and "bis l-alanine lithium chloride" crystals. Spectrochim Acta A 105:623–625

Prelesnik BV, Herak R, Tomic Z, Andjelkovic K, Malinar MJ, Juranic N (1990) Aminocarboxylato chelate ligand conformation influence on the optical activity of cobalt(III) complexes—III. S-valinato ligand conformation in the crystal structure of Δ-cis(NO_2), trans (NH_2)-silver(I)- dinitrobis(S-valinato)cobaltate(III). Polyhedron 9:2149–2153

Price MG, Russell DR (1981) Crystal structure and absolute configuration of an isomer of potassium carbonatobis(L-valinato)cobaltate(III) dihydrate. J Chem Soc Dalton Trans: 1067–1068

Priya SMN, Varghese B, Linet JM, Bhagavannarayana G, Raj CJ, Krishnan S, Dinakaran S, Das SJ (2008) Synthesis, growth, and characterization of novel nonlinear optical active dichlorido-diglycine zinc dihydrate single crystals. Cryst Growth Des 8:1663–1667

Pushilin MA, Gerasimenko AV, Davidovich RL (2007) Bis(DL-valinium) pentafluorido-oxidoniobate(V). Acta Cryst E63:m2086

Qingri J, Xiaoqing W, Tianzhu J, Shiwei Z, Guangxian X (1994) Crystal structure of two neodymium complexes with amino acids: $\{[Nd_2(cystine)_2(OH_2)_8](ClO_4)_6{\cdot}3H_2O_n\}$ and $[Nd_2(phenylalanine)_4(OH_2)_8](ClO_4)_6{\cdot}H_2O$. J Alloys Compd 207:400–405

Rajagopal K, Krishnakumar RV, Subha Nandhini M, Ravikumar K, Natarajan S (2003) Bis(DL-alanine)tetraaquacobalt(II) dinitrate. Acta Cryst E59:m562–m564

Rao JKM, Viswamitra MA (1972) Crystal structure of glycine silver(I) nitrate. Acta Cryst B28:1484–1496

Ravikumar K, Rajan SS, Natarajan S, Ponnuswamy MN, Trotter J (1985) Structure of diaquabis-glycinecobalt(II) bromide. Z Kristallogr 171:201–207

Ravikumar K, Rajan SS, Natarajan S, Ponnuswamy MN, Trotter J (1986) Crystal structure of tris (glycine)calcium(II) dichloride. Z Kristallogr 175:217–225

Ren F, Gao SL, Shi QZ (2000) Xibei Dax.Xue.,Zir.Kex.(Chin.) (J NW Univ Nat Sci Ed) (2000), 30, 303 (Quoted from CSD, code: MOPWUJ)

Reynolds K, Willett RD, Twamley B (2003) (β-Alanine)dibromolead(II). Acta Cryst E59:m402–m404

Rodrigues VH, Beja AM, Paixao JA, Costa MMR (2005a) Crystal structure of hexaaqua-magnesium sulfate bis(N, N, N-trimethylaminoethanoic acid) trihydrate, $[Mg(H_2O)_6]SO_4{\cdot}2(CH_3)_3NCH_2COO{\cdot}3H_2O$. Z Kristallogr NCS 220:620–622

Rodrigues VH, Costa MMR, Klopperpieper A, Chaves MR, Almeida A, Agostinho Moreira J (2005b) Crystal structure of aqua-bis(N, N, N-trimethylammonioacetate) sodium bromide, $[Na(H_2O)\{(CH_3)_3NCH_2COO\}_2]Br$. Z Kristallogr NCS 220:363–364

Rzaczynska Z, Mrozek R, Glowiak C (1997) Crystal structure and properties of catena-poly [manganese(II)-μ-dichloro-μ-L-proline-k2o, o′] monohydrate. J Chem Crystallogr 27:417–422

Rzaczynska Z, Zawadowicz T, Mrozek R, Sikorska M, Glowiak T (1998) The crystal structure and properties of catena-poly[tetraaqua-bis (μ-β-alaninato-K^3O, O', O")-tetrakis (μ-β-alaninato-k^2O, O) dineodymium(III) hexachloride. J Coord Chem 44:343–351

Rzaczynska Z, Mrozek R, Sikorska-Iwan M, Zak Z, Glowiak T (1999) The properties and crystal structure of catena-poly[praseodymium(III)-hexaaqua-μ-L-threonine-κ^3O, O', O'']trichloride and catena-poly(pentaaqua-μ-L-threonine-κ^3O, O', O'') europium(III) and ytterbium(III) trichlorides. Pol J Chem 73:1259–1272

Rzaczynska Z, Mrozek R, Sikorska-Iwan M (2002) Crystal structure, spectroscopy and thermal study of manganese(II) complex with sarcosine. Pol J Chem 76:29–35

Sabo TJ, Dinovic VM, Kaluderovic GN, Stanojkovic TP, Bogdanovic GA, Juranic ZD (2005) Syntheses and activity of some platinum(IV) complexes with *N*-methyl derivate of glycine and halogeno ligands against HeLa, K562 cell lines and human PBMC. Inorg Chim Acta 358:2239–2245

Sangeetha MK, Mariappan M, Madhurambal G, Mojumdar SC (2012) Synthesis growth and characterization of l-valine nickel (II) chloride. J Therm Anal Calor 108:887–894

Saunders CDL, Burford N, Werner-Zwanziger U, McDonald R (2008) Preparation and comprehensive characterization of $[Hg_6(Alanine)_4(NO_3)_4]{\cdot}H_2O$. Inorg Chem 47:3693–3699

Schaak G (1990) Experimental results on phase transitions in betaine compounds. Ferroelectrics 104:147–158

Schaffers KI, Keszler DA (1993) Structure of $CdCl_2.C_3H_7NO_2$. Acta Cryst C 49:1156–1158

Schmidbaur H, Bach I, Riede J, Müller G, Helbig J, Hopf G (1987) Synthese und Struktur des Zink (l-aspartat)chlorids, Zn(L-AspH)Cl. Chem Ber 120:867–869

Schmidbaur H, Wilkinson DL, Schier A, Helbig J, Manniger G (1994) The structure of the pharmacologically active magnesium L-hydrogenaspartate chloride trihydrate. New J Chem 18:507–510

Schobert B, Tschesche H (1978) Unusual solution properties of proline and its interaction with proteins. Biochim Biophys Acta 541:270–277

Schreuer J, Haussühl S (1993) Crystal structure of catena-tri-μ-trimethylammoniumacetato-manganese tetrachloromanganate, $(((CH_3)_3NCH_2COO)_3Mn)MnCl_4$. Z Kristallogr 205:309–310

Schreuer J, Wiehl L, Wagner A, Hofmann P (2006a) Crystal structure of bis(hydrogenbetaine) tetrachlorocuprate(II) monohydrate, $[(CH_3)_3NCH_2COOH]_2[CuCl_4]\cdot H_2O$. Z Kristallogr NCS 221:525–526

Schreuer J, Wiehl L, Biehler J, Hofmann P (2006b) Crystal structure of tetrakis(μ-betaine-O, O') dibromodicopper(II) tetrabromocuprate(II) monohydrate, $[Cu_2\{(CH_3)_3NCH_2COO\}_4Br_2][CuBr_4]\cdot H_2O$, with a propeller-shaped dinuclear copper complex. Z Kristallogr NCS 221:529–531

Shkir M, Alfaify A, Khan MA, Dieguez E, Perles J (2014) Synthesis, growth, crystal structure, EDX, UV–vis–NIR and DSC studies of l-proline lithium bromide monohydrate—a new semiorganic compound. J Cryst Growth 391:104–110

Shtemenko AV, Kozhura OV, Pasenko AA, Domasevitch KV (2003) New octachlorodirhenate (III) salts: solid state manifestation for a certain conformational flexibility of the $[Re_2Cl_8]^{2-}$ ion. Polyhedron 22:1547–1552

Silva MR, Beja AM, Paixao JA, de Veiga LA, Martin-Gil J (1999) Crystal structure of bis (betainium) hexafluorotantalum(IV) dihydrate, $(C_5H_{12}NO_2)2TaF_6.2H_2O$. Z Kristallogr NCS 214:477–478

Silva MR, Beja AM, Paixao JA, de Veiga LA (2001a) Crystal structure of catena diaquadichloro-μ^2-(methylglycine-O, O')manganese(II). Z Kristallogr NCS 216:419–420

Silva MR, Paixao JA, Beja AM, de Veiga LA, Martin-Gil J (2001b) Crystal structure of tetrakis (μ-betaine-O, O′) dichloro-dicopper(II) dichloride tetrahydrated. J Chem Cryst 31:167–171

Silva MR, Beja AM, Paixao JA, Martin-Gil J (2003) Crystal structure of diaqua-monodimethylglycine-hexakis(μ_2-dimethylglycine-*O*:*O*)-μ_3-oxo-triiron(III) heptanitrate—water—dimethylgycine solvate (1:4:0.63), $Fe_3O(C_4H_9NO_2)_7(H_2O)_2(NO_3)_7\cdot 4H_2O\cdot 0.63C_4H_9NO_2$. Z Kristallogr NCS 218:557–560

Silva MR, Beja AM, Paixao JA, Martin-Gil J (2005a) Di-μ-chloro-1:2κ^2*Cl*;3:4κ^2*Cl*-hexachloro-1κ^3*Cl*,4κ^3*Cl*-tetra-μ-dimethylglycine-2:3κ^8*O*:*O*'-tetracopper(II). Acta Cryst C61:m380–m382

Silva MR, Beja AM, Paixao JA, Martin-Gil J (2005b) Weak antiferromagnetic interactions in hydronium diaqua-1κO, 2κO-trichloro-1κCl, 2κ^2Cl-μ-dimethylglycinato-1κO:2κ^2O', N-dimethylglycinato-1κ^2N, O–dicuprate(II). Acta Cryst C61:m507–m509

Silva MR, Beja AM, Paixao JA, Martin-Gil J (2006) Synthesis, crystal structure and magnetic properties of an alternating manganese chain. J Solid State Chem 179:2054–2058

Spek AL (2009) Structure validation in chemical crystallography. Acta Cryst D65:148–155

Steinborn D, Junicke H, Heinemann FW (1997) The crown ether influence on ligand exchange reactions of Na_2PtCl_6 with glycine and D-(+)-alanine; synthesis and characterization of platinum(IV) amino acid complexes. Inorg Chim Acta 256:87–92

Stenzel K, Fleck M (2004) Poly[[[diaquacobalt(II)]-di-μ-glycine] dichloride]. Acta Cryst E60: m1470–m1472

Stephens FS, Vagg RS, Williams PA (1977) The crystal and molecular structure of dichlorobis (δ-*N*-hydro-L-ornithinato)copper(II) dihydrate. Acta Cryst B33:438–443

Subha Nandhini M, Krishnakumar RV (2001) Dichlorobis(sarcosinato)zinc(II). Acta Cryst E57: m435–m437

Subha Nandhini M, Krishnakumar RV, Natarajan S (2001) Dichlorobis(DL-valine)zinc(II). Acta Cryst E57:m498–m500

Subha Nandhini M, Krishnakumar RV, Natarajan S (2002a) Dichlorobis(DL-alanine)zinc(II). Acta Cryst E58:m127–m129

Subha Nandhini M, Krishnakumar RV, Sivakumar K, Natarajan S (2002b) *catena*-Poly[[cadmium (II)]-μ-β-alanine-di-μ-chloro]. Acta Cryst E58:m307–m309

Suga T, Okabe N (1996) Aqua(L-O- phosphato)calcium(II). Acta Cryst C52:1894–1896

Sugandhi K, Verma S, Jose M, Joseph V, Das SJ (2013) Effect of pH on the growth, crystalline perfection, nonlinear optical and mechanical properties of tris-glycine zinc chloride single crystals. Opt Laser Technol 54:347–352

Suresh Kumar MR, Ravindra HJ, Dharmaprakash SM (2007) Synthesis, crystal growth and characterization of glycine lithium sulphate. J Cryst Growth 306:361–365

Taylor NJ, Carty AJ (1977) Nature of mercury(2+) ion-L-cysteine complexes implicated in mercury biochemistry. J Am Chem Soc 99:6143–6145

Tepavitcharova S, Havlicek D, Nemec I, Vojtisek P, Rabadjieva D, Plocek J (2009) Structural and spectral characterization of the compounds nGly·$ZnCl_2$·$m$$H_2O$ ($n=1,2,3$; $m=0,2$). J Mol Struct 918:55–63

Therup N (1977) The crystal structure of D-Histidinato-L-Histidinatocobalt(III) bromide. Acta Chem Scand A31:203–207

Thulasidhass R, Mohana Rao JK (1980) The crystal structure of bisglycine cadmium chloride. Curr Sci 49:349

Tianzhu J, Changqing Y, Qingchuan Y, Jinguang W, Guangxian X (1989) Gaodeng Xuexiao Huaxue Xuebao(Chin.) (Chem J Chin Uni) (1989), 10, 118 (Quoted from CSD, code: VEHJIB)

Tong YX, Chen XM, Ng SW (1997) Synthesis and crystal structures of two polymeric cadmium (II) complexes containing skew-skew carboxylato-bridges. Polyhedron 16:3363–3369

Torres J, Kremer C, Kremer E, Pardo H, Suescun L, Mombru A, Dominguez S, Mederos A (2001) Sm(III) complexation with α-amino acids: X-ray crystal structure of $[Sm_2(Hala)_4(H_2O)_8]$ $(ClO_4)_4(Cl)_2$. J Alloys Compd 323:119–124

Torres J, Kremer C, Kremer E, Pardo H, Suescun L, Mombru A, Dominguez S, Mederos A, Herbst-Irmer R, Arrieta JM (2002) Sm (III) complexation with amino acids. Crystal structures of $[Sm_2(Pro)_6(H_2O)_6](ClO_4)_6$ and $[Sm(Asp)(H_2O)_4]Cl_2$. J Chem Soc Dalton Trans: 4035–4041

Torres J, Kremer C, Pardo H, Suescun L, Mombru A, Castiglioni J, Dominguez S, Mederos A, Kremer E (2003) Preparation and crystal structure of new samarium complexes with glutamic acid. J Mol Struct 660:99–106

Trzebiatowska-Gusowska M, Gagor A, Baran J, Drozd M (2009) Trissarcosine barium(2+) dibromide – a new complex of N-methylglycine. J Raman Spectrosc 40:315–322

Tylczyński Z, Wiesner M (2014) Pyroelectric, piezoelectric, elastic and dielectric properties of triglycine-zinc chloride crystal. J Alloys Compd 588:177–181

Uma Devi T, Lawrence N, Ramesh Babu R, Selvanayagam S, Stoeckli-Evans H, Ramamurthi K (2009) Synthesis, crystal growth and characterization of l-proline lithium chloride monohydrate: a new semiorganic nonlinear optical material. Cryst Growth Des 9:1370–1374

Van Hecke K, Cartuyvels E, Parac-Vogt TN, Gorller-Walrand C, Van Meervelt L (2007) Poly [μ_2-L-alanine-μ_3-nitrato-sodium(I)]. Acta Cryst E63:m2354

Venkatasubramanian K, Saha NN (1984) Crystal and molecular structure of (d-lysinato) (l-lysinato) dichloro copper (II) dihydridate. Curr Sci 53:385–387

Venkatasubramanian K, Suresh E (1993) The crystal and moleculare structure of diaquabis(L-Lysino) nickel(II) dichloride monohydrate. J Coord Chem 29:225–232

Verbist J, Putzeys JP, Piret P, van Meerssche M (1971) Structure cristalline de dérivés d'Acides aminés. V. NaI.$_2$(+NH_3CH_2COO-).H_2O. Acta Cryst B27:1190–1194 (in French)

Via LD, Gia O, Magno SM, Dolmella A, Marton D, Di Noto V (2006) Synthesis, characterization and biological activity of platinum(II) complexes with L- and D-ornithine ligands. Inorg Chim Acta 359:4197–4206

Viera I, Torre MH, Piro OE, Castellano EE, Baran EJ (2005) Structural and spectroscopic characterization of aqua-diargininate-copper(II)-carbonate monohydrate. J Inorg Biochem 99:1250–1254

Vilminot S, Philippot E, Cot L (1974) Structure du sulfate d'ammonium et de glycinium $NH_4NH_3CH_2COOHSO_4$. Acta Cryst B30:2602–2606

Wang ZL, Hu NH, Niu CL, Yang KY, Ni JZ (1991) Chin Chem Lett 2, 961 (Quoted from CSD, code: SUFSUH)

Wang ZL, Hu NH, Niu CJ, Ni JZ (1993) JiegouHuaxue (Chin.) (Chinese J Struct Chem) (1994), 13:5 (Quoted from CSD, code: HEBCOG)

Wang X, Xie Y, Sun J (1996) Synthesis and crystal structure of a novel praseodymium complex with β-alanine. Polyhedron 15:3569–3574

Wang ZM, Lu Z, Liao CS, Yan CH (1998) Tris(glycine)scandium(III) Triperchlorate: a Space-Group Correction. Acta Cryst C54:IUC9800050

Wang R, Liu H, Carducci MD, Jin T, Zheng C, Zheng Z (2001) Lanthanide coordination with α-amino acids under near physiological pH conditions: polymetallic complexes containing the cubane-like $[Ln_4(\mu_3\text{-}OH)_4]^{8+}$ cluster core. Inorg Chem 40:2743–2750

Wang JP, Hu NH, Yang K, Zhang H, Niu CJ (2003) Tetra-μ-α -alanine-bis[tetraaquagadolinium (III)] hexaperchlorate. Acta Cryst C59:m52–m54

Wang S, Qin L, Zhou Z, Wang J (2012) Solubility and solution thermodynamics of betaine in different pure solvents and binary mixtures. J Chem Eng Data 57:2128–2135

Warren RC, McConnell JF, Stephenson NC (1970) The crystal and molecular structure of dichloro-DL-methioninepalladium(II). Acta Cryst B26:1402–1407

Wiehl L, Schreuer J, Stojic A (2006) Crystal structure of tetrakis(μ-betaine-O, O') dibromodicopper(II) dibromide dihydrate, $[Cu_2\{(CH_3)_3NCH_2COO\}_4Br_2]Br_2\cdot 2H_2O$, with a propeller-shaped dinuclear copper complex. Z Kristallogr NCS 221:527–528

Wiehl L, Schreuer J, Haussuhl E, Hofmann P (2008a) Crystal structure of monoaqua(betaine) dichloridocopper(II), $C_5H_{11}NO_2\cdot CuCl_2\cdot H_2O$. Z Kristallogr NCS 223:485–486

Wiehl L, Schreuer J, Haussuhl E, Hofmann P (2008b) Crystal structure of diaquabis(μ-betaine-O, O')tris(copper(II) dichloride), $(C_5H_{11}NO_2)_2\cdot 3CuCl_2\cdot 2H_2O$. Z Kristallogr NCS 223:487–488

Willett RD, Jardine FH, Rouse I, Wong RJ, Landee CP, Numata M (1981) Crystal structure, magnetic susceptibility, and EPR study of bis-(β-alaninium) tetrachlorocuprate(II): spin-diffusion effects in a two-dimensional square planar ferromagnet with anisotropic and anti-symmetric exchange. Phys Rev B24:5372–5381

Willett RD, Wong RJ, Numata M (1983) Magnetic susceptibility and EPR study of bis-(β-alaninium) tetrabromocuprate(II). Inorg Chem 22:3189–3194

Willett RD, Maxcy KR, Twamley B (2002) Polytypism in columnar group 14 halide salts: structures of $(Et_2NH_2)_3Pb_3X_9\cdot nH_2O$ (X = Cl, Br) and $(\beta\text{-alaninium})_2SnI_4$. Inorg Chem 41:7024–7030

Wilson C, Scudder ML, Hambley TW, Freeman HC (1992) Structures of dichloro[(S)-methionine-N, S]platinum(II) and chloro[glycyl-(S)-methioninato-N, N', S]platinum(II) monohydrate. Acta Cryst C48:1012–1015

Wong YS, Carty AJ, Chieh PC (1977) Crystal structure of (DL-methioninato)methylmercury(II). J Chem Soc Dalton Trans: 1157–1160

Xiang SC, Hu SM, Zhang JJ, Wu XT, Li JQ (2005) A new spherical metallacryptate compound $[Na\{Cu_6(Thr)_8(H_2O)_2(ClO_4)_4\}]\cdot ClO_4\cdot 5H_2O$: magnetic properties and DFT calculations. Eur J Inorg Chem: 2706–2713

Xu H, Zhang H, Qu X (2006) Interactions of the human telomeric DNA with terbium–amino acid complexes. J Inorg Biochem 100:1646–1652

Yamada J, Hashimoto H, Inomata Y, Takeuchi T (1994) Preparation and characterization of cadmium(II) halide complexes with N-substituted glycines, and the crystal structures of dichloro(N-methylglycine)cadmium(II) and diaquadichloro(N, N-dimethylglycine)cadmium (II). Bull Chem Soc Jpn 67:3224–3230

Yan CH, Zhang Y, Li B, Jin T, Xu G (1996) Coordination chemistry of scandium—I. Synthesis and crystal structure of scandium six-coordinate glycine complex $[Sc_2(Gly)_6(ClO_4)_6]n$. Polyhedron 15:2895–2899

Yan B, Wang HD, Chen ZD (2001) A novel supramolecular compound with three-dimensional hydrogen-bonded network $[Hbet]_3[Fe(CN)_6]\cdot 4H_2O$: synthesis, characterization and crystal structure. J Mol Struct 569:65–70

Yang G, Chen HA, Zhou ZY, Chen XM (1999) Synthesis and crystal structures of quadruply carboxylate-bridged dimeric lanthanide(III) complexes of betaine. J Chem Cryst 29:309–316

Yifan Z, Kezhen P (1987) Jiegou Huaxue(Chin.) (Chin J Struct Chem) (1987), 6, 132 (Quoted from CSD, code: SASVAK)

Yukawa Y (1992) Structures and properties of (L-prolinato)copper(II) complexes prepared from different solvents. J Chem Soc Dalton Trans: 3217–3221

Yukawa Y, Inomata Y, Takeuchi T (1983) Structure and properties of dichloro(L-proline)cadmium (II) hydrate. Bull Chem Soc Jpn 56:2125–2128

Yukawa Y, Yasukawa N, Inomata Y, Takeuchi T (1985) The structure of dichlorobis(L-proline) zinc(II). Bull Chem Soc Jpn 58:1591–1592

Yusenko K, Fox S, Guni P, Strasdeit H (2008) Model studies on the formation and reactions of solid glycine complexes at the coasts of a primordial salty ocean. Z Anorg Allg Chem 634:2347–2354

Yuzyuk YI, Almeida A, Chaves MR, Pinto F, Klöpperpieper A (2000) Phase transition in betaine potassium bromide dihydrate studied by Raman scattering. J Phys Condens Matter 12:6253–6264

Zemnukhova LA, Davidovich RL, Udovenko AA, Kovaleva EV (2005) Sb(III) fluoride complexes with DL-valine. crystal structure of molecular complex $SbF_3\{(CH_3)_2CHCH(^+NH_3)COO^-\}$. Russ J Coord Chem 31:115–120

Zhang JJ, Xiang SC, Hu SM, Xia SQ, Fu RB, Wu XT, Li YM, Zhang HS (2004) Syntheses and characterizations of a series of heptanuclear octahedral polyhedra $\{LnNi_6\}$ with different amino acids as ligands in aqueous media. Polyhedron 23:2265–2272

Zhang HY, Yu HJ, Xu HX, Ren JS, Qu XG (2007) Structural diversity of lanthanide–amino acid complexes under near physiological pH conditions and their recognition of single-stranded DNA. Polyhedron 26:5250–5256

Zhao JJ, Jin TZ (1996) Gaodeng Xuexiao Huaxue Xuebao(Chin.) (Chem J Chin Uni) (1996), 17, 519 (Quoted from CSD, code: PEVLUX)

Zhou ZH, Wang RJ, Mak TCW, Che CM (1991) Metal-betaine interactions Part V. Molecular and crystal structure of tetra(betaine)dichlorodirhodium(II) dichloride tetrahydrate, $[Rh_2(Me_3NCH_2CO_2)_4Cl_2]Cl_2{\cdot}4H_2O$. Inorg Chim Acta 180:1–4

Zhu ZG, Jin ZS, Liu JT, Liu WM, Ma DG (1990) Synthesis and crystal structure of $[Zn^+\{H_2N(CH_2)_4CH(NH_2)COONa\}2SO_4^-]{\cdot}H_2O$. Chin Sci Bull 35:1521

Chapter 5
Reactions of Amino Acids with Acids

Abstract In this chapter we consider reactions of amino acids with inorganic and organic acids. These reactions may result in salt formation when proton transfer from acid to amino acid takes place. However, in some cases transfer does not occur. In these cases either molecular adducts may be formed or they do not react at all. We divided the salts into three groups. The first group comprises of simple salts, where there is a cation of one type of amino acid and one type of anion. The second group is made up of salts with different types of dimeric cations: (1) $(A\ldots A^+)$, (2) $(A^+\ldots A^+)$, and (3) $(A^+\ldots A^{2+})$. The third group is made up of mixed salts with different cations, different anions, or both.

Keywords Salts of amino acids • Crystal structure • Molecular structure • Vibrational spectra • Strong hydrogen bonds • Nonlinear optical crystals • Mixed salts with different anions

5.1 Simple Salts

Simple salts of amino acids are compounds which consist of singly or doubly protonated cations of one type of amino acid and one type of anion. Compounds being molecular adducts are considered in Sect. 5.1.24. Salts with dimeric cations are considered in Sect. 5.2, while mixed salts with different cations and anions in Sect. 5.3. Because of the limited volume, polyoxometalates of amino acids were not included in this book.

5.1.1 Salts of Glycine

In Table 5.1 salts of glycine with inorganic and organic acids are presented. Out of all the amino acids, except amino acids forming also doubly charged cations, the majority of salts are found in the simplest amino acid glycine. Simple salts of types

M. Fleck and A.M. Petrosyan, *Salts of Amino Acids: Crystallization, Structure and Properties*, DOI 10.1007/978-3-319-06299-0_5,

Table 5.1 Salts of glycine with inorganic and organic anions

#	Crystal	s.g., Z	*a*, *b*, *c* (Å), *α*, *β*, *γ* (°)	Refs
1	(GlyH)F	*Pbca*, 32	15.610(2), 14.001(2), 15.662(3)	Fleck et al. (2010)
2	(GlyH)Cl	$P2_1/c$, 4	7.117(2), 5.234(2), 13.745(3), $\beta = 97.25(1)$	Al-Karaghouli et al. (1975)
3	(GlyH)Br	–	–	Fleck and Weiss (1984)
4	$(GlyH)NO_3$	$P2_12_12_1$, 4	5.607(1), 6.103(1), 16.380(3)	Fleck (n.d.)
5	$(GlyH)BF_4$	–	–	Ghazaryan et al. (2010a)
6	$(GlyH)ClO_4$	–	–	Ghazaryan et al. (2010a)
7	$(GlyH)H_2PO_3$	$P2_1/a$, 4	9.792(1), 8.487(1), 7.411(2), $\beta = 100.43(2)$	Averbuch-Pouchot (1993a)
8	$(GlyH)H_2PO_4$	$P2_1/c$, 4	9.580(2), 7.840(2), 9.249(3), $\beta = 114.7(2)$	Averbuch-Pouchot et al. (1988a)
9	$(GlyH)H_2AsO_4$	$P2_1/c$, 4	7.919(2), 4.7437(10), 17.59(1), $\beta = 91.71(4)$	Andrade et al. (1999)
10	$(GlyH)SbF_4$	$P112_1/b$, 4	8.563, 8.916, 9.52, $\gamma = 112.85$	Davidovich et al. (1991)
11	$(GlyH)ZrF_5{\cdot}2H_2O$	$P2_1/c$, 4	12.633(2), 6.636(2), 10.276(2), $\beta = 92.01(1)$	Davidovich et al. (2007)
12	$(GlyH)ZrF_5$	–	–	Davidovich et al. (2007)
13	$(GlyH)_2ZrF_6$	$P\bar{1}$, 2	7.215(2), 9.108(2), 9.322(1), $\alpha = 73.32(1)$, $\beta = 74.32(1)$, $\gamma = 70.62(1)$	Davidovich et al. (2007)
14	$(GlyH)_2NbOF_5$	$P2_1/c$, 2	5.3532(1), 10.8585(3), 8.9913(2), $\beta = 90.022(1)$	Gerasimenko et al. (2008)
15	$(GlyH)_2SO_4{\cdot}H_2O$	$P2_1/c$, 4	9.717(1), 8.481(1), 13.474(2), $\beta = 105.22(33)$	Cano and Martinez-Carrera (1974)
16	$(GlyH)_2SO_4$	*Pbca*, 8	8.9350(4), 10.2770(3), 21.7640(3)	Cherouana et al. (2002)
17	$(GlyH)_2SeO_4$	*Pbca*, 8	18.163(5), 11.149(3), 10.004(3)	Olejnik et al. (1975)
18	$(GlyH)_2Re_2Cl_8{\cdot}H_2O$	$P\bar{1}$, 2	8.2132(5), 9.6456(6), 12.8504(8), $\alpha = 74.161(1)$, $\beta = 78.217(1)$, $\gamma = 70.455(1)$, at 223 K	Shtemenko et al. (2003)

19	$(GlyH)_2SiF_6$	–	–	Ennan et al. (1981)
20	$(GlyH)_2SiF_6{\cdot}2H_2O$	$P2_1/c$, 2	10.9627(6), 6.7624(4), 9.2693(5), $\beta = 111.108(2)$	Ghazaryan et al. (2013a)
21	$(GlyH)_3P_3O_9$	$P2_1/c$, 4	12.223(8), 14.52(1), 10.229(7), $\beta = 100.47(5)$	Averbuch-Pouchot et al. (1988a)
22	$(GlyH)_4P_4O_{12}$	$P\bar{1}$, 1	7.988(5), 8.449(5), 9.739(5), $\alpha = 111.64(5)$, $\beta = 105.27(5)$, $\gamma = 99.40(5)$	Averbuch-Pouchot et al. (1988b)
23	(GlyH) monophenylphosphate	*Pbca*, 8	26.001(3), 7.423(2), 11.273(2)	Głowiak and Szemik (1986)
24	(GlyH) diphenylphosphate	*Pbca*, 8	11.412(3), 24.624(7), 10.775(3)	Głowiak and Wnęk (1985)
25	$(GlyH)CH_3SO_3$	$P2_1/n$, 4	5.360(2), 8.634(2), 15.703(3), $\beta = 98.06(3)$	Chwaleba et al. (2007)
26	(GlyH) *p*-toluenesulfonate	*Pbca*, 8	9.846(3), 8.119(2), 28.053(2)	Chwaleba et al. (2007)
27	$(GlyH)_2$ 1,5-naphthalenedisulfonate dihydrate	$P\bar{1}$, 1	7.486(4), 7.990(4), 9.214(5), $\alpha = 90.68(4)$, $\beta = 112.65(5)$, $\gamma = 101.51(4)$	Sudbeck et al. (1995)
28	(GlyH) 3-nitrophthalate	$P2_1/c$, 4	12.965(3), 10.966(2), 8.6833(17), $\beta = 107.61(3)$	Li and Chai (2007)
29	$(GlyH)CF_3COO$	$P2_1/c$, 4	4.9598(6), 12.239(2), 12.015(8), $\beta = 106.67(7)$	Rodrigues et al. (2002)
30	$(GlyH)CCl_3COO$	$P4_1$, 8	9.4416(9), 9.4416(9), 20.213(4)	Rodrigues et al. (2006a)
31	(GlyH) hydrogen oxalate	$P2_1/c$, 4	10.5807(15), 5.650(2), 12.093(3), $\beta = 113.83(1)$	Nandhini et al. (2001a)
32	(GlyH) hydrogen oxalate	$P2_1/c$, 8	11.8719(8), 6.1493(2), 20.8247(14), $\beta = 121.180(5)$	Tumanov et al. (2010)
33	$(GlyH)_2$ oxalate	$P2_1/n$, 2	4.9199(18), 9.959(4), 10.859(4), $\beta = 97.513(5)$	Chitra et al. (2006)
34	$(GlyH)_2$ oxalate dimethanol	$P2_1/c$, 2	5.0068(7), 11.3293(15), 12.893(2), $\beta = 106.077(12)$	Tumanov et al. (2010)
35	(GlyH) hydrogen malonate	$P2_1/c$, 4	10.1431(19), 8.1729(11), 9.260(2), $\beta = 101.879(16)$	Losev et al. (2011)
36	(GlyH) hydrogen maleate	$C2/c$, 8	17.689(4), 5.6610(11), 17.328(4), $\beta = 112.30(3)$	Rajagopal et al. (2001a)

Gly^+X^-, $(Gly^+)_2X^{2-}$, $(Gly^+)_3X^{3-}$, and $(Gly^+)_4X^{4-}$ (where Gly^+ is a protonated glycinium cation (GlyH) and X^-, X^{2-}, X^{3-}, and X^{4-} are respective anions) are known for glycine. The greatest number of salts belongs to the (GlyH)X type. Out of the halogenides, (GlyH)F, (GlyH)Cl, and (GlyH)Br are known for glycine. Frost (1942) tried to obtain also (GlyH)I, but could not. Searching for the conditions for the preparation of (GlyH)I led to the discovery of the compound 2Gly·HI (see Sect. 5.2). The fluoride of glycine together with fluorides of other amino acids was patented by Schmid as anticaries preparation (Schmid 1968). In the work of Ennan et al. (1981) this compound was also prepared and characterized by infrared spectrum and thermal analysis. Selvaraju et al. (2006a) and Vijayan et al. (2009a) reported on growth and characterization of this compound as possible NLO material. These papers, however, are questionable, as it was shown in Fleck et al. (2010), because provided IR spectra relate to glycine. The structure of (GlyH)Cl was determined by neutron diffraction method by Al-Karaghouli et al. (1975) and also by X-ray diffraction (Di Blasio et al. 1977a). An asymmetric unit contains one glycinium cation and one chloride anion. Chloride anion forms three hydrogen bonds of N-H. . .Cl type with the NH_3^+ group and one hydrogen bond of O-H. . .Cl type with the COOH group. The crystal of (GlyH)Br was obtained and investigated by nuclear quadrupole resonance (NQR) method (Fleck and Weiss 1984). There is also a paper devoted to the growth and characterization of glycine hydrobromide (Kirubavathi and Selvaraju 2008), which, however, is erroneous. Kirubavathi and Selvaraju (2008) obtained actually the previously known crystal 2Gly·HBr (see Fleck and Petrosyan 2010). Although the structure of (GlyH)Br is not published, one may conclude that it is isostructural with (GlyH)Cl based on great similarity of their IR spectra (Fig. 5.1). For the preparation of (GlyH)Br without the formation of 2Gly·HBr or mixture of (GlyH)Br and 2Gly·HBr, the quantity of HBr should be in excess compared to a stoichiometric 1:1 ratio.

The structure of Gly·HF was determined by Fleck et al. (2010). It differs from the structure of (GlyH)Cl (and hence also (GlyH)Br). There are four crystallographically different cations, (A), (B), (C), and (D). Three of these (A,B,C) form hydrogen bonds O-H. . .F with F1, F2, and F3, respectively, while the fourth cation (D) forms a dimer $[Gly(D)...H...Gly(D)]^+$. Respectively, the fluoride ion F4 forms a dimeric anion $(F4...H...F4)^-$. So, its formula may be presented as $[((GlyH)F)_3(Gly...H...Gly)(F...H...F)]$. In Fleck et al. 2010 the vibrational spectra of Gly·HF are provided and discussed. In Fleck et al. (2010) an interesting peculiarity was observed in that in time HF volatilizes from the crystal with formation of γ-form of glycine.

Glycinium nitrate $(GlyH)NO_3$ may be prepared from an aqueous solution containing equimolar quantities of components. The structure of $(GlyH)NO_3$ was determined by Narayanan and Venkataraman (1975). The non-centrosymmetric structure of the crystal attracted attention as possible NLO material (Martin Britto Dhas and Natarajan 2007a, b). The IR spectrum provided in this work, however, does not correspond to the spectrum of $(GlyH)NO_3$, which was investigated by Khanna et al. (1966), but corresponds to the spectrum of γ-form of glycine. A detailed investigation of the vibrational spectra of $(GlyH)NO_3$ was performed by

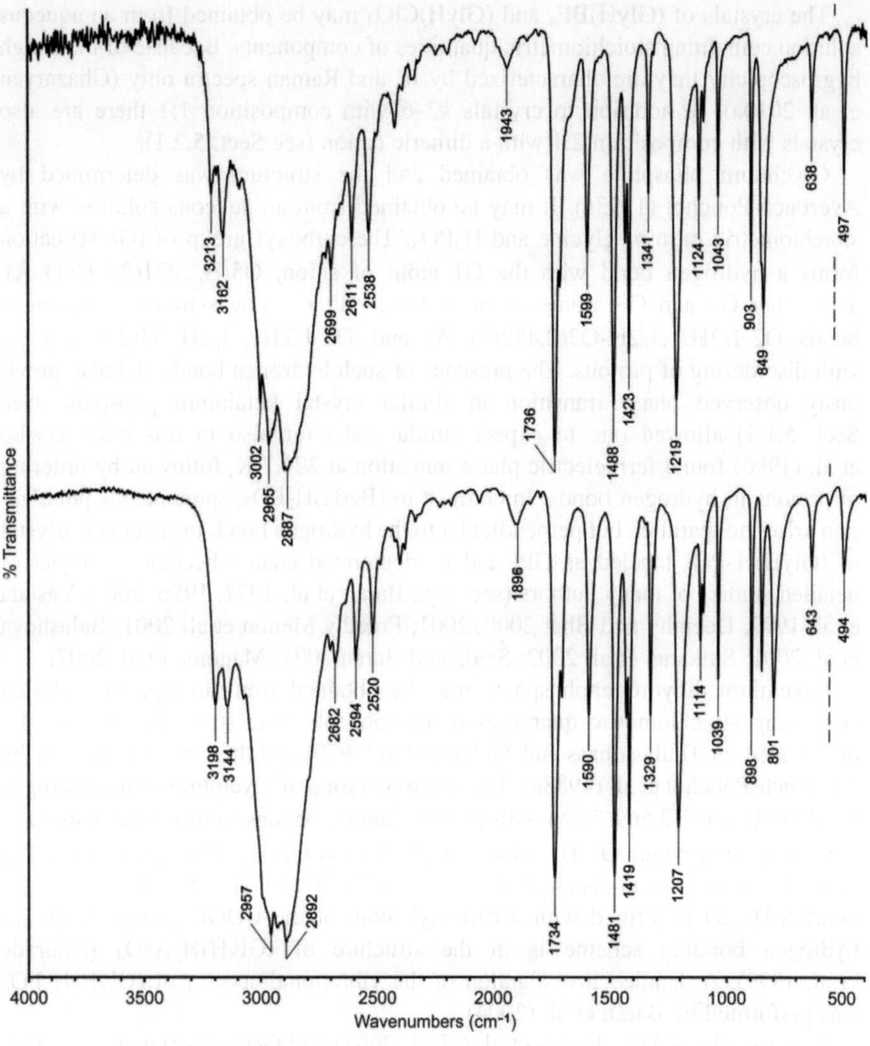

Fig. 5.1 IR spectra of glycinium chloride (*upper*) and glycinium bromide (*lower*)

Baran et al. (2010). More accurate structural data are provided by Fleck (n.d.). An asymmetric unit contains one formula unit. The carboxyl group of (GlyH) cation forms a hydrogen bond with anion, O1-H…O2N(2.641 Å). Here 2.641 Å is the O…O distance. The $N(1)H_3^+$ group forms two hydrogen bonds with nitrate anions, N1-H…O3N(2.793 Å) and N1-H…O2N(2.881 Å), and one hydrogen bond with a carbonyl oxygen atom of symmetry-related cation, N1-H…O2(2.886 Å). So, the atom O2N forms two hydrogen bonds, while the atom O1N does not form a hydrogen bond. As a result, the bond of N-O2N(1.266 Å) is the longest, N-O3N (1.240 Å) is intermediate, and N-O1N(1.225 Å) is the shortest.

The crystals of $(GlyH)BF_4$ and $(GlyH)ClO_4$ may be obtained from an aqueous solution containing stoichiometric quantities of components. Because of their high hygroscopicity they are characterized by IR and Raman spectra only (Ghazaryan et al. 2010a). In addition to crystals #2–6 with composition 1:1 there are also crystals with composition 2:1 with a dimeric cation (see Sect. 5.2.1).

Glycinium phosphite was obtained and the structure was determined by Averbuch-Pouchot (1993a). It may be obtained from an aqueous solution with a stoichiometric ratio of glycine and H_3PO_3. The carboxyl group of (GlyH) cation forms a hydrogen bond with the O1 atom of anion, O5-H. . .O1(2.598(1) Å). Two other O2 and O3 atoms of anion form rather strong symmetric hydrogen bonds O2-1/2H. . .1/2H-O2(2.482(1) Å) and O3-1/2H. . .1/2H-O3(2.518(1) Å) with disordering of protons. The presence of such hydrogen bonds and also previously observed phase transition in similar crystal betainium phosphite (see Sect. 5.1.4) allowed one to expect similar behavior also in this case. Dacko et al. (1996) found ferroelectric phase transition at 224.7 K, followed by ordering of protons in hydrogen bonds. In contrast to $(BetH)H_2PO_3$, spontaneous polarization arises not parallel, but perpendicular to the hydrogen bond. In sequel the crystal of $(GlyH)H_2PO_3$ labeled as GPI and its deuterated analog became a subject of detailed studies of many authors (see, e.g., Baran et al. 1977, 1996, 2002; Yasuda et al. 1997; Deepthy and Bhat 2000, 2001; Preethy Menon et al. 2001; Balashova et al. 2001; Shikanai et al. 2002; Śledź and Baran 2004; Magome et al. 2007).

Glycinium dihydrogenphosphate may be obtained from an aqueous solution containing stoichiometric quantities of components. The structure was initially determined by Thulasidhass and Mohana Rao (1979) and then more accurately by Averbuch-Pouchot et al. (1988a). The carboxyl group of glycinium forms hydrogen bond O6-H. . .O3(2.569(2) Å) with $H_2PO_4^-$ anion. Anions in turn form with each other hydrogen bonds O1-H. . .O4(2.568(2) Å) and O2-H. . .O4(2.596(3) Å). The atoms O3 and O4 form also hydrogen bonds of N-H. . .O type. The third hydrogen bond N-H. . .O is formed with a carbonyl atom of the COOH group. A similar hydrogen bonding scheme is in the structure of $(GlyH)H_2AsO_4$ (Andrade et al. 1999). A detailed investigation of the vibrational spectra of $(GlyH)H_2PO_4$ was performed by Baran et al. (2004)

In the works of Davidovich et al. (1991, 2007) and Gerasimenko et al. (2008), salts with SbF_4^-, ZrF_5^-, ZrF_6^{2-}, and $NbOF_5^{2-}$ anions were reported. In the structure of $(GlyH)SbF_4$ the COOH group forms hydrogen bond O-H. . .F (2.522 Å). The NH_3^+ group also forms N-H. . .F-type hydrogen bonds. The crystal of $(GlyH)SbF_4$ was studied by ^{121}Sb and ^{123}Sb NQR method in (Zemnukhova and Davidovich 1998). The structure of $(GlyH)ZrF_5{\cdot}2H_2O$ was determined by Gerasimenko et al. (2011) at 173 K. The carboxylic group forms hydrogen bond O-H. . .O (2.529(2) Å) with one of the water molecules.

In Gly + H_2SO_4 + H_2O in addition to well-known crystal triglycine sulfate (TGS), the crystals of $(GlyH)_2SO_4{\cdot}H_2O$ and $(GlyH)_2SO_4$ are also formed, which are usual centrosymmetric dielectrics (Srinivasan et al. 1983). Their crystal structures are determined (Cano and Martinez-Carrera 1974; Cherouana et al. 2002).

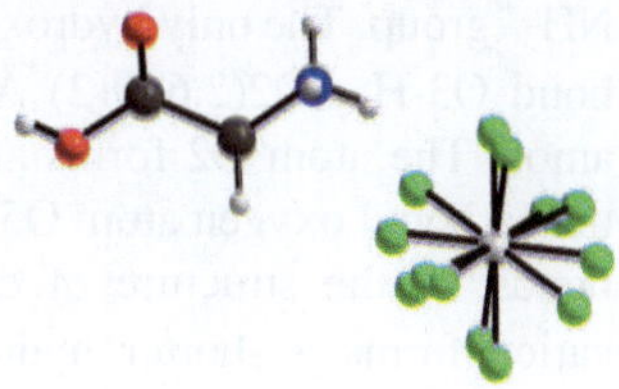

Fig. 5.2 Molecular structure of $(GlyH)_2SiF_6{\cdot}2H_2O$. Note the disorder of the hexafluorosilicate anion

Anhydrous diglycine selenate isostructural with respective sulfate is known as well, the structure of which was determined by Olejnik et al. 1975.

The crystal of $(GlyH)_2Re_2Cl_8{\cdot}H_2O$ is one of three salts of amino acids with $(Re_2Cl_8)^{2-}$ obtained by Shtemenko et al. 2003 and characterized structurally (see Sects. 5.2 and 5.3).

The compound $(GlyH)_2SiF_6$ is the first salt of amino acids with hexafluorosilicate anion obtained by Ennan et al. (1981) and characterized by IR and thermal analysis. The crystals which were obtained from an aqueous solution and dried over phosphorus oxide to constant weight apparently represent noncrystalline mass. In the study by Ghazaryan et al. (2013a) crystals of $(GlyH)_2SiF_6{\cdot}2H_2O$ were obtained and their crystal structure was determined (Fig. 5.2). By dehydration of $(GlyH)_2SiF_6{\cdot}2H_2O$ anhydrous $(GlyH)_2SiF_6$ also was obtained. Attempts to obtain crystals of anhydrous sample from solution at higher temperatures were not successful. However, upon attempting to obtain mixed salts with the supposed composition $(GlyH)_4(ClO_4)_2SiF_6$ from solution with stoichiometric composition instead of mixed salt, crystals with good quality were obtained, which were identical with anhydrous $(GlyH)_2SiF_6$ (Ennan et al. 1981). Their crystal structure will be determined. In addition to salts of $2A^+{\cdot}SiF_6^{2-}$, which is the main type, salts of types $A^+(A\ldots A^+)SiF_6^{2-}$ and $(A\ldots A^+)_2SiF_6^{2-}$ for amino acids forming singly charged cation also are known as well as salts of $A^{2+}{\cdot}SiF_6^{2-}$ for amino acids forming doubly charged cation.

The crystals of $(GlyH)_3P_3O_9$ (Averbuch-Pouchot et al. 1988a) and $(GlyH)_4P_4O_{12}$ (Averbuch-Pouchot et al. 1988b) are the first and yet only known salts of glycine formed by $3A^+{\cdot}X^{3-}$ and $4A^+{\cdot}X^{4-}$ mechanisms, respectively. These salts were obtained by metathesis reaction of $Ag_3P_3O_9{\cdot}H_2O$ and $Ag_4P_4O_{12}$ with (GlyH)Cl in 1:3 and 1:4 molar ratios, respectively. The cyclic anion is in a special position. Two crystallographically independent (GlyH) cations form hydrogen bonds: O10-H…O4A(2.567(2) Å) and O8-H…O1(2.653(1) Å). In case of $(GlyH)_3P_3O_9$ three independent glycinium cations form hydrogen bonds: O11-H…O8(2.702(2) Å), O13-H…O1(2.599(2) Å), and O15-H…O4(2.570(3) Å). Hydrogen atoms of all NH_3^+ groups also form hydrogen bonds with terminal oxygen atoms of cyclic $P_3O_9^{3-}$ anion.

In studies by Głowiak and Szemik (1986) and Głowiak and Wnęk (1985) the crystal structures of monophenylphosphate and diphenylphosphate of glycinium were determined. For monophenylphosphate IR spectrum was also studied. It is interesting to compare hydrogen bonds in the structures of these crystals with the hydrogen bonds of $(GlyH)(H_2PO_4)$. In the structure of monophenylphosphate the carboxyl group of glycinium cation forms hydrogen bond O6-H…O1(2.608(2) Å) with the phenylphosphate group. The atom O1 forms a hydrogen bond also with the

NH_3^+ group. The only hydroxyl group of phenylphosphate anion forms an hydrogen bond O3-H…O2(2.649(2) Å) to the nearest symmetry-related phenylphosphate anion. The atom O2 forms also two hydrogen bonds with the NH_3^+ group, while the carbonyl oxygen atom O5 and also O4 ester oxygen atom do not form hydrogen bonds. In the structure of diphenylphosphate the carboxyl group of glycinium cation forms a shorter hydrogen bond O5-H…O1(2.569(4) Å) with diphenylphosphate anion. The NH_3^+ group forms hydrogen bonds with the O2 atom, the O1 atom, and the carbonyl atom of neighboring glycinium cation. As in previous cases ester atoms O3 and O4 do not form hydrogen bonds.

Chwaleba et al. (2007) obtained and studied the salt of glycine with methanesulfonic and *p*-toluenesulfonic acids. In the structure of (GlyH)CH_3SO_4 the carboxyl group of glycinium forms O-H…O(2.631(2) Å) hydrogen bond with the sulfonate group. The NH_3^+ group also forms a hydrogen bond with oxygen atoms of the sulfonate group. In the structure of the second salt, the carboxyl group also forms O-H…O(2.626(2) Å) hydrogen bond. The NH_3^+ group forms similar hydrogen bonds. Mentioned salts and their deuterated analogs were studied by IR spectroscopic method in detail. The crystals of were studied also by DSC method with the aim to find phase transitions. In the 113–383 K for (GlyH)CH_3SO_3 and in 113–463 K for (GlyH) *p*-toluenesulfonate no phase transitions were observed.

In the structure of (GlyH) 1,5-naphthalenedisulfonate dihydrate (Sudbeck et al. 1995), the anion is in a special position (in the center of symmetry). Unlike the two previous anhydrous salts here the carboxyl groups of two crystallographically equivalent glycinium cations form hydrogen bonds not with the sulfonate group, but with water molecules. The distance O…O in hydrogen bond with water molecule is 2.614(3) Å. The NH_3^+ group forms hydrogen bonds with SO_3^- groups, and water molecule as a proton donor forms hydrogen bonds with the carbonyl oxygen atom and SO_3^- group.

Li and Chai (2007) obtained crystals of glycinium 3-nitrophthalate by heating of an aqueous solution containing equimolar quantities of components up to 378 K with following evaporation. The carboxyl group of glycinium forms O1-H…O5 (2.725(2) Å) hydrogen bond. The atom O5 forms also stronger hydrogen bond O3-H5…O5(2.571(2) Å) and also N1-H3…O5, while the atom O6 forms only one hydrogen bond N1-H2…O6. Probably by this reason the bond length C10-O5 (1.279(2) Å) is noticeably longer than the bond C10-O6(1.225(2) Å). The important role of the nitro group in the formation of salt should be noted. In case of the interaction of glycine with phthalic acid, a molecular adduct is formed instead of salt (see Sect. 5.1.24). In this regard it will be interesting to check the possibility of the formation of salt with 2:1 ratio of glycine and 3-nitrophthalic acid. There are two possibilities of salt formation: (Gly…Gly^+) 3-nitrophthalate$^-$ with dimeric cation and $2Gly^+ \cdot$3-nitrophthalate^{2-}.

Rodrigues et al. (2002, 2006a) obtained salts of (GlyH)CF_3COO and (GlyH) CCl_3COO and determined their structures. Of these the salt (GlyH)CCl_3COO has non-centrosymmetric tetragonal symmetry. This is the second crystal after (GlyH) NO_3 in Table 5.1 having a non-centrosymmetric structure. In the structure of (GlyH)CF_3COO the carboxyl group of glycinium forms hydrogen bond O-H…O

(2.597(2) Å) with the COO^- group of anion. The NH_3^+ group also forms hydrogen bonds with the COO^- group. In the structure of $(GlyH)CCl_3COO$ containing two independent formula units, carboxyl groups of cations form hydrogen bonds O-H...O with COO^- groups of anions with O...O distances 2.644(7) Å and 2.689(7) Å. However, the NH_3^+ groups form hydrogen bonds not only with oxygen atoms of anions, but also with carbonyl oxygen atoms of cations. The crystals of $(GlyH)CF_3COO$ in spite of their centrosymmetric structure were grown and characterized by Singh and Kumar (2011). The crystals decompose at above 101 °C. The crystals of $(GlyH)CCl_3COO$ were also reported by Esthaku Peter and Ramasamy (2012) as possible NLO material. The intensity of SHG is 0.5 compared to the intensity of SHG in KDP. Above 140 °C the crystals decompose. In addition to salts, the existence of which was confirmed by structure determination, there are some reports on obtaining "L-Glycine formate" (Sivanesan et al. 2010) and "triglycine acetate" (Hussani et al. 2009). These reports are, however, erroneous. In the work of Sivanesan et al. (2010) it was supposed that a compound with structure $HC(OH)=N\text{-}CH_2COOH$ was obtained, which was called by authors L-glycine formate. The obtained crystal actually, as it is clear from provided cell parameters, IR spectrum, and data of thermal analysis, is an α-form of glycine. "Triglycine acetate" also is an α-form of glycine, as it is clear from provided cell parameters, IR spectrum, and data of thermal analysis. The α-form of glycine has a centrosymmetric structure. Still, both Sivanesan et al. (2010) and Hussani et al. (2009) reported SHG in their crystals – an impossible physical feature, which can only be explained by erroneous work (e.g., the measurement of a signal arising from impurities or surface effects).

Information on compounds formed in glycine + oxalic acid + H_2O apparently dates back to the nineteenth century. In any event Dubský and Maitner (1936) reported obtaining of compounds $Gly{\cdot}H_2C_2O_4$ and $2Gly{\cdot}H_2C_2O_4$ and quoted earlier works. The structure of $Gly{\cdot}H_2C_2O_4$ was determined by Nandhini et al. (2001a). Asymmetric unit contains one $(GlyH)HC_2O_4$ formula unit. The carboxyl group of glycinium forms O1-H1...O5(2.593(2) Å) hydrogen bond with the carboxylate group of anion. The carboxyl group of anion also forms hydrogen bond O3-H7...O6(2.540(2) Å) with the second oxygen atom of the COO^- group of neighboring anion forming a chain of anions. The NH_3^+ group forms hydrogen bonds with carbonyl atoms of cation and anion as well as with the O6 atom of the carboxylate group of anion. Tumanov et al. (2010) found one more form of (GlyH) HC_2O_4 with the same symmetry, but with two independent formula units. The structure of this form has much common features with the previously known form (Nandhini et al. 2001a). Two independent hydrogen oxalate anions also form infinite chains. The carboxyl group of one anion forms hydrogen bond O4-H4...O11(2.4972(12) Å) with the carboxylate group of the second anion, which in turn forms hydrogen bond O9-H9...O2(2.4744(12) Å) with the carboxylate group of the first anion and so on. Two independent glycinium cations connect two chains of anions by hydrogen bonds O5-H5...O1(2.5923(13) Å) and O7-H7...12(2.6586(14) Å) on one hand and N-H...O hydrogen bonds of NH_3^+

groups on the other hand. Other distinctions of these forms are discussed by Tumanov et al. (2010). The form with $Z = 4(Z' = 1)$ is plate-like, while the form with $Z = 8(Z' = 2)$ is needle-like. Thus, both known forms of (GlyH)HC_2O_4 have monoclinic centrosymmetric $P2_1/c$ symmetry. In spite of this, some authors decided to grow and investigate these crystals and some even as NLO crystal. Sajan et al. (2004) studied both experimentally and theoretically the IR and Raman spectra of (GlyH)HC_2O_4. Arun and Jayalekshmi (2008a) have grown plate-like (GlyH)HC_2O_4 crystals. From TG and DTA curves, it is seen that there is an endoeffect with peak at 179 °C caused by melting, while the second peak at 190 °C is caused by decomposition. However, on the DSC curve before 200 °C there is only one peak at 179 °C. In the work of Arun and Jayalekshmi (2008a) the IR spectrum is also provided. Crystals of (GlyH)HC_2O_4 reported by Mythili et al. (2008) also were plate-like, and unit cell parameters correspond to the form with $Z' = 1$. According to Mythili et al. (2008) and Arun and Jayalekshmi (2011), the grown crystals reveal SHG signal on the level of KDP. On the other hand, crystals obtained by Chandrasekaran et al. (2011a) were not plate-like, but needle-like. Chandrasekaran et al. (2011a) did not identify their crystals; however, thermal curves, powder XRD pattern, and IR spectrum correspond to that provided in previous works for the form with $Z' = 1$. In case of the salt with the composition 2Gly·$H_2C_2O_4$, the formation of at least two types of salts is possible: (Gly. . .GlyH) HC_2O_4 as in the case of 2Bet·$H_2C_2O_4$ (see Sect. 5.2.1.3) and also (GlyH)$_2C_2O_4$ type as in the cases of oxalates of DL-Ser and L-Ser (see Sect. 5.1.8) as well as oxalate of DL-Asp (see Sect. 5.1.18). Crystal structure determination of 2Gly·$H_2C_2O_4$ (Chitra et al. 2006) showed that the second possibility is realized. The asymmetric unit may be expressed as (GlyH)·0.5C_2O_4, because there is one independent glycinium cation and oxalate anion in special position (in center of symmetry). The presence of strong hydrogen bond O3-H6. . .O2(2.4544(15) Å) is noteworthy. The NH_3^+ group forms hydrogen bond both with the oxalate anion (two bonds with the O1 atom) and with the carbonyl O4 atom of the nearest cation. Based on the presence of a strong hydrogen bond, the authors supposed the possibility of phase transitions. However, by DSC method in the interval above 123 K up to melting point (428 K), phase transitions were not observed. Chitra and Choudhury (2007) determined the structure of (GlyH)$_2C_2O_4$ crystal by neutron diffraction method in order to obtain more accurate data about O-H. . .O hydrogen bonds. The distance O3. . .O2 according to neutron diffraction is equal to 2.461(5) Å. It was established that there is no disordering of proton. IR and Raman spectra were studied and also the DSC curves were provided in the 173–473 K interval. In another paper (Choudhury et al. 2009) the structure of (GlyH)$_2C_2O_4$ crystal was determined at 100 K, and the distribution of charge density was determined by the method of maximal entropy. Mishra et al. (2010) and Bhatt et al. (2012) investigated the behavior of (GlyH)$_2C_2O_4$ crystal at higher pressure. According to obtained data the phase existing at atmospheric pressure (phase I) transforms into phase II near 1.7 GPa based on Raman spectra (Mishra et al. 2010) and at 1.6 GPa based on IR spectra (Bhatt et al. 2012). Then phase II transforms into phase III near 5 GPa based on Raman spectra (Mishra et al. 2010) and 4.5 GPa based on IR spectra (Bhatt et al. 2012). It was supposed based on X-ray diffraction studies (Bhatt et al. 2012) that phase II is noncetrosymmetric with space group $P2_1$ and phase III possibly is a disordered low symmetry phase. One more crystal from this family was found by Tumanov et al. (2010) from a solution containing glycine and oxalic acid in 2:1 molar ratio using methanol as an antisolvent, if the solution was not stirred. The volume of added methanol was approximately equal to the volume of the initial saturated aqueous solution. Tumanov et al. (2010) noted that if the same amount of methanol was added rapidly with energetic stirring, a powder sample of known $(GlyH)_2C_2O_4$ was formed. The composition of the crystal is $(GlyH)_2C_2O_4{\cdot}2CH_3OH$. In the structure of this crystal, oxalate anion also is in a special position. So, the asymmetric unit is $(GlyH){\cdot}0.5C_2O_4{\cdot}CH_3OH$. Here a strong hydrogen bond is also formed O2-H2…O3(2.4656(12) Å). There is also O5-H5…O1 hydrogen bond and N-H…O bonds with methanol and oxalate anion.

Glycinium hydrogen malonate was obtained by Losev et al. (2011) from aqueous solution with stoichiometric composition of components. Hydrogen malonate anions form a chain via O6-H6…O3(2.5465(19) Å) hydrogen bond between carboxyl and carboxylate groups. The carboxyl group of glycinium forms O1-H1…O4(2.5289(18) Å) hydrogen bond with the second oxygen atom of the carboxylate group of hydrogen malonate anion. The NH_3^+ group forms a hydrogen bond with the O4 atom and carbonyl atom O5 of hydrogen malonate and a weak hydrogen bond with the O1 atom.

The last salt in Table 5.1 is glycinium hydrogen maleate obtained by Rajagopal et al. (2001a) from an aqueous solution with stoichiometric composition of components. The hydrogen maleate anion usually forms strong intramolecular hydrogen bond O3-H3…O5(2.4465(19) Å). Glycinium cation forms hydrogen bond O1-H1…O6(2.5861(17) Å) with hydrogen maleate anion. The NH_3^+ group forms a hydrogen bond both with the anion and carbonyl atom O2 of the carboxyl group of neighboring cation. In order to further investigate, Singh et al. (2012) and Ruby and Raj (2012) have grown crystals of glycinium hydrogen maleate. In addition, a report was published on "bis-glycine maleate" single crystals having triclinic symmetry (Balasubramanian et al. 2010). In the paper of Petrosyan et al. (2012a) it was shown that this work was erroneous and actually the obtained crystal is glycinium hydrogen maleate obtained by Rajagopal et al. (2001a).

5.1.2 *Salts of Sarcosine*

Known salts of sarcosine (N-methylglycine) are presented in Table 5.2. The preparation of crystals of salt with Sar·2HF composition is described by Schmid (1968). According to this paper, Sar·2HF can be prepared by mixing 17.8 g of sarcosine and 22.2 g of 60 % solution of HF in a platinum dish. After dissolving by

Table 5.2 Salts of sarcosine with inorganic and organic anions

#	Crystal	s.g., Z	*a*, *b*, *c* (Å), *α*, *β*, *γ* (°)	Refs
1	Sar·2HF	–	–	Schmid (1968)
2	(SarH)Cl	$P2_1$, 2	9.00(2), 5.93(1), 5.41(1), $\beta = 96.0$	Bhattacharyya and Saha (1978a)
3	(SarH)Br	–	–	Ghazaryan et al. (2012a)
4	$(SarH)NO_3$	–	–	Fleck et al. (2012a)
5	$(SarH)BF_4$	–	–	Ghazaryan et al. (2012b)
6	$(SarH)ClO_4$	–	–	Ghazaryan et al. (2012b)
7	$(SarH)H_2PO_4$	$P2_12_12_1$, 4	13.180(8), 9.275(3), 6.245(2)	Averbuch-Pouchot et al. (1988c)
8	$(SarH)H_2PO_3$	$P2_12_12_1$, 4	9.405(3), 13.285(5), 6.005(1)	Averbuch-Pouchot (1993b)
9	$(SarH)_2SO_4$	*P(-1)*, 4	7.623(1), 11.538(2), 14.214(3), $\alpha = 71.46(2)$, $\beta = 74.36(2)$, $\gamma = 86.46(2)$ at 140 K	Mostad and Natarajan (1996)
10	$(SarH)_2SiF_6$	$P2_1/n$, 2	5.7240(11), 6.3450(13), 17.453(3), $\beta = 91.99(3)$	Fleck et al. (2012b)
11	$(SarH)_2TeBr_6$	[a]	12.41(2), 11.64(2), 7.91(1), $\gamma = 126.2(2)$	Hempel et al. (1977a)
12	$(SarH)_2TeI_6$	[b]	25.00(3), 25.00(3), 8.26(1)	Dauter et al. (1977)
13	(SarH) *p*-toluenesulfonate	$P2_1/c$, 4	12.432(2), 6.799(2), 15.169(2), $\beta = 98.416(9)$	Ilczyszyn et al. (2008a)
13	(SarH) *p*-toluenesulfonate	$P2_1/n$, 8	15.049(7), 6.660(5), 24.695(2), $\beta = 97.236(4)$, at 100 K	Ilczyszyn et al. (2008a)
14	$(SarH)CF_3COO$	*C2/c*, 8	14.5912(19), 7.0951(11), 15.8414(13), $\beta = 99.138(8)$	Rodrigues et al. (2000)
15	$(SarH)HC_2O_4{\cdot}H_2O$	$P2_1/n$, 4	5.5806(7), 22.641(3), 7.0249(3), $\beta = 99.79(1)$	Krishnakumar et al. (1998)
16	(SarH) hydrogen L-tartrate	*P1*, 1	5.0038(15), 6.442(2), 8.3179(11), $\alpha = 78.60(2)$, $\beta = 80.62(2)$, $\gamma = 79.80(2)$	Krishnakumar et al. (2001)
17	(SarH) hydrogen L-tartrate·Sar	$P2_12_12_1$, 4	9.581(2), 9.809(2), 15.811(3)	Fleck et al. (2013a)
18	(SarH) hydrogen maleate	*C2/c*, 8	22.784(5), 5.9117(12), 13.788(3), $\beta = 96.69(3)$ at 123 K	Rajagopal et al. (2002a)
18	(SarH) hydrogen maleate	*C2/c*, 8	22.754(5), 5.883(1), 13.628(3), $\beta = 96.56(3)$	Ilczyszyn et al. (2003)

[a] *P2/c*, or *Pc*, 2

[b] $P4_2$ or $P4_2/m$, 8

heating the crystals of Sar·2HF are formed in exsiccator with H_2SO_4. Crystals readily dissolve in water. Taken quantities of sarcosine and HF correspond to ca. 1:3.3 molar ratio. The structure of the crystals is not yet determined. Probably the salt is formed by (SarH)(FHF) mechanism.

Crystals of (SarH)Cl were obtained (Bhattacharyya and Saha 1978a) by evaporation of solution of sarcosine in 30 % HCl, and their structure was determined. The molar ratio of sarcosine and HCl was not indicated. The sarcosinium cation forms O-H...Cl hydrogen bond and two N-H...Cl hydrogen bonds. In the paper by Ghazaryan et al. (2012a) it was established that (SarH)Cl and (SarH)Br are possible to obtain from a solution with Sar:HCl(HBr) molar ratio equal to 1:1.3. At stoichiometric ratio 1:1 a mixture of 2Sar·HCl(HBr) and Sar·HCl(HBr) is formed. In the paper by Ghazaryan et al. (2012a) IR and Raman spectra of (SarH)Cl and (SarH)Br are provided, which indicate that they are isostructural. The crystals of (SarH)Cl and (SarH)Br are hygroscopic, which was noted also in Bhattacharyya and Saha (1978a) for (SarH)Cl.

The crystals of (SarH)NO_3 also are hygroscopic. They can be prepared from an aqueous solution containing equimolar ratio of components in a closed vessel in the presence of silica gel as absorber of water vapors (Fleck et al. 2012a). Their IR and Raman spectra are studied. The structure was not determined because of high hygroscopicity.

The crystals of (SarH)BF_4 and (SarH)ClO_4 were obtained by Ghazaryan et al. (2012b) similar to crystals of (SarH)NO_3. They also are very hygroscopic, especially (SarH)BF_4. In the paper, IR and Raman spectra of these crystals are provided. Crystals of sarcosine with HCl, HBr, HNO_3, HBF_4, and $HClO_4$ with 2:1 composition, forming in addition to crystals with 1:1, are considered in Sect. 5.2.

The crystals of (SarH)H_2PO_4 and (SarH)H_2PO_3 were obtained from an aqueous solution containing stoichiometric quantities of components (Averbuch-Pouchot et al. 1988c; Averbuch-Pouchot 1993b). Their structures have some similarity. The anions $H_2PO_4^-$ and HPO_3H^- form chains due to hydrogen bonds: O1-H1...O2 (2.566(2) Å) (Averbuch-Pouchot et al. 1988c) and O3-H2...O2(2.565 Å) (Averbuch-Pouchot 1993b). In both structures, between the chains of anions there are (SarH) cations, which form hydrogen bonds O5-H3...O2(2.539(2) Å) (Averbuch-Pouchot et al. 1988c) and O4-H10...O2(2.567 Å) (Averbuch-Pouchot 1993b). Additionally (SarH) cations form each two N-H...O bonds with neighboring chains of anions. The carbonyl oxygen atom of (SarH) cation in the structure of (SarH)H_2PO_3 does not form a hydrogen bond, while the presence of an additional P-OH group in the structure of (SarH)H_2PO_4 leads to an additional hydrogen bond O4-H2...O6(2.707(2) Å) with carbonyl oxygen atom (Averbuch-Pouchot et al. 1988c). The Raman and IR spectra of (SarH)H_2PO_4 have been published by Hubert Joe et al. (1991). Lemanov et al. (2011) reported about finding a phase transition in (SarH)H_2PO_3; however, later it was shown that observed anomalies are caused not by phase transition but by freezing trapped drops of water solution (Lemanov et al. 2012).

The crystal of $(SarH)_2SO_4$ was obtained by evaporation of an aqueous solution of sarcosine and H_2SO_4 at room temperature (Mostad and Natarajan 1996). An

Fig. 5.3 Molecular structure of $(SarH)_2SiF_6$. The SiF_6 anion is located on a special position, i.e. on an center of inversion

asymmetric unit of a triclinic cell contains two formula units of $(SarH)_2SO_4$. Both sulfate anions are disordered, one in proportion 0.18/0.82 and the second in 0.31/0.69. Each sarcosinium cation forms three hydrogen bonds (one O-H...O and two N-H...O) with the three nearest sulfate anions, and hence each sulfate anion is surrounded by six sarcosinium cations.

The system Sar + H_2SiF_6 + H_2O was studied by Fleck et al. (2012b). In addition to the salt $(SarH)_2SiF_6$ formed by the main mechanism $2A^{+}{\cdot}SiF_6^{2-}$, other salts formed by $A^{+}(A...A^{+})SiF_6^{2-}$ and $2(A...A^{+}){\cdot}SiF_6^{2-}$ comprised of the dimeric cation (Sar...SarH) also were obtained. These salts are considered in Sect. 5.2. In addition to anhydrous salts obtained at 25 °C, it is possible to obtain also hydrate crystal at the same stoichiometric ratio of components at 0 °C. However, crystal hydrate is not stable and in time dehydrates. In the paper by Fleck et al. (2012b) the vibrational spectra of $(SarH)_2SiF_6$ and also the IR spectrum of hydrate sample are shown. The asymmetric unit of $(SarH)_2SiF_6$ contains (SarH) cation in general position and SiF_6^{2-} anion in special position (Fig. 5.3). The sarcosinium cation forms O1-H1...F3(2.615(2) Å) and two N-H...F hydrogen bonds with F1 and F2 atoms with N...F distance ca. 2.80 Å. So, all fluorine atoms form hydrogen bonds, which ensures ordering in the structure. Thermal expansion of the crystal was determined in the 240–290 K interval and thermal expansion parameters were compared with parameters of other crystals in this system. Hempel et al. (1977a) and Dauter et al. (1977) obtained salts of sarcosine with octahedral anions $TeBr_6^{2-}$ and TeI_6^{2-} and determined their crystallographic data.

Crystals of sarcosinium *p*-toluenesulfonate (S-TSA) were obtained by Ilczyszyn et al. (2008a) from an aqueous solution with stoichiometric composition of components. Deuterated samples also were obtained by threefold recrystallization from D_2O. The authors carried out a detailed study of S-TSA by Raman and IR spectroscopy in the 10–400 K interval and by DSC method in the 113–433 K interval (melting temperature is 485 K) and also determined the crystal structure at 293, 200, and 100 K. Determination of the structure at 293 K showed that the sulfonate group is disordered. The sarcosinium cation forms O1-H...O4(2.566(5) Å) and O1-H...O4A(2.662(6) Å) hydrogen bond. According to the DSC data at cooling in the interval 313–288 K nonreversible phase transitions at 302 K (I-order) and 290 K (II-order) occur. Structure determination at 200 K showed that disordering of the sulfonate group disappears and the O1...O4 distance is equal to 2.566 Å. At further cooling reversible phase transition of I-order occurs at ca. 158 K with changing the symmetry from $P2_1/c$ to $P2_1/n$ and doubling the

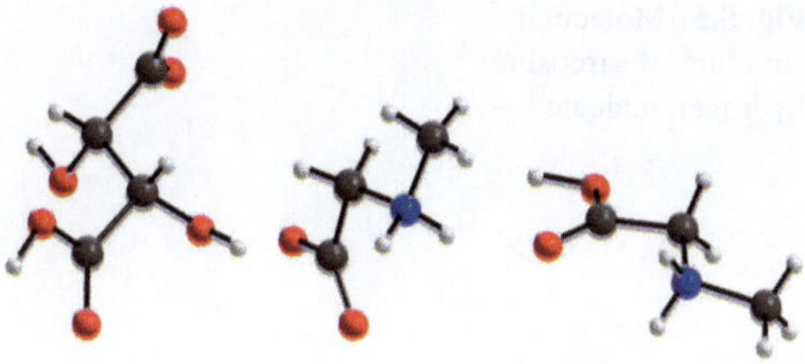

Fig. 5.4 Molecular structure of sarcosine sarcosinium tartrate. The zwitterionic sarcosine molecule is located in the *middle*, the cation on the *right hand side*

volume and Z = 8. Two independent formula units are formed with hydrogen bonds O1-H1...O4(2.599(5) Å) and O11-H11...O14(2.629(6) Å). Dependences of different modes on temperature were studied and interpreted.

The crystals of (SarH)CF_3COO were obtained from a solution of sarcosine in excess of CF_3COOH (Fehst et al. 1993). The asymmetric unit contains one formula unit. The carboxyl group forms O1-H1...O3(2.575(2) Å) hydrogen bond. The NH_2^+ group forms hydrogen bonds both with the carbonyl oxygen atom of neighboring cation and with the other O4 atom of an anion. By taking CF_3COOH in excess, Rodrigues et al. (2000) found another compound in this system (see Sect. 5.2).

The reaction of sarcosine with oxalic acid was investigated by Krishnakumar et al. (1998). The authors found that at equimolar ratio of components, the salt of sarcosinium hydrogen oxalate monohydrate crystallizes from aqueous solution. The carboxyl group cation forms O1-H1...O5(2.525(3) Å) hydrogen bond with the carboxylate group of hydrogen oxalate anion. The anion in turn forms O3-H9...O1W(2.574(3) Å) hydrogen bond with the water molecule. The water molecule as proton donor forms O1W-H10...O6(2.782(2) Å) hydrogen bond with the second atom of the carboxylate group of the anion and also with the carbonyl O4 atom of neighboring cation O1W-H11...O4(2.840 Å). The NH_2^+ group forms N-H...O hydrogen bonds with the O5 and O6 atoms of the carboxylate group of two nearest anions. In a work by Bena Jothy et al. (2006) the crystal was studied by IR and Raman spectroscopy and also by ab initio calculations.

The crystal of sarcosinium hydrogen L-tartrate was obtained by Krishnakumar et al. (2001) from an aqueous solution with stoichiometric composition of components. In the structure hydrogen L-tartrate anions form an infinite chain via strong O-H...O (2.488(2) Å) hydrogen bond between the carboxyl and carboxylate groups. The carboxyl group of the sarcosinium cation forms O-H...O(2.548(2) Å) hydrogen bond with another oxygen atom of the carboxylate group of hydrogen L-tartrate anion. The NH_2^+ group forms N-H...O hydrogen bonds with the carbonyl oxygen atoms of two nearest chains of hydrogen L-tartrate anions. One of two alcoholic C-OH groups forms a hydrogen bond with the oxygen atom of the second alcoholic group, which in turn forms a weaker hydrogen bond with the carbonyl atom of the nearest hydrogen L-tartrate anion. Vibrational spectra of this crystal are discussed by Fleck et al. (2013a). In this paper it was shown that in the Sar + L-tartaric acid + H_2O system, other compounds also exist. At molar ratio 2:1 a compound is formed, which crystallizes in the orthorhombic system with space group $P2_12_12_1$. The asymmetric unit contains a sarcosinium cation, neutral sarcosine in the form of zwitterion, and hydrogen L-tartrate anion (Fig. 5.4).

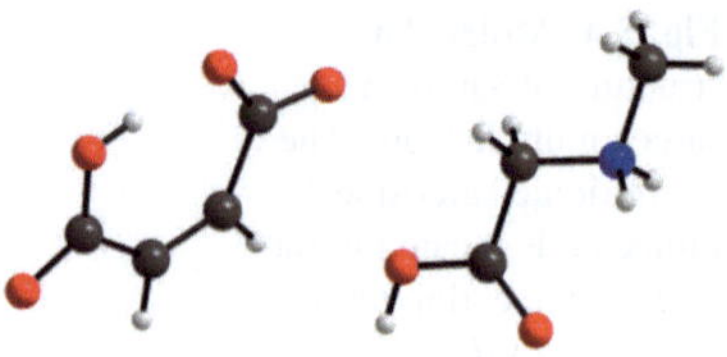

Fig. 5.5 Molecular structure of sarcosine hydrogen maleate

The COOH group of the sarcosinium cation forms a strong hydrogen bond with the carboxylate group of hydrogen L-tartrate anion O1A…H…O5C(2.4553(15) Å), while the carboxyl group of hydrogen L-tartrate anion forms also a rather strong O1C-H…O1B(2.5065(14) Å) hydrogen bond with the carboxylate group of sarcosine zwitterion. The $N(A)H_2^+$ group forms hydrogen bonds with the O2B atom of nearest sarcosine moiety and the carbonyl O2C atom of hydrogen L-tartrate anion, while the $N(B)H_2^+$ group forms hydrogen bonds with the O6C and O2A atoms. Alcoholic C-OH groups form hydrogen bonds with the O1B and O6C atoms. The IR and Raman spectra are discussed in the work of Fleck et al. (2013a). At higher content of sarcosine compared to 2:1 molar ratio (2.5:1, 3:1, 3.2:1) one more crystal with the same composition 2Sar·L-tartaric acid is formed, which has the same space group $P2_12_12_1$ and Z, but different structure: (Sar…SarH) hydrogen L-tartrate (see Sect. 5.2). In addition one more compound in this system, which was characterized by IR spectrum only, was mentioned by Fleck et al. (2013a).

The crystal of sarcosinium hydrogen maleate was obtained from an aqueous solution containing equimolar quantities of components. Its crystal structure was determined at 123 K (Rajagopal et al. 2002a) and at 293 K (Ilczyszyn et al. 2003). The asymmetric unit contains a sarcosinium cation and hydrogen maleate anion (Fig. 5.5). The hydrogen maleate anion usually comprises of a very short hydrogen bond O4-H11…O5(2.418(14) Å) (Ilczyszyn et al. 2003). The carboxyl group of cation forms O1-H1…O(2.563(14) Å) hydrogen bond with anion (Ilczyszyn et al. 2003). The NH_2^+ group in addition forms hydrogen bonds with the O3 atom of anion and carbonyl oxygen atom O2 of the nearest cation. In addition to the structure determination, Ilczyszyn et al. (2003) also investigated the IR and Raman spectra of usual and deuterated crystals. Furthermore, they studied the crystals by ^{13}C solid-state NMR and investigated the thermal behavior by DSC in the 100–446 K interval. However, in this interval phase transitions were not observed. Günay et al. (2013) studied the crystal by theoretical methods including IR and Raman spectra. Ilczyszyn et al. (2003), comparing their own data with those of Rajagopal et al. (2002a), noted that the bond length of intramolecular hydrogen bond in anion (2.418 Å at 293 K) is shorter than the same distance (2.426(2) Å) at 123 K. It should be, however, noted that not only this distance, but also unit cell parameters (see Table 5.2) and volume of unit cell at room temperature ($V = 1812.3(6)$ Å^3) are less than respective parameters and unit cell volume ($V = 1844.5(6)$ Å^3) at 123 K. So, at heating from 123 K to room temperature the crystal does not expand, but in contrast is contracted.

Table 5.3 Salts of dimethylglycine (DMG) with inorganic and organic anions

#	Crystal	s.g., Z	*a*, *b*, *c* (Å), *α*, *β*, *γ* (°)	Refs
1	(DMGH)Cl	$P2_1/c$, 4	7.061(2), 7.194(2), 13.932(3), $\beta = 98.88(2)$	Santarsiero and Marsh (1983)
2	(DMGH) CF_3COO	*Pbca*, 8	10.4286(5), 12.1213(12), 14.6271(18)	Rodrigues et al. (2001a)

5.1.3 Salts of Dimethylglycine (DMG)

In contrast to glycine and its monomethyl and trimethyl derivate (see below), for which a significant number of studied salts are known, for dimethylglycine (DMG) only two salts are known, as presented in Table 5.3. Up to recently even the crystal structure of the amino acid DMG was not known. Minkov and Boldyreva (2012) could for the first time obtain single crystals of DMG and its hemihydrates and determine their crystal structures. The main difficulty was in high hygroscopicity of the amino acid and its salts. Samples of (DMGH)Cl were obtained by Santarsiero and Marsh (1983) from Roger V. Kendall, which were recrystallized from hot acetone/n-butanol solutions. The asymmetric unit contains a cation of DMG and chloride ion. The carboxyl group of dimethylglycinium forms O-H...Cl(3.007 (1) Å) hydrogen bond with chloride ion. The NH^+ group also forms a rather strong hydrogen bond with chloride ion N...Cl(3.087(1) Å). These two strong hydrogen bonds result in a dimeric, centrosymmetric arrangement of cations and anions.

The crystals of (DMGH)CF_3COO were obtained by evaporation of solution containing 1 g DMG and excess of CF_3COOH. The asymmetric unit contains a cation of DMGH and CF_3COO^- anion. The carboxyl group of cation forms hydrogen bond O1-H1...O4(2.593(2) Å) with the carboxylate group of anion. These dimers form zigzag chains. Chains are connected to each other by N1-H4...O4(2.836(2) Å) hydrogen bond. The presence of two hydrogen bonds formed by one oxygen atom of CF_3COO^- anion leads to a significant difference in bond lengths C5-O3(1.210 Å) and C5-O4(1.248 Å).

5.1.4 Salts of Betaine

In Table 5.4 known salts of betaine are presented. Schmid (1968) reported about obtaining of betainium fluoride. Structural data, however, are absent. Crystals of (BetH)Cl and (BetH)Br were reported by Clastre (1964) and their symmetry and structures were determined. More accurately their structures were determined by Fischer et al. (1970) and Rudert et al. (2000) The crystals of (BetH)Cl, (Bet)Br, (BetH)I, (BetH)NO_3, and some others were obtained by Haussühl (1989), and their symmetries and unit cell parameters were determined as well as their elastic and thermoelastic constants. The work of Yip et al. (1990) is of special interest, where a

Table 5.4 Salts of betaine with inorganic and organic anions

#	Crystal	s.g., Z	a, b, c (Å), α, β, γ (°)	Refs
1	(BetH)F	–	–	Schmid (1968)
2	(BetH)Cl	$P2_1/c$, 4	7.428(2), 9.108(5), 11.550(3), $\beta=96.71(3)$	Fischer et al. (1970)
3	(BetH)Cl	Cm, 2	10.578(1), 6.9383(8), 5.7052(8), $\beta=111.45(1)$	Yip et al. (1990)
4	(BetH)Br	$P2_1/c$, 4	7.613(1), 9.236(3), 11.853(4), $\beta=98.08(2)$	Rudert et al. (2000)
5	(BetH)I	$P2_1/c$, 4	12.048, 9.329, 8.016, $\beta=98.9$	Haussühl (1989)
6	(BetH)NO_3	$Pbnm$, 4	10.82, 11.846, 6.703	Haussühl (1989)
7	(BetH)$BF_4 \cdot H_2O$	$P2_1/c$, 4	9.829(3), 9.192(2), 12.105(3), $\beta=110.94(2)$	Jiyang et al. (1986)
8	(BetH)$ClO_4 \cdot H_2O$	$P2_1/c$, 4	9.867(4), 9.246(3), 12.200(5), $\beta=110.93(3)$	Beja et al. (2000)
8	(BetH)$ClO_4 \cdot H_2O$	$P2_1/c$, 4	9.647(2), 9.202(2), 11.999(2), $\beta=110.87(3)$, at 100 K	Ilczyszyn et al. (2002)
9	(BetH)$HSO_4 \cdot H_2O$	$P2_1/a$, 4	14.108(1), 11.702(2), 6.520(3), $\beta=101.00(3)$	Ilczyszyn et al. (1995a)
10	(BetH)$HSeO_4 \cdot H_2O$	$P2_1/c$, 4	6.674(6), 11.912(9), 14.156(14), $\beta=100.61(4)$	Baran et al. (1995a)
11	(BetH)NH_2SO_3	$P2_1/c$, 4	5.3239(1), 11.279(3), 15.675(3), $\beta=99.4(1)$	Haussühl and Haussühl (1995)
12	(BetH)$H_2(IO_3)_2$	$P\bar{1}$, 2	6.232(1), 9.478(1), 12.252(2), $\alpha=87.94(1)$, $\beta=84.59(1)$, $\gamma=85.88(1)$	Haussühl and Schreuer (1996a)
13	(BetH)H_2PO_3	$P2_1/c$, 4	11.267(5), 7.607(5), 12.467(8), $\beta=116.27(4)$	Fehst et al. (1993)

14	$(BetH)_2SiF_6$	*Cmca*, 4	7.7083(3), 13.3512(3), 16.7656(6)	Fleck et al. (2013b)
15	$(BetH)_2SiF_6 \cdot nH_2O$	–	–	
16	$(BetH)_2(F\text{-}H\text{-}F)TaF_6$[a]	*P*2/*a*, 2	12.553(3), 6.531(3), 12.587(3), $\beta = 110.25(1)$	Ramos Silva et al. (1999a)
17	$(BetH)_3Fe(CN)_6 \cdot 4H_2O$	$P\bar{1}$, 2	9.517(2), 11.553(2), 15.145(3), $\alpha = 93.82(3)$, $\beta = 91.00(3)$, $\gamma = 90.11(3)$	Yan et al. (2001)
18	$(BetH)CF_3COO$	$P2_1/c$, 4	6.088(2), 11.9486(15), 15.022(12), $\beta = 107.54(4)$	Rodrigues et al. (2001b)
19	$(BetH)CCl_3COO$	*Pnma*, 4	26.911(7), 7.595(4), 5.954(2) at 243 K	Baran et al. (2000)
20	$(BetH)CHCl_2COO$	–	–	Dega-Szafran et al. (1997)
21	$(BetH)CH_2ClCOO$	–	–	Dega-Szafran et al. (1997)
22	$(BetH)HC_2O_4 \cdot H_2O$	*C*2/*c*, 8	16.238, 11.946, 11.130, $\beta = 98.7$	Haussühl and Wang (1989)
23	(BetH) hydrogen maleate	$P2_1/n$, 4	6.138(3), 10.831(4), 17.244(5), $\beta = 92.85(4)$	Ilczyszyn et al. (1995b)
24	(BetH) hydrogen fumarate	*Pnma*, 4	26.510(5), 6.684(1), 6.376(1)	Haussühl and Schreuer (2001)
25	(BetH) *p*-toluenesulfonate	$P2_1/c$, 4	13.816, 6.434, 17.630, $\beta = 109.3$	Haussühl (1989)
26	(BetH) 2,6-dichloro-4-nitrophenol	$P\bar{1}$, 2	7.177(1), 10.001(2), 11.241(2), $\alpha = 108.81(3)$, $\beta = 100.06(3)$, $\gamma = 106.82(3)$	Dega-Szafran et al. (1997)
27	(BetH) bis(methylsulfonyl)imide	$P2_12_12_1$, 4	7.9820(6), 12.3614(14), 12.9347(12)	Henschel et al. (2002)
28	(BetH) bis(trifluoromethylsulfonyl)imide	*Pbca*, 16	23.4721(8), 10.2027(4), 25.5957(8)	Nockemann et al. (2006)

[a]Ramos Silva et al. (1999a) interpreted the composition of this crystal as $(BetH)_2TaF_6 \cdot 2H_2O$ (see the text)

new form of (BetH)Cl crystal was obtained, which in contrast to previously known isotypic crystals (BetH)Cl, (Bet)Br, and (BetH)I, having centrosymmetric structure with space group $P2_1/c$, has non-centrosymmetric structure with space group *Cm*. The non-cetrosymmetric form of (BetH)Cl (Yip et al. 1990) was obtained in an attempt to obtain a complex of betaine with $SnCl_4$. In 12 ml water 0.8972 g (2.56 mmol) of $SnCl_4{\cdot}5H_2O$ and 1.384 g (10.24 mmol) Bet·H_2O were dissolved. At standing of solution at room temperature crystals were formed after three weeks. In this regard it is interesting to check if it is possible to use this method for obtaining non-centrosymmetric forms for (BetH)Br and (BetH)I as well, using $SnBr_4{\cdot}5H_2O$ and $SnI_4{\cdot}5H_2O$ respectively. In the structures of (BetH)Cl chloride ion forms the only hydrogen bond O-H…Cl with distances O…Cl(2.955(2) Å) (Fischer et al. 1970) and 2.946(3) Å (Yip et al. 1990). In addition to salts of betaine with 1:1 composition with HCl, HBr, HI, and HNO_3, there are also salts with 2:1 composition with dimeric (Bet…BetH) cation (see Sect. 5.2).

The crystals of (BetH)$BF_4{\cdot}H_2O$ (Jiyang et al. 1986) and (BetH)$ClO_4{\cdot}H_2O$ (Beja et al. 2000) have the same symmetry and close cell parameters and apparently are isostructural. In the structure of (BetH)$ClO_4{\cdot}H_2O$ the carboxyl group of betainium forms hydrogen bond O-H…OW(2.597 Å) with a water molecule, which in turn forms hydrogen bonds with nearest perchlorate ions with O…O distances 2.931 Å and 3.024 Å (Beja et al. 2000). Atoms of perchlorate anion are disordered. Ilczyszyn et al. (2002) studied the crystal of (BetH)$ClO_4{\cdot}H_2O$ by XRD at 293 K and at 100 K and also by vibrational spectroscopy in a wide temperature range and by DSC method. Based on DSC data the authors concluded on the existence of phase transition near 180 K. The study of the structure at 100 K allowed to determine that the symmetry and the value of Z did not change; however, disordering of oxygen atoms in perchlorate ion did not allow to refine the structure. It is worth to note that the values of cell parameters at 293 K are lower than at 100 K, which is expressed by the unit cell volumes (990.28 Å3 at 293 K and 995.29 Å at 100 K, Ilczyszyn et al. 2002) and the difference of the parameter “a” at room temperature obtained by Beja et al. (2000) (9.867(4) Å) and by Ilczyszyn et al. (2002) (9.639(2) Å). Obviously additional studies are necessary for the confirmation of the phase transition and to clear up the nature of the phase transition. In addition to (BetH)$BF_4{\cdot}H_2O$ and (BetH)$ClO_4{\cdot}H_2O$ there are salts with 2:1 composition with these anions with dimeric cation (see Sect. 5.2).

The interaction of betaine with sulfuric acid was studied (Haussühl 1989). Crystals with compositions Bet·$H_2SO_4{\cdot}H_2O$ and 2Bet·H_2SO_4 were obtained. The structures were not studied in detail. Ilczyszyn et al. (1995a) studied the structure and vibrational spectra of Bet·$H_2SO_4{\cdot}H_2O$ and Baran et al. (1995a) its selenate analog. Both crystals have the same structure. The asymmetric unit contains protonated cation (BetH), singly charged HSO_4^- anion, and one water molecule. The carboxyl group of cation forms hydrogen bond O-H…O with the oxygen atom of anion with O…O distance 2.630(2) Å in case of (BetH)$HSO_4{\cdot}H_2O$ and 2.625 (5) Å in case of (BetH)$HSeO_4{\cdot}H_2O$. The hydroxyl group of anion forms stronger hydrogen bond O-H…O with a water molecule with O…O distances 2.522(3) Å and 2.501(6) Å for (BetH)$HSO_4{\cdot}H_2O$ and (BetH)$HSeO_4{\cdot}H_2O$, respectively. The water molecule in turn forms hydrogen bonds with two nearest anions with O...O distances 2.711(2) Å and 2.730(2) Å for (BetH)$HSO_4{\cdot}H_2O$ and 2.708(5) Å and 2.710(6) Å for (BetH)$HSeO_4{\cdot}H_2O$. Ilczyszyn et al. (1995a) studied (BetH) $HSO_4{\cdot}H_2O$ and dehydrated (BetH)HSO_4 by vibrational spectroscopy. The investigation of (BetH)$HSO_4{\cdot}H_2O$ in wide temperature range (27–300 K) does not reveal phase transitions in spite of Schaak (1990) reporting on the observation of phase transition at 253.7 K. Baran et al. (1995a) also in detail studied the vibrational spectra of (BetH)$HSeO_4{\cdot}H_2O$ and its deuterated analog and also dehydrated sample. Investigation of the crystal by DSC method in the 100–300 K interval also did not reveal phase transitions. Upon heating over room temperature, it was established that the crystal melts at 346 K without dehydration, that is, it dissolves in own water (Baran et al. 1995a). Then at further heating the anhydrous sample crystallizes. For details about other salts formed in these systems, see Sect. 5.2.

Haussühl and Haussühl (1995) published results on growth and investigation of betainium sulfamate crystals (BetH)NH_2SO_3 and compared these with crystals of other sulfamates. The crystals of have high solubility (223 g in 100 ml water at room temperature) and are hygroscopic. The elastic and thermoelastic constants of crystal are determined.

At the reaction of betaine with iodic acid, well-crystallizing salt Bet·3HIO_3 is formed. The species Bet·3HIO_3 was reported by Haussühl and Schreuer (1996a) and independently by Petrosyan et al. (1995, 1996). The asymmetric unit contains one betainium cation, iodate anion, and two HIO_3 molecules. The (BetH) cation forms a hydrogen bond with the oxygen atom of one of two HIO_3 molecules with O...O distance 2.536(5) Å. The HIO_3 groups form weaker hydrogen bonds. Iodine atoms usually in addition to the three covalent bonds of the first coordination sphere form also weaker I...O bonds with oxygen atoms of the neighboring groups (second coordination sphere). One of such bonds is short enough (I...O 2.455 (3) Å) in order to consider it as intermediate between the first and second coordination sphere, which is in the 2.2–2.5 Å range (Petrosyan and Shishkin 1996). Thermal expansion coefficients and elastic and thermoelastic constants are determined in Haussühl and Schreuer (1996a).

Betaine phosphite (Bet·H_3PO_3) among other salts of betaine was obtained before 1980 (see Haussühl (1989) and references therein). Initially these crystals did not attract much attention because they had centrosymmetric structures. Interest to this crystal renewed (Albers 1988; Albers et al. 1988; Dörffel et al. 1989; Sobiestianskas et al. 1993) after the discovery of phase transition in 1982 in related crystal Bet·H_3PO_4 (see Sect. 5.1.24). Later the crystal was studied in a number of papers (Sobiestianskas et al. 1993; Baran et al. 1997a; Hasebe et al. 1998, 2001; Balashova et al. 1998; Balashova and Lemanov 2003; Machida et al. 2003) and also solid solutions based on Bet·H_3PO_3 and Bet·H_3PO_4. Betainium phosphite contains betainium cation (BetH) and HPO_2OH^- anion, (BetH)H_2PO_3. The crystal undergoes two phase transitions and has three phases. The phase existing above 355 K is labeled as phase I, with $P2_1/m$ symmetry and $Z = 2$. At 355 K a phase

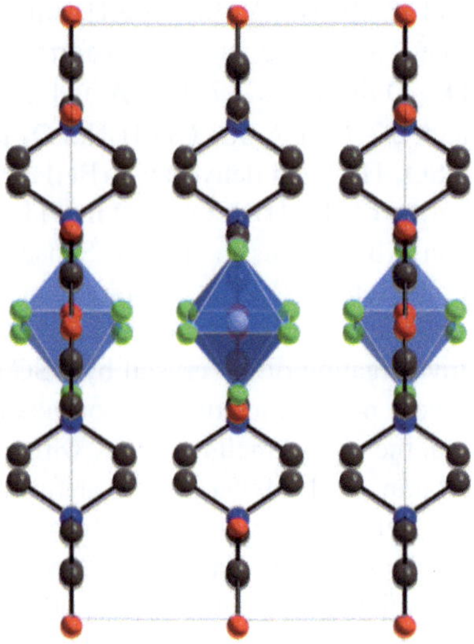

Fig. 5.6 Packing diagram of betaine$_2$SiF$_6$. Note the mirror symmetry (space group Cmca) of the structure. Hydrogen atoms are omitted for clarity

transition occurs, upon which the crystal transforms into phase II with space group $P2_1/c$ and $Z = 4$. Finally, at 216 K, there is a second phase transition to a ferroelectric phase III with space group $P2_1$ and $Z = 4$. The structure of (BetH)H_2PO_3 was determined at room temperature by neutron diffraction (Fehst et al. 1993). The carboxyl group of (BetH) forms O2-H12…O3 hydrogen bond with the O3 atom of phosphite anion. The atom H14 bonded with phosphorus does not participate in hydrogen bonding. The remaining hydrogen atom of the HPO_2OH^- anion is disordered. The atoms H13 and H15 have 0.5 occupation and participate in hydrogen bonding: O4-1/2H13…1/2H13-O4 and O6-1/2H15…1/2H15-O6. Ferroelectric phase transition is caused by loss of symmetry center and ordering of protons in these hydrogen bonds. Detailed investigation of vibrational spectra of (BetH) H_2PO_3 was carried out (Baran et al. 1997a).

In the system Bet + H_2SiF_6 + H_2O different compounds are formed depending on the conditions and molar ratio of components. At 2:1 ratio, which corresponds to the main formation mechanism of amino acids' hexafluorosilicates crystal hydrate (BetH)$_2$SiF$_6$·nH$_2$O is formed, which is not stable and dehydrates out of solution. At ratios of 1:1 and 1:2 at room temperature anhydrous salt (BetH)$_2$SiF$_6$ is formed. At ratios of 3:1 and 4:1 a salt with 4:1 composition is formed, which is considered in Sect. 5.2. The unit cell of (BetH)$_2$SiF$_6$ contains four anions and eight betainium cations in a mirror-symmetric arrangement (Fig. 5.6; Fleck et al. 2013b). The (BetH) cation has mirror symmetry, while the position symmetry of anion is 2/m. The carboxyl group of (BetH) forms a bifurcated hydrogen bond with two crystallographically equivalent F2 atoms with O1-H1…F2(2.9226(18) Å) and H1…F2(2.29(4) Å).

The crystal of "(BetH)$_2$TaF$_6$·2H$_2$O" was obtained by Ramos Silva et al. (1999a) by dissolving pure Ta metal in 40 % solution of HF with addition of a respective amount of betaine. After a few days of slow evaporation small transparent crystals were formed. According to the authors the asymmetric unit contains TaF_6^{2-}

anion in special position, one betainium cation, and one water molecule. According to the authors betainium cation forms hydrogen bond O1-H1. . .O3(2.506 Å) with a water molecule. Such solution of the structure, however, seems to us questionable. The authors found conceivable that tantalum here is tetravalent and there is a doubly charged TaF_6^{2-} anion. This assumption in our opinion is erroneous, because tetravalent tantalum exists only in compounds with Ta-Ta bond and TaF_6^- anion is known exclusively as a singly charged anion. The peak given as "O3" was interpreted by the authors as an oxygen atom of a water molecule, thus two hydrogen atoms were added. These hydrogen atoms, however, do not form any hydrogen bonds, which makes this placement dubious. In addition, which is more important, the atom "O3" is connected with crystallographically equivalent "O3" atom on distance 2.24 Å. This distance is too short to be a distance between oxygen atoms. Based on all these we came to the conclusion that the crystal obtained by Ramos Silva et al. (1999a) actually is $(BetH)_2$(F-H-F)TaF_6. Betainium cation forms hydrogen bond O-H. . .F(2.506 Å) and the F. . .F distance 2.24 Å for symmetric hydrogen bond in F. . .H. . .F anion is typical being short (see, e.g., the anion $(F\text{-}H\text{-}F)^-$ in the structure of L-Arg·2HF (Ramos Silva et al. 2000b).

The crystal of $(BetH)_3Fe(CN)_6{\cdot}4H_2O$ was obtained by Yan et al. (2001) from solution containing (BetH)Cl and $K_3[Fe(CN)_6]$ in ratio 4:1. After 2 days in solution kept in the dark well-shaped blue single crystals were formed. The asymmetric unit contains one formula unit: octahedral $[Fe(CN)_6]^{3-}$ anion, three betainium cations, and four water molecules. Positions of active hydrogen atoms were not determined. The carboxyl groups of cations display disordering. Based on bond lengths it is possible to distinguish between C=O and C-OH bonds, possible hydrogen bonds, as well as hydrogen bonds of water molecules based on their contacts. All three betainium cations form hydrogen bonds O-H. . .O with three water molecules with O. . .O distances: 2.609 Å, 2.563 Å, and 2.518 Å (mean values are provided because of disordering of C-OH bonds). Four water molecules form eight hydrogen bonds as proton donors. Six bonds out of these are O-H. . .N type hydrogen bonds towards the C-N groups of the anion, with O. . .N distances in the 2.754–2.820 Å range. Two bonds extend towards water molecules with O. . .O distances of 2.674 Å and 2.794 Å.

Structures of $(BetH)CF_3COO$ and $(BetH)CCl_3COO$ crystals were determined in (Rodrigues et al. 2001b; Baran et al. 2000). References to the earlier works on these crystals are provided by Baran et al. (2000). Asymmetric unit of $(BetH)CF_3COO$ contains one formula unit, that is, betainium cation and CF_3COO^- anion. Atoms of fluorine and oxygen are disordered. The carboxyl group of cation forms a hydrogen bond with the carboxylate group of anion with O. . .O distances O1-H1. . .O3A (2.565(8) Å) and O1-H1. . .O3B(2.488(14) Å) (Rodrigues et al. 2001b). The crystal of $(BetH)CCl_3COO$ was studied in more detail (Baran et al. 2000). The structure was determined and vibrational spectra were studied in wide temperature range as well as the behavior of crystal in wide temperature range by DSC method. Chlorine and oxygen atoms in anion are disordered as well as oxygen atom of O-H group of cation. Both anion and cation have mirror symmetry. Hydrogen bond O-H. . .O between cation and anion is 2.584(2) Å. By DSC method and by temperature-dependent measurements of IR spectra reversible low-temperature phase transitions were observed at 177 and 187 K. FTIR measurements indicate on the existence of one more phase transition bellow 100 K. In addition irreversible endothermic effects above room temperature were observed by DSC method. Among these, the effect at 338.1 K is connected with a transition into high-temperature phase, while the effects at 365.5 and 376 K are caused by decomposition of sample. Phase transition at 187 K is revealed also on temperature dependence of parameter "b" and unit cell volume determined by X-ray powder diffraction.

Large crystals of $(BetH)HC_2O_4{\cdot}H_2O$ of optical quality were grown from an aqueous solution (Haussühl and Wang 1989). The authors determined the elastic and thermoelastic constants of grown crystals. The structure of crystal was not determined; however, the density of crystal was measured ($\rho_c = 1.396$ g/cm^3), which allows to determine the number of formula units in the unit cell ($Z = 8$). This in turn allows to conclude that for $C2/c$ symmetry the asymmetric unit contains one formula unit ($Z' = 1$). The crystal has a melting point of 337 K. From the behavior of elastic properties and on the basis of investigation by DSC method, phase transitions in the crystal were not observed. The crystals of grown and reported by Haussühl and Wang (1989) were used for investigation of their vibrational spectra (Philip and Aruldhas 1995). Later one more crystal with $2Bet{\cdot}H_2C_2O_4$ composition containing dimeric cation (Bet...BetH) was found in the system $Bet + H_2C_2O_4 + H_2O$ (see Sect. 5.2).

Crystals of betainium hydrogen maleate (BHM) and betainium hydrogen fumarate (BHF) were obtained and studied (Ilczyszyn et al. 1995b; Haussühl and Schreuer 2001). In growing the crystals at higher temperatures, the transformation of maleic acid (*cis*-stereoisomer) into fumaric acid (*trans*-stereoisomer) is observed. Ilczyszyn et al. (1995b) observed growth of BHF crystals when growing crystals of BHM at temperatures above 300 K during an interval of several weeks. In crystal BHF phase transition near 194 K was observed (Unruh et al. 1995). Obviously earlier report about observation of phase transition in BHM (Haussühl 1988) relates to BHF, because subsequent investigations did not reveal phase transitions in BHM. Cell parameters ascribed to BHM (Krummeck et al. 1994) $a' = 6.695$ Å, $b' = 26.528$ Å, and $c' = 12.745$ Å satisfactorily correspond to parameters of BHF (Haussühl and Schreuer 2001) if one assumes $a' = b$, $b' = a$, and $c' = 1/2c$. The temperature of phase transition $T_c = 193 \pm 0.1$ K (Krummeck et al. 1994) corresponds to the temperature of phase transition in BHF (*ca.* 194 K) (Unruh et al. 1995). In the structure of BHM hydrogen maleate anion contains usually short intramolecular hydrogen bond O-H...O(2.429(3) Å). Betainium cation forms hydrogen bond O-H...O(2.527(3) Å) with the second oxygen atom of carboxylate group. In case of BHF hydrogen fumarate anions form head-to-tail chain by means of O-H...O(2.584(3) Å), while betainium cations form strong hydrogen bond O-H...O(2.451(3) Å) with the second oxygen atom of the carboxylate group of anion. Vibrational spectra of BHM are studied in detail (Ilczyszyn et al. 1995b). X-ray diagrams obtained at 160, 180, and 190 K as well as DSC measurement showed no phase transitions in the 100–370 K temperature

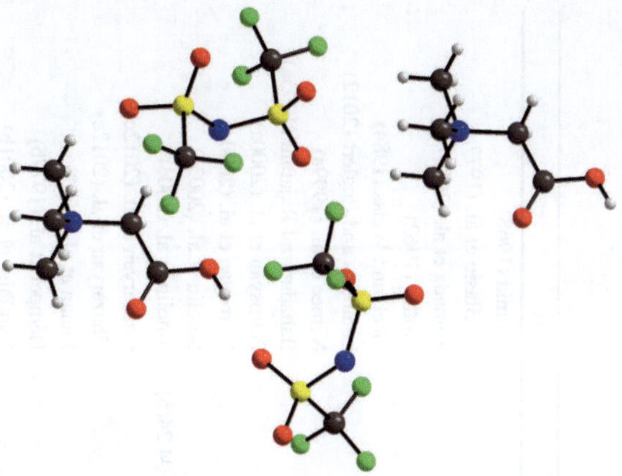

Fig. 5.7 Molecular structure of betainium bis (trifluoromethylsulfonyl) imide

range (Ilczyszyn et al. 1995b). Crystals of (BetH) *p*-toluenesulfonate were obtained by Haussühl (1989), where the unit cell parameters, symmetry, as well as elastic and thermoelastic constants are also reported (but not the structure of the crystal).

The crystals of (BetH) 2,6-dichloro-4-nitrophenol (BET-DCNP) were obtained from methanol solution containing equimolar quantities of components. The melting temperature is 151–152 °C. The presence of two chlorine atoms and nitro group significantly enhances the acid properties of DCNP, which leads to the transition of proton to betaine with the formation of (BetH) cation, and this is revealed in the IR spectrum of the crystal. Crystal structure determination showed the formation of very strong hydrogen bond O-H...O(2.424(3) Å) between the carboxyl group of cation and phenolic oxygen ion.

In the works of Henschel et al. (2002) and Nockemann et al. (2006), the crystalline salts of betainium with similar anions (bis(methylsulfonyl)imide $N(SO_2CH_3)_2^-$, Henschel et al. 2002; and bis(trifluoromethylsulfonyl)imide $N(SO_2CF_3)_2^-$, Nockemann et al. 2006) were obtained. Asymmetric units are shown in Fig. 5.7). As is seen from the figure, in case of $(BetH)N(SO_2CH_3)_2$ the carboxyl group of cation forms hydrogen bond O5-H1...N1(2.647(2) Å), while in case of $(BetH)N(SO_2CF_3)_2$ both crystallographically independent cations form hydrogen bonds O-H...O with SO_2CF_3 group with distances of 2.712 Å and 2.720 Å. One more compound of betaine with $N(SO_2CF_3)_2^-$ with dimeric cation is considered in Sect. 5.2. It is interesting to note that in case of the compound of $N(SO_2CH_3)_2^-$ with cholinium, where unlike of betainium the carboxyl group is replaced by the hydroxyl group, hydrogen bond O-H...O(2.8027(16) Å is formed as well.

5.1.5 *Salts of α-Alanine*

In Table 5.5 salts of the simplest amino acid α-alanine containing a chiral center are presented. The compound L-Ala·HF was obtained and patented by Schmid (1968).

Table 5.5 Salts of α-alanine with inorganic and organic anions

#	Crystal	s.g., Z	a, b, c (Å), α, β, γ (°)	Refs
1	(L-AlaH)F	–	–	Schmid (1968)
2	(L-AlaH)Cl	$P2_12_12_1$, 4	7.148(3), 17.590(8), 5.202(2)	Di Blasio et al. (1977a)
3	(L-AlaH)Cl·H_2O	$P2_12_12_1$, 4	6.1925(13), 9.929(2), 11.759(3)	Yamada et al. (2008)
4	(DL-AlaH)Cl	$P2_1/c$, 4	9.2, 9.04, 7.31, β=95.35	Trotter (1962)
5	(L-AlaH)Br	–	–	Fleck and Weiss (1984)
6	(DL-AlaH)I	$P2_1/n$, 4	7.6975(11), 5.7776(8), 16.034(2), β=98.999(2), at 100 K	Lamberts and Englert (2012)
7	(L-AlaH)NO_3	$P2_12_12_1$, 4	5.6627(6), 7.4705(6), 16.142(1)	Němec et al. (1999a)
8	(DL-AlaH)NO_3	$P2_1/c$, 4	7.648(1), 5.631(1), 16.335(3), β=94.42(1)	Bahadur and Rajaram (1995a)
9	(L-AlaH)IO_3	–	–	Petrosyan et al. (2000a)
10	(DL-AlaH)IO_3·$2H_2O$	$P2_1/c$, 4	13.536(3), 7.3240(15), 9.312(2), β=98.52(3)	Petrosyan et al. (2000a)
11	(L-AlaH)H_2PO_4	$P112_1$, 4	11.918(5), 9.117(5), 7.285(5), β=104.68(5)	Smolin et al. (2003a)
12	(DL-AlaH)$_2SO_4$	$P\bar{1}$, 2	7.431(3), 9.826(3), 10.081(3), α=120.07(5), β=104.73(5), γ=94.24(5)	Smolin et al. (2004)
13	(L-AlaH)$_2SiF_6$·$3H_2O$	$C2$, 2	19.018(6), 5.432(1), 7.908(3), β=104.86(1)	Ghazaryan et al. (2012c)
14	(DL-AlaH)$_2SiF_6$·$2H_2O$	$P2_1$, 2	6.208(1), 9.974(2), 11.840(2), β=101.39(3)	Ghazaryan et al. (2012c)
15	(DL-AlaH)$_2SiF_6$	–	–	Ennan et al. (1983)
16	(DL-AlaH)$_2SeBr_6$	[a]	12.33(2), 7.71(1), 10.26(2), β=105.5(3)	Hempel et al. (1977b)
17	(L-AlaH)HC_2O_4	$P2_12_12_1$, 4	5.6304(15), 7.2353(15), 19.597(3)	Nandhini et al. (2001b)
18	(DL-AlaH)HC_2O_4	$P2_1/n$, 4	5.662(2), 7.342(2), 19.157(6), β=94.48(3)	Nandhini et al. (2001c)
19	(DL-AlaH)HC_2O_4·H_2O	$P2_1/c$, 4	12.4987(19), 6.6017(6), 11.3732(18), β=113.585(12)	Zakharov and Boldyreva (2011)
20	(L-AlaH) hydrogen maleate	$P2_12_12_1$, 4	5.5873(11), 7.3864(17), 23.688(3)	Alagar et al. (2001a)
21	(L-AlaH) hydrogen L-tartrate	$P2_1$, 2	5.1446(8), 13.721(3), 7.4751(9), β=98.09(1)	Rajagopal et al. (2002b)
22	(L-AlaH) X(1)[b]	$P2_12_12_1$, 4	8.806(1), 26.528(9), 5.247	Kimoto et al. (1989)
23	(L-AlaH) X(2)[c]	$P2_1$, 2	11.152(3), 5.373(3), 12.488(6), β=104.85(3)	Sudbeck et al. (1995)
24	(L-AlaH) bnppa[d]	$P2_1$, 2	19.171(2), 6.2961(3), 9.1679(7), β=96.743(6)	Fujii and Hirayama (2002)

[a] *Cm* or *C2/m*, 2
[b] *p*-chlorobenzenesulfonate
[c] 2,4-dinitrobenzenesulfonate·H_2O
[d] bnppa = (+)-(1S)-1,1′-binaphthalene-2,2′-diyl phosphate monohydrate

The structure of the crystal was not determined. The structure of (L-AlaH)Cl is determined by Di Blasio et al. (1977a). However, the conditions of preparation were not described. The asymmetric unit contains one formula unit. The carboxyl group of cation forms a hydrogen bond with chloride ion O...Cl(3.019 Å). The NH_3^+ group forms two hydrogen bonds with chloride ion and one with carbonyl oxygen atom of the nearest cation. Later (L-AlaH)Cl·H_2O also was obtained (Yamada et al. 2008). The crystal of (L-AlaH)Cl·H_2O was obtained from a powder sample after it was left standing in a refrigerator for a few years. Monohydrate has the same symmetry as the anhydrous salt. Here also the carboxyl group forms a hydrogen bond with anion O-H...Cl(3.045 Å). Chloride ion forms three more hydrogen bonds: one N-H...Cl bond and two O-H...Cl bonds with water molecules. The two remaining hydrogen atoms of the NH_3^+ group form N-H...O hydrogen bonds with a water molecule. There is one more compound in L-Ala + HCl + H_2O system, namely, 2L-Ala·HCl with dimeric cation, which is considered in Sect. 5.2. The crystal of (DL-AlaH)Cl was obtained by Trotter (1962). Only symmetry and cell parameters were determined. The crystals of (L-AlaH)Br were obtained and studied by NQR method (Fleck and Weiss 1984). Due to the presence of O-H...Br hydrogen bond spherical distribution of charge bromide ion is disturbed, which creates an electric field gradient and allows detecting nuclear quadrupole resonance. The structure of (DL-AlaH)I was determined by Lamberts and Englert (2012). The crystal was obtained by dissolving DL-Ala and MnI_2·$4H_2O$ in ratio 2:1 in H_2O/MeOH (1:1) solution. After slow evaporation of the solvent yellow oil remained, which was placed in a desiccator. After several weeks colorless needles of (DL-Ala)I were formed. The hydrogen bond scheme is similar to that in the case of (L-AlaH)Cl. The carboxyl group forms hydrogen bond O-H...I(3.391 (2) Å). Iodide ion forms two more hydrogen bonds N-H...I and the third hydrogen atom of the NH_3^+ group forms N-H...O hydrogen bond with the carbonyl atom of the nearest cation. There are also some publications (Lucia Rose et al. 2010, 2011a, b) further investigating (L-AlaH)Cl and (L-AlaH) Cl·H_2O, which are questionable, however. Lucia Rose et al. (2010, 2011a) reported the growth of L-alanine hydrogen chloride at 30 °C, as described by Di Blasio et al. (1977a). However, they identified the grown crystal citing the paper (Yamada et al. 2008), where the crystal of (L-AlaH)Cl·H_2O was described, and gave unit cell parameters which correspond to those published by Yamada et al. (2008). The most striking is that the shape of the grown crystal, IR spectrum, and thermograms clearly show that the obtained crystal actually is L-alanine. In another paper, Lucia Rose et al. (2011b) claim to have grown L-alanine hydrogen chloride monohydrate crystals (both pure and doped by RbCl), providing the unit cell parameters which correspond to parameters given by Yamada et al. (2008). However, provided powder XRD patterns again correspond to L-alanine. Solubility curves of pure and doped with 1 % RbCl crystals are provided, which significantly differ. It should be noted that the solubility curve of pure crystal significantly differs from the curve provided for pure crystal in (Lucia Rose et al. 2010).

The crystal of (DL-AlaH)NO_3 was obtained from aqueous solution containing stoichiometric quantities of components. The carboxyl group of the cation forms O2-H1...O12(2.619 Å) hydrogen bond with nitrate anion. The oxygen atom forms also N1-H6...O12 hydrogen bond, in which the resultant bond length N11-O12 (1.265(6) Å) is the longest among the three bonds of the NO_3^- anion. The second hydrogen atom of the NH_3^+ group forms hydrogen bond N1-H8...O13 also with the NO_3^- anion. The bond length N11-O13(1.245(5) Å) is intermediate between the longest N11-O12 and shortest N11-O11(1.209(6) Å) bonds. Of these acceptors, the atom O11 does not form any hydrogen bonds. The third hydrogen atom of the NH_3^+ group forms hydrogen bond N1-H7...O1 with the carbonyl atom of a neighboring cation. The interaction of L-alanine with nitric acid was studied by Němec et al. (1999a, b). The formation of similar salt (L-AlaH)NO_3 was found; however, due to the presence of optically active form, the crystal of (L-AlaH)NO_3 has non-centrosymmetric space group $P2_12_12_1$. The scheme of hydrogen bonds is similar to that in the previous case. The carboxyl group of cation forms O1-H8...O4(2.617(1) Å) hydrogen bond with the nitrate ion. The atom O4 forms also N1-H2...O4 hydrogen bond and again the bond length N2-O4(1.261 Å) is the longest. Two other hydrogen atoms of the NH_3^+ group form hydrogen bond N1-H7...O1 with the carbonyl atom of the neighboring cation and with the O5 atom of the nitrate ion. The length of the N2-O5 bond, viz.1.243 Å, again is intermediate between 1.261 Å and 1.221 Å. The atom O3, acceptor of the shortest bond, does not form any hydrogen bonds. The crystal of (L-AlaH)NO_3 was studied by vibrational spectroscopic methods in wide temperature range and also by DSC method. Phase transitions were not found in 90–413 K interval. At 415 K the crystal melts. Němec et al. (1999a, b) believed that (L-AlaH)NO_3 is the only compound in the L-Ala + HNO_3 + H_2O system; however, later one more salt with 2 L-Ala·HNO_3 composition was found (see Sect. 5.2). The crystal of (L-AlaH)NO_3 was grown and studied as possible NLO material (Vimalan et al. 2010a). According to this paper, the SHG efficiency estimated by powder method is three times higher than that of KDP.

In a paper by Petrosyan et al. (2000a) iodates of L-alanine DL-alanine ((L-AlaH)IO_3 and (DL-Ala)IO_3·H_2O) were reported. The crystals of (L-AlaH)IO_3 are not stable and in time become reddish. The structure of (DL-AlaH)IO_3·$2H_2O$ crystal was determined and also was investigated by ^{127}I NQR and IR spectroscopy.

Smolin et al. (2003a) reported the synthesis of (L-AlaH)H_2PO_4 from solution containing L-alanine and H_3PO_4 at $pH < 2$ by temperature lowering method in the interval from 40 to -10 °C and determined the structure. Previously it was reported that the crystal displays piezoelectric activity (Lemanov et al. 2002). Asymmetric unit contains two formula units. All active hydrogen atoms of OH and NH_3^+ groups form hydrogen bonds. The strongest hydrogen bonds form carboxyl groups of two crystallographically independent cations: O10-H5...O1(2.512(2) Å) and O12-H13...O5 (2,507(2) Å) with $H_2P(1)O_4^-$ and $H_2P(2)O_4^-$, respectively. Two hydroxyl groups of the anions are forming hydrogen bonds towards each other and the cations. The hydrogen bond O4-H2...O6(2.604(2) Å) connects the OH group of $H_2P(1)O_4^-$ with the O6 atom of $H_2P(2)O_4^-$ anion and hydrogen bond O7-H3...O2(2.592(2) Å) respectively connects the OH group of $H_2P(2)O_4^-$ with the O2 atom of $H_2P(1)O_4^-$ anion. The second OH group of $H_2P(1)O_4^-$ anion forms O3-H1...O11(2.658(1) Å) hydrogen bond with the carbonyl oxygen atom, and the second OH group of $H_2P(2)O_4^-$ anion forms O8-H4...O9(2.628(2) Å) hydrogen bond with another cation. Thus, each anion is connected with both cations and each cation connected with both anions. In addition the $N(1)H_3^+$ group forms two hydrogen bonds N1-H...O2 with $H_2P(1)O_4^-$ anion and one hydrogen bond N1-H...O5 with $H_2P(2)O_4^-$ anion. Respectively, the $N(2)H_3^+$ group forms two hydrogen bonds N2-H...O6 with $H_2P(2)O_4^-$ anion and one N2-H...O1 with $H_2P(1)O_4^-$ anion. It is interesting to note that in case of 2DL-Ala·H_3PO_4 (see Sect. 5.2) alanine molecules are in zwitterionic form.

Crystals of (DL-AlaH)$_2$$SO_4$ were reported by Smolin et al. (2004) by temperature lowering of aqueous solution containing DL-alanine and H_2SO_4 with pH = 1. The asymmetric unit contains one formula unit. Carboxyl groups of cations form hydrogen bonds O6-H1...O4(2.549(3) Å) and O8-H9...O3(2.590(2) Å) with sulfate anion. The $N(1)H_3^+$ group of one of the cations forms two hydrogen bonds N-H...O with O1 and O2 atoms of sulfate anion and one hydrogen bond with the carbonyl oxygen atom of symmetry-related cation with usual N...O distance 2.901(3) Å, while the $N(2)H_3^+$ group of the second cation forms all three hydrogen bonds with sulfate anion. However, Smolin et al. (2004) write that "The Ala1 molecule forms three H bonds to three SO_4^{2-} ions and a strong hydrogen bond to the symmetry-related molecule to yield DL pairs with two hydrogen bonds". It is not clear what the authors have in view under "strong hydrogen bond" and "with two hydrogen bonds."

Crystals of salts of L-alanine and DL-alanine with hexsafluorosilicic acid were reported by Ghazaryan et al. (2012c). Previously anhydrous sample (DL-AlaH)$_2$$SiF_6$, obtained by drying the sample in vacuum exsiccator to constant mass, was obtained and characterized by IR spectra and thermal analysis by Ennan et al. (1983). Crystalline sample is dihydrate with space group $P2_1$ (Ghazaryan et al. 2012c). The structure, however, is close to centrosymmetric and without accounting for hydrogen atoms can be described with space group $P2_1/c$ with $Z = 2$. In case of L-alanine the formation of two salts was established. In addition to the salt formed by main type (L-AlaH)$_2$$SiF_6$·$3H_2O$, a salt with 3L-Ala·$H_2SiF_6$·$H_2O$ composition, comprising dimeric cation, is formed (see Sect. 5.2). In the unit cell of (L-AlaH)$_2$$SiF_6$·$3H_2O$ the cation (L-AlaH) and one water molecule are in general position, while the anion and second water molecule are in special position, which can be presented as (L-AlaH)·0.5SiF_6·1.5H_2O. In addition equatorial fluorine atoms in anion are disordered. The carboxyl group of cation forms O1-H1...OW1 hydrogen bond with water molecule being in general position. In case of the salt with DL-alanine, the carboxyl groups of cations form hydrogen bonds O-H...F with anion. The obtained crystals were studied also by IR and Raman spectroscopy. The crystals are formed from solution containing stoichiometric quantities of components. The molecular structures of (L-AlaH)$_2$$SiF_6$·$3H_2O$ and (DL-AlaH)$_2$$SiF_6$ are shown in Fig. 5.8.

For the crystals of (DL-AlaH)$_2$$SeBr_6$ only preliminary crystallographic data are published (Hempel et al. 1977b).

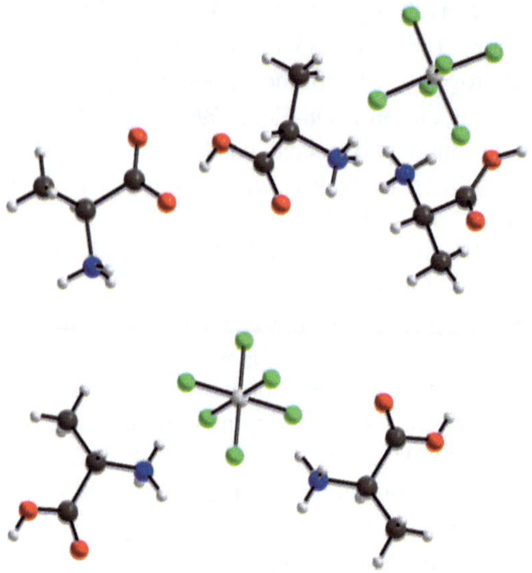

Fig. 5.8 Molecular structure of (L-AlaH)$_2$SiF$_6$·3H$_2$O (*top*) and (DL-AlaH)$_2$SiF$_6$ (*bottom*). Note the chirality of the molecules in the enantiomeric and the racemic structures. Water molecules are omitted for clarity

The crystals of (L-AlaH)HC_2O_4 were obtained from aqueous solution with stoichiometric quantities of components (Nandhini et al. 2001b). The asymmetric unit contains one formula unit. Hydrogen oxalate anions form characteristic head-to-tail sequence, in which the carboxyl group forms O3-H3. . .O6(2.545(2) Å) hydrogen bond. The (L-AlaH) cations by means of the carboxyl group are connected to this sequence via hydrogen bond O1-H1. . .O5(2.575(2) Å) with the second atom of the carboxylate group of anion. The NH_3^+ forms one shorter hydrogen bond N1-H1A. . .O6(2.728 (2) Å). Other bonds are on the level of van der Waals interactions. Stronger hydrogen bond in the chain of anions and additional N1. . .O6 bond led to the lengthening of the bond C5-O6(1.235(2) Å) compared to other C5-O5(1.219(2) Å) bond. Later the crystal of (L-AlaH)HC_2O_4 became a subject for further investigations (Dhanuskodi and Vasantha 2004; Sivabarathy et al. 2004; Krishnakumar and Nagalakshmi 2006; Arun and Jayalekshmi 2008b, 2009; Chandrasekaran et al. 2011b). Some of the provided data badly correspond to each other. IR spectra are provided in the papers (Dhanuskodi and Vasantha 2004; Krishnakumar and Nagalakshmi 2006; Arun and Jayalekshmi 2009; Chandrasekaran et al. 2011b). In the IR spectrum one might expect bands caused by stretching vibration of the C=O bond in the COOH group of cation and anion. Arun and Jayalekshmi (2009) provided the correct IR spectrum, which contains a band at 1,717 cm^{-1} and a shoulder at ca. 1,700 cm^{-1}, while in spectra and tables of Dhanuskodi and Vasantha (2004) and Krishnakumar and Nagalakshmi (2006) bands in the region of 1,700 cm^{-1} are absent. In the spectrum shown by Chandrasekaran et al. (2011a, b) there is a band at 1,724 cm^{-1}. In Dhanuskodi and Vasantha (2004), Arun and Jayalekshmi (2009), and Chandrasekaran et al. (2011b), data of thermal analysis are provided. According to Dhanuskodi and Vasantha (2004) in the interval 444–483 °C decomposition of the residual of sample takes place accompanying with significant exothermic effect with peak at ca. 468 °C, which, however, is absent according to data of Arun and Jayalekshmi (2009) and Chandrasekaran et al. (2011a, b). The value of density 1.35 g/cm^3 measured for crystals obtained by Arun and Jayalekshmi (2009) and calculated value 1.378 g/cm^3 significantly differ from the value 1.49 g/cm^3 determined from X-ray data given by Nandhini et al. (2001b). Chandrasekaran et al. (2011b) obtained a crystal called L-alanine oxalate: $2[C_3H_2NO_2] + C_2H_2O_4 \rightarrow C_8H_{16}N_2O_8$ from a solution containing L-alanine and oxalic acid with 2:1 ratio. For confirmation the authors provide data of chemical analysis, C(45.06 %), H(7.77 %), and N(12.01 %), which in the opinion of the authors correspond to calculated data: C(48.04 %), H(8.06 %), and N(14.00 %). However, these data do not correspond neither to 1:1 composition, for which correct composition was provided by Arun and Jayalekshmi (2009), nor to 2:1 composition, for which calculated composition is C(35.82 %), H(6.01 %), and N(10.44 %). In addition, according to the data given by Sivabarathy et al. (2004) obtained by photoacoustic method, there is phase transition near 100 °C, which was not initially noted at investigation by DTA method. Nandhini et al. (2001c) obtained also the crystal of (DL-AlaH)HC_2O_4. Unlike the crystal of (L-AlaH)HC_2O_4 it crystallizes in centrosymmetric structure. In spite of this obvious difference, they have also significant structural similarity. Hydrogen oxalate anions here also form head-to-tail chain by hydrogen bond O3-H3…O6(2.587(2) Å). The cation also forms a hydrogen bond with the second oxygen atom of the carboxylate group of anion, O1-H1…O5(2.591(2) Å). The atom O5 also forms N1-H1A…O5(2.834(3) Å) hydrogen bond. The atom O6 additionally forms N1-H1C…O6(2.818(2) Å). The third atom of the NH_3^+ group forms N1-H1B…O2(2.863(3) Å) hydrogen bond with the carbonyl oxygen atom of the nearest cation. Thus, in the structure of (DL-AlaH)HC_2O_4 hydrogen bond O-H…O is weaker than in (L-AlaH)HC_2O_4, while N-H…O hydrogen bonds are stronger. Later monohydrate (DL-AlaH)$HC_2O_4{\cdot}H_2O$ was also obtained (Zakharov and Boldyreva 2011). The monohydrate was obtained by slow cooling of an aqueous solution of DL-alanine and oxalic acid in a 1:1 stoichiometric ratio saturated at 323 K. However, it is unclear what determines the formation of the monohydrate, because according to Zakharov and Boldyreva (2013) it may be obtained from aqueous solution of stoichiometric composition at room temperature also by evaporation. As in previous cases hydrogen oxalate anions form the same head-to-tail sequence with O3-H3…O6(2.5732(13) Å) hydrogen bond. However, because of the presence of a water molecule, the carboxyl group of cation forms a hydrogen bond not with anion, but with water molecule O1-H1…O7(2.6234(14) Å). Water molecule in turn as proton donor forms hydrogen bonds O7-H7A…O5(2.7257(14) Å) with the second oxygen atom of the carboxylate group and also with the oxygen atom of the hydroxyl group of neighboring cation O7-H7B…O1(2.8976(16) Å). The NH_3^+ group forms N-H…O hydrogen bonds with water molecule and with O5 and O6 atoms of carboxylate group of anion. There is also a contact with carbonyl atom O4 of anion on the level of van der Waals interactions. When stored in air monohydrate dehydrates within 8–12 h at room temperature transforming into polycrystalline pseudomorph preserving the crystal habit. Recently, Zakharov and Boldyreva (2013) found I-order phase transition in (DL-AlaH)$HC_2O_4{\cdot}H_2O$ in the 1.5–2.4 GPa interval of pressure. The phase transition is not caused with decomposition of sample and has reversible nature. The phase transition is accompanied with an interesting change in hydrogen bonds although without change of symmetry.

Alagar et al. (2001a) found formation of needle-like crystals of L-alaninium hydrogen maleate in aqueous solution containing equimolar quantities of components. Hydrogen maleate anion as usual forms intramolecular hydrogen bond O3-H2...O5(2.425(3) Å). The carboxyl group of the cation forms hydrogen bond O1-H1...O4(2.593(2) Å) with the carbonyl oxygen atom of anion. The NH_3^+ group forms two hydrogen bonds, N1-H1A...O5 and N1-H1C...O6, with the carboxylate groups of two nearest anions and also weak hydrogen bond N1-H1B...O2 with the carbonyl oxygen atom of cation. There is also a contact on the level of van der Waals interactions N1-H1B...O4 with the third anion. As in the case of L-alaninium hydrogen oxalate L-alaninium hydrogen maleate became a subject of growth and investigation as possible NLO material (Vasantha and Dhanuskodi 2004a; Natarajan et al. 2006; Balasubramanian et al. 2009; Charoen-In et al. 2010; Vijayan et al. 2012; Ruby Nirmala and Prakash 2013). Quality crystals were grown, spectral thermal and other properties were studied, and NLO activity was determined by powder method. According to Vasantha and Dhanuskodi (2004a) the SHG efficiency is 0.53 compared to that of KDP and 1.2 according to Balasubramanian et al. (2009) and 0.15 compared to urea according to Natarajan et al. (2006). The crystal melts before decomposition. The melting point is equal to 162 °C according to Vasantha and Dhanuskodi (2004a) and 152 °C according to Vijayan et al. (2012). The crystal was grown also by unidirectional method (Charoen-In et al. 2010).

The crystals of L-alaninium hydrogen tartrate were obtained by Rajagopal et al. (2002b) from aqueous solution containing equimolar quantities of components. These authors as a rule do not indicate which L- or D-form of tartaric acid was used. Crystal structure was determined. From the structure one may conclude that hydrogen tartrate anion is in L-form. Hydrogen L-tartrate anions as usual form head-to-tail sequence by means of strong hydrogen bond O3-H3...O8(2.466(3) Å) between the carboxyl and carboxylate groups. The carboxyl group of the cation forms hydrogen bond O1-H1...O6(2.651(4) Å) with the oxygen atom of the hydroxyl group. The hydroxyl group O(6)-H in turn forms hydrogen bond with the second oxygen atom of the carboxylate group of the neighboring anion O6-H6...O7(2.765(4) Å). The NH_3^+ group forms one more hydrogen bond with O7 atom, N1-H1A...O7, and also N1-H1B...O4 with the carbonyl oxygen atom of the anion. Other contacts are on the level of van der Waals interactions. The crystal attracted the attention of Vimalan et al. (2010b) as possible NLO or piezoelectric material. They have grown high-quality crystal from aqueous solution of L-alanine and tartaric acid (without indication of L- or D-form) at molar ratio 1:1, identified the crystal by cell parameters, registered IR spectrum and thermal properties, and determined NLO and other characteristics. However, some data of this paper and their interpretation are not acceptable. The IR spectrum shown in this paper cannot belong to this crystal. The spectrum does not contain a characteristic absorption band caused by ν(C=O) of carboxyl groups of cation and anion. The authors indicated a transmittance (??) peak at 1,730 cm^{-1} instead of the expected absorption band in this region. Also the authors interpret very weak peaks and at the same time ignore very strong bands in the regions at 2,200 cm^{-1} and 700–1,000 cm^{-1}, which are difficult to expect for the crystal with given composition and structure. In interpreting of thermal data, the authors assign an endothermic peak at 118 °C on DTA curve to mass loss, which, however, occurs above 126 °C. The authors attribute this mass loss with loss of crystal water, which, however, according to Rajagopal et al. (2002b) is absent in the structure.

The next two salts of L-alanine with *p*-chlorbenzensulfonic acid and with 2,4-dinitrobenzensulfonic acid monohydrate (Kimoto et al. 1989; Sudbeck et al. 1995) are useful to consider in comparison. Both salts were obtained from aqueous solution with equimolar ratio of components. The asymmetric units in both cases contain one formula unit. The carboxyl group of the cation in the first case forms hydrogen bond O1-H8...O2(2.646(6) Å) with sulfonate group. The NH_3^+ group also forms hydrogen bonds with the three nearest anions: N1-H1...O3, N1-H2...O2, and N1-H3...O1. Thus, the O2 atom forms two hydrogen bonds, in which the resultant bond length S1-O2(1.469(4) Å) is longer than S1-O1(1.451(5) Å) and S1-O3(1.439(4) Å). Chlorine atom does not form hydrogen bonds. In the second crystal, water molecule plays an important role and significantly changes the scheme of hydrogen bonds. The carboxyl group of the cation forms hydrogen bond O9-H9...O1W(2.554(2) Å) with water molecule. Water molecule in turn forms two hydrogen bonds O1W-H1W...O3(2.762(2) Å) and O1W-H2W...O1(2.903(2) Å) with the sulfonate groups of two nearest anions. The NH_3^+ group forms two hydrogen bonds N4-H1N...O2 and N4-H3N...O1 with the sulfonate-groups of two nearest anions. Of these the first is rather strong with N...O(2.767(2) Å), while the second is on the level of van der Waals interactions. In the third hydrogen bond, the NH_3^+ group forms with the carbonyl oxygen atom of neighboring cation N4-H2N...O8. The nitro groups (similar to the chlorine atom in the first case) do not form hydrogen bonds. The only contact with nitro group N4-H3N...O4 is on the level of van der Waals interactions. Optically active (1S)-(+)-1,1-′-binaphthalene-2,2′-diyl hydrogen phosphate (bnppa) is a good optical resolution agent of amines. Bond lengths P-O1(1.589(2) Å) and P-O2(1.584(2) Å) are significantly longer not only than P-O3(1.453(2) Å), but also P-O4(1.525(2) Å). In bnppa, a hydrogen bond O4-H...O3(2.515(3) Å) is formed. Upon the interaction of L-alanine with bnppa, a proton transfer from PO_4H to L-alanine takes place, resulting in the formation of the salt L-Ala^+·bnppa·H_2O. The carboxyl group of the cation forms hydrogen bond O2A-H...O4(2.551(2) Å) with the PO_4^- group. The water molecule forms one hydrogen bond as proton donor with the second oxygen atom of the PO_4^- group O1W-H...O3(2736(3) Å) and also one hydrogen bond as acceptor with the NH_3^+ group N1A-H...O1W(2.724(3) Å). The NH_3^+ group forms one more hydrogen bond with the PO_4^- group N1A-H...O3(2.812(2) Å). Other contacts are on the level of van der Waals interactions.

In addition to salts of α-alanine provided in Table 5.5, there are also a number of publications on obtaining and investigating L-alanine salts, which actually are L-alanine (Fleck and Petrosyan 2010). The publications may be divided in three groups. The first group comprises of salts which exist and are reliably established in other works. In addition to the salts L-Ala·HCl and L-Ala·HCl·H_2O considered

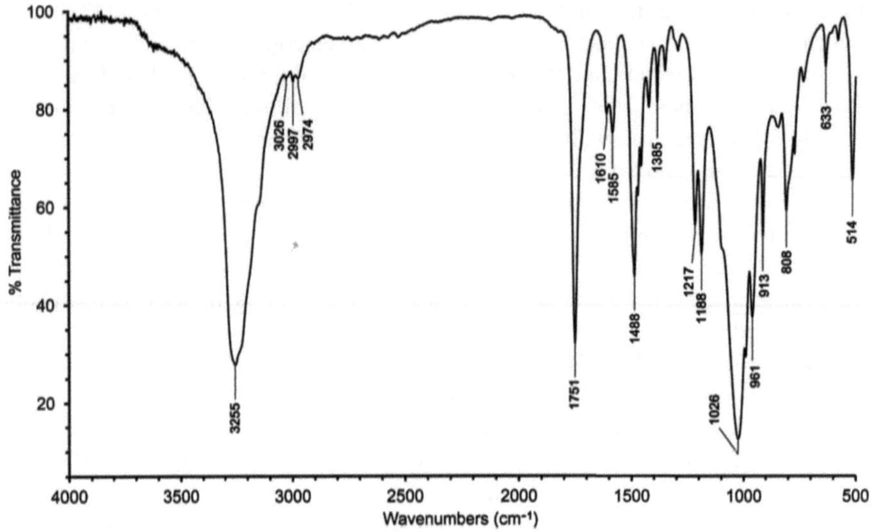

Fig. 5.9 IR spectrum of (L-AlaH)BF_4

above, the existence of the salts 2L-Ala·HNO_3 and 2L-Ala·picric acid was also established. The second group consists of salts the existence of which is not known. It is possible that they do not exist. These salts are L-alanine acetate, urea L-alanine acetate, thiourea L-alanine acetate, L-alanine formate, L-alanine·DL-malic acid, L-alanine fumarate, and L-alanine succinate. The third group comprises of salts which may exist, but no reliable data of their existence is reported. For example there was a report in obtaining L-Ala·HBF_4, which actually was L-alanine (see Fleck and Petrosyan 2010). However, in our group we could find the conditions for obtaining this salt (L-AlaH)BF_4 (see the IR spectrum in Fig. 5.9). The strong absorption band at 1,751 cm^{-1} is caused by ν(C=O) of the COOH group of L-alaninium cation, and the very strong absorption band at 1,026 cm^{-1} is caused by asymmetric stretching vibration of tetrafluoroborate anion.

5.1.6 Salts of β-Alanine

Salts of β-alanine are presented in Table 5.6. Structures of β-alanine halogenides have not been determined yet. Because some salts of β-alanine with 1:1 composition contain dimeric cation of $A^+...A^+$ type (see Sect. 5.2), salts of β-alanine halogenides are placed in Table 5.6 conditionally until the determination of their structures. Ennan et al. (1983) based on the IR spectrum concluded that β-alanine·HF contains the $(F\text{-}H\text{-}F)^-$ group, and also β-alaninium cation and zwitterionic β-alanine, which are connected with anion by hydrogen bonds according to following scheme: $^-OOC\text{-}CH_2\text{-}CH_2\text{-}NH_3^+\cdot(F\text{-}H\text{-}F)^-\cdot{}^+H_3N\text{-}CH_2\text{-}CH_2\text{-}COOH$. In our opinion, the formation of a dimeric cation according to the

Table 5.6 Salts of β-alanine with inorganic and organic anions (Ennan et al. 1983)

#	Crystal	s.g., Z	*a*, *b*, *c* (Å), *α*, *β*, *γ* (°)	Refs
1	(*β*-AlaH)F	–	–	Ennan et al. (1983)
2	(*β*-AlaH)Cl	–	–	Ennan et al. (1983)
3	(*β*-AlaH)Br	–	–	This work
4	(*β*-AlaH)I	–	–	This work
5	(*β*-AlaH)$_2$(IO$_3$)$_2$·H$_2$O	$P2_1/n$, 4	10.324(2), 7.1610(14), 21.178(4), $\beta = 101.11(3)$	Petrosyan et al. (2000a)
6	(*β*-AlaH)IO$_3$·HIO$_3$	–	–	Petrosyan et al. (2000a)
7	(*β*-AlaH)H$_2$PO$_4$	$P2_1/n$, 4	16.115(20), 5.829(2), 8.019(5), $\beta = 93.30(6)$	Averbuch-Pouchot et al. (1988c)
8	(*β*-AlaH)H$_2$PO$_3$	$P2_1/c$, 4	11.667(4), 6.458(2), 10.612(2), $\beta = 112.49(2)$	Averbuch-Pouchot (1993c)
9	(*β*-AlaH)SbF$_4$·H$_2$O	$P2_1/c$, 4	12.1580(5), 7.4167(4), 10.1540(5), $\beta = 108.222(2)$	Udovenko et al. (2005)
10	(*β*-AlaH)SbF$_4$	$P2_1/c$, 4	10.483(1), 7.5906(8), 9.4237(9), $\beta = 90.548(2)$	Udovenko et al. (2005)
11	(*β*-AlaH)ZrF$_5$	$P2_1/c$, 4	10.985(3), 6.626(3), 10.343(3), $\beta = 110.22(1)$	Gerasimenko et al. (2011)
12	(*β*-AlaH)$_2$NbOF$_5$·2H$_2$O	$P2_1/c$, 2	5.3532(1), 10.8585(3), 8.9913(2), $\beta = 90.022(1)$	Gerasimenko et al. (2008)
13	(*β*-AlaH)$_2$SiF$_6$	$P\bar{1}$, 1	5.911(1), 6.881(1), 8.209(2), $\alpha = 75.330(7)$, $\beta = 86.451(8)$, $\gamma = 75.817(7)$	Fleck et al. (2013b)
14	(*β*-AlaH)CCl$_3$OO	$P2_1/n$, 4	6.8049(14), 21.100(4), 6.8968(14), $\beta = 95.75(3)$, at 105 K	Rajagopal et al. (2003a)
15	(*β*-AlaH) maleate	$P2_1/n$, 4	5.4186(11), 22.951(5), 7.4518(15), $\beta = 99.730(16)$	Rajagopal et al. (2001b)
16	(*β*-AlaH)X[a]	$Pna2_1$, 4	22.819(4), 10.771(2), 7.289(1)	Ilyukhin and Davidovich (1999)

[a]X = ethylenediamino-tetrakis(acetato)-antimonate(III)

scheme: (β-Ala. . .β-AlaH)(F-H-F), that is ($^{+}H_3N$-CH_2-CH_2-COO^-. . .HOOC-CH_2-CH_2-NH_3^+).(F-H-F)$^-$ is also possible. The IR spectrum of (β-AlaH)Cl also is provided by Ennan et al. (1983), which differs from the spectrum of β-Ala·HF by the absence of strong and broad absorption band in the 700–2,000 cm^{-1} region. In our group we obtained β-Ala·HF, β-Ala·HCl, β-Ala·HBr, and β-Ala·HI. The IR spectra of β-Ala·HF and β-Ala·HCl are in good agreement with the spectra provided by Ennan et al. (1983), and the spectra of β-Ala·HBr and β-Ala·HI are similar to the spectrum of β-Ala·HCl. We try to obtain single crystals of these halogenides with the aim to determine their structures.

In a study by Petrosyan et al. (2000a) two compounds were obtained from the system β-Ala + HIO_3 + H_2O. At equimolar ratio a salt with β-Ala·HIO_3·0.5H_2O is formed, while at ratios 1:1.5–2 another salt is formed with composition β-Ala·2HIO_3. At ratio 2:1 no new compound is formed. A mixture of β-Ala and β-Ala·HIO_3·0.5H_2O is formed. The obtained salts were studied by ^{127}I NQR method, and also the structure of β-Ala·HIO_3·0.5H_2O was determined. The asymmetric unit contains two (β-AlaH) cations, two IO_3^- anions, and one water molecule. Positions of hydrogen atoms were not determined. Conclusions about structure and hydrogen bonds were made on the basis of bond lengths of nonhydrogen atoms. The water molecule and oxygen atoms of the carboxylate group of one (β-AlaH) are disordered. One of the cations forms hydrogen bond O-H. . .OW with water molecule. The second cation forms O-H. . .O hydrogen bond with one of the IO_3^- anions. Based on the analysis of the IR spectrum of β-Ala·2HIO_3 it was concluded that there are a cation, IO_3^- anion, and HIO_3 molecule. It was observed at heating that the β-Ala·HIO_3·0.5H_2O and β-Ala·2HIO_3 crystals melt at 95 and 100 °C, respectively.

Crystals of β-alaninium dihydrogen phosphate (β-AlaH)H_2PO_4 and phosphite (β-AlaH)HPO_3H were obtained by Averbuch-Pouchot et al. (1988c, 1993c) from aqueous solution with stoichiometric composition. In both cases the asymmetric units contain one formula unit. The $H_2PO_4^-$ anions form a chain by hydrogen bonds. The lengths of P-OH bonds (P-O1 and P-O3) are 1.563(3) Å and the lengths of P-O bonds (P-O2 and P-O4) are 1.498(2) Å and 1.512(2) Å, respectively. The O4 atom as acceptor forms two hydrogen bonds with two neighboring anions O1-H1. . .O4(2.610 (3) Å) and O3-H2. . .O4(2.543(3) Å). The O2 atom forms O6-H3. . .O2(2.657(4) Å) hydrogen bond with the carboxyl group of the cation and also two hydrogen bonds N-H1N. . .O2(2.792(4) Å) and N-H3N. . .O2(3.024(4) Å) with two neighboring cations. The NH_3^+ group forms one more hydrogen bond N-H2N. . .O3(2.933(4) Å).

In the structure of (β-AlaH)H_2PO_3 the anion HPO_3H^- has only one hydroxyl group O3-H9. In the structure of the crystal phosphite anions form cetrosymmetric dimer by means of two hydrogen bonds O3-H9. . .O1(2.567 Å).

This dimeric anion is balanced by two symmetrically oriented (β-AlaH) cations, which form hydrogen bonds: O5-H10. . .O2(2.539 Å). There is a paper by Hubert Joe et al. (1991) in which the vibrational spectra of β-Ala·H_3PO_4 are analyzed. In this work samples were prepared by slow evaporation at room temperature of an aqueous solution of β-alanine and H_3PO_4 in a stoichiometric ratio. However, the obtained samples were not identified. As a result of the analysis of the spectra, the authors came to conclusion that in the structure of the crystal of β-alanine is not in the form of a cation, but in thc form of a zwitterion, which contradicts to the structural data, according to which there is a carboxyl group of cation with bond lengths C=O(1.205(4) Å) and C-OH(1.312(4) Å).

Crystals of (β-AlaH)SbF_4·H_2O and (β-AlaH)SbF_4 were obtained by Udovenko et al. (2005). According to this work, the compounds were prepared by reaction of aqueous solution of β-alanine and SbF_3 with molar ratios of components 1:1 and 2:1, respectively. However, these salts cannot form without the presence in solution of HF, about which the authors forgot to mention. In the structure SbF_4^- anions form a chain by means of Sb…F bond. In the structure of SbF_4^- having the form of bipyramid the F3 and F4 atoms are in axial positions, while the atoms F1 and F2 and lone electron pair of Sb^{3+} occupy equatorial positions. Bond lengths Sb-F1(1.931 (1) Å) and Sb-F2(1.950(1) Å) in monohydrate and in anhydrous salt (1.919(2) Å) and (1.926(2) Å), respectively, significantly differ from the bonds Sb-F3(2.031(1) Å) and Sb-F4(2.187(1) Å) and in anhydrous salt (2.052(2) Å) and (2.101(2) Å), respectively. The chain of anions is formed by means of Sb…F4 bond, which also differ and are Sb…F4(2.669(1) Å) in monohydratc and Sb…F4(2.862(2) Å) in anhydrous salt. Carboxyl groups of cations form hydrogen bond O2-H2…F3(2.563 (2) Å) in monohydrate and O2-H2…F3(2.578(3) Å) in anhydrous salt. In monohydrate the NH_3^+ group forms N-H1B…O3 hydrogen bond with water molecule, which in turn forms hydrogen bond with anion O3-H32…F4 and with the carbonyl oxygen atom of cation O3-H31…O1. The two remaining hydrogen atoms of the NH_3^+ group form hydrogen bonds with anion N-H1A…F4 and N-H1C…F1. In anhydrous salt the NH_3^+ group forms one hydrogen bond N-H1A…O1 with the carbonyl oxygen atom of cation and two bonds with two nearest anions by means of N1-H1B…F4 and N1-H1C…F4. Conformations of cations in monohydrate and anhydrous salts differ significantly. Torsion angle C1C2C3N in monohydrate is 174.5°, while in anhydrous salt it is 77.0°.

Gerasimenko et al. (2011) reported the synthesis of crystals of (β-AlaH)ZrF_5. Zirconium oxide ZrO_2 was dissolved in ca. 40 % solution of HF, and then β-alanine was added in order to obtain equimolar ratio of β-alanine and $HZrF_5$. Crystals were obtained by slow evaporation at room temperature. The asymmetric unit contains one formula unit. Anions form infinite polymeric chain. The carboxyl group of the cation forms the hydrogen bond O2-H1…F5(2.597(2) Å). The NH_3^+ group forms two hydrogen bonds with fluorine atoms N1-H2…F4 and N1-H4…F3 and one with the carbonyl oxygen atom of cation, N1-H3…O1.

The crystal of $(β\text{-AlaH})_2NbOF_5$·$2H_2O$ was obtained by dissolving 1.33 g Nb_2O_5 and 1.78 g β-alanine in 40 ml of 48 % solution of HF in ratio 1:4 and slow evaporation at room temperature (Gerasimenko et al. 2011). The asymmetric unit contains one β-alaninium and one water molecule in general position and anion in special position, which displays disordering. The carboxyl group of cation forms hydrogen bond O3-H3…(F1/O1) with distance O…(F/O) equal to 2.6143(9) Å. Water molecule forms O4-H1…(F1/O1) and O4-H2…F3 hydrogen bonds. The

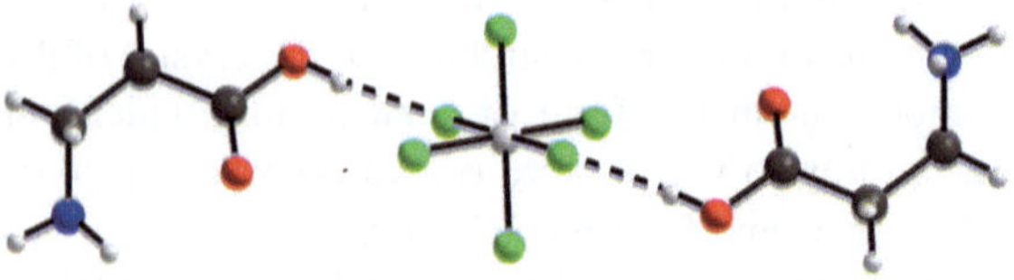

Fig. 5.10 Molecular structure of bis(β-Alaninium) hexafluorosilicate. The centrosymmetric arrangement is similar to that found in the respective sarcosinium salt (compare with Fig. 5.3)

NH_3^+ group forms hydrogen bonds with water molecule N1-H1C...O4 and with anion N1-H1B...F2. The bonds of H1A atom with neighboring O2 and O4 atoms are on the level of van der Waals interactions.

The salt (β-AlaH)$_2$SiF$_6$ was first obtained by Ennan et al. (1983). The labeling DL-β-Ala obviously is misprint. The obtained salt was characterized by IR spectrum and thermal analysis. In the work of Fleck et al. (2013b), the structure of crystal was determined (Fig. 5.10). The asymmetric unit contains a (β-AlaH) cation in general position and SiF_6^{2-} in special position. The carboxyl group of the cation forms hydrogen bond O1-H1...F3(2.6693(13) Å) with anion. The NH_3^+ group forms two hydrogen bonds with anion N1-H12...F2, N1-H13...F1 and one hydrogen bond with the carbonyl oxygen atom of the nearest cation N1-H11...O2. Thus, all proton acceptors form hydrogen bonds.

The crystal of (β-AlaH)CCl$_3$COO was obtained from aqueous solution containing equimolar quantities of components (Rajagopal et al. 2003a). The asymmetric unit of the crystal consists of one β-alaninium cation and one CCl_3COO^- anion. The carboxyl group of cation forms O1-H1...O3(2.649(2) Å) hydrogen bond with anion. There are other hydrogen bonds in addition to this bond. The conformation of the cation is favorable for the formation of an intramolecular weak hydrogen bond N1-H1B...O4. The NH_3^+ group forms hydrogen bonds N1-H1A...O2(2.840(2) Å) and N1-H1C...O3(2.846(3) Å) with the carboxylate group of anion. The atom H1C has also a contact with Cl2 atom (N1-H1C...Cl2) with H...Cl(2.78 Å) and the atom H1B has a weak contact N1-H1B...O4 with H...O(2.32 Å). There are also short contacts C-H...O and C-H...Cl: C1-H...O2 with H...O(2.37 Å) and C1-H...Cl1 with H...Cl(2.88 Å).

The crystal of β-alaninium hydrogen maleate was obtained from aqueous solution containing equimolar quantities of β-alanine and maleic acid (Rajagopal et al. 2001b). In maleate anion as usual there is an intramolecular hydrogen bond O3-H5...O5(2.481(2) Å). This distance is not the longest among known O...O distances for maleate anions; however, it should be considered as one of the longest. The carboxyl group of the cation forms hydrogen bond O1-H1...O6(2.522(3) Å) with the carboxylate group of the anion. The NH_3^+ group forms two normal hydrogen bonds with the anion, N1-H1A...O5 and N1-H1C...O4, and one weak hydrogen bond with the carbonyl oxygen atom of the cation, N1-H1B...O2.

In a study by Ilyukhin and Davidovich (1999) the structure of the salt of β-alaninium with ethylenediaminetetraacetatoantimonate(III) anion was determined. The carboxyl group of the cation forms O10-H1...O6(2.693 Å) hydrogen bond. The NH_3^+ group forms hydrogen bonds: N3-H2...O2(2.886 Å), N3-H3...O8(2.867 Å), and N3-H4...O8(2.748 Å). In Table 5.6 this is the only salt with non-centrosymmetric structure and polar symmetry.

5.1.7 *Salts of Phenylalanine*

In Table 5.7 salts of phenylalanine are presented. Crystalline (L-PheH)Cl was obtained by cooling a boiling saturated solution of L-phenylalanine in 30 % HCl. Then large crystals were obtained by slow evaporation of a saturated solution at room temperature (Al-Karaghouli and Koetzle 1975). The crystal of structure was determined by neutron diffraction. Previously the structure less accurately was determined by XRD (see references in Al-Karaghouli and Koetzle 1975). The asymmetric unit contains L-PheH cation and chloride anion. The carboxyl group of the cation forms hydrogen bond O-H...Cl(2.992(8) Å). Chloride anion additionally forms three more N-H...Cl hydrogen bonds with the three nearest cations.

The crystals of (DL-PheH)NO_3 were obtained from an aqueous solution comprising equimolar quantities of DL-phenylalanine and nitric acid by slow evaporation (Sridhar et al. 2002a). The asymmetric unit contains one formula unit. The carboxyl group of the (PheH) cation forms O1B-H1...O3(2.666(4) Å) with the nitrate anion. The O3 atom forms one more hydrogen bond with the NH_3^+ group, N1-H1B...O3. The NH_3^+ group forms one more hydrogen bond with the nitrate anion N1-H1C...O2. The NH_3^+ group forms one more hydrogen bond with the carbonyl oxygen atom N1-H1A...O1A. Thus, the O1 atom of the nitrate ion does not form any hydrogen bonds. Bond lengths N1-O2(1.256(4) Å) and N1-O3(1.254 (4) Å) are greater than bond length of N1-O1(1.234(4) Å). The crystal of (DL-PheH) NO_3 was studied by IR and Raman spectroscopy (Hubert Joe et al. 1991). The difference of bond lengths in nitrate ion is revealed in vibrational spectra.

The crystal of (DL-PheH)H_2PO_4 was obtained from an aqueous solution containing equimolar quantities of DL-phenylalanine and H_3PO_4 at slow evaporation (Ravikumar et al. 2001a). The salts crystallize in non-centrosymmetric space group *Pc* with $Z' = 2$; however, the structure displays pseudocentrosymmetry (Fig. 5.11), which is often observed when $Z' > 1$. Cations form rather short hydrogen bonds with anions: O1B-H1B...O12(2.473(4) Å) and O2B-H2B...O22(2.497 (4) Å). One of the two hydroxyl groups of $H_2PO_4^-$ anions forms hydrogen bonds with the carbonyl oxygen atom of the same cation: O13-H13...O1A(2.615(4) Å) and O23-H23...O2A(2.684(4) Å). The second hydroxyl group of anions is used for the formation of hydrogen-bonded chain of anions: O14-H14...O12(2.684(3) Å) and O24-H24...O22(2.668(4) Å). The NH_3^+ group of the first cation forms two hydrogen bonds with anions, N11-H11A...O11 and N11-H11C...O21, and one hydrogen bond with the oxygen atom of the carboxyl group of the first cation,

Table 5.7 Salts of phenylalanine with inorganic and organic anions

#	Crystal	s.g., Z	a, b, c (Å), α, β, γ (°)	Refs
1	(L-PheH)Cl	$P2_12_12_1$, 4	27.763(17), 7.059(4), 5.380(5)	Al-Karaghouli and Koetzle (1975)
2	(DL-PheH)NO_3	$P2_1/c$, 4	12.005(7), 5.758(2), 16.250(9), $\beta = 107.48(6)$	Sridhar et al. (2002a)
3	(L-PheH)H_2PO_4·L-Phe	$P2_1$, 2	13.441(1), 4.8760(4), 15.470(2), $\beta = 97.12(1)$	Ravikumar et al. (2002a)
4	(DL-PheH)H_2PO_4	Pc, 4	13.899(7), 9.956(8), 9.051(2), $\beta = 108.726(8)$	Ravikumar et al. (2001a)
5	(L-PheH)$_2SO_4$·H_2O	$P2_1$, 2	13.365(7), 6.735(4), 12.938(7), $\beta = 112.70(6)$	Nagashima et al. (1992)
6	(DL-PheH)$_2SO_4$·H_2O	$P2_1/n$, 4	14.560(2), 6.6000(12), 22.115(4), $\beta = 90.70(2)$	Srinivasan et al. (2001a)
7	(L-PheH)$_2SiF_6$	$P2_1$, 2	11.129(2), 5.690(1), 17.016(3), $\beta = 105.31(3)$	Ghazaryan et al. (2013a)
8	(DL-PheH)$_2SiF_6$	$P2_1/c$, 2	11.183(2), 5.7531(10), 17.000(4), $\beta = 105.59(2)$	Belhouas et al. (2012)
9	(L-PheH)$_2$(CCl_3COO)$_2$·H_2O	$C2$, 4	19.675(3), 6.115(1), 26.584(2), $\beta = 109.6(1)$	Rajagopal et al. (2003b)
10	(DL-PheH)CCl_3COO	$C2/c$, 8	18.838(3), 16.557(2), 10.065(12), $\beta = 111.354(14)$	Rajagopal et al. (2005)
11	(L-PheH) maleate	$P2_1$, 2	11.0560(9), 5.3326(4), 11.4712(7), $\beta = 101.07(1)$	Alagar et al. (2001b)
12	(DL-PheH) maleate	$P2_1/n$, 4	12.308(3), 5.9942(12), 18.061(4), $\beta = 99.15(3)$, at 123 K	Alagar et al. (2003a)

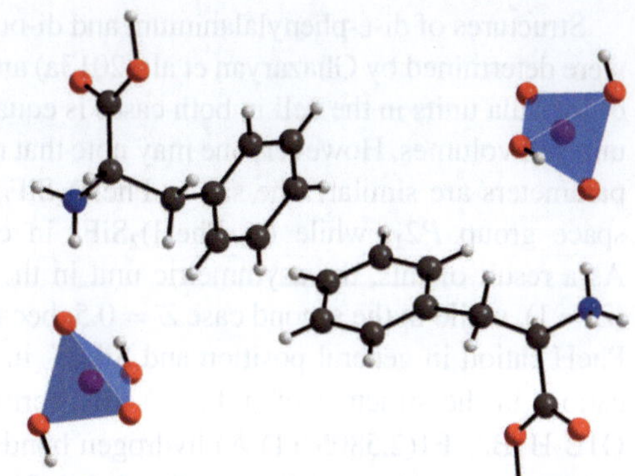

Fig. 5.11 Structure of DL-phenylalaninium dihydrogen phosphate. The symmetry (space group Pc) is pseudocentrosymmetric

N11-H11B…O1B. Similar hydrogen bonds are formed by the NH_3^+ group of the second cation: N21-H21A…O21, N21-H21B…O11, and N21-H21C…O2B.

The interaction of L-phenylalanine with H_3PO_4 with 2:1 molar ratio results in the formation of crystal, the asymmetric unit of which contains one L-PheH cation, one L-Phe in the form of a zwitterion, and one $H_2PO_4^-$ anion (Ravikumar et al. 2002a). There are a number of salts of L-phenylalanine with 2:1 composition with dimeric cation (L-Phe…L-PheH) (see Sect. 5.2). However, the presence of the $H_2PO_4^-$ anion prevents the formation of dimeric cation. In this case the L-phenylalanine dimeric cation is not formed as well. The carboxyl group of (L-PheH) forms hydrogen bond O2B-H2B…O2(2.543(9) Å) with the $H_2PO_4^-$ anion. The anion, in turn, forms a hydrogen bond with the carboxylate group of zwitterion O3-H3…O1A(2.535(9) Å). The second hydroxyl group of the anion forms relatively weak hydrogen bond with the carbonyl group of the cation, O4-H4…O2A (2.938(9) Å). The NH_3^+ group of the zwitterion forms hydrogen bonds with three nearest anions: N11-H11A…O1, N11-H11B…O1, and N11-H11C…O1. The NH_3^+ group of the cation also forms three hydrogen bonds: N21-H21A…O1B, N21-H21B…O2B, and N21-H21C…O2. The vibrational spectra of (DL-PheH) H_2PO_4 and (L-PheH)H_2PO_4. L-Phe were studied by Ravikumar et al. (2006). The crystals of (L-PheH)H_2PO_4. L-Phe were grown from an aqueous solution and characterized as possible NLO material by Sujatha et al. (2010). The SHG efficiency estimated by powder method was 1.2 times higher than in KDP.

In studies by Nagashima et al. (1992) and Srinivasan et al. (2001a), the structures of (L-PheH)$_2SO_4\cdot H_2O$ and (DL-PheH)$_2SO_4\cdot H_2O$ were determined. Both asymmetric units contain each one formula unit. The carboxyl groups of cations form hydrogen bonds with anions. Water molecules as proton acceptors in both cases form hydrogen bonds with the NH_3^+ group, while as proton donors form one hydrogen bond with anion. Hydrogen bonds in the structure of (L-PheH)$_2SO_4\cdot H_2O$ are O1-HO1…O2S(2.585 (6) Å) and O11-O11…O4S(2.535(4) Å), while in the structure of (DL-PheH)$_2SO_4\cdot H_2O$ they are O1B-H1B…O1(2.614(2) Å) and O2B-H2B…O2(2.575(3) Å). Vibrational spectra of (DL-PheH)$_2SO_4\cdot H_2O$ are studied in (Rajkumar and Ramakrishnan 2002).

Structures of di-L-phenylalaninium and di-DL-phenylalaninium hexafluorosilicates were determined by Ghazaryan et al. (2013a) and Belhouas et al. (2012). The number of formula units in the cell in both cases is equal to two, which suggests closeness of unit cell volumes. However, one may note that not only the volumes but also unit cell parameters are similar. The salt (L-PheH)$_2$SiF$_6$ crystallizes in non-centrosymmetric space group $P2_1$, while (DL-PheH)$_2$SiF$_6$ in centrosymmetric space group $P2_1/c$. As a result of this, the asymmetric unit in the first case contains one formula unit ($Z' = 1$), while in the second case $Z' = 0.5$, because the asymmetric unit contains one PheH cation in general position and SiF_6^{2-} in special position. Carboxyl groups of cations in the structure of (L-PheH)$_2$SiF$_6$ form O1A-H1A. . .F5(2.6283(12) Å) and O1B-H1B. . .F1(2.5805(11) Å) hydrogen bonds. The mean value of these distances 2.6044 Å is less than the distance O. . .F(2.6456(14) Å) in hydrogen bond O1-H1. . .F2 in the structure of (DL-PheH)$_2$SiF$_6$. This, however, is compensated by less distances of N. . .F hydrogen bonds (2.810 Å) in the structure of (DL-PheH)$_2$SiF$_6$ compared to the value 2.849 Å for (L-PheH)$_2$SiF$_6$. Mean values of Si-F bond lengths in both structures are the same 1.685 Å. However, the bond lengths Si-F2(1.6952(9) Å) in the structure of (DL-PheH)$_2$SiF$_6$ and Si-F1(1.6892(8) Å) and Si-F5(1.6943(8) Å) in the structure of (L-PheH)$_2$SiF$_6$, which form hydrogen bonds, are the longest. Andreeva et al. (1983) obtained a salt (PheH)$_2$SiF$_6$ and characterized by IR spectrum and thermal analysis. However, from the data of this paper it is unclear if the salt is (DL-PheH)$_2$SiF$_6$, (D-PheH)$_2$SiF$_6$ or (L-PheH)$_2$SiF$_6$.

In the works of Rajagopal et al. (2003b, 2005) crystals of trichloroacetates of L-phenylalaninium and DL-phenylalaninium were obtained from aqueous solutions containing equimolar quantities of components. The salt of L-phenylalanine crystallizes in the form of hemihydrates, while the salt of DL-phenylalanine is anhydrous. The asymmetric unit of the salt of L-phenylalanine contains two L-phenylalaninium cations, two CCl_3COO^- anions, and one water molecule. Chlorine atoms of one of the anions are disordered. The carboxyl group of one of the cations forms hydrogen bond O6-H6. . .O1(2.642(4) Å) with anion in which chlorine atoms are ordered. The second cation forms rather short hydrogen bond with water molecule O8-H8. . .O9(2.507(5) Å). Water molecule as proton donor forms hydrogen bonds with both anions: O9-H9A. . .O1(2.2.760(4) Å) and O9-H9B. . .O3 (2.786(5) Å). The NH_3^+ groups of cations form hydrogen bonds with the carboxylate groups of anions: N1-H1A. . .O2, N1-H1B. . .O2, and N1-H1C. . .O4 and N2-H2A. . .O2 and N2-H2B. . .O3. The asymmetric unit of (DL-PheH)CCl$_3$COO contains one formula unit. The carboxyl group of cation forms O1-H1. . .O4(2.630 (4) Å). The NH_3^+ group forms three N-H. . .O hydrogen bonds with anion. The atom O4 forms also N1-H1C. . .O4, while the atom O3 forms two hydrogen bonds with two nearest cations: N1-1A. . .O3 and N1-H1B. . .O3. The crystal of (L-PheH) CCl$_3$COO·0.5H$_2$O with non-centrosymmetric structure attracted the attention of Prakash et al. (2011a) as possible NLO material. The efficiency of SHG was 0.65 compared to that of KDP. The authors state that the crystal is stable up to 136 °C; however, on the DTA curve there is endothermic effect below 100 °C, which was ignored at discussion of TG and DTA data.

Alagar et al. (2001b, 2003a) obtained crystals of L-phenylalaninium and DL-phenylalaninium maleates from aqueous solutions containing equimolar quantities of components. In both cases asymmetric units contain each one formula unit. Hydrogen maleate anions as usual comprise of intramolecular hydrogen bond. In the structure of (L-PheH) hydrogen maleate, this hydrogen bond O3-H3...O5 (2.412(2) Å) is very short, while in the case of DL-Phe salt the O...O distance of O3-H1...O6 is significantly longer, 2.468(2) Å. The carboxyl group in the structure of L-Phe salt forms hydrogen bond O1-H1...O4(2.547(2) Å) with anion. In the structure of DL-Phe salt this distance is equal to 2.5314(19) Å). The NH_3^+ groups form each two hydrogen bonds N1-H1A...O6 and N1-H1C...O6 with two nearest anions in the structure of L-Phe salt and N1-H1A...O4 and N1-H1B...O6 in the structure of DL-Phe salt. Other contacts are on the level of van der Waals interactions. Anbuchezhiyan et al.(2009a) have grown (L-PheH) hydrogen maleate crystal and characterized it as possible NLO material. It should be noted that almost all salts of phenylalanine have needle-like shape. The crystal obtained in this study had $7 \times 1 \times 0.5$ mm^3 sizes. SHG efficiency estimated by powder method was 0.27 compared to KDP. Anbuchezhiyan et al.(2009a) attributed the absorption band at 3,420 cm^{-1} to the O-H stretching mode. Probably this band is caused by water containing in KBr due to its hygroscopicity, because for hydrogen bonds O3-H3...O5 and O1-H1...O4 having 2.412 Å and 2.547 Å O...O distances, bands caused by OH stretching vibration cannot be in this region. The same crystal was reported as well by Yogam et al. (2012). The IR spectrum provided in this work significantly differs from that provided by Anbuchezhiyan et al.(2009a). Yogam et al. (2012)also assigned the band at 3,423 cm^{-1} to OH stretching of maleate anion. According to their data the efficiency of SHG estimated by powder method is 1.5 times higher than that of KDP. Thermal data also significantly differ from that given by Anbuchezhiyan et al.(2009a)

5.1.8 Salts of Serine

Salts of serine are collected in Table 5.8. The serine molecule $H_2C(OH)$-$CH(NH_2)$ COOH differs from the molecule of α-alanine H_3C-$CH(NH_2)COOH$ in that it contains an OH alcoholic group, which may play the role both of proton donor and proton acceptor in hydrogen bonds.

Crystals of (DL-SerH)Cl were obtained by Jarmelo et al. (2008) from an aqueous solution of DL-serine and excess of concentric HCl by evaporation of solution. The structure was determined and detailed theoretical and experimental investigations of vibrational spectra were performed. The asymmetric unit contains one formula unit. The carboxyl group of the cation forms hydrogen bond not with chloride ion, but with the oxygen atom of alcoholic OH group O-HC...OA(2.6195 Å). Chloride ion as proton acceptor forms three hydrogen bonds with the NH_3^+ group and one hydrogen bond O-HA...Cl with the alcoholic OH group.

Table 5.8 Salts of serine with inorganic and organic anions

#	Crystal	s.g., Z	a, b, c (Å), α, β, γ (°)	Refs
1	(DL-SerH)Cl	$P2_1/c$, 4	9.3040(2), 6.6665(2), 11.7804(3), $\beta = 121.216(2)$	Jarmelo et al. (2008)
2	(DL-SerH)$_2$SO$_4$·H$_2$O	$P2_12_12_1$, 4	10.597(2), 21.337(2), 5.855(1)	Zhang et al. (1997)
3	(L-SerH)H$_2$PO$_4$	$P2_1$, 2	9.134(5), 9.489(5), 4.615(5), $\gamma = 99.54(5)$	Smolin et al. (2005)
4	(L-SerH)·L-Ser·H$_2$PO$_4$·H$_2$O	$P2_1$, 2	4.704(5), 10.512(5), 13.573(3), $\beta = 98.73(5)$	Smolin et al. (2003b)
5	(DL-SerH)SbF$_4$	$P2_1/c$, 4	14.63(1), 6.245(6), 9.070(8), $\beta = 104.88(3)$	Udovenko et al. (2010a)
6	(L-SerH)$_2$SiF$_6$	$C2$, 2	10.455(2), 5.7737(12), 11.239(2), $\beta = 104.84(3)$	Fleck et al. (2013b)
7	(DL-SerH)$_2$SiF$_6$	–	–	This work
8	(L-SerH)HC$_2$O$_4$	$P2_1$, 4	14.3723(8), 6.2614(4), 9.1815(5), $\beta = 92.112(4)$	Braga et al. 2013
9	(L-SerH)$_2$C$_2$O$_4$·2H$_2$O, form I	$P2_1$, 2	4.8832(6), 11.9276(14), 12.4798(15), $\beta = 92.018(10)$	Braga et al. (2013)
10	(L-SerH)$_2$C$_2$O$_4$·2H$_2$O, form II	$P2_1$, 2	12.5711(6), 11.2144(5), 5.2079(2), $\beta = 100.529(3)$	Braga et al. (2013)
11	(DL-SerH)$_2$C$_2$O$_4$·2H$_2$O	$P2_1/c$, 4	4.8685(10), 17.199(7), 17.172(4), $\beta = 91.678(19)$	Alagar et al. (2002a)
12	(L-SerH) maleate	$P2_1$, 2	6.99527(11), 6.5631(10), 10.4284(19), $\beta = 97.094(14)$	Arkhipov et al. (2013)
13	(DL-SerH) maleate	$P\bar{1}$, 2	5.9406(3), 8.9527(4), 9.3796(3), $\alpha = 79.239(3)$, $\beta = 75.528(3)$, $\gamma = 86.771(4)$	Arkhipov et al. (2013)
14	L-(+)-serinium hydrogen squarate	$P2_12_12_1$, 4	5.335(3), 8.769(2), 25.782(5)	Kolev et al. (1998a)
15	L-serinium *p*-toluenesulfonate	$P2_12_12_1$, 4	6.853(0), 27.306(2), 6.783(0)	Kimoto et al. (1989)
16	(L-SerH)X[a]	$P2_1$, 4	11.552(5), 8.698(9), 12.155(3), $\beta = 97.40(3)$	Sudbeck et al. (1995)

[a]X = 4-hydroxybenzenesulfonate

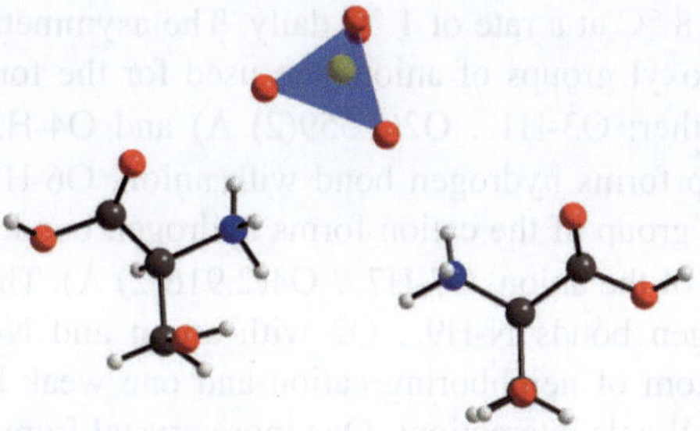

Fig. 5.12 Structure of bis(DL-Serinium) sulfate monohydrate. The structure is non-centrosymmetric, despite the presence of D- and L-molecules and hence the possibility of a centrosymmetric arrangement. The water molecule is omitted for clarity

Crystals of (DL-SerH)$_2$SO$_4$·H$_2$O were obtained by Zhang et al. (1997) from an aqueous solution containing DL-serine and H$_2$SO$_4$ in molar ratio 2:1. The determination of the crystal structure showed that in spite of coexistence of racemic mixture of D- and L-SerH cations, the compound crystallizes in non-centrosymmetric $P2_12_12_1$ space group. The asymmetric unit contains one formula unit (Fig. 5.12). The oxygen atom of water molecule is disordered with occupancy factor 0.5. The O1W...O2W distance is 0.7 Å. The carboxyl groups of both cations form hydrogen bonds with oxygen atoms of alcohol groups: O2-H1...O6(2.613(3) Å) and O5-H9...O3(2.589(3) Å). Thus, specific pairs of DL-cations are formed. The alcoholic group of one of the cations as proton donor forms hydrogen bond with water molecule O3-H8...O1W(O2W) with O...O distances 2.59(1) Å and 2.60(1) Å, respectively, with O1W and O2W, while the alcoholic group of the second cation forms hydrogen bond with anion O6-H16...O4s(2.683(3) Å). Water molecule as proton donor forms hydrogen bonds O1W-H18...O3s(2.82(1) Å) and O1W-H17...O4s(2.76(1) Å) with anion. The NH$_3^+$ groups of both cations form hydrogen bonds with anions: N1-H3...O2s, N1-H4...O3s, and N1-H5...O1s and N2-H11...O1s, N2-H12...O2s, and N2-H13...O3s. There is also a contact N2-H11...O4s on the level of van der Waals bonds with H11...O4s(2.38 Å). Thus, three oxygen atoms of sulfate anion form each two hydrogen bonds, while the O3s atom forms four hydrogen bonds, in which the resultant bond length S1-O3s(1.485(2) Å) is the longest. Smolin et al. (2002) who repeatedly studied the crystal of (DL-SerH)$_2$SO$_4$·H$_2$O apparently were not aware of the work of Zhang et al. (1997). Smolin et al. (2002) concluded that that there is a phase transition at 233 K caused by transition into an incommensurately modulated state. Later Yarmarkin et al. (2006) and Egorov and Lemanov (2006) reported an I-order phase transition in 260–270 K interval caused by freezing of orientational degrees of freedom of crystal water; however, it was then reported that previously observed anomalies are caused by drops of solution entrapped during crystal growth (Lemanov et al. 2012).

Crystals of (L-SerH)H$_2$PO$_4$ were obtained by Smolin et al. (2005) from an aqueous solution containing L-serine and H$_3$PO$_4$ at pH < 2 by cooling method in interval from 28 to 8 °C at a rate of 1 °C daily. The asymmetric unit contains one formula unit. Hydroxyl groups of anions are used for the formation of hydrogen bonds with each other: O3-H1. . .O2(2.559(2) Å) and O4-H2. . .O1(2.537(3) Å). The carboxyl group forms hydrogen bond with anion, O6-H3. . .O1(2.545(2) Å), while the alcoholic group of the cation forms hydrogen bond with the O4 atom of the hydroxyl group of the anion, O7-H7. . .O4(2.916(2) Å). The NH_3^+ group forms two normal hydrogen bonds N-H9. . .O2 with anion and N-H10. . .O5 with the carbonyl oxygen atom of neighboring cation and one weak bond N-H8. . .O2 on the level of van der Waals interactions. One more crystal from this system L-serine + H_3PO_4 + H_2O was obtained previously by Smolin et al. (2003b) from an aqueous solution of L-serine and H_3PO_4 (molar ratio is not indicated) by cooling from room temperature to 8 °C with a rate of 1 °C daily. Earlier this crystal was obtained by Lemanov et al. (2002), in which a piezoelectric effect was observed. The asymmetric unit contains one (L-SerH) cation, one L-serine molecule in the form of zwitterion, one $H_2PO_4^-$ anion, and one water molecule. The structure was determined also at 203 K; however, the structure was the same as at room temperature. In spite of the presence of (L-SerH) cation and L-serine molecule in the structure, here dimeric (L-Ser. . .L-SerH) cation is not observed. The carboxyl group of the cation forms hydrogen bond O6-H3. . .OW(2.562 Å) with water molecule. This hydrogen bond is absent in the table of hydrogen bonds of Smolin et al. (2003b). The alcoholic group of the cation forms hydrogen bond with the oxygen atom of one of the hydroxyl groups of the $H_2PO_4^-$ anion, O7-H7. . .O4(2.840(2) Å). The hydroxyl O4-H2 in turn forms hydrogen bond with neighboring anion, O4-H2. . .O2(2.570 Å). This hydrogen bond also is absent in the table of hydrogen bonds (Smolin et al. 2003b). The hydroxyl group O10-H14 of zwitterionic L-serine also forms hydrogen bond with anion, O10-H14. . .O2(2.845(2) Å). Another hydroxyl group of anion forms the shortest hydrogen bond with the carboxylate group of zwitterion, O3-H1. . .O9(2.481(2) Å). Water molecule as proton donor forms the third hydrogen bond with the O2 atom of the anion, OW-H19. . .O2(2.816 (2) Å), and also with the hydrogen atom of the alcoholic group of the cation, OW-H18. . .O7(2.758(2) Å). The NH_3^+ groups of cation and zwitterion form each three hydrogen bonds: N1-H8. . .O8, N1-H9. . .O1, and N1-H10. . .O1 and N2-H15. . .O10, N2-H16. . .O8, and N2-H17. . .O1.

Crystals of (DL-SerH)SbF_4 were reported by Udovenko et al. (2010a) by the interaction of DL-serine and SbF_3 with molar ratio of 1:1 in aqueous solution of HF (pH = 1) and slow evaporation at room temperature. The SbF_4^- anions form a polymeric chain. Bond lengths in the SbF_4^- anion are Sb-F1(1.945(2) Å), Sb-F2 (1.947(2) Å), Sb-F3(2.054(2) Å), and Sb-F4(2.112(2) Å), while the chain is formed due to Sb. . .F3(2.556(2) Å). The carboxyl group of the (SerH) cation forms O3-H3. . .O1(2.651(3) Å) hydrogen bond with the oxygen atom of the alcoholic group of neighboring cation, which in turn forms O1-H1. . .F4(2.600(2) Å) hydrogen bond. The NH_3^+ group forms hydrogen bonds N-H8. . .F1(2.806(3) Å), N-H7. . .F2(2.817(3) Å), and N-H6. . .F4(2.736(3) Å) with F1, F2, and F4 atoms of the chain.

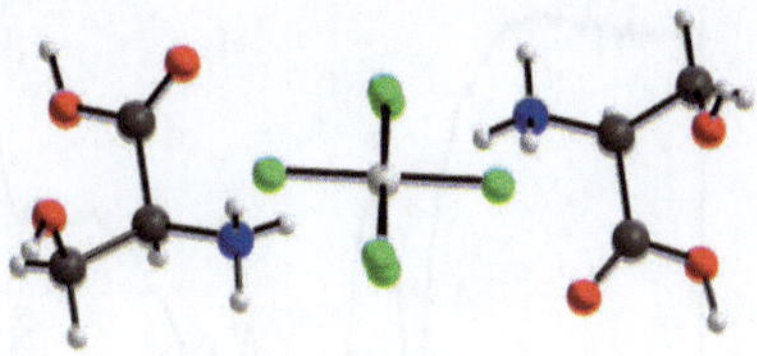

Fig. 5.13 Structure of bis (L-Serinium) hexafluorosilicate. The SiF_6 anion is located on a special position (site symmetry 2)

Fleck et al. (2013b) reported the growth of crystals of (L-SerH)$_2$SiF$_6$. From an aqueous solution of L-serine and H_2SiF_6 irrespective of 2:1, 1:1, and 1:2 molar ratios, the same (L-SerH)$_2$SiF$_6$ phase is formed. However, the quality of crystals is better at 1:1 ratio. The same occurs in the case of (DL-SerH)$_2$SiF$_6$. In this work, the crystal and molecular structure of (L-SerH)$_2$SiF$_6$ was determined. The asymmetric unit contains (L-SerH) cation in general position and SiF_6^{2-} anion in special position (Fig. 5.13). Bond lengths of Si-F in anion are Si-F1(1.6854(9) Å), Si-F2 (1.6682(9) Å), and Si-F3(1.6927(10) Å). The cation does not form O-H…F hydrogen bond with anion, but only N-H…F type does. The carboxyl group of cation forms O1-H1…O3(2.6366(16) Å) hydrogen bond with the oxygen atom of the alcoholic group. The alcoholic group O(3)H forms O3-H3…O2(2.7394(15) Å) hydrogen bond with the carbonyl oxygen atom of the nearest cation. As a result of this each cation is surrounded by four symmetry-related cations, with two as proton donors and two as proton acceptors. The NH_3^+ group forms N1-H11…F1 (2.9075(18) Å), N1-H12…F3(2.8941(16) Å), and N1-H13…F3(2.9054(17) Å). Hexafluorosilicate of serine (SerH)$_2$SiF$_6$ was obtained and characterized by IR spectroscopy and thermal analysis by Andreeva et al. (1983) without indication of what form of L-, D-, or DL-serine was used. In Fig. 5.14 the IR spectra of (L-SerH)$_2$SiF$_6$ and (DL-SerH)$_2$SiF$_6$ are shown. As it is shown in Fig. 5.14, these spectra are rather similar. According to Andreeva et al. (1983) there is a band at 3,395 cm^{-1} in the IR spectrum of their sample assigned to ν(OH) of the alcoholic group. From Fig. 5.14 it is seen that there is a band at 3,384 cm^{-1} in the spectrum of (DL-SerH)$_2$SiF$_6$ and 3,337 cm^{-1} in the spectrum of (L-SerH)$_2$SiF$_6$, which allows supposing that the sample obtained by Andreeva et al. (1983) was (DL-SerH)$_2$SiF$_6$, because the spectrum of (D-SerH)$_2$SiF$_6$ should be similar to that of (L-SerH)$_2$SiF$_6$.

The interaction of serine with oxalic acid was studied by Braga et al. (2013) and Alagar et al. (2002a). Initially the compound (DL-SerH)$_2$C$_2$O$_4$·2H$_2$O was obtained from an aqueous solution with equimolar ratio of components (Alagar et al. 2002a). Braga et al. (2013) and Zakharov et al. (2011) established, however, that it may be obtained also at stoichiometric ratio of components. Alagar et al. (2002a) established – and was confirmed by Braga et al. (2013) and Zakharov et al. (2011) – that the asymmetric unit of (DL-SerH)$_2$C$_2$O$_4$·2H$_2$O contains two cations (L-serinium and D-serinium), two water molecules, and two oxalate $C_2O_4^{2-}$ anions, which, despite being in special positions, lie across the inversion center. Each of the anions forms hydrogen bonds O…H-O with the carboxyl group of one cation and the alcoholic group of another cation, and vice versa. The first anion via O4 atom forms O1′-H1′…O4(2.5374(15) Å) and O3-H3…O4(2.7696(17) Å) hydrogen bonds, and

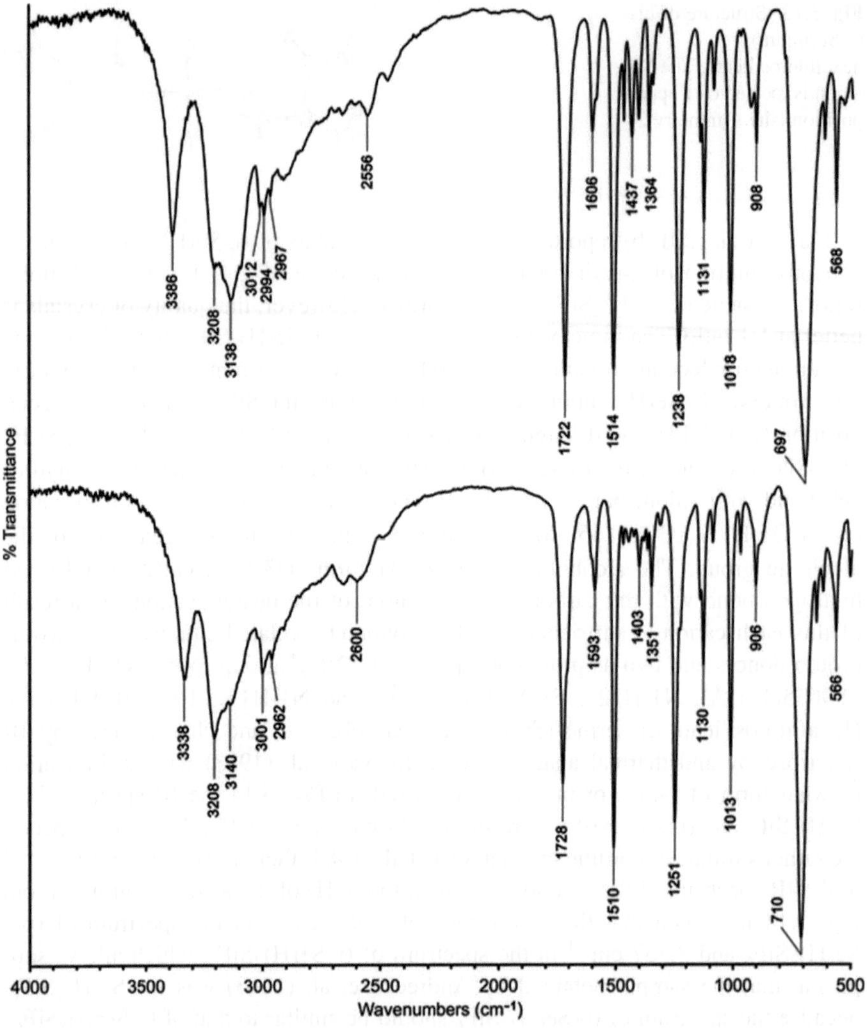

Fig. 5.14 IR spectra of bis(DL-sernium) (*upper*) and bis(L-serinium) hexafluorosilicates

the second anion forms O1-H1…O4′(2.5158(15) Å) and O3′-H3′…O4′(2.7718 (19) Å). The NH_3^+ groups form hydrogen bonds with anions and water molecules: N1-H1A…O1W, N1-H1B…O5, and N1-H1C…O5 and N1′-H1′1…O2W, N1′-H1′2…O5′, and N1-H1′3…O5′. Water molecules form hydrogen bonds both with each other and with the alcoholic group: O1W-H1W1…O3′, O1W-H2W1…O2W, O2W-H1W2…O1W, and O2W-H2W2…O3. Cations are connected with each other only with weak C-H…O bonds: C2-H2…O2 and C2′-H2′…O2′. Further development of these investigations was performed by Braga et al. (2013). The authors established that a similar compound (L-SerH)$_2C_2O_4 \cdot 2H_2O$ may be obtained in the form of a single crystal from a solution with stoichiometric ratio of components.

This salt in the form of a single crystal was labeled as form I, because by grinding of L-serine and oxalic acid ($H_2C_2O_4{\cdot}2H_2O$) it is possible to obtain one more form with the same composition (L-SerH)$_2C_2O_4{\cdot}2H_2O$ (form II). Braga et al. (2013) obtained also a salt with composition (L-SerH)HC_2O_4 by grinding and kneading at 1:1 molar ratio of components. The structure of form I was determined by single-crystal XRD and the structures of form II and (L-SerH)HC_2O_4 by powder XRD. Recently Kulik et al. (2013) could crystallize form II and determine its single-crystal structure. Atomic coordinates of form I were not provided by Braga et al. (2013). So, we consider the structure of form II based on data of Kulik et al. (2013). Crystals of (L-SerH)$_2C_2O_4{\cdot}2H_2O$ (form II) were obtained from an aqueous solution with stoichiometric 2:1 ratio of L-serine and oxalic acid by slow evaporation at 36 °C. Little needle-shaped crystals of form II were formed on bigger needle-shaped crystals of oxalic acid. The crystals of form II melt at 84 °C. The asymmetric unit contains one formula unit, that is, one oxalate anion ($C_2O_4^{2-}$), two (L-SerH) cations, and two water molecules. Carboxyl and alcoholic groups of cations form hydrogen bonds with anions: O2-H2D. . .O6(2.5122(19) Å) and O3-H3. . .O4(2.7887(18) Å) and O12-H12A. . .O5(2.5513(19) Å) and O13-H13. . .O5(2.7637(19) Å). The NH_3^+ groups form each two hydrogen bonds with anion and one hydrogen bond with one of the water molecules: N1-H1B. . .O4(2.809(2) Å), N1-H1C. . .O6(2.779(2) Å), and N1-H1A. . .O9(2.826(2) Å) and N2-H2A. . .O7(3.020(2) Å), N2-H2B. . .O7(2.922 (2) Å), and N2-H2C. . .O8(2.811(2) Å). Water molecules form infinite chain by hydrogen bonds with each other: O8-H8A. . .O9(2.910(2) Å) and O9-H9A. . .O8 (2.911(2) Å). Additionally, water molecules form hydrogen bonds with oxygen atoms of alcoholic groups of cations: O8-H8B. . .O13(2.730(2) Å) and O9-H9B. . .O3(2.7607(19) Å). So, each water molecule forms two hydrogen bonds as acceptor and two as donor of protons. All four oxygen atoms of oxalate anion form each two hydrogen bonds.

Arkhipov et al. (2013) obtained crystals of (L-SerH) hydrogen maleate and (DL-SerH) hydrogen maleate and studied their crystal and molecular structures. The crystals were obtained from an aqueous solution containing equimolar quantities of components by slow evaporation at room temperature. In each case the asymmetric unit contains one formula unit. Hydrogen maleate anions form intramolecular hydrogen bonds: O7-H7. . .O5 with O. . .O distance (2.428(2) Å) for L-serine and 2.4517(19) Å) for DL-serine. Carboxyl groups of cations in each cases form hydrogen bonds with anions: O1-H1. . .O4 with O. . .O distances (2.5338(18) Å) and (2.537(2) Å) for L-serine and DL-serine, respectively. However, apart from this similarity there are also significant differences. The alcoholic group O(3)H in the structure of L-salt forms hydrogen bond with the carbonyl oxygen atom of neighboring cation O3-H3. . .O2(2.7224(18) Å), thus forming infinite chain of cations. In case of DL-salt the alcoholic O(3)H group forms hydrogen bond with anion O3-H3. . .O6(2.706(2) Å). As proton acceptor alcoholic groups in both cases form hydrogen bonds N-H. . .O. Two other hydrogen atoms of the NH_3^+ groups form hydrogen bonds with anions. In the structure of L-salt, two N1-H. . .O6 hydrogen bonds are formed with two nearest anions, while in the case of DL-salt, the hydrogen bonds with two nearest anions are formed with different oxygen atoms: N1-HA...O6 and N1-HB...O5. Conformations of cations in the structures also are different. Torsion angle C1C2C3O3 is equal to −173.31(15)° in the case of L-salt and −69.7(2)° in the case of DL-salt.

Crystals of L-serinium hydrogen squarate were obtained by Kolev et al. (1998a) and their crystal and molecular structure was determined. Another spectroscopic investigation was published later (Kolev et al. 2007). The asymmetric unit contains one formula unit. The (L-SerH) cation has the usual structure; however, the structure of hydrogen squarate anion is not usual. The one expected hydrogen atom of the hydrogen squarate anion is localized neither at O1 nor O3, but disordered in two positions, O1-1/2H1 and O3-1/2H2. As a result, the anions form infinite chains by the bonds O1-1/2H1...1/2H1-O1 and O3-1/2-H2...1/2-O3, with distances O1...O1(2.508 Å) and O3...O3(2.546 Å). As a sequence bond lengths C1-O1 and C3-O3 have no characteristic values (ca. 1.30 Å) for C-OH bond, but have C1-O1(1.267 Å) and C3-O3(1.269 Å) values. Bond lengths C2-O2 and C4-O4 also have no characteristic values (ca. 1.20-1.22 Å) for C=O bonds, but have C2-O2 (1.257 Å) and C2-O4(1.241 Å) values. The C-C distances in the square are also are equalized: C1-C2(1.454 Å), C2-C3(1.441 Å), C3-C4(1.457 Å), and C4-C1 (1.456 Å). The carboxyl group of the cation forms rather strong hydrogen bond O6-H3...O2(2.498 Å) with the anion. The O2 atom forms also O7-H4...O2 (2.867 Å) hydrogen bond with the alcoholic group. The O4 atom also forms two N-H...O-type hydrogen bonds with the NH_3^+ groups of two nearest cations: N1-H5...O4 and N1-H6...O4. The third hydrogen atom of the NH_3^+ group is used for the formation of hydrogen bond with the carbonyl oxygen atom N1-H7...O5. The oxygen atom of the alcoholic group usually forms hydrogen bonds as both acceptor and donor: however, in this structure the O7 atom does not form hydrogen bond as acceptor. Having this in view, one may expect a phase transition caused by the ordering of protons in the chain of anions.

The crystal of (L-SerH) *p*-toluenesulfonate was obtained by Kimoto et al. (1989) from an aqueous solution containing equimolar quantities of components. The asymmetric unit contains one formula unit. The alcoholic group of the cation forms hydrogen bond O1-H10...O5(2.703(6) Å) with the carbonyl oxygen atom of neighboring cation forming chain of cations. The carboxyl group of the cation forms hydrogen bond O6-H12...O4(2.608(7) Å) with the sulfonyl group of the anion. The NH_3^+ group forms three hydrogen bonds with three nearest anions: N1-H2...O3, N1-H3...O3, and N1-H4...O2. Oxygen atom of the alcoholic group does not form hydrogen bonds as acceptor.

Sudbeck et al. (1995) published data on crystals of (L-SerH) 4-hydroxybenzenesulfonate, grown from aqueous solution with stoichiometric composition of components. The asymmetric unit contains two formula units. The presence of a significant number of functional groups capable to form hydrogen bonds as donors and acceptors opens different possibilities of their realization. Both anions form infinite zigzag chain each with its type. The first anion with S(1)O3 sulfonate group forms hydrogen bond O14-H14...O13(2.627(3) Å). The second anion with S(2)O3 sulfonate group forms O24-H24...O21(2.832(4) Å) hydrogen bond. The carboxyl group of the first cation forms O2-H20...O23(2.623(3) Å) hydrogen bond with the sulfonate group of the second anion. At the same time the alcoholic group of this cation forms O3-H30...O12(2.757(4) Å) hydrogen bond with the sulfonate group of the first anion. The carboxyl group of the second cation forms O5-H50...O14(2.630(3) Å) hydrogen bond with the hydroxyl group of the first anion, while the alcoholic group forms O6-H60...O22(2.756(4) Å) hydrogen bond with the sulfonate group of the second anion. The $N(1)H_3^+$ group of the first cation forms two hydrogen bonds N1-H1N...O1 and N1-H2N...O4 with the carbonyl oxygen atoms of the cations and one hydrogen bond N1-H3N...O12 with the sulfonate group of the first anion. The $N(2)H_3^+$ group of second cation forms two hydrogen bonds N2-H5N...O3, N2-H6N...O6 with oxygen atoms of alcoholic groups of cations and one hydrogen bond N2-H4N...O11 with sulfonate group of first anion.

5.1.9 Salts of Threonine

In Table 5.9 known salts of threonine are presented. Threonine has definite chemical similarity with serine. Ghazaryan (2011a) published the growth of crystals of (L-ThrH)Cl and (L-ThrH)Br. The crystal of (L-ThrH)Br was obtained from an aqueous solution containing equimolar quantities of L-threonine and HBr, while the crystal of (L-ThrH)Cl was obtained from a solution containing L-threonine and HCl in molar ratio 1:2, because at molar ratio 1:1 the mixture of 2L-Thr·HCl and L-Thr·HCl is formed. Their vibrational spectra indicate that they are isostructural. Ghazaryan et al. (2011a) reported the structure of (L-ThrH)Cl. Later we have learned that the structure of (L-ThrH)Cl was determined earlier (Chapman and Bryce 2007). The comparison of unit cell parameters of (L-ThrH)Cl with that obtained by Chapman and Bryce (2007) at 217 K, $a = 7.275(4)$ Å), $b = 5.263(3)$ Å), $c = 9.556(5)$ Å), and $\beta = 92.545(5)°$, shows that the phases obtained in both works are identical. In Table 5.9 data of Ghazaryan (2011a) are provided because of their higher accuracy. The carboxyl and alcoholic groups of cation form O1-H1...Cl1(3.0273(12) Å) and O3-H3...Cl1(3.1498(14) Å) hydrogen bonds with chloride anion. The NH_3^+ group forms two hydrogen bonds with chloride anion, N1-H11...Cl1(3.2027(13) Å) and N1-H12...Cl1(3.2531(14) Å), and one hydrogen bond with the carbonyl atom of nearest cation, N1-H13...O2(2.8396(14) Å). So, chloride anion forms four hydrogen bonds. In the model of Chapman and Bryce (2007), the chloride anion forms three hydrogen bonds. Here the hydrogen bond with the alcoholic group is absent.

The crystal of (DL-ThrH)H_2PO_4 was obtained from aqueous solution containing equimolar quantities of DL-threonine and H_3PO_4 (Ravikumar et al. 2002b). The asymmetric unit contains one formula unit. The carboxyl group of the cation forms O1B-H1B...O2(2.470(2) Å) hydrogen bond with the anion. This is the shortest hydrogen bond in the structure. The alcoholic group also forms hydrogen bond with the anion, but with the oxygen atom of the hydroxyl group, O1C-H1C...O3(2.886

Table 5.9 Salts of threonine with inorganic and organic anions

#	Crystal	s.g., Z	*a*, *b*, *c* (Å), *α*, *β*, *γ* (°)	Refs
1	(L-ThrH)Cl	$P2_1$, 2	7.276(1), 5.281(1), 9.587(1), $\beta = 92.464(3)$	Ghazaryan et al. (2011a); Ghazaryan (2011)
2	(L-ThrH)Br	–	–	Ghazaryan (2011)
3	(DL-ThrH)H_2PO_4	*Pbca*, 8	9.0590(6), 19.507(3), 10.1015(9)	Ravikumar et al. (2002b)
4	(L-ThrH)$_2SO_4{\cdot}H_2O$	*C*2, 4	23.096(4), 6.281(9), 11.648(1), $\beta = 116.122(9)$	Sridhar et al. (2001a)
5	(DL-ThrH)HC_2O_4	$P2_1/n$, 4	8.325(5), 10.957(4), 10.363(4), $\beta = 112.39(5)$	Nandhini et al. (2001d)
6	(DL-ThrH)CCl_3COO	$P2_1$, 4	10.3290(11), 10.4271(11), 10.7795(11), $\beta = 103.115(12)$	Rajagopal et al. (2004a)
7	(DL-ThrH) maleate	$P\bar{1}$, 2	5.8050(12), 8.4310(17), 10.740(2), $\alpha = 98.88(3)$, $\beta = 105.17(3)$, $\gamma = 91.68(3)$, at 150 K	Rajagopal et al. (2004b)
8	(L-ThrH) picrate	$P2_1$, 2	9.235(5), 6.265(5), 12.727(5), $\beta = 108.42(5)$	Natarajan et al. (2010a)

(3) Å), which in turn forms hydrogen bond with the neighboring anion, O3-H3P...O2(2.617(2) Å). The second hydroxyl group of the anion forms hydrogen bond with the carbonyl atom of the cation, O4-H4P...O1A(2.685(2) Å). The NH_3^+ group forms two hydrogen bonds with the nearest anions, N11-HA...O1 and N11-HC...O1, while the remaining hydrogen atom is bonded with the neighboring cation via O1B hydroxyl group, N11-HB...O1B. The respective compound of DL-threonine with H_3AsO_4 (DL-Thr·H_3AsO_4) is a molecular adduct, not a salt (see Sect. 5.1.24).

The crystal of (L-ThrH)$_2SO_4$·H_2O was obtained from aqueous solution of a 2:1 stoichiometric ratio of L-threonine and sulfuric acid (Sridhar et al. 2001a). The asymmetric unit contains two (L-ThrH) cations, one SO_4^{2-} anion, and two water molecules, which, however, are on special positions. One of the cations forms a hydrogen bond with the anion via carboxyl and alcoholic groups: O1B-H1B...O1 (2.666(3) Å) and O1C-H1C...O4(2.746(3) Å). The carboxyl group of the second cation forms a hydrogen bond with the alcoholic group of the first cation, O2B-H2B...O1C(2.600(3) Å), while the alcoholic group forms a hydrogen bond with the anion, O2C-H2C...O2(2.764(3) Å). One of the water molecules as proton donor forms hydrogen bonds with two symmetry-related anions, OW1-HW1...O1, and as acceptor forms hydrogen bond, N21-H21B...O1W. The second water molecule forms two hydrogen bonds with symmetry-related cations via OW2-HW2...O2C, not forming, however, hydrogen bonds as acceptor. The NH_3^+ group of the second cation in addition to mentioned hydrogen bond with O1W forms hydrogen bond with anion, N21-HA...O2. The third hydrogen atom has a contact N21-HC...O3 on the level of van der Waals bonds. The NH_3^+ group of the first cation forms hydrogen bond with the anion, N11-HB...O3, and also with the carbonyl atom of the first nearest symmetry-related cation, N11-HC...O1A. The remaining hydrogen atom also has a contact N11-HA...O4 on the level of van der Waals bonds. Vibrational spectra of (L-ThrH)$_2SO_4$·H_2O were studied by Jinnah et al. (2004a).

The crystals of (DL-ThrH)HC_2O_4 were obtained from an aqueous solution containing equimolar quantities of DL-threonine and oxalic acid (Nandhini et al. 2001d). The asymmetric unit contains one formula unit. Hydrogen oxalate anions form usual infinite head-to-tail chain. The carboxyl group of the anion forms a hydrogen bond with the carboxylate group of the next anion in the chain via O4-H4...O6(2.623(4) Å). The carboxyl group of the cation forms the shortest hydrogen bond with the other atom of the carboxylate group of the anion, O2-H2...O7(2.483(4) Å). Such strong bond leads to lengthening of C6-O7 (1.260 Å) compared to C6-O6(1.239 Å). The alcoholic group of the cation forms a hydrogen bond O3-H3...O1(2.854(4) Å) with the carbonyl atom of the neighboring cation. The NH_3^+ group forms N1-H1A...O5(2.810(4) Å) with the carbonyl O5 atom of the cation, N1-H1B...O3(2.895(5) Å) hydrogen bond with alcoholic group, and weaker bond N1-H1C...O7(3.060(4) Å).

Rajagopal et al. (2004a) reported the synthesis of crystals of (DL-ThrH)CCl_3COO from an aqueous solution of DL-threonine and CCl_3COOH with equimolar ratio. In spite of the presence of DL-threonine racemate, the salt crystallizes in non-centrosymmetric space group $P2_1$ with $Z' = 2$. The carboxyl groups of both cations form hydrogen bonds with anions: O1A-H1A…O5A(2.571(3) Å) and O1B-H1B…O4B(2.597(3) Å). Alcoholic groups form hydrogen bonds with carbonyl oxygen atoms of cations: O3A-H3A…O2B(2.824(3) Å) and O3B-H3B…O2A(2.780(3) Å). Each of the NH_3^+ groups forms hydrogen bonds with both anions and one hydrogen bond with the alcoholic group: N1A-H1A2…O4A, N1A-H1A3…O4B, and N1A-H1A1…O3A and N1B-H1B1…O5A, N1B-H1B2…O5B, and N1B-H1B3…O3B. Ramesh Kumar et al. (2006) grew (DL-ThrH)CCl_3COO crystals and characterized them as possible NLO material. SHG efficiency estimated by powder method is equal to 0.62 compared to KDP.

The same group also reported crystals of (DL-ThrH) hydrogen maleate, obtained from an aqueous solution with stoichiometric ratio of components (Rajagopal et al. 2004b). The asymmetric unit contains one formula unit. Hydrogen maleate anion as usual forms an intramolecular short hydrogen bond O5-H5…O6(2.400(1) Å). The carboxyl group of the cation forms hydrogen bond O1-H1…O7(2.598(1) Å) with the second oxygen atom of the carboxylate group of the anion. The alcoholic group forms a hydrogen bond O3-H30…O4(2.771(1) Å) with the carbonyl atom of anion. The NH_3^+ group forms two hydrogen bonds N1-H1N…O4 and N1-H3N…O7 with two nearest anions and one hydrogen bond with the alcoholic group of symmetry-related cation N1-H2N…O3. Thus, the carbonyl atom O2 of the cation does not form hydrogen bonds.

Crystals of (L-ThrH) picrate were obtained from an aqueous solution containing equimolar quantities of L-threonine and picric acid (Natarajan et al. 2010a). The asymmetric unit contains one formula unit. The carboxyl group of the cation forms O2-H2…O4(2.660(3) Å) hydrogen bond with negatively charged O4 atom of the picrate anion. Alcoholic group forms weak bond O3-H3…O2(3.015(5) Å) with the hydroxyl group of the cation. The NH_3^+ group forms N1-H1C…O1(2.868(3) Å) hydrogen bond with the carbonyl atom of the nearest cation and two hydrogen bonds with the NO_2 groups of the anion: N1-H1A…O10(2.884 Å) and N1-H18…O7(3.063 Å). The second hydrogen bond is on the level of van der Waals bonds. In addition, there is a misprint in the paper (2.063 Å is indicated instead of 3.063 Å). The crystal of (L-ThrH) picrate was characterized also as possible NLO material. SHG efficiency was determined by powder method, which was 43 times higher than that of KDP. The IR spectrum and UV–Vis transmittance spectrum are provided. As all picrates the crystal has absorption in violate region of spectrum. In the IR spectrum there is absorption band at 3,400 cm^{-1}, which probably is caused by ν(OH) stretching of the alcoholic group, which forms a weak hydrogen bond. There is a broad absorption band near 3,000 cm^{-1}, which probably is caused by ν(OH) stretching of the carboxyl group. Assignment of the peak at 1,541 cm^{-1} to symmetric stretching is caused probably by misprint. Another point which is not clear is the assignment of the peak at 1,163 cm^{-1} to C=O stretching.

In addition to salts of threonine provided in Table 5.9 there were publications on the existence of L-threonine acetate and L-threonine formate. Fleck and Petrosyan (2010) and Petrosyan et al. (2013a) showed that these papers are erroneous and the obtained crystals actually are L-threonine.

Table 5.10 Salts of tyrosine with inorganic anions

#	Crystal	s.g., Z	a, b, c (Å), α, β, γ (°)	Refs
1	(L-TyrH)Cl	$P2_1$, 2	11.083(5), 9.041(4), 5.099(3), $\beta = 91.82(3)$	Frey et al. (1973)
2	(L-TyrH)Br	$P2_1$, 2	11.353, 9.105, 5.158, $\beta = 91.32$	Moolya and Dharmaprakash (2006)
3	(L-TyrH)I	–	–	Seely (1966a)
4	(DL-TyrH)Cl·$2H_2O$	$P\bar{1}$, 2	5.3330(2), 10.9634(5), 11.2500(4), $\alpha = 113.642(4)$, $\beta = 94.359(3)$, $\gamma = 98.465(3)$, at 100 K	Guenifa et al. (2012)
5	(L-TyrH)ClO_4	$P2_1$, 2	5.335(1), 9.817(2), 12.000(3), $\beta = 94.79(2)$	Yang et al. (1996)
6	$(L\text{-}TyrH)_2SO_4 \cdot H_2O$	$P2_1$, 2	13.473(2), 6.134(1), 13.785(3), 109.51(2)	Sridhar et al. (2002b)

5.1.10 *Salts of Tyrosine*

Known salts of tyrosine are presented in Table 5.10. L-tyrosinium chloride and bromide were obtained by Srinivasan (1956). Based on the symmetry and the unit cell parameters, it was shown that they are isostructural, which was later established when their structures were determined (Srinivasan 1959a, b). Later Frey et al. (1973) determined the structure of (L-TyrH)Cl by neutron diffraction method with accurate determination of the positions of protons. The carboxyl group of the cation forms a hydrogen bond not with anion, but with oxygen atom of the phenolic group O-H...O(2.619(4) Å), which in turn forms rather strong hydrogen bond with chloride anion O-H...Cl(3.045(3) Å). Additionally, the chloride ion forms three more N-H...Cl hydrogen bonds towards the three nearest cations, with N...Cl distances of 3.452(3) Å, 3.329(3) Å, and 3.251(3) Å.

The crystal of (L-TyrH)I was obtained by Oliver Seely Jr. in his dissertation (1966a) and its structure was determined.

The crystals of (L-TyrH)Cl and (L-TyrH)Br having polar symmetry attracted the attention of researchers as possible pyroelectric and NLO materials (Moolya and Dharmaprakash 2006, 2010; Yarmarkin et al. 2005; Natarajan et al. 2008a; Siddheswaran et al. 2009; Anandan et al. 2010a, 2012a; Ilayabarathi et al. 2013). The possibility of growing of crystals from an aqueous solution was shown. Transmittance spectrum in UV–Vis region, IR and Raman spectra, data of thermal analysis, efficiency of SHG by powder method and some other characteristics were determined. Some results and interpretations (Moolya and Dharmaprakash 2006; Siddheswaran et al. 2009; Ilayabarathi et al. 2013) are not acceptable. For example, in the IR spectrum of (L-TyrH)Br the peak at 1,221 cm^{-1} was assigned to bromide ion (Moolya and Dharmaprakash 2006). In the work of Siddheswaran et al. (2009) the peak in the IR spectrum of (L-TyrH)Cl at 1,557 cm^{-1} was assigned to asymmetric stretching of the NH_3^+ group, the peak at 1,481 cm^{-1} is assigned to asymmetric stretching of the COO^- group, and the peak at 639 cm^{-1} is assigned to stretching vibration of the C-Cl bond, while in the structure of (L-TyrH)Cl there are no COO^- group and C-Cl bond. In addition, in the work of Siddheswaran et al. (2009), as if, the structure of (L-TyrH)Cl is shown without indication of the source. Actually this figure relates to the structure of methyl ester of L-tyrosine hydrochloride and not (L-TyrH)Cl. Siddheswaran et al. (2009) used unit cell parameters of methyl ester of L-tyrosine hydrochloride for the interpretation of powder X-ray diffraction pattern of (L-TyrH)Cl. Ilayabarathi et al. (2013), in the interpretation of DSC data of (L-TyrH)Cl, assigned exothermic effect to the melt of the material.

From an aqueous solution of DL-tyrosine and HCl at molar ratio 1:1, dihydrate is formed (DL-TyrH)Cl·$2H_2O$ (Guenifa et al. 2012). The carboxyl group of the cation forms a hydrogen bond with one of the (second) water molecules: O2-H2...O2W (2.5319(15) Å). This water molecule in turn acts as proton donor and forms a hydrogen bond with two nearest (first) water molecules: O2W-H21W...O1W(2.7429(14) Å) and O2W-H22W...O1W(2.8318(15) Å). Phenolic OH group forms a hydrogen bond with chloride ion, O1-H1...Cl1(3.1858(11) Å). The first water molecule as proton donor forms one hydrogen bond with anion, O1W-H12W...Cl(3.1784(11) Å), and one hydrogen bond with the oxygen atom of phenolic OH group, O1W-H11W...O1 (2.9044(13) Å). All three hydrogen atoms of the NH_3^+ group form hydrogen bonds with anion. Thus, chloride ion forms five hydrogen bonds.

Yang et al. (1996) published the structure of (L-TyrH)ClO_4. Coordinates of hydrogen atoms were not provided, so conclusions about hydrogen bonds may be made supposedly. The carboxyl group of the cation forms, as in a number of previous cases, a hydrogen bond with the oxygen atom of the phenolic group, O5...O7(2.659 Å). The phenolic group forms a hydrogen bond O7...O2(2.820 Å) with the anion. The NH_3^+ group forms hydrogen bonds with three nearest anions with N...O distances: 3.009 Å, 2.906 Å, and 3.018 Å.

The crystal of $(L\text{-TyrH})_2SO_4 \cdot H_2O$ was obtained by evaporation of aqueous solution containing stoichiometric ratio 2:1 of L-tyrosine and sulfuric acid (Sridhar et al. 2002b). The asymmetric unit contains one formula unit. Oxygen atom and one hydrogen atom of water molecule are disordered. Two crystallographically independent cations differ by their conformation and by formed hydrogen bonds. The N(11) H_3^+ group in the first cation is in *cis*-position in respect to the phenolic ring, while the N(22)H_3^+ group of the second cation is in *trans*-position. The carboxyl group of the first cation forms a hydrogen bond with the anion, O1B-H1B...O4(2.600(2) Å). The phenolic group also forms a hydrogen bond with the anion, O1C-H1C...O2(2.695 (3) Å). The N(11)H_3^+ group forms two hydrogen bonds with anion, N11-H11B...O4 and N11-H11C...O3, and also with the oxygen atom of the phenolic group, N11-H11A...O1C. The carboxyl group of the second cation forms a hydrogen bond with the oxygen atom of the phenolic group, O2B-H2B...O2C(2.644(3) Å). The phenolic group as proton donor forms a hydrogen bond with water molecule, O2C-H2C...O1W(O2W), with distances O2C...O1W(2.514(6) Å) and O2C...O2W(2.728(7) Å). The N(22)H_3^+ group of the second cation forms all three hydrogen bonds with the anion: N22-H22A...O3, N22-H22B...O1, and N22-H22C...O1. Water molecule as proton donor forms hydrogen bonds with the anion.

5.1.11 *Salts of Cysteine*

Known salts of cysteine are collected in Table 5.11. The feature of cysteine is the presence of the thiol SH group, by which it differs from the similar amino acid serine. The salt (L-CysH)F was described in the patent by Schmid (1968). Its structure is not determined. The crystal of (L-CysH)Cl·H_2O was obtained and studied by Ayyar and Srinivasan (1965) and Ayyar (1968). Later its structure was determined with higher accuracy and was studied by $^{35/37}Cl$ solid-state NMR method (Chapman and Bryce 2007). The carboxyl group of the cation forms rather strong hydrogen bond with chloride ion O1-H5…Cl1(3.032 Å). The thiol group also forms a hydrogen bond with anion, S1-H1…Cl1(3.474 Å). Water molecule also forms as proton donor hydrogen bonds with anion: O3-H9…Cl1(3.208 Å) and O3-H10…Cl1(3.214 Å). The NH_3^+ group forms two hydrogen bonds with water molecule, N-H3…O3 and N-H4…O3, and one hydrogen bond with anion, N-H2…Cl1. So, chloride ion forms five hydrogen bonds. The crystal of (L-CysH) Cl·H_2O attracted the attention of Bhagavannarayana et al. (2010) as possible NLO material. Single crystal of (L-CysH)Cl·H_2O was grown by unidirectional method and characterized by different methods. By means of the powder method, the efficiency of SHG was estimated ($I^{2\omega}/I^{2\omega}{}_{KDP} = 1.2$), also the laser damage threshold was measured, which also is higher than that of KDP. Transmittance spectrum in UV–Vis region, Raman spectrum, and dielectric characteristics are provided. Bhagavannarayana et al. (2010)assigned the Raman peak at 1,145 cm^{-1} to "phenol O," which is absent in the structure of the crystal, and also the peak at 524 cm^{-1} to ν(S-S) stretching. The bond S-S is present in the structure of cystine, but not in the structure of cysteine. By the way, it is worth to note that the authors in the Introduction forewarn the reader "Note that there is another organic compound with a very similar name, L-cystine hydrochloride, but its molecular formula is $C_6H_{12}N_2O_4S_2$·HCl and it crystallizes in a monoclinic structure." In addition, characteristic peak in Raman spectrum at 2,947 cm^{-1} caused by ν(CH) stretching vibration was assigned to NH_3^+ stretching by Bhagavannarayana et al. (2010).

In course of our systematic search for salts of amino acids with dimeric cation (see Sect. 5.2) we studied also salts of cysteine. We found that in system L-Cys + HCl + H_2O in addition to known (L-CysH)Cl·H_2O salt, there is also a salt (L-Cys…L-CysH)Cl with dimeric cation. Similarly, in system L-Cys + HBr + H_2O there are analogous (L-CysH)Br·H_2O and (L-Cys…L-CysH)Br salts. In the systems L-Cys + HNO_3 + H_2O and L-Cys + $HClO_4$ + H_2O, (L-CysH)NO_3 and (L-CysH)ClO_4 salts and respective salts with dimeric cations were found. Further investigation of these newly obtained crystals will be done.

Minkov and Boldyreva (2008, 2011) obtained two forms of (L-CysH)HC_2O_4. The orthorhombic form (Minkov and Boldyreva 2008) was obtained by slow evaporation of a water–ethanol (3:1) solution of an equimolar ratio of L-cysteine and oxalic acid dihydrate, while the monoclinic form (Minkov and Boldyreva 2011) was obtained by slow diffusion of acetonitrile into a saturated aqueous solution of

Table 5.11 Salts of cysteine with inorganic and organic anions

#	Crystal	s.g., Z	*a*, *b*, *c* (Å), *α*, *β*, *γ* (°)	Refs
1	(L-CysH)F	–	–	Schmid (1968)
2	(L-CysH)Cl·H_2O	$P2_12_12_1$, 4	5.4588(9), 7.1570(11), 19.389(3)	Chapman and Bryce (2007)
3	(L-CysH)Br·H_2O	–	–	This work
4	(L-CysH)NO_3	–	–	This work
5	(L-CysH)ClO_4	–	–	This work
6	(L-CysH)HC_2O_4	$P2_12_12_1$, 4	7.0529(11), 10.2407(12), 12.1199(15)	Minkov and Boldyreva (2008)
7	(L-CysH)HC_2O_4	$C2$, 4	18.604(4), 5.5723(6), 11.270(2), $\beta = 124.192(13)$	Minkov and Boldyreva (2011)
8	(DL-CysH)HC_2O_4	$P\bar{1}$, 2	5.6664(2), 9.0149(4), 9.7749(5), $\alpha = 109.349(4)$, $\beta = 102.282(3)$, $\gamma = 100.119(3)$	Minkov and Boldyreva (2009)
9	(DL-CysH)$_2C_2O_4$	$P\bar{1}$, 1	5.2779(6), 6.6526(7), 10.4424(15), $\alpha = 86.840(11)$, $\beta = 76.844(11)$, $\gamma = 70.097(10)$	Drebushchak et al. (2008)
10	(L-CysH) L-tartrate·H_2O	$P2_1$, 2	7.9152(8), 7.7166(6), 10.4111(9), $\beta = 111.0$	Shan and Huang (1999)
11	(L-CysH) picrate	–	–	This work

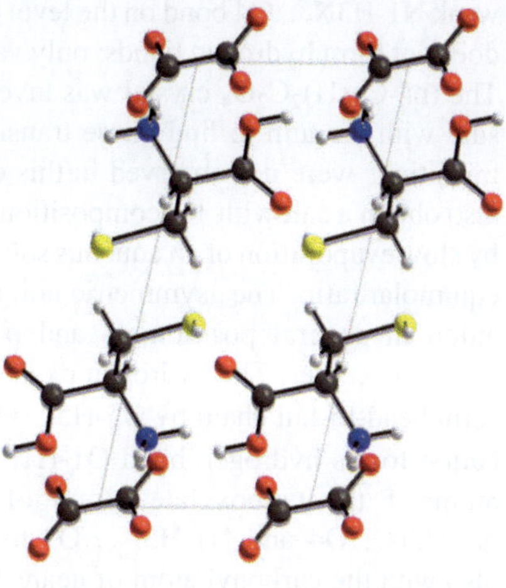

Fig. 5.15 Packing diagram of bis(DL-cysteinium) oxalate. The oxalate anions are located on an inversion centers

L-cysteine and oxalic acid in an equimolar ratio. This form is not stable on storage. The authors assume that the structure of the crystal may comprise strongly disordered water molecules in the voids and the amount of water molecules does not exceed 0.3 molecules per void. In the structures of both forms hydrogen oxalate anions form head-to-tail infinite chain with O5-H5…O4(2.5346(18) Å) hydrogen bond for orthorhombic form and O5-H5…O3(2.523(3) Å) hydrogen bond for monoclinic form. In the structure of monoclinic form anion is almost planar, while in the case of orthorhombic form it is not planar. The angle between the planes of the two groups is 38.6(3)°. The carboxyl group in structures of both forms forms hydrogen bond with the oxygen atom of the carboxylate group of the anion. In the structure of monoclinic form there is O1-H4…O4(2.611(3) Å) and in the structure of orthorhombic form there is O1-H1…O3(2.587(2) Å) hydrogen bonds. In the structure of orthorhombic form, the NH_3^+ group forms three hydrogen bonds with anion: N1-H1N…O4, N1-H2N…O3, and N1-H3N…O6. In the structure of monoclinic form the NH_3^+ group forms one hydrogen bond N1-H1N…O3(2.715 (2) Å). All other contacts are on the level of van der Waals bonds. In the structures of both forms, thiol groups S-H do not form noticeable hydrogen bonds. All contacts are on the level of van der Waals bonds.

In addition to salts of L-cysteine, Minkov and Boldyreva (2009) and Drebushchak et al. (2008) obtained salts with oxalic acid also for DL-cysteine. The salt of (DL-CysH)$_2$C$_2$O$_4$ was reported by Drebushchak et al. (2008). To aqueous solution containing almost equimolar quantities of DL-cysteine (2.1 mmol) and oxalic acid dihydrate (2.0 mmol) propan-2-ol was slowly added at 283 K. The asymmetric unit of (DL-CysH)$_2$C$_2$O$_4$ contains (CysH) cation in general position and $C_2O_4^{2-}$ anion in special position. The $C_2O_4^{2-}$ anion is surrounded by L- and D-(CysH) cations located upon an inversion center (Fig. 5.15). L- and D-(CysH) cations form rather strong hydrogen bond O1-H1…O3(2.4982(13) Å). The NH_3^+ group forms two hydrogen bonds N1-H1N…O3(2.8997(15) Å) and N1-H2N…O4(2.9715(15) Å) and one weak N1-H3N...O4 bond on the level of van der Waals interactions. The thiol group does not form hydrogen bonds; only van der Waals contact with O2 atom is present. The (DL-CysH)$_2$C$_2$O$_4$ crystal was investigated at low temperature and higher pressure with the aim to find phase transitions (Minkov et al. 2012). However, phase transitions were not observed in this crystal. Minkov and Boldyreva (2009) could also obtain a salt with 1:1 composition, (DL-CysH)HC$_2$O$_4$. The crystal was obtained by slow evaporation of an aqueous solution containing DL-cysteine and oxalic acid in equimolar ratio. The asymmetric unit contains (CysH) cation and hydrogen oxalate anion in general position. L- and D-cations are connected with each other by inversion center. The hydrogen oxalate anion is almost planar and as usual anions form head-to-tail chain by O5-H5...O3(2. 6076(12) Å). The carboxyl group of the cation forms hydrogen bond O1-H1...O4(2.5682(12) Å) with the second oxygen atom of the carboxylate group of the anion. The NH_3^+ group forms two N1-H2N...O4 and N1-H3N...O3 hydrogen bonds with two nearest anions and also with the carbonyl atom of nearest cation, N1-H1N...O2. The thiol group also does not form hydrogen bonds. At high pressure (ca. 6 GPa) phase transition was observed in (DL-CysH)HC$_2$O$_4$ crystal (Minkov et al. 2012).

The structure of L-cysteinium hydrogen L-tartrate was determined by Shan and Huang (1999). The asymmetric unit contains one formula unit. The positions of hydrogen atoms in carboxyl and hydroxyl groups of anion were not determined. Bond lengths C1-O1(1.322(4) Å) and C1-O2(1.211(4) Å) show that this is a carboxyl group, while bond lengths C4-O5(1.247(4) Å) and C4-O6(1.255(4) Å) show that this is a carboxylate group. Hydrogen L-tartrate anions form head-to-tail chain via O1...O5(2.702(3) Å) hydrogen bond. The carboxyl group of the cation forms hydrogen bond not with anion, but with water molecule O7-H3...O9 (2.566 Å). Water molecule in turn as proton donor forms hydrogen bonds O9-H11...O6(2.745 Å) and O9-H12...O4(2.925 Å) with two nearest anions. It is very probable that the hydroxyl group O(3)H also forms hydrogen bond with the O5 atom with O...O distance 2.678 Å. The NH_3^+ group forms two hydrogen bonds N1-H8...O6(2.815 Å) and N1-H9...O3(2.836 Å) with two nearest anions and one hydrogen bond N1-H7...O8(2.914 Å) with the carbonyl oxygen atom of nearest cation. It is difficult to say something definite about possible hydrogen bonds of the thiol group, because of suspiciously very short S1-H10(0.58 Å) distance. The crystal was grown and characterized by Martin Britto Dhas and Natarajan (2008a) as possible NLO material. The authors do not indicate which optical isomer of tartaric acid was used as initial reagent. However, they identified the grown crystal by measurement of unit cell parameters and compared with data by Shan and Huang (1999), which allows one to assume that it was L-cysteinium hydrogen L-tartrate. UV–Vis–NIR transmittance and IR spectra were determined, as well as dielectric and microhardness characteristics. NLO activity was estimated by powder SHG method and was found that the efficiency is 1.3 times higher than that of KDP standard. The authors assigned the broad band between 3,500 and 3,000 cm^{-1} to the NH_3^+ and CH_3 asymmetric stretching. In our opinion, it is more probable that this band is caused by stretching vibrations of O-H bonds of water molecule, hydroxyl and carboxyl groups of anion, and stretching vibrations of the NH_3^+ group. The CH_3 group is absent in the structure of the crystal.

In system L-Cys + picric acid + H_2O we found two salts: L-cysteinium picrate and (L-Cys. . .L-CysH) picrate with dimeric cation, which is considered in Sect. 5.2. Structures of these salts will be determined. In the IR spectrum of L-cysteinium picrate the cation is characterized by absorption band ν(C=O) at 1,715 cm^{-1} and also by a band at 2,563 cm^{-1} caused by ν(S-H).

5.1.12 Salts of Methionine

In Table 5.12 known salts of methionine are presented. Obtaining of (L-MetH)F as well as other fluorides is described by Schmid (1968). The structure of the salt was not, however, determined.

The structure of (L-MetH)Cl was determined at room temperature (Di Blasio et al. 1977b), and later at 103 K (Chruszcz et al. 2004). The crystal can be obtained from an aqueous solution with equimolar ratio of components. The crystal was studied also by $^{35/37}$Cl solid-state NMR method (Chapman and Bryce 2007). The asymmetric unit of (L-MetH)Cl contains one formula unit (Fig. 5.16). The carboxyl group of the cation forms a hydrogen bond with chloride ion, O2-H1. . .Cl1 (3.029 Å). The NH_3^+ group also forms hydrogen bonds with anion: N1-H10. . .Cl1(3.184 Å), N1-H11. . .Cl1(3.335 Å), and N1-H12. . .Cl1(3.188 Å). At 103 K the O. . .Cl distance is 3.0239(10) Å and N. . .Cl distances are 3.1683 (10) Å, 3.3054(10) Å, and 3.1709(10) Å (Chruszcz et al. 2004).

The structure of (DL-MetH)NO_3 was determined by Srikrishnan et al. (1984) at room temperature; however, detailed structural data were not reported. Mostad and Natarajan (1985) determined the structure of (DL-MetH)NO_3 at 121 K. The asymmetric unit contains one formula unit. The carboxyl group of the cation forms O2-H12. . .O5(2.602(2) Å) hydrogen bond. The NH_3^+ group forms two hydrogen bonds with two nearest anions, N1-H2. . .O5(2.880(2) Å) and N1-H3. . .O4(2.784 (2) Å), and one hydrogen bond, N1-H1. . .O1(2.841(2) Å). There is a misprint in the value of this bond (2.481(2) Å) in the paper by Mostad and Natarajan (1985). Pandiarajan et al. (2002a) also published on (L-MetH)NO_3. The crystal can be obtained from aqueous solution with stoichiometric composition. The asymmetric unit contains two formula units. In spite of the non-centrosymmetric structure due to the presence of L-form, cations and anions in the structure form a pseudocentrosymmetric arrangement. Carboxyl groups of cations form hydrogen bonds with anions: O1B-H1B. . .O23(2.663(4) Å) and O2B-H2B. . .O12(2.615(4) Å). Each of the two NH_3^+ groups forms two hydrogen bonds towards the nitrate anions and each forms one hydrogen bond towards the carbonyl atom of the same type symmetry-related cation: N11-H11. . .O11, N11-H11B. . .O13, N11-H11C. . .O1A and

Table 5.12 Salts of methionine with inorganic and organic anions

#	Crystal	s.g., Z	a, b, c (Å), α, β, γ (°)	Refs
1	(L-MetH)F	–	–	Schmid (1968)
2	(L-MetH)Cl	$P2_12_12_1$, 4	7.087(4), 24.643(8), 5.297(3)	Di Blasio et al. (1977b)
3	(DL-MetH)Cl	–	–	Shiraiwa et al. (1997); Srimahaprom and Flood (2013)
4	(L-MetH)NO_3	$P2_1$, 4	10.682(2), 5.577(2), 16.829(6), $\beta = 100.92(2)$	Pandiarajan et al. (2002a)
5	(DL-MetH)NO_3	$P2_1/n$, 4	13.663(3), 5.556(1), 13.174(2), $\beta = 103.90(1)$, at 121 K	Mostad and Natarajan (1985)
6	(DL-MetH)H_2PO_4	$P2_1/c$, 8	23.935(4), 9.992(2), 9.026(3), $\beta = 98.25(2)$	Bahadur and Athimoolam (2007)
7	(L-MetH)$_2SO_4$	$P3$, 3	10.281(3), 10.281(3), 14.788(6), $\gamma = 120$	Srinivasan et al. (2001b)
8	(DL-MetH)$_2SO_4$	$R\bar{3}$, 9	10.233(5), 10.233(5), 44.494(5), $\gamma = 120$	Ramaswamy et al. (2004a)
9	(L-MetH)$_2SiF_6 \cdot H_2O$	$P2_1$, 2	5.5287(5), 17.012(1), 10.8999(9), 96.381(4)	Fleck et al. (2013b)
10	(DL-MetH)CCl_3COO	$P\bar{1}$, 2	6.1359(12), 8.9746(18), 12.731(3), $\alpha = 70.54(3)$, $\beta = 85.95(3)$, $\gamma = 85.08(3)$	Rajagopal et al. (2003c)
11	(DL-MetH) maleate	$P2_1/c$, 4	11.070(2), 5.746(5), 19.697(14), $\beta = 102.34(3)$	Alagar et al. (2002b)
12	(*R*)- MetH (*R*)-mandelate·2((*R*)-mandelic acid)	$P2_12_12_1$, 4	9.562(3), 9.830(3), 31.393(5)	Su and Xu (2005)
13	(L-MetH) picrate	–	–	This work

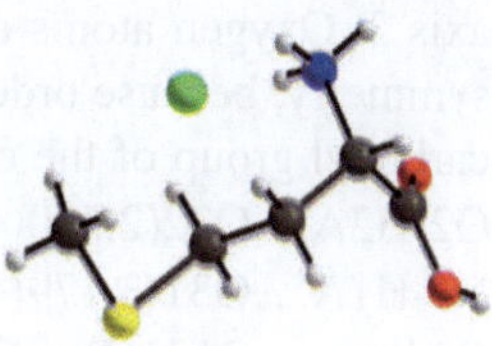

Fig. 5.16 Molecular structure of L-methioninium chloride

N21-H21A...O23, N21-H21B...O22, N21-H21C...O2A. Pandiarajan et al. (2004) studied the IR and Raman spectra of (L-MetH)NO_3. Vasudevan et al. (2013a) have grown the crystal of (L-MetH)NO_3 and characterized it as possible NLO material. Above 143 °C the crystal decomposes with exothermal effect just after melting at 142 °C. According to this study, the intensity of SHG is higher than in L-alanine.

The structure of crystals (DL-MetH)H_2PO_4 is published by Bahadur and Athimoolam (2007). Previously this crystal was obtained and studied by vibrational spectroscopic method (Rajkumar and Ramakrishnan 2001); however, details of the structure were not provided. The asymmetric unit of (DL-MetH)H_2PO_4 contains two formula units ($Z' = 2$) at ($Z = 8$). In one of the cations the atoms C13 and C14 are disordered over two positions with site occupancies of 0.71(1) and 0.29(1). In the structure there are chains formed by $H_2P(1)O_4^-$ and $H_2P(2)O_4^-$. One of the hydroxyl groups of each anion is used for the formation of the chain: O13-H13...O14(2.711(3) Å) in the chain of $H_2P(1)O_4^-$ anions and O23-H23...O24(2.655(3) Å) in the chain of $H_2P(2)O_4^-$ anions. The second hydroxyl group of each anion is used for the formation of a hydrogen bond with carbonyl oxygen atom of respective cation: O11-H11...O1A(2.620(3) Å) between $H_2P(1)O_4^-$ anion and cation (1) and O21-H21...O2A(2.641(3) Å) between $H_2P(2)O_4^-$ anion and cation (2). By doing so, each of the two cations by hydroxyl group forms rather strong hydrogen bond with the same anion: O1B-H1B...O14(2.481(3) Å) and O2B-H2B...O24(2.479(3) Å). The NH_3^+ group of each cation forms each two hydrogen bonds with two nearest anions and each one hydrogen bond with oxygen atom of nearest cation of the same type: N1-H1A...O12, N1-H1C...O12, and N1-H1D...O1B and N2-H2A...O22, N2-H2C...O22, and N2-H2D...O2B.

Srinivasan et al. (2001b) and Ramaswamy et al. (2004a) reported the synthesis and structures of the crystals of (L-MetH)$_2SO_4$ and (DL-MetH)$_2SO_4$. In both cases, DL-methionine and sulfuric acid in 2:1 molar ratio was used. However, in the work of Srinivasan et al. (2001b) actually the crystals of (L-MetH)$_2SO_4$ were formed, instead of (DL-MetH)$_2SO_4$. The asymmetric unit in the case of (L-MetH)$_2SO_4$ contains two (L-MetH) cations and three SO_4^{2-} anions, which are in special positions on 3rd order axes. In one of the cations (cation II) the sulfur atom is disordered. The carboxyl group of cation I forms hydrogen bond with $S(1)O_4^{2-}$ anion O11-H11...O2(2.569(6) Å) and cation II with $S(2)O_4^{2-}$ anion *via* O21-H21...O4(2.576(5) Å). The NH_3^+ group of the first cation forms weak hydrogen bonds: N11-H11A...O3, N11-H11B...O2, and N11-H11C...O6, while the NH_3^+ group of the second cation forms hydrogen bonds: N21-H21A...O4, N21-H21B...O1, and N21-H21C...O6.

The asymmetric unit of (DL-MetH)$_2SO_4$ contains one (MetH) cation in general position and two SO_4^{2-} anions located in special positions. The S1 atom of the cation is disordered. The $S(2)O_4^{2-}$ anion is on $\bar{3}$ axis, while the $S(3)O_4^{2-}$ anion is on axis 3. Oxygen atoms of $S(2)O_4^{2-}$ anion are disordered, which requires center of symmetry, because ordered tetrahedral SO_4^{2-} anion lacks center of symmetry. The carboxyl group of the cation forms hydrogen bonds with ordered $S(3)O_4^{2-}$ anion, O2-H2A...O32(2.570(4) Å). The NH_3^+ group forms one hydrogen bond N1-H1A...O31(3.179(4) Å) on the level of van der Waals bonds; two bonds with O22 atom, N-H1B...O22 with N...O distances 2.772(7) Å) and 2.862(7) Å); and one hydrogen bond N1-H1C...O32(2.895(4) Å). IR and Raman spectra of (DL-MetH)$_2$SO$_4$ were recorded and analyzed (Ramaswamy et al. 2005).

The crystal of (L-MetH)$_2$SiF$_6$·H$_2$O was obtained by Fleck et al. (2013b) from aqueous solution and its structure was determined. Previously (Andreeva et al. 1983) the salt of (MetH)$_2$SiF$_6$ was obtained and characterized by IR spectrum and thermal analysis without, however, indication of methionine form (L, D, DL). The asymmetric unit of (L-MetH)$_2$SiF$_6$·H$_2$O contains one formula unit. The carboxyl group of one cation (cation A) forms a hydrogen bond with water molecule O1A-H1A...O1W(2.642(5) Å). The second cation forms hydrogen bond with anion, O1B-H1B...F1(2.611(3) Å). Water molecule as proton donor forms hydrogen bond with sulfur atom of cation (A), O1W-II1W...S1(3.415(3) Å). The NH_3^+ group of cation (A) forms two hydrogen bonds, N1A-H12A...F6 and N1A-H13A...F3, with anion and also with carbonyl oxygen atom of cation (B), N1A-H11A...O2B, while the NH_3^+ group of cation (B) forms also two hydrogen bonds with anion, N1B-H12B...F4 and N1B-H13B...F5, and one hydrogen bond with carbonyl oxygen atom of cation (A), N1B-H11B...O2A. Due to the significant number of hydrogen bonds with anion, disordering of fluorine atoms of anion is not observed.

The crystal of (DL-MetH)CCl$_3$COO was obtained from an aqueous solution containing DL-methionine and CCl$_3$COOH in a stoichiometric 1:1 ratio (Rajagopal et al. 2003c). The asymmetric unit contains one formula unit. The carboxyl group of cation forms a hydrogen bond with the carboxylate group of anion, O1-H1...O3 (2.589(2) Å). The NH_3^+ group forms hydrogen bonds with two nearest anions, N1-H1C...O3 and N1-H1A...O4, and also with the carbonyl oxygen atom of nearest cation, N1-H1B...O2.

The crystals of (DL-MetH) hydrogen maleate were obtained from an aqueous solution containing DL-methionine and maleic acid in a 1:1 molar ratio (Alagar et al. 2002b). The asymmetric unit contains a methioninium cation and hydrogen maleate anion. The O...O distance of intramolecular hydrogen bond of anion O3-H3...O5 is equal to 2.459(3) Å). The carboxyl group of the cation forms a hydrogen bond with the second oxygen atom of the carboxylate group of the anion O1-H1...O6 (2.552(3) Å). The NH_3^+ group forms two normal hydrogen bonds N1-H1A...O4 and N1-H1C...O5 and one weaker bond N1-H1B...O6. The atom H1B has also a contact with the carbonyl oxygen atom O2 on the level of van der Waals bonds.

The crystal of ((R-MetH)-R-mandelate-2R-mandelic acid was obtained from water/ethanol (1:1) solution. (R)-methionine and racemic mandelic acid in 1:2 ratio were dissolved and refluxed for 2 h and then cooled to room temperature and filtered (Su and Xu 2005). Crystals were obtained from the filtrate after one week. Whereas racemic mixture of mandelic acid is used in the preparation of the

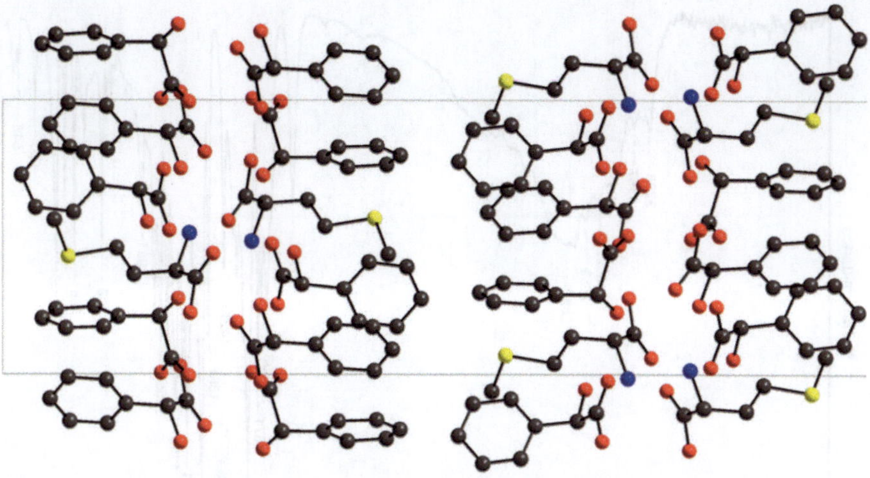

Fig. 5.17 Crystal structure of ((R-MetH)-R-mandelate-2R-mandelic acid. Note the layered arrangement of the building units. Hydrogen atoms are omitted for clarity

crystal, only one enantiomer (R configuration) occurs in the crystal. In the crystal structure the ions and molecules are arranged to form layers, with hydrophilic groups on one side and the hydrophobic groups on the other side (Fig. 5.17). The components in the hydrophilic layers link to each other via O-H. . .O and N-H. . .O hydrogen bonds, whereas in the hydrophobic layers they interact through van der Waals contacts. The asymmetric unit contains one formula unit, that is, methioninium cation, mandelate anion, and two molecules of mandelic acid. One of the molecules of mandelic acid forms strong hydrogen bond with mandelate anion, thus forming a dimeric $(X\ldots HX)^-$ anion with hydrogen bond O11-H11. . .O32(2.495(10) Å). The carboxyl group of the cation forms a hydrogen bond with the carbonyl oxygen atom of the second molecule of mandelic acid O1-H1. . .O22(2.751(10) Å), which in turn forms a hydrogen bond with the carbonyl atom of the first molecule of mandelic acid O21-H21. . .O12(2.640(10) Å). In addition to carboxyl groups of cation and two molecules of mandelic acids there are hydrogen bonds formed by hydroxyl groups O(13)H, O(23)H, and O(33)H of mandelic acids and mandelate anions, which act both as donors and acceptors. The group O(13)H forms hydrogen bond O13-H13. . .O23 and also as acceptor with O(33)H group O13. . .H33-O33. The group O(23)H forms hydrogen bonds with mandelate anion O23-H23. . .O32 as donor and also two hydrogen bonds as acceptor. In addition to the mentioned O23. . .H13-O13, it forms O23. . .H1B-N hydrogen bond with the NH_3^+ group. The group O(33)H in addition to the mentioned O33-H33. . .O13 hydrogen bond also forms O33. . .H1C-N hydrogen bond as acceptor. The group NH_3^+ in addition to two mentioned hydrogen bonds forms also a hydrogen bond with anion N-H1A. . .O31.

For DL-methionine formation of (DL-Met. . .DL-MetH) picrate salt with dimeric cation is known (see Sect. 5.2). In case of L-methionine our attempt to obtain respective salt was not successful; however, we established that the salt with 1:1

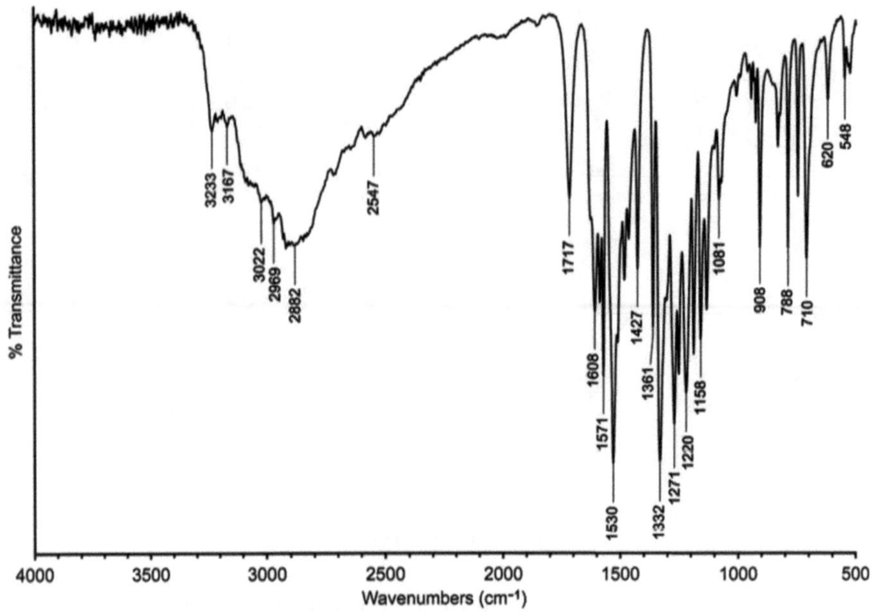

Fig. 5.18 IR spectrum of L-methioninium picrate

composition is formed. The IR spectrum of (L-MetH) picrate contains an absorption band at 1,717 cm^{-1} indicating the presence of (L-MetH) and also characteristic absorption bands of picrate anion (Fig. 5.18). The crystals of (L-MetH) picrate readily grow and give intensive SHG signal. The structure was not yet determined. Its further investigation is in process.

5.1.13 Salts of Valine

In Table 5.13 known salts of valine are presented. Ennan et al. (1983) obtained the (D-ValH)F salt and characterized it by IR spectrum and thermal analysis. The authors note that at significant excess of HF compared to stoichiometric a salt with 1:1 composition is formed. Based on the IR spectrum, Ennan et al. (1983) concluded that the salt has the following structure D-Val.(F. . .HF).(D-ValH); however, this conclusion was not yet confirmed by determination of the structure of the crystal.

The systems L-Val + HCl(HBr) + H_2O were studied in more detail and structurally characterized a number of formed salts. In the system L-Val + HCl + H_2O the following salts are formed: (L-Val. . .L-ValH)Cl, (L-ValH)Cl, and (L-ValH) Cl·H_2O. The first salt with dimeric cation is considered in Sect. 5.2. The structure of (L-ValH)Cl was determined by Parthasarathy (1966) using Weissenberg intensity data, and by comparison of unit cell parameters, it was concluded that the crystal of

Table 5.13 Salts of valine with inorganic and organic anions

#	Crystal	s.g., Z	a, b, c (Å), α, β, γ (°)	Refs
1	(D-ValH)F	–	–	Ennan et al. (1983)
2	(L-ValH)Cl	$P2_1$, 2	10.382(2), 7.066(1), 5.4407(9), $\beta = 91.40(2)$	Koetzle et al. (1974)
3	(L-ValH)Cl·H_2O	$P2_12_12_1$, 4	6.155(1), 6.828(1), 21.099(3)	Li et al. (2006)
4	(DL-ValH)Cl	$P2_1/c$, 4	11.722(4), 7.007(3), 11.143(4), $\beta = 120.53$	Di Blasio et al. (1977c)
5	(L-ValH)Br	$P2_1$, 2	10.154(2), 7.415(2), 5.333(2), $\beta = 91.22(2)$	Moitra and Kar (2008)
6	(L-ValH)NO_3	$P2_1$, 4	5.635(2), 18.478(3), 8.504(4), $\beta = 98.2079$	Srinivasan et al. (1997)
7a	(L-ValH)NO_3	$P2_1$, 4	10.8369(3), 18.459(1), 5.62190(2), $\beta = 129.07(3)$	Němec et al. (2001)
7b	(L-ValH)NO_3	$P2_1$, 8	10.765(2), 18.229(1), 11.109(2), $\beta = 128.64(1)$, at 130 K	Němec et al. (2001)
8	(DL-ValH)NO_3	$P\bar{1}$, 8	9.5695(10), 11.2104(16), 17.537(3), $\alpha = 108.52(3)$, $\beta = 104.14(2)$, $\gamma = 90.25(2)$	Srinivasan et al. (2002)
9	(DL-ValH)ClO_4	$P\bar{1}$, 8	10.3005(6), 13.0905(8), 14.6105(9), $\alpha = 81.288(2)$, $\beta = 88.4473(19)$, $\gamma = 75.9648(19)$, at 105 K	Sridhar et al. (2003)
10	(DL-ValH)H_2PO_4	$P2_1/a$, 4	9.201(1), 10.0035(6), 11.195(2), $\beta = 113.73(1)$	Ravikumar et al. (2002c)
11	(DL-ValH)H_2PO_3[a]	$P2_1/n$, 8	16.3590(3), 6.2540(2), 19.4560(3), $\beta = 109.238(1)$	Bendheif et al. (2003)
12	(DL-ValH)$_2$$NbOF_5$	$P2_1/c$, 4	9.8018(4), 19.4213(8), 10.1979(4), $\beta = 116.190(1)$	Pushilin et al. (2007)
13	(L-ValH)$_2$$SeO_4$·$H_2O$	$P2_1$, 4	6.181(2), 18.675(3), 15.342(4), $\beta = 90.14(2)$	Němec et al. (2002)
14	(DL-ValH)$_2$$TeI_6$	$P2_12_12_1$, 4	18.61(2), 19.88(2), 9.57(1)	Dauter et al. (1977)
15	(D-ValH)$_2$$SiF_6$	–	–	Ennan et al. (1983)
16	(L-ValH)$_2$$SiF_6$	$C2$, 2	17.205(4), 5.682(2), 11.684(4), $\beta = 132.706(9)$	Ghazaryan et al. (2013a)
17	(DL-ValH)CF_3COO	$P\bar{1}$, 2	6.7282(6), 9.0830(4), 9.2148(7), $\alpha = 98.461(1)$, $\beta = 91.990(1)$, $\gamma = 101.052(7)$	Suresh and Natarajan (2006)
18	(DL-ValH)CCl_3COO	$P\bar{1}$, 2	7.2380(14), 8.4150(17), 10.303(2), $\alpha = 106.50(3)$, $\beta = 97.80(3)$, $\gamma = 95.80(3)$, at 123 K	Rajagopal et al. (2002c)
19	(DL-ValH) maleate	$P\bar{1}$, 2	6.1830(12), 9.6480(19), 10.534(2), $\alpha = 104.92(3)$, $\beta = 106.32(3)$, $\gamma = 104.40(3)$, at 150 K	Alagar et al. (2001c)
20	(L-ValH) picrate	$P2_1$, 2	9.9714(13), 6.2930(5), 12.6480(9), $\beta = 110.50(1)$	Anitha et al. (2004a)
21	(DL-ValH)·picrate. DL-Val	$P2_1/n$, 4	5.3822(4), 23.221(2), 16.579(2), $\beta = 94.695(7)$	Anitha et al. (2004b)
22	(L-ValH) bnppa[b]	$P2_12_12_1$, 4	10.7790(7), 37.600(2), 6.0942(5)	Fujii and Hirayama (2002)

[a]The authors considered that the obtained crystal was (L-ValH)H_2PO_3. However, this is not compatible with the $P2_1/n$ symmetry (see the text)
[b]bnppa = (+)-(1S)-1,1′-binaphthalene-2,2′-diyl phosphate monohydrate

(L-ValH)Br has similar structure. In the work of Parthasarathy (1966) the crystal of (L-ValH)Cl was obtained from solution of L-valine in dilute HCl by rapidly evaporating the solution in a desiccator. According to this work, the crystals of (L-ValH) Cl were soft, unstable, and highly hygroscopic. More accurate data on the structure were obtained by Ando et al. (1967). Then the structure (L-ValH)Cl was determined by precision neutron diffraction method (Koetzle et al. 1974). The carboxyl group of the cation and the NH_3^+ group form hydrogen bonds with chloride ion: O-H...Cl (2.986(2) Å), N-H1...Cl(3.374(2) Å), N-H2...Cl(3.187(1) Å), and N-H3...Cl (3.224(2) Å). There is also a short van der Waals contact CA-HA...O1(3.376 Å) with HA...O1(2.368 Å) and C-H-O angle 152°. The structure of (L-ValH)Cl·H_2O was determined by Rao (1969) without determination of hydrogen atom positions and then was refined by Li et al. (2006). The authors note that the crystal studied by Rao (1969) actually was (D-ValH)Cl·H_2O. The crystal of (L-ValH)Cl·H_2O was obtained from an aqueous solution containing equimolar quantities of L-valine and HCl at room temperature. The carboxyl group of cation forms a hydrogen bond with chloride ion O1-H1...Cl1(3.003(2) Å). Water molecule as proton donor also forms hydrogen bonds with chloride ion: O3-H3A...Cl1(3.187(2) Å) and O3-H3B...Cl1(3.171(2) Å). The NH_3^+ group forms one hydrogen bond with chloride ion, N1-H1C...Cl1(3.246(2) Å), and two hydrogen bonds with two nearest water molecules, N1-H1A...O3(2.866(3) Å) and N1-H1B...O3(2.876(3) Å). There are also two short contacts: C2-H11...Cl1(3.758 Å) with H11...Cl1 (2.848 Å) and C4-H14...Cl1(3.794 Å) with H14...Cl1(2.873 Å) which may be classified as intermediate between strong van der Waals and weak hydrogen bonds.

The crystals of (DL-ValH)Cl were grown from a dilute hydrochloric acid solution of the DL-valine (Di Blasio et al. 1977c). As in the case of (L-ValH)Cl in the structure of (DL-ValH)Cl carboxyl NH_3^+ groups form hydrogen bonds with chloride ion: O2-H...Cl(3.006 Å), N-H1...Cl(3.183 Å), N-H2...Cl(3.433 Å), and N-H3...Cl(3.234 Å). There is also a contact N-H3...O1 with H3...O1(2.31 Å) on the level of strong van der Waals bonds. Crystals of isostructural salts (L-ValH) Cl and (L-ValH)Br having polar symmetry attracted the attention of researchers as possible NLO materials (Kirubavathi et al. 2008a; Moitra and Kar 2008; Kar 2012; Hubert Joe 2010; Amalanathan et al. 2011a). Above it was noted that in addition to these salts there are other salts which require the identification of obtained crystals. In some of the mentioned papers apparently there are mistakes because of lack of accurate identification of obtained crystals. In the works of Kirubavathi et al. (2008a) and Moitra and Kar (2008) results on the investigation of (L-ValH) Cl and (L-ValH)Br are presented. One may expect based on their isostructurality that their IR spectra should be similar. However, the IR spectrum of (L-ValH)Cl provided by Kirubavathi et al. (2008a) differs significantly from the spectrum of (L-ValH)Br provided by Moitra and Kar (2008). The authors not only identified the crystal (L-ValH)Br by unit cell parameters, but also determined its structure. So, the results of Kirubavathi et al. (2008a) seem to be questionable. The authors identify their crystal as (L-ValH)Cl based on measured unit cell parameters and consider that they are in good agreement with the reported literature values. However, the value of monoclinic angle β=101.5° differs significantly from the literature value $\beta = 91.5°$ (Parthasarathy 1966). In addition, the interpretation of the IR spectrum contains a number of inacceptable statements. The IR spectrum contains a band at 3,412 cm^{-1}, which may relate either to crystal water, if this is the spectrum of (L-ValH)Cl·H_2O, or to adsorbed water because of hygroscopicity of KBr. Kirubavathi et al. (2008a) assigned this band to C=O stretching, which is not acceptable, while the band at 1,743 cm^{-1} assigned to COO^- stretching, which also is not acceptable because the crystal does not contain the COO^- group and the band at 1,743 cm^{-1} would be assigned to C=O stretching. Interpretation of TG and DTA curves also deserves criticism. The authors ignore mass loss at ca. 53 and 184 °C before decomposition of the sample above 211 °C and consider that the crystal is stable up to 211 °C. Hubert Joe (2010) and Amalanathan et al. (2011a) consider that they obtained the crystal of (L-ValH)Br; however, they have not identified it at all. Both Hubert Joe (2010) and Amalanathan et al. (2011a) carried out calculations for (L-ValH)Br and provided IR spectrum of obtained crystal. Comparison of this spectrum with spectrum of the crystal of 2L-Val·HBr obtained by us (see Sect. 5.2) shows that the crystal obtained in both studies actually is (L-Val...L-ValH)Br and not (L-ValH)Br. Unfortunately, a similar mistake was made also by Kar (2012), where data of (L-Val...L-ValH)Cl (see unit cell parameters and thermal data) are provided as data of (L-ValH)Cl. Interesting results were obtained by Moitra and Kar (2008) for (L-ValH)Br crystal. Quality crystals were obtained and characterized by various methods. The authors found phase transition near 118 °C, which is revealed also on DSC curve (on DTA curve it is not revealed), and also on curve of dielectric constant. Moitra and Kar (2008) consider this phase transition as transition from ferroelectric to paraelectric and the point of phase transition as Curie point. However, data confirming this conclusion were not provided yet. Adhikari et al. (2013) recently have grown and characterized the crystal of (L-ValH)Cl·H_2O as possible NLO material. Crystals were grown by evaporation method from aqueous solution at 42 °C. Solubility curve was determined from 30 to 60 °C. Thermal and mechanical properties, transmittance spectrum in UV–Vis–NIR region, photoluminescence spectrum, and SHG efficiency ($I^{2\omega}/I^{2\omega}{}_{KDP} = 0.3$) by powder method were determined. Dehydration takes place at ca. 68 °C, and then the sample decomposes in the 200–280 °C interval. Crystal structure was redetermined at 120 K (according to CIF (see CCDC #929275) at 150 K). The structure is identical with that determined by Li et al. (2006). One of two hydrogen bonds formed by the water molecule as proton donor is called intramolecular by Adhikari et al. (2013), which is not correct in our opinion.

Srinivasan et al. (1997) published the structure of (L-ValH)NO_3. This crystal was investigated more accurately and in more detail by Němec et al. (2001). By DSC method the authors found a phase transition near 250 K. The structure was redetermined at 293 K and at 130 K. Vibrational spectra were investigated at 298 K and at 90 K. Deuteration and ^{15}N isotope were used as well. Vibrational spectra of (L-ValH)NO_3 were studied also by Rajkumar and Ramakrishnan (2000); however, the quality of IR spectrum was low probably because of low sensitivity of the detector at higher frequencies. Particularly, very strong band at 1,715 cm^{-1} is absent in the spectrum of this work. According to Němec et al. (2001) there is disordering of the isopropyl groups in the structure of (L-ValH)NO_3 at room temperature. Below the phase transition temperature, ordering and thus doubling of unit cell volume take place, although the symmetry does not change. The asymmetric unit at room temperature contains two formula units. The carboxyl groups of cations form hydrogen bonds with nitrate ions: O11-H11. . .O(130)(2.674 (6) Å) and O21-H21. . .O(220)(2.612(5) Å). The NH_3^+ groups form each two hydrogen bonds with nitrate ions and each one hydrogen bond with carbonyl oxygen atom. At 130 K the number of independent formula units is equal to four. Hydrogen bonds of carboxyl groups of cations formed with nitrate ions are O11-H11. . .O(420)(2.658(2) Å), O21-H21. . .O(330)(2.590(2) Å), O31-H31. . .O (120)(2.668(2) Å), and O41-H41. . .O(230)(2.606(2) Å). Hydrogen bonds of NH_3^+ groups are similar.

The crystal of (DL-ValH)NO_3 may be obtained from aqueous solution at equimolar ratio of DL-valine and HNO_3. Initially the structure was determined by S.N. Rao and R. Parthasarathy (CSD code: VALNII) in 1974 in monoclinic space group $P2_1/c$, $Z=4$, $R=0.12$. Later more accurate data showed the correct symmetry is $P\overline{1}$ with $Z=8$ and $Z'=4$ (Srinivasan et al. 2002). The reason of choosing monoclinic symmetry was that valinium cations are related to one another by pseudo-c-glide and pseudo-translation. Carboxyl groups of cations form hydrogen bonds with nitrate anions: O12-H12. . .O1C(2.636(3) Å), O22-H22. . .O4C(2.628(3) Å), O32-H32. . .O2C(2.660 (3) Å), and O42-H42. . .O3C(2.667(3) Å). The NH_3^+ groups form each three hydrogen bonds. One hydrogen bond forms with carbonyl oxygen atom, N11-H11B. . .O21, N21-H21. . .O31, N31-H31B. . .O41, and N41-H41B. . .O11, and two bonds with nitrate anions.

The crystal of (DL-ValH)ClO_4 also may be obtained from aqueous solution at equimolar ratio of DL-valine and $HClO_4$ and also crystallizes in $P\overline{1}$ space group with $Z=8$ and $Z'=4$ (Sridhar et al. 2003). However, there is an interesting distinction of the structure of (DL-ValH)ClO_4 from the structure of (DL-ValH)NO_3. In the structure of (DL-ValH)ClO_4 only three out of four cations form hydrogen bonds with anions: O1B-H1B. . .O34(2.7503(14) Å), O2B-H2B. . .O44(2.7395(12) Å), and O4B-H4B. . .O43(2.7913(13) Å). The remaining cation forms hydrogen bond via carboxyl group not with anion, but with symmetry-related cation: O3-H3B. . .O3A (2.6381(11) Å). This is the second case of formation of the A^+. . .A^+-type dimer, when the A^+ cation is not β-AlaH (see Sect. 5.2). In two cases out of four NH_3^+ groups, the formation of hydrogen bonds with carbonyl oxygen atom is observed: N1-H1C. . .O2A and N3-H3A. . .O1A. In other cases hydrogen bonds N-H. . .O with anions are formed.

The crystal of (DL-ValH)H_2PO_4 was obtained from aqueous solution at equimolar ratio of DL-valine and H_3PO_4 (Ravikumar et al. 2002c). The asymmetric unit contains one formula unit. One of the hydroxyl groups of $H_2PO_4^-$ is used for the formation of infinite chain of anions, O3-H2. . .O2(2.702(2) Å). The second hydroxyl group is used for the formation of hydrogen bond O4-H1. . .O1A(2.6820 (19) Å) with carbonyl oxygen atom of cation. At the same time the carboxyl group of cation forms rather strong hydrogen bond with the O2 atom of anion,

O1B-H1B. . .O2(2.4548(18) Å). The NH_3^+ group forms two hydrogen bonds with two nearest anions, N11-H11A. . .O1 and N11-H11C. . .O1, while the third atom forms weak bond N11-H11B. . .O1B.

Bendheif et al. (2003) claimed the synthesis of the crystal of (L-ValH)H_2PO_3 from aqueous solution containing equimolar quantities of L-valine and H_3PO_3. However, as a result of structure determination of the crystal, the centrosymmetric space group $P2_1/n$ was established, which is not compatible with the presence of optically active form of L-valinium cation. Either space group was not determined correctly or there is DL-ValH instead of L-ValH. Inspection of the structure by CIF (CSD code: HURXEX) showed that the crystal actually is (DL-ValH)H_2PO_3. The asymmetric unit contains two formula units. One of the anions is disordered. The disorder can be described as a rotation of this anion around the axis which bisects the angle O4b-P1b-O3b. The refined model corresponds to a disordered distribution between OH and H, with occupation factors of 0.85 and 0.15. Each of the hydroxyl groups of cations forms O-H. . .O hydrogen bond with anion, while the hydroxyl group of anions forms hydrogen bond O-H. . .O with the carbonyl oxygen atom: O1a-H1a. . .O4a(2.506(2) Å) and O1b-H1b. . .O4b(2.538(2) Å). For anion "a" there is O5a-H14a. . .O2a(2.626(2) Å) hydrogen bond. For anion "b" with disordering there are O5b-H14b. . .O2b(3.026(2) Å) and O13b-H13b. . .O1b(2.681(11) Å). The NH_3^+ groups form N-H. . .O hydrogen bonds with anions.

The crystal of bis(DL-valinium) pentafluorido-oxidoniobate (V) (DL-ValH)$_2NbOF_5$ was obtained by Pushilin et al. (2007) by the reaction of Nb_2O_5 in a solution of HF with DL-valine in molar stoichiometric 4:1 ratio of DL-Val:Nb_2O_5 by slow evaporation of the solution. The asymmetric unit contains one formula unit: discrete anion $(NbOF_5)^{2-}$ and two cations. In the distorted octahedral $(NbOF_5)^{2-}$ anion, the Nb-F bond *trans* to the Nb=O bond is significantly longer than the other four Nb-F bonds: Nb-F1 is (2.2270(7) Å), while Nb-F2, Nb-F3, Nb-F4, and Nb-F5 are (1.9331(9) Å), (1.8906(9) Å), (1.9339(9) Å), and (1.9648(9) Å), respectively. The bond length of Nb-O1 is (1.7172(9) Å). Carboxyl groups of both cations form hydrogen bonds with F1 atom: O2-H2. . .F1(2.5179(11) Å) and O4-H4. . .F1(2.5314(12) Å). The NH_3^+ group of the second cation forms two N-H. . .F type hydrogen bonds, N2-H2B. . .F5 and N2-H2C. . .F4, and one weak bond, N2-H2A. . .O3. The NH_3^+ group of the first cation forms three hydrogen bonds of N-H. . .O type, N1-H1A. . .O3, N1-H1B. . .O5, N1-H1C. . .O1, and one N1-H1B. . .F2 hydrogen bond.

The crystal of (L-ValH)$_2SeO_4{\cdot}H_2O$ was obtained by Němec et al. (2002) from an aqueous solution containing L-valine and H_2SeO_4 in molar ratio 1:1. In this system also the crystal of 3L-Val·H_2SeO_4 containing dimeric cation (L-ValH(L-Val. . .L-ValH)SeO_4) is formed (see Sect. 5.2). The asymmetric unit of (L-ValH)$_2SeO_4{\cdot}H_2O$ contains two formula units. Detailed structural data are not provided in this work, so we used CIF (CSD code: RUXWIQ). Carboxyl groups of all four cations form O-H. . .O hydrogen bonds with anions with O. . .O distances: 2.575 Å, 2.597 Å, 2.601 Å, and 2.609 Å. Water molecules as proton donor also form hydrogen bonds with anions with O. . .O distances: 2.801 Å, 2.803 Å, 2.832 Å, and 2.882 Å. Hydrogen atom of one of the water molecules is not localized. As acceptors water molecules form hydrogen bonds O...H-N with the NH_3^+ groups. In addition, the NH_3^+ groups form hydrogen bonds with anions and also weak bonds with carbonyl atoms of cations. Němec et al. (2002) investigated the IR and Raman spectra of (L-ValH)$_2$SeO$_4$·H$_2$O in detail. The IR spectra were recorded in the temperature range from 298 to 90 K, and also by DSC method in the 95–363 K interval. At 366 K the process of dehydration takes place connected with further decomposition. DSC data and IR spectra show that in investigated temperature interval no phase transitions are observed. According to the authors, the crystal does not reveal high NLO properties. The reagents 2L-Val·H_2SO_4 and L-Val·H_2SO_4 are manufactured by LôôkChem; however, we could not find information about their structures.

At consideration of salts of sarcosine we mentioned about the salt of sarcosine with TeI_6^{2-} anion (Dauter et al. 1977). The authors obtained also the salt (DL-ValH) TeI_6 and determined the symmetry and cell parameters of the crystal. According to the work, despite that valine was used as racemate, the salt crystallizes in non-centrosymmetric $P2_12_12_1$ space group with one asymmetric formula unit.

Ennan et al. (1983) obtained the salt (D-ValH)$_2$SiF$_6$ and characterized with IR spectrum and thermal analysis; however, its structure was not determined. Ghazaryan et al. (2013a) synthesized the salt (L-ValH)$_2$SiF$_6$ and determined its structure. The asymmetric unit contains one (L-ValH) cation in general position and SiF_6^{2-} anion in special position (L-ValH)·0.5SiF$_6$. Some of the fluorine atoms are disordered. The carboxyl group of cation forms hydrogen bonds O1-H1...F2 (2.753 Å) and O1-H1...F3(2.592 Å) with anion. The NH_3^+ group forms one weak bond on the level of van der Waals interactions with carbonyl oxygen atom of neighboring cation, N1-H1B...O2, and two hydrogen bonds, N1-H1A...F1 and N1-H1C...F4, with anion.

The salts of DL-valine with CF_3COOH and CCl_3COOH were obtained by Suresh and Natarajan (2006) and Rajagopal et al. (2002c) from aqueous solution containing DL-valine and respective acid in 1:1 molar ratio. These salts have a number of similar features. The asymmetric unit of both salts contains one formula unit. Carboxyl groups of cations form hydrogen bonds O-H...O with anions. In structures of (DL-ValH)CF$_3$COO and (DL-ValH)CCl$_3$COO the O...O distances are (2.607(4) Å) and (2.601(2) Å), respectively. Fluorine atoms in CF_3COO^- are disordered. The L- and D-forms of cations are connected by center of symmetry and form hydrogen bonds N-H...O between the NH_3^+ group and carbonyl atom, thus forming specific dimer. In the structures of (DL-ValH)CF$_3$COO and (DL-ValH) CCl$_3$COO, the N...O distances are (2.878(5) Å) and (2.871(2) Å), respectively. The remaining two hydrogen atoms of NH_3^+ groups form hydrogen bonds with two atoms of the carboxylate group of anions. In the structures of (DL-ValH)CF$_3$COO and (DL-ValH)CCl$_3$COO the N...O distances are (2.745(5) Å) and (2.827(5) Å) and (2.761(2) Å) and (2.804(2) Å), respectively.

The crystal of (DL-ValH) hydrogen maleate was obtained by Alagar et al. (2001c) from aqueous solution containing equimolar quantities of DL-valine and maleic acid. The asymmetric unit contains one formula unit. Intramolecular hydrogen

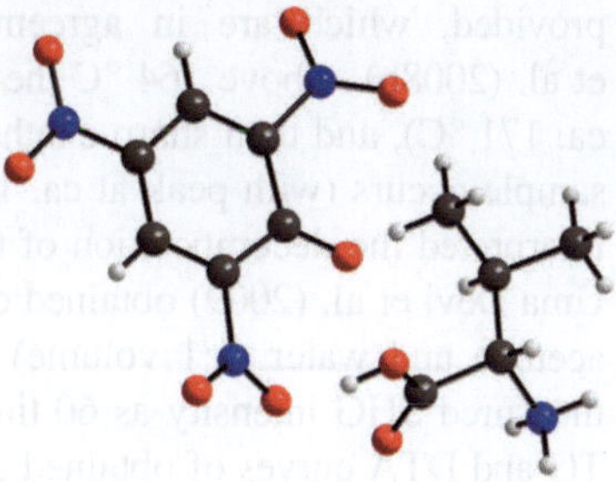

Fig. 5.19 Molecular structure of L-valinium picrate. Note the layered arrangement of the building units. Hydrogen atoms are omitted for clarity

bond in anion O3-H3...O5 is characterized by O...O distance (2.4330(17) Å). The carboxyl group of cation forms hydrogen bond O1-H1...O4(2.5963(15) Å) with the carbonyl atom of anion. The NH_3^+ group forms hydrogen bonds with the carboxylate group of anion: N1-H1A...O6, N1-H1B...O5, and N1-H1C...O6. There is also a contact with carbonyl atom of neighboring cation N1-H1C...O2 on the level of strong van der Waals bonds.

Interaction of L-valine and DL-valine with picric acid was investigated by Anitha et al. (2004a, b). In the case of L-valine a salt with 1:1 composition is formed, (L-ValH) picrate (Anitha et al. 2004a), while in the case of DL-valine the salt has 2:1 composition (Anitha et al. 2004b). In the structure of the salt with 2:1 composition dimeric cation is not formed (see Sect. 5.2). Briget Mary et al. (2006a) studied the vibrational spectra of 2DL-valine·picric acid. Critical notes about this paper are given by Ghazaryan et al. (2012d). The crystal of (L-ValH) picrate was obtained from aqueous solution containing equimolar quantities of L-valine and picric acid. The asymmetric unit contains (L-ValH) cation and picrate anion (Fig. 5.19). The carboxyl group of cation forms hydrogen bond O1B...H1B...O1(2.678(3) Å) with negatively charged oxygen atom of anion. The NH_3^+ group forms one hydrogen bond N11-H11...O5(2.872(4) Å) with nitro group of anion and one with carbonyl oxygen atom of neighboring cation N11-H11B...O1A(2.825(3) Å). Other contacts are intermediate between weak hydrogen bonds and strong van der Waals bonds. Vibrational spectra of (L-ValH) picrate are investigated in detail (Senthilkumar et al. 2007). Crystals of L-valinium picrate became a subject of further investigations as possible NLO material (Kirubavathi et al. 2008b; Martin Britto Dhas and Natarajan 2008b; Srinivasan et al. 2008a; Uma Devi et al. 2009; Koteeswari et al. 2012; Raj and Murugakoothan 2011; Joseph et al. 2011, 2012; Raj et al. 2013). Kirubavathi et al. (2008b) have grown the crystals of (L-ValH) picrate by slow evaporation at room temperature and have characterized them. IR spectrum was provided; however, the authors assigned a number of bands to the COO^- group, which actually is absent in the structure of the crystal. According to this work the intensity of the SHG signal is higher than in KDP two times. The crystal of (L-ValH) picrate was studied as possible NLO material also in (Martin Britto Dhas and Natarajan 2008b; Srinivasan et al. 2008a). However, according to Martin Britto Dhas and Natarajan (2008b) the intensity of the SHG signal is higher than in KDP 60 times, while according to Srinivasan et al. (2008a) the intensity of the SHG signal is higher than in KDP even more than 140 times. In this work, TG and DTA curves are provided, which are in agreement with the curves provided in Kirubavathi et al. (2008b). Above 164 °C the sample starts to melt (DTA endothermic peak at ca. 171 °C), and then sharp exothermic effect connected with decomposition of the sample occurs (with peak at ca. 193 °C). So, Srinivasan et al. (2008a) erroneously interpreted the decomposition of the sample as an endothermic peak at ca. 200 °C. Uma Devi et al. (2009) obtained crystals of (L-ValH) picrate from mixed solvent of acetone and water (1:1 volume) and identified by measured cell parameters and measured SHG intensity as 60 times higher than KDP. However, IR spectrum and TG and DTA curves of obtained crystals do not correspond to previously published data. Particularly, the sample obtained by them decompose without melting with peak on DTA curve at 269 °C assuming this peak as melting, in spite of that decomposition starts above 217 °C, while the IR spectrum does not contain the band related to ν(C=O) vibration of carboxyl group of cation.

Joseph et al. (2011) decided to grow crystals of L-valinium picrate from aqueous solution containing glycine, valine, and picric acid in ratios 0.1:0.9:1. Crystals obtained by the authors were called "glycine mixed L-valine picrate (GVP)." The obtained crystals have no yellow color, which casts doubt on the presence of picrate anion in the structure of crystals. In addition, unit cell parameters and symmetry (orthorhombic) have nothing in common with L-valinium picrate. The IR spectrum of "GVP" has similarity with the spectrum of L-valinium picrate; however, the authors assign the band at 1,739 cm^{-1} not to ν(C=O) COOH of (L-ValH) cation, but to the COO^- group as well as some other bands. We tried to reproduce these results for the determination of the composition and structure of "GVP." From an aqueous solution containing 0.1Gly + 0.9L-Val + picric acid, crystals were obtained by evaporation, which proved to be (L-ValH) picrate.

Raj and Murugakoothan (2011) and Raj et al. (2013) obtained crystals of L-valinium picrate doped by Ba^{2+} and Ni^{2+}. According to these papers, doping with Ni^{2+} reduces the intensity of SHG from 60 times to 45 times compared to KDP, while doping with 10 mol% Ba^{2+} increases the intensity of SHG from 60 times to 160 times. It is not clear what the mechanism of such influence is, and which positions in the structure can be occupied by Ba^{2+} and Ni^{2+} ions.

The crystal of (L-ValH) bnppa as well as (L-AlaH) bnppa (see Sect. 5.1.5) was obtained by evaporation of solvent from solution containing equimolar quantities of components (Fujii and Hirayama 2002). Acetone was used as a solvent. The structure of (L-ValH) bnppa has definite similarities with the structure of (L-AlaH) bnppa; however, there are also some distinctions. The carboxyl group of (L-ValH) cation forms O(2v)-H...O(4) hydrogen bond with the PO_4^- group with O...O distance (2.620(2) Å), which is noticeably longer than in case of the salt with L-alanine (2.551(2) Å). Water molecule as proton donor forms one hydrogen bond O (1 W)-H...O(1v)(2.899(3) Å) with carbonyl oxygen atom and not with PO_4^- group as in the case of the salt of L-alanine. The NH_3^+ group forms three hydrogen bonds, two of them with the PO_4^- group, N(1v)-H...O(3)(2.660(2) Å) and N(1v)-H...O(4)(2.809(2) Å), and the third with water molecule, N(1v)-H...O(1 W) (2.834(4) Å).

5.1.14 *Salts of Leucine*

In Table 5.14 the salts of leucine, which is the homolog of valine, are collected. Halogenides of L-leucine crystallizing in the orthorhombic system (space group $P2_12_12_1$) with one asymmetric unit were obtained. Usually chloride and bromide have similar structures, while fluoride and iodide differ by structure from chloride and bromide. In the case of salts of L-leucine, we have another pattern. Fluoride and chloride crystallize in the form of monohydrate and have similar structure, while bromide and iodide crystallize in the form of anhydrous salt and also have similar structures with each other. Carboxyl groups of (L-LeuH)F·H_2O and (L-LeuH) Cl·H_2O form rather strong hydrogen bonds with anions: O-H...F(2.441(1) Å) and O-H...Cl(3.012 Å). The NH_3^+ groups form each two hydrogen bonds with water molecule and each one hydrogen bond with respective halogenide. Water molecules in both structures as proton donors form hydrogen bonds with respective halogenides. Zeghouan et al. (2012), who obtained (L-LeuH)F·H_2O crystals by evaporation from solution containing equimolar quantities of L-leucine and HF, note that they melt at 618 K. However, it is not clear if it is related to monohydrate or anhydrous salt, because it is improbable that monohydrate can melt at such high temperature without dehydration. It is possible also that there is misprint in number 618 K. In the structures of (L-LeuH)Br and (L-LeuH)I the carboxyl groups of cations form O...Br(3.158 Å) and O...I(3.372 Å) hydrogen bonds. The NH_3^+ groups form each one hydrogen bond with the carbonyl oxygen atom with N...O distances ca. 2.91 Å and each two hydrogen bonds with bromide and iodide anions, respectively. The crystals of (L-LeuH)Br and (L-LeuH)I were investigated by Fleck and Weiss (1984) by ^{79}Br and ^{127}I NQR method. In the crystal of (L-LeuH)Br a phase transition was found at 331.6 K by NQR and DTA methods. The crystal of (L-LeuH) Br attracted the attention of Adhikari and Kar (2013) as possible NLO material, although Chaney et al. (1971) had previously noted that the crystals are soft and hygroscopic. The authors identified the grown crystal by determination of cell parameters based on powder XRD pattern and also determined its various characteristics. According to this study, the intensity of SHG in powder of (L-LeuH)Br is four times higher than in KDP. Quantum chemical calculations of molecular hyperpolarizability were also performed.

Nitrates of L-leucinium and DL-leucinium were obtained by Bahadur et al. (1993) and Berrah et al. (2005) from aqueous solutions containing equimolar quantities of HNO_3 and L-leucine and DL-leucine, respectively. The asymmetric unit of (L-LeuH) NO_3 contains two formula units, while in the case of (DL-LeuH)NO_3 one formula unit. In both cases carboxyl groups of cations form hydrogen bonds with nitrate anions: O8-H14...O1(2.676 Å) and O10-H27...O5(2.574 Å) for (L-LeuH)NO_3 and O2-H6...O3(2.617(1) Å) for DL-LeuH)NO_3. The NH_3^+ groups in structure of (L-LeuH)NO_3 form each two hydrogen bonds with nitrate anions. The third hydrogen bond is formed with carbonyl oxygen atom of other cation. Similarly in the structure of (DL-LeuH)NO_3 the NH_3^+ group forms two hydrogen bonds with nitrate anion and one bond with carbonyl atom of nearest cation. Vibrational spectra of

Table 5.14 Salts of leucine with inorganic and organic anions

#	Crystal	s.g., Z	*a, b, c* (Å), *α, β, γ* (°)	Refs
1	(L-LeuH)F·H_2O	$P2_12_12_1$, 4	5.7058(1), 5.8289(1), 27.3150(4), at 100 K	Zeghouan et al. (2012)
2	(L-LeuH)Cl·H_2O	$P2_12_12_1$, 4	6.22563(13), 6.3951(13), 26.865(5), at 150 K	Atwood et al. (2005)
3	(L-LeuH)Br	$P2_12_12_1$, 4	7.29, 24.51, 5.54	Subramanian (1967)
4	(L-LeuH)I	$P2_12_12_1$, 4	7.693(10), 23.38(2), 5.682(8)	Chaney et al. (1971)
5	(L-LeuH)NO_3	$P1$, 2	5.662(6), 8.445(4), 10.867(6), $\alpha = 75.79(4)$, $\beta = 95.45(6)$, $\gamma = 99.82(6)$	Bahadur et al. (1993)
6	(DL-LeuH)NO_3	$P2_1/c$, 4	11.0324(2), 5.6200(2), 16.4317(3), $\beta = 105.789(2)$	Berrah et al. (2005)
7	(L-LeuH)BF_4	–	–	This work
8	(L-LeuH)ClO_4	$P2_1$, 2	55.640(1), 8.768(2), 10.704(2), $\beta = 92.49(3)$	Janczak and Perptuo (2007)
9	$(L\text{-}LeuH)_2Sb_2F_8$	$P2_12_12_1$, 4	6.1459(6), 14.994(1), 24.789(2)	Udovenko et al. (2010b)
10	$(L\text{-}LeuH)_2SeO_4$	–	–	Němec and Mička (2001a)
11	(L-LeuH)HC_2O_4	$P1$, 2	5.674(3), 9.803(9), 9.906(8), $\alpha = 87.37(7)$, $\beta = 99.64(5)$, $\gamma = 100.35(6)$	Rajagopal et al. (2003d)
12	$(DL\text{-}LeuH)_2TeI_6$	[a]	8.31(1), 12.47(2), 8.25(1), $\alpha = 114.3(2)$, $\beta = 110.2(2)$, $\gamma = 113.2(2)$	Dauter et al. (1977)
13	$(DL\text{-}LeuH)_2SiF_6$	–	–	Ennan et al. (1983)
14	$(L\text{-}LeuH)_2SiF_6 \cdot 2H_2O$	$P2_12_12_1$, 4	5.8221(5), 14.8246(12), 24.9324(18)	Fleck et al. (2013b)
15	(L-LeuH)BS·H_2O[b]	$P2_12_12_1$, 4	9.921(1), 25.741(2), 6.216(0)	Kimoto et al. (1989)
16	(L-LeuH)*o*-DMBS·H_2O[b]	$P2_12_12_1$, 4	9.855(7), 28.479(9), 6.280(1)	Hasegawa et al. (1998)
17	(L-LeuH)*m*-DMBS·H_2O[b]	$P2_12_12_1$, 4	9.839(2), 27.86(1), 6.609(2)	Hasegawa et al. (1998)
18	(L-LeuH)*p*-DMBS·H_2O[b]	$P2_1$, 2	6.163(1), 10.199(9), 14.140(9), $\beta = 94.112(9)$	Hasegawa et al. (1998)
19	(L-LeuH)*p*-TS·H_2O[b]	$P2_12_12_1$, 4	9.866(2), 28.201(9), 6.250(2)	Hasegawa et al. (1998)
20	(L-LeuH)(−)-(PES)[c]	$P2_1$, 2	14.789(1), 5.788(1), 9.830(1), $\beta = 107.41(1)$	Yoshioka et al. (1998)
21	(D-LeuH) (−)-(PES)[c]	$C2$, 4	21.096(3), 5.505(2), 20.667(3), $\beta = 137.35(1)$	Yoshioka et al. (1998)
22	(D-LeuH) (−)-(PES)·H_2O[c]	$P2_12_12$, 4	10.173(1), 27.763(3), 6.137(1)	Yoshioka et al. (1998)

[a] $P1$ or $P\bar{1}$, 1

[b] BS = benzenesulfonate, *o*-DMBS = *3,4*-dimethylbenzenesulfonate, *m*-DMBS = 2,4-dimethylbenzenesulfonate, *p*-DMBS = *2,5*-dimethylbenzenesulfonate, *p*-TS = *4*-methylbenzenesulfonate (*p*-toluenesulfonate)

[c] (−)-(PES) = (S)-(−)-1-phenylethanesulfonate

(L-LeuH)NO_3 were studied by Němec and Mička (1999) and Rajkumar and Ramakrishnan (2000). Němec and Mička (1999) checked possible phase transitions by IR spectroscopic and DSC methods in the 95–425 K interval. However, no phase transitions in this interval were found. Adhikari and Kar (2012) have grown and investigated the crystal of (L-LeuH)NO_3 as possible NLO material. Various characteristics of the crystal were studied as well as hyperpolarizability by calculation, which was rather high. However, the intensity of SHG in powder was on the level of KDP.

Janczak and Perptuo (2007) obtained crystals of (L-LeuH)ClO_4, determined crystal structure, and established its NLO activity by powder method. According to their work, nonlinearity of (L-LeuH)ClO_4 is two times higher than nonlinearity of KDP. The asymmetric unit contains one formula unit. The carboxyl group of (L-LeuH) forms O6-H6...O2(2.746(3) Å) hydrogen bond with anion. This hydrogen bond may be considered as rather weak because of the H...O-distance (2.12 Å) and the OHO angle (133°). The NH_3^+ group forms three also relatively weak hydrogen bonds with three nearest anions: N1-H11...O3(2.939(4) Å), N1-H12...O1(2.954 (3) Å), and N1-H13...O4(3.041(4) Å). Recently, Marchewka and Drozd (2013) theoretically and experimentally studied the vibrational spectra of (L-LeuH)ClO_4 as well as the NLO properties and the thermal behavior by DSC in the 100–350 K interval. Near 200 K the authors found anomalies on heating and cooling curves. According to the authors its origin may be connected with disordered perchlorate anions.

In our group we obtained also the crystals of (L-LeuH)BF_4 which based on similarity of IR spectra are isostructural with (L-LeuH)ClO_4. In addition we established that in the systems L-Leu + $HClO_4$ + H_2O and L-Leu + HBF_4 + H_2O salts 2L-Leu·$HClO_4$ and 2L-Leu·HBF_4 with dimeric cation (L-Leu...L-LeuH) are formed (see Sect. 5.2).

The crystal of (L-LeuH)SbF_4 was obtained by reaction of SbF_3 and L-leucine in HF (1:5) aqueous solution at a 1:1 molar ratio of components by slow evaporation at room temperature (Udovenko et al. 2010b). The unit cell of the crystal structure contains two crystallographically independent Sb atoms (Sb1 and Sb2) and two cations. As asymmetric unit, (L-LeuH)$_2$$Sb_2F_8$ with $Z = 4$ should be given, but not (L-LeuH)SbF_4. The authors consider as formula unit (L-LeuH)SbF_4 with $Z = 4$. In case of (L-LeuH)SbF_4, however, should be $Z = 8$. In its nearest environment Sb1 has three bonds: Sb1-F1(1.931(4) Å), Sb1-F2(1.964(5) Å), and Sb1-F3(1.941(5) Å). In the nearest environment of Sb2 there are five fluorine atoms: Sb2-F4(2.120(5) Å), Sb2-F5(1.927(4) Å), Sb2-F6(2.059(4) Å), Sb2-F7(2.086(4) Å), and Sb2-F8(2.100 (6) Å). Each Sb(1)F_3 group is connected with two nearest Sb(2)F5 groups, and vice versa by Sb1...F4(2.430(5) Å) and (2.366(5) Å), thus forming polymeric anion. Carboxyl groups of cations form O-H...F hydrogen bonds with F6 and F7 atoms: O1-H14...F7(2.588(7) Å) and O3-H28...F6(2.585(7) Å). The NH_3^+ groups form N1-H1A...F1(2.795(7) Å) and N1-H1C...F5(2.943(9) Å) and N2-H2B...F2 (2.836(7) Å) and N2-H2C...F7(2.903(8) Å). There are also some weak contacts. Kavun et al. (2012) studied ^{1}H and ^{19}F NMR spectra and their behavior at heating of the (L-LeuH)$_2$$Sb_2F_8$ sample, and also ionic conductivity. In the range 385–415 K a diffusion phase transition to a relatively high conductive state was found.

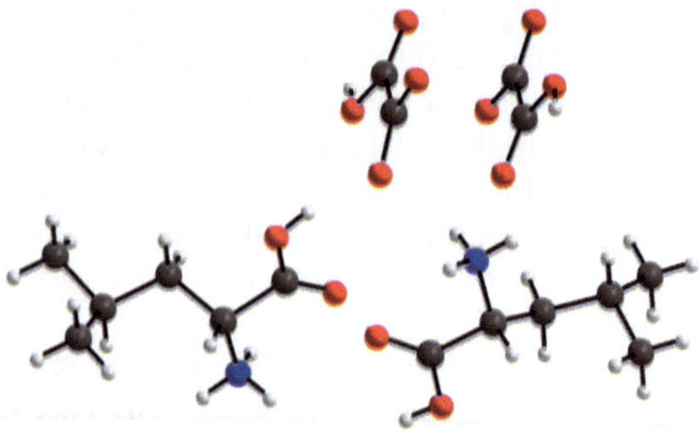

Fig. 5.20 Molecular structure of L-leucinium hydrogen oxalate

The salt (L-LeuH)$_2$SeO$_4$ was obtained by Němec and Mička (2001a, b) from aqueous solution containing equimolar quantities of L-leucine and H$_2$SeO$_4$. Data of chemical analysis showed that (L-LeuH)HSeO$_4$ is not formed as one may expect for 1:1 ratio, but (L-LeuH)$_2$SeO$_4$. The structure was not determined; however, detailed investigation of IR and Raman spectra allowed to establish the presence of cations and together with data of chemical analysis also the formation mechanism. Deuterated samples are studied as well, and also possible phase transitions up to 90 K. No phase transitions were found. Crystals melt at 450 K and then decompose.

The crystals of (L-LeuH) hydrogen oxalate were obtained by Rajagopal et al. (2003d) from aqueous solution containing equimolar quantities of L-leucine and oxalic acid. The triclinic unit of (L-LeuH)HC$_2$O$_4$ contains two formula units. Both cations and anions display pseudocentrosymmetry (Fig. 5.20). Each of the two independent anions forms head-to-tail chain by O31-H31...O34(2.578(3) Å) and O41-H41...O44(2.616(3) Å). The carboxyl groups of cations form hydrogen bonds with the second oxygen atom of the carboxylate group of respective anion: O11-H11...O43(2.593(4) Å) and O21-H21...O33(2.600(4) Å). The NH$_3^+$ groups of one type of cation form hydrogen bonds with the carbonyl oxygen atom of the second type of cation: N11-H11C...O22 and N21-H21B...O12. Two other hydrogen atoms are used for the formation of hydrogen bonds with anions: N11-H11A...O44 and N11-H11B...O33 and N21-H21A...O32 and N21-H21C...O43. In spite of the pseudocentrosymmetry of the structure of (L-LeuH)HC$_2$O$_4$, Anbuchezhiyan et al. (2009b); Anbuchezhiyan (2010) decided to study this crystal as possible NLO material. The IR spectrum was recorded and interpreted, thermal behavior was determined, SHG intensity was estimated by powder method, and some other characteristics were measured. The crystal melts at 162 °C and then decomposes starting at 203 °C. SHG intensity is 0.6 compared to KDP.

Ennan et al. (1983) obtained the salt (DL-LeuH)$_2$SiF$_6$ and characterized it by IR spectrum and thermal analysis. Fleck et al. (2013b) determined the structure of (L-LeuH)$_2$SiF$_6$·2H$_2$O. The asymmetric unit contains one formula unit. Fluorine atoms of SiF$_6^{2-}$ are ordered due to hydrogen bonds. The carboxyl group of cation "A" forms O1A-H1A...F6(2.617(3) Å) hydrogen bond with anion, while the carboxyl group of cation "B" with one of the water molecules, O1B-H1B...O1W (2.583(4) Å). Water molecule 1 W as proton donor form hydrogen bonds with anion,

O1W-H11w…F3(2.755(3) Å) and O1W-H12W…F4(2.738(4) Å). Water molecule 2 W forms one hydrogen bond with anion, O2W-H21W…F4(2.811(4) Å), and one with the carbonyl oxygen atom of cation "B", O2W-H22W…O2B(2.898(4) Å). As an acceptor 2 W molecule forms hydrogen bond with NH_3^+ group, N1B-H13B…O2W(2.759(5) Å). Two other hydrogen bonds of the NH_3^+ group of cation "B" are formed with anion, N1B-H11B…F2 and N1B-H12B…F5. The NH_3^+ group of cation "A" forms one hydrogen bond with the carbonyl oxygen atom of nearest cation "A", N1A-H13A…O2A, and two hydrogen bonds with anion, N1A-H11A…F3 and N1A-H12A…F1.

Interaction of amino acids with various compounds is used for resolving of stereoisomers. In several papers (Kimoto et al. 1989; Hasegawa et al. 1998; Yoshioka et al. 1998) salts of L- and D-leucine with different arenesulfonates were studied with this purpose. L-leucinium benzenesulfonate monohydrate was obtained from aqueous solution containing equimolar quantities of L-leucine and benzenesulfonic acid. The asymmetric unit contains one formula unit. The carboxyl group of cation forms hydrogen bond O5-H6…O1(2.820(7) Å). The NH_3^+ group forms hydrogen bond with anion, N1-H2…O2(2.836(5) Å); with carbonyl oxygen atom of nearest cation, N1-H3…O4(2.868(5) Å); and with water molecule N1-H1…O6(2.850(5) Å). Water molecule as proton donor forms hydrogen bonds, O6-H21…O2(2.836(5) Å) and O6-H20…O3(2.915(8) Å). In the table of hydrogen bonds (Kimoto et al. 1989) the labeling of some atoms is not correct.

Hasegawa et al. (1998) synthesized salts of L-leucine with aromatic sulfonic acid. The crystal structures of L-leucinium 3,4-dimethylbenzenesulfonate monohydrate (Leu·*o*-DMBS·H_2O), 2,5-dimethylbenzenesulfonate monohydrate (Leu·*p*-DMBS·H_2O) and L-leucinium 4-methylbenzenesulfonate monohydrate (Leu·*p*-TS·H_2O) were determined. All salts were obtained from an aqueous solution containing equimolar quantities of components. Hydrogen bond schemes are similar. Carboxyl groups form hydrogen bonds O-H…O with anion. The shortest bond with O…O (2.648 Å) is formed in the case of (L-LeuH) *p*-DMBS·H_2O. The longest bond is formed in the structure of (L-LeuH) p-TS·H_2O with O…O(2.926 Å). The O…O(2.887 Å) bond in the structure of (L-LeuH) *o*-DMBS·H_2O is intermediate. Water molecule as a proton donor forms two hydrogen bonds with two nearest anions and also hydrogen bond N-H…O as an acceptor with the NH_3^+ group. The remaining two hydrogen atoms of the NH_3^+ group form hydrogen bonds with anion and carbonyl oxygen atom of nearest cation.

The (−)-1-phenylethanesulfonic acid (PES) is another efficient resolving agent. Yoshioka et al. (1998) determined structures of three salts of leucine with PES. From acetonitrile–methanol solution containing equimolar quantities of DL-leucine and PES initially anhydrous L-Leu·(−)-PES crystallizes as a less soluble salt, and then more soluble D-Leu·(−)-PES salt is crystallized. If as solvent a mixture of acetonitrile–water is used, then from solution containing equimolar quantities of DL-leucine and PES initially D-Leu·(−)-PES·H_2O is crystallized and then the salt L-Leu·(−)-PES. Attempts to obtain crystal hydrate of L-Leu·(−)-PES at high-humidity condition were not successful, while D-Leu·(−)-PES was easily monohydrated in a 75 % humidity chamber. The salt D-Leu·(−)-PES·H_2O is dehydrated in the 60–90 °C interval. The formed anhydrous product is identical to the salt D-Leu·(−)-PES obtained from solution of acetonitrile–methanol. The D-Leu·(−)-PES melts at 171.5 °C. The crystal of L-Leu·(−)-PES melts at 221.2 °C. At further heating both salts decompose. The structures of L-Leu·(−)-PES and D-Leu·(−)-PES are different. In the structure of L-Leu·(−)-PES the carboxyl group of cation forms O5′-H...O3(2.758(4) Å) hydrogen bond with anion and the NH_3^+ group forms three hydrogen bonds with three nearest anions: N1-H...O3(2.907 (5) Å), N1-H...O4(2.831(5) Å), and N1-H...O4(2.874(4) Å). In the structure of D-Leu·(−)-PES the carboxyl group of cation does not form a hydrogen bond with anion. The (D-LeuH) cations form infinite chain via O5′-H...O4′(2.668(13) Å) hydrogen bond between the hydroxyl group of one cation and the carbonyl oxygen atom of nearest cation and so on. From Fig. 3b of Yoshioka et al. (1998) one may be impressed that cations form dimers via O-H...O and O...H-O hydrogen bonds: however, actually as was mentioned above they form infinite chain. The NH_3^+ group forms three hydrogen bonds with three nearest anions: N1-H...O2(2.923 (7) Å), N1-H...O3(2.836(11) Å), and N1-H...O4(2.821(13) Å). In the structure of D-Leu·(−)-PES·H_2O the carboxyl group of cation forms hydrogen bond with anion, O5′-H...O4(2.633(4) Å). Water molecule as a donor forms hydrogen bonds with two nearest anions: O1″-H...O3(2.890(5) Å) and O1″-H...O4(3.055(5) Å). As an acceptor water molecule forms hydrogen bond with the NH_3^+ group N1-H...O1″ (2.883(5) Å). One more hydrogen bond the NH_3^+ group forms with anion N1-H...O3(2.810(4) Å). The third hydrogen atom of the NH_3^+ group has two contacts with the O4 atom of anion and O4′ carbonyl oxygen atom of nearest cation on the level of van der Waals bonds.

5.1.15 Salts of Isoleucine

In Table 5.15 known salts of isoleucine are provided. The crystals of (D-IleH)Cl·H_2O (I) and (D-IleH)Br·H_2O (II) were obtained by adding concentrated hydrochloric or hydrobromic acid to a suspension in acetone of pure D(−)-isoleucine and allowing the solution to evaporate slowly over soda lime in vacuum. On exposure to air they lose slowly their crystal water. In spite of the same symmetry and similarity of the unit cell parameters crystal structure determination showed that these crystals are not isomorphous. The conformations of (D-IleH) cations are different. Varughese and Srinivasan (1976a, b) labeled the type of (D-IleH)Cl·H_2O as type (I) and that of (D-IleH)Br·H_2O as (II). Trommel and Bijvoet (1954) found that crystals of (D-IleH) Cl·H_2O obtained from a 96 % alcoholic solution are isomorphous with the (D-IleH) Br·H_2O, i.e., is of type (II). As a result of recrystallization (D-IleH)Cl·H_2O (II) transforms into (D-IleH)Cl·H_2O (I). Varughese and Srinivasan (1976a) unexpectedly obtained (L-IleH)Cl·H_2O (II) when they tried to obtain hydrochloride of *t*-BOC-L-isoleucine. The crystals of (L-IleH)Cl·H_2O (II) were obtained by dissolving

Table 5.15 Salts of isoleucine with inorganic anions

#	Crystal	s.g., Z	a, b, c (Å), α, β, γ (°)	Refs
1	(D-IleH)Cl·H_2O (I)	$P2_12_12_1$, 4	6.13(2), 25.01(8), 6.79(2)	Trommel and Bijvoet (1954)
2	(D-IleH)Br·H_2O (II)	$P2_12_12_1$, 4	6.21(4), 24.4(2), 7.00(4)	Trommel and Bijvoet (1954)
3	(L-IleH)Cl·H_2O (II)	$P2_12_12_1$, 4	5.87(1), 24.77(2), 6.85(1)	Varughese and Srinivasan (1976a)
4	(D-alloIleH)Cl·H_2O	$P2_12_12_1$, 4	6.13(1), 24.03(2), 6.80(1)	Varughese and Srinivasan (1976b)
5	(L-IleH)$_3$(SO_4HSO_4)	$P2_12_12_1$, 4	6.743(5), 19.418(5), 22.095(5)	Sridhar et al. (2001b)
6	(L-IleH)$_2$$SiF_6$	–	–	Ennan et al. (1983)
7	(L-IleH)$_2$$SiF_6$·2$H_2O$	$P2_1$, 2	13.555(3), 5.817(1), 13.860(3), β = 105.11(3)	Fleck et al. (2013b)

some *t*-BOC-L-isoleucine in dilute hydrochloric acid by slow evaporation. It is interesting to note that (D-IleH)Cl·H_2O obtained from a solution of D-isoleucine and dilute hydrochloric acid by slow evaporation was type (II). Varughese and Srinivasan (1976b) also determined the structure of (D-*allo*-IleH)Cl·H_2O, which was obtained similarly from D-*allo*-Ile and dilute hydrochloric acid by slow evaporation. In contrast to expectation the crystal of (D-*allo*-IleH)Cl·H_2O proved to be not isomorphous with (D-IleH)Cl·H_2O; however, it has much in common. Carboxyl group in both cases forms O-H...Cl hydrogen bond with chloride ion with O...Cl(2.996 Å) for (L-IleH)Cl·H_2O (II) and O...Cl(2.992 Å) for (D-*allo*-IleH) Cl·H_2O. Water molecule in both cases forms hydrogen bonds with anion with distances O...Cl equal to 3.163 Å and 3.152 Å in the first case and 3.177 Å and 3.150 Å in the second case. The NH_3^+ group forms one N-H...Cl hydrogen bond with chloride anion and two N-H...O hydrogen bonds with two nearest water molecules in both structures.

The next salt of L-isoleucine in Table 5.15 (L-IleH)$_3$(SO_4HSO_4) has unusual composition and structure. The crystal of (L-IleH)$_3$(SO_4HSO_4) may be obtained from aqueous solution containing L-isoleucine and H_2SO_4 in 3:2 molar ratio. The asymmetric unit contains one formula unit, that is, three (L-IleH) cations, sulfate anion (SO_4^{2-}), and hydrogen sulfate anion (HSO_4^-). Sulfate and hydrogen sulfate anions are connected with each other by short hydrogen bond O2-H2...O6(2.502 (7) Å). All three (L-IleH) cations form hydrogen bonds with anions, but not so short: O11-H...O1(2.671(6) Å), O21-H21...O8(2.631(6) Å), and O31-H31...O7(2.588 (7) Å). The NH_3^+ groups of three (L-IleH) cations form N-H...O hydrogen bonds, out of which 8 bonds are formed with oxygen atoms of (SO_4^{2-}) and (HSO_4^-) anions, and only one bond N3-H3C...O32 is formed with the carbonyl oxygen atom of third cation.

Ennan et al. (1983) obtained the salt (L-IleH)$_2$$SiF_6$ and characterized it by IR spectroscopy and thermal analysis. The salt (L-IleH)$_2$$SiF_6$ is characterized by the presence of two absorption peaks in the region of ν(C=O) (1,755 and 1,740 cm^{-1}) and one band in the region of ν_3(Si-F) at 728 cm^{-1}. The authors dried the samples

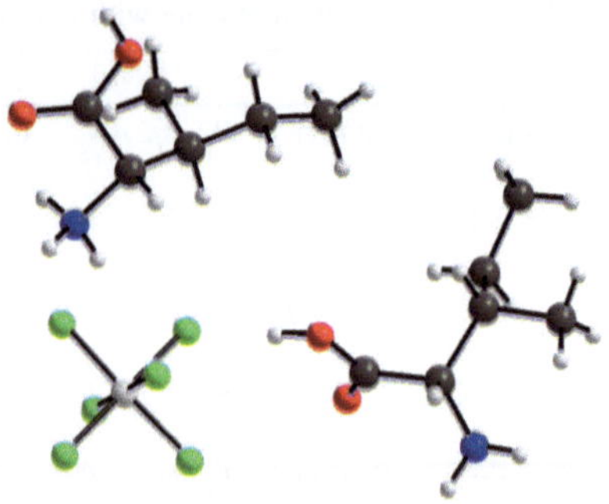

Fig. 5.21 Molecular structure of bis (L-isoleucinium) hexafluorosilicate dihydrate. Water molecules are omitted for clarity

above P_2O_5 to the constant mass. Fleck et al. (2013b) obtained single crystals of (L-IleH)$_2$SiF$_6$·2H$_2$O from aqueous solution containing stoichiometric quantities of components and determined the structure. The asymmetric unit contains one formula unit (Fig. 5.21). The carboxyl group of cation "A" forms O1A-H1A...F4 (2.625(3)). The carboxyl group of cation "B" forms hydrogen bond O1B-H1B...O1W(2.572(3) Å) with one of the water molecules. The second water molecule as an acceptor forms a hydrogen bond with the first water molecule, O1W-H11W...O2W. As proton donor the second water molecule forms a hydrogen bond with anion as well as the second hydrogen atom of the first water molecule. The NH_3^+ group of cation "A" forms one hydrogen bond with the carbonyl oxygen atom of cation "B" and two hydrogen bonds with anion. Similarly, the NH_3^+ group of cation "B" forms one hydrogen bond with the carbonyl oxygen atom of cation "A" and two hydrogen bonds with anion.

At molar ratio 1:2 of L-isoleucine and H_2SiF_6 from aqueous solution, anhydrous salt (L-IleH)$_2$SiF$_6$ is formed. IR and Raman spectra of anhydrous salt are shown in Fig. 5.22. Absorption peaks of ν(C=O) (1,726 and 1,715 cm^{-1}) are significantly lower than in the case of dried sample (Ennan et al. 1983). In addition, strong band of ν_3(Si-F) is split with peaks at 722 and 691 cm^{-1} in contrast to that work, where one band at 728 cm^{-1} was observed. It is possible that anhydrous sample obtained from solution with 1:2 molar ratio differs from anhydrous sample obtained by drying. In the IR spectrum of dihydrate stretching vibrations of water molecule are revealed at 3,623 and 3,534 cm^{-1}. Absorption band caused by ν(C=O) has peaks at 1,731 and 1,703 cm^{-1}, while the vibration ν_3(Si-F) is revealed as non-split band at 702 cm^{-1}.

5.1.16 Salts of Proline

Known salts of proline are collected in Table 5.16. The crystals of (DL-ProH)Cl were synthesized by Mitsui et al. (1969) by dissolving DL-Pro in an excess of concentrated HCl. Salts were obtained after keeping the solution for several days in a desiccator with $CaCl_2$. The authors note that by the same way they could not

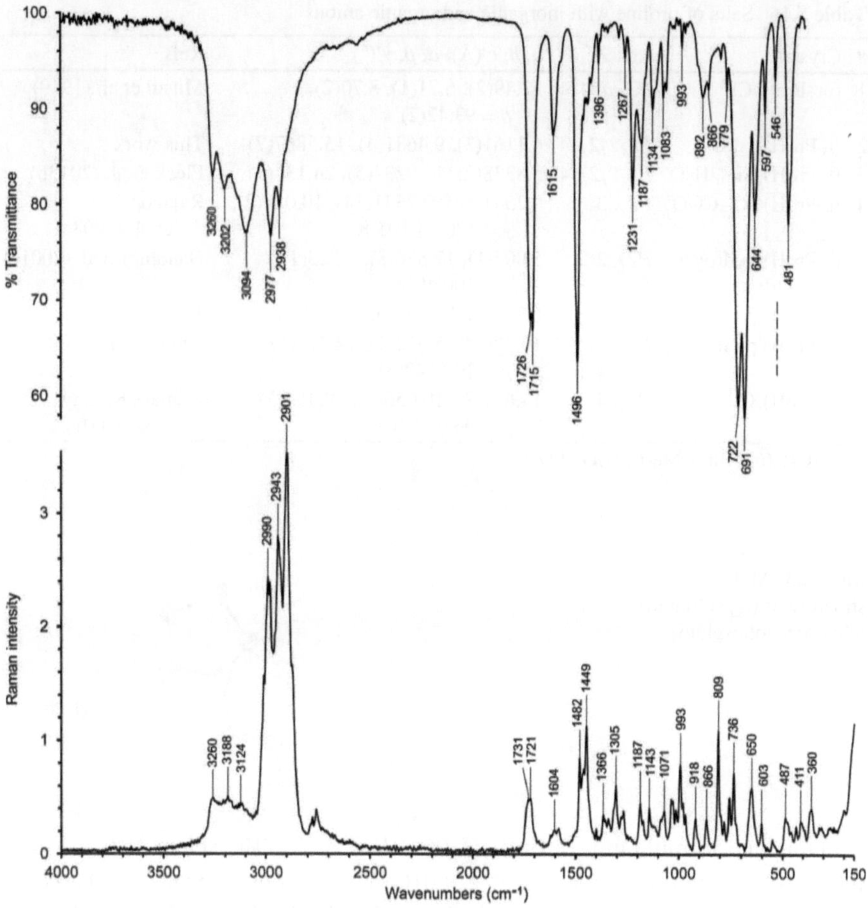

Fig. 5.22 IR and Raman spectra of bis(L-isoleucinium) hexafluorosilicate

obtain crystals with L-Pro. The solution remained viscous during a month. The asymmetric unit of (DL-ProH)Cl contains one formula unit. Positions of hydrogen atoms were not determined; however, bond lengths of C-O bonds show that the O2 atom forms hydroxyl group and forms hydrogen bond O2-H. . .Cl(2.958(7) Å). The NH_2^+ group also forms hydrogen bonds with two nearest chloride ions with N. . .Cl distances 3.135(8) Å) and 3.180(7) Å).

The crystal of (L-ProH)H_2PO_3 was obtained by our group in the course of a search for new salts of amino acids with dimeric (A. . .A^+) type cations. From aqueous solution at 2:1 molar ratio of L-Pro and H_3PO_3 anhydrous L-proline is formed. At molar ratios 1.5:1 and 1:1 of L-Pro and H_3PO_3 identical crystals were formed. The IR spectrum pointed out on the presence of short hydrogen bonds, and we did not rule out the formation of a salt with dimeric cation; however, structure determination showed that its composition is 1:1. The asymmetric unit

Table 5.16 Salts of proline with inorganic and organic anions

#	Crystal	s.g., Z	*a*, *b*, *c* (Å), *α*, *β*, *γ* (°)	Refs
1	(DL-ProH)Cl	$P2_1/a$, 4	12.49(2), 6.71(1), 8.70(2), $\beta = 93.42(2)$	Mitsui et al. (1969)
2	(L-ProH)H_2PO_3	$P2_12_12_1$, 4	6.1161(3), 9.4681(4), 15.5867(7)	This work
3	(L-ProH)$_2SiF_6 \cdot H_2O$	$P2_12_12_1$, 4	5.8338(2), 11.9234(5), 24.134(1)	Fleck et al. (2013b)
4	(L-ProH)CCl_3COO	$P3_1$, 3	9.7541(14), 9.7541(14), 10.089(2), $\gamma = 120$, at 105 K	Rajagopal et al. (2003e)
5	(L-ProH) hydrogen L-tartrate	$P2_1$, 2	5.007(1), 17.676(3), 6.523(1), $\beta = 100.40(2)$	Nandhini et al. (2001e)
6	(L-ProH) picrate	$P2_12_12_1$, 4	6.325(1), 9.269(10, 24.379(4)	Jin et al. (2003)
7	(L-ProH) picrate	$P2_1$, 2	10.909(4), 5.352(3), 12.474(5), $\beta = 109.142(5)$	Anitha et al. (2006a)
8	(L-ProH)X[a]	$P2_1$, 4	14.647(3), 10.056(2), 19.454(3), $\beta = 96.998(10)$	Kumara Swamy et al. (2001)

[a]X = $[CH_2(6\text{-}t\text{-Bu-4-Me-}C_6H_2O)_2PO_2]^-$

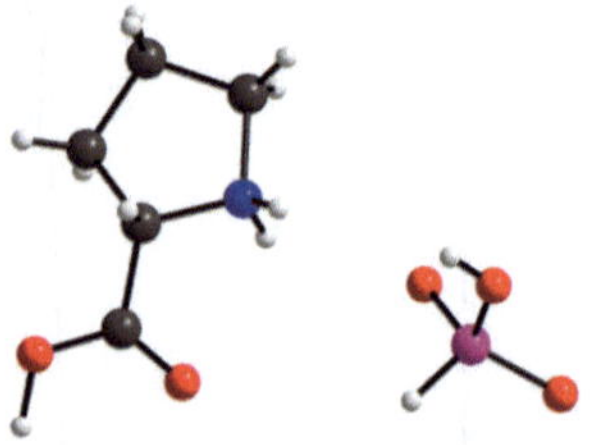

Fig. 5.23 Molecular structure of (L-prolinium) dihydrogenphosphite

contains one formula unit (L-ProH)H_2PO_3 (Fig. 5.23). The anions $H_2PO_3^-$ form infinite chains via the O3-H3...O4(2.5601(15) Å) hydrogen bond. The carboxyl group of cation forms rather strong hydrogen bond O1-H1...O5(2.505(2) Å) with anion. The hydrogen atom H12 of the NH_2^+ group forms relatively short hydrogen bond N1-H12...O4(2.718(2)) with anion, while the atom H11 has three contacts: one is intramolecular with carbonyl oxygen atom O2 with distance H11...O2(2.31 Å); the second also with O2 atom, but intermolecular with nearest cation with distance H11...O2(2.24 Å); and the third contact with O5 atom of anion H11...O5(2.36 Å). All these three contacts are in the 2.15–2.45 Å intermediate interval between weak hydrogen bonds and strong van der Waals bonds.

Two salts (L-ProH)$_2SiF_6 \cdot H_2O$ and (L-ProH)(L-Pro...L-ProH)$SiF_6 \cdot H_2O$ were found in the L-Pro + H_2SiF_6 + H_2O system. The second salt is considered in Sect. 5.2. The asymmetric unit of (L-ProH)$_2SiF_6 \cdot H_2O$ contains one formula unit (Fig. 5.24). The carboxyl group of cation "A" forms rather strong O1-H1A...O1W (2.501(2) Å) hydrogen bond with water molecule, while the carboxyl group of cation "B" forms hydrogen bond with anion O1B-H1B...F5(2.624(3) Å). Water molecule as a proton donor forms hydrogen bonds with the carbonyl oxygen atom of cation "A" O1W-H1W...O2A(2.750(3) Å) and with anion O1W-H2W...F4 (2.746(3) Å). The NH_2^+ groups of both cations form hydrogen bonds with

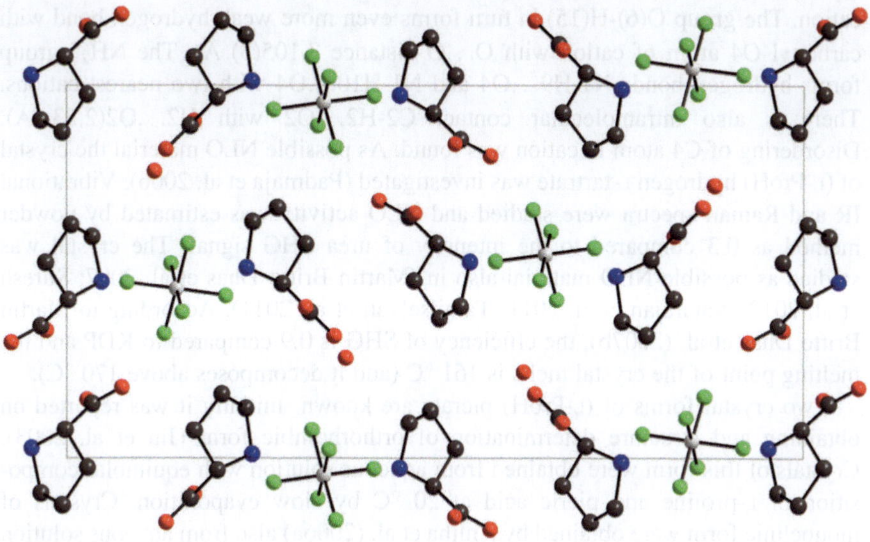

Fig. 5.24 Crystal structure of bis (L-prolinium) hexafluorosilicate monohydrate. Hydrogen atoms are omitted for clarity

nearest anions: N1A-H11A...F2, N1A-H12A...F6, N1B-H11B...F1, and N1B-H12B...F3. In addition, there are intramolecular contacts N1A-H12A...O1A and N1B-H12B...O2B with distances H12A...O1A(2.19 Å) and H12B...O2B (2.24 Å). Thus, all fluorine atoms form hydrogen bonds, which ensure their ordering.

The crystal of (L-ProH)CCl_3COO was obtained from aqueous solution containing equimolar quantities of L-proline and CCl_3COOH. The asymmetric unit of trigonal cell contains one formula unit. The carboxyl group of cation forms O1-H1...O3(2.5775(13) Å) hydrogen bond with anion. The NH_2^+ group forms hydrogen bonds with O4 atom of two nearest anions. There is also intramolecular contact N1-H2...O2 with (H2...O2) distance 2.39 Å. The crystal of (L-ProH)CCl_3COO was grown and characterized by various methods as possible NLO material (Boopathi et al. 2012). The authors assigned the peak at 3,153 cm^{-1} to stretching vibration of O-H bond and concluded that water molecule is present in the structure of crystal, which contradicts to structural data. SHG efficiency is 0.5 compared to the intensity of KDP powder.

Nandhini et al. (2001e) obtained the crystal of (L-ProH) hydrogen L-tartrate from aqueous solution containing stoichiometric quantities of components without indicating the form of tartaric acid. The asymmetric unit contains one formula unit. Anions form head-to-tail chain via O3-H11...O7(2.490(2) Å) hydrogen bond. The carboxyl group of cation forms a hydrogen bond with the second oxygen atom of the carboxylate group of anion O1-H1...O8(2.531(3) Å). The hydroxyl group O(5) H(13) forms weak hydrogen bond O5-H13...O6(2.954(2) Å) with neighboring cation. The group O(6)-H(15) in turn forms even more weak hydrogen bond with carbonyl O4 atom of cation with O...O distance 3.105(3) Å). The NH_2^+ group forms hydrogen bonds N1-H9...O4 and N1-H10...O4 with two nearest cations. There is also intramolecular contact C2-H2...O2 with H2...O2(2.23 Å). Disordering of C4 atom in cation was found. As possible NLO material the crystal of (L-ProH) hydrogen L-tartrate was investigated (Padmaja et al. 2006). Vibrational IR and Raman spectra were studied and NLO activity was estimated by powder method as 0.3 compared to the intensity of urea SHG signal. The crystal was studied as possible NLO material also in (Martin Britto Dhas et al. 2007; Suresh et al. 2010; Natarajan et al. 2011; Tamilselvan et al. 2011). According to Martin Britto Dhas et al. (2007b), the efficiency of SHG is 0.9 compared to KDP and the melting point of the crystal melts is 161 °C (and it decomposes above 170 °C).

Two crystal forms of (L-ProH) picrate are known. Initially it was reported on obtaining and structure determination of orthorhombic form (Jin et al. 2003). Crystals of this form were obtained from aqueous solution with equimolar composition of L-proline and picric acid at 20 °C by slow evaporation. Crystals of monoclinic form were obtained by Anitha et al. (2006a) also from aqueous solution with equimolar composition of L-proline and picric acid at room temperature by slow evaporation. Thus, it is unclear what determines the formation of these two forms, and which form is stable. Structures of these two forms have both similarities and distinctions. Asymmetric units of both forms contain one formula unit. In both cases carboxyl groups form hydrogen bonds O-H...O with negatively charged atom of picrate anion. In the structure of orthorhombic form picrate anion is almost planar, while in the structure of monoclinic form one of the ortho-nitro groups is twisted from the plane of the ring. Cation in the structure of monoclinic form reveals conformational disordering, while in the structure of orthorhombic form ordering was observed. Hydrogen bonds in the structure of monoclinic form are considerably shorter than in the structure of orthorhombic form. The O...O distances are (2.515(4) Å) and (2.658(3) Å), respectively. At the same time the NH_2^+ group forms weak hydrogen bonds N1-H1C...O11 with phenolic oxygen atom with H...O(2.20 Å) and intramolecular contact with carbonyl atom with H...O(2.30 Å), and also N1-H1A...O17 and N1-H1C...O17 with H...O equal to 2.38 Å and 2.37 Å, respectively. In the structure of orthorhombic form the NH_2^+ group forms stronger hydrogen bond with the phenolic oxygen atom of anion N1-H2...O3(2.777 Å) and H...O(2.00 Å), and also weaker hydrogen bond N1-H2...O9 with H...O(2.30 Å). The second hydrogen atom forms both intramolecular bond N1-H3...O2 and intermolecular bond with O2 atom of neighboring cation with H...O 2.29 Å and 2.25 Å, respectively. Monoclinic form was grown and studied as possible NLO material (Martin Britto Dhas et al. 2008a; Uma Devi et al. 2008a). In the first study crystals were obtained from an aqueous solution by evaporation method, while in the second study the crystals were grown by cooling method using as solvent a mixture of 1:1 water and acetone by volume. According to Martin Britto Dhas et al. (2008a), the salt of L-prolinium picrate displays high nonlinearity. SHG efficiency measured by powder method is 74 times higher than in

KDP. Uma Devi et al. (2008a) also found high nonlinearity, 52 times higher than in KDP. As all picrates, L-prolinium picrate has absorption down to 500 nm. According to Uma Devi et al. (2008a), the crystal decomposes near 250 °C. In the IR spectra (Martin Britto Dhas et al. 2008a; Uma Devi et al. 2008a) a broad absorption band is observed in the region 3,400 cm^{-1}, which apparently is caused by the hygroscopicity of KBr and not by absorption of the crystal. In addition, in both papers some absorption peaks are assigned to vibrations of the COO^- group, which is absent in the structure of L-prolinium picrate. Edwin et al. (2010, 2012) studied both experimentally and theoretically vibrational spectra of L-prolinium picrate. In the IR spectrum there is a broad band, which the authors assigned to $\nu_{as}(NH_2^+)$. This band is absent, however, in the Raman spectrum, which points out, in our opinion, that it is caused by water in KBr. The authors assigned the absorption band at 3,235 cm^{-1} to ν(OH) stretching vibration. This assignment, however, is less probable having in view the presence of short O-H…O(2.515 (4) Å) hydrogen bond in the structure of crystal.

For cyclic acid $CH_2(6\text{-}t\text{-}Bu\text{-}4\text{-}Me\text{-}C_6H_2O)_2P(O)OH$ two forms, triclinic and monoclinic, are known. The compound has good solubility in various solvents and is stable against to hydrolysis. Kumara Swamy et al. (2001) obtained various salts, in structures of which very short hydrogen bonds of O-H…O, N-H…O, and C-H…O types are present. The salt of L-proline is obtained from methanol solution containing equimolar quantities of components. Melting point of crystal is 290–292 °C. The values of absorption peaks of IR spectrum and ^1H and ^{31}P NMR signals are provided. The asymmetric unit contains two formula units. The carboxyl groups of cations form hydrogen bonds O9-H78…O4(2.483 Å) and O11-H79…O8(2.487 Å). The NH_2^+ groups form rather short hydrogen bonds with the second oxygen atom of phosphate groups: N1-H77…O3(2.598 Å) and N2-H76…O7(2.631 Å).

5.1.17 Salts of Tryptophan

In Table 5.17 known salts of tryptophan are collected. The distinctive feature of tryptophan is that formally one hydrogen atom of the methyl group of alanine is substituted by an indole group (Stewart 2009). Crystals of (L-TrpH)Cl and (L-TrpH)Br were obtained by Takiagawa et al. (1966) by dissolving L-tryptophan in concentrated hydrochloric and hydrobromic acid, respectively, at room temperature and standing them at 5 °C. By dissolving of L-tryptophan in dilute hydrobromic acid (1 N) a new phase with composition $2\text{L-Trp}\cdot HBr\cdot 1/2H_2O$ was obtained (see Sect. 5.2). The (L-TrpH)Cl and (L-TrpH)Br crystals are isostructural. Asymmetric unit contains one formula unit. The structure of (L-TrpH)Br was determined at room temperature also by Ayyar and Chandrasekharan (1967). Stewart (2009) synthesized crystals of (L-TrpH)Br by dissolving L-tryptophan in concentrated hydrobromic acid followed by slow evaporation at room temperature and then determined the structure at 100 K. Carboxyl groups of cations form hydrogen

Table 5.17 Salts of tryptophan with inorganic and organic anions

#	Crystal	s.g., Z	*a*, *b*, *c* (Å), *α*, *β*, *γ* (°)	Refs
1	(L-TrpH)Cl	$P2_1$, 2	7.45(2), 5.30(1), 14.67(2), $\beta = 98.80(7)$	Takiagawa et al. (1966)
2	(L-TrpH)Br	$P2_1$, 2	7.64(2), 5.41(1), 14.58(2), $\beta = 100.47(10)$	Takiagawa et al. (1966)
3	(L-TrpH)NO_3	–	–	Bouchouit et al. (2005)
4	(L-TrpH)H_2PO_3	$P2_12_12_1$, 4	5.5442(6), 8.3603(4), 27.427(3)	Ramos Silva et al. (2005a)
5	(L-TrpH)$HSeO_3$	$P2_12_12_1$, 4	5.4406(6), 8.2436(6), 27.539(3)	Paixão et al. (2006)
6	$(TrpH)_2SiF_6$	–	–	Andreeva et al. (1983)
7	(DL-TrpH)HCOO	$P2_1/c$, 4	11.237(3), 6.728(2), 16.013(3), $\beta = 96.29(1)$	Bye et al. (1973)
8	(D-TrpH)HC_2O_4	$P2_12_12_1$, 4	5.8773(9), 7.722(1), 28.302(3), at 121 K	Bakke and Mostad (1980)
9	(DL-TrpH) picrate·methanol	$P\bar{1}$, 2	11.733(1), 11.547(1), 7.971(1), $\alpha = 100.34(3)$, $\beta = 81.31(1)$, $\gamma = 97.98(2)$	Gartland et al. (1974)
10	(DL-TrpH) chloranilate·H_2O	$C2/c$, 8	13.1988(5), 8.9661(3), 30.9737 (9), $\beta = 95.401(3)$	Molčanov and Kojić-Prodić (2010)

bonds with chloride and bromide ions, respectively. The NH_3^+ groups also form hydrogen bonds with three nearest chloride and bromide ions, respectively. The N-H group of the indole ring usually does not form a hydrogen bond. In addition to the salt (L-TrpH)Cl another salt (L-Trp...L-TrpH)Cl exists (see Sect. 5.2). In addition to the salts (L-TrpH)Cl (Takiagawa et al. 1966) and 2L-Trp·HCl, the existence of a compound with composition DL-Trp·2HCl with triclinic symmetry (space group $P1$ or $P\bar{1}$ and $Z = 2$) was also reported (Dawson and Mathieson 1951). However, its structure is not determined. Bouchouit et al. (2005) crystallized the salt (L-TrpH)NO_3; however, no structural data were reported.

From aqueous solution of L-tryptophan and H_3PO_3 Ramos Silva et al. (2005a) obtained the crystal of (L-TrpH)H_2PO_3 and determined its structure. Asymmetric unit contains one formula unit. The (L-TrpH) cation via carboxyl group forms very strong hydrogen bond O1-H...O4(2.440 Å) with anion $H_2PO_3^-$. In contrast to phosphite salts of small size amino acids the anions in the structure of given crystal do not form hydrogen bonds with each other. The anion $H_2PO_3^-$ forms a hydrogen bond O3-H...O2(2.713 Å) with the carbonyl oxygen atom. The NH_3^+ group forms three hydrogen bonds: N1...O4, N1...O5, and N1...O1. The strongest bond out of them is N1-H...O5(2.680 Å).

Paixão et al. (2006) obtained one more salt (L-TrpH)$HSeO_3$ from aqueous solution of L-tryptophan and selenious acid. The crystals were investigated by

DSC method in the interval from −170 °C up to +25 °C interval. At temperature −82.5 °C on DSC curves, an anomaly was found both at heating and cooling. For clarification of nature of this anomaly, structures of crystal at room temperature and at 80 K were determined. Asymmetric unit contains one formula unit at both room temperature and 80 K. Bond lengths of the carboxyl group at room temperature C1=O2(1.232(4) Å) and C1-O1(1.278(4) Å) are characteristic for cases of very strong hydrogen bonds. The Se-O bond lengths also are not typical for $HSeO_3^-$ anion. If the bond lengths Se-O5(1.625(3) Å) and Se-O4(1.765(3) Å) are typical for double and single bonds, then the bond length Se-O3(1.711(3) Å) is intermediate between single and double bonds, which is caused by the presence of strong hydrogen bond O1-H1. . .O3(2.482(3) Å). The hydroxyl group of anion forms rather weak hydrogen bond O4-H4. . .O2(2.789(4) Å) with carbonyl oxygen atom of cation. The NH_3^+ group forms one normal hydrogen bond N1-H1B. . .O5(2.806 (4) Å) and two relatively weak hydrogen bonds: N1-H1A. . .O3(2.981(4) Å) and N1-H1C. . .O1(2.945(4) Å). At 80 K the symmetry and number of formula units in cell remained the same. Unit cell parameters changed insignificantly: $a = 5.3910$ (3) Å, $b = 8.2097(4)$ Å, and $c = 27.3216(14)$ Å. According to the interpretation of Paixão et al. (2006) the identified anomaly on DSC curves is caused by the ordering of proton on hydrogen bond O-H. . .O3. The C1-O1 distance became equal to 1.288 (3) Å, which in more extent indicates the presence of the (L-TrpH) cation. However, it should be noted that the bond Se-O3 also increased being equal to 1.7224(16) Å at O1. . .O3(2.474(2) Å). The distance O. . .O (2.800(2) Å) of O4-H4. . .O2 became somewhat longer. Hydrogen bonds of the NH_3^+ group were not changed significantly.

Andreeva et al. (1983) obtained the salt $(TrpH)_2SiF_6$ and characterized it by IR and thermal analysis; however, they did not indicate the form of tryptophan (L-, D- or DL).

The crystals of (DL-TrpH) formate were obtained by diffusion of diethyl ether into a solution of DL-tryptophan in formic acid (Bye et al. 1973). Asymmetric unit contains one formula unit. The carboxyl group forms rather short hydrogen bond O13-H13. . .O18(2.492 Å) with formate anion. The N-H group of the indole group usually does not form a hydrogen bond; however, in given structure there is N1-H1. . .O18(3.003 Å) with the same oxygen atom of formate anion. The bond C16-O18(2.555 Å) is somewhat longer than C16-O17(1.232 Å) possibly because of strong hydrogen bond. The second oxygen atom O17 forms two hydrogen bonds with the NH_3^+ groups of two nearest cations with distances 2.787 Å and 2.811 Å. The third atom of NH_3^+ group forms a hydrogen bond with carbonyl atom N15. . .O14(3.026 Å).

The crystal of (D-TrpH)HC_2O_4 was obtained by Bakke and Mostad (1980) by dissolving DL-Trp in hot aqueous solution of oxalic acid and cooling the solution. Asymmetric unit contains one formula unit (Fig. 5.25). The carboxyl group of (D-TrpH) cation forms hydrogen bond O14-H14. . .O29(2.525(4) Å) with anion. The carboxyl group of the anion forms a hydrogen bond O18-H18. . .O13(2.722(4) Å) to the carbonyl oxygen atom of the cation, i.e., the anions do not form hydrogen bond

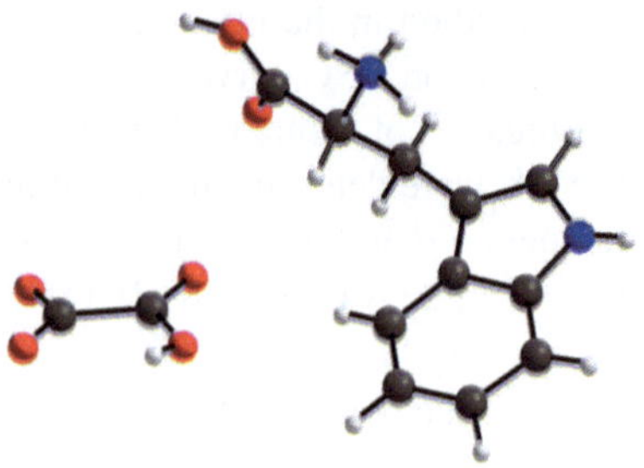

Fig. 5.25 Molecular structure of (L-tryptophanium) hydrogenoxalate

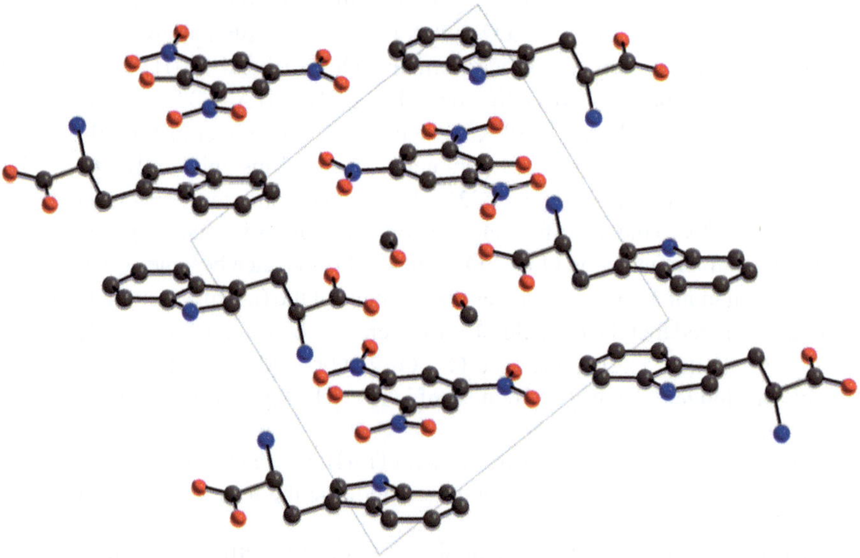

Fig. 5.26 Crystal structure of bis(DL-tryptophanium) picrate·methanol, viewed along [001]. Hydrogen atoms are omitted for clarity

with each other. The NH_3^+ group forms one normal hydrogen bond N15-H…O19 (2.851(4) Å). Other contacts are in intermediate region of weak hydrogen bonds and strong van der Waals bonds.

In the study by Gartland et al. (1974) the structure of crystals of (DL-TrpH) picrate·methanol was determined. Red, needle-like crystals were obtained by slowly evaporating a methanol solution containing equimolar quantities of DL-tryptophan and picric acid. Asymmetric unit contains one formula unit. The molecule of methanol plays an important role in the structure of the crystal. The carboxyl group of cation does not form hydrogen bond with anion. In the structure it is possible to mark out cyclic centrosymmetric grouping consisting of two cations and two methanol molecules, arranged in a layer-like structure (Fig. 5.26). The carboxyl group of cation forms hydrogen bond O12-H…O(2.570 Å) with the oxygen atom of methanol. The hydroxyl group of methanol in turn forms

O-H...O′12(2.795 Å) hydrogen bond with the carbonyl atom of cation. The N1-H1 group of indole ring has a contact with the nitro group of picrate anion N1-H1...O2, which is intermediate between weak hydrogen bonds and strong van der Waals bonds. The NH_3^+ group forms three hydrogen bonds. Two of them are formed with phenolic oxygen atom O1 of two nearest anions and one hydrogen bond is formed with O6 atom of the nitro group of anion.

There are also reports on L-tryptophanium picrate (Ishida et al. 1993; Gowri et al. 2011). These papers are discussed in the paper by Petrosyan et al. (2013b) (see Sect. 5.2).

Molčanov and Kojić-Prodić (2010) discussed the possibility of salt formation in reactions with chloranilic acid as proton donor and proton acceptor (organic base) based on ΔpK_a. In this connection it is worth to note that in addition to the above considered salt (DL-TrpH)HCOO there is also the molecular adduct L-Trp·HCOOH (see Sect. 5.1.24). Molčanov and Kojić-Prodić (2010) particularly obtained crystals of the salt with *R,S*-tryptophan and determined its structure. The asymmetric unit contains (TrpH) cation, singly charged anion, and a water molecule. The carboxyl group of cation forms hydrogen bond O6-H6...O4(2.606(3) Å) with deprotonated oxygen atom O4 of anion. The second hydroxyl group of anion O2-H2 forms hydrogen bond with water molecule, O2-H2...O7(2.694(5) Å). The water molecule accordingly forms two hydrogen bonds: O7-H(A)...O3(2.830(4) Å) and O7-H (B)...O7(2.976(9) Å). It should be noted that at such disposition hydrogen atoms H(B) are at 1.18 Å, which raises doubts. The N2-H2A group of indole ring has a contact N2-H2A...O2 with H2A...O2(2.28(2) Å, which points out that this contact is intermediate between strong van der Waals and weak hydrogen bonds. One of the atoms of the NH_3^+ group forms hydrogen bond N1-H1A...O1(2.798(4) Å) with anion. The atom H1B does not form hydrogen bond, and the H1C atom has two contacts H1C...O3(2.18(4) Å) and H1C...O4(2.15(3) Å).

5.1.18 Salts of Asparagine and Aspartic Acid

In Table 5.18 known salts are collected, where protonated asparagine and aspartic acid act as cations. Crystals of (L-AsnH)NO_3 were obtained from aqueous solution containing equimolar quantities of L-asparagine and HNO_3 by evaporation at room temperature (Aarthy et al. 2005). Triclinic unit cell contains two formula units (Fig. 5.27). Conformations of the cations are different. Carboxyl groups of cations form hydrogen bonds not with anions, but with each other: O1B-H1B...O2C(2.570 (6) Å) and O2B-H2B...O1C(2.490(6) Å). The bond length C21-O2B(1.275(7) Å) is shorter than C11-O1B(1.298(8) Å), which is often observed at strong hydrogen bonds. The NH_3^+ and NH_2 groups form hydrogen bonds with anions and also with carbonyl oxygen atoms of cations.

Crystals of (DL-AsnH)NO_3 were obtained by heating of a mixture of DL-Asn·H_2O and HNO_3 and then by evaporation of the solution at room temperature. Asymmetric

Table 5.18 Salts of asparagine and aspartic acid with inorganic and organic anions

#	Crystal	s.g., Z	a, b, c (Å), α, β, γ (°)	Refs
1	(L-AsnH)NO_3	$P1$, 2	5.357(2), 8.232(3), 10.172(4), $\alpha = 96.88(3)$, $\beta = 104.47(3)$, $\gamma = 102.11(3)$	Aarthy et al. (2005)
2	(DL-AsnH)NO_3	$P2_1/c$, 4	7.923(2), 9.608(2), 10.613(3), $\beta = 107.105(2)$, at 100 K	Slimane et al. (2009)
3	(DL-AsnH)ClO_4	$Pbca$, 8	9.861(5), 10.289(4), 16.700(5), at 100 K	Guenifa et al. (2009)
4	(L-AsnH) hydrogen squarate·0.5H_2O	$P2_12_12$, 4	8.939(2), 19.751(4), 6.060(1)	Kolev et al. (1998b)
5	(L-AsnH) picrate	$P2_1$, 2	10.367(4), 5.1611(7), 13.120(3), $\beta = 93.20(2)$	Anitha et al. (2005a)
6	(L-AspH)Cl	$P1$, 1	5.612(2), 5.647(2), 6.169(2), $\alpha = 114.218(3)$, $\beta = 97.874(4)$, $\gamma = 95.710(4)$	Chapman and Bryce (2007)
7	(DL-AspH)Cl	$P2_1/a$, 4	11.144, 6.314, 11.248, $\beta = 120.20(2)$	Dawson (1977)
8	(DL-AspH)NO_3·H_2O	$P\bar{1}$, 2	5.680(1), 7.331(2), 10.812(1), $\alpha = 76.74(1)$, $\beta = 105.33(1)$, $\gamma = 96.41(1)$	Bahadur and Rajaram (1995b)
9	(DL-AspH)$_2SO_4$	$P2_1/c$, 4	16.774(4), 5.9697(9), 14.5276(18), $\beta = 98.66(2)$	Srinivasan et al. (2001c)
10	(L-AspH)$_2TeBr_6$	$P2_12_12$, 2	8.71(1), 17.62(2), 7.58(1)	Hempel et al. (1977a)
11	(L-AspH)$_2SiF_6$·2H_2O	$C2$, 2	17.9795(8), 5.3803(2), 11.3987(6), $\beta = 125.998(2)$	Fleck et al. (2013b)
12	(DL-AspH)$_2$ oxalate	$P2_1/n$, 2	7.387(1), 5.477(1), 18.522(3), $\beta = 99.36(1)$	Alagar et al. (2003b)

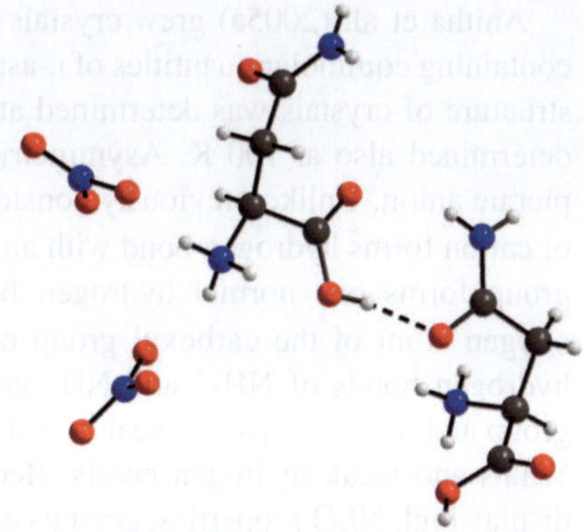

Fig. 5.27 Molecular structure of L-asparaginium nitrate. The O2B-H2B...O1C hydrogen bond is indicated by a *dashed line*

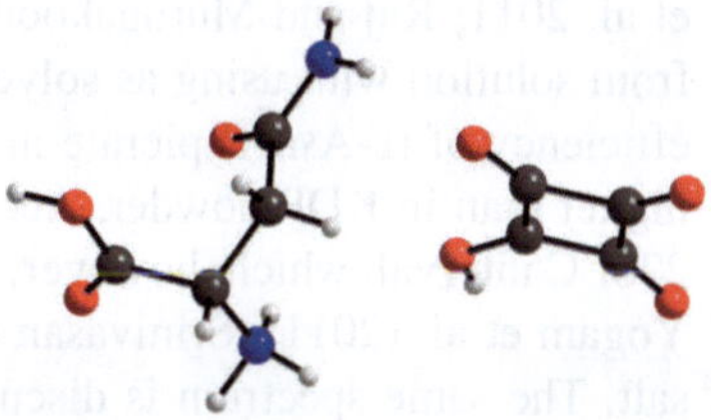

Fig. 5.28 Molecular structure of L-asparaginium hydrogen squarate monohydrate. The water molecule is omitted for clarity

unit contains one formula unit. The carboxyl group as in the case of (L-AsnH)NO_3 forms hydrogen bond with the oxygen atom of amide group O1-H1...O3(2.571 (2) Å), which is close to the value of the first cation in the structure of (L-AsnH)NO_3. The NH_3^+ group forms hydrogen bonds with anions, while the NH_2 group forms one hydrogen bond with the carbonyl atom of carboxyl group.

Crystals of (DL-AsnH)ClO_4 were obtained by Guenifa et al. (2009) from aqueous solution containing equimolar quantities of DL-Asn·H_2O and $HClO_4$ by evaporation at room temperature. Asymmetric unit contains one formula unit. Here also carboxyl group of cation forms hydrogen bond O1-H1...O3 with oxygen atom of amide group of neighboring cation with O...O(2.5485(19) Å). The NH_3^+ group forms hydrogen bonds with three nearest anions. The NH_2 group has contacts with O2 carbonyl atoms on the level of van der Waals bonds.

Kolev et al. (1998b) obtained the crystal of (L-AsnH hydrogen squarate)$_2$·H_2O and determined its structure. The asymmetric unit contains (L-AsnH) cation and hydrogen squarate anion in general position and water molecule in special position (Fig. 5.28). As in previous cases carboxyl group of cation forms O7-H7...O6 (2.551 Å) hydrogen bond with the oxygen atom of the amide group of neighboring cation. Anions form infinite chain via O2-H1...O3(2.620 Å) hydrogen bond. Water molecules serve as bridge between chains of anions forming O8-H11A...O1 (2.701 Å) hydrogen bond. The NH_3^+ group forms two hydrogen bonds N1-H2...O4 and N1-H4...O3 with anion. The NH_2 group forms one hydrogen bond with the carbonyl oxygen atom of carboxyl group N2-H5...O5 and one with anion N2-H6...O4.

Anitha et al. (2005a) grew crystals of (L-AsnH) picrate from aqueous solution containing equimolar quantities of L-asparagine and picric acid by evaporation. The structure of crystals was determined at room temperature. Later the structure was determined also at 100 K. Asymmetric unit contains one (L-AsnH) cation and a picrate anion. Unlike previously considered salts of asparagine the carboxyl group of cation forms hydrogen bond with anion O1B-H1B. . .O1(2.572(2) Å). The NH_3^+ group forms one normal hydrogen bond N11-H11A. . .O1A with the carbonyl oxygen atom of the carboxyl group of nearest cation N. . .O(2.829(2) Å). Other hydrogen bonds of NH_3^+ and NH_2 groups formed with oxygen atoms of amide group and nitro groups are weaker and are in intermediate region of strong van der Waals and weak hydrogen bonds. Because picrates of a number of amino acids display high NLO properties, crystals of (L-AsnH) picrate also attracted attention as possible NLO materials (Thomas et al. 2008; Srinivasan et al. 2006, 2008b; Yogam et al. 2011; Raj and Murugakoothan 2013). Srinivasan et al. (2006) grew crystals from solution with using as solvent a 1:1 mixture of acetone and water. The SHG efficiency of (L-AsnH) picrate measured by powder method was found 66.5 times higher than in KDP powder. According to this study, the melting point is in 223–226 °C interval, which, however, is not in agreement with thermal data provided by Yogam et al. (2011). Srinivasan et al. (2006) also showed the IR spectrum of the salt. The same spectrum is discussed in more detail by Srinivasan et al. (2008b). Yogam et al. (2011) also provide an IR spectrum. Some statements in interpretation of IR spectra in these works are not acceptable, however. Srinivasan et al. (2006, 2008b) assigned the strong band near 3,531 cm^{-1} to stretching vibration of OH group. The OH group of the cation forms an hydrogen bond with picrate anion with O. . .O distance 2.572(2) Å. The expected position of ν(OH) for such hydrogen bond is near the 2,000 cm^{-1} region and hence the band at 3,531 cm^{-1} cannot belong to ν(OH) forming such bond. In our opinion the strong band at 3,531 cm^{-1} is caused by water containing in KBr due to its hygroscopicity. This note concerns also the work of Yogam et al. (2011) where the authors assigned the broad band at 3,434 cm^{-1} to stretching vibration of O-H bond. Srinivasan et al. (2006, 2008b) as well as Yogam et al. (2011) assigned absorption peaks at 1,562 and 1,437 cm^{-1} to vibrations of the COO^- group; meanwhile such group is absent in the structure. Raj and Murugakoothan (2013) studied the influence of cobalt sulfate on morphology and properties of grown crystals.

In Table 5.18 salts aspartic acid are also collected. The aspartium cation contains two carboxyl groups and one NH_3^+ group. So, one may expect that hydrogen bonds scheme, and hence molecular structure of salts with (AspH) cation will significantly differ from salts of (AsnH) cation. In preliminary work (Dawson and Mathieson 1951) the authors reported on obtaining and provided cell parameters and space group of the salt (DL-AspH)Cl·0.5H_2O, however, later this salt was presented as anhydrous (DL-AspH)Cl (Dawson 1977). Later the structure of the salt (L-AspH)Cl was determined and studied by $^{35/37}$Cl solid-state NMR as well. Asymmetric units in both cases contain each one formula unit. The carboxyl group of cation forms hydrogen bond with chloride ion, which in turn is connected with the terminal

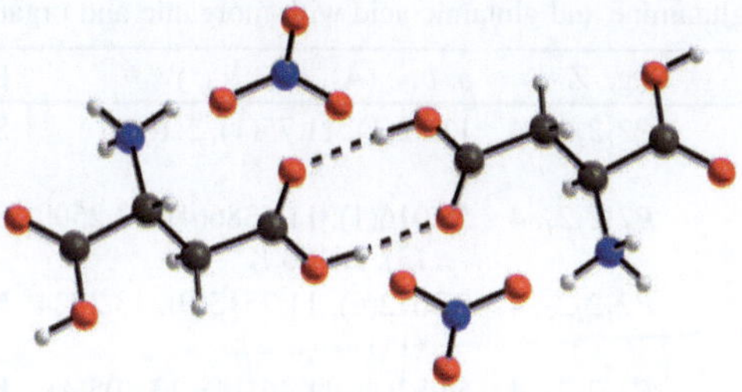

Fig. 5.29 Molecular structure of DL-aspartic acid nitrate monohydrate. Note the formation of a dimeric cation via the O3-H2...O4 hydrogen bonds. This dimeric cation is located around an inversion center. The water molecule is omitted for clarity

carboxyl group of symmetry-related cation and so on. The O...Cl distances for (DL-AspH)Cl are equal to 3.046 Å and 3.060 Å, respectively. The NH_3^+ group forms one hydrogen bond with chloride ion with N...Cl(3.220 Å) and two hydrogen bonds with carbonyl oxygen atoms with N..O equal to 2.878 Å and 2.867 Å. In the structure of (L-AspH)Cl the distances O...Cl are 3.068 Å and 3.096 Å, the N...O are 2.811 Å and 2.833 Å, and for N...Cl is 3.187 Å.

An interesting structure has the crystal of (DL-AspH)$NO_3 \cdot H_2O$ obtained from aqueous solution containing equimolar quantities of DL-aspartic acid and HNO_3 (Bahadur and Rajaram 1995b). Centrosymmetric triclinic cell contains two formula units. The L-AspH and D-AspH cations are connected by center of symmetry and form hydrogen bonds by means of terminal carboxyl groups, O3-H2...O4(2.657 Å) (Fig. 5.29). Thus, a dimeric cation of $A^+ \ldots A^+$ type is formed, which was observed in a number of salts of β-alanine (see Sect. 5.2). The main carboxyl group of (AspH) cation forms hydrogen bond with anion, O1-H1...O11(2.581 Å). The NH_3^+ group forms hydrogen bond N1-H8...OW(2.802 Å) with water molecule, which in turn forms hydrogen bonds with carbonyl atom O2 and O13 atom of anion. The NH_3^+ group has also contacts N1-H6...OW and N1-H7...O12 on the level of strong van der Waals bonds. Vibrational spectra of (DL-AspH)$NO_3 \cdot H_2O$ were studied by Rajkumar et al. (1998).

The crystal of (DL-AspH)$_2SO_4$ was obtained by Srinivasan et al. (2001c) from aqueous solution at 2:1 molar ratio of DL-aspartic acid and H_2SO_4. Asymmetric unit contains one SO_4^{2-} anion and two (L-AspH and D-AspH) cations. Each of four carboxyl groups of two independent cations forms hydrogen bond with one of the oxygen atoms of anion. Thus, sulfate anion is surrounded by four cations by O1B-H1B...O4(2.605(2) Å), O2B-H2B...O3(2.631(2) Å), O1D-H1D...O1 (2.673(2) Å), and O2D-H2D...O2(2.669(2) Å) hydrogen bonds. Six hydrogen atoms of two NH_3^+ groups form four hydrogen bonds with anions and two hydrogen bonds with carbonyl oxygen atoms of terminal carboxyl groups.

Hempel et al. (1977a) reported the unit cell parameters of (L-AspH)$_2TeBr_6$. The value $Z = 2$ points out that the anion is in special position; however, details of hydrogen bonds are not known.

Table 5.19 Salts of glutamine and glutamic acid with inorganic and organic anions

#	Crystal	s.g., Z	a, b, c (Å), α, β, γ (°)	Refs
1	(L-GlnH)Cl[a]	$P2_12_12_1$, 4	13.32(1), 11.75(1), 5.16(1)	Shamala and Venkatesan (1972)
2	(L-GluH)Cl	$P2_12_12_1$, 4	5.1016(1), 11.6386(4), 13.2500 (3), at 153 K	Zhang et al. (2008a)
3	(L-GluH)Br	$P2_12_12_1$, 4	5.3672(6), 11.7515(9), 13.3924 (11)	Natarajan et al. (2008b)
4	(L-GluH)I[b]	$P2_12_12_1$, 4	5.982(1), 11.461(4), 13.405(4)	Kirfel and Wallrafen (1985)
5	$(L\text{-GluH})_2SO_4{\cdot}0.5H_2O$	$P1$, 4	12.536(2), 12.596(2), 13.306 (2), $\alpha = 79.09(1)$, $\beta = 62.05$ (1), $\gamma = 65.88(1)$	Sridhar et al. (2002c)
6	$(L\text{-GluH})_2TeI_6$	$P2_12_12$, 2	8.43(1), 22.05(3), 7.87(1)	Dauter et al. (1977)
7	$(L\text{-GluH})_2SiF_6$	$P2_12_12$, 2	18.5374(7), 5.4721(2), 8.4093 (3)	Ghazaryan et al. (2013a)

[a]In our opinion this crystal actually is (L-GluH)Cl (see the text)
[b]The authors interpreted their results as molecular adduct L-Glu·HI (see the text)

The salt $(L\text{-AspH})_2SiF_6{\cdot}2H_2O$ was obtained by Fleck et al. (2013b) and its crystal structure was determined. Asymmetric unit contains L-aspartium cation and water molecule in general position and SiF_6^{2-} anion in special position, on a twofold axis. The main carboxyl group of the aspartium cation forms hydrogen bond with water molecule O1-H1…O1W(2.569(3) Å). The terminal carboxyl group forms hydrogen bond O3-H3…F2(2.724(2) Å) with anion. The water molecule, in turn, acts as proton donor and forms hydrogen bonds with carbonyl oxygen atom of terminal carboxyl group O1W-H1W…O4 and with anion, O1W-H2W…F3. The NH_3^+ group forms one hydrogen bond with the carbonyl oxygen atom of the main carboxyl group N1-H1A…O2, whereas two bonds extend towards the two nearest anions, i.e., N1-H1B…F4 and N1-H1C…F4.

The salt $(DL\text{-AspH})_2C_2O_4$ was obtained by Alagar et al. (2003b) from aqueous solution containing DL-Asp and oxalic acid in a 1:1 molar ratio. Asymmetric unit contains AspH cation in general position and $C_2O_4^{2-}$ in special position, in center of symmetry. The main carboxyl group of the cation forms a hydrogen bond with anion O1-H1…O6(2.530(3) Å). The terminal carboxyl group also forms a hydrogen bond with anion O3-H3…O5(2.565(2) Å). The NH_3^+ group forms one hydrogen bond with anion, N1-H1C…O6(2.733(3) Å), and also two hydrogen bonds with the carbonyl oxygen atom of two nearest cations, N1-H1A…O4 and N1-H1B…O4. Thus the $C_2O_4^{2-}$ anion is surrounded by six cations.

5.1.19 Salts of Glutamine and Glutamic Acid

In Table 5.19 known salts of glutamine and glutamic acid are collected. It is well known that in acid medium glutamine is easily hydrolyzed into glutamic acid.

There are a number of reports on obtaining of salts of glutamine; however, only for (L-GlnH)Cl (Shamala and Venkatesan 1972) the structure of crystal was determined, although conditions of obtaining the crystal were not provided. From aqueous solution of L-glutamine and HCl it is possible to obtain (L-GluH)Cl instead of (L-GlnH)Cl owing to hydrolysis of L-glutamine in acid medium. The unit cell parameters, space group, and atomic coordinates of nonhydrogen atoms at $R = 0.096$ are provided. In our opinion at this accuracy it is difficult to distinguish the NH_2 group of the amide group of (L-GlnH) from the OH group of the terminal group of the carboxyl group of (L-GluH). Shamala and Venkatesan (1972) noticed that the supposed NH_2 group of the amide group forms only one hydrogen bond. Comparison of unit cell parameters, symmetry, structure, and hydrogen bond scheme allows to conclude that actually the crystal of (L-GluH)Cl was obtained and the crystal structure of (L-GluH)Cl was determined and not that of (L-GlnH)Cl.

Amalanathan et al. (2010) obtained crystals of L-glutamine picrate from aqueous solution containing equimolar quantities of L-glutamine and picric acid by slow evaporation at room temperature during two weeks and then repeatedly recrystallized. The authors show IR and Raman spectra of obtained compound and also results of calculations by DFT method without determination of crystal structure. It is worth to note that IR and Raman spectra of obtained crystal do not contain any bands in the region of 1,700 cm^{-1}.

About one more salt of glutamine was reported by Andreeva et al. (1983). The authors reported on obtaining of glutamine hexafluorosilicate $(GlnH)_2SiF_6$ without pointing out the form of glutamine (L-, D- or DL) and characterized the obtained product by IR spectroscopy and thermal analysis also without structure determination. Using non-aqueous solvent would allow avoiding hydrolysis of glutamine.

The salt (L-GluH)Cl easily grows in the form of large crystals from aqueous solution. Large crystals were obtained from an aqueous solution containing excess of 20 % HCl, and the structure was determined by neutron diffraction (Sequeira et al. 1972). Apparently earlier reports on obtaining and investigating (DL-GluH)Cl and DL-GluH)Br relate to the salts of optical isomers and not to the racemates (see Sequeira et al. 1972). Later the structure of (L-GluH)Cl was determined more accurately by Zhang et al. (2008a) and that of (L-GluH)Br by Natarajan et al. (2008b). Kirfel and Wallrafen (1985) determined the structure of the respective iodide but, however, interpreted obtained results as L-$C_5H_9NO_4 \cdot HI$, that is, as molecular adduct. Actually all three compounds exist as isostructural crystals of the type (L-GluH)X, where X stands for chloride, bromide and iodine anions. Asymmetric units of all three salts contain (L-GluH) and chloride, bromide, and iodide anions, respectively. Cations in all three structures are connected with each other by O1-H1…O3 hydrogen bond, where O3 is the carbonyl oxygen atom of terminal carboxyl group, while O1 is the hydroxyl oxygen atom of the main carboxyl group, and form head-to-tail zigzag chain. In the structures of chloride, bromide, and iodide, the O…O distances make (2.6208(13) Å), (2.658(4) Å), and (2.717(5) Å), respectively. Hydroxyl groups of terminal carboxyl groups form O4-H4…X hydrogen bonds, where X is Cl, Br, I with distances O…Cl(3.0389(9) Å), O…Br(3.206 (3) Å), and O…I(3.410(5) Å), respectively. The NH_3^+ groups form each two hydrogen bonds with respective anions and one with carbonyl O2 atom of the main carboxyl group. The distances N1...O2 make 2.8700(13) Å, 2.841(4) Å, and 2.805(5) Å, respectively, for the structures of chloride, bromide, and iodide. The reason for the interpretation of the structure of (L-GluH)I as L-Glu·HI was the impossibility of finding hydrogen atom nearby O4 atom. However, the presence of iodide ion does not allow safe fixation of hydrogen atom. More trusted are bond lengths of heavy atoms. Bond lengths of C=O and C-OH of terminal carboxyl groups in structures of chloride and bromide make 1.2183(15) Å and 1.3139(14) Å for chloride and 1.198(5) Å and 1.320(5) Å for bromide. Comparison of these bond lengths with distances C5-O3(1.199(6) Å) and C5-O4(1.306(5) Å) for iodide proves that the O4 atom is bonded with the hydrogen atom and the presence of O4-H4...I hydrogen bond.

Several papers (Aleksandrov and Haussühl 1975; Delfino et al. 1976, 1977, 1978) attracted the attention of researchers to the crystals of halogenides of L-glutamic acid as materials with NLO and other physical properties (Recker et al. 1986). Elastic, thermoelastic, electrooptic, piezoelectric properties, thermal expansion, IR spectra, UV–Vis transmittance spectra, etc. were studied. Later, interest in the growth and investigation of (L-GluH)Cl and (L-GluH)Br resumed (Selvaraju et al. 2006b, 2009; Sathyalakshmi et al. 2007; Ganesh et al. 2007; Rathyalakshmi et al. 2009; Natarajan et al. 2008b). According to Selvaraju et al. (2009) the efficiency of SHG of (L-GluH) Cl is 1.2 times higher than in KDP, while in (L-GluH)Br it is 0.25 compared to KDP (Natarajan et al. 2008b). However, thermal stability of (L-GluH)Br is higher than that of (L-GluH)Cl. The onset of decomposition of (L-GluH)Br is at 219 °C (Natarajan et al. 2008b) and 195 °C for (L-GluH)Cl (Selvaraju et al. 2009). Some assignments of absorption peaks in the IR spectra are not acceptable. Particularly, some peaks were assigned to vibrations of the COO^- group, which is absent in structures of (L-GluH)Cl and (L-GluH)Br.

The crystals of $(L\text{-GluH})_2SO_4{\cdot}0.5H_2O$ were obtained by Sridhar et al. (2002c) by evaporation of aqueous solution containing L-glutamic acid and sulfuric acid in 2:1 molar ratio. Triclinic cell contains eight (L-GluH) cations, four sulfate anions, and two water molecules. The attempt of authors to solve the structure in a C-centered monoclinic cell was not successful. Six main carboxyl groups of eight cations form hydrogen bonds with carbonyl oxygen atoms of terminal groups of other cations: O11B-H1B...O21C(2.644(4) Å), O31B-H3B...O51C(2.733(4) Å), O41B-H4B...O71C (2.720(4) Å), O61B-H6B...O41C(2.657(4) Å), O71B-H7B...O11C(2.591(4) Å), and O81B-H8B...O31C(2.625(5) Å). Two other carboxyl groups form hydrogen bonds with water molecules: O21B-H2B...O2W(2.525(5) Å) and O51B-H5B...O1W(2.495(4) Å). As is seen hydrogen bonds with water molecules are the shortest. The remaining two carbonyl atoms of terminal carboxyl groups form hydrogen bonds with water molecules: O1W-H1WB...O61C(2.803(5) Å) and O2W-H2WB...O81C(2.725(6) Å). All hydroxyl groups of terminal carboxyl groups form hydrogen bonds with anions: O11D-H1H...O31(2.585(4) Å), O21D-H2F...O12 (2.555(4) Å), O31D-H3F...O14(2.604(4) Å), O41D-H4F...O32(2.577(4) Å), O51D-H5F...O41(2.595(4) Å), O61D-H6F...O42(2.659(4) Å), and O71D-H7F...O33

(2.625(4) Å). The hydroxyl group of the last cation forms bifurcated weak hydrogen bonds: O81D-H8E...O22(2.958(6) Å) and O81D-H8E...O23(3.081(6) Å). Water molecules also form each one hydrogen bond with anions: O1W-H1WA...O21 (2.732(4) Å) and O2W-H2WA...O11(2.706(5) Å). The NH_3^+ groups form hydrogen bonds mainly with anions, but also with carbonyl atoms of carboxyl groups.

Dauter et al. (1977) obtained the crystals of (L-GluH)$_2$TeI$_6$ and determined their crystallographic data. It is worth to note that in CSD there is CIF (refcode: LGLTEI) of bis(L-Glutaminium) hexaiodotellurium (IV), which, however, actually relates to (L-GluH)$_2$TeI$_6$, as it is evident from the formula $(C_5H_{10}NO_4)_2TeI_6$ provided by Dauter et al. (1977). In a table of the paper apparently there is a misprint. Instead of "L-Glutamic acid," "L-Glutamic and" is written, which possibly was the reason of including in CSD as a salt of L-glutamine. The value $Z=2$ at space group $P2_12_12_1$ points out that TeI_6^{2-} anion is in special position.

The crystals of (L-GluH)$_2$SiF$_6$ were obtained by Ghazaryan et al. (2013a), who also determined the structure. As in the previous case the anion is in special position, on twofold axis. So, asymmetric unit may be expressed as (L-GluH)·0.5SiF$_6$. The main carboxyl group forms hydrogen bond O1-H1...O4(2.683(2) Å) with the carbonyl oxygen atom of the terminal carboxyl group, while the terminal carboxyl group forms hydrogen bond with anion, O3-H3...F2(2.602(2) Å). The NH_3^+ group forms two hydrogen bonds with anion: N1-H11...F1(2.790(2) Å) and N1-H12...F3 (2.914(2) Å), and one weak bond N1-H13...O2(3.019(2) Å) with carbonyl oxygen atom. Previously Andreeva et al. (1983) obtained the salt of glutamic acid with H_2SiF_6 and characterized it by IR spectroscopy and thermal analysis without indication of the form of glutamic acid (L, D, or DL).

5.1.20 Salts of Ornithine

Above we considered salts of amino acids which at protonation could form singly charged cations. Now we will set out to consider salts of amino acids which are capable to form both singly and doubly charged cations. These are salts of ornithine, lysine, arginine, and histidine. The possibility of forming doubly charged cation is caused by the presence of one more amino group in the case of ornithine and lysine, and also guanidine and imidazole groups in the case of arginine and histidine, respectively. In Table 5.20 salts of ornithine are presented.

The structure of (L-OrnH)Cl was determined by Chiba et al. (1967), who showed also that (L-OrnH)Br has similar structure. The structure of (L-OrnH)Cl was determined also by Guha et al. (1969) and redetermined at 100 K (Dittrich et al. 2007). The asymmetric unit contains one (L-OrnH) cation and one chloride anion. The (L-OrnH) cation contains carboxylate COO^- group and two protonated NH_3^+ groups. The N(1)H_3^+ group forms one hydrogen bond with chloride anion, N1-H1...Cl (3.161 Å), and two hydrogen bonds with O1 atom of carboxylate groups of two nearest cations: N1-H2...O1(2.850 Å) and N1-H3...O1(2.859 Å). Second the N(2)

Table 5.20 Salts of ornithine with inorganic and organic anions

#	Crystal	s.g., Z	a, b, c (Å), α, β, γ (°)	Refs
1	(L-OrnH)Cl	$P2_1$, 2	10.005(10), 7.992(6), 5.000(10), $\beta = 96.98(5)$	Chiba et al. (1967)
2	(L-OrnH)Br	$P2_1$, 2	10.27, 8.04, 4.98, $\beta = 97$	Chiba et al. (1967)
3	(DL-OrnH)Br	$P2_1/c$, 4	9.3880(94), 7.901(8), 11.663(13), $\beta = 109.83(25)$	Kalyanaraman and Srinivasan (1971)
4	(L-OrnH)NO_3	$P2_1$, 2	5.1944(6), 7.803(1), 11.050(1), $\beta = 98.75(1)$	Ramaswamy et al. (2002)
5	(DL-OrnH_2)$SeBr_6$	$P1$ or $P\bar{1}$, 2	10.89(2), 7.45(1), 11.84(2), $\alpha = 101.8(3)$, $\beta = 100.5(3)$, $\gamma = 103.3(3)$	Hempel et al. (1977b)
6	(L-OrnH_2)$SO_4 \cdot H_2O$	$P2_12_12_1$, 4	6.1589(2), 9.682(6), 18.714(6)	Ravikumar et al. (2004)
7	(L-OrnH_2)$SO_4 \cdot HF$	$P2_12_12_1$, 4	6.1234(5), 10.655(1), 16.018(2)	Fleck et al. (2014)
8	(L-OrnH) $_2SO_4 \cdot H_2O$	–	–	Kyriakos and Karlheinz (1995)
9	(L-OrnH)$_2SO_4$	–	–	Levintow and Greenstein (1951)
10	(DL-OrnH)$_2SO_4$	–	–	Levintow and Greenstein (1951)
11	(L-OrnH_2)$SiF_6 \cdot H_2O$	$P2_12_12_1$, 4	6.1310(1), 9.8651(2), 19.4339(3)	Ghazaryan et al. (2013a)
12	(L-OrnH)$_2SiF_6$	$P2_1$, 6	9.846(2), 20.641(4), 13.789(3), $\beta = 110.67(3)$	Fleck et al. (2013b)
13	(L-OrnH_2) dipicrate	$P2_1$, 2	11.692(2), 14.058(3), 7.364(1), $\beta = 104.65(2)$	Nagata et al. (1995)
14	(L-OrnH)L-aspartate·0.5H_2O	$C2$, 4	21.858(2), 4.718(1), 18.046(2), $\beta = 137.4(1)$	Salunke and Vijayan (1983)
15	(L-OrnH)D-aspartate·H_2O	$P2_1$, 2	5.118(1), 7.881(2), 16.025(2), $\beta = 91.78(2)$	Soman and Vijayan (1988)

H_3^+ group forms two hydrogen bonds with chloride anion and one with O2 atom: N2-H4…Cl(3.258 Å), N2H5…Cl(3.278 Å), and N2-H6…O2(2.879). Balakrishnan and Ramamurthi (2009) and Senthil et al. (2009b) have grown and characterized the crystal of (L-OrnH)Cl as possible NLO material. IR spectra are provided. The IR spectrum provided by Balakrishnan and Ramamurthi (2009) contains a strong band at 3,442 cm^{-1}, which obviously is caused by water containing in KBr. The crystal is stable up to 260 °C according to Balakrishnan and Ramamurthi (2009) and 240 °C according to Senthil et al. (2009b). The powder SHG efficiency is 1.25 (Balakrishnan and Ramamurthi 2009) or 1.7 (Senthil et al. 2009b) times higher than that of KDP.

The structure of (DL-OrnH)Br was determined as well (Kalyanaraman and Srinivasan 1971; Schaffrin and Trotter 1970). In contrast to (L-OrnH)Cl here the $N(1)H_3^+$ group forms two hydrogen bonds with bromide anion and one with O1, N1…Br(3.46 Å), N1…Br(3.40 Å), and N1…O1(2.86 Å), while the $N(2)H_3^+$ group forms one hydrogen bond with bromide anion and two hydrogen bonds with O1 and O2 atoms: N2…Br(3.33 Å), N2…O1(2.88 Å), and N2…O2(2.86 Å).

The crystals of (L-OrnH)NO_3 were obtained by evaporation of aqueous solution containing equimolar quantities of L-ornithine and HNO_3 (Ramaswamy et al. 2002). Asymmetric unit contains one formula unit. The $N(1)H_3^+$ group forms one hydrogen bond with nitrate anion, N1-H1B…O1(2.818(3) Å), and two with O1B atom of the carboxylate group of two nearest cations: N1-H1A…O1B(2.813(3) Å) and N1-H1C…O1B(2.911(3) Å). The terminal $N(2)H_3^+$ group forms one hydrogen bond with the second atom of the carboxylate group of neighboring cation, N2-H2C…O1A(2.900(3) Å), and two hydrogen bonds with nitrate anion: N2-H2A…O2(2.899(4) Å) and N2-H2B…O3(2.911(4) Å). IR and Raman spectra of (L-OrnH)NO_3 were studied in detail (Ramaswamy et al. 2003a). Hempel et al. (1977b) obtained (DL-$OrnH_2$)$SeBr_6$ crystals and determined their symmetry and unit cell parameters.

The system L-Orn + H_2SO_4 + H_2O is studied not enough. Different salts are formed in the system: $(L\text{-}OrnH)_2SO_4$ (Levintow and Greenstein 1951), $(L\text{-}OrnH)_2SO_4{\cdot}H_2O$ (Kyriakos and Karlheinz 1995), and $(L\text{-}OrnH_2)SO_4{\cdot}H_2O$ (Ravikumar et al. 2004) and according to our preliminary data also $(L\text{-}OrnH_2)(HSO_4)_2$; however, the structure was determined only for $(L\text{-}OrnH_2)SO_4{\cdot}H_2O$ (Ravikumar et al. 2004). The crystal was obtained by evaporation from aqueous solution containing equimolar quantities of L-ornithine and H_2SO_4. Asymmetric unit contains one formula unit. Carboxylic acid of (L-$OrnH_2$) cation forms hydrogen bond with anion, O1B-H1…O1(2.546(7) Å). The $N(1)H_3^+$ group forms the shortest hydrogen bond with water molecule, N1-H1B…OW (2.790(7) Å). Water molecule in turn forms hydrogen bond with anion, OW-H1W1…O4(2.726(7) Å). The contacts of the second hydrogen atom of the water molecule are on the level of strong van der Waals interactions. The $N(1)H_3^+$ group forms one more hydrogen bond with anion, N1-H1C…O3(2.810(7) Å). The contacts of H1A are on the level of strong van der Waals interactions. The $N(2)H_3^+$ group forms hydrogen bonds with three nearest anions: N2-H2A…O3(2.729(7) Å), N2-H2B…O2(2.781(8) Å), and N2-H2C…O2(2.945(10) Å).

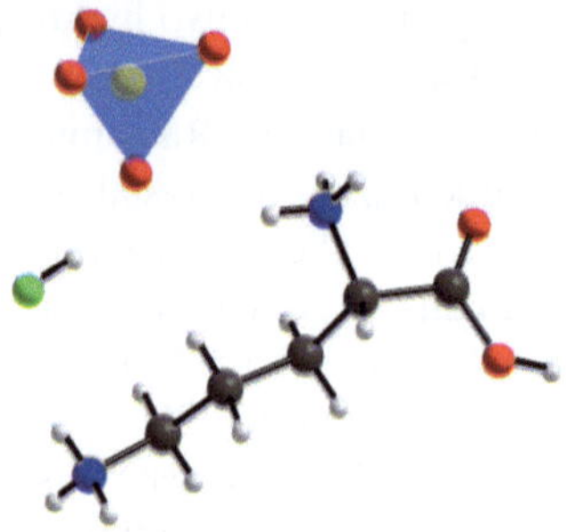

Fig. 5.30 Molecular structure of L-ornithinium sulfate hydrogenfluoride. SO_4 anions are displayed as tetrahedra

The crystal similar by structure with (L-$OrnH_2$)SO_4·H_2O was obtained by us during searching for new mixed salts with different anions. The crystal of (L-$OrnH_2$)SO_4·HF was obtained from aqueous solution containing L-ornithine, HF, HNO_3, and H_2SO_4 in molar ratios 2:1:1:1 and also containing L-ornithine, HF, and H_2SO_4 in molar ratios 2:2:1. Asymmetric unit contains one formula unit (Fig. 5.30). Here the HF molecule plays the role of water molecule in the structure of (L-$OrnH_2$)SO_4·H_2O. The carboxylic group of cation forms hydrogen bond with anion, O1-H1...O13(2.542(4) Å). The $N(1)H_3^+$ group forms two hydrogen bonds with anion: N1-H1B...O14(2.846(4) Å) and N1-H1C...O12(3.027(4) Å), and one with HF molecule, N1-H1A...F1(2.872(5) Å). The $N(2)H_3^+$ group forms two hydrogen bonds with anion, N2-H2A...O11(2.879(4) Å) and N2-H2B...O14 (2.779(4) Å), and one with HF molecule, N2-H2C...F1(3.022(4) Å). The HF molecule forms rather strong hydrogen bond F1-H1F...O12(2.503(4) Å). Thus, the atoms O11, O12, O13, O14 of the anion in the present structure correspond to the atoms O2, O4, O1 and O3 in the structure of (L-$OrnH_2$)SO_4·H_2O. IR and Raman spectra of (L-$OrnH_2$)SO_4·HF are provided and discussed in Fleck et al. (2014). As expected these spectra are similar to that of (L-$OrnH_2$)SO_4·H_2O.

One more crystal which may be compared with (L-$OrnH_2$)SO_4·H_2O is (L-$OrnH_2$) SiF_6·H_2O (Ghazaryan et al. 2013a), in which the role of sulfate anion plays the hexafluorosilicate anion. Here also there are the same symmetry and close unit cell parameters. Asymmetric unit here also contains one formula unit. The carboxylic group of cation also forms a hydrogen bond O1-H1...F6(2.5724(16) Å) with anion. The $N(1)H_3^+$ group also forms one hydrogen bond with water molecule, N1-H1C...O1W(2.7743(18) Å); one with anion, N1-H1A...F1(2.8876(17) Å); and one with carbonyl oxygen atom of neighboring cation, N1-H1B...O2(2.9665 (16) Å). The group $N(2)H_3^+$ group also forms hydrogen bonds with anion, N2-H2A...F3(2.7998(18) Å) and N2-H2B...F2(2.8899(19) Å), and two weaker bonds with H2C atom, N2-H2C...F1(3.0344(18) Å) and N2-H2C...F2(3.0677 (19) Å). Water molecule as proton donor forms one relatively strong hydrogen bond O1W-H1W...F5(2.7039(18) Å) and one relatively weak hydrogen bond O1W-H2W...F1(3.013(2) Å). In the system L-Orn + H_2SiF_6 + H_2O in addition to the salt (L-$OrnH_2$)SiF_6·H_2O with doubly charged cation, another salt with singly charged cation is formed (L-OrnH)$_2SiF_6$ (Fleck et al. 2013b). Asymmetric unit contains three formula units. In two of three anions disordering of some fluorine atoms is observed. Twelve NH_3^+ groups of six cations form 36 hydrogen bonds with carboxylate groups of neighboring cations, and also with anions. The determination of the structure of (L-OrnH)$_2SO_4$ will show whether the structures of (L-OrnH)$_2SiF_6$ and (L-OrnH)$_2SO_4$ are similar.

Several picrates of DL-arginine, L-arginine, L-lysine and L-ornithine were obtained from aqueous solutions containing equimolar quantities of respective amino acids and picric acid (Nagata et al. 1995). In all cases (except L-ornithine) salts with composition 1:1 were obtained. Only in the case of L-ornithine a compound with 1:2 composition was formed. Based on C-O bond lengths (C1-O1(1.172 (6) Å) and C1-O1(1.311(6) Å), Nagata et al. (1995) fairly concluded on the presence of doubly charged (L-OrnH$_2$) cation and two picrate anions in the structure of crystal. The position of hydrogen atom in the COOH group of (L-OrnH$_2$) cation was calculated based on its stereochemistry. According to the authors hydrogen atom occupies *trans*-position in respect to O1 atom. As a result of this it turned out that the COOH group does not form hydrogen bond. In our opinion hydrogen atom occupies cis-position and forms hydrogen bond O1′-H…O6′ with the nitro group with O…O distance 2.868(5) Å. Phenolic oxygen atoms of both picrate anions form each two hydrogen bonds with both NH_3^+ groups: N2-H…O1A(2.789(5) Å) and N6-H…O1A(2.786(5) Å) and N2-H…O1B(2.924(5) Å) and N6-H…O1B(2.767 (5) Å). In addition N(2)H_3^+ and N(6)H_3^+ groups form each one hydrogen bond with nitro group N2…O6′B(3.065(5) Å) and N6…O6′B(2.827(5) Å).

Salunke and Vijayan (1983) and Soman and Vijayan (1988) published the structures of (L-OrnH) L-aspartate·0.5H_2O and (L-OrnH) D-aspartate·H_2O. The first crystal was grown by the slow diffusion of n-propanol into an aqueous solution of the compounds, while in the second case acetone was used as the precipitant. The water molecule in the first case is in special position, while in the second case it is in general position. In spite of that cations and anions have the same charge states; the (L-OrnH) cation consists of two protonated NH_3^+ groups and COO^- carboxylate group, and aspartate anion one NH_3^+ group and two COO^- carboxylate groups. The schemes of hydrogen bonds in these structures have significant distinctions. In the structure of L-L salt water molecule as acceptor does not form hydrogen bonds, while in the structure of L-D salt water molecule forms as acceptor hydrogen bond with NH_3^+ groups of anion, N11-H1…W1(2.955(6) Å), and cation, N6-H1…W1 (2.857(5) Å). The α-amino group of cation in the structure of L-L salt forms hydrogen bonds with oxygen atoms of carboxylate groups of three nearest anions, N1-H1…O12(2.965(3) Å), N1-H2…O12(2.789(3) Å), and N1-H3…O11(2.765 (2) Å), while the N(1)H_3^+ group of cation in the structure of L-D salt forms two hydrogen bonds with carboxylate groups of two nearest cations and one hydrogen bond with anion, N1-H1…O2(2.929(4) Å), N1-H3…O2(2.945(4) Å), and N1-H2…O11(2.844(4) Å). In contrast, the terminal NH_3^+ group of the cation in the structure of L-L salt forms two hydrogen bonds with the carboxylate group of nearest cations and one hydrogen bond with the terminal carboxylate group of anion: N6-H1…O2(2.851(4) Å), N6-H3…O2(2.802(4) Å), and N6-H2…O16 (2.757(3) Å). The terminal NH_3^+ group of the cation in the structure of L-D salt in addition to hydrogen bonds with water molecule forms one hydrogen bond with nearest cation, N6-H3...O1(2.811(4) Å), and one with anion, N6-H2...O12 (2.778 Å). The hydrogen bonds formed by NH_3^+ groups of anions also have significant differences. In the structure of L-L salt, the $N(11)H_3^+$ group forms one hydrogen bond with cation, N11-H3...O1(2.783(2) Å), and two hydrogen bonds with two nearest anions, N11-H1...O15(2.783(5) Å) and N11-H2...O15(2.874 (5) Å). In the structure of L-D salt, the NH_3^+ group of anion in addition to one hydrogen bond with water molecule forms one hydrogen bond with nearest anion, N11-H3...O12(2.624(5) Å), while the third hydrogen atom forms intramolecular hydrogen bond, N11-H2...O15(2.808(5) Å). Water molecules in both structures form hydrogen bonds with anions. In the structure of L-L salt there is O16...H1-W1-H1...O16 hydrogen bond with O...O distance (2.908(5) Å), while in the structure of L-D salt water molecule connects two anions: W1-H1...O16(2.719(5) Å) and W1-H2...O15(2.713(5) Å). In addition to salts provided in Table 5.20 other salts are patented (Chibata et al. 1983), which also are not studied sufficiently.

5.1.21 *Salts of Lysine*

The largest number of salts is known for lysine, arginine, and histidine. Therefore, we separately considered salts of these amino acids with inorganic and organic acids. In Table 5.21 salts of lysine with inorganic acids are provided. The preparation of fluoride of L-lysine is described by Schmid (1968). In a study by Sukiasyan et al. (2014), a detailed investigation of the structure of L-lysinium fluoride was performed and results of vibrational spectra, UV–Vis–NIR transmittance, thermal behavior, and NLO activity estimated by SHG powder method are provided. Quality single crystals of (L-LysH)F were grown. The crystal is stable up to 250 °C and the efficiency of SHG is $I^{2\omega}/I^{2\omega}{}_{KDP} = 0.65$. Asymmetric unit contains one formula unit. The $N(1)H_3^+$ and $N(2)H_3^+$ groups form hydrogen bonds with fluoride anion and also with carboxylate group of symmetry-related cations: N1-H1A...O2 (2.823(2) Å), N1-H1B...F(2.630(2) Å), N1-H1C...F(2.722(2) Å), N2-H2A...O1 (2.745(3) Å), N2-H2B...O2(2.917(2) Å), and N2-H2C...F(2.628(2) Å). From the systems L-Lys + HCl + H_2O and DL-Lys + HCl + H_2O, the salt of monochloride of L-form is crystallized at room temperature in the form of dihydrate, while in the case of racemate in the form of anhydrous salt. It was also known that the anhydrous salt L-Lys·HCl can be obtained by dehydration of the dihydrate salt.

Takayanagi et al. (1981) found that there are two forms of anhydrous L-Lys·HCl. They labeled known form as β-form and new form as α-form. The α-form is formed when dehydration is carried out at 115 °C and higher temperature. According to this work, the new α-form is metastable and at high humidity transforms into β-form. The β-form is formed when dehydration is carried out at lower temperatures (100 °C or lower) under atmospheric pressure or preferably under reduced pressure. The structure of (L-LysH)Cl·$2H_2O$ was determined by X-ray (Wright and Marsh

Table 5.21 Salts of lysine with inorganic anions

#	Crystal	s.g., Z	a, b, c (Å), α, β, γ (°)	Refs
1	(L-LysH)F	$P2_1$, 2	5.4641(11), 7.472(2), 10.252(2), $\beta = 99.17(3)$	Sukiasyan et al. (2014)
2	(L-LysH)Cl (alpha)	–	–	Takayanagi et al. (1981); Petrosyan and Ghazaryan (2009)
3	(L-LysH)Cl (beta)	–	–	Takayanagi et al. (1981); Petrosyan and Ghazaryan (2009)
4	(L-LysH)Cl·$2H_2O$	$P2_1$, 2	7.492(1), 13.320(4), 5.879(1), $\beta = 97.79(1)$	Bugayong et al. (1972); Petrosyan and Ghazaryan (2009)
5	(DL-LysH)Cl	$P2_1/c$, 4	9.157(1), 11.166(2), 8.547(1), $\beta = 105.83(3)$, at 100 K	Flaig et al. (2002)
6	(L-LysH)(IO_3)	–	–	Petrosyan et al. (1999)
7	(L-LysH_2)$(IO_3)_2$	–	–	Petrosyan et al. (1999)
8	(L-LysH_2)$(IO_3)_2$·HIO_3	$P2_1$, 2	8.360(2), 6.330(1), 14.750(3), $\beta = 92.86(3)$	Petrosyan et al. (1999)
9	(L-LysH_2)$(IO_3)_2$·4HIO_3	$P2_12_12_1$, 4	7.366(1), 14.400(3), 22.171(4)	Petrosyan et al. (2000a)
10	(L-LysH)(H_2AsO_3)	–	–	Marchewka et al. (2003a)
11	(DL-LysH)($TiNbO_5$)	$Pnma$, 8	6.487(34), 3.804(6), 47.32(11)	Grandin et al. (1985)
12	(L-LysH_2)SO_4	$P2_12_12_1$, 4	5.573(1), 11.536(1), 16.594(2)	Capasso et al. (1983)
13	(L-LysH_2)$SeBr_6$	$C2$, 8	34.00(4), 7.88(1), 12.62(2), $\beta = 93.0(0)$	Hempel et al. (1977b)
14	(L-LysH_2)SiF_6·H_2O	$P2_12_12_1$, 4	6.1888(3), 10.0473(5), 19.817(1)	Ghazaryan et al. (2013a)
15	(L-LysH_2)($PtCl_6$)	$P2_12_12_1$, 4	6.868(3), 10.111(5), 22.49(2)	L'Haridon et al. (1978)

1962) and neutron diffraction (Koetzle et al. 1972; Bugayong et al. 1972) methods. The symmetry and cell parameters were determined earlier (Srinivasan 1956). Asymmetric unit contains one formula unit. The terminal NH_3^+ group forms one hydrogen bond with chloride anion N...Cl(3.173 Å) and two bonds with the carboxylate group of two nearest cations with N...O distances 2.788 Å and 2.887 Å (Koetzle et al. 1972). The main NH_3^+ group forms one hydrogen bond with chloride anion (N...Cl(3.217 Å)), one with the carboxylate group of nearest cation (N...O(2.795 Å)), and one hydrogen bond with one of the water molecules (N...O(2.824 Å)). This water molecule in turn forms hydrogen bonds with another water molecule (O...O(2.718 Å)) and with anion (O...Cl(3.254 Å)). The second water molecule forms a hydrogen bond with anion (O...Cl(3.220 Å)) and with carboxylate group (O...O(2.805 Å)). Later the crystal of (L-LysH)Cl·2H_2O became a subject for growth and characterization for revealing NLO and other properties (Ramesh Babu et al. 2006a, b, c; Krishnakumar et al. 2008, 2010; Kalaiselvi et al. 2008a; Ozga et al. 2008; Senthil et al. 2009a; Vasudevan et al. 2011a). In apaper by Petrosyan and Ghazaryan (2009) vibrational spectra of (L-LysH) Cl·2H_2O and two anhydrous forms of (L-LysH)Cl were discussed and critically considered earlier published papers. The structure of (DL-LysH)Cl was determined by Bhaduri and Saha (1979) and later more accurately by Flaig et al. (2002). Asymmetric unit contains one formula unit. The terminal NH_3^+ group forms two hydrogen bonds with chloride anion, N2-H13...Cl1(3.1705 (4) Å) and N2-H15...Cl(3.1770(5) Å), and one with carboxylate group of nearest cation, N2-H14...O2(2.8658(6) Å). The main NH_3^+ group forms one hydrogen bond with chloride anion N1-H2...Cl1(3.3818(5) Å) and two hydrogen bonds with carboxylate group of two nearest cations: N1-H1...O2(2.8128(7) Å) and N1-H3...O1(2.8512(7) Å).

Suzuki and Matsuoka (1991) reported the preparation of L-lysine iodate as NLO material; however, its composition was not clear. More detailed investigation of compounds formed in L-Lys + HIO_3 + H_2O system was performed in (Petrosyan et al. 1998, 1999, 2000a; Nalbandyan et al. 2000). The compounds (L-LysH)IO_3 and (L-$LysH_2$)$(IO_3)_2$ were prepared by evaporation of aqueous solution with stoichiometric composition in the form of powder. The presence of water crystallization is not ruled out. The presence of (L-LysH) and (L-$LysH_2$) and also iodate anions is confirmed by IR spectra. The compound (L-$LysH_2$)$(IO_3)_2$ has internal crystalline structure, which is evident from the observation of NQR signals. The salts (L-$LysH_2$)$(IO_3)_2$·HIO_3 and (L-$LysH_2$)$(IO_3)_2$·4HIO_3 were obtained in the form of single crystals, and their crystal structures were determined. In the structure of (L-$LysH_2$)$(IO_3)_2$·HIO_3 one of the IO_3 groups (I(3)O_3) forms short bond O31...I2 with atom I(2) of the HIO_3 group with distance 2.429(6) Å, which is intermediate between the bond of the first coordination sphere (1.77–1.99 Å) and the second coordination sphere (2.6–3.2 Å). This bond leads to elongation of I3-O31 bond (1.845 Å) compared with I-O distances in I(1)O_3 (1.804–1.821 Å). This allows to consider $[I(3)O_3...I(2)O_3H]^-$ as a complex anion $(I_2O_5H)^-$. The HIO_3 group forms hydrogen bond O22-H. . .O21(2.586(8) Å), while the carboxylic group of cation forms hydrogen bond O2-H. . .O32(2.549(8) Å). The NH_3^+ groups also form hydrogen bonds with the IO_3^- groups. The crystal of L-Lys·3HIO_3 is formed from aqueous solution with L-Lys:HIO_3 ratios in the interval from 1:2.5 to 1:4.75. In the interval 1:5–1:6 a mixture of L-Lys·3HIO_3 and L-Lys·6HIO_3 is formed. In the interval 1:6–1:8.5 L-Lys·6HIO_3 is formed, while at higher than 1:8.5 ratio HIO_3 crystal is formed. One might suggest that the crystal of L-Lys·6HIO_3 may be present as (L-LysH_2)$(IO_3)_2$·4HIO_3; however, the IR and ^{127}I NQR spectra are not possible to interpret based on mentioned assumption. Crystal structure determination showed that this salt has a complex and interesting structure. L-lysine as expected presents in the form of doubly charged cation (L-LysH_2), which is balanced by two complex anions: $[HO_3I(5)...OI(1)O_2...I(3)O_3H]^-$ and $[I(6)O_3...H...OI(4)O_2...I(2)O_3H]^-$. Hydrogen atoms were not localized. On their presence one might judge based on bond lengths of nonhydrogen atoms. The presence of long bond I5-O51 (1.894(7) Å) allows to conclude that the I(5) atom forms HIO_3 group. The same may be concluded on I(3) atom based on bond length I3-O31(1.941(7) Å). The atoms I(5) and I(3) form short bonds I5. . .O13(2.498(6) Å) and I3. . .O12(2.403 (6) Å), which have intermediate values between I-O covalent bonds (first coordination sphere) and usually observing I. . .O bonds in the interval 2.6–3.2 Å (second coordination sphere), which allows to consider I(2)O_3 together with HI(5)O_3 and HI (3)O_3 as one complex anion. In the second complex anion there is a short I(2). . . O41(2.491(6) Å). However, the HI(6)O_3 group forms with the I(4)O_3 group not an I. . .O bond, but a strong hydrogen bond O62. . .H. . .O42(2.482(7) Å). Such strong bond reduces the length of I(6)O62 bond up to 1.833(7) Å and respectively increase the bond length I(4)-O42(1.829(6) Å). So, typical HIO_3 groups are HI(2)O_3, HI(3) O_3, and HI(5)O_3 groups, which form weaker hydrogen bonds O21-H. . .O53(2.654 (7) Å), O31-H. . .O1(2.812(7) Å), and O52-H. . .O63(2.620(7) Å). The carboxylic group of cation forms hydrogen bond O2-H. . .O22(2.608(7) Å). The N(1)H_3^+ and N(2)H_3^+ groups have in their surrounding four oxygen atoms, which may be considered as probable hydrogen bonds. Determination of crystal structure allowed interpreting of the IR and NQR spectra (Petrosyan et al. 2000a). The crystals of (L-LysH_2)$(IO_3)_2$·HIO_3 display pronounced NLO properties. So, conditions for growing large crystals were studied in detail (Nalbandyan et al. 2000). Unfortunately, these crystals are not stable. With the lapse of time they became yellow and then (after several years) became black due to liberation of iodine. Other salts of L-lysine also were studied and identified (Sukiasyan 2005). The majority of their structures were not determined because of hygroscopicity and instability of obtained salts.

Another species, (L-LysH)(H_2AsO_3) was reported (Marchewka et al. 2003a). The composition was confirmed by chemical analysis. The IR spectrum is discussed and NLO activity determined by powder method ($d_{eff} = 0.97 d_{eff}$(KDP)).

Intercalation compound of DL-Lys and $HTiNbO_5$ was obtained by Grandin et al. (1985) in the form of powder and unit cell parameters and symmetry are determined from the pattern taken with an X-ray powder diffractometer.

Capasso et al. (1983) obtained crystals of (L-LysH_2)SO_4 from aqueous alcoholic solution of L-lysine and sulfuric acid. Crystal and molecular structure has been determined. Carboxylic group of cation forms hydrogen bond O5-H...O3(2.634 (5) Å) with anion. Conformation of cation has unusual bent nature, which allows forming hydrogen bond of both (the main NH_3^+ and terminal N(Z)H_3^+) groups with the same anion: N-H...O2(2.764(5) Å) and NZ-H...O3(2.909(6) Å). These groups form also each two hydrogen bonds with symmetry-related anions: N-H...O1 (2.774(5) Å) and N-H...O2(2.835(5) Å) and NZ-H...O1(2.894(6) Å) and NZ-H...O4(2.777(7) Å). Petrosyan et al. (2011a) provided and discussed the IR and Raman spectra of (L-LysH_2)SO_4 and indicated an erroneous paper on the synthesis and characterization of this crystal as well.

Three salts of L-lysine with octahedral anions, $SeBr_6^{2-}$ (Hempel et al. 1977b), SiF_6^{2-} (Ghazaryan et al. 2013a), and $PtCl_6^{2-}$ (L'Haridon et al. 1978), are known. Details of the structure of (L-LysH_2)$SeBr_6$ were not determined. For (L-LysH_2) ($PtCl_6$) positions of nonhydrogen atoms are determined. Asymmetric unit contains one formula unit. The carboxylic group of cation forms hydrogen bond O1-H...Cl4 (3.123 Å). Carbonyl oxygen atom forms hydrogen bond with main N(1)H_3^+ group, N1-H...O2(2.781 Å). Two other hydrogen atoms form probably hydrogen bonds with anion: N1...Cl3(3.283 Å) and N1...Cl6(3.275 Å). The structural data of L'Haridon et al. (1978) contains longer contacts of the N(2)H_3^+ group with chlorine atoms. Asymmetric unit of (L-LysH_2)SiF_6·H_2O also contains one formula unit. The carboxylic group of cation forms rather strong hydrogen bond with anion, O1-H1...F6(2.555(3) Å). The main N(1)H_3^+ group forms hydrogen bonds with water molecule, N1-H13...O1W(2.811(3) Å), and also has weak contacts with carbonyl oxygen atom O2 and with F2 and F5 atoms of anion. The terminal N(2) H_3^+ group forms stronger hydrogen bonds with anion: N2-H21...F3(2.935(3) Å), N2-H22...F1(2.858(3) Å), and N2-H23...F5(2.847(3) Å). The water molecule forms hydrogen bonds with anion: O1W-H11W...F2(2.768(3) Å) and O1W-H12W...F5(3.044(3) Å).

More salts of lysine are known with organic acids. They are collected in Table 5.22.

The structures of salts of L- and DL-lysine with simplest organic acid (L-LysH) HCOO and (DL-LysH)HCOO were determined by Suresh and Vijayan (1995a). The crystal is obtained by the slow diffusion of acetonitrile into aqueous solutions of the components. In the structure of (L-LysH)HCOO the N(1)H_3^+ group forms two hydrogen bonds, N1-H1...O1(2.760(2) Å) and N1-H3...O2(2.744(3) Å), with carboxylate groups of two nearest cations and one hydrogen bond with anion, N1-H2...O11(2.897(3) Å). The terminal N(7)H_3^+ group forms two normal hydrogen bonds with two nearest anions: N7-H1...O11(2.823(3) Å) and N7-H2...O12 (2.751(3) Å). The third hydrogen atom of the N(7)H_3^+ group forms two contacts with O11 and O12 atoms of the same third anion, N7-H3...O11(3.081 Å) and N7-H3...O12(3.035 Å) with H...O (2.3 Å) and (2.4 Å), respectively. In the structure of (DL-LysH)HCOO the main N(1)H_3^+ group similarly forms two

Table 5.22 Salts of lysine with organic anions

#	Crystal	s.g., Z	a, b, c (Å), α, β, γ (°)	Refs
1	(L-LysH)HCOO	$P2_1$, 2	5.431(1), 7.546(1), 12.095(2), $\beta=93.42(1)$	Suresh and Vijayan (1995a)
2	(DL-LysH)HCOO	$P2_1/c$, 4	10.205(2), 11.152(2), 8.491(1), $\beta=97.51(1)$	Suresh and Vijayan (1995a)
3	(L-LysH)CH_3COO	$P2_1$, 2	5.411(1), 7.562(1), 12.635(2), $\beta=91.7(1)$	Suresh and Vijayan (1983a)
4	(DL-LysH)CH_3COO	$P\bar{1}$, 2	4.571(2), 7.656(2), 12.841(2), $\alpha=94.48(1)$, $\beta=94.59(2)$, $\gamma=98.83(2)$	Soman et al. (1989)
5	(L-LysH)CF_3COO	$P2_1$, 4	5.6985(2), 23.5430(8), 8.5007(3), $\beta=91.630(2)$	Sun et al. (2008a)
6	(L-LysH)CCl_3COO	$P2_1$, 2	5.7283(2), 9.0412(3), 12.4994(4), $\beta=97.2510(9)$	Raghavalu et al. (2007)
7	(L-LysH)L-aspartate	$P2_1$, 2	5.539(3), 7.848(4), 15.324(15), =99.1(1)	Bhat and Vijayan (1976)
8	(L-LysH) D-aspartate·H_2O	$P2_12_12_1$, 4	5.458(1), 7.152(2), 36.022(5)	Soman et al. (1988)
9	(L-LysH) D-glutamate	$P2_1$, 4	4.902(1), 30.719(2), 9.679(5), 90.00(3)	Soman et al. (1988)
10	(L-LysH) D-pantothenate	$P2_1$, 2	5.883(2), 16.218(5), 10.024(2), $\beta=106.6(2)$	Salunke and Vijayan (1984)
11	(L-LysH) picrate	$P1$, 2	10.367(3), 15.962(6), 5.081(3), $\alpha=99.07(5)$, $\beta=99.13(4)$, $\gamma=74.70(3)$	Nagata et al. (1995)
12	(L-LysH) 4,5-IDC	$P2_1$, 4	6.634(2), 28.629(7), 7.057(2), $\beta=97.53(2)$	Görbitz and Husdal 1998
13	(L-LysH)HSq·H_2O[b]	$P1$, 1	4.900(1), 8.2000(16), 8.7700(18), $\alpha=102.32(3)$, $\beta=101.46(3)$, $\gamma=104.98(3)$	Kolev et al. (2009)
14	(L-LysH_2)$_2$(H_2Sq)(HSq)$_2$(Sq)·2H_2O[b]	$P1$, 1	8.578(3), 10.0295(17), 10.7089(10), $\alpha=77.917(9)$, $\beta=77.649(12)$, $\gamma=74.29(2)$	Kolev et al. (2009)
15	(L-LysH)X(1)[a]	$C2$, 4	32.91(2), 5.91(2), 17.64(2), $\beta=112.52(6)$	Tani et al. (2002)
16	(L-LysH_2)$_4$X(2)·17.5H_2O[a]	$P1$, 1	13.599(4), 14.369(2), 15.169(3), $\alpha=111.02(1)$, $\beta=100.02(2)$, $\gamma=99.02(2)$	Selkti et al. (2000)
17	(DL-LysH_2)(X(3))$_2$·H_2O[a]	$P\bar{1}$, 2	8.282(1), 12.513(2), 17.044(2), $\alpha=72.626(1)$, $\beta=77.647(1)$, $\gamma=84.358(1)$	Ojala et al. (1996)
18	(DL-LysH_2)X(4)·2H_2O[a]	$P\bar{1}$, 2	10.398(2), 15.508(2), 8.704(2), $\alpha=97.16(1)$, $\beta=99.63(1)$, $\gamma=99.75(1)$	Ojala et al. (1996)
19	(L-LysH_2)X(4)·1.5H_2O[a]	$P2_1$, 4	8.973(2), 28.989(2), 10.539(1), $\beta=113.93(1)$	Ojala et al. (1996)
20	(L-LysH)$_3$X(5)·2C_2H_5OH·7H_2O[a]	$P2_1$, 2	10.7934(2), 29.3481(5), 21.8847(4), $\beta=91.974(1)$	Lazar et al. (2006)
21	(L-LysH_2)(HC_2O_4)$_2$·H_2O	$P2_12_12_1$, 4	5.6396(11), 15.125(3), 17.787(4)	Sukiasyan et al. (2008)
22	(L-LysH_2)(HC_2O_4)$_2$	–	–	Sukiasyan et al. (2008)
23	(L-LysH_2)$_2$(HC_2O_4)$_2$(C_2O_4)	$P2_1$, 2	4.906(1), 20.145(4), 12.455(1), $\beta=92.5(1)$	Venkatraman et al. (1997)
24	(DL-LysH)(HC_2O_4)·2H_2O	$P\bar{1}$, 2	5.540(1), 10.764(2), 12.056(2), $\alpha=77.8(1)$, $\beta=80.6(1)$, $\gamma=75.6(1)$	Venkatraman et al. (1997)

(continued)

Table 5.22 (continued)

#	Crystal	s.g., Z	a, b, c (Å), α, β, γ (°)	Refs
25	(L-LysH)($C_4H_5O_4$)	$P1$, 1	5.125(2), 8.087(1), 8.689(1), $\alpha = 112.06(1)$, $\beta = 99.08(2)$, $\gamma = 93.77(2)$	Prasad and Vijayan (1991)
26	(L-LysH)$_4$($C_4H_4O_4$)($C_4H_5O_4$)$_2$·($C_4H_6O_4$)	$P1$, 1	10.143(4), 10.256(2), 12.916(3), $\alpha = 105.00(2)$, $\beta = 99.09(3)$, $\gamma = 92.78(3)$	Prasad and Vijayan (1991)
27	(DL-LysH)$_2$($C_4H_4O_4$)·($C_4H_6O_4$)	$P\bar{1}$, 2	5.506(4), 8.070(2), 14.089(2), $\alpha = 92.02(1)$, $\beta = 100.69(3)$, $\gamma = 95.85(3)$	Prasad and Vijayan (1991)
28	(L-LysH)($C_5H_7O_4$)	$P3_1$, 3	9.044(1), 9.044(1), 14.943(3), $\gamma = 120$	Saraswathi et al. (2001)
29	(DL-LysH)($C_5H_7O_4$)	Ia, 4	10.398(6), 8.662(2), 15.393(6), $\beta = 93.06(2)$	Saraswathi et al. (2001)
30	(L-LysH)($C_6H_9O_4$)	$P2_1$, 2	10.532(3), 7.2834(17), 10.599(3), $\beta = 113.352(3)$	Sharma et al. (2006)
31	(DL-LysH)$_2$($C_6H_8O_4$)	$P\bar{1}$, 1	5.4730(15), 7.773(2), 13.011(4), $\alpha = 100.112(4)$, $\beta = 93.292(4)$, $\gamma = 100.744(4)$	Sharma et al. (2006)
32	(DL-LysH)$_2$($C_7H_{10}O_4$)·0.4C_2H_6O·1.06H_2O	$C2/c$, 4	32.811(17), 11.012(6), 8.552(4), $\beta = 91.333(9)$	Saraswathi et al. (2003)
33	(L-LysH)(DL-$C_4H_4O_5$)	–	–	Marchewka et al. (2003a)
34	(L-LysH)(L-$C_4H_5O_6$)	$P2_1$[c], 2	5.103(1), 17.445(2), 7.538(1), $\beta = 97.94(1)$	Debrus et al. (2005)
35	(L-LysH)(D-$C_4H_5O_6$)	$P2_1$, 2	5.1849(11), 16.667(4), 7.6701(17), $\beta = 96.366(4)$	Selvaraj et al. (2007)
36	(DL-LysH)$_2$(DL-$C_4H_4O_6$)·H_2O	$P2_1/c$, 4	10.087(3), 23.003(8), 9.414(3), $\beta = 90.820(7)$	Selvaraj et al. (2007)
37	(L-LysH) maleate	$P2_1$, 2	9.854(4), 7.136(2), 9.7388(1), $\beta = 115.887(12)$	Pratap et al. (2000)
38	(L-LysH) fumarate	–	–	Marchewka et al. (2003a)

IDC imidazoledicarboxylate, *H_2Sq* squaric acid, *$H_2C_2O_4$* oxalic acid, *$C_4H_6O_4$* succinic acid, *$C_5H_8O_4$* glutaric acid, *$C_6H_{10}O_4$* adipic acid, *$C_7H_{12}O_4$* pimelic acid, *$C_4H_5O_5$* malic acid, *$C_4H_6O_6$* tartaric acid

[a]X(1) = (16S)-9-Deoxy-9β-chloro-15-deoxy-16-hydroxy-17,17-trimethylene-20-norPGF$_2$, X(2) = 2($C_{28}H_{20}O_{16}S_4$), *p*-sulfonatocalix[4]arene, X(3) = 4-((4-hydroxyphenyl)azo)benzenesulfonate, X(4) = 7-oxo-8-(phenylhydrazo)-7,8-dihydro-1,3-naphthalenedisulfonate, X(5) = 4($C_{28}H_{22}O_{10}P_2$)

[b]We do not agree with interpretation of hydrogen bond motif proposed by the authors for this species and offered another formula, i.e. (L-LysH)(L-LysH$_2$)(HSq)(Sq)(H_2Sq)$_2$·2H_2O (see text)

[c]Incommensurate phase with super space group (α0γ)

hydrogen bonds with carboxylate groups of two nearest cations, N1-H3...O1(2.843 (3) Å) and N1-H2...O2(2.754(3) Å), and one hydrogen bond with anion, N1-H1...O12(2.775(3) Å), while the terminal $N(7)H_3^+$ group forms two hydrogen bonds with two nearest anions, N7-H1...O11(2.792(3) Å) and N7-H2...O11(2.762 (3) Å), and one with carboxylate group of nearest cation, N7-H3...O2(2.876(3) Å).

The structures of (L-LysH)CH_3OO and (DL-LysH)CH_3COO were determined (Suresh and Vijayan 1983a; Soman et al. 1989). In both cases asymmetric units contain one formula unit. Schemes of hydrogen bonds are similar in both cases. The main $N(1)H_3^+$ groups in both structures form two hydrogen bonds with carboxylate groups of two nearest cations (with N...O distances 2.740 Å and 2.752 Å in (L-LysH)CH_3OO, 2.749 Å and 2.817 Å in (DL-LysH)CH_3COO and each one hydrogen bond with anion (with N...O distances 2.867 Å and 2.758 Å for L- and DL-salts, respectively). The terminal $N(7)H_3^+$ groups in both structures form three hydrogen bonds with three nearest anions. Two of them are formed with the same type oxygen atom of the $N(1)H_3^+$ group, with N...O distances 2.983 Å and 2.998 Å for L-salt and 2.966 Å, 3.051 Å for DL-salt. The third hydrogen atom of $N(7)H_3^+$ groups forms a hydrogen bond with the second type of oxygen atom of anion with N...O distances 2.762 Å for L-salt and 2.764 Å for DL-salt. Non-centrosymmetric salt (L-LysH)CH_3OO attracted the attention of researchers (Sun et al. 2008b; Neelam et al. 2013) as possible NLO material. Sun et al. (2008b) grew single crystals of (L-LysH)CH_3OO from aqueous solution by evaporation at 40 °C. The solubility curve was determined, IR and Raman spectra registered and discussed, and transmittance spectrum in UV–Vis region, thermal behavior, and the presence of SHG confirmed by powder method. Neelam et al. (2013) reported crystals grown by evaporation method at 34 °C and measured different characteristics. In addition to (L-LysH)CH_3OO, crystals of (L-LysH)CCl_3OO and (L-LysH)CF_3OO also were studied. The crystal of (L-LysH)CCl_3OO (Raghavalu et al. 2007) was obtained by evaporation at room temperature from aqueous solution containing L-lysine and trichloroacetic acid in equimolar ratio. Asymmetric unit contains one formula unit. The main $N(1)H_3^+$ group forms one hydrogen bond with anion N1-H1A...O4 (2.905(2) Å) and one with carboxylate group of nearest cation, N1-H1C...O1 (2.823(2) Å). It is interesting to note that this second cation with its terminal N(2) H_3^+ group in turn forms hydrogen bond N2-H2A...O2(2.894(2) Å) with carboxylate group of the first cation. The remaining two hydrogen atoms form hydrogen bonds with nearest anions, N2-H2B...O3(2.756(2) Å) and N2-H2C...O4(2.881 (2) Å). The $N(1)H_3^+$ group has also a contact N1-H1B...O2 on the level of van der Waals interactions with H1B...O2(2.31 Å). The crystal of (L-LysH)CF_3OO was obtained by Mathivanan et al. (2008) from aqueous solution containing equimolar quantities of L-lysine and CF_3COOH. The symmetry and structure were determined as well as IR and UV–Vis spectra, thermal behavior, and NLO activity ($d_{eff} = 0.96 \cdot d_{eff}$(KDP)) was estimated. However, some statements of Mathivanan et al. (2008) on interpretation of structure and IR spectrum are not acceptable. Asymmetric unit contains two formula units. Fluorine atoms of one of the anions are disordered. The authors assumed that instead of the usual structure of singly charged cation

$^{+}H_3N\text{-}(CH_2)_4\text{-}CH(NH_3^+)COO^-$ the L-LysH cation has the following structure $H_2N\text{-}(CH_2)_4\text{-}CH(NH_3^+)COOH$. The bond lengths C-O in L-LysH cations are equal to C6-O1(1.237 Å) and C6-O2(1.249 Å) and C12-O3(1.253 Å) and C12-O4 (1.256 Å). These bond lengths are characteristic for COO^- groups. The authors decided that O2 and O4 are hydroxyl groups. The authors interpreted the contacts N2…O2 and N4…O4 as O2-H…N2 and O4-H…N4 not as N2-H…O2 and N4-H…O4 hydrogen bonds, which is very improbable, considering that the nitrogen atoms N2 and N4 have positive charges. The presence of absorption band at 1,710 cm^{-1} probably served as a basis for such interpretation. Possibly this band is caused by short C14-O5(1.229 Å) and C14-O6(1.222 Å) bonds in anion $C(13)F_3\text{-}C(14)OO^-$. In the IR spectrum the authors ignored more pronounced peaks at 3,870 and 3,750 cm^{-1}, but interpreted a peak at 3,620 cm^{-1} being the result of noise as O-H stretching vibration. It is difficult to agree with assignment of weak peak at 1,350 cm^{-1} to C-F stretching vibration as well as assignment of peaks at 2,337 and 2,364 cm^{-1} to symmetric stretching vibration of NH_2 groups, which in our opinion are caused by the absorption of atmospheric CO_2 gas. The crystal of (L-LysH) CF_3OO was obtained also by Sun et al. (2008a) and the structure was determined. Sun et al. (2009a, b) provided the results of further investigation of the crystal and conditions of its crystallization. According to the model by Sun et al. (2008a, 2009a)] L-LysH cations contain each two NH_3^+ groups and each one carboxylate group as it is usual for singly charged cations of lysine. Four NH_3^+ groups of two independent cations form hydrogen bonds with four COO^- groups (two of L-LysH cations and two of anions). The $C(6)OO^-$ group as proton acceptor forms three hydrogen bonds: N4-H4E…O7(2.838(4) Å), N3-H3E…O8(2.866(4) Å), and N1-H1C…O8(2.901(4) Å). The $C(12)OO^-$ group also forms three hydrogen bonds: N3-H3C…O5(3.064(4) Å), N1-H1E…O5(2.870(4) Å), and N2-H2E…O6 (2.853(4) Å). The $C(14)OO^-$ group of anion forms three hydrogen bonds: N1-H1D…O1(2.886 Å), N4-H4C…O1(3.034 Å), and N4-H4C…O2(2.986 Å). The hydrogen atom H4C forms bifurcated hydrogen bonds. The H4C atom has also two weak bonds with O1 and O2. None of these three hydrogen bonds of the $C(14)OO^-$ group are included in the table of hydrogen bonds in Sun et al. (2008a and 2009a). Only the contact N4-H4D…O1 (with H…O(2.63 Å)) is included, which cannot be considered as a hydrogen bond. The $C(15)OO^-$ group of the second anion forms four hydrogen bonds: N4-H4D…O3(2.842(4) Å), N2-H2D…O3(2.800 (4) Å), N3-H3D…O4(2.864(4) Å), and N2-H2C…O4(2.934(4) Å). Sun et al. (2009a) provided and discussed the IR and Raman spectra. Comparison of the IR spectrum shown by Sun et al. (2009a) with that provided by Mathivanan et al. (2008) shows that the band at 3,421 cm^{-1} probably is caused by adsorbed water because of the hygroscopicity of KBr. From thermal curves, Sun et al. (2009a) concluded that before decomposition the sample melts near 215 °C. The authors estimated the hardness of the crystal in Mohs scale in the range from 2.0 to 2.5.

The structure of (L-LysH) L-aspartate was determined by Bhat and Vijayan (1976). The crystals were obtained by recrystallization at room temperature of the compound obtained commercially. Asymmetric unit contains one formula unit. The $N(1)H_3^+$ group of the (L-LysH) cation forms two hydrogen bonds with carboxylate groups of nearest cations N1-H13...O1(2.833(7) Å) and N1-H11...O2(2.730(6) Å) and one hydrogen bond with main carboxylate group of anion, N1-H12...O12 (2.843(7) Å). The terminal $N(7)H_3^+$ group forms two hydrogen bonds with the main carboxylate groups of anions N7-H71...O11(2.789(7) Å) and N7-H73...O12 (2.874(8) Å) and one hydrogen bond with terminal carboxylate group N7-H72...O15(2.795(8) Å). The terminal $N(11)H_3^+$ group of anion forms two normal hydrogen bonds with the main and terminal carboxylate groups of two nearest anions, N11-H113...O11(2.810(7) Å) and N11-H112...O16(2.705(8) Å), and two contacts on the level of strong van der Waals interactions, N11-111...O15 (2.953(8) Å) and N11-H111...O16(2.930(8) Å).

The crystal of (L-LysH) D-aspartate monohydrate was obtained by Soman et al. (1988) by vapor diffusion of n-propanol into aqueous solution containing L-Lys and D-aspartic acid in 1:1 proportion. Soman et al. (1988) solved the structure in space group $P2_12_12_1$ but, however, considered that the crystal has strong pseudo-orthorhombic symmetry and that the true symmetry is monoclinic. Asymmetric unit contains one formula unit. The $N(1)H_3^+$ and $N(7)H_3^+$ groups of cation form hydrogen bonds similar to the previous case. The $N(1)H_3^+$ group forms hydrogen bonds, N1-H2...O1(2.782(3) Å), N1-H1...O2(2.719(3) Å), and N1-H3...O12 (2.959(3) Å), and $N(7)H_3^+$ group forms hydrogen bonds, N7-H2...O12(2.745 (4) Å), N7-H3...O11(2.838(4) Å), and N7-H1...O16(2.797(4) Å). The presence of water molecule, however, significantly alters the scheme of hydrogen bonds of anion. The $N(11)H_3^+$ group forms hydrogen bond with water molecule, N11-H3...W1(2.835(4) Å), which in turn forms hydrogen bonds with two nearest anions W1-H1...O15(2.929(4) Å) and W1-H2...O16(2.793(4) Å). Two other atoms of the $N(11)H_3^+$ group form hydrogen bonds with two nearest anions: N11-H1...O15(2.915(4) Å) and N11-H2...O15(2.914(4) Å).

The crystal of (L-LysH) D-glutamate was also obtained by Soman et al. (1988) by vapor diffusion of methanol into aqueous solution containing L-lysine and D-glutamic acid in 1:1 proportion. Asymmetric unit contains two formula units. Due to the presence of an optically active amino acids the structure is non-centrosymmetric, although it was found that it is actually pseudocentrosymmetric. The structure was determined with lesser accuracy ($R = 0.125$). The $N(1)H_3^+$ group of the first cation forms one hydrogen bond with the carboxylate group of the second cation, N1-H3...O21(2.76(2) Å), and two hydrogen bonds with the second anion, N1-H1...O32(2.86(2) Å) and N1-H2...O32(2.77(2) Å). The $N(7)H_3^+$ group forms hydrogen bonds with the terminal carboxyl groups of anions: N7-H1...O32(2.77 (2) Å), N7-H2...O17(2.84(2) Å), and N7-H3...O17(2.86(2) Å). The $N(21)H_3^+$ group of the second cation similarly forms one hydrogen bond with the carboxylate group of the first cation, N21-H3...O1(2.77(2) Å), and two hydrogen bonds with the first anion: N21-H1...O12(2.86(2) Å) and N21-H2...O12(2.81(2) Å). The $N(11)H_3^+$ group of the first anion forms two hydrogen bonds with the carboxylate groups of two nearest second type cation and one hydrogen bond with the second anion: N11-H2...O22(2.86(2) Å), N11-H3...O22(2.93(2) Å), and N11-H1...O31(2.80 (2) Å). Similarly the $N(31)H_3^+$ group of the second anion forms two hydrogen bonds with the carboxylate groups of first type cation, N31-H2...O2(2.88(2) Å) and N31-H3...O2(2.90(2) Å), and one hydrogen bond with the main carboxylate group of the first anion, N31-H1...O11(2.81(2) Å).

Salunke and Vijayan (1984) obtained the crystal of (L-LysH) D-pantothenate and determined its structure. Pantothenic acid also called vitamin B_5 is the amide between pantoic acid ($OHCH_2C(CH_3)_2CH(OH)COOH$) and β-alanine, $OHCH_2C(CH_3)_2CH(OH)C(O)NHCH_2CH_2COOH$. Free pantothenic acid was prepared by the addition in molar proportion of oxalic acid to an aqueous solution of commercially available hemicalcium salt of pantothenic acid and consequent precipitation of calcium oxalate. L-lysine was then added in molar proportion to the supernatant. This solution on slow evaporation yielded the crystals of the salt. Asymmetric unit contains one formula unit. Pantothenate anion has rather bent conformation, which insures the formation of intramolecular hydrogen bonds: N16-H1...O12(2.853 (4) Å) and N16-H1...O20(2.545(4) Å). In the table of hydrogen bonds given by Salunke and Vijayan (1984) these values are mixed up. The anions form a chain due to hydrogen bond O25-H1...O11(2.654(3) Å). The other hydroxyl group of the anion provides hydrogen bonding of chains with each other, via the hydrogen bond O20-H1...O18(2.821(3) Å). The main $N(1)H_3^+$ group of the cation forms a hydrogen bond with nearest cation via N1-H2...O2(2.714(3) Å) and also with two nearest anions via N1-H1...O25(2.820(3) Å) and N1-H3...O18(2.956(3) Å). The terminal NH_3^+ group (labeled in Fig. 1 of Salunke and Vijayan 1981 as N(6), but in the table of hydrogen bonds (Salunke and Vijayan 1981) as N(7) Å) forms hydrogen bonds with the carboxylate group of nearest cation, N7-H2...O1(2.798(3) Å) (in Salunke and Vijayan 1984, this value in the table of hydrogen bonds indicated as 2.789(3) Å) and also with two carboxylate groups of two nearest anions: N7-H1...O11(2.823 (4) Å) and N7-H3...O12(2.801(4) Å).

Picrates of amino acids used long ago for isolation of amino acids from hydrolyzate (see Robson and Selim 1952, and references on earlier works). Nagata et al. (1995) determined the crystal structure of (L-LysH) picrate. The crystals were obtained from aqueous solution containing equimolar quantities of components. Triclinic cell (space group *P*1) contains two formula units of (L-LysH) picrate. The structure is pseudosymmetric. This means that the properties of the crystal caused by molecular hyperpolarizability will in essentially lesser extent be revealed in macroscopic nonlinearity due to pseudosymmetry. Twelve hydrogen atoms of four NH_3^+ groups of two independent cations will act as hydrogen bond donors. So, it is probable that in addition to two carboxylate groups of cations and two phenolic oxygen atoms also oxygen atoms of nitro groups will be involved. Each of the two main NH_3^+ groups of cation forms one hydrogen bond with phenolic oxygen atom of anions: N(2)LA-H...O(1)PB(2.789(7) Å) and N(2)LB-H...O(1)PA(2.859(7) Å). The rest two hydrogen atoms form hydrogen bonds with carboxylate groups of cations: N(2)LA-H...O(1′)LA(2.845(6) Å) and N(2)LA-

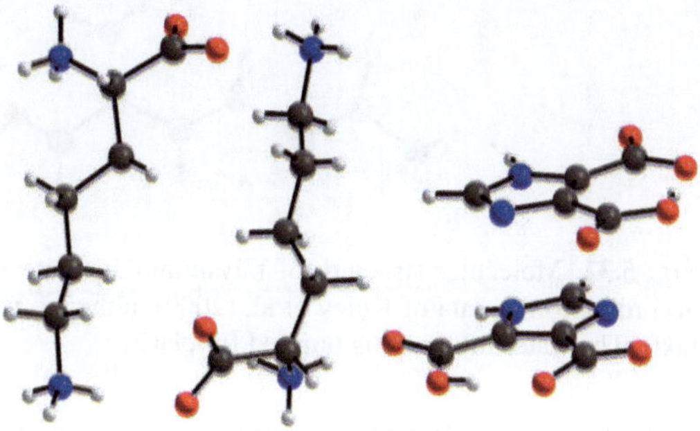

Fig. 5.31 Molecular structure of (L-LysH) 4,5-imidazoledicarboxylate. Note the different orientation of the lysinium cations and the imidazoledicarboxylate anions

H...O(1)LB(2.789(6) Å) and N(2)LB-H...O(1)LA(2.764(6) Å) and N(2)LB-H...O(1′)LB(2.728(7) Å). Here LA and LB are A and B cations of (L-LysH) and PA and PB are A and B picric acids. Each of the terminal NH_3^+ groups forms hydrogen bonds with the carboxylate group of cation: N(7)LA-H...O(1)LB(2.893 (7) Å) and N(7)LB-H...O(1′)LA(2.798(7) Å). In addition each of them forms two hydrogen bonds with nitro groups of two nearest anions: N(7)LA-H...O(61)PB (2.951(7) Å) and N(7)LA-H...O(6′)PB(2.954(7) Å) and N(7)LB-H...O(6′)PA (3.062(8) Å) and N(7)LB-H...O(6′)PA(3.057(8) Å).

The crystals of (L-LysH) 4,5-imidazoledicarboxylate were grown by slow diffusion of ethanol into an aqueous solution containing equimolar amounts of the components (Görbitz and Husdal 1998). The asymmetric unit contains two formula units (Fig. 5.31). The 4,5-IDC forms intramolecular hydrogen bonds: O2C...O4C(2.465 (3) Å) and O2D-O4D(2.473(3) Å). The authors consider that in hydrogen bond O2C...O4C the hydrogen atom is covalently bonded with O4C atom. In our opinion it is more probable that it covalently bonded to O2C atom based on bond lengths C6C-O2C(1.306(4) Å) and C7C-O4C(1.281(4) Å). In case of hydrogen bond O2D...O4D these bond lengths are equal to (1.305(4) Å) for (C6D-O2D) and (1.287(4) Å) for (C7D-O4D). In this case the authors fairly consider that hydrogen atom covalently bonded with O2D atom. Anions form a chain via N1C-H1C...O1D (2.774(3) Å) and N1D-H1D...O1C(2.723(3) Å). In addition nitrogen atoms N3C and N3D as acceptors form hydrogen bonds with terminal NH_3^+ groups of (L-LysH) cations: N2A-H22A...N3C(3.062(4) Å) and N2B-H22B...N3D(3.017(4) Å). The main NH_3^+ group of the cation "A" forms two hydrogen bonds with the carboxylate group of the cation "B", N1A-H11A...O1B(2.783(3) Å) and N1A-H13A...O2B (2.772(3) Å), and one hydrogen bond with anion "D", N1A-H12A...O4D(2.812 (3) Å). The terminal $N(2A)H_3^+$ group in addition to the mentioned N2A-H22A...N3C forms hydrogen bond N2A-H23A...O1B(2.868(4) Å) and weak hydrogen bond N2A-H21A...O3D(3.091(4) Å). The main NH_3^+ group of the cation "B" forms two hydrogen bonds with carboxylate groups of both cations, N1B-H11B...O1B(2.731 (3) Å) and N1B-H12B...O2A(2.811(3) Å), and one hydrogen bond with anion "C", N1B-H12B...O3C(2.718(3) Å). The terminal $N(2B)H_3^+$ group in addition to the mentioned N2B-H22B...N3D forms two weak hydrogen bonds: N2B-H21B...O1A (2.970(3) Å) and N2B-H23B...O3C(3.182(4) Å).

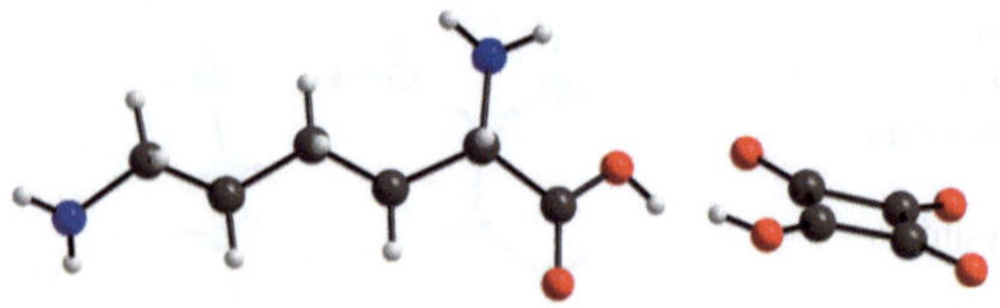

Fig. 5.32 Molecular structure of L-lysinium squarate monohydrate. The protonation is shown according to the data of Kolev et al. (2009), although we consider this state to be incorrect (see text). The water molecule is omitted for clarity

Kolev et al. (2009) obtained two new salts of L-lysine by interaction with squaric acid. The crystal of (L-LysH) HSq·H_2O was obtained by evaporation from aqueous solution containing equimolar quantities of L-lysine and squaric acid. Asymmetric unit contains one formula unit: (L-LysH) cation, HSq^-, and one water molecule. The hydrogen atoms were constrained to calculated positions and refined using riding models in all cases. However, the model suggested for (L-LysH) cation in our opinion is not acceptable. Kolev et al. (2009) came to the conclusion that the carboxylic group and terminal amino group are in neutral states (COOH and NH_2) and only the main amino group is protonated: $H_2N(CH_2)_4CH(NH_3^+)COOH$. All available structural data on singly charged cation (L-LysH) show that both amino groups are protonated, while the carboxylic group is in the state of negatively charged carboxylate group (COO^-): $^+H_3N(CH_2)_4CH(NH_3^+)COO^-$. So, we will consider the structural data of Kolev et al. (2009) on the basis of this model of cation (L-LysH) (Fig. 5.32). The hydroxyl group of anion forms strong hydrogen bond with the carboxylate group of cation, O3-H3′...O11(2.515 Å). The bond length C1-O11(1.259(3) Å) is close to the typical value of carboxylate group (ca. 2.5 Å), while the bond length C1-O12(1.224(3) Å) is lesser than the expected value. The main $N(2)H_3^+$ group forms hydrogen bonds with three nearest anions: N2-H21...O4(2.778 Å), N2-H22...O2(2.853 Å), and N2-H23...O2(2.825 Å). The positions of hydrogen atoms of the water molecule were not determined; however, based on distances between nonhydrogen atoms, it will be possible to determine probable hydrogen bonds. The terminal $N(6)H_3^+$ group forms hydrogen bond with water molecule, N6-H64...OW1(2.803 Å). Water molecule in turn forms hydrogen bonds with anion and the carboxylate group of cation: OW1...O1(2.754 Å) and OW1...O12(2.706 Å). In addition, the $N(6)H_3^+$ group forms weak hydrogen bond N6-H63...O4(2.962 Å). The composition of the crystal of (L-LysH)(HSq)·H_2O in Table 1 by Kolev et al. (2009) is indicated incorrectly ($C_{11}H_{16}N_2O_6$). The correct composition is $C_{10}H_{18}N_2O_7$.

The crystal of $(L\text{-}LysH_2)_2(H_2Sq)(HSq)_2(Sq)\cdot 2H_2O$ was obtained from aqueous solution of L-lysine and squaric acid at molar ratio 1:2. Asymmetric unit contains two lysine moieties, four squaric acid moieties, and two water molecules. Not high enough X-ray data and uncertainty of hydrogen atom positions forced the authors to suggest different structural models of the structure of hydrogen bond schemes. The authors came to the conclusion that there is one neutral molecule of squaric acid H_2Sq, two singly charged HSq^- anions, and one doubly charged Sq^{2-} anion. For ensuring neutrality the authors had to assume the presence of two doubly charged (L-$LysH_2$) cations. In our opinion the suggested model is not convincing and more accurate determination of the structure and positions of hydrogen atoms is desirable. In our opinion another model may be suggested based on the distances between nonhydrogen atoms. We agree with authors that the lysine moiety with C7-O4(1.229(8) Å) and C7-O3(1.300(8) Å) is doubly charged cation. For the second lysine moiety the following bond lengths were obtained: C1-O2(1.256(12) Å), C1-O2′(1.218(16) Å), and C1-O1(1,257(8) Å), where the two values of C1-O2 and C1-O2′ are caused by disordering of oxygen atom. In our opinion these data indicate that in this case there is carboxylate group COO^- and hence singly charged cation (L-LysH). This forces to reconsider the conclusion about states of squarate moieties. We agree with Kolev et al. (2009) that the moiety with C13-O13(1.257(8) Å), C14-O14(1.231(8) Å), C15-O15(1.279(8) Å), and C16-O16(1.270(8) Å) is doubly charged Sq^{2-} anion; the moiety with two short C21-O21(1.209(7) Å) and C22-O22(1.215(8) Å) bonds is neutral H_2Sq molecule; and the moiety with C25-O25(1.274(8) Å), C26-O26(1.218(8) Å), and C27-O27(1.245(8) Å) bonds and one bond with longer bond C28-O28(1.299(8) Å) is singly charged HSq^- anion. However, we assume that the moiety with two short bonds C19-O19(1.217 (8) Å) and C20-O20(1.220(8) Å) and two longer bonds C17-O17(1.292(8) Å) and C18-O18(1.290(8) Å) also is neutral H_2Sq molecule. So, the composition of the salt may be presented as (L-LysH)(L-$LysH_2$)(HSq)(Sq)(H_2Sq)·$2H_2O$. In order to establish the details of the hydrogen bonds, an objective determination of the hydrogen atom position is necessary.

In several works (Tani et al. 2002; Selkti et al. 2000; Ojala et al. 1996; Lazar et al. 2006) a series of salts with L- and DL-Lys with singly (LysH) and doubly ($LysH_2$) charged cations with various complex anions were reported.

Interaction of L- and DL-lysine with oxalic acid was studied by Venkatraman et al. (1997). The crystals of (L-$LysH_2$)$_2$(HC_2O_4)$_2$(C_2O_4) and (DL-LysH)(HC_2O_4) ·$2H_2O$ were obtained and their crystal structures were determined. The crystals of (DL-LysH)(HC_2O_4)·$2H_2O$ were obtained by the diffusion of acetonitrile into the aqueous solution containing equimolar proportion of components. In case of more strong acid H_2SO_4 the crystal of (L-$LysH_2$)SO_4 is formed, while in this case (DL-LysH)(HC_2O_4)·$2H_2O$ is formed. Asymmetric unit contains one formula unit. The hydrogen oxalate anions form usual head-to-tail chain by O22-H…O25(2.513 Å) hydrogen bond. The main N(1)H_3^+ group forms hydrogen bonds with carboxylate groups of nearest cations N1-H1…O1(2.901 Å) and N1-H3…O2(2.904 Å) and with anion N1-H2…O26(2.805 Å). The terminal N(7)H_3^+ group forms hydrogen bonds with one of the water molecules, N7-H1…OW1(2.766 Å), with carboxylate group of nearest cation, N7-H2…O1(2.767 Å), and with anion, N7-H3…O25 (2.787 Å). The second water molecule is disordered. Oxygen atoms of this disordered water molecule are at 2.583 Å and 3.032 Å distances and form a chain. The first water molecule forms hydrogen bonds with cation OW1-H2…O2(2.802 Å) and anion OW1-H1…O26(2.889 Å). In case of L-lysine, Venkatraman et al. (1997)

could not obtain respective crystals with molar ratio 1:1 under the same conditions. However, at molar ratio of L-lysine and oxalic acid 1:5 crystals were obtained, which had unusual composition and charge states of cation and anion. Asymmetric unit contains two doubly charged cations (L-LysH_2), two hydrogen oxalate $HC_2O_4^-$ anions, and one doubly charged oxalate anion $C_2O_4^{2-}$. These three anions form a trimer $C_2O_4^{2-}$...$HC_2O_4^-$...$HC_2O_4^-$ by rather strong hydrogen bonds: O22...H1-O42(2.574 Å) and O45...H1-O32(2.543 Å). This trimer from two sides is closed by (L-LysH_2) cations by hydrogen bonds O1-H1...O25(2.497 Å) and O35...H1-O12 (2.577 Å). Four NH_3^+ groups of two cations form additional N-H...O bonds with this trimer. The O26 of $C_2O_4^{2-}$ anion forms three hydrogen bonds: N1-H2...O26 (3.159 Å), N7-H1...O26(2.888 Å), and N7-H2...O26(2.898 Å). The atoms O25 and O21 each form one hydrogen bond: N17-H3...O25(2.905 Å), N11-H2...O21 (2.812 Å). The atoms O41, O46, and O36 form each two hydrogen bonds, N11-H1...O41(3.094 Å) and N11-H3...O41(2.910 Å), N1-H1...O46(2.913 Å) and N1-H3...O46(2.681 Å), and N17-H1...O36(2.946 Å) and N17-H2...O36 (2.946 Å), and one more hydrogen bond with O35, N7-H3...O35(2.876 Å). Venkatraman et al. (1997) here noted that there is pseudosymmetry in the structure of the crystal. Approximate symmetry is orthorhombic $P2_12_12_1$. Further investigation of the L-Lys + $H_2C_2O_4$ + H_2O system was performed (Sukiasyan 2005). At 0, 20, and 40 °C solutions with different ratios of L-Lys and oxalic acid were prepared. As a result of evaporation of these solutions, at least seven different phases were identified by IR spectra. Among them the highest SHG activity showed the crystal with composition L-Lys·$2H_2C_2O_4$·H_2O. Therefore, this crystal was studied in more detail (Sukiasyan et al. 2008). Asymmetric unit contains one formula unit. Unlike the crystal studied by Venkatraman et al. (1997), here two crystallographically different anions do not interact with each other. Each of them forms usual chain by hydrogen bonds: O3-H17...O5(2.535(3) Å) and O8-H18...O10(2.608(3) Å). Carboxyl group of cation interacts with one of these anions, O1-H1...O9(2.559(3) Å). The water molecule as proton acceptor forms hydrogen bonds with both NH_3^+ groups, N1-H5...O11(2.841(4) Å) and N2-H14...O11(2.966(5) Å), while as proton donor forms hydrogen bonds with both anions, O11-H19...O6(2.798(3) Å) and O11-H20...O10(2.876(4) Å). The rest of the hydrogen atoms of NH_3^+ groups also form hydrogen bonds with carboxylate groups of anions. The shorter of them are N1-H3...O6(2.835(4) Å), N1-H4...O9(2.972(4) Å), and N2-H16...O5(2.772(4) Å). Thermal properties of (L-LysH_2)$(HC_2O_4)_2$·H_2O also were examined (Sukiasyan et al. 2008). Conditions of dehydration were determined and anhydrous salt (L-LysH_2)$(HC_2O_4)_2$ was obtained as well. IR and Raman spectra of anhydrous and hydrated salts were studied. It turned out that one of the phases obtained from solution in the form of crystals at room temperature at ratio of L-lysine and oxalic acid in the interval 1.1–1.5 is identical with anhydrous salt. Monohydrate is formed at stoichiometric ratio 1:2.

The group of Vijayan systematically investigated interaction of lysine with other dicarboxylic acids. No salts with malonic acid (HOOC-CH_2-COOH) were found, while with next member in the series of dicarboxylic acids, namely with succinic acid (HOOC-$(CH_2)_2$-COOH), interesting salts were obtained (Prasad and Vijayan 1991). Crystals of salts of DL- and L-lysine were grown by diffusion of ethanol into aqueous solutions of amino acid and succinic acid in 1:2 molar ratio. All three salts, however, have the composition 1:1. In the case of L-lysine two salts were obtained at different concentrations of used solutions. In one case the obtained salt has the expected structure (from a chemical point of view). Unit cell contains one (L-LysH) cation and one hydrogen succinate anion HOOC-$(CH_2)_2$-COO^-. The anions form usual head-to-tail sequence by hydrogen bond O18-H18...O12(2.527 (4) Å). The main $N(1)H_3^+$ group forms one hydrogen bond with carboxylate group of nearest cation N1-H1...O2(2.797(3) Å) and two hydrogen bonds with anion, N1-H2...O17(2.946(3) Å) and N1-H3...O11(2.854(4) Å). The terminal $N(7)H_3^+$ group forms two hydrogen bonds with carboxylate groups of two nearest cations, N1-H1...O1(2.946(5) Å) and N7-H2...O2(2.694(5) Å), and one hydrogen bond with anion, N7-H3...O12(2.818(3) Å). In the case of salt of DL-lysine asymmetric unit of centrosymmetric triclinic cell contains (L-LysH) cation in general position, succinate anion in special position, and neutral succinic molecule also in special position. The succinate anion ^-OOC-$(CH_2)_2$-COO^- and the molecule of succinic acid form a sequence by hydrogen bond O22-H22...O12(2.465(4) Å) in such manner that each anion is surrounded by two succinic acid molecules and each molecule of succinic acid is surrounded by two succinate anions. The main $N(1)H_3^+$ group forms two hydrogen bonds with carboxylate groups of two nearest cations, N1-H3...O1(2.804(3) Å) and N1-H2...O2(2.835(4) Å), and one weak hydrogen bond with anion, N1-H1...O11(3.013(4) Å). The terminal $N(7)H_3^+$ group forms hydrogen bonds with nearest cation, N7-H1...O1(2.904(4) Å), with molecule of succinic acid, N7-H2...O21(2.892(4) Å), and succinate anion, N7-H3...O11(2.714 (4) Å). The second form of salt of L-lysine contains in unit cell four (L-LysH) cations, one doubly charged anion (Succ. 2), two hydrogen succinate anions (Succ. 1 and Succ. 4), and one neutral molecule of succinic acid (Succ. 3). Positions of hydrogen atoms are not determined, the structure have pseudosymmetry. So, all conclusions were made based on bond lengths and valence angles of nonhydrogen atoms. There are two crystallographically independent ribbons. In the one involving (Succ 2) and (Succ 3), succinic acid molecules and succinate ions alternate along the ribbon. The second involves (Succ 1) and (Succ 4). There are two O...O-type hydrogen bonds between succinate anion and molecule of succinic acid: O62-H...O52(2.477(8) Å) and O68-H...O58(2.464(9) Å). In the case of (Succ. 1) and (Succ. 4) there is a hydrogen bond O72-H...O42(2.533(9) Å) and one more shorter hydrogen bond between O48 and O78 with O...O distance 2.445(10) Å), which according to Prasad and Vijayan (1991) is symmetric hydrogen bond. Eight NH_3^+ groups of four (L-LysH) cations form 24 hydrogen bonds with carboxylate groups of cations (12 hydrogen bonds) and succinate anions and molecule of succinic molecule (12 hydrogen bonds).

Saraswathi et al. (2001) reported salts of L- and DL-lysine with glutaric acid (HOOC-$(CH_2)_3$-COOH). The crystals of (L-LysH) hydrogen glutarate were obtained by the diffusion of isobutyl alcohol into an aqueous solution of L-lysine and glutaric acid in a 1:3 molar ratio. The crystals of (DL-LysH) hydrogen glutarate were obtained by the diffusion of 2-propanol into an aqueous solution of DL-lysine and glutaric acid mixed in a 1:2 molar ratio. In spite of chemical similarity of obtained salts (in both cases there are singly charged (LysH) cation and hydrogen glutarate anion), their structures are quite different. The structure of (DL-LysH) hydrogen glutarate salt is similar to the structure of (L-LysH) hydrogen succinate. Here also anions form head-to-tail sequence connected with each other by rather strong hydrogen bond O19-H19…O12(2.506(4) Å). The main $N(1)H_3^+$ group forms one hydrogen bond with carboxylate group of nearest cation, N1-H1B…O2(2.780(4) Å), and two hydrogen bonds with anion: N1-H1A…O11 (2.849(4) Å) and weak hydrogen bond N1-H1C…O18(3.054(4) Å). The terminal N(7)H_3^+ group also forms one hydrogen bond with nearest cation, N7-H7A…O2 (2.779(4) Å), and two hydrogen bonds with anion: N7-H7B…O12(2.807(4) Å) and weak hydrogen bond N7-H7C…O11(2.934(4) Å). In the structure of (L-LysH) hydrogen glutarate having somewhat unusual trigonal symmetry there is no expected chain of anions. Instead this carboxyl group of anion forms rather strong hydrogen bond with carboxylate group of cation: O19-H19…O1(2.502(7) Å). The main $N(1)H_3^+$ group forms one weak hydrogen bond with nearest cation N1-H1B…O2(3.096(6) Å) and two hydrogen bonds with anion: N1-H1A…O12 (2.729(6) Å) and N1-H1C…O11(2.683(7) Å). The terminal $N(7)H_3^+$ group forms also one hydrogen bond with nearest cation N7-H7B…O2(2.923(6) Å) and two hydrogen bonds with anion: N7-H7A…O18(2.781(8) Å) and N7-H7C…O12 (2.685(6) Å).

Interaction of L- and DL-lysine with adipic acid was studied by Sharma et al. (2006). Crystals were obtained by the method of diffusion of acetonitrile into an aqueous solution of L- or DL-lysine and adipic acid in 1:1 molar ratio in case of L-Lys, the expected salt (L-LysH) hydrogen adipate (L-LysH)(OOC-$(CH_2)_4$-COOH is formed, while in case of DL-lysine obtained salt has the composition (DL-LysH)$_2$(OOC-$(CH_2)_4$-COO), that is, 2:1. Asymmetric unit contains one formula unit. Anions form usual chain …OOC-$(CH_2)_4$-COOH…OOC-$(CH_2)_4$-COOH… by hydrogen bond O20-H20…O12(2.467(3) Å). Rather strong O20…O12 hydrogen bond led to significant difference in bond lengths C13-O13(1.195 Å) and C13-O12(1.263 Å). The main $N(1)H_3^+$ group forms hydrogen bonds only with anions: N1-H1C…O11(2.761(4) Å) and N1-H1B…O20(2.796(4) Å) and weak hydrogen bond N1-H1A…O19(2.935(5) Å). By hydrogen bond N7-H7B…O2 (2.918(4) Å) cations form head-to-tail chain. One more hydrogen bond N7-H7A…O2(2.770(5) Å) connects two such chains with each other. Hydrogen bonds formed by H7C (N7-H7C…O12 and N7-H7C…O1) are on the level of strong van der Waals interactions. There is one formula unit in the cell of (DL-LysH)$_2$(OOC-$(CH_2)_4$-COO). Asymmetric unit contains one (LysH) cation and half of –OOC-$(CH_2)_4$-COO^- anion, because anion is across inversion center. The main $N(1)H_3^+$ group forms two hydrogen bonds with two nearest cations by

N1-H1B...O2(2.815(2) Å) and N1-H1C...O1(2.762(2) Å) and one hydrogen bond with anion N1-H1A...O11(2.730(2) Å). The terminal $N(7)H_3^+$ group forms two hydrogen bonds with two nearest anions N7-H7A...O12(2.838(2) Å) and N7-H7B...O12(2.764(2) Å). The third H1C atom has two contacts on the level of strong van der Waals interactions.

An investigation of lysine with the next homolog, namely, pimelic acid (HOOC-$(CH_2)_5$-COOH), was performed (Saraswathi et al. 2003). Crystals of salt of DL-lysine with pimelic acid were obtained by the diffusion of ethanol into an aqueous solution of DL-lysine and pimelic acid mixed in 1:1 molar ratio. Extensive attempts to crystallize the respective salt of L-lysine were not fruitful. As in previous case from solution containing the components with 1:1 ratio a salt with 2:1 composition was formed being a solvate. The refinement of the structure was complicated due to the disordering of one carbon atom in the anion as well as the presence of also disordered water and ethanol molecules. The authors explain this by the looseness of the structure. The main $N(1)H_3^+$ group forms two hydrogen bonds with two nearest cations, N1-H1A...O1(2.868(3) Å) and N1-H1C...O2(2.802(2) Å), and one hydrogen bond with anion, N1-H1B...O11(2.769(3) Å), while the terminal $N(7)H_3^+$ group forms two hydrogen bonds with anions, N7-H7B...O12(2.817 (3) Å) and N7-H7C...O12(2.807(3) Å), and one hydrogen bond with cation, N7-H7A...O2(2.899(3) Å). Also, the presence of hydrogen bonds of water molecule to the anion, OW1...O11(2.707(12) Å), water to ethanol, OW1...O19(2.921 (13) Å) and ethanol to anion, O19...O11(2.790(12) Å) is supposed.

Marchewka et al. (2003a) obtained a series of salts of L-lysine, determined their composition by chemical analysis, characterized by IR spectroscopy, and estimated their NLO activity by powder SHG method. Among them was previously known (L-LysH) acetate salt (Suresh and Vijayan 1983a) and also salts of L-lysine with L-tartaric, DL-malic, and fumaric acids. The SHG intensities compared to KDP were 0.80((L-LysH) acetate), 0.34((L-LysH) hydrogen L-tartrate), 0.84((L-LysH) DL-malate), and 1.01((L-LysH) hydrogen fumarate). Later the crystal of (L-LysH) hydrogen L-tartrate was investigated in more detail (Debrus et al. 2005; Drozd and Marchewka 2006). Selvaraj et al. (2007) studied compounds of L-lysine with D-tartaric acid and L-tartaric acid as well as DL-lysine with DL-tartaric acid. The crystal of (L-LysH) hydrogen L-tartrate was obtained by evaporation at room temperature of an aqueous solution containing equimolar quantities of components (Debrus et al. 2005). By DSC method two phase transitions were observed. At heating with rate 20 K/min the first peak on DSC curve appears at ca. 297 K and second at 302 K. At cooling with the same rate second peak appears near 297 K and the first peak at ca. 290 K. The phase above 303 K was labeled as phase I. The symmetry and structure of this phase was determined at 320 K. The phase at temperatures lower than 290 K was labeled as phase III. The symmetry and structure of this phase were determined at 260 K. Intermediate phase was labeled as phase II. According to Debrus et al. (2005) phase II is an incommensurate phase. The average structure of this phase has been determined at 298 K for unit cell parameters shown in Table 5.22. The symmetries of all three phases are the same (space group $P2_1$); however, the parameter "c" of phase III and hence the volume are trebled ($Z = 6$, $Z' = 3$). The [101] direction in phase III corresponds to the "c" direction in phases II and I. The difference of phases I and III is clear from dependence of unit cell parameters on temperature shown in Fig. 13 of Debrus et al. (2005). Asymmetric units of phases I and II contain each one formula unit, while phase III contains three formula units. Cations (L-LysH) are highly disordered. Hydrogen L-tartrate anions form usual chain by means of O-H...O hydrogen bond between carboxyl and carboxylate groups. In phase I the O...O distance is equal to 2.542 Å. In phase III there are three different hydrogen L-tartrate anions with three O...O distances: 2.548 Å, 2.532 Å, and 2.517 Å. Selvaraj et al. (2007) grew crystals by liquid diffusion of methanol or propanol into aqueous solutions of the components, mostly in equimolar ratio. It was verified that L-lysine and D-tartaric acid and D-lysine and L-tartaric acid form crystals with the same space group and cell parameters. The authors determined the structure of (L-LysH) hydrogen L-tartrate crystal at 298 K without suspicion about existence of incommensurate phase. Here the disordering of the cation was also found. Hydrogen bond O-H...O in chain is characterized by O...O(2.532 Å) distance. The authors also provided data on (L-LysH) hydrogen D-tartrate. Here also there is the chain of hydrogen tartrate anions with O...O distance 2.527 Å of O-H...O hydrogen bond. Only oxygen atoms of the COO^- group of the (L-LysH) cation are disordered. The crystal has the same space group and similar unit cell parameters with slightly greater density (1.494 g/cm^3) than in the case of L-·L- salt (1.478 g/cm^3). In this regard in our opinion it will be interesting to investigate the crystal of (L-LysH) hydrogen D-tartrate for checking the presence of phase transitions.

Interaction of DL-lysine and DL-tartaric acid leads to formation of centrosymmetric $(DL\text{-}LysH)_2$ DL-tartrate·H_2O salt (Selvaraj et al. 2007). One of the cations displays disordering as well as the water molecule. Four NH_3^+ groups of two cations form hydrogen bonds with two carboxylate groups of cations, with two carboxylate groups of anion and also with water molecule. Alcoholic groups of anion form the most strong hydrogen bonds with carboxylate groups of two nearest anions O25-H25...O29(2.708) Å) and O27-H27...O22(2.702(4) Å).

Pratap et al. (2000) prepared the salt of L-lysine with maleic acid. Crystals of (L-LysH) hydrogen maleate were crystallized by the slow diffusion of acetonitrile into solution of L-lysine and maleic acid in a 1:3 molar ratio. Extensive attempts to crystallize the corresponding salt of DL-lysine were not fruitful. Asymmetric unit contains one formula unit. Hydrogen maleate anion as usual forms intramolecular hydrogen bond O17-H...O12(2.427(6) Å). The bond lengths C13-O12 and C16-O17 are 1.268 Å and 1.278 Å respectively. The main $N(1)H_3^+$ group forms hydrogen bonds with three nearest anions: N-H1...O11(2.798(4) Å), N-H2...O18 (2.835(5) Å), and N1-H3...O11(2.875(5) Å), while the terminal $N(7)H_3^+$ group forms hydrogen bonds with three nearest cations: N7-H1...O2(2.784(4) Å), N7-H2...O1(2.802(3) Å), and N7-H3...O2(2.846(4) Å).

5.1.21.1 Crystals of Salts of Amino Acids as Possible NLO Materials

Salts of L-arginine, particularly (L-ArgH)(H_2PO_4)·H_2O, played an important role in attracting the attention of researchers to the salts of amino acids as possible NLO materials. So, before discussing the salts of arginine let us shortly consider how and why salts of amino acids attracted attention as possible NLO materials.

Before the creation of lasers it was considered that in all phenomena involving light the frequency of light remains unchanged. Isaac Newton in his book of "Optics" in 1704 even considered it as an axiom. The rapid development of quantum radiophysics in the 1950s led to the inevitable creation of lasers. The year 1960 marked as the beginning of the laser era, when Theodore Maiman (1960) built the first (ruby) laser, which had radical influences in almost all areas of science and technology, as well as in everyday life. In the following year already, Franken et al. (1961) published results regarding an experiment discovering of the second harmonic generation in quartz crystal, announcing the birth of nonlinear laser optics.

The material equation of linear optics is a linear relationship between the polarization and the electric field, $P_i = \chi_{ij}E_j$, reflecting the state of affairs in the pre-laser optics. In general case one should also take into account the nonlinear terms in the power series expansion of the field.

$$P_i = P_0 + \chi_{ij}E_j + 1/2\chi_{ijk}E_jE_k + 1/6\chi_{ijkl}E_jE_kE_l + \ldots \quad (5.1)$$

The first term P_0, which does not depend on the field, is nonzero only for pyroelectric crystals. The χ_{ijk} quadratic coefficient of the field is substantially smaller than χ_{ij} term, just as χ_{ijkl} is substantially smaller than χ_{ijk}. Therefore, for the electric fields produced by non-laser sources, you can neglect the nonlinear terms in (5.1). However, laser sources can create fields that are comparable to intra-atomic fields. What are the new effects that may occur due to the existence of the quadratic nonlinear term? For simplicity let us omit the indices and write the quadratic term $P_i^{(NL)} = 1/2\chi_{ijk}E_jE_k$ in the form of

$$P^{(NL)} = dE^2 \quad (5.2)$$

For terms with even powers, and in particular for the quadratic term, coefficients are nonzero only for crystals exhibiting non-centrosymmetric structure. Suppose two waves are incident on a crystal: $E_1 = E_{o1}\cos(\omega_1 t + k_1 r)$ and $E_2 = E_{o2}\cos(\omega_2 t + k_2 r)$. Substituting the values of E_1 and E_2 in the expression (5.2) we will have $P^{(NL)} = d(E_1 + E_2)^2 = d(E_{o1}\cos(\omega_1 t + k_1 r) + E_{o2}\cos(\omega_2 t + k_2 r))^2 = d(E_{o1}{}^2\cos^2(\omega_1 t + k_1 r) + 2E_{o1} E_{o2}\cos(\omega_1 t + k_1 r)\cdot\cos(\omega_2 t + k_2 r) + E_{o2}{}^2\cos^2(\omega_2 t + k_2 r))$. And by using trigonometric transformations we have $P^{(NL)} = d[(E_{o1}{}^2 + E_{o2}{}^2)/2 + 1/2E_{o1}{}^2\cos(2\omega_1 t + 2k_1 r) + 1/2E_{o2}{}^2\cos(2\omega_2 t + 2k_2 r) + E_{o1}E_{o2}\cos[(\omega_1 + \omega_2)t + (k_1 + k_2)r] + E_{o1}E_{o2}\cos[(\omega_1 - \omega_2)t + (k_1 - k_2)r]]$.

Thus, the nonlinear quadratic term results in a number of nonlinear effects: the appearance of constant polarization, the second harmonic waves with frequencies

$2\omega_1$ and $2\omega_2$, and waves with the sum $(\omega_1+\omega_2)$ and difference $(\omega_1-\omega_2)$ of frequencies. In the particular case of a single wave with a frequency ω, there will be constant polarization and second harmonic wave 2ω. Following the discovery of second harmonic generation (Franken et al. 1961), other nonlinear effects have also been identified: the appearance of a wave with the sum frequency $(\omega_1+\omega_2)$ (Bass et al. 1962a), the occurrence of constant polarization also known as optical rectification (Bass et al. 1962b), the appearance of waves with a difference frequency $(\omega_1-\omega_2)$ (Wang and Racette 1965; Zernike and Berman 1965), and the creation of optical parametric oscillator, allowing continuously tuning of frequency, the principle of which is the opposite to appearance of sum frequency (Giordmaine and Miller 1965; Akhmanov et al. 1966).

Given the importance of the quadratic coefficient NLO, let's state its symmetrical properties. Expression (5.2), taking into account the indices, will have this form

$$P_i^{(\mathrm{NL})} = d_{ijk}E_jE_k \tag{5.2'}$$

This expression shows a fundamental property of the d_{ijk} coefficient. It doesn't change when interchanging second and third indices, $d_{ijk}=d_{ikj}$. This indicates the fact that the number of components d_{ijk}, as third-rank tensor, in the general case when crystal lacks any symmetry should decrease from 27 to 18. This allows us to use two-index notation instead of a three-index notation. If we replace the second and third indices with one index according to the substitution rule: $11\rightarrow1$, $22\rightarrow2$, $33\rightarrow3$, $23,32\rightarrow4$, $13,31\rightarrow5$, $12,21\rightarrow6$, then in the two-index notation index i will still have values of 1,2,3 while index j will go from 1 to 6.

$$d_{ij} = \begin{pmatrix} d_{11} & d_{12} & d_{13} & d_{14} & d_{15} & d_{16} \\ d_{21} & d_{22} & d_{23} & d_{24} & d_{25} & d_{26} \\ d_{31} & d_{32} & d_{33} & d_{34} & d_{35} & d_{36} \end{pmatrix} \tag{5.3}$$

Another symmetrical property of d_{ijk}, pointed out by Kleinman (1962) consists of the fact that when there is no absorption in the relevant frequency range and dispersion of d is weak, then d is symmetric with respect to exchange of all three indices. This property, called Kleinman symmetry, is not fundamental in character, but is very useful in practice. Thus, for example, $d_{121}(d_{16})=d_{211}(d_{21})$, $d_{123}(d_{14})=d_{231}(d_{25})=d_{312}(d_{36})$, and so on. By finding all equal components, one can see that there are only 10 independent components:

$$d_{ij} = (d_{11}, d_{12}, d_{13}, d_{14}, d_{15}, d_{16}, d_{22}, d_{23}, d_{24}, d_{33}) \tag{5.3'}$$

Further decrease in the number of independent components comes from symmetry of the crystal. The principle of the impact of crystal's symmetry on the number of independent components is described clearly by Zernike and Mitwinter (1970). For example, for α-HIO_3 crystal with orthorhombic symmetry 222, only three components are different from zero, d_{14}, d_{25}, and d_{36}. Also, according to the Kleinman condition, all three components are equal and therefore the crystal has only one independent d_{ijk} coefficient.

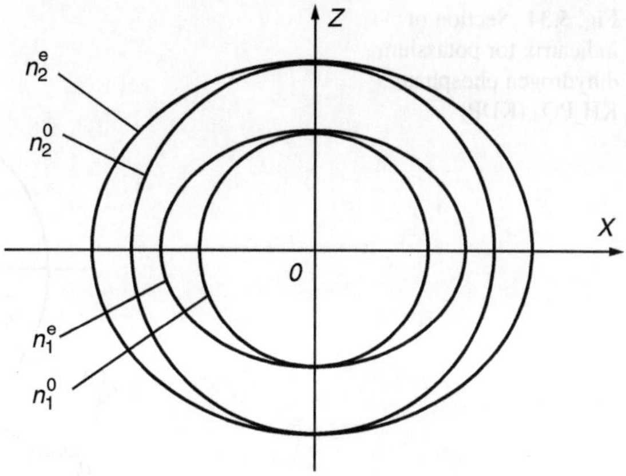

Fig. 5.33 Section of indicatrix for quartz

Even with great scientific importance of detection of second harmonic generation (SHG), this phenomenon is unlikely to have had practical significance, if in other crystals second harmonic intensity would have been as small (approx. 10^{-8} of the power of the original wave) as in quartz. Actually the choice of quartz appeared to be unfortunate.

Despite the fact that quartz is birefringent, the birefringence is less than the variation of refractive index due to dispersion (Fig. 5.33).

Because of this, quartz crystal has no such direction where the refractive indices and hence the velocity of propagation of main wave and second harmonic wave are equal. And this leads to the fact that even at small distances dephasing between the main wave and the second harmonic wave occurs, which is the main reason for the low intensity of second harmonic generation in quartz crystal. It was comprehended by Giordmaine (1962) and Maker et al. (1962), who almost simultaneously reported the possibility of a substantial increase in the intensity of the second harmonic wave, with propagation of main and second harmonic waves in the direction where their refractive indices are equal (phase-matching method). Analysis of indicatrix of optically negative uniaxial potassium dihydrogen phosphate KH_2PO_4 (KDP) crystal (Fig. 5.34) showed that in this case, the birefringence is sufficient to ensure the phase-matching condition (synchronization). At a certain angle with respect to the optical axis (called the phase-matching angle) $n_o^{(\omega)}$ and $n_e^{(2\omega)}$ are equal.

The use of phase-matching method led to huge increase in second harmonic generation efficiency compared with quartz, and reached 10–50 %. This indicates that a suitable nonlinear optical crystal cut in the right direction can produce coherent emission with comparable power at a different frequency, i.e., substitutes an entire laser.

Of particular interest is the case when the phase-matching angle is 90° (Fig. 5.35). In this case an appreciable variation of the phase-matching angle does not lead to a significant reduction in conversion efficiency. Therefore, the phase-matching at 90° degree is also called noncritical.

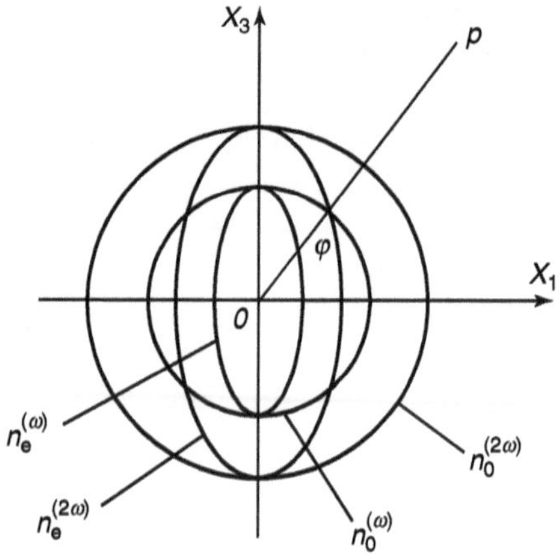

Fig. 5.34 Section of indicatrix for potassium dihydrogen phosphate KH_2PO_4 (KDP)

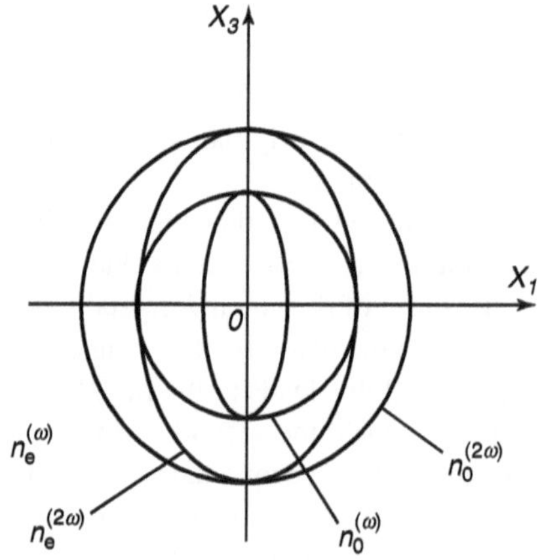

Fig. 5.35 Noncritical phase-matching condition

Actually since that time (1962) the search for effective nonlinear optical (NLO) crystals for frequency conversion of laser radiation began, and a new direction in materials science has been founded. Since the transparent crystals grown for piezoelectric transducers meet the basic requirements of non-centrosymmetricity, it was expected to investigate them as well as NLO materials. On this path a number of attractive crystals have been found.

Examination of available at that time crystals allowed Miller (1964) to formulate a constructive empirical rule for a targeted search of NLO materials. According to Miller NLO coefficient is

$$d_{ijk}{}^{2\omega} = \delta_{ijk}{}^{2\omega} \chi_{ii}{}^{2\omega} \chi_{jj}{}^{\omega} \chi_{kk}{}^{\omega}, \tag{5.4}$$

where the $\delta_{ijk}{}^{2\omega}$ coefficient, later called as Miller's delta, changed slightly from crystal to crystal, while the linear susceptibilities $\chi_{ii}{}^{2\omega}, \chi_{jj}{}^{\omega}, \chi_{kk}{}^{\omega}$ were proportional to refractive indices. Therefore, according to the Miller's rule, the larger the refractive index, the bigger the d_{ijk} coefficient.

Garrett and Robinson based on the anharmonic oscillator model derived relation (5.4) and found that Miller's delta is due to anharmonicity in the crystal (Garret and Robinson 1966).

Although SHG was also observed in powders, and comparative measurements of effectiveness were carried out (Rieckhoff and Peticolas 1965), it was considered that to determine eligibility of materials for practical applications, in particular for determining the presence of phase-matching, it is necessary to grow of single crystals, which is a very challenging task. Kurtz and Perry (1968) developed a powder method, which allows you to determine the availability of phase-matching. Kurtz and Perry investigated a large number of materials in the form of crystal and powder (about 100 inorganic and organic) and divided them into 5 classes. The first class included materials with high nonlinear properties, having phase-matching. The second class included materials with low nonlinear properties (higher or equal to quartz), but exhibiting phase-matching. The third class of nonlinear materials included materials with higher nonlinear coefficient than quartz, but without phase-matching. The forth class included materials with nonlinearity not higher than that of quartz and having no phase-matching. Finally, the fifth class included centrosymmetric materials with no nonlinear properties. They were the first to discover SHG signal in 56 materials, 27 of which exhibited a phase-matching. Several crystals from the first class ($LiNbO_3$, $KNbO_3$, HIO_3, $LiIO_3$, KIO_3, and urea) attracted attention, and some of them have found practical applications (Avdienko et al. 1980; Kuzminov 1987).

Powder method promoted wide search for new materials among the huge number of organic compounds, single crystals of which is very difficult to obtain. As a result a number of molecular organic materials with very high nonlinear properties have been identified. Initially, it seemed that these materials do not conform to the rule of Miller as their refractive indices were low. A comparison of these compounds with each other and with similar compounds that do not exhibit high nonlinearity allowed establishing that the high nonlinearity is due to intramolecular charge transfer from donor to acceptor groups through a system of conjugated double bonds (Davydov et al. 1970, 1971).

Thus, it appears that along with the mechanism of the anharmonicity there is another mechanism typical for some organic molecular crystals (charge-transfer mechanism) for nonlinearity. And molecular nonlinear coefficient $\beta_{c.t.}$ is proportional to the alteration of the dipole moment due to the charge transfer under the influence of light $\beta = \beta_{anh} + \beta_{c.t.}$. This discovery led to further direction for search and targeted design of new molecular organic crystals. The results are summarized in several monographs (see Koreneva et al. 1975, 1985; Chemla and Zyss 1987; Dmitriev et al. 1999). A very good example is *para*-nitroaniline (*p*NA), in which an electron-donating group is amino (NH_2) and an electron-acceptor group is nitro (NO_2). *Para*-position provides maximum alteration in the dipole moment, but due to the centrosymmetric structure of *p*NA, there is no high molecular nonlinearity in the crystal.

mNA
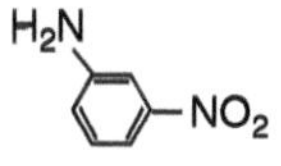

pNA
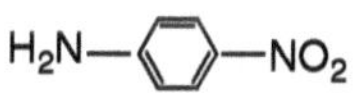

Meta-nitroaniline (mNA), however, crystallizes in non-centrosymmetric structure and, despite un-optimal location of donor and acceptor groups, exhibits pretty high NLO properties (Dmitriev et al. 1999). Numerous other organic molecular crystals have been synthesized (Koreneva et al. 1975, 1985; Chemla and Zyss 1987; Dmitriev et al. 1999). Besides high nonlinearity organic molecular crystals have another advantage in comparison with inorganic: they have smaller temporal response because nonlinear processes involve the movement of fast electrons and not heavy ions. However, it became clear that they all have some inherent drawbacks that limit their practical application. Since the molecules in molecular crystals are interconnected by weak van der Waals forces, this leads to the fact that they have low mechanical hardness and low melting temperature. The organic molecular crystals are generally not water soluble, necessitating the use of organic solvents, expensive and at times toxic. In addition, it was found that the increase of nonlinearity leads to narrowing of the transparency region.

This prompted researchers in the late 1970s to look for such materials, which could to some extent combine the positive qualities of both inorganic (high hardness and high melting point, wide window of transparency, the possibility of growing single crystals from an aqueous solution or melt) and organic (the possibility of increasing the nonlinearity through targeted changes in chemical composition and performance) crystals. The search for new crystals remains among inorganic and organic crystals as well. In addition to KDP and its analogs, lithium niobate and other niobates and lithium iodate (α-$LiIO_3$), potassium titanyl phosphate, KTP ($KTiOPO_4$) and its analogs, barium borate, BBO (β-BaB_2O_4), lithium borate, LBO (LiB_3O_5), and many others (Dmitriev et al. 1999) have been discovered. A fundamental solution to the problem would be the discovery of inorganic crystals in which the charge-transfer mechanism is realized. However, until now, implementation of the mechanism of charge transport through the conjugated bonds has not been seen among the inorganic crystals.

Important for the progress in studies looking for materials that can unite the positive qualities of inorganic and organic crystals was the work of Davydov et al. (1977). Although the work has been dedicated to identify new molecular organic NLO materials, it emphasized the importance of the investigation of crystals of organic salts. The organic salts were the most important materials, with the ability to combine positive qualities of both inorganic and organic crystals. It was pointed out to the importance of the investigation of optically active amino acids and their salts, although amino acids and their salts have previously been examined as NLO materials (Rieckhoff and Peticolas 1965; Kurtz and Perry 1968). Rieckhoff and Peticolas (1965) examined all the essential amino acids and/or salts thereof, including L-arginine, L-lysine hydrochloride, and L-histidine hydrochloride; Kurtz and Perry (1968) examined, among other crystals, L-histidine (with orthorhombic symmetry), L-glutamic acid, and L-lysine hydrochloride too, which were allocated to the second class. Davydov et al. (1977) investigated L-arginine hydrochloride, L-tryptophan, L-asparagine, potassium salt of L-asparagine, and dinitrophenyl derivatives of L-alanine and L-serine.

Extremely important for the further development of research in this direction was the work of Chinese researchers (Xu et al. 1983a, b), in which it was found that L-arginine phosphate monohydrate L-Arg·H_3PO_4·H_2O (LAP) has several advantages over the widely used already by this time crystal KDP. LAP can be also easily grown to large size and high quality from the aqueous solution. Compared with KDP its nonlinear coefficient is more than three times higher, and it has several times higher laser damage threshold and is less hygroscopic (Eimerl et al. 1989a; Sasaki et al. 1989). Moreover, a comparative study of the characteristics of the LAP and the most effective crystals used for SHG Nd:YAG laser showed that LAP exceeds in number of parameters the crystals KTP, BBO, and LBO (Bhar et al. 1995). It is important to note on the example of LAP that the properties of the hybrid may be better than the individual components, particularly, crystals of L-arginine and crystals of phosphoric acid salts.

All this stimulated the search for new crystals of salts of protonated amino acids. Crystals of salts of a number of amino acids (lysine, valine, leucine, threonine, and tryptophan) with phosphoric, boric, and iodic acids were obtained (Suzuki and Matsuoka 1991). In many of them the SHG signal exceeds the signal from the KDP. Successful search for new crystalline salts of L-arginine has been done by Monaco et al. (1987) and Petrosyan et al. (2000b). In numerous reports in subsequent years L-arginine crystalline salts obtained by Monaco et al. (1987) and Petrosyan et al. (2000b) have been more thoroughly investigated. Discovery of nonlinear optical properties of L-arginine phosphate monohydrate, L- Arg·H_3PO_4·H_2O (LAP) (Xu et al. 1983a, b), had a significant importance not only to attract attention to the salts of protonated amino acids, but also for the development of the entire direction of hybrid organic–inorganic crystals. Warren (1990), based on analysis of previous works, identified ways to obtain materials which can combine the positive qualities of the inorganic and organic materials. He called such materials as hybrid or semi-organic materials. They are salts in which the cations or anions are organic or metal complex salts of organic ligands. Due to the nature of ionic bonds of these crystals, they often can be grown from an aqueous solution (similar to inorganic crystals), have sufficiently good mechanical properties, thermal stability and thermal conductivity. At the same time, as in case of organic compounds, a broad, purposeful change of properties is possible (by modifying the chemical composition and crystalline structure). Among the semi-organic crystals in which inorganic part is the cation and organic is anion, there are such previously known classes of compounds as formates, oxalates, and tartrates. However, these anions were not selected taking into account of the intramolecular charge-transfer requirements. Higher nonlinearity is found in crystals in which the anions contain conjugated bonds and favorable groups. An example is the dihydrate of the sodium para-nitrophenolate (NPNa) (Minemoto et al. 1993), first proposed in (Davydov et al. 1977), an effective nonlinear coefficient of which is higher than that of KTP, and the thermal conductivity is significantly higher than that of the organic molecular crystals. Deuteration of crystal water (DNPNa) significantly reduces the absorption loss and increases the SHG efficiency, and DNPNa can be used successfully in the intracavity mode (Minemoto et al. 1993). Out of complex salts containing polarizable organic molecules as ligands, the chlorides, nitrates, sulfates, etc. of metals such as Zn, Cd, Hg, Pb can be named as examples. Zinc tris(thiourea) represents the greatest interest of this class (Marcy et al. 1992). From ionic compounds with organic cations, besides salts of amino acids, salts of styrylpyridinium and aminopyridinium should be mentioned as well. Warren (1990) reported the synthesis of 4-aminopyridinium salts with different anions: dihydrogen phosphate, tetrafluoroborate, hydrogen sulfate, sulfonate, oxalate, tosylate, etc. The most promising from pyridinium salts is 2-amino-5-nitropyridine salts, demonstrating optimal disposition of the donor and acceptor groups. Crystals of dihydroarsenate and dihydrophosphate of 2-amino-5-nitropyridine were obtained and studied for nonlinear optical and electro-optical properties (Kotler et al. 1992; Salvestrini et al. 1998), which demonstrate that these crystals have promising potential for further examination.

Another class of salts with organic cations is a styrylpyridinium (stilbazolium) salts with different anions, which as in the case of stilbene have an extended system of conjugated double bonds (Marder et al. 1989). From this class, tosylate of 4′-dimethylamino-N-methyl-4-stilbazolium (DAST) attracted the most attention. From the aqueous solution, DAST crystallizes in the form of monohydrate with centrosymmetric structure. However, anhydrous DAST, obtainable from methanol, crystallizes with non-centrosymmetric structure and exhibiting great nonlinearity (Pan et al. 1996; SHG signal of fundamental wave with a wavelength $\lambda = 1{,}907$ nm is greater than the signal from urea about 1,000 times).

As mentioned above, this kind of large nonlinearity leads to a limited transparency in the visible region of the spectrum (the absorption edge of 700 nm). For this reason DAST can be applied in electro-optics and the difference frequency generation and for SHG in the IR region.

In 1995, the discovery of L-histidine tetrafluoroborate (L-His·HBF_4) (Marcy et al. 1995) as a new NLO crystal was reported, the significance of which can be compared with the finding of LAP's NLO properties. The L-His·HBF_4 crystal, having monoclinic symmetry (space group $P2_1$), can easily be crystallized from the aqueous solution and is transparent in a wide range from 250 to 1,300 nm, non-hydroscopic, and thermally stable up to 205 °C. Its effective nonlinearity is 5 times higher than KDP's and phase-matchable. The importance of this study is the fact that previous works in which L-His·HCl (Rieckhoff and Peticolas 1965) and L-histidine (Kurtz and Perry 1968) has been investigated did not indicate at the potential of L-histidine crystalline salts. One would expect that this crystal will attract the attention of researchers investigating new NLO materials based on amino acid salts. Indeed, the works devoted to the more detailed study of the growth conditions and the various characteristics of L-His·HBF_4 crystal shortly appeared (Aggarwal et al. 1999, 2003; Rajendran et al. 2001, 2003a, b; Ramajothi and Dhanuskodi 2003), continuing until now (Dhanuskodi and Ramajothi 2004; Gokul Raj et al. 2005, 2006a, 2007).

5.1.22 Salts of Arginine

In Table 5.23 salts of arginine with inorganic acids are collected. The salt (L-ArgH)F was obtained by Schmid (1968) and patented as a means for prevention of caries. Crystals of (L-ArgH)F as possible NLO materials were obtained and investigated by Monaco et al. (1987). The obtained data on NLO activity and presence of nearly noncritical phase-matching showed its availability of this crystal and later attracted the attention of various researchers. Monaco et al. (1987) did not publish details of the structure. Detailed investigation of the structure of (L-ArgH)F was performed (Ramos Silva et al. 2000a). The authors obtained the crystal of (L-ArgH)F in an attempt to synthesize L-arginine dysprosium fluoride. Asymmetric unit contains one formula unit. Five N-H bonds of guanidyl group and three N-H bonds of NH_3^+ group form eight hydrogen bonds. Out of these, four N-H...O hydrogen bonds are formed with carboxylate group and four with fluoride anion: N1-H1B...O1(2.819(2) Å), N1-H1C...O2(2.8252(19) Å), N3-HA...O2(2.835(2) Å), N3-HB...O1(2.916(2) Å) and N1-HA...F(2.5895(19) Å), N2-H2...F(2.720(2) Å), N4-HA...F(2.677(2) Å), N4-HB...F(2.797(2) Å). In several studies, crystals of (L-ArgH)F were grown from aqueous solution by temperature lowering and evaporation methods solubility curve was determined, and IR and UV–Vis spectra, defect structure, and thermal and other characteristics were determined (Pal and Kar 2002, 2005a; Haja Hameed et al. 2003a; Packiam Julius et al. 2004a; Mohandoss et al. 2012). The presence of fluoride anions in solution prevents generation of microbes in solution, which is a serious problem in crystal growing of some salts of amino acids. In the system L-Arg + HF + H_2O in addition to (L-ArgH)F salt with 1:1 ratio a salt with 1:2 ratio also is formed. Petrosyan et al. (2000b) obtained this salt from aqueous solution containing L-arginine and HF in molar ratio 1:2. The crystal has orthorhombic symmetry and contains as independent part a (L-ArgH) cation and also fluoride ion (F^-) and HF molecule, which form short hydrogen bond $(F...H\text{-}F)^-$.

Ramos Silva et al. (2000b) reported another form of this salt, which has triclinic symmetry and two formula units in unit cell. Conditions of obtaining of this form are not clear. Ramos Silva et al. (2000b) indicated that the crystals were prepared by reacting fluoric acid (40 %) with a dilute aqueous solution of L-arginine without indication of molar ratio. The C-O bond lengths of carboxylate groups of cations (1.255(3) Å), (1.242(3) Å), and (1.252(3) Å), (1.246(3) Å) of triclinic form (Ramos Silva et al. 2000b), and (1.251(3) Å) and (1.240(3) Å) of orthorhombic form (Petrosyan et al. 2000b) are characteristic for singly charged (L-ArgH) cations. The distances F...F in anions $(F...H\text{-}F)^-$ of triclinic form are (2.248(3) Å) and (2.233(2) Å), while in orthorhombic form the distance is (2.243(3) Å). All other active hydrogen atoms form hydrogen bonds with oxygen atoms of the COO^- group and fluoride atoms of $(F...H\text{-}F)^-$ anions. In doing so, each oxygen atom and each fluoride atom in the structures of both forms form each two hydrogen bonds. The volume per one formula unit for triclinic form ($V/Z = 244.6$ Å^3) is noticeably lesser than in the case of orthorhombic form ($V/Z = 262.3$ Å^3), which leads to

Table 5.23 Salts of arginine with inorganic anions

#	Crystal	s.g., Z	a, b, c (Å), α, β, γ (°)	Refs
1	(L-ArgH)F	$P2_1$, 2	5.4475(4), 8.5133(6), 10.2195(7), $\beta = 93.475(6)$	Ramos Silva et al. (2000a)
2	(L-ArgH)(F·HF)	$P2_12_12_1$, 4	7.145(1), 8.258(2), 17.782(4)	Petrosyan et al. (2000b)
3	(L-ArgH)(F·HF)	$P1$, 2	5.1813(11), 9.2173(18), 10.6278(17), $\alpha = 87.878(14)$, $\beta = 74.948(16)$, $\gamma = 86.653(17)$	Ramos Silva et al. (2000b)
4	(L-ArgH)Cl	$P2_1$, 4	5.330, 9.460, 20.070, $\beta = 90.50$	Mazumdar et al. (1969)
5	(L-ArgH)Cl	$P1$, 2	5.1263(8), 9.461(1), 10.322(2), $\alpha = 88.138(5)$, $\beta = 76.447(4)$, $\gamma = 89.745(5)$	Sridhar et al. (2002d)
6	(L-ArgH)Cl·H_2O	$P2_1$, 4	11.044(2), 8.481(1), 11.214(2), $\beta = 91.31(1)$	Dow et al. (1970)
7	(L-ArgH_2)Cl_2·H_2O	$P2_12_12_1$, 4	7.500(2), 7.518(2), 21.710(4), at 193 K	Petrosyan et al. (2004a)
8	(L-ArgH)Br·H_2O	$P2_1$, 4	11.213(2), 8.630(3), 11.243(9), $\beta = 91.64(5)$	Nabakhtiani et al. (1992)
9	(L-ArgH_2)Br_2·H_2O	$P1$, 1	6.7700(14), 7.2183(14), 7.594(2), $\alpha = 69.08(3)$, $\beta = 83.39(3)$, $\gamma = 68.17(3)$, at 233 K	Petrosyan et al. (2004b)
10	(L-ArgH_2)I_2	$P2_12_12_1$, 4	26.039(13), 6.995(10), 7.319(7)	Seely (1966b)
11	(L-ArgH)ClO_3	$P2_12_12_1$, 4	5.1928(10), 13.852(3), 15.73493)	Petrosyan et al. (2005a)
12	(L-ArgH)BrO_3	$P1$, 2	8.426(2), 8.737(2), 9.301(2), $\alpha = 111.05(3)$, $\beta = 99.43(3)$, $\gamma = 110.90(3)$, at 233 K	Petrosyan et al. (2005a)
13	(L-ArgH)IO_3	–	–	Petrosyan et al. (1998)
14	(L-ArgH_2)$(IO_3)_2$	$P2_12_12_1$, 4	6.9554(14), 7.998(2), 25.038(5)	Petrosyan et al. (1998)
15	(L-ArgH)$_2$$(NO_3)_2$·$H_2O$	$P1$, 1	5.142(10, 10.439(2), 10.557(2), $\alpha = 88.43(3)$, $\beta = 79.57(3)$, $\gamma = 81.28(3)$	Terzyan et al. (2004)
16	(L-ArgH_2)$(NO_3)_2$	$P2_1$, 2	7.754(2), 7.286(1), 11.673(2), $\beta = 92.60(3)$	Terzyan et al. (2004)

17	(L-ArgH)($HSeO_3$)[a]	*C2*, 4	22.493(5), 5.1624(13), 9.730(4), $\beta = 95.68(3)$	de Matos et al. (2001)
18	(L-ArgH)BF_4	$P2_12_12_1$, 4	5.050(2), 13.619(6), 16.470(3)	Wang et al. (1994)
19	(L-ArgH)ClO_4	$P2_12_12_1$, 4	5.0883(2), 13.8332(3), 16.5612(4)	Srinivasan and Rajaram (1997a)
20	(L-ArgH_2)(ClO_4)$_2$	*P1*, 1	5.6099(1), 6.3496(1), 10.2927(2), $\alpha = 94.582(1)$, $\beta = 93.904(1)$, $\gamma = 92.071(1)$, at 120 K	Suponitsky et al. (in prep.)
21	(L-ArgH)(H_2PO_4)·H_2O	$P2_1$, 2	7.319(1), 7.912(2), 10.779(3), $\beta = 98.05(2)$, at 130 K	Espinosa et al. (1996)
22	(L-ArgH)(C_2H_5)$_2PO_4$	$P3_121$, 6	9.251(3), 9.251(3), 34.25(2), $\gamma = 120$	Furberg and Solbakk (1973)
23	(L-ArgH_2)(H_2PO_4)$_2$	$P2_12_12_1$, 4	9.681(2), 21.756(4), 7.000(14)	Haussühl et al. (2003)
24	(L-ArgH)(H_2AsO_4)·H_2O	–	–	Marchewka (2006)
25	(L-ArgH_2)(H_2AsO_4)$_2$	$P2_12_12_1$, 4	9.787(2), 22.024(2), 7.134(3)	Zalkin et al. (1989)
26	(L-ArgH)$C_2H_5SO_4$	$P2_12_12_1$, 4	9.1504(18), 12.519(3), 12.551(3)	Karapetyan (2008)
27	(L-ArgH)ReO_4	$P2_12_12_1$, 4	5.0901(3), 13.9239(6), 16.8224(8)	Isakov et al. (2012)
28	(L-ArgH)X(1) ·CH_3COOH·H_2O[b]	$P2_1$, 2	9.602(5), 7.636(6), 16.661(5), $\beta = 92.16(3)$	Ramos Silva et al. (1999b)
29	(L-ArgH)$_3VF_6$·3H_2O	$P2_1$, 2	10.805(2), 9.050(2), 17.030(2), $\beta = 96.96(1)$	Ramos Silva et al. (2000c)
30	(L-ArgH)$_3CrF_6$·3H_2O	$P2_1$, 2	10.8194(15), 9.0434(15), 16.998(2), $\beta = 96.890(12)$	Ramos Silva et al. (2005b)
31	(L-ArgH_2)$SeBr_6$, needles	*P1*, 2	11.66(2), 7.78(1), 12.43(2), $\alpha = 106.5(3)$, $\beta = 114.3(3)$, $\gamma = 96.7(3)$	Hempel et al. (1977b)
32	(L-ArgH_2)$SeBr_6$, plates	$P2_1$, 2	7.43(1), 10.74(2), 11.77(2), $\beta = 102.2(3)$	Hempel et al. (1977b)

[a]The authors believe that this crystal is hydrate. We do not agree with this assumption (see the text)

[b]X(1) = [(UO_2(CH_3COO)$_3$]

respective difference in values of densities (1.454 and 1.356 g/cm^3 respectively). The reason of this difference is the difference in conformations of cations in the structures of these forms. Single crystals of the orthorhombic form were obtained and thermal properties were studied (Petrosyan et al. 2002). On heating crystal of orthorhombic form loses the HF molecule before 160 °C and transforms into (L-ArgH)F. Sankar et al. (2010) write that they have grown crystals of orthorhombic form (L-ArgH)(F...HF) from aqueous solution of stoichiometric composition and investigated it as possible NLO material. It should be, however, noted that the IR spectrum registered by KBr pellet technique differs from the spectrum showed in the work of Petrosyan et al. (2000b). Thermal properties also differ. In addition, to confirm the agreement of unit cell parameters, Sankar et al. (2010) did not quote the paper by Petrosyan et al. (2000b) or (2002), but an article in which fluoride and dihydrofluoride of L-histidine were studied. Search for possible explanation of differences of data in Petrosyan et al. (2000b, 2002) with data of Sankar et al. (2010) led us to the conclusion that the crystal obtained by Sankar et al. (2010) is (L-ArgH)Cl·H_2O or (L-ArgH)Br·H_2O or (L-ArgH)($Cl_{0.5}Br_{0.5}$)·H_2O (see references below). Mazumdar and Srinivasan (1964) reported the synthesis (L-ArgH)Cl, (L-ArgH)Cl·H_2O, and (L-ArgH)Br·H_2O and determined their symmetry and structure. All three crystals have the symmetry $P2_1$ and $Z = 4$, $Z' = 2$. Unfortunately the authors do not report conditions of obtaining particularly (L-ArgH)Cl and (L-ArgH)Cl·H_2O. More detailed structural results are provided by Mazumdar and Srinivasan (1966) for (L-ArgH)Br·H_2O and Mazumdar et al. (1969) for (L-ArgH)Cl. More accurate data with determination of positions of hydrogen atoms are reported, Dow et al. (1970) and Xian and Lu (2007) for (L-ArgH)Cl·H_2O and Nabakhtiani et al. (1992) for (L-ArgH)Br·H_2O. Later (Sridhar et al. 2002d) another triclinic form of anhydrous (L-ArgH)Cl salt was obtained by slow evaporation from an aqueous solution of 1:1 stoichiometric ratio of L-arginine and hydrochloric acid. The unit cell of triclinic form also contains two independent formula units (L-ArgH)Cl. The crystals of (L-ArgH)Cl·H_2O and (L-ArgH)Br·H_2O are isostructural. The main NH_3^+ groups of the first cation in both structures form two N1...O2 and N1...O4 hydrogen bonds with carboxylate groups of both cations. The third atom of the first cation forms hydrogen bond with one (second) of the halogenide ions, while the N(2)H group of the first cation forms hydrogen bond with the other (first) halogenide ion. One of the NH_2 groups of the guanidyl group of the first cation forms hydrogen bonds with the first halogenide ion and carboxylate group of the first cation. The second NH_2 group of the first cation forms two N4...O6 bonds with two nearest water molecules. The second cation similarly forms N5...O2 and N5...O5 hydrogen bonds with carboxylate groups and hydrogen bonds with first halogenide ion, while the N(6)H group forms hydrogen bond with the second halogenide ion. One of the NH_2 groups of the second cation forms hydrogen bonds with the second halogenide ion and carboxylate group of second cation. The second NH_2 group forms two hydrogen bonds N8...O3 with two nearest water molecules. The first water molecule forms hydrogen bond with the first halogenide ion and the second respectively with second halogenide ion. In addition they form each weak O3...O1 and O6...O4 hydrogen bonds with carboxylate groups.

Thus each oxygen atom of the carboxylate groups forms two hydrogen bonds and each halogenide ion forms three N-H...Cl(Br) hydrogen bonds and each one O-H...Cl(Br) with water molecule. Anhydrous (L-ArgH)Cl salts of monoclinic and triclinic forms in spite of some similarity have also differences in conformations of cations and details of hydrogen bonds. In the structure of the monoclinic form, the main NH_3^+ groups of the cations each form one N1...Cl1 and N5...Cl2 hydrogen bonds towards the chloride ions and each two hydrogen bonds with carboxylate groups: N1...O1, N1...O2 and N5...O3, N5...O4. The N(2)H group of the first cation and N(6)H group of the second cation form N2...Cl1 and N6...Cl2 hydrogen bonds. The guanidyl group of the first cation forms hydrogen bonds with Cl2 chloride ion (N3...Cl2) and carboxylate groups of second cation (N3...O4 and N4...O3). Respectively the $N(5)H_3^+$ group of the second cation forms N5...Cl2, N5...O3, and N5...O4 hydrogen bonds and guanidyl group forms N7...Cl1, N7...O2, and N8...O1 hydrogen bonds, that is, with other chloride ion and carboxylate groups of the first type cation. In case of triclinic form the main NH_3^+ groups also form each one hydrogen bond with ion N11-HA...Cl1 and N21-H...Cl2. However, other two hydrogen atoms form hydrogen bonds with carboxylate groups of both cations N11-HB...O2A and N11-HC...O1B and N21-HB...O1A and N21-HC...O2B. The N(12)H and N (22)H groups form hydrogen bonds with other chloride ions: N12-H12...Cl2 and N22-H22...Cl1. The guanidyl group of each cation forms hydrogen bonds with both ions N13-HC...Cl2 and N4-HC...Cl1 and N23-HC...Cl2 and N24-HD...Cl1 as well as with carboxylate groups of the same type of cations: N13-HD...O1A and N14-HD...O1B and N23-HD...O2B and N24-HC...O2A. After discovering of promising NLO properties of (L-ArgH)$H_2PO_4 \cdot H_2O$ (LAP) (see below) the first work on systematic search of new analogs of LAP was the article of Monaco et al. (1987), where an attempt was made to obtain about 20 new salts of L-arginine and to investigate their NLO properties. Among the obtained crystals were (L-ArgH)F and also previously studied (L-ArgH)Cl (monoclinic), (L-ArgH)Cl·H_2O, and (L-ArgH)Br·H_2O, NLO efficiency of which was higher than in quartz. These results attracted the attention of researchers for growing their crystals and further investigation. In the works of Evlanova et al. (1990) and Rashkovich and Shekunov (1991) crystallization conditions of (L-ArgH)Cl·H_2O were studied and also showed that above 60 °C anhydrous (L-ArgH)Cl salt crystallizes. Haussühl et al. (1990) have grown large crystals of (L-ArgH)Cl·H_2O and (L-ArgH)Br·H_2O and measured some physical parameters (elastic, thermoelastic, dielectric, thermal expansion, indices of refraction, pyroelectric, electrooptic, etc.) and solubility of these crystals. Indices of refraction of (L-ArgH)F, (L-ArgH)Cl·H_2O, (L-ArgH)Br·H_2O, and some other crystals depending on wavelengths of light in the range 365–1,050 nm as well as phase-matching conditions were determined (DeLoach 1994). Moreover, crystals of (L-ArgH)Cl·H_2O and (L-ArgH)Br·H_2O were grown and studied by IR spectroscopy and thermal analysis (Mukerji and Kar 1998a, b). Dislocation structure of these crystals revealed by etching technique was studied (Mukerji and Kar 1999a, b). Vickers microhardness (Mukerji and Kar 1999c, 2000a) as well as Knoop microhardness (Mukerji and Kar 2000a, b) of these crystals were investigated as well. Isostructuality of (L-ArgH)Cl·H_2O and (L-ArgH)Br·H_2O allowed obtaining also crystals of solid solutions. Crystals of solid solutions were obtained from aqueous solution containing equimolar quantities of (L-ArgH)Cl·H_2O and (L-ArgH)Br·H_2O (Pal et al. 2002). Unit cell parameters of (L-ArgH)($Cl_{0.5}Br_{0.5}$)·H_2O ($a = 11.158(2)$ Å, $b = 8.579(3)$ Å, $c = 11.235(3)$ Å, $\beta = 91.55(4)°$) have values intermediate between values of pure components (see Table 5.23). The temperature of dehydration of solid solution equal to 92 °C also has intermediate value of dehydration temperature for (L-ArgH) Cl·H_2O (65 °C) and (L-ArgH)Br·H_2O (110 °C). Similar situation is in the case of melting temperatures of dehydrated samples: 208 °C for (L-ArgH)($Cl_{0.5}Br_{0.5}$), 221 °C for (L-ArgH)Cl, and 200 °C for (L-ArgH)Br. The efficiencies of powder SHG for (L-ArgH)Cl·H_2O, (L-ArgH)Br·H_2O, and (L-ArgH)($Cl_{0.5}Br_{0.5}$)·H_2O are 3.41, 2.80, and 2.64 times that of quartz powder respectively. These values are higher than that measured in Monaco et al. (1987). The Nd^{3+}:YAG laser damage threshold of (L-ArgH)Cl·H_2O, (L-ArgH)Br·H_2O, and (L-ArgH)($Cl_{0.5}Br_{0.5}$)·H_2O is about 27.72, 16.37, and 29.84 GW/cm^2, respectively. Interesting results were obtained for mechanical properties. For the (100) cleavage plane the maximum value of hardness is 62.29 kg/mm^2 for (L-ArgH)Cl·H_2O and 47.98 kg/mm^2 for (L-ArgH)Br·H_2O, while for the crystal of solid solution is equal to 114.46 kg/mm^2. Transmittance spectra in UV–Vis region for all three crystals do not differ practically. More detailed results on mechanical properties and surface micromorphology obtained by chemical etching technique are provided by Pal and Kar (2003, 2005b, 2006a). Interesting results were obtained by XRD structural analysis of crystals of solid solution (Pal et al. 2004). Crystals of solid solution were obtained by mixing an equimolar solution of (L-ArgH)Cl·H_2O and (L-ArgH)Br·H_2O. As in the case of pure crystals asymmetric unit contains two (L-ArgH) cations, two anions, and two water molecules. According to obtained data the composition of the crystal corresponds to the formula: (L-ArgH) ($Cl_{0.44}Br_{0.56}$)·H_2O, which agrees with expected composition. However, in two crystallographically different positions relation of Cl and Br anions differs noticeably: $Cl_{0.63547}Br_{0.36453}$ in one position and $Cl_{0.24306}Br_{0.75694}$ in the second position. Curves of solubility, IR spectrum, and specific heat data at different temperatures are provided. The data of the specific heat were used to explain the higher optical damage threshold of the crystal of solid solution, compared to KDP crystal. Another study (Thomas et al. (2006)) also is devoted to growing and investigation of crystals of solid solution. This paper, however, contains a series of contradicting statements. From the Introduction of Thomas et al. (2006) one may conclude that the aim was to obtain crystals of solid solution (L-ArgH) ($Cl_{0.5}Br_{0.5}$)·H_2O, obtained by Pal et al. (2002). Unit cell parameters are close to that obtained by Pal et al. (2002). However, from the other hand according to data of Thomas et al. (2006) the crystal is stable at heating with rate 10°/min up to 124.3 °C, while according to Pal et al. (2002) dehydration of the crystal of solid solution starts from 92 °C. In order to obtain crystals of solid solution (L-ArgH)($Cl_{0.5}Br_{0.5}$)·H_2O one should take L-arginine, HCl, and HBr in molar ratio 2:1:1. However, Thomas et al. (2006) took L-arginine, HCl, and HBr in molar ratio 1:1:1. Moreover, Thomas et al. (2006) state that they obtained the crystal of (L-ArgH_2)Cl·Br·H_2O, which corresponds to hypothetic salt with different anions and not to solid solution (L-ArgH)($Cl_{0.5}Br_{0.5}$)·H_2O. Interpreting the Raman spectrum of obtained crystal Thomas et al. (2006) write: "Based on the –OH vibrations in the higher energy region, it can be said that both HCl and HBr are used up in the formation of argininium salts with the stoichiometry 1:1:1, L-arginine:HCl:HBr." One more similar work is the paper by Siddheswaran et al. (2006), the authors of which declared on obtaining a new crystal L-arginine hydrochlorofluoride monohydrate. The authors write that the crystal was synthesized by an equimolar ratio of (L-ArgH)Cl·H_2O and (L-ArgH)F·H_2O. Actually the crystal with composition (L-ArgH)F·H_2O is not known, while known anhydrous crystal (L-ArgH)F is not isostructural with (L-ArgH)Cl·H_2O. So, the statement of a question is dubious. Siddheswaran et al. (2006) do not indicate the composition of supposed crystal. For example, it may be a solid solution (L-ArgH)($Cl_{1-x}F_x$)·H_2O or mixed salt with different anions (L-Arg$H)_2$F·Cl·H_2O. Siddheswaran et al. (2006) indicate that obtained crystals belong to monoclinic system (space group $P2_1$) and provide unit cell parameters ($a = 11.0289$ Å, $b = 8.4752$ Å, $c = 11.1955$ Å, $\alpha = 90.0504°$, $\beta = 91.3670°$, $\gamma = 89.9717°$, $Z = 4$). Clearly, only a crystal with two angles of exactly 90° (usually given as α and γ) belongs to the monoclinic system. If the angles deviate from 90° (even slightly) the crystal is triclinic. We suppose the refinement was carried out in the triclinic system. The authors show also the IR spectrum of the crystal and thermal data. Discussing the IR spectrum the authors assume the existence of covalent bonds C-Cl and C-F, which is erroneous in our opinion. Unit cell parameters and IR spectrum in our opinion showed that the crystal obtained by Siddheswaran et al. (2006) is (L-ArgH)Cl·H_2O. The IR spectrum of obtained crystal perfectly corresponds to the spectrum of (L-ArgH)Cl·H_2O provided by Kalaiselvi et al. (2008b) and also to the spectrum of "(L-ArgH)(F…HF)" given by Sankar et al. (2010). Lima et al. (2002) studied Raman spectra of (L-ArgH)Cl·H_2O in wide temperature region from 295 K to 10 K. As a result of this investigation the authors came to the conclusion that there is phase transition in the region 100–110 K. Meera et al. (2004) have grown the crystals of (L-ArgH)Cl·H_2O from aqueous solution and also from mixture of 70 % ethanol and 30 % water by temperature lowering method in the interval 45 °C to room temperature. A large crystal of (L-ArgH)Cl was grown by evaporation method at 60 °C. According to the data of the authors (L-ArgH)Cl·H_2O crystals obtained from water and water–ethanol solutions have Vickers hardness 178 kg/mm^2 and 190 kg/mm^2 respectively, which are significantly higher not only than that of previously obtained crystal (L-ArgH)Cl·H_2O (62.29 kg/mm^2), but also for crystal of solid solution (114.47 kg/mm^2) (Pal and Kar 2003). Almeida et al. (2003, 2006) used the X-ray multiple diffraction method to the single crystal (L-ArgH)Cl·H_2O for determination of all piezoelectric coefficients. It should be, however, noted that the values of piezoelectric coefficients obtained in these works are three orders higher than values of piezoelectric coefficients measured by traditional method. For example, d_{11} for quartz is 2.3 pC/N, while the values measured in these works have nC/N orders. Recently three more papers were published devoted to these crystals (Sangeetha et al. 2011a, b, c). Sangeetha et al. (2011a) studied the influence of irradiation by Ag^{8+} ions on the structure and spectral properties of (L-ArgH)Cl·H_2O and (L-ArgH)Br·H_2O, while Sangeetha et al. (2011b) investigated the influence of doping by Cu^{2+} and glycine on structural, optical, and dielectric properties of (L-ArgH)Cl·H_2O.

Sangeetha et al. (2011c) grew (L-ArgH)Cl·H_2O crystals by conventional and unidirectional solution growth methods and characterized them.

Unidirectional method is a simple and effective method for crystal growth from solution. This method was developed in 1961 by the outstanding Soviet scientist and practitioner A. A. Shternberg (AIP 1993) and named his method as "growth in forms" (Shternberg 1961). Another modification is described by Nepomnyashaya et al. (1961). This method is described in a book for beginners (Petrov et al. 1967, 1969, 1983) and in a book for students and researchers (Kozlova 1967, 1972, 1980). Later this method was reinvented by Sankaranarayanan and Ramasamy (2005) and further developed by these and other authors.

In the systems L-Arg + HCl + H_2O and L-Arg + HBr + H_2O in addition to the abovementioned salts with singly charged cation (L-ArgH) salts with doubly charged cation (L-ArgH_2) were obtained: (L-ArgH_2)Cl_2·H_2O and (L-ArgH_2)Br_2·H_2O (Petrosyan et al. 2000b). The first salt of arginine ((L-ArgH_2)I_2 with doubly charged [$^+(H_2N)_2CNH(CH_2)_3CH(NH_3{}^+)COOH$] cation was discovered by Seely (1966b). Seely obtained the crystal when he tried to find crystallization conditions of (L-ArgH)I and determined the crystal structure of (L-ArgH_2)I_2. This important work, however, remained essentially unnoticed. Petrosyan et al. (2000b) determined the structures of (L-ArgH_2)Cl_2·H_2O and (L-ArgH_2)Br_2·H_2O at room temperature and IR spectra are provided. Petrosyan et al. (2004a) determined structures at low temperatures as well as pyroelectric and piezoelectric properties. The main difference of (L-ArgH) and (L-ArgH_2) cations is in that in the case of (L-ArgH) there is negatively charged carboxylate group COO^-, which serves as hydrogen bond acceptor, while in the case of (L-ArgH_2) there is neutral carboxylic group COOH, which serves as hydrogen bond donor. This leads to significant difference in hydrogen bond schemes in the structures of salts of arginine with singly and doubly charged cations. In spite of that asymmetric units of both (L-ArgH_2)Cl_2·H_2O and (L-ArgH_2)Br_2·H_2O salts contain one formula unit, the symmetries and structures of these salts are different. The first salt crystallizes in the orthorhombic system, while the second has triclinic symmetry. In the structure of (L-ArgH_2)Cl_2·H_2O carboxylic group of cation forms hydrogen bond O2-H…O3W (2.635(2) Å) with water molecule, which in turn forms hydrogen bond with anion O3W-H1…Cl2(3.189(2) Å) and O3W-H2…Cl2(3.232(2) Å). All active hydrogen atoms of $NH_3{}^+$ and guanidyl group form hydrogen bonds with anions. In the structure of (L-ArgH_2)Br_2·H_2O carboxylic group of cation forms hydrogen bond with one of the anions O2-H…Br1(3.108(6) Å). Water molecule as hydrogen bond acceptor forms hydrogen bond with the $NH_3{}^+$ group N1-H1…O3W(2.732 Å), which in turn does not form noticeable hydrogen bonds as proton donor, which explain the difference in the IR spectra of these salts in the region of absorption of stretching vibrations of water molecules. Thus, in the (L-ArgH_2)Cl_2·H_2O, (L-ArgH_2)Br_2·H_2O, (L-ArgH_2)I_2 series one can see the weakening of water molecule bonding up to its absence in the last term. In accordance with its polar symmetry the crystal of (L-$ArgH_2$)$Br_2{\cdot}H_2O$ exhibits pyroelectric effect. Along with the direction perpendicular to (100) crystallographic plane pyroelectric coefficient reaches 44μC/m^{2}K, which is considerably higher than in the case of corresponding monohalide monohydrates (−0.66 and −0.92 μC/m^{2}K for (L-ArgH)$Cl{\cdot}H_2O$ and (L-ArgH)$Br{\cdot}H_2O$ respectively, see Haussühl et al. 1990). The value of dielectric constant ε in this direction is equal to 6.5. So, the figure of pyroelectric merit of the crystal γ/ε is equal to 6.8 μC/m^{2}K, which is comparable with the corresponding values of the best pyroelectric materials. In this regard it should be noted that the existence of triclinic form of (L-$ArgH_2$)$Cl_2{\cdot}H_2O$ cannot be ruled out, which also may be interesting as pyroelectric material. Taking into account that in the structure of (L-$ArgH_2$)$Br_2{\cdot}H_2O$ carboxylic group forms hydrogen bond with bromide ion, one may assume that excess of HCl in solution compared to stoichiometry will stimulate the formation of triclinic form of (L-$ArgH_2$)$Cl_2{\cdot}H_2O$. The first in the series of salts of L-arginine with doubly charged cation was (L-$ArgH_2$)$(IO_3)_2$ (see Petrosyan et al. 1998 and references therein). Monaco et al. (1987) reported that the salt (L-ArgH)IO_3 as well as some others is formed in the form of fine powder and does not crystallize. Petrosyan et al. (1998) provided IR and ^{127}I NQR spectra of (L-ArgH)IO_3. It was not possible to obtain crystals of (L-ArgH)IO_3; however, the possibility of registration of NQR spectrum indicates that the powder is not amorphous and has to some extent an ordered structure. An attempt was made to obtain the salt (L-ArgH)(IO_3...HIO_3) taking into account the propensity of HIO_3 for formation of acid salts. However, the IR spectrum of crystals obtained from aqueous solution containing L-arginine and HIO_3 in molar ratio 1:2 indicated on the absence of the HIO_3 group in the structure of the crystal. Determination of the structure by XRD method confirmed the presence of (L-$ArgH_2$) cation and two IO_3^- anions. Then other salts with doubly charged (L-$ArgH_2$) cation were obtained, (L-$ArgH_2$)$(NO_3)_2$, (L-$ArgH_2$)$(H_2PO_4)_2$, (L-$ArgH_2$)$Cl_2{\cdot}H_2O$, and (L-$ArgH_2$)$Br_2{\cdot}H_2O$; their crystals were obtained and structures were determined (Petrosyan et al. 2000b). By analogy an attempt was made to obtain also salts with ClO_3^- and BrO_3^- anions. In this case, however, salts with doubly charged cation were not possible to obtain. The crystals of salts (L-ArgH)ClO_3 and (L-ArgH)BrO_3 were obtained. The salts were obtained from aqueous solution containing L-arginine and respective acid in molar ratio 1:1. The solution of $HClO_3$ was obtained by passing an aqueous solution of $KClO_3$ through an exchange column with a sulphocationite in H^+ form. The solution of $HBrO_3$ was obtained after precipitation and isolation of barium sulfate by the reaction of $Ba(BrO_3)_2{\cdot}H_2O$ with H_2SO_4. Petrosyan et al. (2005a) studied IR and NQR spectra of obtained crystals, determined the structures, grew crystals, and measured piezo- and pyroelectric characteristics. In spite of the same formation mechanism of these salts their structures have significant differences: the number of independent formula units, conformation of cations, packing, and hydrogen bonds. Salts with doubly charged cation do not form obviously because of instability of respective acids at higher concentrations.

The crystals of (L-$ArgH_2$)$(IO_3)_2$ were studied also (Preema et al. 2006, 2007; Sankar et al. 2008a, b; see also Petrosyan 2008a). One more system from which a powdered "L-arginine nitrate" salt was obtained was the system L-Arg + HNO_3 + H_2O.

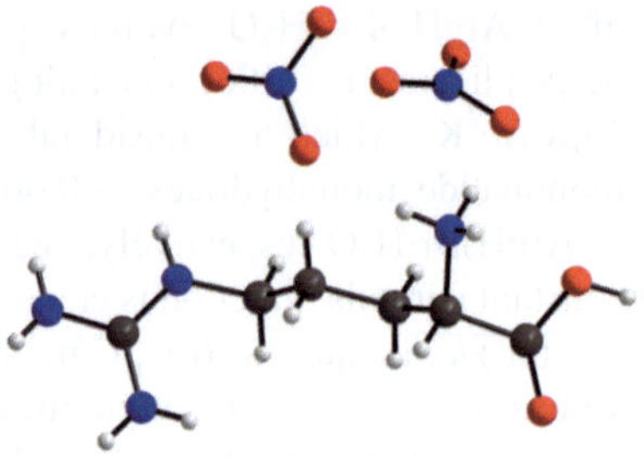

Fig. 5.36 Molecular structure of L-argininium^{2+} dinitrate

Initially plate-like crystals of (L-ArgH$_2$)(NO$_3$)$_2$ were obtained (Petrosyan et al. 2000b), then larger bulk crystals with sizes $20 \times 10 \times 4$ mm^3 were obtained (Petrosyan et al. 2002), and later crystallization conditions of "L-arginine nitrate" and its crystal structure were determined (Terzyan et al. 2004), which turned out hydrate (L-ArgH)$_2$(NO$_3$)$_2$·H$_2$O or formally (L-ArgH)NO$_3$·0.5H$_2$O. Needle-like crystals (L-ArgH)$_2$(NO$_3$)$_2$·H$_2$O were possible to obtain from aqueous solution with molar ratio 1:1.25 of L-Arg:HNO$_3$. At equimolar ratio powdered sample is formed. Triclinic cell contains one formula unit, i.e., two (L-ArgH) cations, two nitrate anions, and one water molecule. Water molecule as acceptor forms hydrogen bonds with both cations: N1A-H3A…O1W(2.824 Å) and N4B-H14B…O1W(2.945 Å). As a donor water molecule forms hydrogen bonds with two symmetrically equivalent cations of the same type: O1W-H2W…O1B(2.727 Å) and O1W-H1W…O2B(2.688 Å). The NH$_3^+$ groups of cations form hydrogen bonds with carboxylate groups: N1A-H1A…O1A (2.837 Å) and N1B-H1B…O1B(2.844 Å). The asymmetric unit of (L-ArgH$_2$)(NO$_3$)$_2$ contains one formula unit. The carboxylic group of cation forms hydrogen bond O2-H1…O4(2.655 Å) with one of the nitrate anions. The remaining active hydrogen atoms form hydrogen bonds N-H…O with nitrate anions. The main distinction of (L-ArgH) and (L-ArgH$_2$) cations is that the (L-ArgH$_2$) cations do not form hydrogen bonds with each other.

The structure and vibrational spectra of (L-ArgH$_2$)(NO$_3$)$_2$ are studied also (Ramaswamy et al. 2001, 2003b; the molecular structure is shown in Fig. 5.36). The IR spectrum of (L-ArgH$_2$)(NO$_3$)$_2$ registered by KBr pellet technique (Ramaswamy et al. 2003b) differs from the spectrum shown by Petrosyan et al. (2000b). Petrosyan and Sukiasyan (2008) studied IR and Raman spectra of (L-ArgH)$_2$(NO$_3$)$_2$·H$_2$O and (L-ArgH$_2$)(NO$_3$)$_2$ and the distinction of the spectra in Petrosyan et al. (2000b) and Ramaswamy et al. (2003b) explained by possible partly decomposition and ionic exchange during preparation of pellet with KBr under high pressure. The crystals of (L-ArgH$_2$)(NO$_3$)$_2$ were grown and characterized as possible NLO material (Thomas et al. 2005a; Preema et al. 2007). Vibrational spectra of (L-ArgH)$_2$(NO$_3$)$_2$·H$_2$O were studied theoretically and experimentally (Vijayakumar et al. 2009).

De Matos et al. (2001) obtained crystals of (L-ArgH)(HSeO$_3$) from aqueous solution containing L-arginine and H$_2$SeO$_3$ in molar ratio 1:2. Obtained crystals had small sizes (less than 0.5 mm). The SHG efficiency measured by powder method was two times higher than that of urea. The authors determined the structure of obtained crystals. Asymmetric unit contains one formula unit. The authors assumed also (in our opinion erroneously) the presence of water crystallization. In our opinion in the structure there was some quantity of Se, which usually is formed as a result of decomposition of selenite salts. This is evident from the light red color of obtained crystals. The anions form infinite chain by means of O1-H1...O2(2.696 (3) Å) hydrogen bond. Active hydrogen atoms form hydrogen bonds with oxygen atoms of the carboxylate group and anion. De Matos et al. (2001) found an anomaly on DSC curve at 289 K and were going to determine the structure of the crystal at temperatures lower than this temperature for elucidation of the nature of this anomaly. However, we could not find in literature results of this investigation.

Crystals of salts (L-ArgH)BF_4 and (L-ArgH)ClO_4 were obtained by Monaco et al. (1987). The symmetries, unit cell parameters, and linear and nonlinear characteristics were determined. Intensities of SHG signals according to this work were 0.07 and 0.14 compared to quartz standard for (L-ArgH)BF_4 and (L-ArgH)ClO_4 respectively. Structures of these crystals were determined (Wang et al. 1994, Srinivasan and Rajaram 1997a). The structure of (L-ArgH)BF_4 is determined not with high accuracy. In the structure of (L-ArgH)ClO_4 oxygen atoms of anion are disordered, and in guanidyl group there is an excess hydrogen atom attached to N2 nitrogen atom. Owens et al. (2001) determined the solubility of (L-ArgH)BF_4 in water and some other solvents and have grown crystals from aqueous solution by temperature lowering method. Transmittance spectrum in the UV–Vis region and thermal behavior by DSC method were determined. According to this work the crystal of (L-ArgH)BF_4 melts at 242 °C. The crystals have been grown and studied also by Rajan Babu et al. (2003) and Jeyakumari et al. (2006). The solubility curve given by Rajan Babu et al. (2003) noticeably differs from that provided by Owens et al. (2001). For confirmation of the presence of functional groups, the authors provide the IR spectrum, which as indicated by Petrosyan (2006a), actually is the spectrum of L-alanine. The crystal of (L-ArgH)BF_4 obtained by Jeyakumari et al. (2006) has melting temperature 225 °C, which significantly differs from the value indicated by Owens et al. (2001). Solubility curves in water and ethanol are shown by Jeyakumari et al. (2006). The solubility in ethanol is significantly higher than that given by Owens et al. (2001). The IR spectrum and the TG and DTA curves also are provided by Jeyakumari et al. (2006). The efficiency of SHG $I^{2\omega}/I^{2\omega}_{KDP} = 0.8$ was determined, which is significantly higher than that obtained by Monaco et al. (1987). Jeyakumari et al. (2006), also showed the presence of phase-matching by Kurtz–Perry method. They also observed that the presence of HBF_4 acid in solution prevents the generation of microbes during one month. Growth and further investigation of (L-ArgH)ClO_4 crystals after Monaco et al. (1987) was performed by Mallik and Kar (2005a). The solubility was determined, crystals were grown from aqueous solution by evaporation method, UV–Vis and IR spectra were provided, thermal behavior was studied, and efficiency of SHG was measured by powder method. Crystals melt at 230 °C and then decompose at 250 °C with exothermic effect. According to the data of this work, the efficiency of SHG is 3.12 times higher than in quartz and 0.13 that of urea, which is significantly higher than that reported by Monaco et al. (1987). Later (L-ArgH)ClO_4 crystals were studied also (Aruna et al. 2007a, b; Kalaiselvi et al. 2008c). Aruna et al. (2007a) prepared (L-ArgH)ClO_4 crystals from aqueous solution by evaporation at constant temperature and provided IR and Raman spectra. However, discussion of IR spectrum is contradictory. Aruna et al. (2007a) assigned some bands in the spectrum to various types of OH and COOH groups, which are actually absent in the structure of crystal (see Petrosyan 2006a). Thermal data also are provided, which are in agreement with data of Mallik and Kar (2005a). The efficiency of SHG by powder method is also measured as 0.2 of urea (Aruna et al. 2007a, b). Kalaiselvi et al. (2008c) grew crystals by temperature lowering method, the solubility was determined, and UV–Vis and IR spectra are provided, as well as thermal curves and efficiency of SHG by powder method, which was 0.17 that of urea. Recently one more paper was published, where theoretically and experimentally vibrational spectra and NLO characteristics were studied (Manimaran et al. 2013). The efficiency of SHG measured by powder method was 0.18 that of KDP and 0.02 of urea, which is significantly lower than that measured earlier (Mallik and Kar 2005a; Aruna et al. 2007a, b; Kalaiselvi et al. 2008c). Wang et al. (1994) overestimated the calculated data and strong absorption band in the IR spectrum at 1,121 and 1,076 cm^{-1}, which in our opinion relate to asymmetric stretching vibration of perchlorate anion, assigned to other vibrations. To obtain trustworthy data on NLO properties of this crystal, measurements on single crystal are necessary in our opinion. One more compound (L-$ArgH_2$)$(ClO_4)_2$ was obtained in the L-Arg + $HClO_4$ + H_2O system (Sukiasyan et al. 2007). Previous attempt to prepare this crystal (Petrosyan et al. 2000b) was not successful. The reason of the previous unsuccessful attempt became clear. It turns out that the crystal of (L-$ArgH_2$)$(ClO_4)_2$ is extremely hygroscopic. At usual laboratory conditions the crystal absorbs moisture from air and rapidly turns into liquid. The crystal is possible to obtain from aqueous solution containing stoichiometric quantities of components in a tightly closed vessel in the presence of water absorber. The formed crystals also should be kept in close vessel in the presence of moisture absorber. In Fig. 5.37 IR and Raman spectra of (L-$ArgH_2$)$(ClO_4)_2$ are shown. Comparison of spectra of (L-ArgH)ClO_4 (Petrosyan 2006a) with the spectra shown in Fig. 5.37 allows to make the following conclusions. The presence of the absorption band at 1,734 cm^{-1} and respective Raman line at 1,739 cm^{-1} shows the presence of carboxyl COOH group and hence the (L-$ArgH_2$) doubly charged cation. The most intensive line in the Raman spectrum relates to symmetric stretching vibration of perchlorate anion. This line in Fig. 5.37 is split into two components with wavenumbers 943 cm^{-1} and 933 cm^{-1}, which indicates on the presence of two crystallographically different perchlorate anions. In the IR spectrum in this region in addition to very strong band caused by asymmetric stretching vibration there are weak peaks at 940 cm^{-1} and 928 cm^{-1}. For ideal tetrahedral perchlorate ion symmetric stretching vibration is not active in IR spectrum. However, in real crystal as a result of inequivalence of Cl-O bonds, this vibration may reveal also in IR spectrum. Therefore, we believe that peaks at 940 cm^{-1} and 938 cm^{-1} also reflect the presence of two crystallographically different perchlorate ions. Recently the structure of the (L-$ArgH_2$)$(ClO_4)_2$ crystal

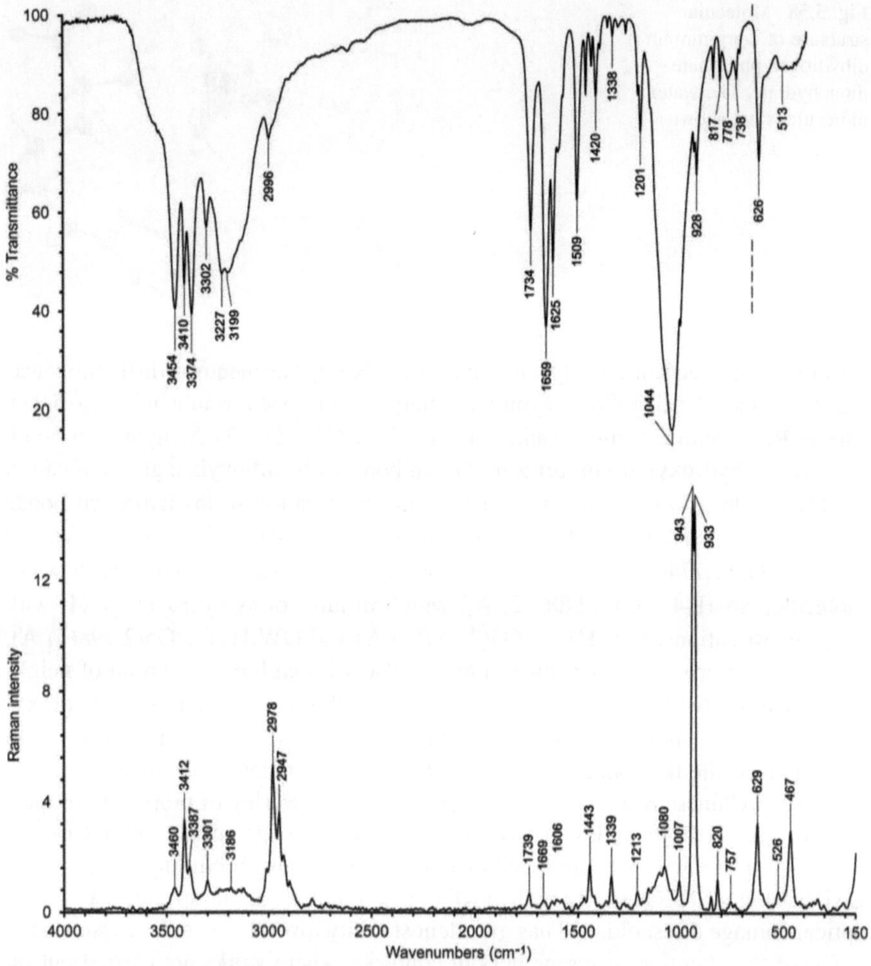

Fig. 5.37 IR and Raman spectra of L-argininium^{2+} di-perchlorate

was determined at 120 K (Suponitsky et al. in prep.). Unit cell contains one (L-ArgH_2) cation and two perchlorate anions. Some oxygen atoms of perchlorate anions are disordered. The carboxylic group of cation forms O2-H2. . .O7 (2.733 Å) hydrogen bond with one of the anions. The NH_3^+ group forms two hydrogen bonds with $Cl(1)O_4^-$ anion and one hydrogen bond with $Cl(2)O_4^-$ anion. Hydrogen bonds formed by guanidyl group are weak.

The crystal of (L-ArgH)(H_2PO_4)·H_2O (Table 5.23) greatly stimulated further investigation of salts of arginine and other amino acids. Crystals of (L-ArgH) (H_2PO_4)·H_2O were obtained by Aoki et al. (1971) and Saenger and Wagner (1972) and its structure was determined. Later the structure of $[(D_2N)_2CND(CH_2)_3CH(ND_3)COO]^+ \cdot D_2PO_4^- \cdot D_2O$ partly deuterated crystal by neutron diffraction was determined (Cheng et al. 1995, 1997). More accurate determination of the

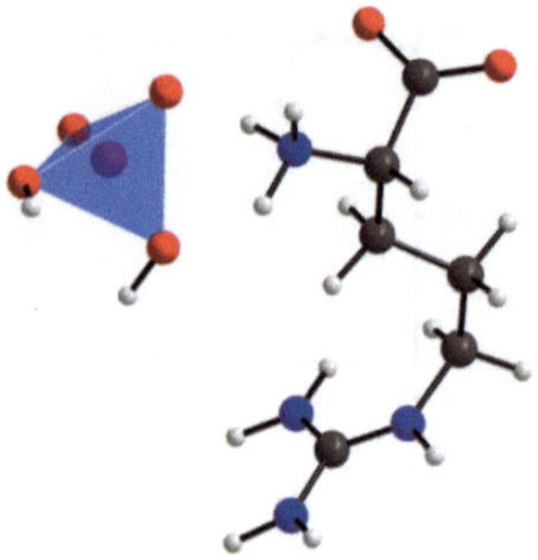

Fig. 5.38 Molecular structure of L-argininium dihydrogen phosphate monohydrate. The water molecule is not shown

structure and electron density distribution by X-ray and neutron diffraction was made (Espinosa et al. 1996). Asymmetric unit contains one formula unit (Fig. 5.38). The $H_2PO_4^-$ anions form a chain via O4-H19...O1(2.568(3) Å) hydrogen bond. The second hydroxyl group forms hydrogen bond with carboxylate group of cation O3-H18...O6(2.598(3) Å). The $N(1)H_3^+$ group of cation forms hydrogen bonds with three nearest anions: N1-H2...O1(2.829(2) Å), N1-H3...O2(2.801(2) Å), and N1-H4...O2(2.774(2) Å). The guanidyl group forms hydrogen bonds with water molecule, N4-H14...OW(2.887(2) Å), which in turn forms hydrogen bonds with two nearest cations: OW-H16...O5(2.751(3) Å) and OW-H17...O5(2.798(3) Å). The guanidyl group forms hydrogen bonds also with carboxylate group of neighboring cation, N3-H12...O6(2.926(2) Å), and with oxygen atom O3 of hydroxyl group of anion N4-H15...O3(2.852(2) Å). There is also weak bond N2-H11...O6 (3.070(3)), while the contact N3-H13...O1 is usual van der Waals bond.

In 1983 Chinese researchers have reported on discovering of promising properties of (L-ArgH)(H_2PO_4)·H_2O (LAP) as NLO material (Jiang et al. 1983). Crystals of LAP were grown by cooling and evaporating methods. According to this work, LAP has higher NLO efficiency than KDP, i.e., 3.5 times of this species, has higher optical damage threshold, and has excellent stability toward moisture. These results attracted the attention of researchers in countries where works are carried out on laser nuclear fusion program. For this purpose, KDP crystals were intended to be used in this program. These are the USA (Velsko and Eimerl 1986, 1988; Eimerl et al. 1989a, b, 1990; Fuchs et al. 1989; Langhoff 1990; Barker et al. 1991; Robertson and Dunn 1993), Japan (Sasaki et al. 1989; Kitaoka et al. 1989; Yokotani et al. 1989, 1990), the USSR (Andreev et al. 1990; Silvestrova et al. 1992), India (Dhanaraj et al. 1991a, b, 1992; Venkatararamanan et al. 1994; Bhat 1994; Bhar et al. 1995) which have a long tradition on the study of amino acids and their salts, and other countries as well. Crystals of LAP have a little absorption in the region of fundamental wavelength of Nd:YAG laser (1.06 μ) caused by overtones of vibrations of N-H and O-H bonds. In addition it turned out that during growth of the crystal microorganisms may appear in solution, which are undesirable for growth process and properties of crystal. The problem of undesirable absorption was solved by deuteration of NH and OH bonds by recrystallization of crystals from D_2O (Eimerl et al. 1989a; Barker et al. 1991; Chen et al. 1988). To prevent the appearance of microorganisms it was suggested to add into solution H_2O_2, $CHCl_3$ (Yokotani et al. 1990), cover the solution by a layer of n-hexane (Dhanaraj et al. 1991b), add sodium azide (Carvalho et al. 1997), etc. Investigation of this problem continued also later (see below). Growth conditions of LAP and deuterated LAP (D-LAP) crystals were investigated; linear and nonlinear optical properties and dielectric, elastic, piezoelectric, thermal, and other characteristics were studied in detail and compared with KDP and other NLO crystals. Temperature-insensitive phase-matching for second harmonic generation in D-LAP was found (Barker et al. 1991); it was pointed out on possible application of D-LAP as optical parametric oscillator (Robertson and Dunn 1993) and high-energy operation of stimulated Brillouin scattering mirror due to much lower stimulated Brillouin scattering threshold (Yoshida et al. 1997). The crystals of LAP (and D-LAP) surpass now-in-use crystal of KDP by nonlinearity and damage threshold; however, they are inferior by thermal stability and by transmittance in UV region. Thermal behavior was studied also by Mazumder et al. (1995). Vibrational spectra were studied (Dhanaraj et al. 1991a; Chen et al. 1988; Mazumder et al. 1995; Espinosa et al. 1994). The method of EPR with using VO^{2+} and Cu^{2+} as paramagnetic ions was used to determine the locations of these ions in lattice of LAP crystal (Angeli Mary and Dhanuskodi 2001; Dhanuskodi and Angeli Mary 2002; Santana et al. 2002). Significant attention was paid to the problems of investigation of solution, growth mechanism, nucleation, revealing, and investigation of crystal defects by various methods (Kim and Lal 1998; Kim 2000; Arunmozhi et al. 1997; Mohan Kumar et al. 1999; Arunmozhi et al. 2004; Haja Hameed and Rohani 2007; Rao et al. 1990; Sangwal et al. 1994, 1995a, b, 1996, 1997a, b, c, Sangwal and Sanz 1997; Liu et al. 2000; Geng et al. 2004a, b, 2005a, b).

Influence of dopants was studied in order to suppress microorganisms as well as improve the properties of grown crystal (Ravi et al. 1994; Haja Hameed et al. 1999; Li et al. 2000; Haja Hameed et al. 2001, 2002a, 2003b, 2011; Kavitha et al. 2004; Shanmugavadivu et al. 2004a, b; Dhanuskodi et al. 2003, 2005; Singh and Lal 2008; Machado et al. 2010; Arjunan et al. 2010, 2012a; Anandan et al. 2011, 2014). The investigation of LAP crystals also continued (Su et al. 2000a, b; Herreros-Cedrés et al. 2003). The main sources of nonlinearity of LAP crystal based on consideration of isolated L-ArgH cation were studied theoretically by Langhoff (1990). Significantly more realistic information was obtained by Wu et al. (2007) (see also Fonseca et al. 2010). Impact of cation, anion, and water molecule was considered. The most interesting result in our opinion was the showing of possibility of charge transfer from carboxylate group to guanidyl group via hydrogen bonds. Riscob et al. (2012) grew crystals of LAP by unidirectional method and characterized by various methods.

Furberg and Solbakk (1973) obtained (L-ArgH)$(C_2H_5)_2PO_4$ and determined its symmetry and structure. Unfortunately the accuracy of structure determination is low. Guanidyl and amino groups of the cation form hydrogen bonds with two oxygen atoms of the anion, which are not bonded to ethyl groups. The carboxylate group forms two hydrogen bonds with the guanidyl group and one hydrogen bond with the amino NH_3^+ group.

The crystal of (L-$ArgH_2$)(H_2PO_4)$_2$ is one of the salts of L-arginine with doubly charged cation obtained by Petrosyan et al. (2000b). Details of structure and results of measurements of a number of physical properties are provided by Haussühl et al. (2003). Solubility curve was determined showing the possibility of crystal growing by both lowering of temperature and evaporation. The grown crystals are transparent from 237 nm to at least 900 nm. From DTA and DTG curves as well as by thermomicroscopic observations it was found that melting starts at about 174 °C and decomposition above 185 °C occurs. Crystals with dimensions 15 × 20 40 mm^3 for physical measurements were grown at 30 °C by evaporation. Asymmetric unit contains one formula unit, i.e., doubly charged L-argininium(2+) cation and two $H_2PO_4^-$ anions. The dihydrogenphosphate anions $H_2P(1)O_4^-$ and $H_2P(2)O_4^-$ form chains with rather strong hydrogen bonds: O2-H16...O8 (2.605 Å), O7-H19...O3(2.643 Å), and O4-H17...O6(2.519 Å). These chains build up ribbons via O5-H18...O8(2.576 Å) bonds. The carboxyl group of the cation forms hydrogen bond with $H_2P(1)O_4^-$ anion, O9-H20...O1(2.562 Å). The N (1)H_3^+ group forms hydrogen bonds with two nearest $H_2P(1)O_4^-$ anions, N1-H9...O3(2.915 Å) and N1-H10...O3(2.899 Å), and one with the second anion, N1-H8...O8(2.733 Å). The guanidyl group forms two normal hydrogen bonds with both anions, N3-H13...O4(2.978 Å) and N4-H14...O6(2.928 Å), and two weak bonds, N2-H11...O2(3.077 Å) and N4-H15...O1(2.978 Å), while the group N3-H12 is not involved in any hydrogen bonds. For grown crystals of (L-$ArgH_2$)(H_2PO_4)$_2$ coefficients of thermal expansion, dielectric constants as well as piezoelectric, elastic, thermoelastic, and piezoelastic constants were measured at 293 K and compared with respective values of LAP crystals. The crystals of (L-$ArgH_2$)(H_2PO_4)$_2$ attracted also the attention of other authors (Haja Hameed et al. 2002b; Ittyachan and Sagayaraj 2002, 2003a; Pragasam et al. 2005a, b, 2006). Haja Hameed et al. (2002b) reported growth and characterization of (L-$ArgH_2$)(H_2PO_4)$_2$ crystals. However, this paper is questionable. The authors determined unit cell parameters, which are in agreement with previously published values (Petrosyan et al. 2000b). However, powder X-ray diffraction pattern, IR spectrum, and thermoanalytic characteristics show that the crystal actually was LAP and not (L-$ArgH_2$)(H_2PO_4)$_2$. In contrast, Ittyachan and Sagayaraj (2002) have grown (L-$ArgH_2$)(H_2PO_4)$_2$ crystals, recored the IR spectrum, which is in agreement with the spectrum shown by Petrosyan et al. (2000b), confirmed by the determination of crystal structure (Ittyachan et al. 2003) and the reported TG and DTA curves and UV–Vis spectrum. The later, however, is questionable (compare with spectra shown by Pragasam et al. (2005a, b, 2006). Pragasam et al. (2005a, b) studied the influence of Cu^{2+} and Mg^{2+} dopants on some properties of (L-$ArgH_2$) (H_2PO_4)$_2$ crystals. Although Ittyachan and Sagayaraj (2002) determined the crystal structure, Pragasam et al. (2005a, b) assume the presence of water molecule, C (NH_2)$_3$ as well as COO^- groups, which are actually absent in the structure. In another paper (Pragasam et al. 2006) the influence of glycine and valine (without indication the form, L-, D-, or DL-valine) on properties of (L-$ArgH_2$)(H_2PO_4)$_2$ crystals was studied. Here the presence of water molecule $C(NH_2)_3$ as well as

COO^- groups in the structure of crystal also was suggested. In addition, Pragasam et al. (2006) assume the formation of peptides as a result of the interaction of L-arginine with glycine and valine, which is not reflected in the IR spectra.

The crystal of (L-$ArgH_2$)(H_2AsO_4)$_2$ was obtained (Monaco et al. 1987) when the authors attempted to prepare arsenate analog of LAP. The structure was determined (Zalkin et al. 1989). The crystal of (L-$ArgH_2$)(H_2PO_4)$_2$ considered above is isostructural with (L-$ArgH_2$)(H_2AsO_4)$_2$. Optical properties of (L-$ArgH_2$)(H_2AsO_4)$_2$ were determined (Monaco et al. 1987). According to this work, the powder SHG efficiency is 0.12 compared to quartz. Later this crystal was not studied. However, it should be noted that such low values were obtained also for other crystals (except (L-ArgH)F), which, nevertheless, was not a hindrance for their further investigation. Particularly, the salt (L-ArgH)(H_2AsO_4)·H_2O obtained by Marchewka MK, 2006, personal communication was not studied in detail. The IR spectrum of this salt (Marchewka MK, 2006, personal communication) shows that it is isotypic with LAP and is worth for further study. Monaco et al. (1987) attempted to obtain crystals of salts (L-ArgH)(HSO_4) and (L-ArgH)$_2SO_4$; however, these salts had very high solubility and formed viscous solutions. Petrosyan (2006a) synthesized crystals of the salt (L-ArgH)($C_2H_5SO_4$) by reaction (L-ArgH)BF_4 + $KC_2H_5SO_4$ → (-L-ArgH)($C_2H_5SO_4$) + KBF_4↓. The obtained crystals were characterized by IR and Raman spectra and by observation of SHG activity. This indicates its non-centro-symmetric structure, which was confirmed by the result of structure determination (Karapetyan 2008). Asymmetric unit contains one formula unit. The N(1)H_3^+ group forms two hydrogen bonds with two nearest anions, N1-HB...O5(2.823(5) Å) and N1-HC...O6(2.896(4) Å), and one with the carboxylate group of nearest cation, N1-HA...O1(2.786(4) Å). Guanidyl group also forms hydrogen bonds with anion and carboxylate group of cation, N3-HA...O2(2.771(4) Å), N4-HA...O4(2.945 (5) Å), and N4-HB...O1(2.847(4) Å), and also weak bonds on the level of van der Waals interaction, N2-H2...O3(3.098(6) Å) and N3-HB...O6(3.196(5) Å).

A very interesting crystal of (L-ArgH)ReO_4 was obtained recently by Isakov et al. (2012). The crystal was obtained from aqueous solution containing equimolar quantities of L-arginine and $HReO_4$ at 40 °C. According to data of thermal analysis there are no phase transitions from 80 K to melting point (518 K). Asymmetric unit contains one formula unit. The N(1)H_3^+ group forms rather strong hydrogen bonds with carboxylate group of neighboring cation, N1-H1C...O5(2.737(8) Å), and also with anion, N1-H1B...O4(2.900(11) Å). The third contact N1-HA...O1(2.774 (10) Å) in spite of short N...O distance is very weak (H1A...O1)(2.44 Å). The guanidyl group forms hydrogen bond with the second oxygen atom of carboxylate group of nearest cation, N3-H3C...O6(2.842(8) Å). Other hydrogen bonds are weak (N4-H4C...O2, N4-H4D...O5) or very weak (N2-H2A...O6, N3-H3D...O5, N3-H3D...O6). The authors determined NLO properties of the crystal by Maker-fringe method. From three nonzero NLO coefficients (d_{14}, d_{25}, d_{36}) the authors have determined $d_{14} = 1.1 d_{36}^{KDP}$ and $d_{36} = 4.6 d_{36}^{KDP}$. The crystal of (L-ArgH)ReO_4 has high optical damage threshold and very good optical transmittance spectrum.

The crystal of L-arginine tris(acetato-*O*,*O′*)dioxouranium(VI)-acetic acid monohydrate was prepared by Ramos Silva et al. (1999b). The authors indicated uranium (III) in the title of the article, which is not correct and probably is a misprint. The crystal was synthesized by mixing uranyl acetate dihydrate with L-arginine in water–acetic acid solution and slow evaporation. Asymmetric unit contains one formula unit. In complex anion $[UO_2(CH_3COO)_3]^-$ carboxylate groups of acetate ions are located in plane perpendicular to central ion $[UO_2]^{2+}$. The CH_3 group of one acetate ion is disordered. Neutral molecule of acetic acid forms hydrogen bond with carboxylate group of L-argininium cation, O12-H14…O9(2.565(6) Å), which is the shortest among all hydrogen bonds in the structure of crystal. The water molecule as an acceptor forms hydrogen bond N2-H2…O13(2.840(7) Å), while as a donor forms hydrogen bond O13-H13B…O7(2.762(6) Å) with one of the acetate ions and weak bond O13-H3A…O2(2.965(9) Å) with one of the oxygen atoms of $[UO_2]^{2+}$ cation. The amino group $N(1)H_3^+$ and guanidyl group of (L-ArgH) cation are also involved in hydrogen bonds. The second oxygen atom of the carboxylate group of (L-ArgH) cation forms two hydrogen bonds N1-H82A…O10(2.820(6) Å) and N3-H85A…O10(2.946(5) Å). There are also three hydrogen bonds with acetate ions: N1-H82C…O3(2.899(8) Å), N3-H85B…O4(2.917(6) Å), and N4-H4D…O5(2.937(6) Å). The remaining contacts are on the level of strong van der Waals interactions.

Ramos Silva et al. (2000c, 2005b) obtained and determined the structures of two isostructural salts: $(L\text{-}ArgH)_3VF_6\cdot 3H_2O$ and $(L\text{-}ArgH)_3CrF_6\cdot 3H_2O$. The first salt was obtained by reacting of fluoric acid (40 %) with powdered vanadium, and then this solution was added to an aqueous solution of L-arginine in a 1:2 molar ratio of vanadium and L-arginine. Obtained powder was recrystallized in a 1:1 ethanol/water solution. The second salt also was obtained in similar way. However, based on the reaction of the first salt the molar ratio of chromium and L-arginine was 1:3. Because these salts are isostructural, we will consider the structural features on example of the second salt. One of the three (L-ArgH) cations (in both structures) is disordered. One of the three water molecules ($H_2O(8)$) forms hydrogen bonds with $(CrF_6)^{3-}$ anion and carboxylate group of disordered cation: O8-H82…F3(2.787(3) Å) and O8-H81…O5(2.804(3) Å). Another water molecule $H_2O(7)$ forms hydrogen bonds with anion and third water molecule $H_2O(9)$: O7-H72…F5(2.687(2) Å) and O7-H71…O9(2.875(3) Å). The third water molecule forms two hydrogen bonds with two other cations: O9-H91…O2(2.718(3) Å) and O9-H92…O3(2.996(3) Å). The water molecule $H_2O(7)$ as an acceptor forms N12-H12C…O7(2.756(3) Å) with the $N(12)H_3^+$ group of disordered cation. The water molecule $H_2O(9)$ as an acceptor forms one more hydrogen bond N4-H4B…O9(2.946(3) Å) with the $N(4)H_3^+$ group, while the $H_2O(8)$ water molecule as an acceptor forms very weak hydrogen bonds on the level of van der Waals interactions with the guanidyl group of disordered cation. The remaining N-H groups of amino and guanidyl groups form hydrogen bonds with carboxylate groups of cations, N8-H8C…O1(2.884(2) Å), N8-H8A…O5(2.860(3) Å), N6-H6B…O3(2.846(3) Å), N5-H5B…O4(2.929(3) Å), N4-H4A…O4(2.901

(3) Å), N2-H21...O2(2.818(3) Å), and N10-H10A...O6(2.740(5) Å), and also with anions. Some of these bonds are rather strong: N8-H8B...F6(2.640(2) Å), N7-H7...F2(2.723(2) Å), N5-H5A...F1(2.747(2) Å), N4-H4C...F4(2.696(2) Å), N3-H3...F5(2.726(2) Å), N1-H1B...F1(2.732(2) Å), and N11B-H11B...F2(2.744 (5) Å).

Two crystalline forms are known for the salt (L-ArgH_2)$SeBr_6$ (Hempel et al. 1977b). The one in the form of needles crystallizes in triclinic system with two formula units in the cell, while the monoclinic form crystallizes in the form of plates. This salt of L-arginine is the first and yet the only salt with doubly charged arginine cation of $A^{2+}Y^{2-}$ type.

In Table 5.24 salts of arginine with organic acids are collected. The salt of simplest organic acid, namely, formic acid, was obtained in the form of fine powder (Monaco et al. 1987). The IR spectrum of this powder form was published (Haussühl et al. 2006). Later, Suresh et al. (1994a) could obtain monoclinic crystals of (L-ArgH)HCOO by slow diffusion of acetonitrile into aqueous solution of the L-arginine and formic acid and determined its crystal structure. By the same procedure also crystals (DL-ArgH)HCOO·$2H_2O$ were obtained. It was observed that from aqueous solution containing L-arginine and formic acid in molar ratio L-Arg: HCOOH 1:2–5 crystals of (L-ArgH)HCOO may be obtained, which, however, have orthorhombic symmetry (Petrosyan et al. 2000b). The obtained crystals had $3 \times 5 \times 7$ mm^3 sizes. The structure of crystal was determined and IR spectrum was shown. It was found also that white noncrystalline precipitate obtained from solution with equimolar ratio can be recrystallized by using seed crystals of (L-ArgH)HCOO obtained from more acidic solution. Later the sizes of orthorhombic form of (L-ArgH)HCOO were brought to $40 \times 30 \times 29$ mm^3 sizes (Petrosyan et al. 2002) and more detailed results of investigation are provided by Haussühl et al. (2006), where the sizes were brought to $45 \times 60 \times 60$ mm^3. Crystals of orthorhombic form were obtained and studied also (Packiam Julius et al. 2004b; Janarthanan et al. 2009). Haussühl et al. (2006) investigated conditions of growth of orthorhombic form. The solubility of crystals was determined and also the dependence of pH on molar ratio of HCOOH and L-arginine. Abrupt change of pH in the interval from 1:1 to 1:2 for molar ratio of L-Arg:HCOOH was observed. An occurrence of disturbing microorganisms was not observed during crystal growth due to the bactericidal properties of formic acid. Crystal structure was considered and compared with structure of monoclinic form. Asymmetric unit of orthorhombic form contains one formula unit, while the monoclinic form contains two formula units and has pseudo-orthorhombic structure with $Pna2_1$ approximate space group (Suresh et al. 1994a). The most noticeable structural differences are observed in the hydrogen bond interactions with anion. In the monoclinic form, in contrast to the expectation of Suresh et al. (1994a) the guanidyl group interacts primarily with formate ion, while the protonated α-amino group interacts primarily with the carboxylate group. In the orthorhombic form the guanidyl group interacts primarily with the carboxylate group. These distinctions lead to significant distinction in densities (1.308 and 1.422 g/cm^3 in orthorhombic and monoclinic forms

Table 5.24 Salts of arginine with organic anions

#	Crystal	s.g., Z	a, b, c (Å), α, β, γ (°)	Refs
1	(L-ArgH)HCOO, powder form	–	–	Monaco et al. (1987); Haussühl et al. (2006)
2	(L-ArgH)HCOO, monoclinic	$P2_1$, 4	8.490(1), 6.020(1), 20.207(2), $\beta = 92.29(1)$	Suresh et al. (1994a)
3	(L-ArgH)HCOO, orthorhombic	$P2_12_12_1$, 4	7.706(2), 7.808(2), 18.594(4)	Haussühl et al. (2006)
4	(DL-ArgH)HCOO·$2H_2O$	$P2_1/c$, 4	14.587(3), 5.027(1), 18.647(3), $\beta = 111.53(1)$	Suresh et al. (1994a)
5	(L-ArgH)CH_3COO	$P2_1$, 2	9.229(2), 5.178(3), 13.271(4), $\beta = 111.4(1)$	Suresh and Vijayan (1983b)
6	(DL-ArgH)CH_3COO·H_2O	$P2_1/c$, 4	13.552(2), 5.048(2), 18.837(3), $\beta = 101.34(2)$	Soman et al. (1989)
7	(L-ArgH)CF_3COO	$P2_1$, 2	10.581(2), 5.710(1), 10.861(2), $\beta = 106.81$	Xu et al. (2003)
8	(L-ArgH_2)$(CF_3COO)_2$	$P2_1$, 2	9.8457(9), 5.7677(13), 14.5035(12), $\beta = 95.178(7)$	Ramos Silva et al. (2003)
9	(L-ArgH) diphenylacetate	$P2_12_12_1$, 8	9.266(3), 14.659(5), 30.70(2)	Ramos Silva et al. (1999c)
10	(L-ArgH) hydrogen squarate	$P1$, 2	5.113(2), 8.279(2), 14.860(5), $\alpha = 93.01(2)$, $\beta = 96.27(3)$, $\gamma = 99.87(3)$	Angelova et al. (1996)
11	(L-ArgH) L-ascorbate	$P2_1$, 2	5.060(8), 9.977(9), 15.330(13), $\beta = 97.5(2)$	Sudhakar and Vijayan (1980)
12	(L-ArgH) 4-nitrophenolate·H_2O	$P2_12_12_1$, 4	5.1254(2), 9.3683(3), 31.5145(10)	Srinivasan et al. (2008c)
13	(L-ArgH) 4-nitrophenolate 4-nitrophenol·$2H_2O$	$P2_1$, 2	7.8669(5), 10.3764(2), 13.8301(2), $\beta = 98.2128(9)$	Wang et al. (2011)
14	(L-ArgH) picrate·$2H_2O$	$P2_1$, 2	17.798(3), 5.005(1), 10.732(2), $\beta = 98.32(3)$	Nagata et al. (1995)
15	(DL-ArgH) picrate	$P2_1/n$, 4	32.833(3), 9.989(1), 5.051(1), $\beta = 91.25(3)$	Nagata et al. (1995)
16	(L-ArgH) 2,5-dihydroxy benzoate·H_2O	$P2_1$, 2	5.3104(3), 18.446(2), 8.1639(6), $\beta = 94.319(5)$	Kinsel et al. (2004)
17	(L-ArgH) *p*-nitrobenzoate·H_2O	$P2_1$, 2	8.566(3), 5.817(2), 17.131(7), $\beta = 101.223(5)$	Wang et al. (2014)
18	(*S*-ArgH) *S*-naproxen[a]	$C2$, 4	14.662(5), 9.225(5), 17.129(5), $\beta = 113.865(5)$	Vayá et al. (2005)
19	(*S*-ArgH) *R*-naproxen	$P3_221$, 6	9.145(5), 9.145(5), 43.307(5), $\gamma = 120$	Vayá et al. (2005)
20	(*R*-ArgH) *S*-naproxen	–	–	Vayá et al. (2005)
21	(L-ArgH)X(1).$(CH_3)_2OS$[a]	$C2$, 8	51.897(14), 9.101(2), 14.864(2), $\beta = 97.46(1)$	Ebdrup et al. (2003)

22	(L-ArgH) L-aspartate	$P2_1$, 2	5.511(3), 8.438(4), 15.265(9), $\beta=97.9(1)$	Salunke and Vijayan (1982)
23	(L-ArgH) D-aspartate	$P1$, 2	5.239(1), 9.544(1), 14.064(2), $\alpha=85.58(1)$, $\beta=97.9(1)$, $\gamma=84.35(1)$	Suresh et al. (1986)
24	(DL-ArgH) DL-aspartate	$P\bar{1}$, 2	5.402(3), 9.933(3), 13.881(2), $\alpha=99.24(2)$, $\beta=99.73(3)$, $\gamma=97.28(3)$	Soman et al. (1990)
25	(L-ArgH) L-glutamate·H_2O	$P2_12_12_1$, 4	5.047(8), 9.849(8), 32.500(12)	Bhat and Vijayan (1977)
26	(L-ArgH) D-glutamate·$3H_2O$	$P2_1$, 2	9.968(2), 4.652(1), 19.930(2), $\beta=101.20(1)$	Suresh et al. (1986)
27	(L-ArgH) D-glutamate·H_2O	$P2_1$, 2	9.941(1), 4.668(2), 17.307(1), $\beta=95.27(1)$	Soman and Vijayan (1989)
28	(DL-ArgH) DL-glutamate·H_2O	$P\bar{1}$, 2	5.139(2), 10.620(1), 14.473(2), $\alpha=101.34(1)$, $\beta=94.08(2)$, $\gamma=91.38(2)$	Soman et al. (1990)
29	(L-ArgH)$_8$(oxalate)$_4$.10H_2O	$P2_1$, 2	9.9733(2), 36.7922(5), 12.1965(2), $\beta=91.035(1)$, at 120 K	Petrosyan et al. (2005b)
30	(L-ArgH) hydrogen oxalate	$P1$, 2	5.0668(14), 9.7572(13), 13.141(3), $\alpha=111.11(2)$, $\beta=92.75(2)$, $\gamma=91.97(3)$	Chandra et al. (1998)
31	(L-ArgH) hydrogen oxalate	$P2_1$, 2	5.0697(10), 7.1977(14), 16.595(3), $\beta=92.18(3)$	Petrosyan et al. (2005b)
32	(L-ArgH) hydrogen oxalate·oxalic acid	$P2_1$, 2	5.0237(10), 22.292(5), 6.8851(14), $\beta=107.91(3)$	Petrosyan et al. (2005b)
33	(DL-ArgH) hydrogen oxalate	$P2_1/c$, 4	5.1107(12), 23.937(3), 10.947(2), $\beta=116.91(2)$	Chandra et al. (1998)
34	(L-ArgH) hydrogen malonate	$P1$, 1	5.353(6), 6.931(8), 9.922(11), $\alpha=69.530(16)$, $\beta=89.035(17)$, $\gamma=71.128(16)$	Saraswathi and Vijayan (2002a)
35	(DL-ArgH) hydrogen malonate·H_2O	$C2/c$, 8	19.559(2), 5.0296(6), 28.407(4), $\beta=90.553(2)$	Saraswathi and Vijayan (2002a)
36	(L-ArgH)$_2$ succinate·succinic acid·$2H_2O$	$P1$, 1	5.099(1), 10.222(1), 14.626(1), $\alpha=77.31(9)$, $\beta=89.46(9)$, $\gamma=78.42(10)$	Prasad and Vijayan (1990)
37	(DL-ArgH)$_2$ succinate·$4H_2O$	$P2_1/c$, 2	5.292(1), 16.296(3), 15.203(2), $\beta=92.89(8)$	Prasad and Vijayan (1990)
38	(L-ArgH)$_2$ glutarate·$2H_2O$	$C2$, 2	19.008(3), 5.073(2), 13.6723(19), $\beta=101.109(13)$	Saraswathi and Vijayan (2001)
39	(L-ArgH)$_2$ adipate	$P2_1$, 2	12.494(4), 5.9510(17), 16.719(5), $\beta=105.977(5)$	Roy et al. (2005)
40	(DL-ArgH)$_2$ adipate·$2H_2O$	$P2_1/c$, 2	13.825(4), 5.0531(16), 18.804(6), $\beta=102.900(6)$	Roy et al. (2005)
41	(L-ArgH)$_2$ pimelate·$2H_2O$	$C2$, 4	30.278(7), 5.1414(12), 19.355(5), $\beta=115.776(7)$	Saraswathi et al. (2003)
42	(L-ArgH) hydrogen maleate·$2H_2O$	$P1$, 1	5.2641(11), 8.0388(8), 9.7860(11), $\alpha=106.197(9)$, $\beta=97.275(14)$, $\gamma=101.64(2)$	Ravishankar et al. (1998)
43	(DL-ArgH) hydrogen maleate	$Pca2_1$, 4	15.9829(13), 5.413(5), 16.188(18)	Ravishankar et al. (1998)
44	(L-ArgH) hydrogen malate	$P2_1$, 2	10.514, 12.633, 5.07, $\beta=100.76$	Monaco et al. (1987)
45	(L-ArgH)$_2$L-tartrate	$P2_1$, 2	9.907(2), 8.7428(18), 14.076(3), $\beta=109.412(3)$	Selvaraj et al. (2007)
46	(DL-ArgH) hydrogen DL-tartrate	$P\bar{1}$, 2	5.4876(16), 10.021(3), 13.868(4), $\alpha=109.079(5)$, $\beta=94.378(5)$, $\gamma=94.538(5)$	Selvaraj et al. (2007)

[a]Naproxen = mono[6-methoxy-α-methyl-2-naphthalene acetate], X(1) = (S)-2-ethoxy-3-(4-(2-(phenoxazin-10-yl)ethoxy)phenyl)propionate (ragaglitazar)

respectively). IR and Raman spectra, transmittance in UV–Vis–NIR region, thermal stability, thermal expansion, and elastic and thermoelastic constants were measured. Asymmetric unit of (DL-ArgH)·HCOO·2H_2O contains one formula unit. The N(1)H_3^+ forms two hydrogen bonds with carboxylate groups of two nearest cations, N1-H1...O2(2.942(8) Å) and N1-H3...O2(2.822(10) Å), and one hydrogen bond with anion, N1-H2...O11(2.713(10) Å). Guanidyl group forms hydrogen bonds mainly with anion, N8-H1...O12(2.883(10) Å), N8-H2...O12(2.829(10) Å), and N9-H1...O11(2.834(10) Å), and also one hydrogen bond with carboxylate group of nearest cation, N9-H2...O1(2.892(11) Å), and one hydrogen bond with one of the water molecules, N6-H...W1(2.905(9) Å). One of the water molecules forms hydrogen bond as proton donor with cation, W1...O2(2.751(9) Å), while the second molecule with anion, W2...O12(2.884(10) Å). In addition water molecules form hydrogen bonds with symmetry-equivalent molecules: W1...W1(2.859(8) Å) and W2...W2(2.789(11) Å).

Interaction of L- and DL-arginine with acetic acid was investigated. The crystals of (L-ArgH)CH_3COO (Suresh and Vijayan 1983b) and (DL-ArgH)CH_3COO·H_2O (Soman et al. 1989) were obtained by slow diffusion of acetone vapor into an aqueous solution of the components in equimolar proportion. Asymmetric unit in both cases contains one formula unit. In the structure of (L-ArgH)CH_3COO the N(1) H_3^+ group forms two hydrogen bonds with carboxylate groups of two nearest cations: N1-H2...O1(2.907(5) Å) and N1-H3...O2(2.688(4) Å). Guanidyl group forms three hydrogen bonds with anions, N8-H1...O12(2.830(5) Å), N8-H2...O12 (2.827(3) Å), and N9-H1...O13(2.824(6) Å), and two hydrogen bonds with carboxylate groups of two nearest cations, N9-H2...O1(2.937(5) Å) and N6-H1...O2 (2.848(3) Å). In doing so, the hydrogen bonds N8-H1...O12 and N9-H1...O13 are formed with carboxylate group of the same anion. In case of (DL-ArgH) CH_3COO·H_2O the N(1)H_3^+ group forms two hydrogen bonds with carboxylate groups of two nearest cations and one hydrogen bond with anion: N1-H1...O2 (2.825(6) Å), N1-H3...O2(2.926(7) Å), and N1-H2...O11(2.726(6) Å). The guanidyl group also forms three hydrogen bonds with anions: N8-H1...O12(2.844 (6) Å), N8-H2...O12(2.812(7) Å), and N9-H2...O11(2.844(8) Å). Here also the bonds N8-H2...O12 and N9-H2...O11 are formed with the carboxylate group of the same anion. There is also a hydrogen bond N1-H1...O1(2.846(6) Å). However, the group N6-H forms a hydrogen bond not with the carboxylate group of the cation, but with the water molecule: N6-H...W1(2.933(6) Å). The water molecule in turn as a proton donor forms hydrogen bonds with symmetry-equivalent water molecule, W1-H2...W1(2.901(7) Å), and with carboxylate group of cation, W1-H1...O2 (2.756(7) Å). Monaco et al. (1987) obviously independently of Suresh and Vijayan (1983b) also obtained the crystal of (L-ArgH)CH_3COO and determined its symmetry, unit cell parameters, and linear and nonlinear optical characteristics. Particularly the authors found phase-matching of both I and II types, and also SHG efficiency, which was 1.2 times higher than in quartz standard. The crystal attracted the attention of other researchers for further growing and investigation

(DeLoach 1994; Pal et al. 2003; Muralidharan et al. 2003; Pal and Kar 2005c, 2006b; Geng et al. 2005c; Gulam Mohamed et al. 2007; Selvaraju et al. 2007; Meena and Mahadevan 2008, 2010; Natarajan et al. 2009a; Gnanasekaran and Madhavan 2010; Renuka et al. 2012). More detailed optical characteristics and phase-matching conditions were studied (DeLoach 1994), the solubility was determined, crystals were grown, and the morphology was determined, IR spectrum and thermal characteristics (Pal et al. 2003; see also Muralidharan et al. 2003), transmittance in UV–Vis region, mechanical properties (Pal and Kar 2005c, 2006b), surface morphology by AFM (Geng et al. 2005c), the influence of doping by Cu^{2+}, Mg^{2+} (Gulam Mohamed et al. 2007), Li^{+} (Natarajan et al. 2009a), Na^{+}, K^{+}, glycine, urea (Meena and Mahadevan 2010), La^{3+} (Gnanasekaran and Madhavan 2010) on various characteristics, nucleation kinetics (Selvaraju et al. 2007), dielectric (Meena and Mahadevan 2008) and other properties (Renuka et al. 2012) were studied.

Xu et al. (2003) discovered one more crystal of (L-ArgH)CF_3COO, determined its symmetry and structure, have grown single crystal with $28 \times 17 \times 3$ mm^3 sizes, and measured a number of its characteristics showing that it is promising as NLO material. Asymmetric unit contains one formula unit. The $N(4)H_3^+$ group forms two hydrogen bonds with carboxylate groups of two nearest cations, N4-H41…O2(2.836(3) Å) and N4-H42…O1(2.845(4) Å), and one hydrogen bond with carboxylate group of anion, N4-H43…O1′(2.824(3) Å). The guanidyl group forms two hydrogen bonds with cations, N3-H31…O1(2.909(3) Å) and N2-H21…O2(2.806(3) Å), and three hydrogen bonds with anions, N1-H11…O2′ (2.866(3) Å), N1-H12…O2(2.905(5) Å), and N2-H22…O1(2.905(5) Å). The crystal (L-ArgH)CF_3COO was obtained from aqueous solution containing equimolar quantities of L-arginine and CF_3COOH acid. Sometime later Ramos Silva et al. (2003) obtained the crystal of (L-$ArgH_2$)$(CF_3COO)_2$ by adding an excess of CF_3COOH to an aqueous solution of L-arginine, which crystallizes in the same space group with one formula unit in asymmetric part of unit cell. Fluorine atoms of one of the anions are disordered. The carboxyl group of the (L-$ArgH_2$) cation forms hydrogen bond with carboxylate group of the second anion O6-H6…O3 (2.581(3) Å). The $N(4)H_3^+$ group forms two hydrogen bonds with the first anion, N4-H4B…O1(2.899(4) Å) and N4-H4C…O2(2.772(4) Å), and one hydrogen bond with carbonyl oxygen atom of symmetry-related cation, N4-H4A…O5(2.803(4) Å). The guanidyl group forms hydrogen bonds with anions: N1-H1A…O4(2.874(3) Å), N1-H1B…O4(2.889(4) Å), N2-H2A…O3(2.943(4) Å), N2-H2B…O1(2.990(4) Å), and N3-H3…O1(3.058(3) Å). The last two hydrogen bonds are very weak and are in the intermediate region of weak hydrogen bonds and strong van der Waals interactions. Suresh et al. (2006), being familiar with the paper from Ramos Silva et al. (2003) and evidently believing that the crystal of (L-ArgH)CH_3COO was not obtained, repeatedly obtained this crystal and determined its structure. Xu et al. (2003) provided IR and UV–Vis–NIR spectra (UV transparency cut off is 232 nm); determined SHG efficiency by powder method, which was 2.5 times higher than in KDP; determined also optical damage threshold, which was higher than in LAP; and determined the melting point (ca. 212 °C), which is much higher than in LAP (140 °C). In addition, the presence of fluorine atoms prevents the appearance of microorganisms during crystal growth. During following years this crystal was a subject of more detailed investigation of this group and other authors (Liu et al. 2007a, b, c, d, e, 2008a, b, c, 2009, 2010a, b, c, 2011, 2013; Sun et al. 2008c, 2010; Arjunan et al. 2008a, b).

Liu et al. (2007a) determined by microcrystallization method that the optimal value of pH is 3. It should be, however, noted that alteration of pH was made by the addition not of CF_3COOH, but HCl, i.e., in addition to alteration of acidity a new ion (Cl^-) was added to solution. Probably using the CF_3COOH acid would allow to discover also the salt (L-$ArgH_2$)$(CF_3COO)_2$ in this system. In papers by Sun et al. (2007a) and Liu et al. (2008a) high-quality crystals of (L-ArgH)CF_3COO were grown and IR and Raman spectra and TG and DSC curves are provided. Liu et al. (2008c) investigated the morphology of crystal and various physical properties. Dielectric properties, specific heat, thermal expansion, refractive indices, and their dispersion were investigated. Moreover, Arjunan et al. (2008a) provided results of investigation of dielectric properties but, however, obtained values of dielectric constants and dielectric loss significantly higher than that reported by Liu et al. (2008c), probably because of the presence of systematic errors. Sun et al. (2008c) provided more detailed analysis of vibrational spectra. In a number of works growth morphology of {101} surfaces by AFM (Liu et al. 2007b, c, d) as well as nucleation growth mechanism (Liu et al. 2008a, b, 2009, 2010b, c, 2011; Arjunan et al. 2008b), including nucleation kinetics of deuterated crystal (Sun et al. 2010), are investigated. Liu et al. (2010a) grew crystals of (L-ArgH)CF_3COO doped with Cu^{2+} and Mg^{2+} and their influence on a number of characteristics of the crystal was investigated. In order to obtain doped crystals 2 mol% each of $CuCl_2{\cdot}2H_2O$ and $MgCl_2$ was added in solution. Concentrations of dopants were 0.06 % Cu^{2+} and 0.15 % Mg^{2+} in crystals. Different amounts of dopants in crystals are explained by the authors by the difference of the ionic radii of Mg^{2+}(0.65 Å) and Cu^{2+}(0.72 Å). The influence of doping on solubility was studied. The highest solubility is found in the pure (L-ArgH)CF_3COO crystal, followed by crystals doped with Mg^{2+} and Cu^{2+}. Influence of doping on cell parameters, thermal behavior, IR and UV–Vis–NIR spectra, NLO property, and surface morphology was studied. Crystals doped by Cu^{2+} were blue tinted and had additional absorption in region of 500 nm. According to Liu et al. (2010a), doping leads to increasing of SHG intensity in 1.2 and 1.5 times for Cu^{2+} and Mg^{2+}, respectively. Sun et al. (2011a) have grown deuterated crystal and investigated the influence of deuteration on its various physical characteristics. The deuterated crystal has been grown at pH 4.5–5.0 by temperature reducing method. The structure of deuterated crystal is determined at 273 K, as well as solubility in D_2O, X-ray powder diffraction pattern, IR and Raman spectra, thermal (TG, DTA) curves, dependence of specific heat on temperature, UV–Vis–NIR transmittance spectrum, intensity of SHG, and optical damage threshold. The obtained results show that the properties of deuterated crystal are better than nondeuterated. Liu et al. (2013) theoretically considered optical and electronic properties of (L-ArgH)CF_3COO. Hartree–Fock and second-order

Möller–Plesset perturbation methods were used for determination of molecular dipole moment, polarizability, and first hyperpolarizability. The oriented-gas model predicted the following values of nonlinear optical coefficients: $d_{22} = 3.67$, $d_{21} = 1.86$, $d_{23} = 4.25$, and $d_{25} = 1.39$ relative to d_{36} of KDP.

Sun et al. (2009c) grew and investigated also the crystal of (L-ArgH$_2$)(CF$_3$COO)$_2$. The crystal with $32 \times 21 \times 4$ mm^3 sizes were obtained from aqueous solution by temperature lowering method. The solubility was determined. The structure of crystal was determined at 93 K. It was found that disordering of fluorine atoms of one of the anions disappear at 93 K. The IR and Raman spectra are provided (in more detail the spectra are discussed also by Sun et al. (2011b)) as well as UV–Vis–NIR transmittance spectrum, thermal (TG, DTG, DTA, DSC) curves, temperature dependence of specific heat, thermal expansion and dielectric properties. Sun et al. (2009c) estimated NLO properties by powder SHG method and also optical damage threshold. As a result of obtained data Sun et al. (2009c) came to the conclusion that the crystal of (L-ArgH$_2$)(CF$_3$COO)$_2$ also is a potential candidate for NLO material. We cannot agree, however, with the assignment of stretching vibration of O-H bond of COOH group assumed by Sun et al. (2009c, 2011b). The authors assigned a relatively narrow absorption peak at 3,457 cm^{-1} to stretching vibration of O-H bond. According to the authors this peak is absent in the spectrum of (L-ArgH)CF$_3$COO and this assignment corresponds to the calculated value. It should be noted, however, that in calculations only intramolecular hydrogen bonds are considered. Actually in the structure of (L-ArgH$_2$)(CF$_3$COO)$_2$ the carboxyl group forms rather strong hydrogen bond O-H…O with O…O distance (2.581(3) Å) (Ramos Silva et al. 2003). Respective ν(OH) value of this distance according to Novak's correlation is near 2,500 cm^{-1}. Actually in this region there is a band with two peaks, one of which may relate to ν(OH). As to the peak at 3,457 cm^{-1}, this peak probably relates to ν(NH) vibration of N-H bonds forming the weakest hydrogen bonds. In the structure of (L-ArgH$_2$)(CF$_3$COO)$_2$ there are contacts N2-H2B…O1, N3-H…O1, and N3-H3…F1 with distances H…O(2.21 Å, 2.35 Å) and H…F(2.54 Å), which are in the intermediate region of very weak hydrogen bonds and strong van der Waals interactions. In the structure of (L-ArgH)CF$_3$COO there are no such weak bonds, and this may be the reason of the presence of the peak at 3,457 cm^{-1} in the spectrum of (L-ArgH$_2$)(CF$_3$COO)$_2$ and absence in the spectrum of (L-ArgH)CF$_3$COO.

The crystal of L-argininium diphenylacetate was obtained by Ramos Silva et al. (1999c) from 1:1 water–ethanol solution containing equimolar amounts of components by slow evaporation. Asymmetric part of unit cell contains two formula units. The structure is pseudocentrosymmetric. Two (L-ArgH) cations form a cycle where the carboxylate group of one cation is connected with the guanidyl group of another cation (the A type according to Salunke and Vijayan 1981): N7-H42…O5 (2.793 Å), N8-H44…O6(2.871 Å), N3-H27…O7(2.797 Å), and N4-H29…O8 (2.860 Å). Additionally N(1)H$_3^+$ and N(5)H$_3^+$ groups form each two hydrogen bonds with anions and one with carboxylate group of cation: N1-H23…O6(2.819 Å), N1-H24…O2(2.941 Å), and N1-H25…O4(3.008 Å) and N5-H38…O8(2.804 Å),

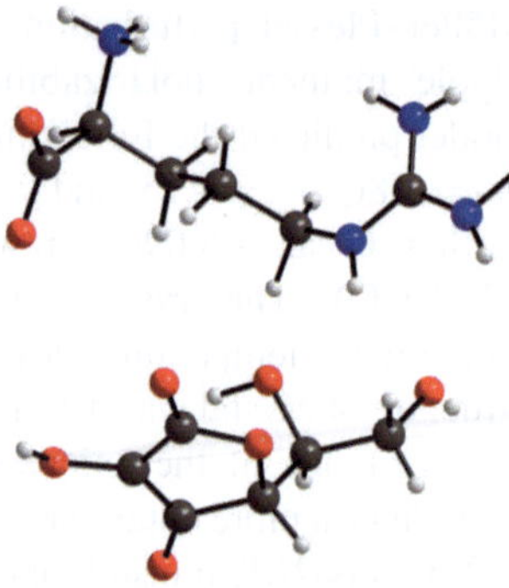

Fig. 5.39 Molecular structure of L-argininium L-ascorbate

N5-H39...O4(2.797 Å), and N5-H40...O2(2.923 Å). The rest of the hydrogen atoms of guanidyl groups form hydrogen bonds with anions: N3-H28...O1 (2.892 Å), N4-H30...O2(2.821 Å), N7-H43...O3(2.815 Å), N8-H45...O4 (2.816 Å), N2-H26...O1(3.008 Å), and N6-H41...O3(3.045 Å). The last two bonds are quite weak.

The crystals of L-argininium hydrogen squarate were obtained by Angelova et al. (1996). The salt was obtained from aqueous solution containing L-arginine and squaric acid in equimolar ratio by slow evaporation at room temperature. Triclinic unit cell contains two formula units. Two hydrogen squarate anions form pseudocentrosymmetric dimer via strong hydrogen bonds: O11-H110...O12′(2.494 (2) Å) and O11′-H110′...O12(2.499(2) Å). Cations also form pseudocentrosymmetric dimers via guanidyl-carboxylate bonds. In contrast to the previous case, in this case hydrogen bonds form NH and NH_2 groups with carboxylate groups (the type B; Salunke and Vijayan 1981): N2-HN21...O1′(2.895(3) Å) and N3-HN3...O2′(2.952(2) Å) and N2′-HN21′...O1(2.992(2) Å) and N3′-HN3′...O2 (2.957(2) Å). The groups $N(4)H_3^+$ and $N(4')H_3^+$ form each two hydrogen bonds with carboxylate groups of nearest cations and each one hydrogen bond with anions: N4-HN41...O2′(3.073(2) Å), N-HN43...O2(2.859(2) Å), and N4-HN42...O14 (2.721(2) Å) and N4′-HN41′...O2(3.071(2) Å), N4′-HN43′...O2′(2.888(2) Å), and N4′-HN42′...O14′(2.744(2) Å). The rest of the hydrogen atoms of guanidyl groups form hydrogen bonds with anions: N1-HN11...O12′(2.933(3) Å) and N2-HN22...O 13(3.185(2) Å) and N1′-HN11′...O12(2.849(3) Å), N1′-HN12′...O13(3.000(2) Å), and N2′-HN22′...O13′(3.025(2) Å).

The crystal of (L-ArgH)L-ascorbate was obtained by Sudhakar and Vijayan (1980) by slow evaporation in nitrogen atmosphere of an aqueous solution containing equimolar quantities of components. Asymmetric unit of unit cell contains one formula unit (Fig. 5.39). The authors mention that this is the first complex between an amino acid and a vitamin. The $N(1)H_3^+$ group forms one normal hydrogen bond with the carboxylate group of a nearest cation, N1-H2 (N1)...O2(2.891(6) Å). Other contacts are very weak: N1-H1(N1)...O2(3.061 (6) Å), N1-H3(N1)...O11(2.904(6) Å), and N1-H3(N1)...O12(3.065(5) Å). The guanidyl group forms normal hydrogen bonds with the anions: N6-H(N6)...O15

(2.846(6) Å), N8-H2(N8)...O13(2.815(6) Å), N9-H1(N9)...O16(2.769(6) Å), and N9-H2(N9)...O13(2.830(6) Å), except for one weak hydrogen bond (N8-H1(n8)... O1(3.025(6) Å). Additionally three OH groups of ascorbate anion form four hydrogen bonds. The group O(12)H forms hydrogen bond with cation, O12-H(O12)...O1 (2.676(6) Å). The group O(15)H forms bifurcated hydrogen bonds, O15-H(O15)... O11(2.913(5) Å) and O15-H(O15)...O14(2.921(5) Å), and the group O(16)H forms hydrogen bond with negatively charged oxygen atom O13 of ascorbate anion, O16-H (O16)...O13(2.622(5) Å). Thus, the negatively charged oxygen O13 atom forms three hydrogen bonds.

Srinivasan et al. (2008c) obtained a new crystal (L-ArgH) 4-nitrophenolate monohydrate from aqueous solution containing equimolar quantities of L-arginine and 4-nitrophenol, determined its symmetry and crystal structure, studied vibrational spectra, and also calculated NLO properties and estimated by powder SHG method. Another similar crystal (L-ArgH) 4-nitrophenolate 4-nitrophenol dihydrate was obtained by Wang et al. (2011), who determined the solubility in water; determined its symmetry and structure, IR absorption spectrum, UV–Vis–NIR transmittance spectrum, and TG curve; and calculated NLO properties and measured SHG efficiency and phase-matching by Kurtz–Perry method. Srinivasan et al. (2008c) have grown a crystal with $15 \times 8 \times 4$ mm^3 sizes by slow evaporation method at 30 °C. Asymmetric part of unit cell contains one formula unit. The N(2) H_3^+ group forms two hydrogen bonds with carboxylate groups of two nearest symmetry-equivalent cations: N2-H2A...O5(2.8779(17) Å) and N2-H2B...O4 (2.7763(15) Å). The third atom forms two very weak hydrogen bonds with nitro group of anion and water molecule: N2-H2C...O1(3.0482(18) Å) and N2-HC...O1W (3.0974(19) Å). The water molecule forms as proton donor hydrogen bonds with nitro group of anion and carboxylate group of cation: O1W-H1A...O1(2.8839 (19) Å) and O1W-H1B...O5(2.8001(17) Å). The N(5)H group of guanidyl group forms hydrogen bond with water molecule, N5-H5N...O1W(2.8765(18) Å). The N(3)H_2 group of guanidyl group forms two weak hydrogen bonds with nitro group and phenolate oxygen atom of the anion: N3-H3A...O2(3.094(2) Å) and N3-H3B...O3(2.981(2) Å). The other N(4)H_2 group forms two normal hydrogen bonds with phenolate oxygen atom of two nearest anions: N4-H4A...O3(2.8226 (17) Å) and N4-H4B...O3(2.8093(19) Å). Thus, negatively charged phenolate oxygen atom forms three hydrogen bonds. Calculated value of the hyperpolazability β_{tot} is equal to $4.20.10^{-29}$ esu. Experimental SHG efficiency measured by powder method was very high compared to KDP and urea. Said paper (Srinivasan et al. 2008c), however, contains some contradictory statements. The authors write: "The internal vibrations of argininium cation can be classified into those arising from functional groups NH_3^+, CH_3 and COOH." One may suppose that CH_3 and COOH are misprints. However, the group COOH is present also in reaction scheme in Fig. 1 and also in Table 6, where one may meet assignments to C-OH vibrations. Actually (L-ArgH) cation does not contain CH_3 and COOH groups as it is evident from Fig. 4 of the paper (Srinivasan et al. 2008c).

Asymmetric unit of (L-ArgH) 4-nitrophenolate 4-nitrophenol dihydrate contains one formula unit (Wang et al. 2011). The presence of neutral molecule of nitrophenol significantly changed the scheme of hydrogen bonds compared to the structure reported by Srinivasan et al. (2008c). The group $N(4)H_3^+$ forms two hydrogen bonds with water molecules and one rather strong hydrogen bond with negatively charged oxygen atom of nitrophenolate anion: N4-HC...O10(2.784 (3) Å), N4-HD...O9(3.089(3) Å) and N4-HE...O6(2.757(2) Å), that is, does not form hydrogen bonds with carboxylate group in contrast to previous case. Carboxylate group forms strong hydrogen bond with nitrophenol, O3-H3C...O1 (2.592(2) Å), which is the strongest hydrogen bond in the structure. Additionally carboxylate group forms hydrogen bonds with one water molecule and guanidyl group: O10-H101...O1(2.733(2) Å), N2-H2D...O2(2.865(2) Å). The difference in bond lengths of carboxylate group (C6-O1(1.261(2) Å) and C6-O2(1.233(2) Å) may be caused by formed hydrogen bonds. The second hydrogen bond water molecule $H_2O(10)$ forms with nitro group of anion:O10-H102...O7(2.925(3) Å). Additionally this nitro group forms hydrogen bonds with guanidyl group: N1-H1B...O8(3.009(3) Å), N3-H3...O7(3.098(3) Å). All three hydrogen bonds formed by nitro group $N(6)O_2$ are very weak, while the nitro group of nitrophenol group $N(5)O_2$ does not form any hydrogen bonds. The water molecule $H_2O(9)$ in addition to mentioned weak hydrogen bond with $N(4)H_3^+$ group forms as proton acceptor also with guanidyl group, N2-H2C...O9(2.960(2) Å). As proton donor the water molecule $H_2O(9)$ forms hydrogen bonds with nitrophenol and nitrophenolate anion: O9-H92...O3(2.885(3) Å), O9-H91...O6(2.918(2) Å). Additionally O6 atom forms weak hydrogen bond with guanidyl group, N1-H1A...O6 (2.917(3) Å). Thus, the negatively charged oxygen atom of nitrophenolate anion forms three hydrogen bonds. Wang et al. (2011) have grown a crystal with sizes $14 \times 8 \times 7$ mm^3. In the interpretation of the IR spectrum the authors write: "In addition to this, the stretching vibration of C=O falls to 1,687 cm^{-1}. As has been proved by XRD analysis, carboxyl group acts as a proton donor in the molecule." This is not correct statement. The authors assigned the strong absorption band at 1,583 cm^{-1} to phenyl stretching vibration. In our opinion it is more probable that this band is caused by characteristic vibration $\nu_{as}(NO_2)$. Thermogravimetric curve shows that above 80 °C the process of dehydration takes place. Calculated value for hyperpolarizability β_{tot} is equal to $3.11.10^{-29}$ esu. Measured value of SHG efficiency is about of efficiency of urea, while in case of (L-Arg) 4-nitrophenolate monohydrate (Srinivasan et al. 2008c) the SHG efficiency is much higher. Wang et al. (2011) explain this by that neutral 4-nitrophenol and negatively charged 4-nitrophenolate are located coincidently inversely to each other in molecular arrangements. The presence of water crystallization in the structures of both crystals leads to reduced thermal stability of crystals compared to urea. In addition, the presence of nitrophenolate anions leads to worsening of transmittance in UV region compared to urea.

The interaction of picric acid with L- and DL-arginine was studied by Nagata et al. (1995). The crystals of (L-ArgH) picrate·$2H_2O$ and (DL-ArgH) picrate were prepared from aqueous solution containing an equimolar amount of arginine and picric acid by slow evaporation at room temperature. In each case asymmetric unit

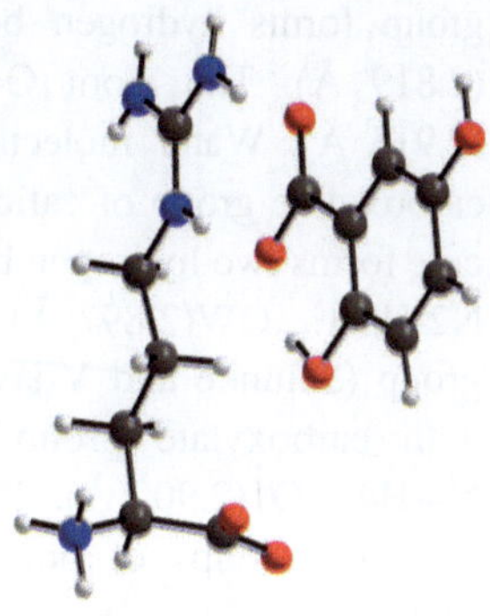

Fig. 5.40 Molecular structure of L-argininium L-2,5-dihydroxybenzoate monohydrate. The water molecule is not shown

contains one formula unit. In the structure of (L-ArgH) picrate·$2H_2O$ the $N(2)H_3^+$ group forms hydrogen bonds with carboxylate group of nearest cation, N2-H...O1′A(2.740(7) Å) and weak hydrogen bond with nitro group of anion, N2-H...O4′P(2.948(8) Å). The guanidyl group forms hydrogen bonds of type D (Salunke and Vijayan 1981) with phenolate oxygen atom of picrate anion: N8-H...O1P(2.823(9) Å), N6-H...O1P(2.823(8) Å). The guanidyl group forms also a hydrogen bond with carboxylate group of nearest cation, N7-H...O1A (2.814(9) Å) and two hydrogen bonds with water molecules: N7-H...O2W(2.949 (9) Å), N8-H...O1W(2.928(8) Å). Water molecules in turn as proton donors form hydrogen bonds with symmetry-equivalent molecules and with carboxylate group of cation: O1W...O1W(2.850(7) Å), O1W...O1′A(2.911(7) Å), and O2W...O2W (2.912(8) Å), O2W...O1′A(2.835(8) Å). Thus, there are two chains of water molecules in the structure of crystal. The oxygen atom O1′A of carboxylate group forms three hydrogen bonds, while the atom O1A only one hydrogen bond. The difference in bond lengths C1-O1(1.219 Å), C1-O1′(1.266 Å) may be caused by difference of formed hydrogen bonds. In the structure of (DL-ArgH) picrate L-ArgH and D-ArgH form centrosymmetric cyclic dimer via hydrogen bonds formed between guanidyl group of one isomer with carboxylate group of other: N6-H...O1′A(2.886(3) Å), N8-H...O1A(2.865(4) Å), that is, by specific interaction of type B (Salunke and Vijayan 1981). Three other hydrogen bonds of guanidyl group are weak: N7-H...O1P(2.924(3) Å), N7-H...O2P(3.052(4) Å), N8-H...O1A (2.994(4) Å). The $N(2)H_3^+$ group forms hydrogen bonds with negatively charged oxygen atom of picrate anion, N2-H...O1P(2.863(3) Å), with carboxylate group of nearest cation, N2-H...O1′A(2.793(3) Å), and weak bond with nitro group of anion, N2-H...O4′P(2.921(4) Å). So, the phenolate atom O1P and oxygen atoms of carboxylate group form each two hydrogen bonds.

The crystal of (L-ArgH) 2,5-dihydroxybenzoate·H_2O was obtained from water/acetonitrile 50:50(v/v) solution (Kinsel et al. 2004). Asymmetric unit contains one formula unit (Fig. 5.40). Carboxylate group of the anion has intramolecular hydrogen bond O3-H3A...O1(2.508 Å). The atom O3 in turn as proton acceptor forms hydrogen bond with water molecule OW...O3(2.835 Å). The second hydroxyl group forms hydrogen bond with carboxylate group of cation, O4-H4A...O6 (2.819 Å). The atom O4 in turn forms weak hydrogen bond N2-H2B...O4 (2.915 Å). Water molecule as proton donor forms second hydrogen bond with carboxylate group of cation, OW...O6(2.893 Å). As proton acceptor water molecule forms two hydrogen bonds with guanidyl group: N1-H1B...OW(2.881 Å) and N2-H2A...OW(2.897 Å). This is the type C of specific interactions of guanidyl group (Salunke and Vijayan 1981). Guanidyl group forms also hydrogen bonds with carboxylate group of anion of type B: N1-H1A...O2(2.826 Å) and N4-H4...O1(2.900 Å). The group $N(3)H_3^+$ forms two hydrogen bonds with carboxylate groups of both cation and anion: N3...O5(2.690 Å) and N3...O2 (2.774 Å).

A new salt L-argininium *p*-nitrobenzoate monohydrate has been obtained from aqueous solution containing equimolar quantities of components by slow evaporation at 40 °C (Wang et al. 2014). Single crystal with $15 \times 8 \times 2$ mm^3 sizes was grown by slow cooling method. The solubility in water was determined from 25 to 45 °C. The crystal structure was determined as well as X-ray diffraction powder pattern, IR and Raman, ^1H, ^{13}C NMR, UV–Vis. NIR spectra were provided. Thermal properties were characterized by TG and DTA curves. The efficiency of SHG by Kurtz–Perry method was determined and compared with KDP. Asymmetric unit contains one formula unit. The $N(4)H_3^+$ group forms a hydrogen bond with water molecule, N4-H18...O1(2.802(4) Å) and two hydrogen bonds with carboxylate group of two nearest cations, N4-H16...O6(2.717(4) Å), N4-H17...O7(2.913 (4) Å). Water molecule in turn forms hydrogen bonds with carboxylate groups of anion and cation: O1-H1A...O3(2.838(4) Å) and O1-H1B...O6(2.764(4) Å). Guanidyl group forms two types of specific interactions. The groups N1-H1C and N2-H2B form specific hydrogen bonds (type B); (Salunke and Vijayan 1981) with carboxylate group of anion: N1-H1C...O3(3.026 Å) and N2-H2B...O2(2.736 (4) Å). The first hydrogen bond (N1...O3) is absent in the table of hydrogen bonds given in Wang et al. (2014). The groups N2-H2A and N3-H4C form hydrogen bonds with the same oxygen atom O2, that is, specific interaction of type C (Salunke and Vijayan 1981): N2-H2A...O2(2.821(4) Å) and N3-H4C...O2 (2.957(4) Å). In the discussion of vibrational spectra Wang et al. (2014) mention characteristic bands of COOH group and assign the peak at 3,454 cm^{-1} to stretching vibration of O-H bond and bands at 1,691 and 1,667 cm^{-1} to C=O of L-arginine and *p*-nitrobenzoic acid. Actually there are carboxylate COO^- groups in the structures of L-argininium(+) cation and *p*-nitrobenzoate anion and in our opinion the band at 3,454 cm^{-1} may relate to stretching vibration of water molecule, while the bands at 1,691, 1,667, 1,639, 1,576 cm^{-1} may, in addition to mentioned $\nu_{as}(COO^-)$ and $\nu_{as}(NH_3^+)$, be caused by deformation vibration of water molecule, which is not mentioned in discussion of the vibrational spectra. Dehydration of the crystal starts at 130 °C with peak in DTA curve at 152 °C. The efficiency of SHG is about four times that of KDP.

Interaction of arginine with mono[6-methoxy-α-methyl-2-naphthalene acetate], naproxen (NPX), having anti-inflammatory properties, has been studied (Vayá et al. 2005). (S-ArgH) S-naproxen was synthesized by mixing equimolar amount of arginine and naproxen in water. After removal of the solvent, the residue was recrystallized from methanol/acetonitrile. The obtained salts were characterized by ^{1}H, ^{13}C NMR, IR and fluorescence spectra. Crystal structures of (S-ArgH) S-naproxen and (S-ArgH) R-naproxen have been determined. In both cases asymmetric unit contains one formula unit. In both cases argininium cations form cyclic dimers by hydrogen bonds between guanidyl and carboxylate groups by type A (Salunke and Vijayan 1981). In the structure of S-S salt hydrogen bonds are: N2-H23…O4(2.790 Å), N3-H25…O5(2.856 Å). In addition to this type A interaction between guanidyl and carboxylate group of cations there is also type D interaction between guanidyl group and carboxylate group of anion: N2-H22…O3 (2.892 Å), N1-H21…O3(2.898 Å). The rest H24 atom of $N(3)H_2$ group forms hydrogen bond N3-H24…O1(2.804 Å) with symmetry-equivalent anion. The $N(4)H_3^+$ group forms hydrogen bonds N4-H26…O5(2.807 Å), N4-H27…O1 (2.817 Å), N4-H28…O1(2.836 Å) with carboxylate groups of nearest cation and anions. In case of S-R salt hydrogen bonds of type A are: N2-H23…O4(2.919 Å), N4-H27…O5(2.782 Å). As in previous case the rest atoms of $N(2)H_2$ and $N(4)H_2$ groups form hydrogen bonds with carboxylate groups of nearest anions: N2-H22… O3(2.839 Å), N4-H28…O1(2.841 Å). The main distinction from the previous S-S salt is that in this case N1-H21 group does not form hydrogen bond. The $N(3)H_3^+$ group as in case S-S salt forms hydrogen bonds: N3-H24…O1(2.848 Å), N3-H25…O1(2.955 Å), N3-H26… O5(2.772 Å).

The structure of (L-ArgH) ragaglitazar ((S)-2-ethoxy-3-(4-(2-(phenoxazin-10-yl) ethoxy) phenol) propionate salvate has been determined (Ebdrup et al. 2003). Crystals were obtained by dissolving L-arginine and ragaglitazar in dimethylsulfoxide and slow evaporation at 40 °C. Asymmetric unit contains two formula units. One of dimethylsulfoxide molecules is disordered. Each of two L-argininium cation forms cyclic dimer with its symmetry-equivalent by type A interaction (Salunke and Vijayan 1981) between guanidyl and carboxylate groups. Hydrogen bonds in one dimer are: N5-H60…O11(2.804 Å), N6-H62…O12(2.819 Å), and in second dimer are: N9-H75…O13(2.801 Å), N10-H78…O14(2.826 Å). The rest hydrogen bonds of guanidyl groups form hydrogen bonds with carboxylate groups of anion: N5-H61…O4(2.867 Å), N6-H63…O3(2.867 Å), and N9-H76…O9 (2.830 Å), N10-H77…O8(2.862 Å). The N-H groups of guanidyl group form hydrogen bonds with oxygen atoms of ethoxy groups: N4-H59…O5(2.975 Å), N8-H74…O10(2.961 Å). The NH_3^+ groups of cations form each two hydrogen bonds with carboxylate groups of anions and each one hydrogen bond with carboxylate groups of cations: N3-H51…O8(2.860 Å), N3-H52…O9(2.882 Å), N3-H50…O14(2.778 Å) and N7-H65…O3(2.880 Å), N7-H66…O4(2.854 Å), N7-H67…O12(2.756 Å). Dimethylsulfoxide molecules do not form hydrogen bonds.

Interaction of arginine with aspartic and glutamic acids is investigated in the series of papers (Salunke and Vijayan 1982; Suresh et al. 1986; Soman et al. 1990; Bhat and Vijayan 1977; Soman and Vijayan 1989). The crystal of (L-ArgH)

L-aspartate was grown by the slow diffusion of 1-propanol into an aqueous solution of the commercially obtained compound (Salunke and Vijayan 1982). Asymmetric unit contains one formula unit. Guanidyl group of the (L-ArgH) cation forms hydrogen bonds with carboxylate group of the L-aspartate anion by type A (Salunke and Vijayan 1981): N8-H2…O11(2.896(4) Å), N9-H2…O12(2.889(4) Å). Two other hydrogen atoms of these groups form hydrogen bonds with terminal carboxylate group of nearest anion, N8-H1…O15(2.814(4) Å) and carboxylate group of nearest cation, N9-H1…O1(2.852(4) Å). The N6-H1…O1(2.852(4) Å). The N6-H1 group also forms hydrogen bond with carboxylate group of other nearest cation, N6-H1…O2(2.960(4) Å). The $N(1)H_3^+$ group of the (L-ArgH) cation forms two hydrogen bonds with carboxylate groups of two nearest cations: N1-H1…O1 (2.819(3) Å), N1-H2…O2(2.877(3) Å) and with main carboxylate group of anion, N1-H3…O12(2.797(3) Å). The $N(11)H_3^+$ group of the anion forms hydrogen bonds with carboxylate groups of three nearest anions: N11-H1…O16(2.758 (5) Å, N11-H2…O11(2.928(4) Å) and N11-H3…O15(2.819(3) Å). The crystal of (L-ArgH) D-aspartate was grown from aqueous solution containing equimolar amounts of components by the vapor diffusion technique and 1-propanol as precipitant (Suresh et al. 1986). Triclinic unit cell contains two formula units. Both cations form hydrogen bonds via guanidyl groups with terminal carboxylate groups of anions by type B (Salunke and Vijayan 1981): N6-H…O15(2.872(3) Å), N9-H1…O16(2.778(3) Å) and N26-H…O35(2.856(3) Å), N29-H1…O36(2.734 (3) Å). The $N(8)H_2$ group forms hydrogen bonds with carboxylate group of second anion: N8-H1…O36(2.837(3) Å) and weak bond N8-H2…O32(3.103(3) Å). In case of second cation N(28)H2 group forms hydrogen bonds with carboxylate groups of first cation and anion: N28-H2…O2(2.882(3) Å), N28-H1…O16(2.808 (3) Å). The $N(1)H_3^+$ group of first cation forms hydrogen bonds with main carboxylate groups of anions: N1-H1…O11(2.725(2) Å), N1-H2…O12(2.752 (3) Å), N1-H3…O31(2.809(3) Å). The $N(21)H_3^+$ group of second cation forms two hydrogen bonds with main carboxylate groups of anions and one hydrogen bond with carboxylate group of nearest second cation: N21-H3…O12(3.132(3) Å), N21-H3…O11(2.897 Å), N21-H2…O32(2.831(3) Å), N21-H1…O22(2.920 (2) Å). The $N(11)H_3^+$ and $N(31)H_3^+$ groups of anions form each two hydrogen bonds with carboxylate groups of cations and each one hydrogen bond with carboxylate groups of anions: N11-H1…O21(2.983(2) Å), N11-H2…O2(2.789 (2) Å), N11-H3…O12(2.910(1) Å), and N31-H1…O1(2.799(3) Å), N31-H2… O22(2.767(2) Å), N31-H3…O35(2.827(3) Å).

The crystal of (DL-ArgH) DL-aspartate was obtained in (Soman et al. 1990) from aqueous solution containing equimolar quantities of components by vapor diffusion technique. As precipitant propanol was used. Centrosymmetric triclinic cell contains one formula unit. Guanidyl group of the cation forms hydrogen bonds with main carboxylate group of anion by type A (Salunke and Vijayan 1981): N8-H2…O12(2.73(1) Å), N9-H2…O11(2.93(1) Å). The rest atoms of $N(8)H_2$ and $N(9)H_2$ groups form hydrogen bonds with neighboring terminal carboxylate groups of anion: N8-H1...O16(2.86(1) Å), N9-H1...O15(2.92(1) Å). The N6-H1 group forms hydrogen bond with carboxylate group of nearest cation, N6-H1...O2(2.77 (1) Å). The N(1)H_3^+ group forms two hydrogen bonds with carboxylate groups of two nearest cations, N1-H1...O1(2.84(1) Å), N1-H2...O1(2.76(1) Å) and one with terminal carboxylate group of anion, N1-H3...O16(2.78(1) Å). The N(11)H_3^+ group of the anion forms intramolecular hydrogen bond N11-H1...O15(2.91 (1) Å), and also with both carboxylate groups of nearest anions: N11-H2...O15 (2.83(1) Å), N11-H3...O11(2.81(1) Å).

Salts of arginine with glutamic acid in contrast to the salts with aspartic acid exist in form of crystal hydrates (Suresh et al. 1986; Soman et al. 1990; Bhat and Vijayan 1977; Soman and Vijayan 1989). The crystal of (L-ArgH) L-glutamate monohydrate was grown by slow diffusion of acetone into an aqueous solution of commercially obtained compound. Asymmetric unit contains one formula unit. Guanidyl group of the cation forms hydrogen bonds with terminal carboxylate group of anion by type A of specific interactions (Salunke and Vijayan 1981) and with carboxylate group of symmetry-equivalent cation by type B (Salunke and Vijayan 1981): N8-H82...O17(2.83 Å), N9-H92...O16(2.85 Å) and N9-H91...O1 (2.95 Å), N6-H6...O1(3.06 Å). The rest hydrogen atom forms hydrogen bond with main carboxylate group of anion, N8-H81...O12(3.04 Å). The N(1)H_3^+ group of cation forms two hydrogen bonds with carboxylate groups of two nearest cations and one with water molecule: N1-H12...W1(2.81 Å). Water molecule in turn forms hydrogen bonds with terminal and main carboxylate groups of two nearest anions: W1-H1...O16(2.72 Å), W1-H2...O12(2.84 Å). The N(11)H_3^+ group of anion forms three hydrogen bonds with carboxylate groups of three nearest anions: N11-H111...O11(2.92 Å), N11-H112...O17(2.70 Å), N11-H113...O11(2.88 Å). Crystals of (L-ArgH) D-glutamate trihydrate (Suresh et al. 1986) and (L-ArgH) D-glutamate monohydrate (Soman and Vijayan 1989) were obtained from aqueous solutions containing equimolar quantities of L-arginine and D-glutamic acid by diffusion of acetone. Two types crystals with different morphology were formed. The crystal of (L-ArgH) D-glutamate monohydrate was needle-like in shape. Surprisingly both crystals crystallize in the same space group and Z and with close *a* and *b* parameters. In both cases asymmetric unit consists of one formula unit. In the structure of trihydrate guanidyl group forms hydrogen bonds with two of three water molecules and with terminal carboxylate group of anion: N8-H1...W3 (2.822(2) Å), the contact N8-H2...W2(3.166(3) Å) is usual van der Waals contact, N9-H1...W2(2.977(3) Å), N9-H2...W2(2.954(3) Å), and N6-H...O17(2.799 (2) Å). The N(1)H_3^+ group forms two hydrogen bonds with main carboxylate groups of two nearest anions and one with carboxylate group of nearest cation: N1-H1...O12(2.913(2) Å), N1-H2...O12(2.877(2) Å), N1-H3...O1(2.748(2) Å). In contrast, the N(11)H_3^+ group forms two hydrogen bonds with carboxylate groups of two nearest cations and one with main carboxylate group of nearest anion: N11-H2...O2(2.765(3) Å), N11-H3...O2(2.813(3) Å), N11-H1...O11(2.813 (2) Å). Water molecules as proton donor form hydrogen bonds with terminal carboxylate group of anions and with each other: W1-H1...W1(2.849(3) Å),

W1-H2...O16(2.762(2) Å), W2-H1...O17(2.695(3) Å), W2-H2...W1(3.008 (3) Å), W3-H1...O16(2.746(3)), W3-H2...O16(2.833(3) Å). The structure of monohydrate (Soman and Vijayan 1989) in spite of lesser number of water molecules has some similarity with hydrogen bonds scheme of trihydrate (Suresh et al. 1986). Guanidyl group of cation forms hydrogen bonds with terminal carboxylate group of anions and with water molecule: N6-H1...O16(2.83(1) Å), N8-H1...W1(2.89(1) Å), N8-H2...W1(2.98(1) Å), N9-H1...O16(2.99(1) Å), N9-H2...O16(3.12(1) Å). The same oxygen atom forms hydrogen bond with N9-H2 and N6-H1 groups. So, there is type B specific interaction in this case. The $N(1)H_3^+$ group also forms two hydrogen bonds with main carboxylate groups of two nearest anions and one with carboxylate group of nearest cation: N1-H1... O12(2.83(1) Å), N1-H2...O12(2.81(1) Å), N1-H3...O1(2.77(1) Å). The $N(11)H_3^+$ group of anion as in case of trihydrate forms two hydrogen bonds with carboxylate group of two nearest cations and one with main carboxylate group of nearest anion: N11-H2...O2(2.79(1) Å), N11-H3...O292.81(1) Å), N11-H1...O11(2.80(1) Å). Water molecule forms hydrogen bonds with terminal carboxylate group of anions: W1-H1...O17(2.73(1) Å), W1-H2...O17(2.77(1) Å).

The crystal of (DL-ArgH) DL-glutamate monohydrate was grown from aqueous solution, containing equimolar quantities of DL-arginine and DL-glutamic acid by vapor diffusion technique using acetone as precipitant (Soman and Vijayan 1989). The crystals of grew as thin long, double crystals. Asymmetric unit contains one formula unit. Cations form cyclic dimers via type B specific interaction: N9-H1...O1(2.961(9) Å), N6-H1...O2(2.875(9) Å). In addition, the guanidyl group forms type A specific hydrogen bonds with terminal carboxylate group of anion: N8-H2...O17(2.843(8) Å), N9-H2...O16(2.865(8) Å). The rest hydrogen atom of guanidyl group forms hydrogen bond with terminal carboxylate group of anion, N8-H1...O17(2.853(9) Å). The $N(1)H_3^+$ group of cation forms two hydrogen bonds with main carboxylate group of anions and one hydrogen bond with carboxylate group of nearest cation: N1-H1...O12(2.887(7) Å), N1-H2...O11 (2.745(8) Å) and N1-H3...O2(2.849(8) Å). The $N(11)H_3^+$ group of anion forms one hydrogen bond with main carboxylate group of nearest anion, a contact with carboxylate group of nearest cation, and with water molecule: N11-H1...O12 (2.865(8) Å), N11-H2...W1(2.739(9) Å), and N11-H3...O2(3.142(8) Å). Water molecule in turn forms hydrogen bond with terminal carboxylate group of anion, W1-H1...O16(2.690(9) Å).

In the group of Prof. Vijayan interaction of arginine with dibasic acids was also investigated. Interaction of L- and DL-arginine with oxalic (HOOC-COOH) (Chandra et al. 1998), malonic ($HOOC-CH_2-COOH$) (Saraswathi and Vijayan 2002a), succinic ($HOOC-(CH_2)_2-COOH$ (Prasad and Vijayan 1990), glutaric ($HOOC-(CH_2)_3-COOH$) (Saraswathi and Vijayan 2001), adipic ($HOOC-(CH_2)_4-COOH$) (Roy et al. 2005) and pimelic ($HOOC-(CH_2)_5-COOH$ (Saraswathi et al. 2003) acids was investigated. Chandra et al. (1998) obtained crystals of (L-ArgH)HC_2O_4 and (DL-ArgH)HC_2O_4 by slow diffusion of ethanol and acetone into aqueous solutions containing equimolar quantities of L- or DL-arginine and oxalic acid. Triclinic cell of (L-ArgH)HC_2O_4 contains two formula units, while in case of (DL-ArgH)HC_2O_4 asymmetric unit contains one formula unit. Hydrogen bonds scheme in these structures are quite different. In the structure of (DL-ArgH) HC_2O_4 cations form cyclic centrosymmetric dimers via type B hydrogen bonds: N9-HB...O1(2.911(3) Å), N6-H6...O2(2.849(2) Å). These cyclic dimers form chain via N9-HA...O1(2.2.844(3) Å) hydrogen bond. Hydrogen oxalate anions form infinite chain via O12-H12...O16(2.578(2) Å) hydrogen bond. Anions form additionally hydrogen bonds with cations, N8-HA...O15(2.932(3) Å), N8-HB... O11(3.008(3) Å), N1-HA...O15(2.812(2) Å), N1-HB...O16(2.897(2) Å). The $N(1)H_3^+$ group forms also N1-HC...O2(2.766(2) Å) hydrogen bond. The conformation of hydrogen oxalate anion deviates significantly from planarity. The torsion angle O-C-C-O is 95.2(1)°. In case of triclinic (L-ArgH)HC_2O_4 each cation is surrounded by two anions and each anion is surrounded by two cations. Carboxyl groups of anions form hydrogen bonds with carboxylate groups of cations: O22-H22...O1(2.563(3) Å), O32-H32...O11(2.511(1) Å), while carboxylate groups of anions form hydrogen bonds with guanidyl groups of cations by type A: N8-HB...O35(2.900(4) Å), N9-HA...O36(2.794(4) Å), and N18-HB... O25(2.944(4) Å), N19-HB...O26(2.901(4) Å). Guanidyl groups additionally form hydrogen bonds: N8-HA...O12(2.906(5) Å), N6-H6...O25(2.969(4) Å) and N18-HA...O2(2.969(5) Å), N16-H16...O35(2.912(4) Å), while the contacts N9-HB...O25(3.384(4) Å) and N19-HA...O35(3.315(3) Å) are usual van der Waals interactions. The $N(1)H_3^+$ group forms hydrogen bonds with anions and one hydrogen bond with carboxylate group of the same type cation: N1-HA...O21 (2.936(3) Å), N1-HB...O36(2.726(4) Å) and weak bonds N1-HB...O32(3.066 (4) Å), N1-HC...O2(2.799(4) Å). The $N(11)H_3^+$ group forms hydrogen bonds only with anions: N11-HA...O26(3.082(4) Å), N11-HB...O26(2.871(4) Å), and N11-HB...O22(2.956 Å), N11-HC...O31(2.902(3) Å).

The system L-Arg + $H_2C_2O_4$ + H_2O was investigated also by Petrosyan et al. (2002, 2005b). By variation of molar ratio of L-arginine and oxalic acid in 0–100 °C interval several phases were identified by their IR spectra. One phase was obtained at equimolar ratio at 40–45 °C, which was labeled as α-form. Structure determination showed that α-phase is identical with triclinic crystal obtained in (Chandra et al. 1998). The molecular structure is shown in Fig. 5.41. The solubility of α-phase at 45 °C is 86 g in 100 ml water. It melts at 150 °C with decomposition. Another phase was obtained at equimolar ratio at 20 °C by slow evaporation, which was labeled as β-form. Determination of its crystal structure showed that it crystallizes in monoclinic system and is a new form of (L-ArgH)HC_2O_4 salt. Asymmetric unit contains one formula unit. As in case of α-form here also hydrogen oxalate anions do not form chain with each other, but carboxyl group of anion forms hydrogen bond with carboxylate group of cation, O6-H...O2(2.544 Å). The $N(1)H_3^+$ group forms two hydrogen bonds with carboxylate group of two nearest cations and one hydrogen bond with carbonyl oxygen atom of anion: N1-HA...O1 (2.796 Å), N1-HB...O1(2.814 Å), N1-HC...O5(2.849 Å). Guanidyl group forms

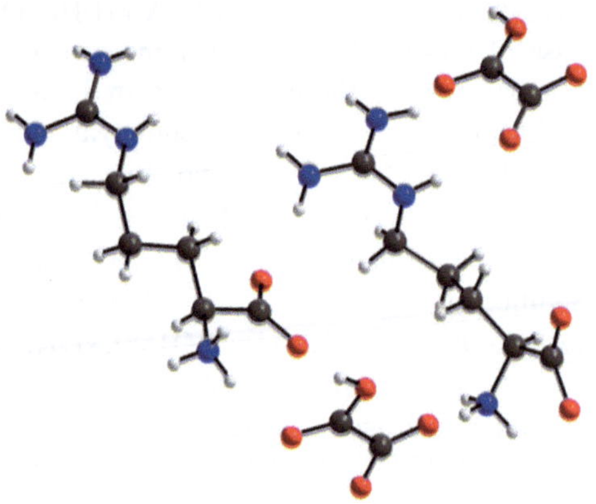

Fig. 5.41 Molecular structure of the α-form of L-argininium hydrogen oxalate

type A hydrogen bonds with carboxylate group of anion: N3-HA...O4(2.984 Å), N4-HA...O3(2.975 Å). Additionally two hydrogen bonds are formed with two nearest anions: N3-HB...O3(2.904 Å) and N4-HB...O4(2.917 Å). The group N(2)-H has two contacts on the level of strong van der Waals interactions. The β-form has very high solubility (140 g in 100 ml water at 20 °C). Crystals of the β-form are stable up to 165 °C.

Recrystallization of the α-form at room temperature leads to the β-form. One more phase was identified by IR spectrum and labeled as γ-form (Petrosyan et al. 2005b) Later structure determination of this phase (Antipin et al. 2006) showed that it probably was formed in result of partial racemization at high temperatures and is identical with (DL-ArgH)HC_2O_4 (Chandra et al. 1998). In addition to these phases obtained at equimolar ratios at different temperatures, two more salts were obtained. From aqueous solution with 2:1 molar ratio of L-arginine and oxalic acid needle-like crystals are formed. The crystal structure was determined at 120 K. Asymmetric unit contains eight (L-ArgH) cations, four oxalate ($C_2O_4^{2-}$) anions and ten water molecules (L-ArgH)$_8$(C_2O_4)$_4$.10H_2O. The solubility of the crystal at room temperature is 87 g in 100 ml water. Numerous hydrogen bonds connect (L-ArgH) cations with anions, with each other and with water molecules. Water molecules as proton donors and acceptors form hydrogen bonds with cations, anions and in some cases with each other. Dehydration of the crystal takes place at 143 °C (Petrosyan et al. 2005b). One more salt in this system is formed at 1:2 molar ratio of L-arginine and oxalic acid. The crystal of L-Arg·2$H_2C_2O_4$ is monoclinic. Crystal structure determination showed that asymmetric unit contains one (L-ArgH) cation, one hydrogen oxalate ($HC_2O_4^-$) anion and one neutral molecule of oxalic acid. The crystal was investigated in more detail in (Antipin et al. 2006). The structure and electron density distribution was determined at 100 K. IR and Raman spectra were provided and discussed. The neutral molecule and hydrogen oxalate anion are bonded with short O10-H11...O5(2.500(2) Å) hydrogen bond, which allows to consider the fragment (HOOC-COO...HOOC-COOH)$^-$ as a dimeric anion. These anions form weaker hydrogen bond O8-H10...O4(2.661(3) Å) with each other.

Carboxylic group of hydrogen oxalate anion forms very strong hydrogen bond with carboxylate group of cation, O3-H9...O2(2.440(2) Å). This strong hydrogen bond probably the reason of difference in bond lengths of carboxylate group of cation: C1-O1(1.227(3) Å) and C1-O2(1.263(3) Å). The $N(1)H_3^+$ group forms a hydrogen bond with nearest hydrogen oxalate group, N1-H3...O6(2.846(3) Å), while guanidyl group forms a hydrogen bond with carboxylate group of nearest cation, N3-H5...O1(2.855(4) Å) and with carbonyl oxygen atom of neutral oxalic molecule, N4-H8...O7(2.980(3) Å). Other N-H...O type hydrogen bonds are weak. The solubility of the crystal is 28 g in 100 ml water. It is stable up to 160 °C. Thus in the system L-Arg + $H_2C_2O_4$ + H_2O, the following salts are formed: $(L\text{-}ArgH)_8(C_2O_4)\cdot 10H_2O$, two forms (triclinic (α) and monoclinic (β) of (L-ArgH) HC_2O_4 and (L-ArgH)(HC_2O_4...$H_2C_2O_4$).

One more compound was discovered when a mixture of equal volumes of water and ethanol was used as a solvent to reduce the high solubility of the β-form of (L-ArgH)HC_2O_4 (Petrosyan et al. 2005b). At 32 °C brilliant and well-shaped crystals in form of hexagonal plates were formed. Crystal structure determination showed that they are formed according to the reaction: L-arginine + $H_2C_2O_4$ → N^{α}-oxalyl-L-arginine (NOLA) + H_2O. NOLA crystallizes in trigonal system with unit cell parameters: $a = b = 9.1700(13)$ Å, $c = 25.710(5)$ Å, $\gamma =$ 120, $Z = 6$, $\rho_c = 1.310$ g/cm^3, space group $P3_221$. In the molecule of NOLA there are positively charged protonated guanidyl group, neutral carboxylic group and negatively charged deprotonated oxalyl group. In the structure molecules are bonded with rather strong O2-H...O5(2.548 Å) hydrogen bond. The solubility of NOLA is very low compared to the above mentioned salts (0.15 g in 100 ml water at 20 °C). NOLA is stable up to 216 °C.

Recently two papers were published (Vasudevan et al. 2013b, c), where the crystal (L-ArgH)HC_2O_4 was grown and characterized by different methods. Particularly powder SHG intensity is 1.5 times higher than that of KDP. However, it is not clear which form of (L-ArgH)HC_2O_4 was obtained and studied. From one hand the authors identified their crystal by measuring the unit cell parameters. According to these data obtained crystal is α-form, that is triclinic form. From the other hand the crystal was grown at room temperature and recrystallized. According to Petrosyan et al. (2005b) at room temperature β-form is formed. In addition recrystallization of α-form at room temperature leads to formation of β-form. Indeed, comparison of the IR spectrum shown by Vasudevan et al. (2013b) with spectra of α- and β-forms shows its agreement with the spectrum of β-form and not α-form.

The crystal of (L-ArgH) hydrogen malonate was obtained by the diffusion of butanol into an aqueous solution containing L-arginine and malonic acid in a 1:1.5 molar ratio (Saraswathi and Vijayan 2002a). Triclinic unit cell contains one formula unit (Fig. 5.42). Hydrogen malonate anion forms very short intramolecular hydrogen bond O11-H11...O16(2.423(3) Å). Guanidyl group forms two hydrogen bonds with two nearest cations and three nearest anions: N6-H6...O12(3.033 (4) Å), N8-HA...O2(2.853(4) Å), N8-HB...O16(3.056(4) Å), N9-HA...O1

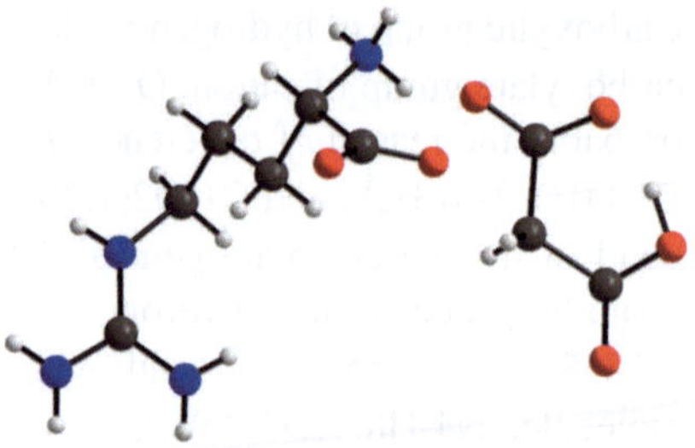

Fig. 5.42 Molecular structure of L–argininium hydrogen malonate

(2.865(4) Å), N9-HB...O17(2.900(3) Å). Of these three are weak: N6...O12, N8...O16, N9...O1. The $N(1)H_3^+$ group forms hydrogen bonds with two nearest anions and one cation: N1-HA...O17(2.778(4) Å), N1-HB...O1(2.724(4) Å), N1-HC...O12(2.811(3) Å). The crystal of (DL-ArgH) hydrogen malonate monohydrate was obtained by the diffusion of isopropyl alcohol into an aqueous solution containing DL-arginine and malonic acid in a 1:2 molar ratio (Saraswathi and Vijayan 2002a). Asymmetric unit contains one formula unit. Cations form centrosymmetric cyclic dimers by type B hydrogen bonds: N6-H6...O2(2.926 (4) Å), N9-HA...O1(2.905(4) Å). At the same time guanidyl group forms hydrogen bonds with carboxylate group of anion by type B: N8-HB...O16(2.918(4) Å), N9-HB...O17(2.823(3) Å). Intramolecular hydrogen bond of anion is O11-H11...O16(2.443(5) Å). The $N(1)H_3^+$ group forms two hydrogen bonds with two nearest cations and with water molecule: N1-HA...O2(2.828(4) Å), N1-HC...O2(3.006(3) Å), and N1-HB...OW(2.746(4) Å). The water molecule in turn forms hydrogen bonds with two nearest anions: OW-HA...O17(2.788(5) Å), OW-HB...O12(2.710(5) Å).

The crystal of $(L\text{-ArgH})_2$ succinate succinic acid dihydrate was obtained by the diffusion of ethanol into an aqueous solution containing equimolar quantities of L-arginine and succinic acid (Prasad and Vijayan 1990). Triclinic cell contains one formula unit, that is, two (L-ArgH) cations, succinate $(OOC\text{-}(CH_2)_2\text{-}COO)^{2-}$ anion, neutral succinic acid molecule and two water molecules. Prasad and Vijayan (1990) considered this salt as L-arginine hemisuccinate hemisuccinic acid monohydrate with $Z = 2$ based on pseudosymmetry. The structure of the crystal is very interesting. From one hand two cations form hydrogen bonds of type A with succinate anion: N8-H2...O32(2.797(3) Å), N9-H2...O31(2.943(3) Å) and N28-H2...O38 (2.826(3) Å), N29-H2...O37(2.853(3) Å). From the other hand succinate anion and succinic acid form infinite chain such that each succinate anion is surrounded by two succinic acids and each succinic acid is surrounded by two succinate anions: O42-H1...O38(2.479(4) Å), O48-H1...O32(2.517(4) Å). Water molecules as acceptors form hydrogen bonds with $N(1)H_3^+$ and $N(21)H_3^+$ groups: N1-H2... W1(2.773(4) Å), N21-H2...W2(2.771(3) Å). As proton donors water molecules form hydrogen bonds with nearest succinate anion and succinic acid: W1-H1... O42(3.067(3) Å), W1-H2...O31(2.816(4) Å), W2-H1...O37(2.834(3) Å), W2-H2... O48(2.940(3) Å). The $N(1)H_3^+$ and $N(21)H_3^+$ groups form additionally each two hydrogen bonds with carboxylate groups of nearest cations: N1-H1...O2(2.895(3) Å),

N1-H3...O21(2.755(4) Å), N21-H1...O22(2.789(3) Å), N21-H3...O1(2.987(3) Å). The rest hydrogen bonds formed by guanidyl groups are: N6-H1...O21(2.949 Å) (in Prasad and Vijayan 1990, the value (2.984 Å) is provided for this bond), N8-H1...O41(2.910(3) Å), N9-H1...O22(3.070(3) Å), and N26-H1...O2(2.873 (4) Å), N28-H1...O47(2.970(3) Å), N29-H1...O1(2.837(4) Å)(in Prasad and Vijayan 1990, by misprint N28 is pointed out).

The crystal of (DL-ArgH)$_2$ succinate tetrahydrate was obtained by the same way as previous crystal. Asymmetric unit consists of one (ArgH) cation, two water molecules and succinate anion, which is in special position (across the inversion center). The succinate anion is symmetrically surrounded by L- and D-ArgH cations via the same type A specific hydrogen bonds interaction: N8-H2...O12(2.820 (6) Å), N9-H2...O11(2.802(6) Å). Other hydrogen bonds formed by guanidyl group are N8-H1...W2(2.968(6) Å), N9-H1...O1(2.890(6) Å), N6-H1...O2 (3.101(6) Å). The N(1)H_3^+ group forms hydrogen bonds with nearest anion, cation and W1 water molecule: N1-H1...O12(2.715(6) Å), N1-H2...O2(2.892(5) Å), N1-H3...W1(2.823(5) Å). The water molecule W1 forms hydrogen bonds with second water molecule and anion: W1-H1...W2(2.812(5) Å), W1-H2...O11(2.745 (5) Å), while the molecule W2 forms hydrogen bonds with carboxylate group of two nearest cations: W2-H1...O1(2.829(5) Å), W2-H2...O2(2.861(5) Å).

The crystals of (L-ArgH)$_2$ glutarate dihydrate were obtained by diffusion of isobutyl alcohol into an aqueous solution of L-arginine and glutaric acid mixed in an equimolar ratio (Saraswathi and Vijayan 2001). Asymmetric unit consists of one (L-ArgH) cation, one water molecule both in general position and one glutarate anion $^-$OOC-(CH$_2$)$_3$-COO$^-$ in special position. Glutarate anion is surrounded symmetrically by two (L-ArgH) cations by type A hydrogen bonds: N8-HA...O11 (2.797(6) Å), N9-HA...O12(2.857(7) Å). Guanidyl group forms additionally hydrogen bonds with nearest anion and cation and with water molecule: N8-HB...O11 (2.881(4) Å), N9-HB...O1(2.955(4) Å) and N6-H6...OW(2.948(5) Å). Water molecule forms hydrogen bonds with nearest cation, OW-H1...O2(2.738(4) Å) and with symmetrically equivalent water molecule, OW-H2...OW(2.940(6) Å). The N(1)H_3^+ group forms hydrogen bonds with two nearest cations, N1-HA...O2(2.733(5) Å), N1-HC...O1(2.753(5) Å) and with nearest anion, N1-HB...O12(2.687(4) Å).

The crystals of (L-ArgH)$_2$ adipate were obtained by the diffusion of ethanol into an aqueous solution of L-arginine and adipic acid mixed in equimolar ratio (Roy et al. 2005). Asymmetric unit consists of one adipate anion $^-$OOC-(CH$_2$)$_4$-COO$^-$ and two (L-ArgH) cations. Adipate anion is surrounded by two cations by type A hydrogen bonds interaction: N8-HA...O30(2.740(3) Å), N9-HB...O29 (2.902(3) Å), and N18-HB...O22(2.731(4) Å), N19-HA...O21(2.904(4) Å). In addition, cations form pseudocentrosymmetric cyclic dimers by type B specific interaction: N6-HA...O11(2.970(3) Å), N9-HAO11(3.010(3) Å) and N16-HA...O1(3.007(3) Å), N19-HB...O1(3.069(3) Å). The N(1)H_3^+ and N(11) H_3^+ groups form each one hydrogen bond with anion and two hydrogen bonds with two nearest cations: N1-HA...O21(2.738(3) Å), N1-HB...O2(2.782(3) Å),

N1-HC...O1(2.803(3) Å), and N11-HB...O29(2.902(3) Å), N11-HA...O12(2.755 (3) Å), N11-HC...O11(3.026(3) Å).

The crystal of (DL-ArgH)$_2$ adipate dihydrate was obtained by the diffusion of ethanol into an aqueous solution of DL-Arg and adipic acid mixed in equimolar ratio. Asymmetric unit consists of one argininium cation, one water molecule and half of adipate anion $^-$OOC-(CH$_2$)$_4$-COO$^-$ anion. Adipate anion is surrounded by symmetrically located L- and D-argininium cations by type A specific interaction: N8-HB...O11(2.809(3) Å), N9-HA...O12(2.828(3) Å). The groups N(8)H$_2$ and N(9)H$_2$ form additionally hydrogen bonds with neighboring anion and cation: N8-HA...O11(2.847(2) Å), N9-HB...O1(2.885(3) Å). The rest hydrogen bond guanidyl group forms with water molecule, N6-H6...OW(2.933(3) Å). Water molecule forms hydrogen bonds with carboxylate group of cation and symmetry-equivalent water molecule: OW-HB...O2(2.751(3) Å), OW-HA...OW(2.885 (3) Å). The N(1)H$_3^+$ group forms hydrogen bonds with two nearest cations and one with anion: N1-HB...O2(2.947(3) Å), N1-HC...O2(2.815(2) Å), N1-HA...O12 (2.742(2) Å).

The crystals of (L-ArgH)$_2$ pimelate dihydrate were obtained by the diffusion of methanol into an aqueous solution of L-arginine and pimelic acid in a 1:3 molar ratio (Saraswathi et al. 2003). Asymmetric unit consists of two (L-ArgH) cations, two water molecules and two half of pimelate anions. Pimelate anions are located on twofold axis. The structure of (L-ArgH)$_2$ pimelate dihydrate has similarity with the structure of (L-ArgH)$_2$ glutarate dihydrate. Pimelate anions are surrounded by symmetrically equivalent (L-ArgH) cations. By the way, in CIF of CCDC (Refcode AKOFUB, CCDC #223328) the crystal is indicated as bis(DL-argininium) pimelate dihydrate. Inspection of the structure shows that all four (ArgH) cations are in L-form. Guanidyl groups of both inequivalent (L-ArgH) cations form hydrogen bonds with carboxylate groups of anions by type A interaction: N8-HB...O31(2.890(3) Å), N9-HB...O32(2.730(4) Å), and N18-HB...O22 (2.970(3) Å), N19-HA...O21(2.914(3) Å). These NH$_2$ groups additionally form hydrogen bonds with neighboring anion and cation: N8-HA...O22(2.834(3) Å), N9-HA...O1(2.953(3) Å), and N18-HA...O31(2.878(3) Å), N19-HB...O11(2.926 (3) Å), while NH groups form hydrogen bonds with water molecules: N6-H6...OW2 (2.958(3) Å), N16-H16...OW1(2.953(3) Å). The N(1)H$_3^+$ and N(11)H$_3^+$ groups form hydrogen bonds with two nearest cations and one anion: N1-HA...O2(2.770 (3) Å), N1-HB...O21(2.733(3) Å), N1-HC...O11(2.781(2) Å), and N11-HA...O12 (2.774(3) Å), N11-HB...O32(2.664(3) Å), N11-HC...O1(2.778(2) Å). Water molecules form each one hydrogen bonds with carboxylate group of cation and the same type crystallographically equivalent water molecule: OW1-HA...O12(2.721(4) Å), OW1-HB...OW1(2.945(4) Å), and OW2-HA...O2(2.808(4) Å), OW2-HB...OW2 (2.965(4) Å).

The crystals of (L-ArgH) hydrogen maleate dihydrate were obtained by Zheng et al. (1994a). Large crystal (25 × 40 × 13 mm^3) was grown by a slow cooling method. Second harmonic generation was observed and crystal structure was determined. Triclinic cell contains one formula unit. Later, Ravishankar et al. (1998) investigated interaction of L- and DL-arginine with maleic acid seemingly being unaware that the structure of the salt of L-arginine was determined previously. The structure of (L-ArgH) hydrogen maleate dihydrate was determined two more times (Mallik and Kar 2005b; Sun et al. 2007b). We will consider the structures of L- and DL-salts on the basis of Ravishankar et al. (1998). The salts were obtained by the slow diffusion of acetonitrile and acetone into solution of equimolar quantities of L-arginine (DL-arginine) and maleic acid respectively. The salt of L-arginine is identical with previously and later obtained salts by evaporation. Hydrogen maleate anion as usual has intramolecular short hydrogen bond O12-H...O17(2.405(2) Å). The bond lengths C13-O12 and C16-O17 are 1.289 Å and 1.269 Å respectively, while O12-H and H...O17 are equal to 1.12(5) Å and 1.33(5) Å accordingly. Guanidyl group of cation has two specific interactions. By type A interaction cations form infinite chain: N8-HA...O2(2.944(3) Å), N9-HA...O1(2.930(3) Å). At the same time by type B guanidyl group forms hydrogen bonds with hydrogen maleate anion: N6-H6...O18(2.952(3) Å), N9-HB...O17(2.845(3) Å). The rest hydrogen atom of guanidyl group forms one more hydrogen bond with anion, N8-HB...O11(2.852(3) Å). The $N(1)H_3^+$ group forms hydrogen bonds with one water molecule, cation and anion: N1-HA...OW1 (2.798(3) Å), N1-HB...O2(2.842(2) Å), N-HC...O11(3.076(3) Å). The rest water molecule forms hydrogen bonds with second molecule and cation, OW2-H1...O18 (2.862(3) Å), OW2-H2...O2(2.764(3) Å). The salt (DL-ArgH) maleate also crystallizes in non-centrosymmetric polar space group. Guanidyl group also has two specific interactions, but in this case by type A it interacts with anion and with type B with cation: N8-HA...O18(2.837(6) Å), N9-HA...O17(2.934(7) Å), and N6-H6...O1(2.856(6) Å), N9-HB...O2(2.814(6) Å). However, in this case cations form zigzag and not cyclic dimers as in some other structures. The rest hydrogen bond is formed with O11 atom of anion, N8-HB...O11(3.102(6) Å). The $N(1)H_3^+$ group forms two hydrogen bonds with anions and one with cation: N1-HA...O11 (2.803(6) Å), N1-HB...O2(2.729(5) Å), N-HC...O18(2.757(5) Å). Intramolecular hydrogen bond of anion in this case is equal to 2.419(7) Å.

Various authors have grown and characterized the crystal of (L-ArgH) hydrogen maleate dihydrate (Zheng et al. 1994a; Mallik and Kar 2005b, c; Sun et al. 2007b, 2008d; Vasantha and Dhanuskodi 2004b; Kalaiselvi et al. 2008a; Baraniraj and Philominathan 2010; Charoen-In et al. 2011; Vasudevan et al. 2013d). Solubility curves in water, acetone/water and ethanol/water have determined (Mallik and Kar 2005b). Unusual solubility curve in water may indicate on possibility to obtain a new phase at crystallization in the interval 45–55 °C. Infrared, UV–Vis–NIR spectra, thermal behavior and NLO activity were studied (Vasantha and Dhanuskodi 2004b; Mallik and Kar 2005c). The crystal is phase-matchable and SHG efficiency is 1.5 times higher than that of KDP (Vasantha and Dhanuskodi 2004b), 0.28 times that of urea (Mallik and Kar 2005c) and 6.8 times of quartz (Mallik and Kar 2005c). Sun et al. (2008d) discussed IR and Raman spectra, based on structure as well as thermal properties. According to this work, the sample starts losing one water molecule from 90 to 105 °C. Endothermic peak at 99 °C corresponds to the melting point, then the rest of water liberates between 140 and 160 °C. According to the authors, the intensity of SHG signal is three times higher than that of KDP. In connection with low thermal stability caused by crystal water it should be noted that the crystal of (DL-ArgH) hydrogen maleate may be more promising due to the absence of crystal water and presence of non-centrosymmetric polar symmetry. One more interesting material was obtained by Mallik and Kar (2007). The authors tried to obtain solid solution of (L-ArgH) hydrogen maleate dihydrate and (L-ArgH) formate by the reaction: L-Arg + (formic acid)$_x$ + (maleic acid)$_{(1-x)}$, where $x = 0.2$–0.8. High-quality single crystals were obtained at $x = 0.30$, which was called L-arginine formomaleate. Actual composition according to C, H, N test was $x = 0.25$. According to the IR spectrum, thermal curves and unit cell parameters the crystal is formed on the basis of (L-ArgH) hydrogen maleate dihydrate structure. Powder SHG efficiency was 1.2 times that of KDP.

Crystals of (L-ArgH) hydrogen malate were obtained by Monaco et al. (1987). The authors did not indicate the form of malic acid. Probably it is (L-ArgH) hydrogen L-malate, because an attempt to obtain (L-ArgH) hydrogen DL-malate was not successful.

Interaction of L- and DL-arginine with L-, D-, and DL-tartaric acid was investigated by Selvaraj et al. (2007). In case of L-arginine a salt with 2:1 ratio with L-tartaric acid was obtained, while in case of DL-arginine a salt with 1:1 ratio was obtained only with DL-tartaric acid. The crystal of (L-ArgH)$_2$L-tartrate was obtained by diffusion of butanol into an aqueous solution containing equimolar ratio of components, while in case of crystals (DL-ArgH) hydrogen DL-tartrate acetonitrile was used as precipitant. In both cases asymmetric units contain one formula unit. In the structure of (L-ArgH)$_2$L-tartrate the anion forms two hydrogen bonds as proton donor with carboxylate groups of both cations: O27-H27...O11(2.698(4) Å) and weaker O25-H25...O1(2.840(3) Å). Cations form hydrogen bonds with each other and with anion. Guanidyl group of one cation forms type D specific hydrogen bonds with other cation: N16-H16...O2(2.781(3) Å), N19-HA...O2(2.841(4) Å). The other NH_2 group forms hydrogen bonds with nearest cation and anion: N18-HA...O1(2.863(3) Å), N18-HB...O29(2.840(3) Å). The rest contact N19-HB...O29 is usual van der Waals interaction. Guanidyl group of the first cation forms hydrogen bond with one hydroxyl group of anion, N6-H6...O27 (2.785(3) Å), two weak hydrogen bonds with nearest cation and anion, N8-HA...O12(2.961(4) Å), N8-HB...O21(2.934(3) Å) and two more weaker hydrogen bonds, N9-HA...O29(3.130(4) Å), N9-HB...O21(3.127(4) Å). The NH_3^+ groups form each one hydrogen bond with carboxylate group of nearest other cation and each two hydrogen bonds with anion: N1-HB...O11(2.759 (3) Å), N11-HA...O11(3.046(3) Å), N1-HA...O21(2.892(3) Å), N-HC...O22 (2.770(3) Å) and N11-HB...O30(2.732(3) Å), N11-HC...O29(2.906(4) Å).

The main feature of the structure of (DL-ArgH) hydrogen DL-tartrate is the chain of hydrogen L- and D-tartrate anions connected by short centrosymmetric hydrogen bonds: O12-H12-O12(2.451(4) Å) and O20-H20-O20(2.466(4) Å). The C(3)H_2 and

$C(4)H_2$ groups of the cation are disordered. Hydroxyl groups of anion form hydrogen bonds with nearest cation and anion: O15-H15...O1(2.813(3) Å), O17-H17...O20(2.953(4) Å). Guanidyl group forms one hydrogen bond with nearest cation, others with anion. There is N6-H6...O19(2.892(4) Å) hydrogen bond. If one consider the contact N9-HB...O19(3.111(5) Å) as weak hydrogen bond, then it may be concluded on existence of type B specific interaction with the anion. The hydrogen atom HB has one more contact with O20 atom, N9-HB...O20 (2.976(4) Å), which also may be considered as weak hydrogen bond. If one considers the contact N9-HA...O12(3.099(4) Å) as weak hydrogen bond, then one may conclude on existence of one more specific interaction with second group of anion, N8-HB...O11(2.801(4) Å). The rest hydrogen bond of guanidyl group is N8-HA...O2(2.807(4) Å. The $N(1)H_3^+$ group forms one normal hydrogen bond, N1-HA...O1(2.788(4) Å). The rest hydrogen bonds are very weak: N1-HB...O17(3.001(4) Å), N1-HC...O2(3.300(4) Å), N1-HC...O1(3.171(4) Å).

5.1.23 Salts of Histidine

In Tables 5.25 and 5.26 salts of histidine with inorganic and organic acids are shown respectively. The review by Petrosyan (2010a) was used as a background for writing this section. Like the previously considered amino acids ornithine, lysine and arginine capable to form doubly charged cation, histidine also can form doubly charged cation due to presence of imidazole ring in its molecule. However, unlike these amino acids, which in zwitterionic state have protonated more basic terminal amino groups (in case of ornithine and lysine) or guanidyl group (in case of arginine) histidine in zwitterionic state has protonated α-amino group and not lesser basic imidazole group. Singly charged (OrnH) and (LysH) have protonated both amino groups and negatively charged carboxylate group, as well as in (ArgH) protonated guanidyl and amino groups and negatively charged carboxylate group. In case of (HisH) these are protonated amino and imidazole groups and negatively charged carboxylate group. Doubly charged $(OrnH_2)$, $(LysH_2)$, $(ArgH_2)$, $(HisH_2)$ cations are characterized by presence of neutral COOH carboxylic group. So, the presence of doubly charged $(HisH_2)$ cation as in previous cases may be established by vibrational spectroscopic feature of COOH group, that is, by presence of ν(C=O) vibration at 1,700 cm^{-1} or higher wavenumbers. Spectroscopic confirmation of the presence of $(HisH_2)$ cation was made by Garfinkel and Edsall (1958) on example of $(HisH_2)Cl_2$. These authors also indicated on the important vibrational spectroscopic feature of presence of histidine, namely, very characteristic narrow peak caused by stretching vibrations of C-H bonds in imidazole ring at about 3,150 cm^{-1}, which later was confirmed. In majority of published papers these vibrations incorrectly are assigned to vibrations of other bonds. The conformation of histidine may be defined by torsion angles χ (C1C2C3C4) and χ^1 (N1C2C3C4), which define the disposition of carboxy and amino groups in respect to imidazole ring. Kistenmacher et al. (1972) suggested using the torsion angle χ and calling

Table 5.25 Salts of histidine with inorganic anions

#	Crystal	s.g., Z	*a*, *b*, *c* (Å), *α*, *β*, *γ* (°)	Refs
1	(L-HisH)F·2H_2O	$P2_12_12_1$, 4	8.436(2), 8.560(2), 13.892(3)	Petrosyan et al. (2005c)
2	(L-HisH_2)F_2	$P2_12_12$, 4	13.142(3), 6.5829(13), 9.650(2)	Petrosyan et al. (2005c)
3	(L-HisH)(F-H. . .F. . .HF)	–	–	Karapetyan et al. (2003)
4	(L-HisH)Cl·H_2O	$P2_12_12_1$, 4	15.301(3), 8.921(2), 6.846(2)	Oda and Koyama (1972)
5	(L-HisH_2)Cl_2	$P2_12_12_1$, 4	6.970(4), 16.165(7), 8.964(6)	Kistenmacher and Sorrell (1974)
6	(L-HisH_2)Cl_2	$P2_1$, 2	5.243(2), 7.257(1), 13.350(1), $\beta = 92.65(1)$	Mostad and Natarajan (1995)
7	(DL-HisH)Cl·2H_2O	$P2_1/a$, 4	8.87(1), 15.30(2), 8.48(1), $\beta = 114.5(2)$	Bennett et al. (1970)
8	DL-His·2HCl	$P2_1/c$, 4	8.38, 15.42, 8.81, $\beta = 113.55$	Chacko et al. (1966)
9	(L-HisH)Br·H_2O	$P2_12_12_1$, 4	7.0480(14), 9.032(2), 15.257(3)	Petrosyan et al. (2005c)
10	(L-HisH_2)Br_2	$P2_12_12_1$, 4	7.1983(14), 9.153(2), 16.523(3)	Petrosyan et al. (2005c)
11	(L-HisH_2)Br_2	$P2_1$, 2	5.334(1), 7.483(2), 13.700(3), $\beta = 91.70(3)$	Petrosyan et al. (2005c)
12	(L-HisH)NO_3	$P2_12_12_1$, 4	5.243(2), 7.116(1), 25.030(2)	Srikrishnan et al. (1984)
13	(L-HisH)NO_3·H_2O	$P2_12_12_1$, 4	8.257(2), 11.286(2), 11.518(2)	Petrosyan et al. (2006)
14	(L-HisH_2)$(NO_3)_2$	$P2_12_12_1$, 4	5.4810(2), 8.2860(5), 25.5350(1)	Benali-Cherif et al. (2002)
15	(DL-HisH_2)$(NO_3)_2$	$P2_1/a$, 4	8.370(2), 14.973(3), 9.342(2), $\beta = 100.69(2)$	Bahadur et al. (1991)
16	(L-HisH)IO_3	–	–	Petrosyan et al. (2010a)
17	(L-HisH)BF_4	$P2_1$, 2	5.022(2), 9.090(1), 10.216(2), $\beta = 93.484(8)$	Marcy et al. (1995)

18	(L-HisH$_2$)(BF$_4$)$_2$ (α)	–	–	Petrosyan et al. (2010a)
19	(L-HisH$_2$)(BF$_4$)$_2$ (β)	–	–	Petrosyan et al. (2010a)
20	(L-HisH)ClO$_4$	$P2_1$, 2	5.052(1), 9.194(2), 10.388(2), $\beta = 92.34(1)$	Román et al. (1987)
21	(DL-HisH)ClO$_4$	$Pbca$, 4	9.5281(6), 9.7227(9), 20.5794(12)	Bahadur and Athimoolam (2009)
22	(L-HisH)H$_2$PO$_3$	$P2_1$, 2	8.497(7), 11.999(10), 4.967(3), $\beta = 104.15(6)$	Averbuch-Pouchot et al. (1988d)
23	(L-HisH)(H$_2$PO$_4$)·H$_2$O	$P2_12_12_1$, 4	8.923(5), 14.383(9), 8.362(5)	Averbuch-Pouchot et al. (1988d)
24	(L-HisH)(H$_2$PO$_4$)·H$_3$PO$_4$	$P2_1$, 2	9.16(1), 8.90(1), c = 8.81(1), $\beta = 111.38(5)$	Averbuch-Pouchot et al. (1988d)
25	(L-HisH)(H$_2$AsO$_4$)·H$_3$AsO$_4$	$P2_1$, 2	9.264(2), 8.929(2), 8.874(2), $\beta = 108.61(3)$	Ratajczak et al. (2000)
26	(L-HisH)NH$_2$SO$_3$	$P2_12_12_1$, 4	–	Rajkumar et al. (1999)
27	(DL-HisH)NH$_2$SO$_3$	–	–	Bahadur (1992)
28	(L-HisH)$_2$SO$_4$·H$_2$O	$P2_1$, 2	5.3316(11), 24.287(5), 6.8190(14), $\beta = 94.77(3)$	Petrosyan et al. (2010a)
29	(L-HisH)$_2$SO$_4$	–	–	Petrosyan et al. (2010a)
30	(L-HisH$_2$)$_2$(SO$_4$)$_2$·3H$_2$O	$P1$, 1	8.3400(13), 8.9291(11), 10.724(2), $\alpha = 102.576(13)$, $\beta = 107.704(15)$, $\gamma = 119.667(11)$	Srinivasan et al. (2001d)
31	(L-HisH$_2$)SO$_4$	$P2_1$, 4	6.8616(6), 5.8084(6), 25.6901(18), $\beta = 90.733(8)$	Kumar et al. (2005)
32	(L-HisH$_2$)(HSO$_4$)$_2$	$P2_1$, 2	8.1354(3), 7.4734(2), 10.2251(3), $\beta = 91.540(1)$, at 120 K	Ghazaryan et al. (in prep.)
33	(L-HisH)HgCl$_3$	$P2_12_12_1$, 4	8.116(30), 7.206(30), 19.77(4)	Adams et al. (1970)
34	(L-HisH)$_2$SiF$_6$·2H$_2$O	$P4_12_12$, 4	8.308(1), 8.308(1), 29.740(6)	Ghazaryan et al. (2013a)
35	(L-HisH$_2$)SiF$_6$·2H$_2$O	$P3_1$, 6	8.892(1), 8.892(1), 27.865(4), $\gamma = 120$	Ghazaryan et al. (2013a)

Table 5.26 Salts of histidine with organic anions

#	Crystal	s.g., Z	a, b, c (Å), α, β, γ (°)	Refs
1	(L-HisH)HCOO	–	–	Petrosyan (2010a)
2	(DL-HisH)HCOO·H_2O	$P2_1/c$, 4	9.309(1), 9.138(1), 11.741(1), $\beta=92.10(1)$	Suresh and Vijayan (1995b)
3	(L-HisH)HCOO·HCOOH	$P2_1$, 2	7.261(1), 7.441(1), 11.062(2), $\beta=104.93(1)$	Suresh and Vijayan (1995b)
4	(L-HisH)CH_3COO	$P2_12_12_1$, 4	5.027(1), 11.126(1), 17.473(3)	Suresh et al. (1994b)
5	(L-HisH)CH_3COO·$2H_2O$	$C2$, 4	15.649(2), 9.276(1), 8.566(1), $\beta=94.65(1)$	Suresh et al. (1994b)
6	(L-HisH)CH_3COO·$2H_2O$	$P1$, 2	8.519(3), 9.056(4), 9.026(3), $\alpha=61.72(3)$, $\beta=86.63(3)$, $\gamma=86.26(3)$, at 123 K	Mostad et al. (1995)
7	(L-HisH)CH_2(OH)COO	$P2_12_12_1$, 4	5.084(1), 11.012(1), 17.899(3)	Suresh and Vijayan (1996)
8	(DL-HisH)CH_2(OH)COO	$P2_1/c$, 4	5.248(1), 26.485(4), 7.248(1), $\beta=95.47(1)$	Suresh and Vijayan (1996)
9	(L-HisH)CF_3COO	$P1$, 2	5.1724(6), 8.8183(12), 12.481(3), $\alpha=96.193(17)$, $\beta=99.853(13)$, $\gamma=102.106(13)$	Gokul Raj et al. (2006b)
10	(L-HisH)CCl_3COO	$P2_1$, 4	5.4505(18), 25.769(8), 9.210(2), $\beta=99.98(2)$	Gokul Raj et al. (2006c)
11	(L-HisH)HC_2O_4	$P2_12_12_1$, 4	5.535(4), 6.809(4), 26.878(3)	Prabu et al. (1996)
12	(L-HisH)HC_2O_4 powdery	–	–	Petrosyan (2010a)
13	(L-HisH)$_2C_2O_4$·$3H_2O$ (α)	–	–	Petrosyan (2010a)
14	(L-HisH)$_2C_2O_4$·$3H_2O$ (β)	$P2_12_12$, 2	7.7978(16), 22.313(5), 5.6582(11)	Petrosyan (2010a)
15	(L-HisH)$_2C_2O_4$·$3H_2O$ (γ)	$C2$, 2	22.005(4), 5.7740(12), 7.8760(16), $\beta=96.20(3)$	Petrosyan (2010a)
16	(L-HisH_2)(HC_2O_4)$_2$·$2H_2O$	–	–	Petrosyan (2010a)
17	(L-HisH_2)(HC_2O_4)$_2$	–	–	Petrosyan (2010a)
18	(DL-HisH_2)(HC_2O_4)$_2$	$P2_1/c$, 4	6.750(7), 10.139 (2), 19.352(2), $\beta=90.8(1)$	Prabu et al. (1996)
19	(L-HisH) hydrogen malonate	$P2_1$, 4	5.3287(15), 25.534(3), 8.2421(15), $\beta=97.00(2)$	Saraswathi and Vijayan (2002b)
20	(DL-HisH) hydrogen malonate	$P2_1/n$, 4	5.3643(10), 25.3538(6), 8.3043(14), $\beta=96.202(15)$	Saraswathi and Vijayan (2002b)
21	(L-HisH) hydrogen succinate·$3H_2O$	$P2_12_12_1$, 4	7.030(1), 8.773(1), 24.322(3)	Prasad and Vijayan (1993a)
22	(DL-HisH)$_2$ succinate·$4H_2O$	$P\bar{1}$, 1	7.654(1), 8.723(1), 9.260(1), $\alpha=77.23(1)$, $\beta=72.37(1)$, $\gamma=82.32(1)$	Prasad and Vijayan (1993a)
23	(L-HisH_2) tetrafluorosuccinate	$C2$, 4	21.8863(16), 5.1411(4), 15.5619(12), $\beta=132.601(5)$	Akkurt et al. (2004)
24	(L-HisH) hydrogen glutarate·H_2O	$P2_12_12_1$, 4	8.335(1), 8.856(1), 19.522(2)	Saraswathi and Vijayan (2001)
25	(L-HisH) hydrogen glutarate·L-His	$P1$, 2	5.791(1), 13.110(3), 14.360(3), $\alpha=110.46(2)$, $\beta=89.78(1)$, $\gamma=95.62(2)$	Saraswathi and Vijayan (2001)
26	(L-HisH) hydrogen maleate	$P1$, 2	5.3889(12), 9.3851(13), 12.3681(9), $\alpha=107.054(9)$, $\beta=98.569(12)$, $\gamma=97.303(14)$	Pratap et al. (2000)
27	(L-HisH) hydrogen maleate·H_2O	$P2_1$, 2	5.4225(11), 6.8274(14), 17.098(3), $\beta=96.50(3)$	Fleck et al. (2013c)

28	(L-HisH)$_2$ (hydrogen maleate)$_2$·3H_2O	$P2_1$, 2	11.4656(7), 8.0530(5), 14.9705(9), $\beta = 101.657(2)$	Gonsago et al. 2012a)
29	(L-HisH$_2$) (hydrogen maleate)$_2$	$P1$, 2	7.2108(4), 9.8648(4), 12.9405(6), $\alpha = 104.006(2)$, $\beta = 90.174(2)$, $\gamma = 109.045(2)$	Fleck et al. (2013c)
30	(L-HisH) hydrogen L-malate	$P2_12_12_1$, 4	5.2148(7), 13.5259(18), 17.5369(19), at 90 K	de Matos Gomes et al. (2006)
31	(L-HisH)$_2$ D-tartrate	–	–	Petrosyan (2009)
32	(L-HisH)$_2$ (hydrogen L-tartrate)$_2$·H_2O	$C2$, 2	23.002(5), 7.676(2), 7.657(2), $\beta = 96.91(3)$	Marchewka et al. (2003b)
33	(L-HisH)$_2$ (hydrogen D-tartrate)$_2$·H_2O	$C2$, 2	23.020(5), 7.6813(15), 7.6587(15), $\beta = 96.84(3)$	Petrosyan (2009)
34	(L-HisH) hydrogen L-tartrate	$P1$, 1	5.3712(18), 7.637(3), 8.460(3), $\alpha = 72.025(5)$, $\beta = 73.872(5)$, $\gamma = 81.144(5)$	Johnson and Feeder (2004a)
35	(D-HisH) hydrogen D-tartrate	$P2_12_12_1$, 4	5.0207(9), 12.109(2), 20.797(4)	Johnson and Feeder (2004b)
36	(DL-HisH) hydrogen DL-tartrate	$P2_1/c$, 4	4.9695(5), 13.4392(12), 19.2749(18), $\beta = 90.253(2)$	Johnson and Feeder (2004c)
37	(L-HisH) L-aspartate·H_2O	$P2_1$, 2	5.131(1), 6.881(1), 18.277(2), $\beta = 97.26(1)$	Suresh and Vijayan (1987)
38	(L-HisH) 4,5-imidazoledicarboxylate	$P2_1$, 2	8.921(2), 6.656(1), 12.040(2), $\beta = 110.19(1)$	Görbitz and Husdal (1998)
39	(L-HisH) tetrafluorophthalate[a]	–	15.602(7), 5.150(6), 16.152(6), $\beta = 92.328(7)$	Ramajothi and Dhanuskodi (2006)
40	(L-HisH) 4-nitrophenolate·4-nitrophenol	$P2_1$, 2	8.8475(7), 8.8161(8), 12.4684(9), $\beta = 102.623(2)$	Dhanalakshmi et al. (2010)
41	(L-HisH) 2-nitrobenzoate	$P2_1$, 2	5.147(3), 7.228(5), 18.887(1), $\beta = 92.72(1)$	Natarajan et al. (2012)
42	(L-HisH) α-ketoisocaproate·C_2H_5OH	$C2$, 4	17.675(10), 5.146(6), 20.510(15), $\beta = 105.77(7)$	Nagashima et al. (1993)
43	(L-HisH) trimesate·1/3[$(CH_3)_2CO$]	$P2_12_12_1$, 4	18.097(7), 13.353(5), 6.737(4)	Herbstein and Kapon (1979)
44	(DL-HisH) trimesate·H_2O	$Pna2_1$, 4		Herbstein and Kapon (1979)
45	(L-HisH)$_2$ squarate	$C2$, 2	19.282(3), 8.105(2), 5.888(1), $\beta = 91.13(2)$	Karle et al. (1996)
46	(L-HisH) hydrogen squarate	$P2_12_12_1$, 4	4.936(2), 14.923(4), 15.068(3)	Karle et al. (1996)
47	(L-HisH) p-toluenesulfonate	$P2_12_12_1$, 4	5.2700(2), 7.3691(3), 38.2042(14)	Muralidharan et al. (2013)
48	(L-HisH) 5-nitrouracilate	$P2_1$, 2	6.1911(16), 7.332(2), 13.729(4), $\beta = 99.990(13)$	Domingos et al. (2012)
49	(L-HisH$_2$) (picrate)$_2$·2H_2O	$P2_1$, 2	6.6060(4), 25.7003(13), 7.9627(5), $\beta = 107.532(7)$	Sethuram et al. (2013)

[a]The authors have proposed (L-HisH$_2$) tetrafluorophthalate molecular structure without structure determination and proper substantiation (see Petrosyan 2010a and the text)

open or extended conformation if the angle χ is near ±180 ° and closed or folded conformation if the angle χ is near $\pm60°$. Bhat and Vijayan (1978) suggested to use χ^1 and call open I conformation if $\chi^1 = -60°$, open II if $\chi^1 = 180°$ and closed if $\chi^1 = +60°$. Krause et al. (1991) agree with this suggestion. The literature analysis shows that there is confusion in using the terms "close" and "open." We suggest to use both angles and call "closed" if (χ^1, χ) are $\pm60°$ and "open" if χ^1 or χ is near 180°, that is, if amino or carboxy groups are trans in respect to imidazole ring. So, we suggest calling "closed" in case of gauche-gauche and "open" in cases gauche-trans and trans-gauche.

The first compound in the system L-His + HF + H_2O, viz. L-His·2HF, was reported by Schmid (1968). This work was not known to Petrosyan and Petrosyan (2002) and Karapetyan et al. (2003) as well as Ramos Silva et al. (2004), who obtained the same compound. In the works of Petrosyan and Petrosyan (2002) and Karapetyan et al. (2003), four phases were identified by IR spectra, the structure of two of them (L-HisH)F·$2H_2O$ and (L-HisH_2)F_2 were determined as well (Petrosyan et al. 2005c; Karapetyan et al. 2003). The structure of (L-HisH_2)F_2 was independently determined also by Ramos Silva et al. (2004). Asymmetric unit of (L-HisH) F·$2H_2O$ contains one formula unit. The (L-HisH) cation has open conformation with carboxylate group trans in respect to imidazole group. Imidazolium group forms rather strong hydrogen bonds with anion and with carboxylate group of nearest cation: N2-H8…F(2.529(2) Å), N3-H10…O1(2.645(2) Å). The $N(1)H_3^+$ group forms two hydrogen bonds with anion and one with nearest carboxylate group: N1-H2…F1(2.721(3) Å), N1-H4…F1(2.798(2) Å), N1-H3…O2(2.776 (2) Å). Water molecules form chain with each other and also with anion and cation: O3-H11…O2(2.812(3) Å), O3-H12…O4(2.785(3) Å), O4-H13…F1(2.720(3) Å), O4-H14…O3(2.790(3) Å). The asymmetric unit of (L-HisH_2)F_2 also contains one formula unit. The (L-HisH_2) cation also has open conformation with carboxylic group trans in respect to imidazolium group. Carboxylic group of cation forms very strong hydrogen bond with one of fluoride anions: O2-H1…F1(2.3839(13) Å) (Petrosyan et al. 2005c). According to Ramos Silva et al. (2004) this O…F distance is equal to 2.3818(10) Å. The N-H groups of imidazolium ring form hydrogen bonds with second fluoride anion: N2-H9…F2(2.5421(16) Å), N3-H11…F2 (2.5269(15) Å). One of C-H group of imidazolium group has a short contact C6-H10…O2(3.266(2) Å) with H…O(2.28 Å), which may be classified as weak hydrogen bond. The $N(1)H_3^+$ group forms hydrogen bonds with anions: N1-H3…F1(2.712(2) Å), N1-H4…F2(2.690(2) Å), N1-H5…F1(2.6967(16) Å). Madhavan et al. (2006a) have grown and characterized the crystal of (L-HisH) F·$2H_2O$. Solubility curve was determined as well as IR and UV–Vis–NIR spectra, SHG efficiency, microhardness, dielectric properties, dark and photoconductivity. Dehydration occurs at 108 °C. Interpretation of the IR spectrum is contradictory. From one hand the authors write about carboxylate COO^- group and from the other hand absorption band of water molecule at 3,366 cm^{-1} assign to O-H stretching of carboxylic COOH group. Sharp peak at 3,138 cm^{-1}, which is caused by stretching vibration of C-H bonds of imidazolium group, is assigned to N-H group.

The authors found high SHG efficiency, which is 6.15 times higher than that of KDP. Petrosyan (2007a) found that $I^{2\omega}/I^{2\omega}_{KDP}$ is 1.7 and 2.0, for (L-HisH)F·$2H_2O$ and (L-HisH_2)F_2, respectively, while the Ramos Silva et al. (2004) write that no SHG signal for (L-HisH_2)F_2 was detected. The compound (L-HisH)(F-H...F...HF) is not stable. In result of recrystallization the (L-HisH_2)F_2 crystal is formed (Karapetyan et al. 2003). Because of volatility of HF the conditions of obtaining of crystals in the system L-His + HF + H_2O depend not only on molar ratios of L-His and HF.

Three compounds were found in the system L-His + HCl + H_2O. First the crystal of (L-HisH)Cl·H_2O was known. Its crystal structure was determined (Donohue et al. 1956; Donohue and Caron 1964; Oda and Koyama 1972; Fuess et al. 1977). Then two forms (orthorhombic, Kistenmacher and Sorrell 1974, and monoclinic, Mostad and Natarajan 1995) of (L-HisH_2)Cl_2 were obtained and their crystal structures were determined. Asymmetric unit in the structure of (L-HisH)Cl·H_2O contains one formula unit. The cation has closed conformation. In imidazolium ring N-H groups form hydrogen bonds with carboxylate group of two nearest cations: N2-H9...O2(2.642(6) Å), N3-H10...O1(2.829(4) Å). The N(1)H_3^+ group forms two hydrogen bonds with anion and one with water molecule: N1-H6...Cl(3.193 (4) Å), N1-H8...Cl(3.198(6) Å), N1-H7...O3(2.790(5) Å) (Fuess et al. 1977). Water molecule in turn forms hydrogen bonds with anion and with carboxylate group of cation: O3-H11...Cl(3.200(7) Å), O3-H12...O2(2.764(4) Å). One of C-H groups of imidazolium ring forms weak hydrogen bond, C5-H4...Cl(3.548 Å). The structures of (L-HisH_2)Cl_2 (again, one orthorhombic form, Kistenmacher and Sorrell 1974, and a monoclinic one, Mostad and Natarajan 1995) were determined. Conditions of preparation of these forms as well as interconnection between them was not known. Asymmetric unit in both forms contain one formula unit. Conformations of (L-HisH_2) cations and hydrogen bonding schemes are different in these forms. Conformation of cation in orthorhombic form is open with carboxyl group trans in respect to imidazolium group. The N(1)H_3^+ group forms one hydrogen bond with carbonyl oxygen atom of nearest cation and with both types of anions: N1-H8...O1(2.90 Å), N1-H9...Cl1(3.15 Å), N1-H10...Cl2(3.14 Å). Carboxyl group forms O2-H11...Cl1(2.95 Å) hydrogen bond. Imidazolium group forms hydrogen bonds: N2-H1...Cl1(3.11 Å), N3-H3...Cl2(3.10 Å) and also C5-H2...Cl2 (3.52 Å) with H2...Cl2(2.60 Å). In monoclinic form conformation of cation is closed. Carboxyl group as well as N(1)H_3^+ group form hydrogen bonds with one type of anion: O2-H2...Cl1(3.057(2) Å), N1-H1...Cl1(3.193(1) Å), N1-H2...Cl1(3.243 (1) Å), N1-H3...Cl1(3.318(1) Å), while imidazolium group forms hydrogen bonds with second type anion: N2-H2...Cl2(3.060(2) Å), N3-H3...Cl2(3.027(3) Å), C6-H6...Cl2(3.505(2) Å) with H6...C2(2.56(2) Å).

The crystal of (DL-HisH)Cl·$2H_2O$ was obtained by recrystallization of commercial (DL-HisH)Cl from aqueous alcohol (1:1) (Bennett et al. 1970). Asymmetric unit consists of one formula unit. The (HisH) cation has open conformation in contrast to the case of (L-HisH)Cl·H_2O. Carboxylate group is trans with respect to imidazolium ring. The N(1)H_3^+ group forms hydrogen bonds with anion, one of water molecules and carboxylate group of nearest cation: N1...Cl(3.29 Å), N1...OW1(2.79 Å), N1...O2(2.84 Å). Imidazolium group forms hydrogen bonds with anion and carboxylate group of nearest cation: N2...O1(2.73 Å), N3...Cl (3.11 Å). Water molecule OW1 forms hydrogen bonds OW1...OW2(2.75 Å) and with carboxylate group, OW1...O2(2.80 Å)(this bond is absent in the table of hydrogen bonds shown by Bennett et al. 1970), while OW2 forms hydrogen bonds with anion and carboxylate group: OW2...Cl(3.19 Å), OW2...O2(3.00 Å).

The crystal of DL-His·2HCl was obtained and studied by Chacko et al. (1966). One may suggest that it is formed similarly to L-crystals, that is, (DL-HisH$_2$)Cl$_2$, however, the crystal structure was not determined. Crystals formed in system L-His + HBr + H_2O, namely, (L-HisH)Br·H_2O, (L-HisH$_2$)Br$_2$ (monoclinic and orthorhombic) are isostructural with respective crystals formed in system L-His + HCl + H_2O (Petrosyan et al. 2005c; Karapetyan et al. 2003; Petrosyan and Petrosyan 2002). It was observed that the monoclinic forms of (L-HisH$_2$)Cl$_2$ and (L-HisH$_2$) Br$_2$ upon recrystallization transform to the respective orthorhombic forms. In addition, it was found that strong SHG signals give monoclinic forms of (L-HisH$_2$)Cl$_2$ and (L-HisH$_2$)Br$_2$. The crystal of (L-HisH)Br·H_2O was independently obtained and characterized as well (Ittyachan and Sagayaraj 2003a; Rajendran et al. 2003c). Unfortunately, the authors called the crystal of L-histidinium bromide in the title of their papers without any indication that it is a monohydrate, which led to confusion later. Ittyachan and Sagayaraj (2003a) provided the correct formula $C_6H_{12}N_3O_3Br$ and assigned absorption bands of water molecule to vibrations of NH_3^+ group, while Rajendran et al. (2003c) indicated the molecular formula $C_6H_{10}O_2N_3Br$ of anhydrous sample, but in interpretation of the IR spectrum indicated absorption band of water crystallization. Ittyachan and Sagayaraj (2003a) and Rajendran et al. (2003c) report growth of crystals by evaporation and temperature cooling methods, provided unit cell parameters, IR spectra, UV–Vis–NIR transmittance spectrum. In the work of Rajendran et al. (2003c), the thermal behavior is also determined and SHG efficiency is estimated (*ca.* 0.84 that of KDP). Out of crystals formed in system L-His + HCl + H_2O the crystal of (L-HisH)Cl·H_2O was investigated mainly. (L-HisH)Cl·H_2O crystals doped by Cu(II) (Colaneri and Peisach 1992), Ni(II) (Remédios et al. 2008) and Cr^{3+} (Kripal and Pandey 2011) were studied by ESR method. The crystal was investigated also by ^{1}H, ^{2}H (Wei and McDermott 1999; Zhao and Harbison 2006) ^{15}N, ^{13}C (Harbison et al. 1981; Song et al. 2001; Strohmeier et al. 2003) and $^{35/37}$Cl NMR method. Faria et al. (2004) studied the (L-HisH)Cl·H_2O crystal by Raman spectroscopy in wide region of wavenumbers (20–3,400 cm^{-1}) and temperature (40–295 K). On the basis of the obtained data, the authors came to conclusion on existence of two structural phase transitions between 140 and 110 K as well as between 80 and 60 K. It should be noted, however, that the symmetry and structure of (L-HisH)Cl·H_2O are the same at 100 K (Andreas et al. 2007). De Sousa et al. (2011) also studied the (L-HisH)Cl·H_2O crystal by Raman spectroscopy method under high pressure. Based on their data they came to conclusion that a reversible structural phase transition is between 2.7 and 3.1 GPa.

De Menezes et al. (2007) have determined piezoelectric coefficients of (L-HisH) Cl·H_2O crystal by electric field induced strain with the angular shift in the multiple diffraction peak position based on synchrotron radiation. Kannan et al. (2006) have grown the (L-HisH)Cl·H_2O crystal and monoclinic (L-$HisH_2$)Cl_2 and characterized the obtained crystals. Provided solubility curves obviously are mixed up. Also Raman spectra, thermal curves, spectra in UV–Vis–NIR region as well as values of damage threshold for (L-HisH)Cl·H_2O (10.1 J/cm^2), (L-$HisH_2$)Cl_2 (14 J/cm^2) and KDP (6.8 J/cm^2) are provided. Several papers (Madhavan et al. 2007a; Ahmed et al. 2010; Robert et al. 2010; Anandan et al. 2012b) also are devoted to the study of (L-HisH)Cl·H_2O. Especially the work of Madhavan et al. (2007a) was considered critically (see comments in Petrosyan et al. 2010a, b), some mistakes were found. Anandan et al. (2012b) sadly repeated the same mistakes in interpretation of vibrational spectra. The authors based on the spectra substantiate the presence of COOH, COO^- and also water molecule. Ahmed et al. (2010) considered the structure by DFT method as well as vibrational spectra and NLO properties. The crystal structure also experimentally repeatedly was determined. The peak in the IR spectrum at 3,157 cm^{-1} is assigned to N-H asymmetric stretching, which actually is caused by ν(CH) of imidazolium group and is characteristic indication of presence of histidine. Robert et al. (2010) grew the (L-HisH)Cl·H_2O crystal by unidirectional method and characterized it. Anandan and Jayavel (2011) obtained pure (L-HisH)Cl·H_2O as well as a crystal doped by tartaric acid (without indication what form of tartaric acid was used). According to the authors, doping with tartaric acid slightly reduces NLO and thermal stability of crystal. Powder SHG for pure crystal is three times higher than that of KDP, while for doped is 2.5 times higher.

Out of crystals formed in L-His + HBr + H_2O system later the crystal of (L-HisH) Br·H_2O only was studied called "L-histidinium bromide" without indication that it is monohydrate. Ittyachan et al. (2005a) studied the thermal properties of (L-HisH) Br·H_2O, the morphology of grown crystals, photoconductivity and microhardness. According to TG, DTG and DTA curves the crystal losses water crystallization near 130 °C. However, in discussion of DSC curve the authors assume the existence of more than one water molecule, which contradicts to the structural data. Ramajothi and Dhanuskodi (2007) have grown and characterized by various methods. However, some statements in this work are not acceptable. The solubility curve in water is not in agreement with previously published curves (Ittyachan and Sagayaraj 2003a; Rajendran et al. 2003c). Ramajothi and Dhanuskodi (2007) write that space group $P2_12_12_1$ allows maximal contribution of the molecular nonlinearity of the macroscopic crystal nonlinearity without indication of reference. Actually this statement is erroneous, because this space group allows minimal contribution (see e.g. Petrosyan et al. 2000b, and references therein). Ramajothi and Dhanuskodi (2007) erroneously assigned the absorption band at 3,443 cm^{-1} to vibration of NH_3^+ group, while in interpretation of thermal data consider that these data confirm the absence of crystal water. In discussion of ^{1}H NMR spectra in solution of D_2O Ramajothi and Dhanuskodi (2007) assigned the signal with chemical shift at 8.30 ppm to N-H group of imidazole, which is erroneous because of rapid exchange of protons at NH and NH_3^+ groups with deuterium. Actually this signal as well as the signal at 7.09 ppm are caused by CH groups of imidazolium ring. The ^{13}C spectrum contains additional signal at 182.762 ppm, which was ignored by the authors. Some more papers have been published on this species: Ahmed et al. (2008) redetermined the structure of (L-HisH)Br·H_2O and also theoretically investigated the geometric parameters, vibrational spectra and hyperpolarizability. Vijayan et al. (2009b) obtained a crystal of (L-HisH)Br·H_2O with very high quality and characterized it. Anandan et al. (2010b) grew the crystal of (L-HisH)Br·H_2O ("L-histidinium hydrobromide"), moreover, IR and Raman spectra, data of thermal analysis are provided and again nothing about water of crystallization in interpretation of vibrational spectra and curves of thermal analysis. On the contrary they mention about COOH group, which is actually absent in the structure and absorption band of water at 3,443 cm^{-1} is assigned to N-H asymmetric stretching. Sajan et al. (2011) also investigated theoretically vibrational spectra and NLO properties of (L-HisH)Br·H_2O. Joema et al. (2011a, b) reported growth of crystals of (L-HisH)Br·H_2O doped by ammonium sulfate (Joema et al. 2011a) and by urea (Joema et al. 2011b). According to the authors, doping by ammonium sulfate leads to improvement of SHG efficiency (0.91 for pure and 1.2 for doped crystals compared to KDP). Doping by 0.5 % of urea is not reflected noticeably on cell parameters, but significantly on thermal properties. According to Joema et al. (2011b) in pure (L-HisH)Br·H_2O crystal the first stage of dehydration starts at room temperature and ends at 165 °C. In doped crystal the first stage of dehydration starts at 30 °C and ends at 100 °C. In conclusion the authors noted that doped crystals are found to be thermally stable than pure crystals.

Having in view that the crystals of (L-HisH)Cl·H_2O and (L-HisH)Br·H_2O are isostructural, Rani et al. (2013) decided to study the properties of (L-HisH)Cl_xBr_{1-x}·H_2O in whole interval from 0 to 1 with incremental step 0.1. Unit cell parameters were determined based on powder diffraction pattern, IR spectra were registered, the solubility was determined and also SHG efficiency. The highest solubility had crystals with composition (L-HisH)$Cl_{0.5}Br_{0.5}$·H_2O. In spite of increasing cell volume on portion of Br the value of "c" parameter had maximal value at $Br_{0.5}$ and then decreased. The values of "a" and "b" increased on increasing of portion of Br. According to this study, the maximal value of SHG efficiency had the crystal with (L-HisH)$Cl_{0.5}Br_{0.5}$·H_2O composition (($I^{2\omega}/I^{2\omega}{}_{KDP} = 3.53$), when it is ($I^{2\omega}/I^{2\omega}{}_{KDP} = 3.14$) for (L-HisH)Cl·$H_2O$ and ($I^{2\omega}/I^{2\omega}{}_{KDP} = 1.2$) for (L-HisH)Br·$H_2O$.

Three salts are formed in L-His + HNO_3 + H_2O system: (L-HisH)NO_3, (L-HisH)NO_3·H_2O and (L-$HisH_2$)$(NO_3)_2$. All three salts crystallize in the orthorhombic system with space group $P2_12_12_1$. The crystal structure of (L-HisH)NO_3 was determined (Srikrishnan et al. 1984). However, the authors did not indicate the conditions of its obtaining and did not provide the coordinates of atoms. So, some important details of the structure, particularly, hydrogen bonds are not known. Based on existence of the (DL-$HisH_2$)$(NO_3)_2$ (Bahadur et al. 1991), Petrosyan and Petrosyan (2002) decided to investigate the possibility of obtaining also (L-$HisH_2$)$(NO_3)_2$. The crystal of (L-$HisH_2$)$(NO_3)_2$ was obtained and characterized spectroscopically and also was found that in addition to anhydrous salt (L-HisH)NO_3 a hydrate also exists, turned out monohydrate (Petrosyan et al. 2005c).

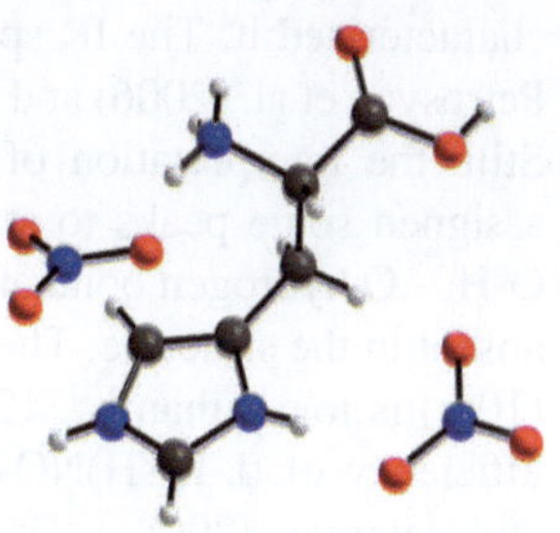

Fig. 5.43 Molecular structure of L-histidinium^{2+} dinitrate

The crystal of (L-HisH$_2$)(NO$_3$)$_2$ was also obtained by Benali-Cherif et al. (2002) and its crystal structure was determined (Fig. 5.43). Later it became known that the (L-HisH$_2$)(NO$_3$)$_2$ crystal was obtained by Bahadur (1992) and its crystal structure was determined, however, these results were not published. Petrosyan et al. (2006) reported crystals of all three salts, conditions of their obtaining were determined, crystal structure of (L-HisH)NO$_3$·H$_2$O was determined and also IR and Raman spectra of all three salts were investigated. The NLO activity of these salts was estimated by powder SHG method (Petrosyan 2007a). These crystals attracted the attention of other researchers as possible NLO materials: Aruna et al. (2007c) obtained needle-like crystals of (L-HisH$_2$)(NO$_3$)$_2$ with $23 \times 2 \times 2$ mm^3 sizes, described the morphology of crystals, determined their quality by high resolution X-ray diffraction, transmittance in UV–Vis–NIR region, thermal behavior and NLO properties by Kurtz–Perry powder method. The (L-HisH$_2$)(NO$_3$)$_2$ crystals are stable up to 165 °C and then decompose. According to this study, the efficiency of SHG is 2.8 times higher than in KDP. Based on dependence of efficiency of SHG on sizes of particles the authors concluded on possibility of phase-matching, however, it should be noted that the plateau is absent in the graph. According to Petrosyan (Petrosyan 2007a), the efficiency of powder SHG for (L-HisH$_2$)(NO$_3$)$_2$ is ($I^{2\omega}/I^{2\omega}{}_{KDP} = 0.52$), while for (L-HisH)NO$_3$·H$_2$O and (L-HisH)NO$_3$ are ($I^{2\omega}/I^{2\omega}{}_{KDP} = 0.90$ and ($I^{2\omega}/I^{2\omega}{}_{KDP} = 0.87$) respectively. The crystal of (L-HisH) NO$_3$ was studied also in several works: Zhang et al. (2008b) determined the solubility of crystals in water and have grown quality (L-HisH)NO$_3$ crystal by cooling method and determined its morphology. The authors also determined crystal structure of (L-HisH)NO$_3$, however, did not provide coordinates of atoms and did not deposit its CIF in CSD. The authors only showed unit cell packing. They also determined thermal behavior, provided IR and UV–Vis–NIR spectra, microhardness and estimated the SHG efficiency by powder method. According to the authors, the crystal of (L-HisH)NO$_3$ melts near 234 °C and then decomposes with exothermic effect with peak of DTA at 254 °C. The IR spectrum contains a peak at 3,443 cm^{-1}, which the authors assigned to asymmetric stretching vibrations of NH$_3^+$ group, which, however, is absent in the spectrum given by Petrosyan et al. (2006). In the text the authors write that the efficiency of the crystal is about 2.5 times as large as that of KDP, while in Abstract and Conclusion they write that SHG intensity is 1.5 times greater than that of KDP. Martin Britto Dhas and Natarajan (2008a) have grown the (L-HisH)NO$_3$ crystal by evaporation and characterized it. The IR spectrum showed in their work agrees with that given by Petrosyan et al. (2006) and also does not contain the absorption peak at 3,443 cm^{-1}. Still, the interpretation of the IR spectrum is rather contradictory. The authors assigned some peaks to stretching and out of plane deformation of O-H bond in O-H...O hydrogen bond and to CH_3 asymmetric stretching, while these groups are absent in the structure. The microhardness of crystal 73.0 kg/mm^2 at the same load (10 g) is lower than *ca.* 125 kg/mm^2 obtained by Zhang et al. (2008b). The SHG efficiency of (L-HisH)NO_3 is ($I^{2\omega}/I^{2\omega}{}_{KDP} = 0.5$) according to Martin Britto Dhas and Natarajan (2008a). The crystal (L-HisH)NO_3 was grown by evaporation method at 35 °C and characterized also by Riscob et al. (2011a). According to this work, the (L-HisH)NO_3 crystal has very high birefringence, which is two times higher than that in calcite. The authors determined the presence of phase-matching by Kurtz–Perry method and determined the SHG efficiency ($I^{2\omega}/I^{2\omega}{}_{KDP} = 0.8$).

So, the structures of all three crystals formed in L-His + HNO_3 + H_2O system are determined, and in spite of the fact that the structure of (L-HisH)NO_3 was determined twice, details of its structure are not published. Asymmetric units of (L-HisH)NO_3·H_2O (Petrosyan et al. 2006) and (L-HisH_2)$(NO_3)_2$ (Benali-Cherif et al. 2002; Bahadur 1992) contain each one formula unit. Cations in both structures have open conformation. Judging by figure of the structure of (L-HisH)NO_3 the cation here also has open conformation. In the structure of (L-HisH)NO_3·H_2O the $N(1)H_3^+$ group forms two hydrogen bonds with nearest anions and one hydrogen bond with carboxylate group of cation: N1-H2...O3(2.863(2) Å), N1-H3...O5(2.875(2) Å), N1-H4...O2(2.765(2) Å). The imidazolium group forms hydrogen bonds with carboxylate group of nearest cation and with water molecule: N3-H7...O1(2.653 (2) Å) and N2-H9...O6(2.669(2) Å). The water molecule in turn forms hydrogen bonds with anion and carboxylate group of cation: O6-H12...O5(2.807(2) Å), O6-H11...O2(2.787(2) Å). In the structure of (L-HisH_2)$(NO_3)_2$ (Benali-Cherif et al. 2002) the carboxyl group forms hydrogen bond with one of anions: O1-H1...O5(2.621(3) Å). The $N(1)H_3^+$ group forms two hydrogen bonds with two nearest the same type anions and with carbonyl oxygen atom of nearest cation: N1-H1...O3(2.957(3) Å), N1-H2...O3(2.900(3) Å), N1-H3...O2(2.876(3) Å). At presence of oxygen containing anion doubly charged cations usually do not interact with each other. In this case there is such bond. The imidazolium group forms hydrogen bonds with second type two symmetry-equivalent anions: N2-H...O7(2.767(3) Å), N3-H...O6(2.836(3) Å). There is also a weak contact N3-H...O6(2.836(3) Å) with H...O(2.36 Å), which is intermediate between strong van der Waals and weak hydrogen bonds. The table of hydrogen bonds published by Benali-Cherif et al. (2002) contains many other contacts, which are not hydrogen bonds. Asymmetric unit of (DL-HisH_2)$(NO_3)_2$ also contains one formula unit (Bahadur et al. 1991). In one of the anions, one oxygen atom is disordered. Cation also has open conformation. Carboxyl group forms hydrogen bond with anion, which has disordered oxygen atom. One contact O2-H11...O21(2.824 Å) with H11...O21(2.276 Å) is weak. There is one more contact with one of disordered position of oxygen O2-H11...O(231)(2.667 Å). The $N(1)H_3^+$ group forms hydrogen bond with both types of anions: N1-H1...O12(2.885 Å), N1-H2...O22 (3.026 Å) (there is misprint in Bahadur et al. 1991, the value 3.306 Å is indicated instead of 3.026 Å based on CIF), N1-H3...O21(2.876 Å). So, in this case cations do not interact with each other in contrast to the (L-HisH$_2$)(NO$_3$)$_2$. The imidazolium group forms hydrogen bonds with two the same type symmetry-related anions: N2-H7...O13(2.863 Å), N3-H9...O11(2.856 Å). In addition there are short contacts C5-H8...O with disordered oxygen atoms with H8...O(2.222 Å and 2.274 Å), which can be considered as weak C-H...O hydrogen bond. Vibrational spectra of (DL-HisH$_2$)(NO$_3$)$_2$ were investigated by Rajkumar et al. (1999).

In the L-His + HIO_3 + H_2O system there was found only one compound (L-HisH) IO_3, which is formed in form of powder (Petrosyan 2010a). In the interval from 2:1 to 1:7 ratios of L-His:HIO_3 no other compounds were found. At 2:1 ratio L-histidine is forming and at ratios from 1:1 to 1:7 mentioned powdered (L-HisH)IO_3 is formed. The compound was characterized by IR and Raman spectra.

Let us now consider the system L-His + HBF_4 + H_2O. Marcy et al. (1995) found that the (L-HisH)BF_4 crystal grows readily from aqueous solution, is stable and nonhygroscopic, displays favorable mechanical and thermal characteristics, and has adequate birefringence to permit phase-matched nonlinear parametric processes over much of its transparency range (250–1,300 nm). Coefficients in Sellmeier equation for refractive indices have been determined. They have measured Type I and Type II 1,064 nm SHG phase-matching loci and found that the (L-HisH)BF_4 crystal has an effective nonlinear coefficient five times that of KDP. As it would be expected the (L-HisH)BF_4 crystal attracted great attention. Solubility in water and other solvents has been determined. Crystallization conditions were studied and rather large crystals with good quality have been grown and characterized by determination of thermal properties, microhardness, transmittance in UV–Vis–NIR range, IR and Raman spectra, and SHG efficiency measured by powder method. In addition, an attempt was made to obtain new salt with doubly charged (L-HisH$_2$) cation (Petrosyan and Petrosyan 2002). Aggarwal et al. (1999, 2003) have determined solubility in water, acetone and ethanol, have grown (L-HisH)BF_4 crystals with $20 \times 20 \times 10$ mm^3 dimensions and determined transmittance in UV–Vis–NIR range. They, however, could not observe high NLO properties of grown crystal. Solubility in water was determined also in several works Rajendran et al. 2003a, b; Ramajothi and Dhanuskodi 2003; Dhanuskodi and Ramajothi 2004), the results of which are in satisfactory agreement. Solubility in acetone (Rajendran et al. 2001; Gokul Raj et al. 2005), methanol (Ramajothi and Dhanuskodi 2003; Gokul Raj et al. 2005) and ethanol (Aggarwal et al. 1999; Dhanuskodi and Ramajothi 2004) is significantly lower than in water, therefore crystals are grown from aqueous solutions. Determination of growth parameters and influence of pH on growth and properties have been carried out (Rajendran et al. 2003a, b). Measurements of SHG efficiency by powder method did not give expected (on the basis of Marcy et al. 1995) high values. The following values of SHG efficiency relative to KDP were obtained: 1.3 (Zhang et al. 2008b), 1.70 (Gokul Raj et al. 2007), *ca.* 2.2 (Rajendran et al. 2003a), 0.23 relative to urea (Ramajothi and Dhanuskodi 2003), which is *ca.* 2 relative to KDP, *ca.* 2 (Syamala et al. 2007), 5.13 (Madhavan et al. 2006a). The value around 2 seems more realistic. Gokul Raj et al. (2007) on the basis of relative efficiency concluded that $d_{eff} = 1.70 d_{eff}$(KDP), which is not correct (see *e.g.* Marcy et al. 1995). Rajendran et al. (2003a) observed that decreasing of pH increases solubility, however, the accuracy of measurements of SHG efficiency is insufficient to determine the influence of pH on SHG efficiency. Thermal properties were studied in several works: The analysis of these results shows that the (L-HisH)BF_4 crystal is stable up to 250 °C and at further heating decomposes. The conclusion (Ramajothi and Dhanuskodi 2003; Dhanuskodi and Ramajothi 2004) that the (L-HisH)BF_4 crystal melts at 235 ± 1 °C should be considered as incorrect. This conclusion contradicts not only to TG, DTA and DSC showed in (Rajendran et al. 2003a; Syamala et al. 2007; Gokul Raj et al. 2005), but also to TG and DTA data given by Dhanuskodi and Ramajothi (2004). From TG curve it can be seen that above 250 °C the sample begins to decompose. The DTA peak of this is at 275 °C. The temperature of decomposition 205 °C specified by Marcy et al. (1995) probably is a misprint (should be 250 °C). The higher decomposition temperature (270 °C) indicated in the works of Syamala et al. (2007) and Gokul Raj et al. (2005) is probably caused by higher heating rate (20°/min). Results of microhardness measurements (Syamala et al. 2007; Rajendran et al. 2001; Ramajothi and Dhanuskodi 2003) are not even in qualitative agreement. The ^{1}H NMR spectrum recorded (Dhanuskodi and Ramajothi 2004) to our opinion is not adequate because the choice of solvent is not appropriate (compare with Petrosyan et al. 2009a). It was already discussed by Petrosyan (2007b) that the IR spectra shown by Rajendran et al. (2001) and Ramajothi and Dhanuskodi (2003) and have noted that these spectra are distorted because of partial ionic exchange at registration of the spectra by KBr pellet technique. In the reports by Rajendran et al. (2003a) and Syamala et al. (2007) practically complete ionic exchange took place and the IR spectrum represents the spectrum of mixture of (L-HisH)Br·H_2O and KBF_4. In the study of Marcy et al. (1995) the structure of (L-HisH)BF_4 has been determined, but details of hydrogen bonding have not been showed. Sinha et al. (2009) grew the crystal of (L-HisH)BF_4 from aqueous solution by two methods. Leaving without consideration these methods, we shall note that allegedly different IR spectra (recorded by KBr pellet technique) of the (L-HisH)BF_4 crystals grown with different methods, in fact are the spectra with complete ionic exchange (Fig. 8a in Sinha et al. 2009) as in case of the studies by Syamala et al. (2007) and Rajendran et al. (2003a) and partly ionic exchange (Fig. 8b in Sinha et al. 2009) as in cases shown by Rajendran et al. (2001) and Ramajothi and Dhanuskodi (2003). Gokul Raj et al. (2006a) have repeatedly determined the structure of (L-HisH)BF_4, however, it is a pity that the paper by Marcy et al. (1995), in which the structure of (L-HisH)BF_4 has been determined for the first time, is not mentioned. Even more noteworthy is that Gokul Raj et al. (2007), who presented the structure of (L-HisH)BF_4 in more detail for interpretation of vibrational (IR and Raman) spectra, noted that crystal structure of (L-HisH)BF_4 has not been solved so far, though the paper by Marcy et al. (1995) is mentioned. The hydrogen bonding scheme in the structure of (L-HisH)BF_4,

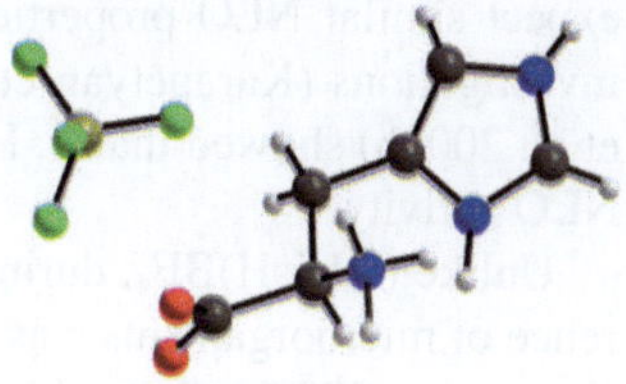

Fig. 5.44 Molecular structure of L-histidinium perchlorate

as was expected (Petrosyan 2007b), is similar to the scheme of isostructural (L-HisH)ClO_4 (Román et al. 1987). Asymmetric unit contains one formula unit. (L-HisH) cation has open conformation. The $N(3)H_3^+$ group forms two hydrogen bonds with carboxylate groups of two nearest cations and one hydrogen bond with anion: N3-HA…O2(2.722(3)Å), N3-HC…O1(3.010(3)Å), N3-HB…F3(2.945 (3)Å). Imidazolium group forms one hydrogen bond with anion and one with nearest cation: N1-H1…F1(2.856(3)Å) and N2-H2…O1(2.953(3)Å). Gokul Raj et al. (2007) used results of structure determination for interpretation of IR and Raman spectra. The IR spectrum of (L-HisH)BF_4 (Gokul Raj et al. 2005, 2007) recorded by KBr pellet method is not distorted and well corresponds to the IR-ATR spectrum (Petrosyan 2007b). Raman spectra recorded by Gokul Raj et al. (2007); Syamala et al. (2007) and Petrosyan (2007b) also are in good agreement. We will not consider the discussion of the spectra of Syamala et al. (2007) because, as it was noted already, the IR spectrum completely distorted and also there is assumed that monoclinic form of L-histidine is obtained. Gokul Raj et al. (2005) noted that the characteristic intense broad band at 1,025 cm^{-1} is due to BF_4 vibrations. However, in their 2007 paper the authors write that it is not possible to isolate the bands originating only due to tetrafluoroborate ions and in the Table 6 of this work of bands assignments for IR and Raman spectra one cannot find bands related to vibrations of BF_4 anion. Gokul Raj et al. (2007) tried to confirm assignments of observed stretching vibrations of N-H bonds by calculation of vibrational frequencies on the basis of correlation between frequencies and N…O, N…F distances. However, the Raman lines at 3,180, 3,148 and 2,950 cm^{-1}, which correspond to 3,184, 3,150 and 2,951 cm^{-1} in our spectrum (Petrosyan 2007b) are caused by C-H bonds of imidazolium (3,180 and 3,148 cm^{-1}) and aliphatic C-H bonds (2,950 cm^{-1}), but not by N-H bonds. In addition, it is worth to note that noticeable absorption in the transmittance curve near 970 nm (Gokul Raj et al. 2005, 2007) seems to be an artifact, because this absorption is absent in the curves recorded by other authors (Aggarwal et al. 1999, 2003; Rajendran et al. 2001; Dhanuskodi and Ramajothi 2004). In the system L-His + HBF_4 + H_2O in addition to the crystal of (L-HisH)BF_4 a salt with doubly charged (L-$HisH_2$) cation also is formed, which has two (α and β) forms. Crystal structures of these forms were not determined. They were characterized by infrared and Raman spectra (Petrosyan 2010a).

The crystal of (L-HisH)ClO_4 was known before (L-HisH)BF_4. Its crystal structure was determined (Román et al. 1987, Fig. 5.44). As it was mentioned (L-HisH) ClO_4 and (L-HisH)BF_4 have similar structure. Solid-state NMR (^1H and ^{15}N) of (L-HisH)ClO_4 was studied (Wei and McDermott 1999; Song et al. 2001). One might expect similar NLO properties also in case of (L-HisH)ClO_4. Indeed preliminary investigations (Karapetyan et al. 2003; Petrosyan and Petrosyan 2002; Petrosyan et al. 2004b) showed that (L-HisH)ClO_4 crystal has the same morphology and high NLO activity.

Unlike (L-HisH)BF_4, during crystal growth of (L-HisH)ClO_4 undesirable occurrence of microorganisms was observed (Petrosyan et al. 2004b). Later it was found (Petrosyan 2007b) that addition of small (1 %) amount of L-His·HBF_4, having strong bactericide property, prevents occurrence of microorganisms in solution during crystal growth. The ATR IR and Raman spectra of (L-HisH)BF_4 and (L-HisH)ClO_4 are discussed (Petrosyan 2007b). SHG efficiency for (L-HisH)ClO_4 is $I^{2\omega}/I^{2\omega}_{KDP} = 2.38$ (Petrosyan 2007a) The (L-HisH)ClO_4 crystals attracted also attention of other authors: High-quality crystals with $5 \times 4 \times 1$ mm^3 (Ittyachan et al. 2005b) and $5 \times 6 \times 2.5$ mm^3 (Aruna et al. 2007d) dimensions were grown and their morphology, UV–Vis–NIR transmittance, microhardness and photoconductivity (Ittyachan et al. 2005b) as well as IR spectrum, thermal properties, nonlinearity and dielectric studies (Aruna et al. 2007d) have been performed. According to this work, the crystal is stable up to 270 °C, whereas Román et al. (1987) reported that the crystal decomposes above 195 °C. This may be caused by impurities, because the (L-HisH)ClO_4 crystals in the older work were obtained from the MoO_3-histidine-$HClO_4$ system. The UV–Vis–NIR curve shown by Ittyachan et al. (2005b) disagrees with that given by Aruna et al. (2007d). The IR spectrum of (L-HisH)ClO_4 recorded by KBr pellet technique in the second work is in agreement with our spectrum (Petrosyan 2007b), however, interpretation of the spectrum in our opinion is not correct. All peaks from 3,252 to 3,038 cm^{-1} are assigned to vibrations of N-H bonds. The peak at 1,336 cm^{-1} is assumed to be superimposed of deformational vibration of CH_2 group and ClO_4^- anion. Meanwhile the ClO_4^- ion has no absorption in this region. Wavenumbers of fundamental vibrations of free ClO_4^- ion are 935 cm^{-1} (ν_1), 1,102 cm^{-1} (ν_3), 360 cm^{-1} (ν_2), 533 cm^{-1} (ν_4). At last, it is worth to note that the most intensive band at 1,068 cm^{-1} (ν_3 of ClO_4^- ion) was assigned to the vibration of C-N bond. SHG efficiency of (L-HisH)ClO_4 measured in (Aruna et al. 2007d) is $I^{2\omega}/I^{2\omega}_{KDP} = 3.19$.

The crystal of (DL-HisH)ClO_4 was crystallized from aqueous solution containing equimolar quantities of DL-histidine and perchloric acid (Bahadur and Athimoolam 2009). Crystal structure was determined at room temperature. Asymmetric unit contains one formula unit. The (HisH) cation has open conformation with carboxylate group located *trans* in respect to imidazolium group. The authors incorrectly call monoprotonated cation as diprotonated. The $N(1)H_3^+$ group forms two hydrogen bonds with carboxylate groups of two nearest cations and two weak bonds with anion: N1-HB...O1B(2.847(4) Å), N1-HC...O1B(2.774(4) Å) and N1-HA...O2 (2.970(4) Å), N1-HA...O3(3.142(4) Å). Interpretation of hydrogen bonds formed by imidazolium group is erroneous. The authors mixed up carbon and nitrogen atoms of imidazolium group. Actually nitrogen atoms labeled in this work as carbon atoms form hydrogen bonds with anion and carboxylate group of nearest cation with N...O distances 2.856(4) Å and 2.742(4) Å respectively.

From the system L-His + H_3PO_4 + H_2O initially easier forming salt of L-His·2H_3PO_4 composition was obtained. Preliminary results were reported by Blessing and McGandy (1972), then Blessing (1987) reported a full determination of the structure. The L-His·2H_3PO_4 crystals are formed easily from aqueous solution containing stoichiometric quantities of components. Crystal structure determination showed that L-histidine exists as singly charged cation (L-HisH)(H_2PO_4)·H_3PO_4. From the work of Rajendran et al. (2007) it is clear that this crystal may be obtained also from solution containing equimolar quantities of components as it is shown by Petrosyan (2008b). Asymmetric unit contains one formula unit. The (L-HisH) cation has open conformation. The H_3PO_4 molecule forms short hydrogen bond with $(H_2PO_4)^-$ anion: O22-H…O11(2.554 Å). The two rest O-H groups of the H_3PO_4 molecule also form hydrogen bonds with nearest anions: O23-H…O12(2.573 Å), O24-H…O11(2.583 Å). However, the $(H_2PO_4)^-$ anions form with each other shorter hydrogen bond: O13-H…O12(2.460 Å). Second OH group forms hydrogen bond O14-H…O″(2.569 Å) with carboxylate group of cation. The NH_3^+ group forms one hydrogen bond with carboxylate group of nearest cation and also with (H_3PO_4) molecule and $(H_2PO_4)^-$ anion: N-H2…O″(2.854 Å), N-H1…O21 (2.777 Å) and N-H3…O13(2.954 Å). The imidazolium group forms one hydrogen bond with carboxylate group of nearest cation, $N^{\delta1}$-H…O′(2.658 Å) and with H_3PO_4 molecule, $N^{\varepsilon2}$-H…O21(2.733 Å). There is also a contact $C^{\varepsilon1}$-H…O22 (3.222 Å) with H…O(2.30 Å). From the system L-His + H_3PO_4 + H_2O Averbuch-Pouchot et al. (1988d) could prepare another salt with composition (L-HisH) (H_2PO_4)·H_2O. The crystals have been prepared by mixing equimolar quantities of dilute solutions of L-histidine and H_3PO_4. The solution is kept at room temperature until a dry solid gel is obtained (2–3 months). Then a few drops of water are added every day over several weeks, until some small crystalline areas dispersed in the gel appear. At this step, daily addition of larger quantities of water accelerates the crystallization. Asymmetric unit contains one formula unit. The (L-HisH) cation has open conformation. One of two O-H groups of anion is used for creating a chain of anions: O2-H2…O4(2.562(3) Å). The second group forms hydrogen bond with carboxylate group of cation: O1-H1…O6(2.598(4) Å). Water molecule as proton donor forms one hydrogen bond with cation and one with anion: OW-H1…O5(2.861(5) Å), OW-H2…O4(2.852(4) Å). The N(1)H_3^+ group forms two hydrogen bonds with water molecule and with anion: N1-H1…OW(2.862 (5) Å), N1-H2…O3(2.791(4) Å). The imidazolium group forms one hydrogen bond with carboxylate group of nearest cation and with anion: N2-H…O5(2.672 (4) Å), N3-H…O3(2.684(4) Å). There is also a short contact C6-H…O4(3.081 Å) with H…O distance 2.35 Å. Later Espinosa et al. (1995) apparently independently prepared by another procedure the same crystal and determined its structure.

Averbuch-Pouchot (1993d) has also prepared L-histidine phosphite L-His·H_3PO_3 ((L-HisH)H_2PO_3), from aqueous solution containing L-histidine and phosphorous acid in 1:1 proportion. Asymmetric unit contains one formula unit. The (L-HisH) cation has closed conformation. The anions do not form hydrogen bonds with each other. The only OH group of anion forms hydrogen bond with carboxylate group of cation, O3-H...O4(2.622(3) Å). (L-HisH) cations form hydrogen bonds with each other and with anion. The $N(1)H_3^+$ group forms hydrogen bonds with three nearest anions: N1-H1...O2(2.744(2) Å), N1-H2...O1(2.873(3) Å), N1-H3...O1 (2.833 Å). Imidazolium group forms hydrogen bonds with anion and cation: N2-H...O2(2.709(3) Å), N3-H...O5(2.702(3) Å). The crystals of (L-HisH) $H_2PO_4{\cdot}H_2O$ and (L-HisH)H_2PO_3 have not been considered as NLO materials yet, while the (L-HisH)$H_2PO_4{\cdot}H_3PO_4$ has been studied by several authors. Espinosa et al. (1997) have grown comparatively large crystals and performed vast spectroscopic investigations in wide range of frequencies and temperature. Electron density study was performed in this crystal as well (Mata et al. 2006).

Ratajczak et al. (2000) have prepared the L-His·$2H_3AsO_4$ crystal, which is formed by the same (L-HisH)$H_2AsO_4{\cdot}H_3AsO_4$ mechanism and is isostructural with respective diphosphate. Crystal structure has been determined as well as IR, Raman spectra, SHG activity and thermal properties were studied. The crystal is thermally stable up to 170 °C and shows SHG efficiency 0.46 relative to KDP. Characteristic peak caused by ν(CH) of imidazole ring at 3,137 cm^{-1} in IR spectrum and respective Raman line at 3,138 cm^{-1} were assigned to vibration of C(2)H bond. The same peak in the IR spectrum of L-His·$2H_3PO_4$ at 3,130 cm^{-1} was assigned to vibration of N-H bond (Espinosa et al. 1997). The crystal of L-histidine diphosphate was reported as a new crystal (Ittyachan and Sagayaraj 2003b). Solubility curve in water has been determined and comparatively large crystal was grown. IR spectrum and unit cell parameters prove that the crystal is identical with that obtained by Blessing (1987) and Espinosa et al. (1997). However, as was noted (Petrosyan et al. 2005c), in result of incorrect interpretation of X-ray data, authors came to conclusion that the crystal is formed by mechanism of doubly charged cation. Therefore interpretation of IR spectrum is contradictory. Further work on growth and characterization of the crystal carried out (Madhavan et al. 2006b). UV–Vis–NIR transmittance curve of grown crystal has noticeable absorption peak at about 1,200 nm, which is absent in curves shown in (Espinosa et al. 1997; Ittyachan and Sagayaraj 2003b). TG/DTA curves (Madhavan et al. 2006b) show that the crystal is stable up to 200 °C, hence is more stable than its arsenate analog. However, interpretation of thermogravimetric data is incorrect. Authors write that "A small weight loss of about 3 % is observed between 217 and 264 °C, which may be attributed to the adsorption of water molecules." Meanwhile the loss of weight is about 30 % as can be seen from Fig. 4 (Madhavan et al. 2006b) caused by decomposition of the sample. Apparently, authors used DTA scale instead of TG scale. Authors reported also data on crystal morphology, microhardness measurements, dielectric and photoconductivity as well as SHG efficiency. The authors found that SHG efficiency of L-His·$2H_3PO_4$ nearly four times higher than that of KDP. This value is much higher than that of L-His·$2H_3AsO_4$ (0.46) (Ratajczak et al. 2000) and values for the same crystal of other authors (0.41 (Petrosyan 2007a) and 1.1 (Rajendran et al. 2007), where, as was mentioned, the L-His·$2H_3PO_4$ crystal was in fact grown and studied). The work of Ahmed et al. (2009) also deals with this crystal. Unfortunately, out of all papers concerning L-His·$2H_3PO_4$ the authors were apparently familiar only with (Ittyachan and Sagayaraj 2003b). Therefore they decided to determine once again the crystal structure and recorded the Raman spectrum. Structure determination is in good agreement with previous data. The optimized geometric parameters, vibrational frequencies, dipole moment, polarizability and hyperpolarizability were computed using DFT/B3LYP/6-31G(d) method. Unfortunately, Ahmed et al. (2008) assigned the characteristic peaks of stretching vibrations of C-H bonds of imidazolium ring at 3,154 and 3,142 cm^{-1} in Raman spectrum to N-H...O and O-H...O stretching. This is hardly probable. Particularly all O-H...O type hydrogen bonds are strong with O...O distances shorter than 2.6 Å. According to the correlation between ν(OH) and O...O distances (Novak 1979) all (OH) frequencies should be below 2,500 cm^{-1}. It is worth noting that one may expect that arsenate analog of the (L-HisH)H_2PO_4·H_2O, that is (L-HisH)H_2AsO_4·H_2O can exist as well.

The crystals of (L-HisH)NH_2SO_3 and (DL-HisH)NH_2SO_3 were studied by Bahadur (1992). Needle-shaped colorless crystals of (L-HisH)NH_2SO_3 were obtained from aqueous solution containing equimolar quantities of L-histidine and sulfamic acid by slow evaporation and investigated by infrared and Raman spectroscopic methods (Rajkumar et al. 1999).

The system L-His + H_2SO_4 + H_2O was studied by Srinivasan et al. (2001d) and Kumar et al. (2005). In the older work, the crystal of (L-HisH_2)$_2$(SO_4)$_2$·$3H_2O$ was obtained by slow evaporation from aqueous solution containing L-histidine and sulfuric acid. The temperature was not, however, indicated. The authors indicated that the crystal "was crystallized from an aqueous solution of DL-histidine and sulfuric acid," however, actually L-histidine was used, according to the letter of Prof. Rajaram to author of these lines. Asymmetric unit contains one formula unit, that is, two (L-HisH_2) cations, two $(SO_4)^{2-}$ anions and three water molecules. It was indicated $Z = 2$, however, this number relates to the composition (L-HisH_2)(SO_4)·$1.5H_2O$, as it is clear from the value of molecular mass $M_r = 280.26$ g/mol. One oxygen atom of one of anions is disordered. Both cations have open conformation. Carboxylic groups of both cations form very strong hydrogen bonds with two water molecules: O1B-H1B...OW1(2.474(6) Å) and O2B-H2B...OW2(2.485(7) Å). Water molecules H_2O(W1) and H_2O(W2) in turn also form rather strong hydrogen bonds with anions: OW1-H1W...O5(2.629(9) Å), OW1-HW2...O3(2.684(8) Å) and OW2-HW3...O2(2.666(7) Å), OW2-HW4...O8(2.588(14) Å). The water molecule H_2O(W3) as proton acceptor does not form hydrogen bonds, while as proton donor also form hydrogen bonds with anions, in spite that they are relatively weak: OW3-HW5...O1(2.775(13) Å), OW3-HW6...O7(2.909(15) Å). Hydrogen bonds of N-H...O type also are mainly strong, especially hydrogen bonds formed by imidazolium groups. The N(11)H_3^+ group forms hydrogen bond with carbonyl oxygen atom O2A and with anions: N11-H11C...O2A(2.989(6) Å), N11-H11A...O7(2.795(8) Å), N11-H11B...O2(3.046(6) Å), N11-H11B...O4(3.055(8) Å). The N(21)H_3^+ group also forms hydrogen bond with carbonyl oxygen O1A and with anions: N21-H21C...O1A(2.895(6) Å), N21-H21A...O4(2.714(6) Å), N21-H21B...O6(2.767(7) Å). The imidazolium groups form hydrogen bonds with anions: N12-H12A. . .O3(2.662(7) Å), N13-H13. . .O1(2.794(8) Å) and N22-H22A. . . O8(2.665(11) Å), N22-H22a. . .O8′(2.713(19) Å), N23-H23. . .O6(2.821(7) Å). There are also two short C-H. . .O contacts of imidazolium groups: C16-H. . .OW1(3.214 Å) with H. . .O(2.31 Å) and C15-H. . .O1A(3.076 Å) with H. . .O(2.20 Å). Kumar et al. (2005) obtained anhydrous crystals (L-HisH_2)SO_4 also from aqueous solution containing equimolar quantities of L-histidine and sulfuric acid by slow evaporation at room temperature. The salt (L-HisH_2)SO_4 crystallizes in monoclinic system with $Z = 4$. Asymmetric unit contains two formula units and there is pseudosymmetry. Both cations have open conformation. Carboxyl groups of cations form rather short hydrogen bonds with anions: O1B-H1B. . .O11(2.521(7) Å) and O2B-H2B. . .O24 (2.554(7) Å). The group $N(11)H_3^+$ forms one hydrogen bond with carbonyl oxygen atom of symmetry-related cation and two hydrogen bonds with anions: N11-H11A. . .O1A(2.834(8) Å), N11-H11B. . .O13(2.721(8) Å), N11-H11C. . .O12 (2.757(10) Å). The group $N(21)H_3^+$ also forms one hydrogen bond with carbonyl oxygen atom of symmetry-equivalent cation and two hydrogen bonds with second anion: N21-H21A. . .O2A(2.790(8)), N21-H21B. . .O22(2.708(8) Å), N21-H21C. . . O23(2.679(8) Å). The imidazolium groups form hydrogen bonds with both anions: N12-H12A. . .O21(2.719(8) Å), N13-H13. . .O14(2.729(8) Å) and N22-H22A. . .O13 (2.694(8) Å), N23-H23. . .O24(2.811(7) Å). There is also one relatively short contact C15-H. . .O12(3.228 Å) with H. . .O(2.35 Å). Investigation of the system L-His + H_2SO_4 + H_2O was continued by Petrosyan et al. (2010a). In addition to salts obtained by Srinivasan et al. (2001d) and Kumar et al. (2005) it seems possible also existence of salts with singly charged cation (L-HisH) and also with HSO_4^- anion. From aqueous solution containing L-histidine and H_2SO_4 in molar ratio 2:1 by evaporation at different temperatures samples in form of powder were obtained. Infrared spectra indicated on formation of (L-HisH)$_2SO_4$ salt and its hydrate (Petrosyan et al. 2010a). The hydrate being monohydrate was possible to obtain in form of single crystal and determine its crystal structure. Asymmetric unit of (L-HisH)$_2SO_4$·H_2O contains one formula unit. The (L-HisH) cations have different conformations. One of them has closed conformation and the second has open conformation. In *trans*-position in respect to imidazolium ring is NH_3^+ group and not COO^- group. The NH_3^+ group of first cation forms hydrogen bond with water molecule, sulfate anion and with carboxylate group of symmetry-equivalent cation: N1-H2. . .O992.724(4) Å), N1-H3. . .O5 (2.669(4) Å), N1-H4. . .O2(2.911(4) Å). The $N(4)H_3^+$ group of second cation forms similar hydrogen bonds: N4-H12. . .O9(2.971(4) Å), N4-H14. . .O8(2.783(4) Å), N4-H13. . .O4(2.998(4) Å). The imidazolium groups of cations form one hydrogen bond with anion and one with carboxylate group of other cation: N2-H7. . .O7(2.734 (4) Å), N3-H9. . .O4(2.759(4) Å) and N5-H17. . .O6(2.696(4) Å), N6-H19. . .O2 (2.724(4) Å). Water molecule forms hydrogen bonds with anion: O9-H21. . .O7 (3.035(5) Å), O9-H21. . .O8(2.863(4) Å), O9-H22. . .O6(2.765(4) Å). There are also short C-H. . .O contacts of imidazolium group with carboxylate groups: C5-H8. . .O1 (3.089 Å), C11-H18. . .O3(3.089 Å), C12-H20. . .O1(3.088 Å) with H. . .O distances 2.26 Å, 2.16 Å and 2.31 Å respectively. At least the contact C11-H18. . .O3 is probably

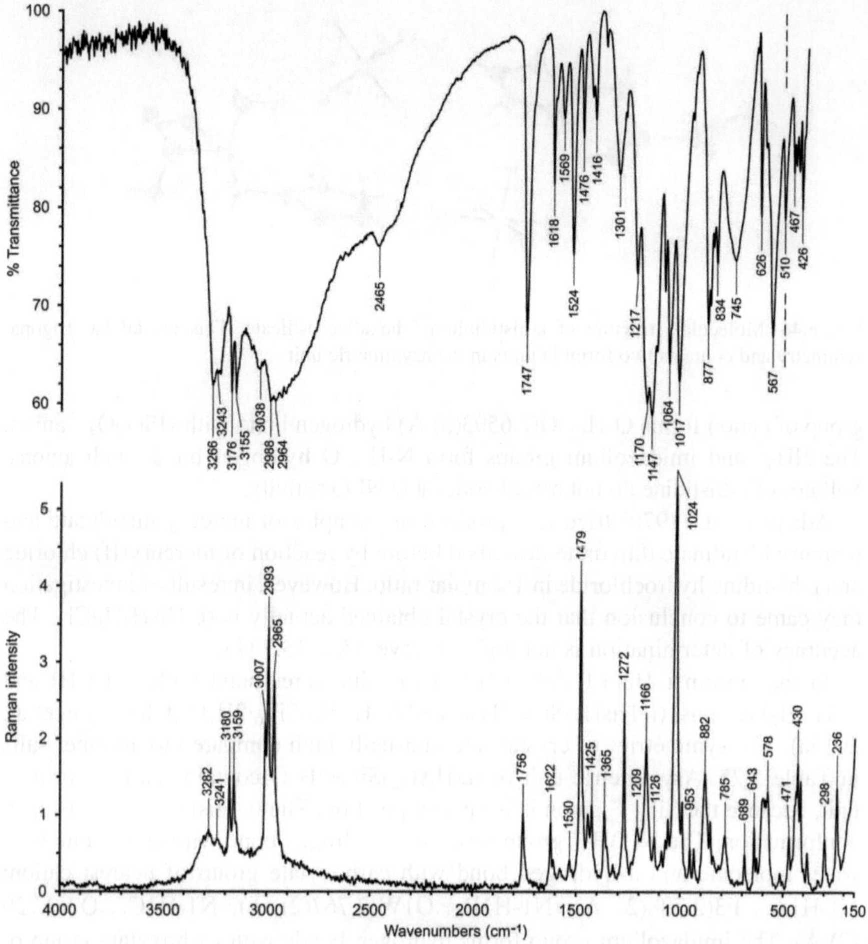

Fig. 5.45 IR and Raman spectra of (L-HisH$_2$)(HSO$_4$)$_2$

a real hydrogen bond. The crystal of (L-HisH$_2$)(HSO$_4$)$_2$ is extremely hygroscopic Ghazaryan et al. (in prep). The crystals were possible to obtain from aqueous solution with stoichiometric composition in presence of silica gel as water absorber in closed vessel. In Fig. 5.45 IR and Raman spectra of obtained crystal are provided. In the Raman spectrum peaks at 3,180 cm and 3,159 cm^{-1} relate to ν(CH) stretching vibration of imidazolium group. More intensive peaks at 3,008–2,965 cm^{-1} relate to ν(CH) of aliphatic CH, CH$_2$ groups. The peak at 1,756 cm^{-1} indicates on presence of (L-HisH$_2$) cation, while intensive peak at 1,024 cm^{-1} indicates on presence of HSO$_4^-$ anion. Structure determination confirmed the conclusions made based on vibrational spectra. Asymmetric unit contains one formula unit. Both HS(1)O$_4^-$ and HS(2)O$_4^-$ anions form infinite chains *via* O-H…O(2.6043(3) Å) and O-H…O(2.6175(4) Å) hydrogen bonds respectively. The (L-HisH$_2$) cation has close conformation. Carboxyl

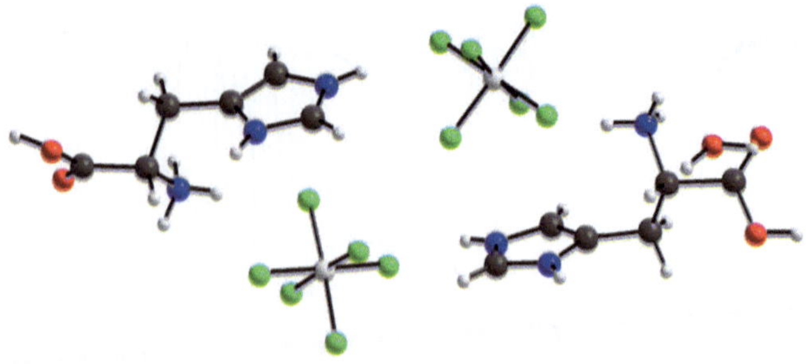

Fig. 5.46 Molecular structure of L-histidinium^{2+} hexafluorosilicate. This crystal has trigonal symmetry and contains two formula units in the asymmetric unit

group of cation forms O-H...O(2.6593(3) Å) hydrogen bond with $HS(1)O_4^-$ anion. The NH_3^+ and imidazolium groups form N-H...O hydrogen bonds with anions. Sulfates of L-histidine do not reveal noticeable NLO activity.

Adams et al. (1970) tried to reproduce the samples of mercury histidinate and mercury histidinate dihydrate described before by reaction of mercury(II) chloride and L-histidine hydrochloride in 1:2 molar ratio. However, in result of investigation they came to conclusion that the crystal obtained actually is (L-HisH)$HgCl_3$. The accuracy of determination is not high, however ($R = 15.7$ %).

In the system L-His + H_2SiF_6 + H_2O two salts were found with (L-HisH) and (L-$HisH_2$) cations: (L-HisH)$_2SiF_6{\cdot}2H_2O$ and (L-$HisH_2$)$SiF_6{\cdot}2H_2O$ (Ghazaryan et al. 2013a). The symmetries of crystals are unusually high compared to all other salts in Table 5.25. Asymmetric unit of (L-$HisH_2$)$SiF_6{\cdot}2H_2O$ contains half of formula unit, because the SiF_6^{2-} anion is in special position. The (L-HisH) cation has open conformation. The $N(1)H_3^+$ group forms one hydrogen bond with anion, one with water molecule and a hydrogen bond with carboxylate group of nearest cation: N1-H1A...F3(2.788(2) Å), N1-H1B...O1W(2.767(2) Å), N1-H1C...O2(2.729 (2) Å). The imidazolium group forms hydrogen bonds with carboxylate group of nearest cation and with anion: N2-H2...O1(2.658(2) Å), N3-H3...F2(2.747(2) Å). Water molecule as proton donor forms hydrogen bonds with cation and anion: O1W-H1W...O2(2.719(3) Å), O1W-H2W...F1(2.808(3) Å). In contrast to the previous case asymmetric part of unit cell of (L-$HisH_2$)$SiF_6{\cdot}2H_2O$ contains two formula units (Fig. 5.46). Both salts are formed from aqueous solution containing stoichiometric quantities of components that is 2:1 and 1:1 respectively by slow evaporation at room temperature.

In Table 5.26 salts of histidine with organic anions are shown. Interaction of histidine with simplest HCOOH (formic acid) organic acid has been studied by the laboratory of Vijayan. The majority of known salts of histidine with organic acids have been studied by his group. The salt with formic acid was obtained by vapor diffusion of acetone into aqueous solution of L-histidine containing an excess of formic acid. The obtained salt (L-histidine formate formic acid) is formed by (L-HisH)HCOO·HCOOH mechanism (Suresh and Vijayan 1995b). The (L-HisH) has open conformation. Asymmetric unit contains one formula unit. The neutral

HCOOH molecule does not interact with formate ion, but forms strong hydrogen bond with carboxylate group of (L-HisH) cation, O22-H22...O2(2.507(3) Å). The NH_3^+ group forms hydrogen bonds with nearest cation, N1-H2...O1(2.790(3) Å) and formate anion, N1-H3...O12(2.738(2) Å). There are also contacts with HCOOH molecule, N1-H1...O21(2.929(3) Å) and N1-H1...O22(2.941(3) Å). Imidazolium group forms hydrogen bonds with two nearest formate ion: N5-H1N5...O11(2.696 (3) Å), N7-H1N7...O11(2.810(4) Å). Further investigation of the interaction of L-histidine with formic acid at various molar proportions of the components showed that there are three different phases identified by IR spectra (see Petrosyan 2010a, and references therein). The first phase is formed at 1:1 and 1:2 ratios, the second at 1:3 and the third at from 1:4 to 1:7. The first two phases are formed in form of polycrystals, while the third phase is formed in form of single crystal. Later, measurement of unit cell parameters of the third phase showed that it is identical with the known salt (Suresh and Vijayan 1995b). The phase obtained at 1:3 ratio probably is another crystalline form of (L-HisH)HCOO. All three phases show high nonlinearity, however, the difficulty of growing the first two in form of single crystal caused with volatility of the formic acid. The third phase can be prepared in form of single crystal because it is formed in wider range of molar ratios.

The salt (DL-HisH)HCOO·H_2O also was obtained by vapor diffusion of acetone into aqueous solution of DL-histidine containing an excess of formic acid (Suresh and Vijayan 1995b). Asymmetric unit also contains one formula unit. The HisH cation has open conformation as well. The NH_3^+ group forms hydrogen bonds with nearest cation, anion and water molecule: N1-H1N1...O2(2.736(2) Å), N1-H1N1...O11(2.871(2) Å), N1-H3N1...W1(3.001(3) Å). Imidazolium group forms hydrogen bonds with nearest cation and anion: N5-H1N5...O1(2.659(2) Å), N7-H1N7...O12(2.646(2) Å). Water molecule forms hydrogen bonds with two nearest anions: W1-H1W1...O12(2.846(3) Å), W1-H2W1...O11(2.809(3) Å).

Interaction of L-histidine with acetic acid also have been studied by Suresh et al. (1994b). Two salts with acetic acid (anhydrous and dihydrate) were obtained. Anhydrous (L-HisH)CH_3COO is orthorhombic, while dihydrate (L-HisH)CH_3COO·$2H_2O$ is monoclinic. Crystals were obtained by dissolving L-histidine in an excess of acetic acid, then lyophilized until dry. The dry powder was dissolved in a minimal amount of water, and crystals were grown from aqueous solution by the slow diffusion of acetonitrile. Two crystal forms, one needle-like and second chunky were obtained in these experiments. Compounds with DL-histidine could not be made despite several attempts. Asymmetric unit in crystal of anhydrous salt contains one formula unit. Conformation of (L-HisH) cation is closed. The $N(1)H_3^+$ group forms three hydrogen bonds with three nearest anions: N1-H1N1...O11(2.935(5) Å), N1-H2N1...O11(2.840(5) Å), N1-H3N1...O11(2.873(4) Å). Imidazolium group forms hydrogen bonds with one nearest anion and cation: N5-HN5...O12(2.687 (4) Å), N7-HN7...O2(2.585(5) Å). Asymmetric unit of (L-HisH)CH_3COO·$2H_2O$ also contains one formula unit. Conformation of (L-HisH) cation is open. The $N(1)H_3^+$ group forms one hydrogen bond with nearest symmetry-related cation and two hydrogen bonds with two nearest anions: N1-H1N1...O2(2.760(5) Å),

N1-H2N1...O11(2.780(4) Å), N1-H3N1...O12(2.814(5) Å). Imidazolium group forms one hydrogen bond with cation and one with anion: N5-HN5...O1(2.694 (5) Å), N7-HN7...O11(2.685(4) Å). Hydrogen atoms of water molecules could not be located experimentally. In addition one water molecule is disordered. Possible hydrogen bonds are: W1...W1(2.797(9) Å), W1...W2(2.789(10) Å), W2...W2 (2.804(12) Å), W1...O2(2.836(8) Å), W2...O12(2.776(9) Å).

Another (triclinic) form of L-histidine acetate dihydrate (L-HisH) $CH_3COO{\cdot}2H_2O$ was obtained by Mostad et al. (1995b) by slow evaporation of an aqueous solution containing L-histidine and acetic acid. The crystal structure has been determined at 123 K. Asymmetric unit contains two formula units; calculated density $\rho_c = 1.36$ g/cm^3 at 123 K, measured density at room temperature is equal to 1.32 g/cm^3. Both (L-HisH) cations have open conformation. The NH_3^+ groups form each one hydrogen bond with carboxylate group of symmetry-related cation (with N...O distances 2.752 Å and 2.778 Å) and each two hydrogen bonds with anions (with N...O distances 2.765 Å, 2.808 Å and 2.817 Å, 2.771 Å respectively). Imidazolium groups form one hydrogen bond with carboxylate group of other cation and one with anion (with N...O distances 2.685 Å, 2.676 Å and 2.685 Å, 2.687 Å). Water molecules form one hydrogen bond with carboxylate group of cation (O...O(2.831 Å)), two hydrogen bonds with anions (with O...O distances 2.771 Å and 2.804 Å). The rest hydrogen bonds they form with each other.

Monoclinic (L-HisH)$CH_3COO{\cdot}2H_2O$ has been grown from aqueous solution by cooling method (with dimensions $20 \times 12 \times 8$ mm^3) and characterized as NLO material (Mathivanan et al. 2007). The crystal was identified by measuring of unit cell parameters. The morphology, IR and Raman spectra, transmittance in UV–Vis–NIR range, thermal properties and SHG efficiency ($I^{2\omega}/I^{2\omega}_{KDP} = 2.20$) were reported.

Madhavan et al. (2007b) reported their results on obtaining and investigation of triclinic form of (L-HisH)$CH_3COO{\cdot}2H_2O$ and Kumar et al. (2008) reported their results on the same triclinic (L-HisH)$CH_3COO{\cdot}2H_2O$ as well as on crystals (L-HisH) $CH_3COO{\cdot}2H_2O$ doped by 2 mol% $MgSO_4$ and $CuSO_4$. Unfortunately, analysis of this paper shows that this paper is questionable. Madhavan et al. (2007b) consider that they have grown the triclinic form of (L-HisH)$CH_3COO{\cdot}2H_2O$ and as a proof show unit cell parameters ($a = 8.520$ Å, $b = 9.059(4)$ Å, $c = 9.023(3)$ Å, $\alpha = 61.70(2)°$, $\beta = 86.60(3)°$, $\gamma = 86.30(4)°$, $V = 611.6(4)$ Å^3, $Z = 2$, $M_r = 251.24$ and $\rho_c = 1.33$ g/cm^3) and note that these values are in good agreement with the reported work (Mostad et al. 1995). Unfortunately, these values almost coincide and just this causes suspicion that these parameters are really measured, because it is hardly probable that unit cell parameters measured at 123 K and at room temperature can be so close. From the values $V = 611.6(4)$ Å^3, $Z = 2$, $M_r = 251.24$ the value $\rho_c = 1.36$ g/cm^3 one may calculate but not $\rho_c = 1.33$ g/cm^3, which is close to the density at room temperature $\rho_c = 1.32$ g/cm^3 (Mostad et al. 1995). Analysis of the shown IR spectrum also does not help. Madhavan et al. (2007b) expecting absorption bands of stretching and deformation vibration of water molecules in the spectrum assigned noise at 3,948 cm^{-1} to stretching vibration of water and small peak at 1,788 cm^{-1} to deformation vibration. In reality the shown spectrum is exactly the spectrum of L-histidine. However, the solubility curve shown in Fig. 1 of the paper of Madhavan et al. (2007b) can not belong to L-histidine because the solubility of the crystal at 40 °C is *ca.* 80 g/100 ml H_2O, while the solubility of the L-histidine at the same temperature is *ca.* 4 g/100 ml H_2O (Kitamura 1993). Comparison with data shown in (Kumar et al. 2008), where pure and doped (L-HisH)$CH_3COO{\cdot}2H_2O$ crystals were studied, also have not helped. All curves (except TG curve, which satisfactory agrees with that of pure (L-HisH)$CH_3COO{\cdot}2H_2O$) including solubility, microhardness, dielectric constant, dielectric loss and dark- and photoconductivity disagree with respective curves of undoped (L-HisH)$CH_3COO{\cdot}2H_2O$. So, there is something amiss with this paper. However, the Raman spectrum of pure (L-HisH)$CH_3COO{\cdot}2H_2O$ in the paper of Kumar et al. (2008) is in agreement with Raman spectrum of monoclinic (L-HisH)$CH_3COO{\cdot}2H_2O$ (Mathivanan et al. 2007).

L-histidine glycolate and DL-histidine glycolate were prepared by the slow diffusion of acetone into aqueous solution of the components in molar proportions of amino acid and glycolic acid (Suresh and Vijayan 1996). As can be seen the unit cell parameters of anhydrous (L-HisH)CH_3COO and glycolate (L-HisH)$CH_2(OH)COO$ are close to each other. Asymmetric units of both crystals contain each one formula units. The L-HisH cations in both structures have closed conformation. However, the schemes of hydrogen bonds are different. In the structure of (L-HisH)$CH_2(OH)COO$ the $N(1)H_3^+$ group forms three hydrogen bonds with three nearest anions: N1-H1N1…O11(3.088(2) Å, N1-H2N1…O11(2.788(2) Å, N1-H333…O11(3.088(2) Å. Imidazolium group forms one hydrogen bond with nearest anion and one with cation: N5-H2N5…O12(2.704(2) Å, N7-H2N7…O2 (2.651(2) Å. Hydroxyl group of glycolate anion forms hydrogen bond with nearest cation, O13-H2O13…O1(2.676(2) Å. In the structure of (DL-HisH)$CH_2(OH)COO$ the $N(1)H_3^+$ group forms one hydrogen bond with oxygen atom of hydroxyl group of anion, one with carboxylate group of anion and one with carboxylate group of nearest cation: N1-H1N1…O13(2.792(1) Å, N1-H2N1…O12(2.715(2) Å, N1-H3N1…O2(2.866(1) Å. Imidazolium group forms one hydrogen bond with carboxylate groups of anion and cation: N5-H2N5…O11(2.751(2) Å, N7-H2N7…O2(2.748(2) Å. Hydroxyl group of glycolate anion forms hydrogen bond with carboxylate group of anion, O13-H2O13…O11(2.643(1) Å. Solid-state ^1H NMR spectrum of L-histidine glycolate is shown by Wei and McDermott (1999).

The crystal of (L-HisH)CF_3COO was obtained and crystal structure was reported by Gokul Raj et al. (2006b). The crystal was obtained by slow evaporation at room temperature from solution containing equimolar quantities of L-histidine and trifluoroacetic acid. Triclinic cell contains two formula units. Both (L-HisH) cations have open conformation, however, the authors called the conformation as closed. Crystal structure, IR and Raman spectra and SHG efficiency was determined also by Petrosyan et al. (2005d). In contrast to the data of (Gokul Raj et al. 2006b) it was found that there is disorder of some atoms in one of two CF_3COO anions (Petrosyan 2010a,). For this reason the structure was determined also at 120 K and was found that disordering disappears at this temperature (Petrosyan 2009, 2010a).

Structural data of (L-HisH)CF_3COO at room temperature are provided by Gokul Raj et al. (2006b). The structure at 120 K is similar to that at room temperature (Petrosyan 2009). The $N(1)H_3^+$ group forms two hydrogen bonds with carboxylate groups of cations and one hydrogen bond with carboxylate group of anion: N1-HA…O2(2.863(3) Å, N1-HB…O2A(2.803(3) Å, N1-HC…O4(2.781(3) Å, while the $N(1A)H_3^+$ group forms two hydrogen bonds with carboxylate groups of anions and one with cation: N1A-HD…O4(2.836(3) Å, N1A-HE…O2A(2.725(3) Å, N1A-HF…O3(2.727(3) Å. Each imidazolium group of cations forms one hydrogen bond with nearest anion and one with cation: N2-H…O4(2.806(3) Å, N3-H…O1(2.844(3) Å and N2A-H…O4A(2.681(3) Å, N3A-H…O2(2.656(3) Å. The efficiency of SHG of the crystal measured by powder method is $I^{2\omega}/I^{2\omega}_{KDP} = 0.84$ (Gokul Raj et al. 2007) and $I^{2\omega}/I^{2\omega}_{KDP} = 0.86$ according to Petrosyan (2007a). The crystals of (L-HisH)CF_3COO were grown from water–ethanol solution, containing stoichiometric quantities of components and characterized also (Martin Britto Dhas et al. 2008b), the authors apparently being unaware about the work of Gokul Raj et al. (2006b). Results of thermal analysis, SHG efficiency, IR and UV–Vis–NIR spectra are provided. The crystal decomposes above 200 °C and $I^{2\omega}/I^{2\omega}_{KDP} = 1.2$ according to Martin Britto Dhas et al. (2008b). Sun et al. (2009d) also grew and characterized the crystal of (L-HisH)CF_3COO. The solubility curve was determined in aqueous solution in the interval 35–55 °C. Unit cell parameters, powder X-ray diffraction pattern, IR, UV–Vis–NIR transmittance spectra, TG, DTA curves, SHG efficiency as well as hyperpolarizability calculated by DFT method were reported. For particles sizes *ca.* 130 μm the efficiency of SHG was found as $I^{2\omega}/I^{2\omega}_{KDP} = 2.6$. Sun et al. (2011c) studied nucleation parameters of crystal growth of (L-HisH)CF_3COO. The morphology of crystals depending on the pH was studied using the microcrystallization method. Optical and dielectric properties of (L-HisH)CF_3COO grown from water–ethanol solution were studied as well (Suresh 2012). In Fig. 5.47 the IR, registered by ATR method, and Raman spectra are shown taken from Petrosyan (2009). Characteristic sharp peak at 3,138 cm^{-1} in the IR spectrum and rather intensive line with peaks at 3,160, 3,148 and 3,137 cm^{-1} in the Raman spectrum are assigned to stretching vibration of C-H bonds in imidazolium rings. Respective peaks at 3,144 cm^{-1} (Martin Britto Dhas et al. 2008b) and 3,142 cm^{-1} (Sun et al. 2009d) in the IR spectra the authors assigned to stretching vibration of NH_3^+ group. Characteristic strong peaks at 2,966 and 2,944 cm^{-1} in the Raman spectrum are caused by stretching vibrations of aliphatic CH and CH_2 groups. The peak at 2,941 cm^{-1} in the IR spectrum caused by these vibrations is overlapped with broad band (3,300–2,200 cm^{-1}) caused by hydrogen-bonded N-H groups. Similar broad band is seen also in the Raman spectrum. Three very strong absorption peaks at 1,678, 1,660 and 1,628 cm^{-1} and respective Raman lines at 1,681, 1,653 and 1,623 cm^{-1} are caused by asymmetric stretching vibration of COO^- groups of (L-HisH) cations and CF_3COO anions and also asymmetric deformation vibration of NH_3^+ group. The peak at 1,515 cm^{-1} and respective line in the Raman spectrum at 1,507 cm^{-1} are probably caused by symmetric deformation vibration of NH_3^+ group. Absorption band with peaks

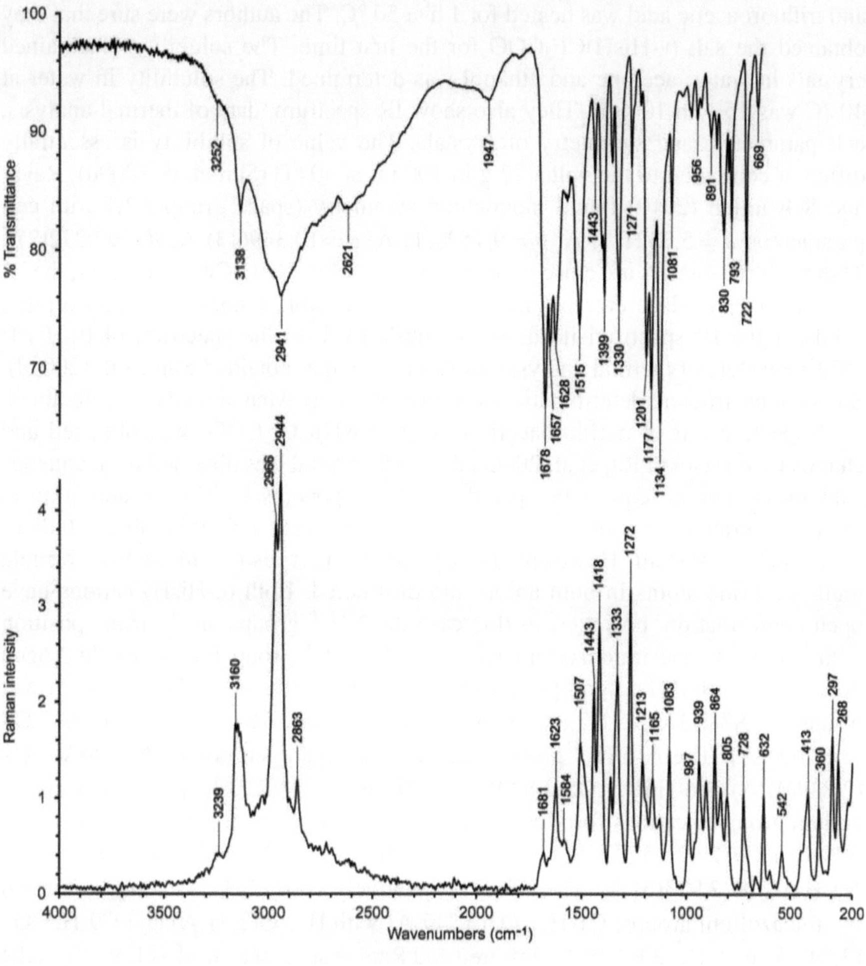

Fig. 5.47 IR and Raman specta of (L-HisH)CF_3COO

at 1,410 and 1,399 cm^{-1} and respective Raman line at 1,418 cm^{-1} are caused by symmetric stretching vibration of COO^- groups. Martin Britto Dhas et al. (2008b) the peak at 1,406 cm^{-1} in this region assigned to O-H deformation. Actually the group O-H is absent in the structure of (L-HisH)CF_3COO. Stretching vibrations of C-F bonds in CF_3 group are characterized by strong absorption peaks. Peaks at 1,201, 1,177 and 1,134 cm^{-1} and respective relatively weak line at 1,213, 1,165 and 1,131 cm^{-1} in the Raman spectrum are assigned to stretching vibration of CF_3 group. This is in agreement with assignment of Sun et al. (2009d). Martin Britto Dhas et al. (2008b) additionally assigned the peaks at 1,265, 1,334 cm^{-1} to this vibration.

There is one more publication devoted to the (L-HisH)CF_3COO crystal (Rayar and Selvarajan 2010). The solution containing equimolar quantities of L-histidine and trifluoroacetic acid was heated for 1 h at 50 °C. The authors were sure that they obtained the salt (L-HisH)CF_3COO for the first time. The solubility of obtained crystals in water, acetone and ethanol was determined. The solubility in water at 40 °C was 15 g in 100 ml. They also show IR spectrum, data of thermal analysis, cell parameters and symmetry of crystals. The value of solubility is essentially different compared to the value 77 g in 100 ml at 40 °C (Sun et al. 2009d). Rayar and Selvarajan (2010) found monoclinic symmetry (space group $P2_1$) with cell parameters: $a=5.0321(1)$ Å, $b=9.1590(1)$ Å, $c=10.3496(3)$ Å, $\beta=92.272(2)°$. These cell parameters are close to the parameters of (L-HisH)ClO_4, however, the IR spectrum is quite different. In spite of some unacceptable assignments of absorption bands of the IR spectrum it has some similarity with the spectrum of (L-HisH)CF_3COO. Data of thermal analysis also similar to that obtained Sun et al. (2009d). So, without structure determination it is difficult to say what actually they obtained.

Respective salt of trichloroacetic acid ((L-HisH)CCl_3COO) was obtained and characterized (Gokul Raj et al. 2006c, d, e). The crystal was obtained from aqueous solution containing equimolar quantities of components by slow evaporation at room temperature. In contrast to the salt of trifluoroacetic acid this salt crystallizes in monoclinic system. However, the asymmetric unit also contains two formula units. Chlorine atoms in both anions are disordered. Both (L-HisH) cations have open conformation, however, in this case the NH_3^+ groups are in *trans*-position with respect to the imidazolium ring. The N(3)H_3^+ group forms two hydrogen bonds with carboxylate groups of cations and one with carboxylate group of one of anions: N3-H3A...O1(2.815(5) Å), N3-H3C...O6(2.787(5) Å), N3-H3B...O8 (2.681(5) Å). The N(6)H_3^+ group forms similar hydrogen bonds: N6-H6A...O2 (2.839(4) Å), N6-H6B...O5(2.839(4) Å), N6-HC...O4(2.713(5) Å). Each imidazolium group of cations form one hydrogen bond with anion and one with cation: N1-H1...O7(2.671(5) Å), N2-H2...O2(2.761(5) Å) and N5-H5...O3(2.715(5) Å), N4-H4...O6(2.923(6) Å). There are two relatively short C-H...O contacts formed by imidazolium groups: C1-H...O3(3.217 Å) with H...O(2.31 Å) and C9-H...O7 (3.247 Å) with H...O(2.38 Å). Infrared and Raman spectra as well as UV–Vis–NIR spectrum are registered and discussed (Gokul Raj et al. 2006d). The efficiency of SHG measured by powder method is $I^{2\omega}/I^{2\omega}_{KDP}=0.33$. The crystal is stable up to 200 °C.

Interaction of L-histidine and DL-histidine with oxalic acid was studied by Vijayan with coworkers (Prabu et al. 1996). The crystals were obtained by slow diffusion of acetone into aqueous solution of the components in equimolar proportions. In case of L-histidine the crystal of (L-HisH)HC_2O_4 was formed, while in case of DL-histidine the crystal of (DL-HisH_2)$(HC_2O_4)_2$ was formed. Asymmetric units in both crystals contain one formula unit. Conformations of cations in both cases are closed. In the structure of (L-HisH)HC_2O_4 hydrogen oxalate anions form usual chain HC_2O_4...HC_2O_4...HC_2O_4 via hydrogen bond O12-HO12...O15 (2.523(3) Å). The N(1)H_3^+ group forms two hydrogen bonds with carboxylate group of anion and one with carboxylate group of nearest cation: N1-H1...O16 (2.802(3) Å), N1-H2...O15(2.923(3) Å), N1-H3...O2(2.949(3) Å). The imidazolium group forms one relatively short hydrogen bond with carboxylate group of nearest cation, N7-H…O1(2.624(3) Å) and two weak hydrogen bonds with anion: N5-H…O11(2.910(3) Å) and N5-H…O15(2.996(3) Å).

In the structure of (DL-$HisH_2$)$(HC_2O_4)_2$ both hydrogen oxalate anions form chains: O12-H12…O16(2.525(2) Å), O22-H22…O26(2.528(2) Å). The carboxyl group of cation forms hydrogen bond with second anion, O2-H…O25(2.707(2) Å). There is also a weak bond, O2-H…O21(2.898(2) Å). The $N(1)H_3^+$ group forms hydrogen bonds with anions: N1-H1…O12(2.919(2) Å), N1-H1…O16(2.761 (2) Å), N1-H2…O26(2.922(2) Å). There is also a weak contact N1-H3…O11 (3.168(2)Å). The imidazolium group forms hydrogen bonds with both anions: N5-H…O15(2.784(2) Å), N7-H…O25(2.841(2) Å). Deuterated salts of (L-HisH) HC_2O_4, (DL-$HisH_2$)$(HC_2O_4)_2$ and some other compounds were used (Hiramatsu et al. 2010) for establishing a correlation between Raman spectra and histidinium's conformation and hydrogen bonding. Further studies of the system L-His + oxalic acid + H_2O were performed (Petrosyan 2007a, 2009, 2010a; Petrosyan and Petrosyan 2002; Petrosyan et al. 2007; Dammak et al. 2007; Ahmed et al. 2011; Chimpri et al. 2013). In the study by Petrosyan and Petrosyan (2002), in addition to known (L-HisH)HC_2O_4 crystal, three more phases (2L-His·$H_2C_2O_4$·nH_2O, powdered L-His·$H_2C_2O_4$ and L-His·$2H_2C_2O_4$) were obtained and identified by IR spectra and was observed qualitatively high SHG activity of 2L-His·$H_2C_2O_4$·nH_2O crystals. Later this crystal was designated as α-form, because after recrystallization it transforms into another β-form and also a crystal with L-His·$2H_2C_2O_4$ composition (Petrosyan et al. 2005c). Moreover, the structure and composition of β- form ((L-HisH)$_2C_2O_4$·$3H_2O$) was determined and also a new γ-form (L-HisH)$_2C_2O_4$·$3H_2O$ was obtained at above room temperature (Petrosyan et al. 2007). In addition to the L-His·$2H_2C_2O_4$ crystal the crystal of L-His·$2H_2C_2O_4$·$2H_2O$ also was obtained from aqueous solution at below room temperature and characterized by IR and thermal analysis methods. It is interesting to note that fine, amorphous-like powdery L-His·$H_2C_2O_4$ is not, however, amorphous because it shows strong enough SHG signals ($I^{2\omega}/I^{2\omega}{}_{KDP} = 2$). Later the structure of the crystal of γ-form (L-HisH)$_2C_2O_4$·$3H_2O$ also was determined (Petrosyan 2009, 2010a). In spite of different symmetry they have similar structure, which confirms the conclusion made based on similarity of the IR spectra (Petrosyan et al. 2007). Asymmetric units in both structures can be presented as (L-HisH)$(C_2O_4)_{0.5}$·$1.5H_2O$, because (L-HisH) cation and one water molecule are in general position, while the C_2O_4 anion and second water molecule are in special position. Cations have open conformation in both structures. The $N(1)H_3^+$ group in the structure of β- form forms hydrogen bonds with one water molecule and carboxylate group of nearest cations: N1-H2…O5(2.802(2) Å), N1-H3…O2(2.971(2) Å), N1-H4…O2(2.835 (2) Å), and also a weak bond N1-H4…O1(3.100(2) Å). The imidazolium group forms hydrogen bonds with nearest cation and anion: N2-H8…O1(2.764(2) Å), N3-H10…O4(2.684(2) Å). Water molecules form hydrogen bonds with anions: O5-H11…O3(2.768(2) Å), O5-H12…O4(2.762(2) Å), O6-H13…O3(2.805(2) Å). In the structure of γ-form the $N(1)H_3^+$ group also forms hydrogen bonds with one water molecule and carboxylate group of nearest cations: N1-H3...O5(2.808(3) Å), N1-H2...O2(2.877(3) Å), N1-H4...O1(2.714(3) Å). The imidazolium group has two weak contacts, N2-H8...O5(3.014(3) Å), N2-H8...O2(2.847(3) Å) and one normal hydrogen bond, N3-H10...O4(2.699(3) Å). Water molecules also form hydrogen bonds with anions: O5-H11...O3(2.782(3) Å), O5-H12...O4(2.762 (3) Å), O6-H13...O3(2.812(3) Å).

Dammak et al. (2007) and Ahmed et al. (2011) obtained the same L-His·$H_2C_2O_4$ crystal, which was reported by Prabu et al. (1996), apparently not aware of this paper. In both studies, the crystal structure was determined, IR and Raman spectra were studied experimentally and theoretically as well as dipole moment, isotropic polarizability and hyperpolarizability were calculated for isolated molecule (only in the work of Ahmed et al. 2011). Characteristic vibrational peaks at 3,137 cm^{-1} (R) and 3,132 cm^{-1} (IR) were assigned to NH_3^+ and O-H...O stretching respectively (Dammak et al. 2007). Raman and IR peaks at 1,715 cm^{-1} were assigned not to COOH group of anion, but asymmetric stretching of COO^- group (Ahmed et al. 2011). The crystal of (L-HisH)HC_2O_4 attracted the attention of Chimpri et al. (2013) to determine accurate electron density based on structure determination at 100 and 110 K. More accurate structural data and hydrogen bonding were obtained. Refractive indices were calculated and compared with experimentally determined value. In addition to synthetic work with L-histidine the authors tried also to obtain respective salt with DL-histidine. They observed spontaneous resolution of a racemate of (L-HisH)HC_2O_4 and (D-HisH)HC_2O_4 with the same structure. However, when crystals were grown from a solution stirred above 50 °C a new oxalate compound with divalent cation (L-or D-) was obtained ($HisH_2$)C_2O_4 with orthorhombic symmetry (space group $P2_12_12_1$, $Z=4$) and cell parameters: $a=6.6800(2)$ Å, $b=9.2116(2)$ Å, $c=16.3534(4)$ Å at 253 K.

The highest SHG efficiency among oxalates of L-histidine show crystals (L-HisH)$_2C_2O_4$·$3H_2O$ ($I^{2\omega}/I^{2\omega}{}_{KDP}=4.3$) (Petrosyan et al. 2007).

The next homolog of oxalic acid is malonic acid HOOC-CH_2-COOH. Interaction of L- and DL-histidine with malonic acid was studied by Saraswathi and Vijayan (2002b). The crystals of (L-HisH) hydrogen malonate were obtained by the diffusion of isopropyl alcohol into an aqueous solution of L-histidine and malonic acid mixed in a 1:2.5 molar ratio, while crystals of DL-HisH hydrogen malonate were obtained with the same method, but with 1:6 molar ratio of DL-His and malonic acid and ethanol as precipitant. Asymmetric unit of L-HisH hydrogen malonate contains two formula units. Both (L-HisH) cations have open conformation with trans disposition of NH_3^+ group in respect to imidazolium ring. Both hydrogen malonate anions form very short intramolecular hydrogen bonds: O21-H21...O26(2.427 (8) Å), O31-H31...O36(2.427(8) Å). The N(1)H_3^+ group forms two hydrogen bonds with carboxylate groups of both cations and one hydrogen bond with anion: N1-H1A...O12(2.862(6) Å), N1-H1B...O1(2.703(6) Å), N1-H1C...O37 (2.827(7) Å). The imidazolium group forms hydrogen bonds with two anions of the same type: N5-H5...O27(2.798(8) Å), N7-H7...O22(2.751(8) Å). The second cation forms similar hydrogen bonds: N11-H11A...O2(2.768(6) Å),

N11-H11B...O11(2.725(7) Å), N11-H11C...O27(2.848(7) Å). Imidazolium group forms hydrogen bonds with two anions of the second type: N15-H15...O37(2.866 (8) Å), N17-H17...O32(2.760(8) Å). There are also short contacts C6-H...O21 (3.186 Å) with H...O(2.29 Å) and C16-H...O31(3.208 Å) with H...O(2.32 Å), which may be considered as weak hydrogen bonds.

Asymmetric unit of DL-HisH hydrogen malonate contains one formula unit. The $N(1)H_3^+$ group forms two hydrogen bonds with carboxylate group of two nearest cations and one hydrogen bond with anion: N1-H1A...O2(2.815(2) Å), N1-H1B...O1(2.733(2) Å), N1-H1C...O37(2.859(2) Å). The imidazolium group forms hydrogen bonds with two anions: N5-H5...O37(2.859(2) Å), N7-H7... O32(2.761(2) Å). Here also there is a short contact C6-H...O31(3.200 Å) with H...O(2.30 Å).

Interaction of L- and DL-histidine with next homolog succinic acid HOOC-$(CH_2)_2$-COOH was studied by Prasad and Vijayan (1993a). Crystals of L-HisH hydrogen succinate trihydrate and $(\text{DL-HisH})_2$ succinate tetrahydrate were obtained from concentrated *aqueous* solution of the L-His or DL-His and succinic acid in 1:3 molar ratio by liquid diffusion using acetonitrile. Asymmetric unit of L-HisH hydrogen succinate trihydrate contains one formula unit. The conformation of the L-HisH cation is open. Hydrogen succinate anions are not hydrogen bonded with each other. Carboxylic group forms hydrogen bond with one water molecule, O12-H...W1(2.601(4) Å). The $N(1)H_3^+$ group forms one hydrogen bond with carboxylate group of anion, one with carbonyl oxygen atom of anion and one with carboxylate group of nearest cation: N1-H1...O18(2.822(4) Å), N1-H2...O11 (2.791(4) Å), N1-H3...O2(2.896(4) Å). The imidazolium group forms hydrogen bonds with carboxylate groups of nearest cation and anion: N5-H...O1(2.701 (4) Å), N7-H...O17(2.701(4) Å). Water molecules form hydrogen bonds with each other and with carboxylate groups of cation and anion: W1-H2...W3(2.785(5) Å), W2-H1...W3(2.761(5) Å), W3-H2...W2(2.761(5) Å), W1-H1...O18(2.653(4) Å), W3-H1...O17(2.732(4) Å), W2-H2...O2(2.826(4) Å).

Asymmetric unit of $(\text{DL-HisH})_2$ succinate tetrahydrate contains one HisH cation, two water molecules and succinate anion in special position. So, it may be presented as DL-HisH$(C_2H_4O_4)_{0.5}\cdot 2H_2O$. The $N(1)H_3^+$ group forms two hydrogen bonds with succinate anion and one with carboxylate group of nearest cation: N1-H1...O11(2.838(2) Å), N1-H2...O11(2.850(3) Å), N1-H3...O1(2.793(2) Å). The imidazolium group forms one hydrogen bond with one water molecule and one with carboxylate group of nearest cation: N5-H...W1(2.673(4) Å), N7-H...O2 (2.674(3) Å). The water molecule W1 in turn forms a hydrogen bond with anion and with second water molecule: W1-H1...O11(2.850(3) Å), W1-H2...W2(2.687 (4) Å). Second water molecule forms hydrogen bonds with anion and cation: W2-H1...O12(2.668(4) Å), W2-H2...O2(2.782(3) Å).

Akkurt et al. (2004) obtained L-$HisH_2$ tetrafluorosuccinate by evaporation from aqueous solution and determined its crystal structure. Asymmetric unit contains one formula unit (Fig. 5.48). The L-$HisH_2$ has closed conformation. However, the authors write that "...the cation adopts the sterically most favorable open

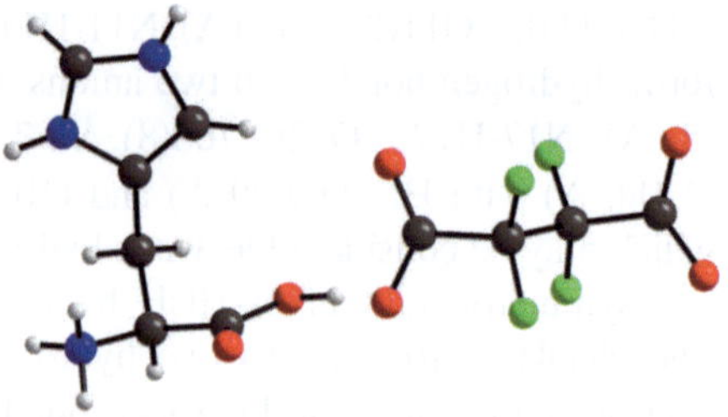

Fig. 5.48 Molecular structure of L-histidinium^{2+} tetrafluorosuccinate

conformation....” Carboxyl group of L-HisH_2 cation forms rather strong hydrogen bond with anion, O2-H22. . .O6(2.544(3) Å). The $N(3)H_3^+$ group forms hydrogen bonds with three nearest anions: N3-HA. . .O4(2.958(3) Å), N3-HB. . .O4(3.079 (3) Å), N3-HC. . .O6(2.775(2) Å). The imidazolium group also forms hydrogen bonds with anions: N1-H1. . .O5(2.796(2) Å), N2-H2. . .O3(2.690(3) Å). There are two relatively short C-H. . .O contacts: C3-H3. . .O3(3.158(3) Å) with H. . .O (2.26 Å) and C5-H5. . .O1(3.040(3) Å) with H. . .O(2.29 Å). Ramajothi et al. (2008) grew the crystal of L-HisH_2 tetrafluorosuccinate by slow evaporation technique and characterized this species. The solubility of the crystal in water and ethanol was determined. However, we think that the units of concentration (g/cm^3) is not provided correctly (probably should be (g/100 cm^3). In addition to solubility curve the results of powder XRD pattern, IR, UV–Vis and ^{1}H NMR spectra, thermal analysis (TG and DTA curves), powder SHG efficiency (0.21 times that of urea) and laser damage threshold are provided. The crystal melts at 195 °C and then decomposes. Some points of interpretations of IR and NMR spectra are not acceptable. The characteristic sharp peak of stretching vibration of C-H bonds of imidazolium ring at 3,140 cm^{-1} is assigned to stretching vibration of N-H bonds. The authors write that the C=O stretching of carboxylate group appears at 1,633 cm^{-1} in L-histidine is shifted to 1,650 cm^{-1} in the formation of L-histidinium tetrafluorosuccinate. Actually the L-HisH_2 cation contains COOH carboxyl group, which is characterized by presence of stretching of C=O bond and indeed is present at 1,720 cm^{-1} (Akkurt et al. 2004), while the band at 1,650 cm^{-1} probably relates to asymmetric vibration of COO^- groups of anion. Characteristic signals at 7.30 ppm and 8.74 ppm in the ^{1}H NMR spectrum, which are caused by protons at C-H bonds of imidazole group, are assigned to carboxyl acid (OH) group and N-H group.

Interaction of L-histidine and DL-histidine with the next homolog glutaric acid ($HOOC(CH_2)_3COOH$) was studied by Saraswathi and Vijayan (2001). They could not obtain crystalline salts with DL-histidine, while with L-histidine they obtained two crystalline salts: L-HisH hydrogen glutarate monohydrate and another, rather unusual compound L-HisH hydrogen glutarate L-histidine. The first salt was obtained by the diffusion of acetonitrile into an aqueous solution of L-histidine and glutaric acid mixed in a 1:6 molar ratio. The second salt was obtained by the same method and with the same 1:6 molar ratio, but isobutyl alcohol was used instead of acetonitrile. Asymmetric unit of L-HisH hydrogen glutarate monohydrate contains one formula unit. The cation L-HisH has open conformation with trans disposition of carboxylate group in respect to imidazolium ring. Hydrogen glutarate anions form usual chain by O19-H19...O12(2.574(5) Å) hydrogen bond. The $N(1)H_3^+$ group forms two hydrogen bonds with carboxylate groups of two nearest anions and one with carboxylate group of nearest cation: N1-HB...O18(2.963 (4) Å), N1-HC...O11(2.2.726(4) Å), N1-HA...O2(2.765(5) Å). The imidazolium group forms hydrogen bonds with carboxylate groups of cation and anion: N5-H5...O1(2.694(5) Å), N7-H7...O11(2.699(4) Å). Water molecule does not form hydrogen bond as proton acceptor. As proton donor it forms hydrogen bonds with carboxylate groups of anion and cation: OW-H1...O12(2.924(5) Å), OW-H2...O2(2.804(5) Å).

Triclinic cell of L-HisH hydrogen glutarate L-histidine contains two formula units. Both L-HisH cations have open conformation with trans disposition of carboxylate groups in respect to imidazolium rings. One of L-histidine molecules has the same conformation, while the second molecule having open conformation has NH_3^+ group in trans disposition in respect to imidazole group. This crystal has an interesting structure. Hydrogen glutarate anions do not form chains similar to the previous case. Carboxyl groups of both anions form hydrogen bonds with nitrogen atoms of imidazole groups of both L-histidine molecules: O49-H49...N27 (2.707(4) Å), O59-H59...N37(2.576(5) Å). The $N(1)H_3^+$ and $N(11)H_3^+$ groups of both (L-HisH) cations form each two hydrogen bonds with carboxylate groups of cations and each one hydrogen bond with carboxylate group of L-histidine molecules: N1-HA...O1(2.617(3)), N1-HB...O2(2.866(3)), N1-HC...O32(2.662(4) Å) and N11-HB...O12(2.718(3) Å), N11-HC...O2(3.152(5) Å), N11-HA...O22 (2.770(4) Å). The imidazolium groups of both cations form each two hydrogen bonds with carbonyl oxygen atom and carboxylate group of anions: N5-H5...O48 (2.710(4) Å), N7-H7...O51(2.557(5) Å), and N15-H15...O58(2.840(4) Å), N17-H17...O41(2.573(4) Å). The $N(21)H_3^+$ and $N(31)H_3^+$ groups of both zwitter-ionic L-His molecules form each two hydrogen bonds with carboxylate groups of L-His molecules and each one hydrogen bond with anion: N21-HA...O31 (2.784(3) Å), N21-HB...O22(2.813(3) Å), N21-HC...O42(2.812(4) Å), and N31-HA...O21(2.805(3) Å), N31-HB...O32(2.792(4) Å), N31-HC...O52(2.680 (4) Å). The imidazole groups of L-His molecules form each one hydrogen bond with carboxylate group of cations: N25-H25...O11(2.883(3) Å) and N35-H35...O1 (2.777(3) Å). In the structure of orthorhombic (Madden et al. 1972a; Lehmann et al. 1972) and monoclinic forms of L-histidine (Madden et al. 1972b; Averbuch-Pouchot 1993d) there are intramolecular hydrogen bonds between NH_3^+ groups and nitrogen atoms of imidazole ring. However, in this structure L-histidine molecules do not form intramolecular hydrogen bonds.

Four salts were obtained in the system L-His + maleic acid + H_2O (Pratap et al. 2000; Fleck et al. 2013c; Gonsago et al. 2012a). Interaction of L-histidine with maleic acid HOOCCH=CHCOOH was first studied by Pratap et al. (2000). They obtained the crystal of (L-HisH) hydrogen maleate by the slow diffusion of acetonitrile or methanol into solutions of equimolar quantities of L-histidine and maleic acid. Triclinic cell contains two formula units. Both (L-HisH) cations have closed conformation. Hydrogen maleate anions form short intramolecular hydrogen bonds: O22…O27(2.406(9) Å) and O32…O37(2.429(8) Å). The N(1)H_3^+ group forms two hydrogen bonds with carboxylate groups of two nearest cations: N1-H2…O12(2.773(5) Å), N1-H3…O11(2.812(6) Å). There is also a weak hydrogen bond with one of anion, N1-H1…O32(3.162(5) Å) with H…O(2.28 Å). The N(11)H_3^+ group of the second cation forms two hydrogen bonds with carboxylate groups of two nearest cations: N11-H2…O2(2.696(5) Å), N1-H3…O12(2.923 (5) Å). The imidazolium groups of both cations form hydrogen bonds each with both anions: N5-H…O21(2.866(6) Å), N7-H…O38(2.717(7) Å) and N15-H…O31 (2.809(6) Å), N17-H…O28(2.688(6) Å). Gokul Raj et al. (2007) reported the efficiency of SHG of L-histidine maleate ($I^{2\omega}/I^{2\omega}{}_{KDP} = 1.3$), however, without providing any data confirming identification of the crystal. One more salt (L-HisH)$_2$ (hydrogen maleate)$_2$·3H_2O from this system was obtained by Gonsago et al. (2012a, b) from aqueous solution containing equimolar quantities of L-histidine and maleic acid by slow evaporation at 32 °C. The authors apparently were not aware of the existence of anhydrous salt (L-HisH) hydrogen maleate (Pratap et al. 2000) and obtained and characterized the crystal from point of view of possible NLO material. Asymmetric unit contains one formula unit, that is, two (L-HisH) cations, two hydrogen maleate anions and three water molecules. One of cation has open conformation, while the second has closed conformation. Two hydrogen maleate anions have intramolecular hydrogen bonds: O3-H…O2(2.412 Å), O9-1/2H…1/2H…O8 (2.424 Å). Consideration of hydrogen bonding in the structure of crystal is incomplete (see Fleck et al. 2013c). The N(3)H_3^+ group forms one hydrogen bond with one of water molecules, one with carboxylate group of symmetry-related cation and one with anion: N3-HA…O3W(2.815(4) Å), N3-HB…O6(2.673(3) Å), N3-HC…O7 (2.796(3) Å). The N(6)H_3^+ group also forms one hydrogen bond with another of water molecules, one with carboxylate group of symmetry-related cation and one with anion: N6-HB…O1W(2.809(3) Å), N6-HA…O11(3.049(3) Å), N6-HC…O1 (2.785(3) Å). Hydrogen bonds formed by imidazolium groups and water molecules as proton donors was not considered by Gonsago et al. (2012a). These hydrogen bonds were considered by Fleck et al. (2013c). In one of imidazolium group carbon and nitrogen atoms as it was shown by Fleck et al. (2013c) are not assigned correctly in the paper of Gonsago et al. (2012a). Actually, C5 and C6 are nitrogen atoms (should be N2 and N1) and respectively N1 and N2 are carbon atoms (should be C6 and C5). Imidazolium groups form hydrogen bonds with carboxylate group of cation and with anion: N1-H…O12(2.662 Å), N2-H…O9(2.815 Å), and N4-H…O5 (2.781 Å), N5-H…O4(2.826 Å). The water molecules form the following hydrogen bonds: O1W-H…O2W(2.782 Å), O2W-H2…O11(2.799 Å), O2W-H3…O6 (2.759 Å) and probably O3W…O1(2.767 Å). Gonsago et al. (2012a, b) determined the solubility of the crystal in water, IR and UV–Vis spectra, curves of thermal analysis and estimated SHG efficiency by powder method ($I^{2\omega}/I^{2\omega}{}_{KDP} = 0.65$).

Fleck et al. (2013c) obtained two more salts in the system L-His + maleic acid + H_2O. From aqueous solution containing L-histidine and maleic acid in

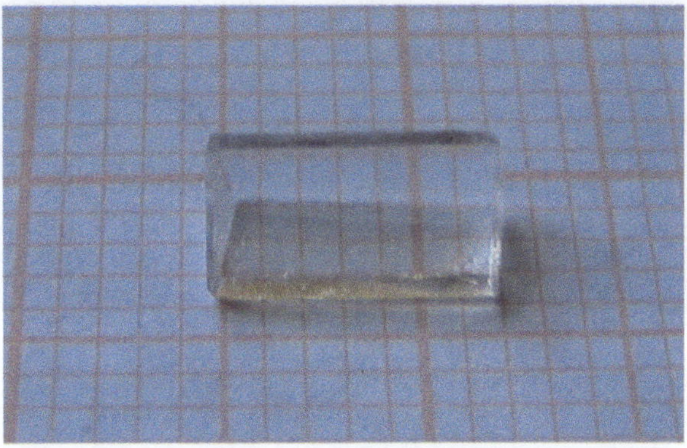

Fig. 5.49 Grown crystal of L-histidinium^{2+} bis (hydrogenmaleate)

molar ratios 1:1 and 1:2, we grew crystals of (L-HisH) hydrogen maleate monohydrate and (L-$HisH_2$) (hydrogen maleate)$_2$, respectively, by evaporation at 25C. A photograph of the grown crystal of L-histidinium^{2+} bis(hydrogen maleate is shown in Fig. 5.49. The asymmetric unit of monoclinic cell of (L-HisH) hydrogen maleate monohydrate contains one formula unit. The (L-HisH) cation has close conformation. Hydrogen maleate anion forms intramolecular hydrogen bond O1B-H1B…O4B(2.437(2) Å). The $N(1)H_3^+$ group forms two hydrogen bonds with two nearest anions and one with carboxylate group of nearest cation: N1A-H11A…O2B(2.767(2) Å), N1A-H12A…O3B(2.890(3) Å), N1A-H13A… O1A(2.920(2) Å). Imidazolium group forms hydrogen bonds with water molecule and carboxylate group of nearest cation: N2A-H21A…O1W(2.719(2) Å), N3A-H3A…O1A(2.680(2) Å). Water molecule in turn forms two hydrogen bonds with two nearest anions: O1W-H1W..O4B(2.874(2) Å), O1W-H2W…O3B (2.742(2) Å).

Triclinic cell of (L-$HisH_2$) (hydrogen maleate)$_2$ contains two formula units, that is, two (L-$HisH_2$) cations and four hydrogen maleate anions (Fig. 5.50). Both cations have open conformation. The hydrogen maleate anions may be divided into two groups. In first group, intramolecular hydrogen bonds are longer (O1C…O4C(2.439(4) Å), O1E…O4E(2.430(5) Å)) than in second group (O1D…O4D(2.400(5) Å), O1F… O4F(2.390(5) Å)). Two (L-$HisH_2$) cations form hydrogen bonds via carboxyl groups with anions having longer intramolecular O…O hydrogen bonds: O1A-H1A…O3C (2.659(4) Å), O1B-H1B…O4E(2.635(4) Å). The NH_3^+ groups as well as imidazolium groups of both cations form hydrogen bonds with anions: N1A-H11A…O4D(2.869 (5) Å), N1A-H12A…O2F(2.818(4) Å), N1A-H13A…O3F(2.955(4) Å) and N2A-H21A…O3E(2.713(5) Å), N3A-H3A…O1D(2.841(5) Å), N3A-H3A…O2D (2.997(6) Å) for one cation and N1B-H11B…O2E(2.798(6) Å), N1B-H12B…O3D (2.859(4) Å), N1B-H13B…O2D(2.985(5) Å), N2B-H21B…O3C(2.746(4) Å), N3B-H3B…O3F(2.880(5) Å) for second cation. Vibrational spectra (IR and Raman) of both crystals are provided and discussed.

Crystals of (L-HisH) hydrogen L-malate were obtained and investigated as possible NLO material by de Matos Gomes et al. (2006). The crystal was obtained

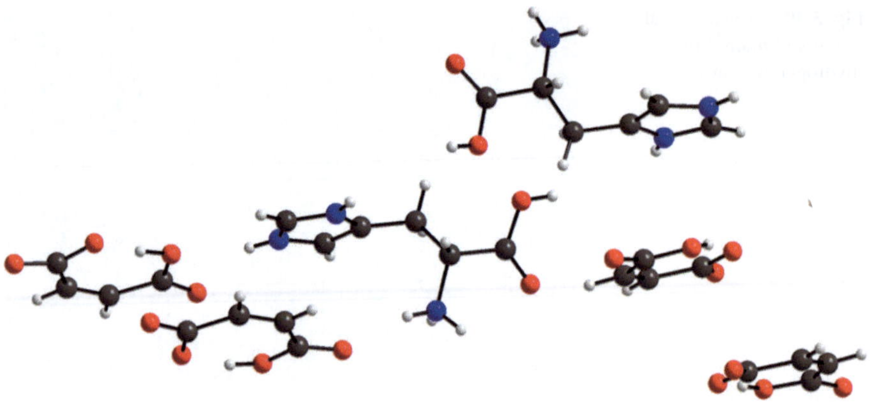

Fig. 5.50 Molecular structure of L-histidinium^{2+} bis(hydrogen maleate). The triclinic unit cell (space group P1) contains four monovalent hydrogen maleate anions, charge-counterbalanced by two L-histidinium^{2+} cations

from aqueous solution containing equimolar quantities of L-histidine and L-malic acid ($HOOCCH_2CH(OH)COOH$) by evaporation at ambient temperature. Crystal structure was determined at 90 K. Asymmetric unit contains one formula unit. The (L-HisH) cation has open conformation with trans disposition of NH_3^+ group in respect to imidazolium group. In contrast to expectation of formation of head-to-tail chain of anions, the carboxyl group of hydrogen L-malate forms hydrogen bond with carboxylate group of cation, O14-H14...O1(2.568(4) Å). Hydroxyl group of anion forms hydrogen bond with carboxylate group of symmetry-related anion, O13-H13...O11(2.709(4) Å). The $N(1)H_3^+$ group forms two hydrogen bonds with two nearest cations and one with carboxylate group of nearest anion: N1-H1A...O2(2.832(4) Å), N1-H1C...O2(2.724(5) Å), N1-H1B...O12 (2.746(4) Å). The imidazolium group forms hydrogen bonds with two nearest anions, one with oxygen atom of hydroxyl group and with carboxylate group: N5-H5...O13(2.807(4) Å), N7-H7...O11(2.701(4) Å). The transmittance curve in UV–Vis–NIR region was measured and IR spectrum is shown. The melting point is between 171 and 174 °C, followed by decomposition. The powder SHG efficiency is ($I^{2\omega}/I^{2\omega}{}_{KDP} = 0.70$).

Interesting results were obtained in result of studies of histidine with tartaric acids, HOOCCH(OH)CH(OH)COOH. Reaction of L-histidine with L-tartaric acid was studied by Marchewka et al. (2003b). From the aqueous solution containing equimolar quantities of components a salts was obtained, which according to chemical analysis had the composition (L-HisH)(OOCCH(OH)CH (OH)COOH)·$0.5H_2O$. This composition was confirmed by structure determination. The (L-HisH) cation and hydrogen L-tartrate anion are in general position, while water molecule is in special position, on axis of symmetry. So, the formula may be expressed as (L-HisH)(OOCCH(OH)CH(OH)COOH)·$0.5H_2O$ with $Z = 4$ or $(\text{L-HisH})_2(\text{OOCCH(OH)CH(OH)COOH})_2\cdot H_2O$ with $Z = 2$. Hydrogen L-tartrate anions form usual head-to-tail chain by O4-H15...O7(2.5479(15) Å) hydrogen bond. Two such symmetry-related chains are connected by water molecules, O9-H16...O8(2.9461(17) Å). Both hydroxyl groups of anion form hydrogen bonds with the same oxygen atom of carboxylate group of cation: O5-H11...O2 (2.8065(16) Å), O6-H13...O2(2.9446(17) Å). The (L-HisH) cation has open conformation. The $N(1)H_3^+$ group forms one hydrogen bond with carboxylate group of symmetry-related cation and two hydrogen bonds with anions: N1-H7...O2(2.7320 (18) Å), N1-H8...O6(2.8718(17) Å), N1-H9...O3(2.7573(18) Å). There is also an additional contact N1-H9...O7(3.1196(18) Å), which may be considered as weak hydrogen bond. The imidazolium group forms hydrogen bonds with carboxylate groups of cation and anion: N2-H3...O1(2.7225(16) Å), N3-H1...O8(2.7062 (17) Å). Infrared and Raman spectra of the crystal are given with assignment of the bands. The efficiency of SHG determined by powder method is $I^{2\omega}/I^{2\omega}_{KDP} = 0.79$.

Johnson and Feeder obtained three more salts: (DL-HisH) hydrogen DL-tartrate (Johnson and Feeder 2004c), (D-HisH) hydrogen D-tartrate (Johnson and Feeder 2004b), (L-HisH) hydrogen L-tartrate (Johnson and Feeder 2004a). The salts were prepared by mixing of equivolume (5 ml) saturated aqueous solutions of respective forms of histidine and tartaric acid. Every mixture was placed in a glass vial containing 25 ml of methanol and sealed. Crystals were obtained after three weeks at room temperature. Asymmetric units of all three salts contain one formula unit. The (HisH) cation in the structure of (DL-HisH) hydrogen DL-tartrate has open conformation with trans disposition of NH_3^+ group in respect to imidazolium group. The carboxyl group of anion forms hydrogen bond with carboxylate group of cation, O3-H3D...O1(2.5856(17) Å). Hydroxyl groups of anion form hydrogen bonds with nearest anions: O5-H5A...O4(2.8174(19) Å), O6-H6...O8(2.7152 (18) Å). The $N(3)H_3^+$ group forms one hydrogen bond with carboxylate group of symmetry-related cation and two hydrogen bonds with carboxylate groups of two nearest anions: N3-H3A...O2(2.7532(18) Å), N3-H3B...O7(2.7937(18) Å), N3-H3C...O7(2.7178(18) Å). There is also a contact, N3-H3B...O6(3.1374 (18) Å). The imidazolium group forms hydrogen bonds with carboxylate groups of both cation and anion: N1-H1...O2(2.6871(19) Å), N2-H2A...O8(2.7689 (19) Å). In the structure of (D-HisH) hydrogen D-tartrate the cation (D-HisH) has closed conformation. Carboxylic group of anion as in previous case forms rather strong hydrogen bond with carboxylate group of cation, O3-H3D...O2(2.565 (2) Å). Hydroxyl groups of anion also form hydrogen bonds with nearest anions: O5-H5A...O4(2.713(2) Å), O6-H6...O7(2.754(2) Å). The $N(3)H_3^+$ group forms one hydrogen bond with carboxylate group of symmetry-related cation and one hydrogen bond with oxygen atom of hydroxyl group of another nearest anion: N3-H3B...O8(2.743(3) Å), N3-H3C...O6(2.901(3) Å). There is also a contact, N3-H3C...O8(3.004(3) Å). The imidazolium group forms hydrogen bonds with carboxylate groups of both cation and anion: N1-H1...O2(2.747(3) Å), N2-H2A...O7(2.683(3) Å). The imidazolium group via C3-H group forms two more contacts: C3-H2...O5(3.224 Å) with H...O(2.40 Å) and C3-H2...O4 (2.987 Å) with H...O(2.21 Å). The last contact can be considered as weak C-H...O hydrogen bond. The structure and symmetry of (L-HisH) hydrogen L-tartrate are quite different. In this case hydrogen L-tartrate anions form head-to-tail chain with hydrogen bond O3-H3...O8(2.533(2) Å). Hydroxyl groups of anion form hydrogen bonds with carboxylate groups of nearest cation and anion: O5-H5...O1(2.741(3) Å), O6-H6...O7(2.744(2) Å). The (L-HisH) cation has open conformation with trans disposition of NH_3^+ group in respect to imidazolium group. The $N(3)H_3^+$ group forms one hydrogen bond with carboxylate group of anion, one with carbonyl oxygen atom of another nearest anion and one with carboxylate group of symmetry-related cation: N3-H3A...O7(2.799(2) Å), N3-HB...O4 (2.866 Å), N3-HC...O2(2.822(3) Å). Johnson and Feeder (2004a) indicated only two hydrogen bonds formed by $N(3)H_3^+$ group. The imidazolium group forms hydrogen bonds with carboxylate groups of both anion and cation: N1-H1...O8 (2.760(2) Å), N2-H2A...O1(2.810(3) Å). There is also a contact, C3-H2...O5 (2.922 Å) with H...O(2.37 Å). It is not clear, is it possible to obtain a crystal form of (L-HisH) hydrogen L-tartrate with the symmetry and structure of (D-HisH) hydrogen D-tartrate and vice versa.

Interaction of L-histidine with D-tartaric acid was also studied (Petrosyan 2007a, 2009, 2010a). Two salts were obtained from aqueous solutions containing 2:1 and 1:1 molar ratios by evaporation. The crystal obtained at equimolar ratio had the same composition, symmetry, unit cell parameters, structure and similar infrared spectrum (Fig. 5.51) as in case of respective salt with L-tartaric acid (Marchewka et al. 2003b). Peaks at 3,183 and 3,156 cm^{-1} are assigned to stretching vibration of C-H bonds of imidazolium group, while Marchewka et al. (2003b) assigned these peaks to vibrations of NH_3^+ group. Marchewka et al. (2003b), in addition, observed strong absorption peaks at 1,408, 1,422, 1,254 and 1,261 cm^{-1}, which are absent in Fig. 5.51. We obtained the crystal of (L-HisH)(hydrogen L-tartrate)·$0.5H_2O$ and its infrared spectrum was identical to the spectrum of (L-HisH)(hydrogen D-tartrate) ·$0.5H_2O$ shown in Fig. 5.51 registered by ATR method. So, the difference probably is caused by that the spectrum of Marchewka et al. (2003b) was registered by the method of suspension in oil. It should be noted that the structures of L·L and L·D salts may be close or different. As an example of salts with close structures salts of L-lysine with L- and D-tartaric acids may be indicated, which have the same symmetry $P2_1$ and unit cell parameters (Selvaraj et al. 2007), while the salts of L-arginine with L- and D-aspartic acids have different structures and symmetries: L-Arg·L-Asp (with space group $P2_1$) (Salunke and Vijayan 1982) and L-Arg·D-Asp (with space group $P1$ (Suresh et al. 1986). The spectrum of sample obtained from solution with 2:1 molar ratio corresponds to that expected for $(L\text{-}HisH)_2$D-tartrate (Fig. 5.52) (Petrosyan 2009). The absence of any bands in the region of 1,700 cm^{-1} corresponds to absence of carboxyl groups in L-HisH cation and D-tartrate anion (OOCCH(OH)CH(OH)COO). Sharp peak at 3,114 cm^{-1} is caused by characteristic stretching vibration of C-H bonds of imidazolium group. Very broad absorption band in the region 3,400–2,500 cm^{-1} is caused by hydrogen bonds N-H...O formed

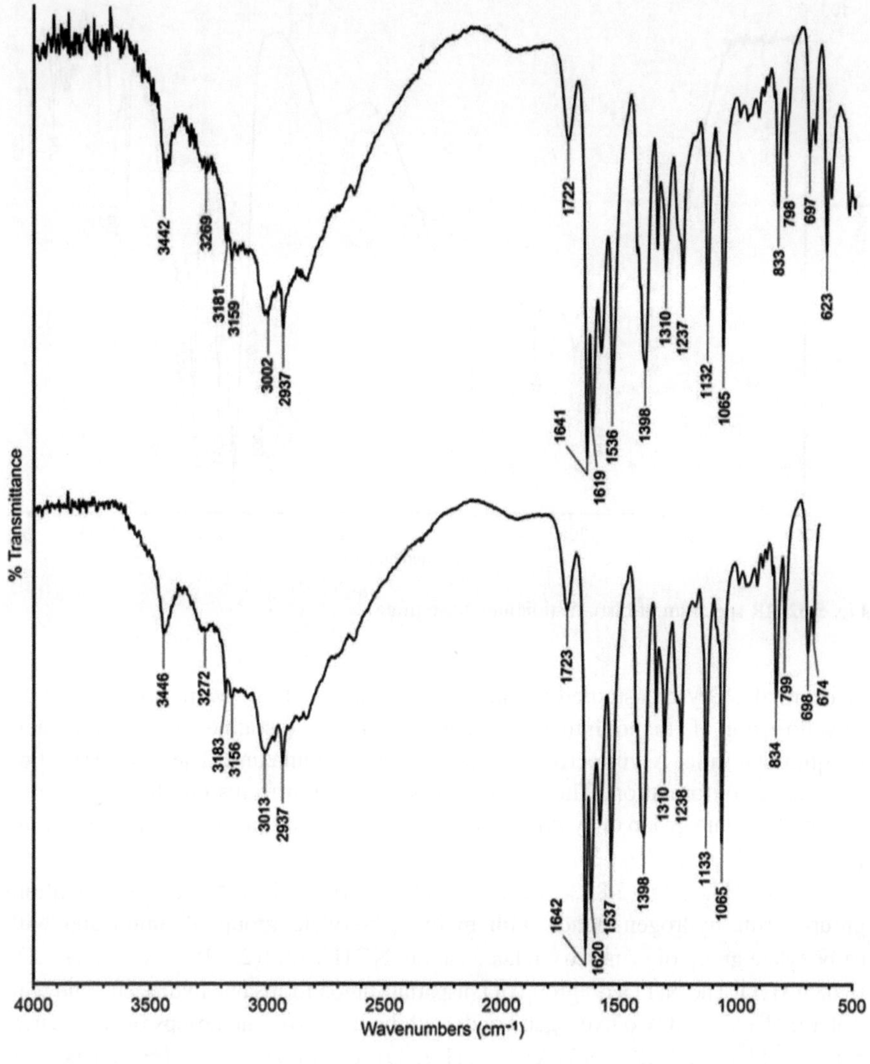

Fig. 5.51 IR spectra of (L-histidinium$^+$)$_2$(hydrogen L-tartrate)$_2$H$_2$O (top) and (L-histidinium$^+$)$_2$ (hydrogen D-tartrate)$_2$H$_2$O (bottom)

by cations and O-H…O formed by anions. The efficiency of SHG of this salt determined by powder method is $I^{2\omega}/I^{2\omega}{}_{KDP} = 0.54$ (Petrosyan 2007a). Respective salt with L-tartaric acid ((L-HisH)$_2$L-tartrate) was also obtained by the same way. The IR spectrum of (L-HisH)$_2$L-tartrate) was identical to the spectrum of (L-HisH)$_2$D-tartrate).

Crystals of (L-HisH) L-aspartate monohydrate were obtained by Suresh and Vijayan (1987). Previously another paper was published on structure determination of (L-HisH) L-aspartate monohydrate (Bhat and Vijayan 1978), where, however, because of twinning, pseudosymmetry and disorder the structure was determined

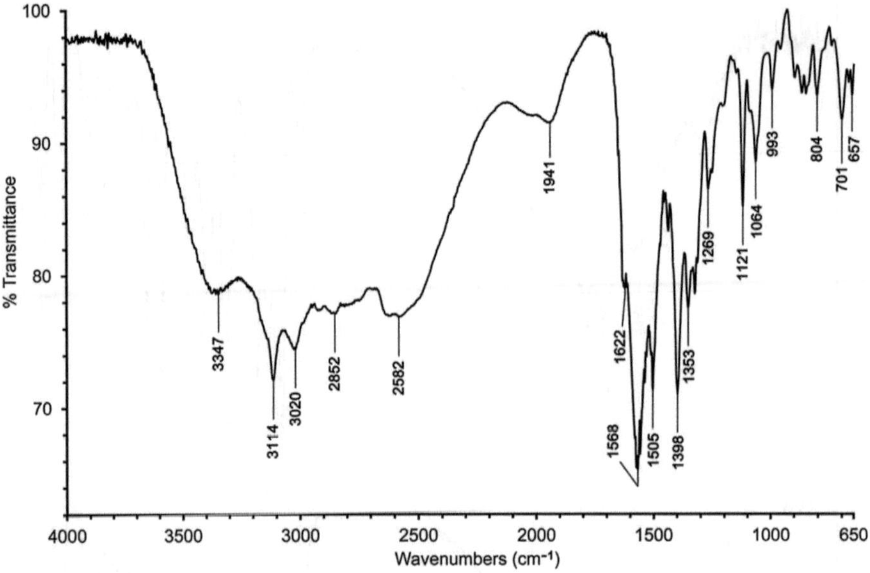

Fig. 5.52 IR spectrum of bis(L-histidinium^{+})(D-tartrate)

inaccurately. Crystals studied by Suresh and Vijayan (1987) were obtained by the slow diffusion of ethanol into an aqueous solution of L-histidine and L-aspartic acid in equimolar ratio. Asymmetric unit contains one formula unit. The (L-HisH) cation has closed conformation. The $N(1)H_3^+$ group of cation forms one hydrogen bond with carboxylate group of symmetry-related cation as well as with both carboxylate groups of three nearest anions: N1-H1...O2(2.851(3) Å), N1-H2...O15(2.906(5) Å), N1-H3...O12(2.833(5) Å), N1-H3...O11(2.935(5) Å). The imidazolium group forms hydrogen bonds with main carboxylate group of anion and with carboxylate group of symmetry-related cation: N5-H...O11(2.800(4) Å), N7-H...O2(2.688(3) Å). The $N(11)H_3^+$ group of L-aspartate anion forms one hydrogen bond with water molecule and two hydrogen bonds with both carboxylate groups of two nearest anions: N11-H2...W(2.845(5) Å), N11-H1...O12(2.774(4) Å), N11-H3...O16(2.936(4) Å). Water molecule in turn forms hydrogen bonds with terminal carboxylate group of two nearest anions: W-H1...O16(2.743(5) Å), W-H2...O16(2.972(5) Å). The crystal was studied also by solid-state ^{1}H NMR method (Wei and McDermott 1999).

The salt of L-histidine with 4,5-imidazoledicarboxylic (IDC) acid was prepared by slow diffusion of 2-propanol into an aqueous solution containing equimolar quantities of components (Görbitz and Husdal 1998). Asymmetric unit contains one formula unit, that is, one (L-HisH) cation and one hydrogen 4.5-imidazoledicarboxylate anion (Fig. 5.53). Carboxyl group of anion forms strong intramolecular hydrogen bond with carboxylate group, O2B-H2B...O4B(2.492(2) Å). Strong intramolecular hydrogen bond in this anion is usual and the value 2.492(2) Å is a new upper limit for the observed *d*(O...O) range. The (L-HisH)

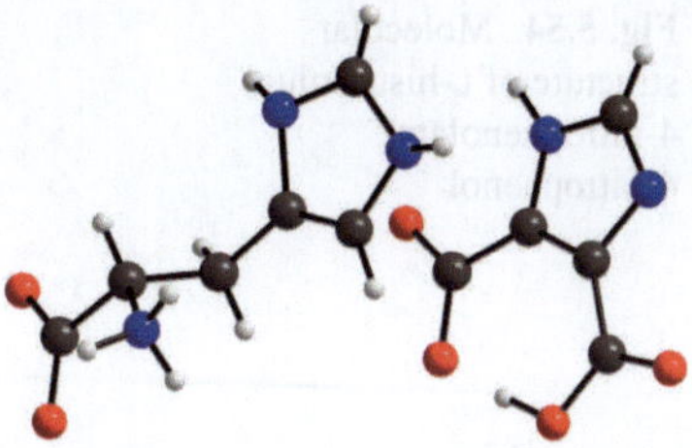

Fig. 5.53 Molecular structure of L-histidinium$^+$ 4,5-imidazoledicarboxylate

cation has open conformation. The $N(1A)H_3^+$ group forms three hydrogen bonds, one with carboxylate group of anion, N1A-H11A…O4B(2.821(2) Å), one with carbonyl oxygen atom of carboxyl group of anion, N1A-H12A…O1B(2.820 (2) Å), and one with carboxylate group of nearest cation, N1A-H13A…O2A (2.718(2) Å). The imidazolium group forms three hydrogen bonds. In addition to usual hydrogen bonds formed by N-H bonds there is also one C-H…O bond. There is a hydrogen bond with carboxylate group of nearest anion, N2A-H2A… O3B(2.806(2) Å), a hydrogen bond N3A-H3A…N3B(2.758(2) Å) with imidazole group of another nearest anion, and with carboxylate group of nearest cation, C6A-H61A…O1A(2.912(2) Å) with (H…O)(2.11(2) Å). Imidazole group of anion also forms a hydrogen bond with carboxylate group of cation, N1B-H1B…O2A (2.608(2) Å).

Ramajothi and Dhanuskodi (2006) obtained and crystallized a salt from the L-His + tetrafluorophthalic acid + H_2O system. The salt was synthesized by evaporation from aqueous solution containing L-histidine and tetrafluorophthalic acid in equimolar ratio. Solubility in water and ethanol was determined. However, measuring unit of solubility probably should be (g/100 ml) and not (gm/cc). The authors have grown single crystals with $12 \times 6 \times 3\ mm^3$ dimensions and characterized by various methods as NLO material. The efficiency of SHG is equal to 0.20 relative to urea. Interpretation of data of thermal analysis as well as IR and ^1H NMR spectra is contradictory. From one hand the authors write that the crystal melts at 138 °C and from other hand write that the material is very stable and there is no phase transition up to 146 °C. Interpretation of ^1H NMR spectrum is incorrect. The authors write that the singlet due to C-H of the imidazole ring is observed at $\delta = 4.1$ ppm, the signal $\delta = 8.8$ ppm is due to the proton of the carboxyl (OH) group, etc. Actually signals at 7.5 and 8.8 ppm are caused by protons of two C-H bonds in imidazolium group. The authors determined unit cell parameters without structure determination. They suggested the molecular structure of the salt, according to which, there are doubly charged (L-$HisH_2$) cation and $C_6F_4(COO^-)_2$ anion, however, without any substantiation. The shown IR spectrum contains a band at about 1,740 cm^{-1}, which is not discussed in the text and is not included in the Table 1 of the paper by Ramajothi and Dhanuskodi (2006). This band may be caused by stretching vibration of C=O bond of (L-$HisH_2$) cation, if the suggested (L-$HisH_2$) $C_6F_4(COO)_2$ structure is correct or $C_6F_4COOH(COO^-)$ anion, if one suggests the (L-HisH)$C_6F_4COOH(COO)$ mechanism. In our opinion the structure (L-HisH)C_6F_4

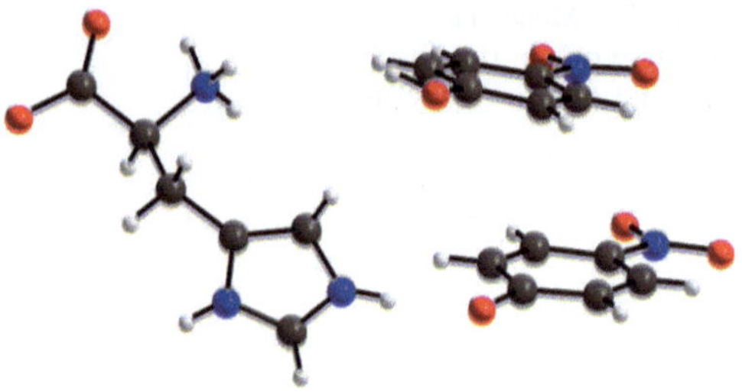

Fig. 5.54 Molecular structure of L-histidinium 4-nitrophenolate 4-nitrophenol

COOH(COO) is more probable because hydrogen-phthalate anions form stable structure with very strong intramolecular hydrogen bond between COOH and carboxylate group (Görbitz and Husdal 1998). The authors assigned the sharp characteristic peak at 3,140 cm^{-1} caused by stretching vibration of C-H bonds of imidazolium group to N-H symmetric stretching and a peak at 1,519 cm^{-1} assigned to N-H asymmetric stretching, which is incorrect as well.

Dhanalakshmi et al. (2010) obtained a new crystal L-histidinium 4-nitrophenolate 4-nitrophenol from aqueous solution containing L-histidine and 4-nitrophenol in equimolar ratio. Probably it was expected to obtain 1:1 salt similar to previously known L-argininium 4-nitrophenolate monohydrate. Later the crystal of L-histidinium 4-nitrophenolate 4-nitrophenol was obtained also from aqueous solution containing stoichiometric (1:2) ratio of components (Chen et al. 2012). The crystal was grown by evaporation method with sizes 7 × 5 × 2 mm^3 and had yellow color. Crystal structure was determined. Asymmetric unit contains one formula unit (Fig. 5.54). The (L-HisH) cation has open conformation. The $N(5)H_3^+$ group forms three hydrogen bonds, one with carboxylate group of symmetry-related cation, one with nitro group of nitrophenol, and one with phenolate oxygen atom: N5-H5A. . .O7(2.8448(16) Å), N5-H5B. . .O2(2.8787(17) Å), N5-H5C. . .O5(2.7645(16) Å). Imidazolium group forms one hydrogen bond with carboxylate group of nearest cation and one with nitrophenolate anion: N3-H3A. . .O8(2.7383(15) Å), N4-H4A. . .O5(2.6083(15) Å). The O-H group of nitrophenol molecule forms hydrogen bond with carboxylate group of nearest cation, O6-H14A. . .O7(2.6520(15) Å). The IR and UV–Vis–NIR spectra were registered and thermal properties are studied by TG, DTA and DSC methods. The efficiency of SHG measured by powder method is $I^{2\omega}/I^{2\omega}{}_{KDP} = 3.425$. The crystal decomposes above 154 °C. According to the authors the crystal is transparent in the region between 209 and 1,200 nm. However, this is not compatible with yellow color of the crystal. Some assignments of peaks in the IR spectrum are questionable. Characteristic peak of stretching vibration of C-H bonds of imidazolium group at 3,154 cm^{-1} is assigned to symmetric stretching of N-H bonds, and also the peak at 1,602 and 712 cm^{-1} are assigned to $\nu_{as}(NH_3)$ and $\nu(CH_2)$ respectively. Further investigation of this crystal was carried out by Chen et al. (2012). As mentioned above the salt of L-histidinium 4-nitrophenolate 4-nitrophenol was obtained from aqueous solution containing L-histidine and 4-nitrophenol in 1:2 molar ratio. The solubility of the crystal in water was determined. Bigger, with 45 × 25 × 10 mm^3 sizes, and higher quality crystals were grown by temperature lowering method. Transmittance spectrum of the crystal is shown. The UV transparency cutoff is 480 nm, which is compatible with yellow color of crystal. The SHG efficiency was studied by the Kurtz–Perry powder method. The presence of phase-matching was established and the efficiency was $I^{2\omega}/I^{2\omega}{}_{KDP} = 3.55$. Thermal expansion coefficients were determined using thermal dilatometer. Specific heat in the 300–380 K were measured as well as laser damage threshold, which is 3.02 GW/cm^2.

Natarajan et al. (Natarajan et al. 2012) obtained a new salt (L-HisH) 2-nitrobenzoate from aqueous solution containing L-histidine and 2-nitrobenzoic acid in equimolar proportion by slow evaporation. Crystal structure was determined. Asymmetric unit contains one formula unit. The (L-HisH) has open conformation with $N(1)H_3^+$ group trans in respect to imidazolium ring. The $N(1)H_3^+$ group forms two hydrogen bonds with carboxylate group of two nearest cations and one with carboxylate group of anion: N1-H1...O2(2.847(2) Å), N1-H2...O2 (2.738(2) Å), N1-H3...O3(3.020(2) Å). The imidazolium group forms hydrogen bonds with carboxylate group of two nearest anions: N2-H4N...O3(2.685(3) Å), N3-H5N...O4(2.712(3) Å). There are also C-H...O contacts of imidazolium group with carboxylate group of cation and nitro group: C5-H5...O1(3.052(2) Å) with H...O(2.41 Å) and C6-H6...O5(2.931(3) Å) with H...O(2.36 Å). The efficiency of SHG measured by powder method is about two times that of KDP. Results of further investigation of this crystal reported in (Moovendaran et al. 2013). The IR and UV–Vis–NIR spectra and thermal properties are reported. Two characteristic sharp peaks caused by stretching vibration of C-H bond at 3,173 and 3,138 cm^{-1} are assigned to N-H symmetric stretching and NH_3^+ stretching. The crystal melts near 189 °C and then decomposes.

Nagashima et al. (1993) prepared a solvate salt (L-HisH) α-ketoisocaproate· C_2H_5OH of L-histidine with α-ketoisocaproic acid ($CH_3CH(CH_3)CH_2C(O)$ COOH) from an aqueous ethanol solution. Asymmetric unit contains one formula unit. The (L-HisH) cation has open conformation. The $N(1)H_3^+$ group forms three hydrogen bonds, one with carboxylate group of symmetry-related cation, one with carboxylate group of anion and with oxygen atom of ethanol molecule: N1-H...O2 (2.761(11) Å), N-H′...O1′(2.838(12) Å), N1-H′...O1S(2.850(15) Å). The imidazolium group forms hydrogen bonds with carboxylate groups of anion and cation: N2-H...O1′(2.675(14) Å), N3-H...O1(2.642(11) Å). The ethanol molecule in turn forms hydrogen bond with carboxylate group of anion, O1S-H...O2′(2.764 (10) Å). The crystal easily decomposed by loss of solvent of crystallization.

One more solvated salt is (L-HisH) trimesate one-third acetone prepared by Herbstein and Kapon (1979) by dissolving L-histidine and trimesic acid in water–acetone solution. Asymmetric unit contains one (L-HisH) cation and singly charged trimesate anion. In addition the authors found two peaks, which they treated as C=O group of a solvent acetone with occupancy of 1/3. Non-appearance of the methyl groups was attributed to rotational disorder of the acetone molecule. The (L-HisH) cation has open conformation. On the basis of H-N-H angles the authors came to conclusion that the position of one hydrogen atom (H3) of the $N(1)H_3^+$

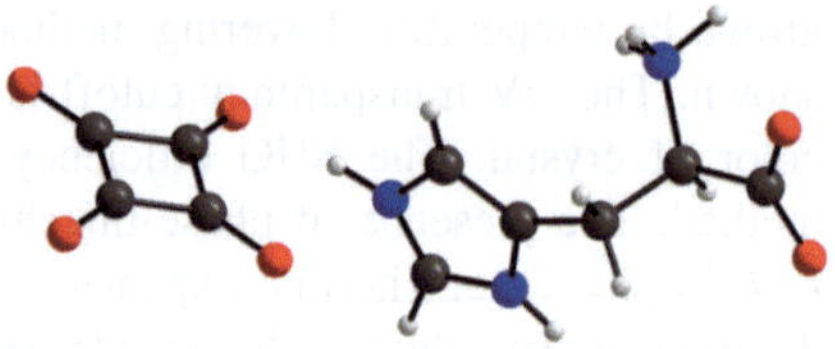

Fig. 5.55 Molecular structure of L-histidinium$^+$ hydrogen squarate

group was not determined correctly. In addition, the authors used for oxygen atoms of carboxylate groups of anion and cation the same numbering O1 and O2. In order to avoid confusion we will use O7 and O8 for carboxylate group of cation instead of original O1 and O2 labeling. The N1 atom has four intermolecular contacts, which may be hydrogen bonds: N1...O2(2.786 Å), N1...O2(2.860 Å), N1...O6 (2.805 Å), N1...O8(2.926 Å). The authors indicated only the first two contacts as hydrogen bonds. In our opinion at least one of two other contacts (with carbonyl oxygen atom O6 of one of COOH group of anion and oxygen atom O8 of carboxylate group of cation) may be a hydrogen bond depending on correct positions of hydrogen atoms of $N(1)H_3^+$ group. The imidazolium group forms hydrogen bonds with carboxylate groups of anion and cation: N2-H...O1 (2.567 Å) and N3-H...O7(2.741 Å). Carboxylic groups of anion form hydrogen bonds with carboxylate group of two nearest cations: O4-H...O7(2.601 Å) and O5-H...O8(2.572 Å). The authors reported some results on the salt formed by DL-histidine. The space group of the salt of DL-histidine is $Pna2_1$. In their abstract, Herbstein and Kapon (1979) write that the salt has the same composition and have essentially identical cell dimensions, while in Experimental section they write that a similar complex with DL-histidine, in which one water molecule replaced each one-third molecule of acetone, was prepared from water alone. The crystal of (L-HisH) trimesate one-third acetone was studied also by solid-state ^{15}N NMR method (Song et al. 2001).

Squaric acid ($H_2C_4O_4$, SQ) is not usual carboxylic acid, but it is rather strong dibasic acid. Numerous salts of squaric acid are known, including salts of amino acids. Reaction of L-histidine with squaric acid was studied by Karle et al. (Karle et al. 1996). Two salts were obtained, both with (L-HisH) cation ((L-HisH)$_2C_4O_4$ and (L-HisH)HC_4O_4). The salts were prepared from aqueous solution containing stoichiometric amounts of components by slow evaporation. The (L-HisH)HC_4O_4 salt crystallizes in small needles, while (L-HisH)$_2C_4O_4$ as large plates (15–20 mm in length). Melting points of (L-HisH)$_2C_4O_4$ and (L-HisH)HC_4O_4 (250–255 and 225–230 °C respectively), wavenumbers of IR spectra and λ_{max} in UV region were determined. Crystal structures of grown crystals were determined. Asymmetric unit of (L-HisH)HC_4O_4 contains one formula unit (Fig. 5.55). Monohydrogen squarate anions in the structure do not form expected head-to-tail chains, but are hydrogen bonded with carboxylate group of cations, O3s...O1(2.510(7) Å). The position of hydrogen in HC_4O_4 anion was not determined, however, comparison of C-O distances (C1-O1s(1.247 Å), C2-O2s(1.237 Å), C3-O3s(1.294 Å), C4-O4s (1.258 Å) in the anion and short hydrogen bond O3s...O1 shows that hydrogen

Fig. 5.56 Molecular structure of bis(L-histidinium$^+$) squarate

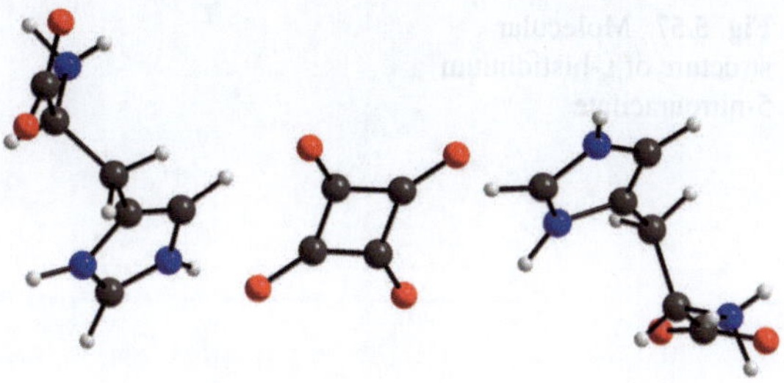

atom is attached to O3s atom. The L-HisH cation has open conformation. The N(1) H_3^+ group forms three hydrogen bonds with three nearest anions: N1-H1...O4s (2.812(7) Å), N1-H2...O4s(2.941(7) Å), N1-H3...O2s(2.757(7) Å), while the imidazolium group forms hydrogen bonds with carboxylate group of nearest cation and anion: N11-H...O2(2.644(7) Å) and N12-H...O1s(2.768(7) Å). The imidazolium group has also C-H...O contacts: C8-H8...O2s(3.019 Å) with H...O (2.33 Å) and C9-H7...O1s(3.108 Å) with H...O(2.17 Å). Asymmetric unit of $(L\text{-HisH})_2C_4O_4$ contains half of molecular unit (Fig. 5.56). The L-HisH cation is in general position, while the squarate anion is in special position. Twofold axis passes through O1s-C1s-C3s-O3s. The (L-HisH) cation has open conformation. The $N(1)H_3^+$ group forms two hydrogen bonds with carboxylate group of two nearest cation and with anion: N1...O2(2.891(7) Å), N1...O2s(2.700(7) Å), N1...O2 (2.884(7) Å). The imidazolium group forms N-H...O and C-H...O types hydrogen bonds: N11...O1(2.716(10) Å), N12...O3s(2.711(11) Å) and C11...O1(2.973 (11) Å) with H...O(2.06(11) Å). The second C...O2s contact is characterized by C...O(3.076 Å) and H...O(2.17 Å) parameters. Both crystals display NLO activity with SHG efficiency 0.18 relative to urea. The crystal of $(L\text{-HisH})_2C_4O_4$ was studied by ^{1}H (Wei and McDermott 1999) and ^{15}N (Song et al. 2001) solid-state nuclear magnetic resonance method.

Muralidharan et al. (2013) obtained crystals of L-histidinium *p*-toluenesulfonate by slow evaporation of aqueous solution at room temperature containing equimolar quantities of L-histidine and 4-toluenesulfonic acid and determined its crystal structure. Asymmetric unit contains one formula unit. The (L-HisH) cation has open conformation with NH_3^+ group disposited trans in respect to imidazolium group. The $N(3)H_3^+$ group forms two hydrogen bonds with carboxylate group of two nearest cations and one with anion: N3-H3B...O4(2.755(2) Å), N3-H3C...O4(2.2.896 (2) Å), N3-H3D...O3(3.000(2) Å). The imidazolium group forms hydrogen bonds with two nearest anions: N1-H1...O2(2.828(2) Å) and N2-H2A...O3(2.746(2) Å). Due to two hydrogen bonds formed by O4 oxygen atom of carboxylate group the length of C13-O4(1.260(2) Å) bond is longer than C13-O5(1.231(2) Å). The same is in case of sulfonate group: the bond S1-O3(1.4652(17) Å) is longer compared to two other bonds, S1-O1(1.4361(17) Å) and S1-O2(1.4490(18) Å).

Domingos et al. (2012) obtained pink transparent single crystals of L-histidinium 5-nitrouracilate with prismatic habit from aqueous solution containing equimolar

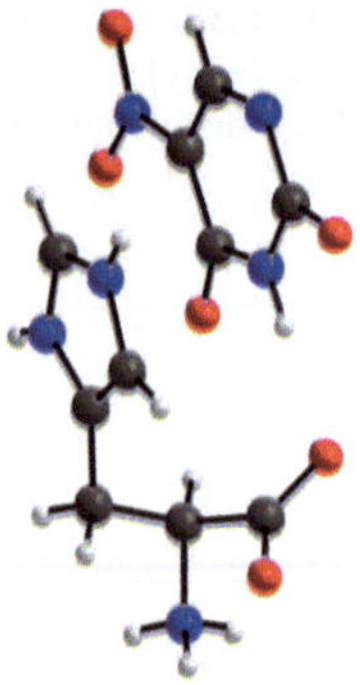

Fig. 5.57 Molecular structure of L-histidinium^{+} 5-nitrouracilate

quantities of L-histidine and 5-nitrouracil by evaporation. One might expect non-centrosymmetric structure of this salt due to the presence of optically active L-histidine and high nonlinearity due to the presence of 5-nitrouracil, which itself has high NLO properties (Puccetti et al. 1993). Previously known two salts with 5-nitrouracilate anion (Pereira Silva et al. 2008; Portalone 2010) and possibly 3-aminopyridine 5-nitrouracil (Thomas et al. 2005b) had centrosymmetric structures. Asymmetric unit of L-histidinium 5-nitrouracilate contains one formula unit (Fig. 5.57). The (L-HisH) cation in the structure has open conformation with NH_3^+ group in *trans*-position in respect to imidazolium group. The anion is deprotonated at N1 position similar to other cases (Pereira Silva et al. 2008; Portalone 2010). The $N(4)H_3^+$ group forms two hydrogen bonds with carboxylate group of two nearest cations and one with anion: N4-H4A…O6(2.914(2) Å), N4-H4B…O5(2.767(2) Å), N4-H4C…O2(2.752(2) Å). The imidazolium group forms one hydrogen bond with carboxylate group of nearest cation and one with deprotonated N1 atom of anion: N5-H5…O6(2.852(2) Å), N6-H6A…N1(2.726(2) Å). There is also a short contact C12-H12…O1(2.989 Å) with H…O(2.27 Å). Thus the atom O5 forms one hydrogen bond, while O6 forms three hydrogen bonds. Probably this is the reason of difference in bond lengths C7-O5(1.233 Å) and C7-O6(1.261 Å). The remaining N-H group of anion forms hydrogen bond with neighboring carboxylate group of cation, N3-H3…O6(2.880(2) Å) (there is a misprint in the paper of Domingos et al. (2012), this hydrogen bond is indicated as N1-H3…O6). There is also a contact C6-H6…O1(3.212 Å) with H…O(2.41 Å). Domingos et al. (2012) performed detailed theoretical investigation of microscopic and macroscopic nonlinear properties of the L-histidinium 5-nitrouracilate as well as investigated vibrational circular dichroism. The efficiency of SHG determined by powder method showed that $I^{2\omega}/I^{2\omega}_{KDP} = 10.5$, thus confirming high expected value.

Sethuram et al. (2013) recently obtained yellowish crystals of (L-$HisH_2$) dipicrate dihydrate from water and ethanol mixture containing L-histidine and picric acid in 1:2 molar ratio by slow evaporation. Asymmetric unit contains one formula unit. The (L-$HisH_2$) cation has closed conformation. Carboxyl group of (L-$HisH_2$) cation forms hydrogen bond with phenolate oxygen atom of one of picrate anion, O2-H2…O3(2.657(3) Å). The $N(1)H_3^+$ group forms one rather strong hydrogen bond with one of water molecules, N1-HB…O18W(2.700(5)) and also three more weak hydrogen bonds: N1-HA…O10(2.909(3) Å), N1-HA…O12(2.841(4) Å), N1-HC…O15(3.007(4) Å). The imidazolium group forms one rather strong hydrogen bond with phenolate oxygen atom of second picrate anion, N2-H2B…O10 (2.709(4) Å) and one more weak hydrogen bond with nitro group, N3-H3…O14 (3.077(4) Å). The water molecule H_2O(18 W) forms rather strong hydrogen bond with other water molecule, O18W-HA…O17W(2.664(5) Å). Other hydrogen bonds formed by water molecules are weak: O18W-HB…O7(3.005(5) Å), O17W-HA…O5(3.082(4) Å), O17W-HB…O3(2.864(4) Å). Further investigation of this crystal was performed (Sethuram et al. 2014). The IR, Raman and UV–Vis spectra as well as ^{1}H, ^{13}C NMR spectra were studied. Thermal analysis curves (TG and DTA) show that dehydration takes place in the interval between 81 and 101 °C. Experimental mass loss is about 5 %, which agrees with calculated value 5.5 %. Above 160 °C anhydrous sample starts to decompose. The optimal geometry, spectral characteristics and hyperpolarizability were studied using the DFT method. According to their calculations hyperpolarisability is 39 times higher than that of urea. However, SHG efficiency measured by powder method showed that it is higher than in KDP only 2.5 times. For picrate salts of amino acids this value is not very high and is much low than that expected from calculations. The authors write that there is no significant absorption in the range of 190–800 nm. However, this is not in agreement with that expected for colored crystals containing picrate anions. In addition, interpretation of vibrational spectra based on calculated data is not acceptable. For example, the authors assigned the bands at 3,599 and 3,321 cm^{-1} to stretching vibration of N-H and C-H bonds respectively. In our opinion it is more probable that these bands are caused by stretching vibrations of water molecules. The band at 1,741 cm^{-1} the authors assigned to stretching vibration of C-O bond of picrate anion (to C7-O3 bond). In our opinion this is the characteristic band caused by stretching vibration of C=O bond of COOH group, etc.

5.1.24 Adducts of Amino Acids with Inorganic and Organic Acids

Above we considered salts of amino acids, which are formed due to interaction between amino acids (as bases) with different (inorganic and organic) acids in result of proton (or protons) transfer from acid to amino acid. However, interaction of amino acids with acids may occur also without proton transfer with formation of molecular adducts. This type of compounds is widely called also cocrystals. However, this term in our opinion is not felicitous. It means coexistence of different crystals, but actually there is coexistence of different molecules and not different crystals. One may imagine a crystal which consists of two molecules of substances one of which (or even both) is liquid in usual conditions. Actually we have in one crystal different molecules of different substances, but not different crystals. This type of crystals is called also multicomponent crystals, which is not more felicitous, because in definite sense crystals of salts also are multicomponent crystals. In our opinion the terms adducts or comolecules better reflect the nature of these compounds than the term cocrystal. The term cocrystal has sense, e.g. when two substances crystallize in eutectic or eutonic points in process of cocrystallization. Although the salts and adducts may be distinguished by different methods, transition between them occurs smoothly. We collected compounds of amino acids which represent molecular adducts in Table 5.27. Formation of salts may be predicted if the ΔpK_a ($\Delta pK_a = pK_a$(base)-pK_a(acid)) is greater than 3 and formation of adducts when ΔpK_a is less than zero (Majerz et al. 1997). In case of $0 < \Delta pK_a < 3$ it is difficult to predict what will form. Obviously other factors are comparable with ΔpK_a in this range. For example in case of $\Delta pK_a = 0.77$ formation of salt or adduct depends on temperature for the same crystal (Steiner et al. 2001; see also Asaji et al. 2007; Globelny et al. 2011). In this regard it is interesting to mention the salt (DL-TrpH)HCOO (see Sect. 5.1.17) with hydrogen bond O-H...O (2.492 Å) between (TrpH) cation and HCOO anion, while in case of L-Trp·HCOOH there is an adduct between HCOOH and zwitterionic L-Trp with hydrogen bond O-H...O (2.487(2) Å) (Hübschle et al. 2002) at 183 K and O-H...O(2.4870(4) Å) (Scheins et al. 2004) at 100 K.

More detailed consideration of compounds with molecular ratio 2:1 of amino acids and acids (## 2, 9, 10, 19, 28, 46–48, 56, 57) were performed in Sect. 5.2. It is interesting to note that in the system Gly + H_2SeO_3 + H_2O two compounds were discovered. The compound Gly·H_2SeO_3 is an adduct (Ondráček et al. 1992), while the second is a salt (Gly...GlyH)$HSeO_3$ (Němec et al. 1998) with dimeric cation (see Sect. 5.2). The compound Gly·H_2SeO_3 is also studied in the work of Němec et al. (1998). The compounds Gly·*m*-nitrobenzoic acid (Zheng et al. 1994b) and Gly·*3,5*-dihydroxybenzoic acid·H_2O (Chen et al. 2006) have non-centrosymmetric, polar symmetry and display SHG. Anandha Babu et al. (2011) also obtained and investigated the crystal of Gly·*3,5*-dihydroxybenzoic acid·H_2O. Interaction of glycine with trimesic acid results in (Gly·trimesic acid·H_2O) adduct, while interaction of L-histidine and DL-histidine with trimesic acid results in salts (see Sects. 5.1.22 and 5.1.23). The crystal of Gly·glutaric acid was obtained and single crystal was grown and investigated by (Riscob et al. 2011b). Crystal structure was solved by Anandha Babu et al. (2011) in space group $P2_1$/n. Further results on the investigation of this crystal are given by Zakharov et al. (2012, 2013). However, L-arginine having strong basic properties forms (L-ArgH) glutarate dihydrate salt (see Sects. 5.1.22 and 5.1.23). There is also (L-HisH) hydrogen glutarate monohydrate salt (see Sects. 5.1.22 and 5.1.23).

Mohandas et al. (2013) obtained needle crystals by evaporation of aqueous solution containing glycine and D-tartaric acid in equimolar ratio. Crystal structure determination showed that the obtained crystal crystallized in centrosymmetric $P2_1/n$ space group, which is not compatible with the presence of optically active D-tartaric acid in the structure. Inspection of the structure showed that the obtained crystal actually is Gly·DL-tartaric acid. There are several tartrates salts of amino acids, however, in case of glycine and DL-tartaric acid the formed compound is

Table 5.27 Adducts of amino acids with inorganic and organic acids

#	Crystal	s.g., Z	*a*, *b*, *c* (Å), *α*, *β*, *γ* (°)	Refs
1	Gly·H_2SeO_3	$P2_1/n$, 4	5.049(1), 13.677(2), 8.664(1), $\beta = 94.92(1)$	Ondráček et al. (1992)
2	2Gly·$Te(OH)_6$·$2H_2O$[a]	$P2_1/n$, 2	7.117(4), 12.201(5), 7.936(4), $\beta = 110.46(4)$	Andersen et al. (1983)
3	Gly·*m*-nitrobenzoic acid	$Pca2_1$, 4	22.701(5), 5.852(2), 7.815(2)	Zheng et al. (1994b)
4	Gly·*3,5*-dihydroxybenzoic acid·H_2O	$Pca2_1$, 4	25.513(10), 4.668(2), 9.150(4)	Chen et al. (2006)
5	Gly·phthalic acid	*Pbca*, 8	7.9657(5), 11.3470(7), 23.513(2)	Balakrishnan et al. (2013)
6	Gly·trimesic acid·H_2O	$Pna2_1$, 4	7.971(5), 24.615(10), 6.476(5)	Herbstein et al. (1981)
7	Gly·glutaric acid	$P2_1/c$, 4	4.8954(4), 20.8944(14), 10.8462(8), $\beta = 120.648(6)$	Losev et al. (2011)
8	Gly·DL-tartaric acid[b]	$P2_1/n$, 4	4.8387(2), 9.2913(4), 20.0273(8), $\beta = 90.171(1)$	Mohandas et al. (2013)
9	2Sar·$Te(OH)_6$	$P2_1$, 2	10.045(8), 13.080(9), 5.106(4), $\beta = 96.09(1)$	Averbuch-Pouchot (1988)
10	2Sar·pyromellitic acid	$P2_1/c$, 2	8.8894(3), 5.4118(2), 20.2205(7), $\beta = 104.388(2)$	Domingos et al. (2008)
11	Sar·sucrose·H_2O	$P2_1$, 2	7.175(1), 18.898(6), 7.777(4), $\beta = 110.3(2)$	Krishnakumar and Natarajan (1996)
12	Sar·ellagic acid	$P\bar{1}$, 2	7.6115(2), 9.4592(2), 11.5037(3), $\alpha = 105.294(2)$, $\beta = 109.068(2)$, $\gamma = 91.848(2)$	Kavuru et al. (2010)
13	Sar·L-ascorbic acid	$P2_1$, 2	5.223(2), 8.486(3), 12.871(4), $\beta = 95.220(60$	Kavuru et al. (2010)
14	Dimethylglycine·ellagic acid	$C2/c$, 8	20.7781(4), 12.4012(2), 13.0557(2), $\beta = 107.333(1)$	Kavuru et al. (2010)
15	Bet·H_3PO_4	$P2_1/c$, 4	11.076(4), 7.852(3), 12.944(6), $\beta = 119.62(2)$	Schildkamp and Spilker (1984)
16	Bet·H_3AsO_4	$P2_1/n$, 4	9.556(4), 12.897(6), 8.128(3), $\beta = 91.23(2)$	Schildkamp et al. (1984)
17	Bet·H_3BO_3	*Pmcn*, 4	7.769(1), 9.873(2), 11.974(2)	Zobetz and Preisinger (1989)
18	Bet·H_2SeO_3	*Fdd2*, 16	28.042(2), 21.815(2), 6.073(2)	Paixão et al. (1997)
19	2Bet·$Te(OH)_6$	$P\bar{1}$, 1	7.164(5), 8.325(5), 10.035(6), $\alpha = 108.58(5)$, $\beta = 123.82(4)$, $\gamma = 96.02(5)$	Ilczyszyn et al. 1992)
20	Bet·bis(imidazolium) dipicrate	$P\bar{1}$, 2	7.604(8), 13.403(4), 16.366(8), $\alpha = 109.47(3)$, $\beta = 98.36(6)$, $\gamma = 99.17(5)$	MacDonald et al. (2005)
21	Bet·bis(imidazolium) dipicrate	$C2/c$, 8	33.908(7), 7.765(1), 25.389(6), $\beta = 114.74(2)$	MacDonald et al. (2005)
22	Bet·citric acid	$P\bar{1}$, 2	7.175(3), 10.065(4), 11.099(3), $\alpha = 103.85(4)$, $\beta = 97.50(3)$, $\gamma = 105.32(4)$	Andrade et al. (2002)

(continued)

Table 5.27 (continued)

#	Crystal	s.g., Z	a, b, c (Å), α, β, γ (°)	Refs
23	Bet·4-ethoxyphenylboronic acid	$P2_1/n$, 8	19.222(2), 9.1788(11), 19.337(2), $\beta = 118.755(12)$	Ragowska et al. (2006)
24	Bet·L-ascorbic acid	$P2_1$, 2	7.2174(3), 9.1125(4), 10.4125(4), $\beta = 96.479(3)$	Kavuru et al. (2010)
25	Bet·catechol	$P2_1/c$, 8	9.88686(4), 12.0700(4), 19.9224(7), $\beta = 97.460(2)$	Kavuru et al. (2010)
26	Bet·gallic acid	$P2_1/c$, 4	6.9856(1), 14.7063(2), 12.7854(2), $\beta = 97.184(1)$	Kavuru et al. (2010)
27	Bet·resveratrol monohydrate	$P2_12_12_1$, 4	9.5079(3), 9.6920(3), 26.9227(8)	Kavuru et al. (2010)
28	2DL-Ala·H_3PO_4	$C2/c$, 4	14.442(7), 10.352(6), 9.062(4), $\beta = 119.08(4)$	Averbuch-Pouchot et al. (1988c)
29	(S)-Ala·(R)-2-Phenoxypropionic acid	$P2_12_12_1$, 4	7.3300(4), 34.587(2), 5.2367(4)	Takahashi and Fujii (2004)
30	((S)-Ala)$_2$·((R)-mandelic acid)$_2$·H_2O	$P1$, 1	6.0193(11), 8.2285(10), 12.55811(11), $\alpha = 82.545(9)$, $\beta = 86.000(14)$, $\gamma = 89.752(13)$	Hu et al. (2002)
31	(S)-Ala.(S)-mandelic acid	$C2$, 4	17.795(4), 5.394(2), 12.431(2), $\beta = 100.65(1)$	Hu et al. (2004a)
32	L-Phe·benzoic acid	$P2_1$,2	5.419(2), 7.438(2), 17.814(3), $\beta = 92.94(2)$	Suresh et al. (2005)
33	L-Phe·4-nitrophenol	$P2_1$, 2	5.8327(2), 7.0099(9), 17.8751(4), $\beta = 94.59(10)$	Rodrigues et al. (2006b)
34	(R)-Phe·(S)-mandelic acid	$C2$, 4	19.839(2), 5.557(1), 17.095(2), $\beta = 123.83(1)$	Okamura et al. (1997)
35	(S)-Phe·(S)-mandelic acid	$C2$, 4	19.389(2), 5.725(1), 15.748(2), $\beta = 115.25(1)$	Okamura et al. (1997)
36	(R)-Phe·(R)-mandelic acid	$C2$, 4	19.391(3), 5.715(4), 15.755(3), $\beta = 115.23(1)$	Hu et al. (2004b)
37	L-Phe·fumaric acid	$P1$, 2	5.7016(10), 11.4864(15), 11.5542(17), $\alpha = 67.953(11)$, $\beta = 81.158(13)$, $\gamma = 79.379(15)$	Alagar et al. (2003c)
38	DL-Phe·fumaric acid	$P\bar{1}$, 2	5.499(3), 11.661(5), 11.909(5), $\alpha = 64.67(4)$, $\beta = 78.00(4)$, $\gamma = 84.17(4)$	Klussmann et al. (2007)
39	L-Ser·L-ascorbic acid	$P2_12_12_1$, 4	5.335(3), 8.769(2), 25.782(5)	Sudhakar et al. (1980)

40	L-Ser·2(inosine-5′-phosphate)·$12H_2O$	$P2_1$, 2	8.695(7), 21.898(6), 12.374(3), $\beta = 110.59(3)$	Mukhopadhyay et al. (1995)
41	L-Ser·pyridine-2,4-dicarboxylic acid	$P1$, 1	4.4941(13), 6.4512(18), 10.123(3), $\alpha = 81.273(3)$, $\beta = 87.060(3)$, $\gamma = 86.247(3)$	Liang (2008)
42	DL-Thr·H_3AsO_4	$P2_1/c$, 4	10.1195(6), 9.8062(6), 8.9643(6), $\beta = 99.098(1)$	Wilkinson and Harrison (2005)
43	L-(*R*)-cysteine D-(*R*)-mandelic acid	$P2_12_12_1$, 4	7.0358(8), 34.272(8), 5.1868(8)	Fujii et al. (2005a)
44	L-(*R*)-cysteine L-(*S*)-mandelic acid	$P2_12_12_1$, 4	7.0514(6), 33.870(2), 5.1688(4)	Fujii et al. (2005a)
45	(*S*)-Val (*R*)-2-Phenoxypropionic acid	$P2_12_12_1$, 4	7.6493(4), 34.436(1), 5.4074(3)	Fujii et al. (2005b)
46	2DL-Val·fumaric acid	$C2/c$, 4	24.417(4), 7.5713(10), 10.013(2), $\beta = 109.268(10)$	Alagar et al. (2003d)
47	2L-Val·fumaric acid	$C2$, 2	22.308(1), 7.2294(4), 5.3070(2), $\beta = 91.205(4)$	Klussmann et al. (2007)
48	2DL-Val·succinic acid	$C2/c$, 4	24.373(2), 7.5744(6), 10.0284(9), $\beta = 106.134(2)$	Alagar et al. (2004)
49	(*S*)-Pro·X(1)[c]	$P2_12_12_1$, 4	8.436(8), 18.9881(12), 12.6937(11)	Ramanathan and Periasamy (1998)
50	(*S*)-Pro·2X(2)[c]	$P2_12_12_1$, 4	9.061(1), 13.943(2), 27.480(4)	Periasamy et al. (1997)
51	(*S*)-(−)-Pro·X(3)[c]	$P2_1$, 4	11.650(5), 8.530(7), 26.837(3), $\beta = 97.60(1)$	Fu et al. (1997)
52	L-Pro·X(4)	$P2_1$, 2	7.745(2), 8.028(3), 12.561(4), $\beta = 90$	Marchand AP, Watson WH, 2006, private communication to CSD. CSD code: VESCUS
53	L-Pro·2,5-dihydroxybenzoic acid	$P2_12_12_1$, 4	5.896(2), 11.486(1), 18.016(3)	Aakeröy et al. (1995)
54	L-Pro·X(5)	$P2_1$, 4	14.755(2), 5.283(1), 18.856(2), $\beta = 111.45(1)$, at 110 K	Timofeeva et al. (2003)
55	L-Pro·*p*-ethoxyphenylboronic acid	$P1$, 2	7.376(1), 7.830(1), 12.959(1), $\alpha = 84.11(1)$, $\beta = 78.72(1)$, $\gamma = 73.91(1)$, at 100 K	Ragowska et al. (2006)
56	2L-Pro·4-aminobenzoic acid·H_2O	$P2_1$, 2	9.8171(6), 10.4167(11), 10.1713(12), $\beta = 112.078(9)$	Athimoolam and Natarajan (2007)
57	2DL-Pro·succinic acid	$P2_1/c$, 2	5.254(1), 17.480(1), 10.230(1), $\beta = 119.60(6)$	Prasad and Vijayan (1993b)
58	L-Trp·HCOOH	$P2_12_12_1$, 4	5.3163(3), 8.1348(4), 27.259(2), at 183 K	Hübschle et al. (2002)
59	L-Trp·CH_3COOH·H_2O	$C2$, 4	11.277(1), 6.8760(8), 18.527(2), $\beta = 102.680(2)$	Li et al. (2009)

(continued)

Table 5.27 (continued)

#	Crystal	s.g., Z	a, b, c (Å), α, β, γ (°)	Refs
60	L-Trp·3(4-nitrophenol)	$P2_1$, 2	13.0321(9), 6.7332(4), 17.3091(10), $\beta = 104.479(3)$	Rodrigues et al. (2012)
61	L-Trp·$D(R)$-mandelic acid·1.5H_2O	$C2$, 4	32.997(2), 5.8823(8), 9.8717(9), $\beta = 103.483(6)$	Fujii (2009)
62	L-Trp·X[d]	$P1$, 1	7.0320(14), 7.7590(16), 9.5800(19), $\alpha = 85.44(3)$, $\beta = 81.89(3)$, $\gamma = 71.84(3)$	Di (2010)
63	L-Asn·L-tartaric acid	$P2_1$, 2	5.0860(4), 9.6720(6), 11.8340(8), $\beta = 95.311(8)$	Natarajan et al. (2010b)
64	L-Glu·L-pyroglutamic acid·H_2O	$P2_1$, 2	18.473(5), 7.190(1), 5.113(1), $\beta = 96.77(2)$	Taira and Watson (1977)

[a]The authors indicated for this crystal the composition 2Gly·Te$(OH)_6$·H_2O. Actually the correct formula should be 2Gly·Te$(OH)_6$·2H_2O (more detailed discussion see in Sect. 5.2)

[b]The authors indicated for this crystal Gly·D-tartaric acid. However, centrosymmetric space group $P2_1/n$ is not compatible with the presence of optically active D-tartaric acid in the structure of this crystal. Actually the crystal is Gly·DL-tartaric acid

[c]X(1) = (11R,12R)-(+)-9,10-Dihydro-9,10-ethanoanthracene-11,12-dicarboxylic acid; X(2) = (S)-(−)-(1,1′-Binaphth-2,2′-diol); X(3) = 4-[2,4,6,-tris (1-methylethyl)benzoyl]benzoic acid; X(4) = Pentacyclo(5.3.0.0^2,5^.0^3,9^.0^4,8^)decane-2,5-dicarboxylic acid; X(5) = 1,1-Dicyano-2-(4-hydroxyphenyl)-ethene

[d]X = pyridine-2,4-dicarboxylic acid·ethanol

molecular adduct. There are molecular adducts of L-ascorbic acids with L-serine (Sudhakar et al. 1980) as well as sarcosine and betaine (Kavuru et al. 2010). Again, L-arginine having strong basic properties forms the salt (L-ArgH) L-ascorbate (see Sects. 5.1.22 and 5.1.23). Similarly the salts (L-ArgH) 2,5-dihydroxybenzoate monohydrate may be mentioned in contrast to above considered molecular adduct Gly·*3,5*-dihydroxybenzoic acid·H_2O as well as (L-ArgH) 4-nitrophenolate monohydrate in contrast to L-Phe·4-nitrophenol and L-Trp·3(4-nitrophenol). Betaine forms molecular adducts with H_3PO_4 and H_3AsO_4 acids, DL-Thr·H_3AsO_4 also may be mentioned, while glycine forms (GlyH)(H_2PO_4) and (GlyH)(H_2AsO_4) salts as well as (DL-ThrH)(H_2PO_4). The compound Bet·H_2SeO_3 crystallizes in non-centrosymmetric polar space group (Paixão et al. 1997). Further investigations of this crystal (Arjunan et al. 2012b; Ilczyszyn et al. 2013) showed that it displays SHG signal ($I^{2\omega}/I^{2\omega}{}_{KDP} = 0.97$) (Arjunan et al. 2012b) and the structure at 270 and 170 K is the same as at room temperature (Ilczyszyn et al. 2013). L-phenylalanine and DL-phenylalanine form molecular adducts with fumaric acid (Alagar et al. 2003c; Klussmann et al. 2007), while in case of betaine the salt (BetH) hydrogen fumarate is known (see Sect. 5.1.4).

Natarajan et al. (2010b) obtained the crystal of L-asparagine·L-tartraric acid from aqueous solution containing equimolar quantities of components. Structure determination showed that it is a molecular adduct in structure of which there are hydrogen bonds between carboxylic groups of L-tartraric acid and oxygen atom of amide group and carboxylate group of zwitterionic L-asparagine: O4-H4...O1 (2.500(6) Å), O9-H9...O2(2.570(7) Å). Hydrogen bonds formed by NH_3^+ and NH_2 groups as well as O-H groups are also present. Moovendaran et al. (2012) have grown and characterized the crystal as possible NLO material. Three more papers also were published (Vivek and Murugakoothan 2013; Tamilselvan et al. 2014; Shkir and Abbas 2014). Vivek and Murugakoothan (2013) call the crystal L-asparaginium L-tartrate presuming formation of salt, which is not in agreement with the result of structure determination (Natarajan et al. 2010b). The morphology of crystals obtained by Tamilselvan et al. (2014) differs from morphologies of crystals obtained in other papers (Moovendaran et al. 2012; Vivek and Murugakoothan 2013; Shkir and Abbas 2014). In addition the band in the IR spectrum at 3,450 cm^{-1} the authors assigned to stretching vibration of water, while the crystal does not contain water molecules. The authors use the formula $C_4H_8N_2O_3{}^+ \cdot C_4H_6O_6{}^-$ for the crystal, while the formulas $C_4H_8N_2O_3$ and $C_4H_6O_6$ concern to neutral asparagine and tartaric acid. Shkir and Abbas (2014) provided results of their experimental and theoretical investigations. Strangely they consider that the structure of the crystal determined by X-ray diffraction (particularly, the zwitterionic structure of L-asparagine molecule) is not correct, while calculated structure by DFT method for gas phase (particularly, neutral structure of L-asparagine molecule) is correct. This is evident from their statement: "We strongly believe that this structure cannot be a true ground state equilibrium structure." The efficiencies of SHG $I^{2\omega}/I^{2\omega}{}_{KDP}$ of the L-asparagine·L-tartraric acid crystal measured by powder method by Moovendaran et al. (2012), Vivek and Murugakoothan (2013) and Tamilselvan et al. (2014): 0.31, 1.5 and 3, respectively, differ significantly.

5.2 Salts of Amino Acids with Dimeric Cations

In this section we will consider salts with dimeric cations of three types. The first type is the (A...A^+) type, where A is an amino acid in zwitterionic state comprising carboxylate (COO^-) group and A^+ is a cationic, protonated amino acid (AH) comprising neutral carboxyl (COOH) group except for ornithine, lysine, arginine and histidine. The dimeric cation of this type is formed due to O-H...O type short hydrogen bond between COOH and COO^- groups. The majority of salts with this type of dimeric cations have 2A·HX composition [(A...A^+)·X^-], and few of 4A·H_2Y [2(A...A^+)·Y^{2-}]. However, not all compounds with 2A·HX composition form salts with this type of dimeric cation. Some of them do not form salts and are molecular adducts, others form salts, but without formation of dimeric cation. At the same time there are salts with (A...A^+) type dimeric cation, but not with 2A·HX composition. We will consider all these compounds in Sect. 5.2.1.

The second type is (A^+...A^+) type dimeric cation. In this case, hydroxyl group of one protonated amino acid forms O-H...O type hydrogen bond with carbonyl oxygen atom of second cation, which in turn forms O-H...O hydrogen bond with carbonyl oxygen atom of the first cation. Almost all known salts of this type are salts of β-alanine. Salts with this type of dimeric cations will be considered in Sect. 5.2.2.

Amino acids capable forming singly charged and doubly charged cations, namely ornithine, lysine, arginine and histidine, cannot form salts with dimeric cations of mentioned (A...A^+) and (A^+...A^+) types, because A and A^+ of these amino acids comprise negatively charged COO^- carboxylate groups and only doubly charged A^{2+} cation comprises neutral COOH group. For these amino acids (A...A^{2+}), (A^+...A^{2+}) and (A^{2+}...A^{2+}) types dimeric cations are possible *a priori*. However, only salts with (A^+...A^{2+}) type dimeric cations are known to date for these amino acids. Salts with this type of cation will be considered in Sect. 5.2.3.

5.2.1 Salts with (A...A^+) Type Dimeric Cations

The chronology of discovering and investigation of salts with (A...A^+) type dimeric cations can be divided into four stages.

The first stage started from the year of discovering of the first amino acid asparagine (1806) until 1912. During the nineteenth century almost all amino acids (except threonine) were isolated and salts of glycine with 2Gly·HCl and 2Gly·HNO_3 were obtained. In 1912 it was shown that the crystal "glycine picrate" previously obtained from the solution containing glycine and picric acid in 1:1 molar ratio actually has 2:1 composition. This stage can be called as the stage of obtaining the first salts of amino acids with 2:1 composition, because at that time it was early to talk about (Gly...Gly^+) dimeric cation. For this the concept of hydrogen bond and the technique of X-ray diffraction analysis would have to be developed. Interestingly, that year according to L. Pauling was the year of appearance of hydrogen bond concept and at the same time was the year of discovery of X-ray diffraction.

The second stage can be called the period until 1956. Even though by that time, both the concept of hydrogen bond and the technique of X-ray diffraction have been developed, it took more than 40 years for them to get applied to crystals of glycine salts with 2:1 composition. The most important event in this period should be considered the work of Walter Frost (1942). Frost has obtained the previously known crystal 2Gly·HCl, but also the 2Gly·HBr and 2Gly·HI, and also tried to obtain the 2Gly·HF. Moreover, Frost supposed that similar compounds with 2:1 composition may exist also for other amino acids and set a goal of synthesizing them, particularly for alanine. It is worth noting that the crystals of 2Gly·HBr and 2Gly·HCl obtained by Frost were used for determination of their structures.

The third stage includes the period since 1956 till 1990. Since 1956 a number of important papers concerning salts of amino acids with dimeric cation were published. The structures of 2Gly·HBr (1956) and 2Gly·HCl (1957) crystals were determined, where the existence of dimeric Gly...Gly$^+$ cations was established. The ferroelectricity in triglycine sulfate and triglycine selenate (1956) was discovered, as well as in triglycine tetrafluoroberyllate (1957), which also comprise dimeric Gly...Gly$^+$ cations, while in 1958 the ferroelectricity in 2Gly·HNO_3 was discovered. These discoveries attracted significant interest to such crystals, however, before 1990 search for and revealing of new salts of amino acids with dimeric cation had unsystematic character and sometimes was accidental. Nevertheless, during this period seven crystals were discovered of six new amino acids, and a new anion. In 1963 Rolski et al. established the existence of 2L-Leu·HCl. In 1967 based on IR spectrum it was concluded about existence of (L-Leu...L-Leu$^+$) dimeric cation, which was confirmed in 1972 by crystal structure determination. The importance of this discovery is that previously salts with 2:1 composition with dimeric cation were known only for glycine. In 1966 Takiagawa et al. obtained L-tryptophanium chloride and bromide by dissolving L-tryptophan in concentrated hydrochloric and hydrobromic acids respectively, while in case of dilute hydrobromic acid they obtained a new phase with 2L-Trp·HBr·0.5H_2O composition according to elemental analysis. The structure of this crystal remained undetermined, however, the IR spectrum of the crystals obtained by us allows to suppose that the structure contains (L-Trp...L-Trp$^+$) dimeric cation. In addition, we found based on the IR spectra that there are anhydrous 2L-Trp·HCl and 2L-Trp·HBr with (L-Trp...L-Trp$^+$) dimeric cation. The first structurally characterized salt with (L-Trp...L-Trp$^+$) dimeric cation and a new anion was (L-Trp...L-Trp$^+$)·$HgCl_3^-$ (1981). In 1978 Bhattacharyya and Saha published (preliminary report was in 1974) a paper on obtaining a new salt with new (Sar...Sar$^+$) dimeric cation, which was the beginning of significant contribution of Indian researchers in the area of discovery and study of such crystals. One more salt with dimeric cation of a new amino acid is 2DL-alloThr·HBr (1975) in spite of that the authors consider it as DL-allothreonine hydrobromide. The first salt, which comprises dimeric cation of proline, was 2DL-Pro·HCl (1981). At last in 1984 the salt (L-Phe...L-Phe$^+$)·NO_3^-

was found when the authors tried to obtain L-Phe·HNO_3 out of solution with equimolar quantities of components.

The last fourth stage began in 1990. Chen and Mak based on their previous work on bis(pyridine betaine) hydrochloride monohydrate assumed the existence of the similar salt of betaine and indeed found the crystal of 2Bet·HCl·H_2O with (Bet...Bet^+) dimeric cation (1990). Even though accidental findings were encountered also in this stage, this stage can be called the stage of systematical search and investigation of salts of amino acids with dimeric cations. If during the previous three stages (during 184 years) 12 such salts with 7 amino acids and 6 different anions were found, then during 20 odd years the number of amino acids with dimeric cations reached 16, the number of anions reached 25, and the number of salts with dimeric cations and 2:1 composition reached almost 90(89). Below we will consider all known such salts. They are listed in Tables 5.28, 5.29, 5.30, 5.31, 5.32, 5.33, 5.34, 5.35, 5.36, 5.37, and 5.38. The known salts are distributed by amino acids and anions very nonuniformly. Out of 25 known anions 7 anions (chloride, bromide, iodide, nitrate, tetrafluoroborate, perchlorate, and picrate) form 66 salts (*ca.* 74 %). Out of 16 amino acids, which form salts with dimeric cation, 7 amino acids (glycine, sarcosine, betaine, α-alanine, β-alanine, phenylalanine, cysteine) form 66 salts (*ca.* 74 %). At the same time 15 anions out of 25 form only each one salt.

5.2.1.1 Salts of Glycine with (A...AH) Type Dimeric Cations

Salts of glycine are shown in Table 5.28 together with symmetry, crystallographic data, the values of O...O distance in dimeric cations, and respective references.

Frost (1942) prepared previously known 2Gly·HCl salt and also two new similar salts: 2Gly·HBr and 2Gly·HI. The crystals of 2Gly·HBr (Buerger et al. 1956) and 2Gly·HCl (Hahn and Buerger 1957) were the first crystals in which the existence of Gly...Gly^+ dimeric cations have been established. Piret et al. (1972) determined also the structure of 2Gly·HI and showed that in this case also there are Gly...Gly^+ dimeric cations. Later the structures of these crystals were reinvestigated (Natarajan et al. 1992; Natarajan and Zangrando 1992; Ghazaryan et al. 2010a; Faamau and Tiekink 1993) and results of previous works were improved. The value of O...O distance (2.562(3) Å) in case of 2Gly·HBr (Ghazaryan et al. 2014a) is the longest among O...O distances of salts with compositions 2:1. All three these crystals have non-centrosymmetric structure, but only 2Gly·HCl and 2Gly·HBr were considered as possible nonlinear optical crystals (Moolya and Darmaprakash 2006; Ambujam et al. 2006, 2007; Moolya and Dharmapraksh 2007; Surekha et al. 2014a, b). Detailed vibrational spectroscopic investigation of 2Gly·HCl was performed by Baran et al. (2012). In 1958 the ferroelectricity was discovered in 2Gly·HNO_3 (Pepinsky et al. 1958). At room temperature the crystal has centrosymmetric space group $P2_1/a$, while below order–disorder type phase transition ($T_c = 206$ K) the symmetry is Pa. As one may expect, later the crystal became the subject of numerical studies (see e.g. Hoshino et al. 1963; Sato 1966, 1968; Easwaran 1966;

Table 5.28 Salts of *glycine* with (A…AH) type dimeric cations

#	Crystal	s.g., Z	a, b, c (Å), α, β, γ (°)	O…O, (Å)	Refs
1	(Gly…GlyH)Cl	$P2_12_12_1$, 4	5.2992(6), 8.083(2), 17.973(2)	2.552(5)	Natarajan et al. (1992)
2	(Gly…GlyH)Br	$P2_12_12_1$, 4	5.3895(6), 8.212(1), 18.424(2)	2.562(3)	Ghazaryan et al. (2014a)
3^{i}	(Gly…GlyH)I	$C2$, 8	18.589(4), 5.161(1), 20.389(4)	2.451(4)	Ghazaryan et al. (2010a)
3^{ii}	(Gly…GlyH)I		$\beta = 110.67(3)$	2.494(4)	
4	(Gly…GlyH)NO_3	$P2_1/a$, 2	9.460(1), 5.1728(5), 9.224(1), $\beta = 97.217(15)$	2.454(1)	Łukaszewicz et al. (1996a, b)
5	(Gly…GlyH)$HSeO_3$	$P2_1/c$, 4	12.2651(7), 4.8079(6), 19.955(1), $\beta = 122.745(4)$	2.556(6)	Němec et al. (1998)
6^{i}	(Gly…GlyH)BF_4	$P\bar{1}$, 6	8.6069(2), 11.8626(3), 14.7597(4),	2.453(1)	Ghazaryan et al. (2010a)
6^{ii}	(Gly…GlyH)BF_4		$\alpha = 104.451(1)$, $\beta = 106.417(1)$, $\gamma = 94.918(1)$	2.462(1)	
6^{iii}	(Gly…GlyH)BF_4			2.465(1)	
7^{i}	(Gly…GlyH)ClO_4	$P\bar{1}$, 10	8.700(2), 14.842(2), 19.396(2),	2.457(2)	Ghazaryan et al. (2010a)
7^{ii}	(Gly…GlyH)ClO_4		$\alpha = 98.748(7)$, $\beta = 93.056(9)$, $\gamma = 106.053(9)$	2.464(2)	
7^{iii}	(Gly…GlyH)ClO_4			2.464(2)	
7^{iv}	(Gly…GlyH)ClO_4			2.466(2)	
7^{v}	(Gly…GlyH)ClO_4			2.474(2)	
8a	(Gly…GlyH)ReO_4	$P2_1/c$, 4	15.7095(5), 8.1826(3), 8.2909(3), $\beta = 103.715(2)$	2.445(2)	Rodrigues et al. (2009)
$8b^{i}$	(Gly…GlyH)ReO_4	$P1$, 2	5.2688(2), 9.4982(3), 10.5374(3),	2.450(5)	Dekola et al. (2009)
$8b^{ii}$	(Gly…GlyH)ReO_4		$\alpha = 79.458(2)$, $\beta = 78.811(2)$, $\gamma = 76.383(2)$	2.456(6)	
9	(Gly…GlyH) hydrogen fumarate·H_2O	$P2_1/n$, 4	13.058(1), 6.8251(7), 15.326(1), $\beta = 112.65(2)$	2.487(2)	Natarajan et al. (2009b)
10	(Gly…GlyH) picrate	$P2_1/c$, 4	15.184(3), 6.7276(13), 14.987(3), $\beta = 93.64(3)$	2.480(1)	Ghazaryan et al. (2011b)

Table 5.29 Salts of sarcosine with (A...AH) type dimeric cations

#	Crystal	s.g., Z	*a, b, c* (Å), *α, β, γ* (°)	O...O, (Å)	Refs
1^{i}	(Sar...SarH)Cl	$Pca2_1$, 8	23.5139(8), 6.0609(2), 15.2695(6)	2.463(2)	Ghazaryan et al. (2012a)
1^{ii}	(Sar...SarH)Cl			2.469(2)	
2^{i}	(Sar...SarH)Br	$Pca2_1$, 8	23.771(1), 6.1868(4), 15.405(1)	2.453(3)	Ghazaryan et al. (2012a)
2^{ii}	(Sar...SarH)Br			2.464(3)	
3	(Sar...SarH)I	$P2_1/c$, 4	9.5168(8), 6.2610(6), 20.301(1), $\beta = 100.929(5)$	2.496(4)	Ghazaryan et al. (2013b)
4	(Sar...SarH)NO_3	$P\bar{1}$, 2	5.320(1), 6.578(1), 15.957(2), $\alpha = 95.448(6)$, $\beta = 95.944(6)$, $\gamma = 93.410(6)$	2.524(1)	Fleck et al. (2012a)
5	(Sar...SarH)BF_4	$P2_12_12_1$, 4	6.0083(2), 9.7640(3), 20.5822(7)	2.506(3)	Ghazaryan et al. (2012b)
6	(Sar...SarH)ClO_4	$P2_12_12_1$, 4	6.0410(2), 9.8005(4), 20.7882(7)	2.527(4)	Ghazaryan et al. (2012b)
7	(Sar...SarH)CH_3SO_3	$C2/c$, 8	30.448(3), 6.0453(7), 16.120(2), $\beta = 121.863(19)$	2.457(2)	Chwaleba et al. 2006)
8	(Sar...SarH)CF_3COO	–	–	–	Ghazaryan et al. (in prep.)
9	(Sar...SarH) hydrogen L-tartrate	$P2_12_12_1$, 4	6.0518(5), 13.8637(11), 17.4678(12)	2. 512(3)	Fleck et al. (2013a)
10	(Sar...SarH)$_2SiF_6$	$P2_1/c$, 2	11.485(2)Å, 10.606(1)Å, 9.768(1)Å, $\beta = 114.63(1)°$	2.506(2)	Fleck et al. (2012b)

Table 5.30 Salts of betaine with (A. . .AH) type dimeric cations

#	Crystal	s.g., Z	*a*,*b*,*c* (Å), α, β, γ (°)	O. . .O, (Å)	Refs
1	(Bet. . .BetH)Cl·H_2O	*Pnma*, 4	11.904(1), 22.454(5), 5.624(1)	2.454(4)	Chen and Mak (1990)
2	(Bet. . .BetH)Br	–	–	<2.454[a]	Godzisz et al. (2002)
3	(Bet. . .BetH)I	–	–	<2.454[a]	Godzisz et al. (2002)
4	(Bet. . .BetH)NO_3	$P2_1/c$, 4	5.668(1), 11.259(2), 23.073(5), $\beta=93.65(3)$	2.449(2)	Baran et al. (1995c)
5	(Bet. . .BetH)BF_4	$C2/c$, 4	24.1762(9), 5.7806(2), 11.8445(4), $\beta=110.541(1)$	2.454(1)	Ghazaryan et al. (2011a)
6	(Bet. . .BetH)ClO_4	–	–	2.440(2)	Godzisz et al. (2002)
6	(Bet. . .BetH)ClO_4	$P2_1$, 2	5.8308(11), 23.086(4), 6.2130(11), $\beta=107.624(6)$	2.455(4)	This work
7	(Bet. . .BetH)HSO_4	*Pbca*, 8	11.442(2), 22.710(5), 11.941(2)	2.532(3)	Ratajczak et al. (1994)
8	(Bet. . .BetH)$HSeO_4$	*Pbca*, 8	11.591(2), 22.930(5), 12.045(2)	2.551(9)	Baran et al. (1997b)
9	(Bet. . .BetH)$_2SeO_4$	*Fddd*, 8	13.243(8), 19.077(9), 21.96(2)	2.424(6)	Baran et al. (1995d)
10^{i}	(Bet. . .BetH)NH_2SO_3	$P2_1/c$, 4	13.503(1), 8.435(1), 14.839(1),	2.442(2)[b]	Ilczyszyn et al. (2000)
10^{ii}	(Bet. . .BetH)NH_2SO_3		$\beta=112.24(1)$	2.446(2)[b]	
11	(Bet. . .BetH)HC_2O_4	$P2_1/c$, 4	11.7729(7), 5.5841(4), 24.549(3), $\beta=97.520$	2.457(3)	Rodrigues et al. (2001c)
12	(Bet. . .BetH)$_2SiF_6$	*Fddd*, 8	13.237(1), 19.311(1), 22.057(2)	2.432(3)	Ghazaryan et al. (2013a)

[a]Estimated from asymmetric stretching vibration of the hydrogen-bonded O-H group
[b]At 250 K

Table 5.31 Salts of β-alanine with (A. . .AH) type dimeric cations

#	Crystal	s.g., Z	a, b, c (Å), α, β, γ (°)	O. . .O, (Å)	Refs
1	(β-Ala. . .β-AlaH)Cl	$C2/c$, 4	19.938(4), 5.199(1), 10.719(2), $\beta = 118.03(3)$	2.4729(17)	Godzisz et al. (2003a)
2	(β-Ala. . .β-AlaH)Br	$C2/c$, 4	20.3113(4), 5.2250(1), 10.8983(2), $\beta = 117.971(1)$	2.472(2)	Ghazaryan et al. (2011a)
3	(β-Ala. . .β-AlaH)I	–	–	–	This work
4	(β-Ala. . .β-AlaH)NO_3	$C2/c$, 4	19.791(1), 5.3220(3), 10.974(1), $\beta = 113.923(6)$	2.467(2)	Sridhar et al. (2001c)
5	(β-Ala. . .β-AlaH)ClO_4	–	–	2.457	Ilczyszyn et al. (2006)
6	(β-Ala. . .β-AlaH)BF_4	$C2/c$, 4	21.2844(8), 5.2987(2), 11.1557(4), $\beta = 116.651(3)$	2.460(2)	Ghazaryan et al. (2011a)
7	(β-Ala. . .β-AlaH) L-tartrate	–	–	–	This work
8^{i}	(β-Ala. . .β-AlaH) picrate	$P\bar{1}$, 2	4.9515(4), 11.725(2), 14.931(2),	2.472(4)	Anitha et al. (2004c)
8^{ii}	(β-Ala. . .β-AlaH) picrate		$\alpha = 78.00(1)$, $\beta = 83.75(1)$, $\gamma = 82.45(1)$	2.477(4)	
9	(β-Ala. . .β-AlaH)$_2SiF_6$	–	–	–	This work

Table 5.32 Salts of α-alanine with (A...AH) type dimeric cations

#	Crystal	s.g., Z	a, b, c (Å), α, β, γ (°)	O...O, (Å)	Refs
1[i]	(L-Ala...L-AlaH)Cl	$P2_1$, 4	5.084(1), 19.371(4), 10.958(2),	2.414(5)	Ghazaryan et al. (2011a)
1[ii]	(L-Ala...L-AlaH)Cl		$\beta = 90.217(4)$	2.437(5)	
2[i]	(L-Ala...L-AlaH)Br	$P2_1$, 4	5.1853(10), 18.950(4), 11.281(2),	2.381(12)	Ghazaryan et al. (2011a)
2[ii]	(L-Ala...L-AlaH)Br		$\beta = 90.20(3)$	2.451(10)	
3	(D-Ala...D-AlaH)Br	$P2_1$, 2	9.485(4), 5.1678(6), 11.258(2), $\beta = 92.97(3)$	2.431(8)	Fischer (2006)
4[i]	(L-Ala...L-AlaH)I_3	$P2_12_12_1$, 8	8.366(2), 8.912(2), 41.889(2)	2.475(3)	Fleck et al. (2008)
4[ii]	(L-Ala...L-AlaH)I_3			2.505(3)	
5	(L-Ala...L-AlaH)NO_3	$P2_1$, 2	7.8578(5), 5.4516(6), 12.8276(7), $\beta = 94.73(4)$	2.455(2)	Ramos Silva et al. (2001)
6	(L-Ala...L-AlaH)·H_2PO_3·H_2O	$P2_1$, 2	8.592(3), 6.475(3), 12.004(3), $\beta = 104.8(5)$	2.530(2)	Lemanov et al. (2002)
7	(L-Ala...L-AlaH) picrate·H_2O	$P2_1$, 2	8.2678(17), 7.5097(15), 15.540(3), $\beta = 105.23(3)$	2.553(2)	Ghazaryan et al. (2012d)
8	(L-Ala...L-AlaH) picrate	–	–	–	Ghazaryan et al. (2011a)

Table 5.33 Salts of phenylalanine with (A. . .AH) type dimeric cations

#	Crystal	s.g., Z	*a*, *b*, *c* (Å), *α*, *β*, *γ* (°)	O. . .O, (Å)	Refs
1	(L-Phe. . .L-PheH)Cl	–	–	–	Ghazaryan et al. (2011a)
2[i]	(L-Phe. . .L-PheH)Br	$P2_1$, 4	5.3041(3), 31.218(2), 11.3346(8), β = 90.105(2)	2.444(5)	Ghazaryan et al. (2011a)
2[ii]	(L-Phe. . .L-PheH)Br			2.436(6)	
3	(L-Phe. . .L-PheH)NO_3	$P2_1$, 2	12.536(16), 5.378(4), 14.962(9), β = 107.83(8)	2.426(8)	Srinivasan et al. (2001e)
4	(L-Phe. . .L-PheH)BF_4	$P2_12_12_1$, 4	5.3668(11,) 12.556(3), 29.373(6)	2.444(5)	Ghazaryan et al. (2011a)
5	(L-Phe. . .L-PheH)ClO_4	$P2_12_12_1$, 4	5.3907(11), 12.680(2), 29.152(5)	2.440(4)	Srinivasan and Rajaram (1997b)
6[i]	(L-Phe. . .L-PheH)$HSeO_3$	$P1$, 4	11.0507(4), 12.8611(5), 15.8820(6),	2.420(2)	Ghazaryan et al. (2011a)
6[ii]	(L-Phe. . .L-PheH)$HSeO_3$		α = 107.867(2),	2.429(2)	
6[iii]	(L-Phe. . .L-PheH)$HSeO_3$		β = 110.392(2),	2.439(2)	
6[iiii]	(L-Phe. . .L-PheH)$HSeO_3$		γ = 90.017(2)	2.476(2)	
7	(L-Phe. . .L-PheH)HCOO	$P2_1$, 2	11.507(6), 5.638(3), 14.610(5), β = 100.65(4)	2.496(3)	Görbitz and Etter (1992)
8	(L-Phe. . .L-PheH)HC_2O_4	$P1$, 2	5.555(1), 13.530(3), 14.646(3), α = 71.98(3), β = 88.04(3), γ = 86.75(3)	-	Ghazaryan et al. (2011a)
9	(L-Phe. . .L-PheH) hydrogen malonate	$P2_1$, 2	14.031(3), 5.5082(10), 14.601(3), β = 107.412(3)	2.429(4)	Alagar et al. (2005)
10	(DL-Phe. . .DL-PheH) picrate	$P\bar{1}$, 2	7.333(3), 13.737(3), 15.381(6), α = 113.04(2), β = 94.93(4), γ = 105.23(3)	2.473(5)	Anitha and Rajaram (2005)

Table 5.34 Salts of proline with (A...AH) type dimeric cations

#	Crystal	s.g., Z	a, b, c (Å), α, β, γ (°)	O...O, (Å)	Refs
1	(DL-Pro...DL-ProH)Cl	$C2/c$, 4	19.589(12), 6.571(6), 10.563 (20), $\beta=105.38(4)$	2.438(1)	Swaminathan and Chacko (1981)
2	(L-Pro...L-ProH) Cl	$P4_12_12$, 8	$a=b=12.8188(4)$, $c=16.0876(7)$	2.471(2)	Ghazaryan et al. (2011a)
3	(L-Pro...L-ProH) Br	$P4_12_12$, 8	$a=b=12.8482(6)$, $c=16.4126(7)$	2.452(2)	Ghazaryan et al. (2011a)
4	(L-Pro...L-ProH) NO_3	$P2_12_12_1$, 4	7.2006(6), 7.711(1), 24.060(3)	2.414(3)	Pandiarajan et al. (2002b)
5	(L-Pro...L-ProH)ClO_4	$P2_1$, 2	7.848(6), 7.3951(14), 12.834 (5), $\beta=90.95(4)$	2.474(6)	Pandiarajan et al. (2002c)
6	(L-Pro...L-ProH) BF_4	$P2_1$, 2	7.7860(9), 7.3654(4), 12.6730 (9), $\beta=90.088(1)$	2.483(3)	Ghazaryan et al. (2011a)

Table 5.35 Salts of leucine with (A...A⁺) type dimeric cations

#	Crystal	s.g., Z	a, b, c (Å), α, β, γ (°)	O...O, (Å)	Refs
1	(L-Leu...L-LeuH)Cl	$P2_1$, 2	11.152(1), 5.116(1), 15.405 (2), $\beta=108.94(2)$	2.429(7)	Golič and Hamilton (1972)
2	(L-Leu...L-LeuH)Br	$P2_1$, 2	11.529(1), 5.247(1), 15.105 (1) $\beta=108.032(4)$	2.454(5)	Ghazaryan et al. (2011a)
3	(L-Leu...L-LeuH)BF_4	–	–	–	This work
4	(L-Leu...L-LeuH)ClO_4	–	–	–	Ghazaryan et al. (2011a)
5[i]	(L-Leu...L-LeuH) picrate	$P1$, 2	7.1470(5), 11.8540(8), 15.456(1),	2.454(5)	Anitha et al. (2005b)
5[ii]	(L-Leu...L-LeuH) picrate		$\alpha=106.45(2)$, $\beta=95.17$ (1), $\gamma=91.02(2)$	2.471(5)	

Table 5.36 Salts of leucine with (A...A⁺) type dimeric cations

#	Crystal	s.g., Z	a, b, c (Å), α, β, γ (°)	O...O, (Å)	Refs
1	(L-Val...L-ValH) Cl	$P2_1$, 2	11.144(2), 5.179(9), 13.028(2), $\beta=111.85(3)$	2.460(4)	Moitra et al. (2010)
2	(L-Val...L-ValH) Br	$P2_1$, 2	11.480(1), 5.222(1), 12.062 (1) $\beta=96.776(4)$	2.479(8)	Ghazaryan et al. (2011a)
3	(L-Val...L-ValH) $ClO_4{\cdot}H_2O$	$P2_1$, 2	11.235(3), 5.8312(14), 13.0948 (19), $\beta=91.282$ (16)	2.562(4)	Pandiarajan et al. (2001a)

Table 5.37 Salts of methionine with (A...AH) type dimeric cations

#	Crystal	s.g., Z	a, b, c (Å), α, β, γ (°)	O...O, (Å)	Refs
1	(L-Met...L-MetH) $ClO_4{\cdot}H_2O$	$P2_1$, 2	11.3720(15), 5.5876(8), 15.323(2), $\beta=107.782$ (5), at 105 K	2.500(1)	Sridhar et al. (2002e)
2	(L-Met...L-MetH) $BF_4{\cdot}H_2O$	–	–	–	This work
3	(L-Met...L-MetH) hydrogen maleate	$P2_1$, 2	12.981(8), 5.326(3), 15.124 (9), $\beta=114.09(12)$	2.437(3)	Natarajan et al. (2008c)
4	(DL-Met...DL-MetH) picrate	$P\overline{1}$, 2	7.076(1), 12.102(1), 14.895 (2), $\alpha=103.90(1)$, $\beta=97.04(2)$, $\gamma=101.73(2)$	2.457(5)	Anitha et al. (2006b)

Table 5.38 Salts of other amino acids with (A...AH) type dimeric cations

#	Crystal	s.g., Z	*a*, *b*, *c* (Å), *α*, *β*, *γ* (°)	O...O, (Å)	Refs
1	(D-Ser...D-SerH)NO_3	$P2_1$, 2	5.929(1), 8.195(7), 11.503(2), $\beta = 100.06(1)$	2.462(3)	Ravikumar et al. (2001b)
2	(DL-alloThr...DL-alloThrH)Br	$P2_1/c$, 4	8.91(1), 7.79(1), 18.91(1), $\beta = 92.0(2)$	2.442(2)	Swaminathan and Srinivasan (1975)
3	(L-Thr...L-ThrH)Cl	–	–	–	Ghazaryan et al. (2011a)
4	(L-Cys ...L-CysH)Cl	–	–	–	This work
5	(L-Cys ...L-CysH)Br	–	–	–	This work
6	(L-Cys ...L-CysH)I·nH_2O				This work
7	(L-Cys ...L-CysH)NO_3·nH_2O	–		–	This work
8	(L-Cys ...L-CysH)BF_4	–		–	This work
9	(L-Cys ...L-CysH)ClO_4	–		–	This work
10	(L-Cys ...L-CysH) picrate	–	–	–	This work
11	(L-Asp...L-AspH)NO_3	$P2_12_12_1$, 4	5.5840(7), 11.491(3), 21.043(5)	2.512(5)	Sridhar et al. (2002e)
12	(L-Trp...L-TrpH)Cl	–	–	–	This work
13	(L-Trp...L-TrpH)Br	–	–	–	This work
14	(L-Trp...L-TrpH)Br·0.5H_2O	$P2_12_12_1$, 4	31.30, 13.86, 5.73	–	Takiagawa et al. (1966)
15	(L-Trp...L-TrpH)$HgCl_3$	$P2_1$, 2	16.688(5), 5.626(1), 14.022(5), $\beta = 108.17(3)$	2.41(2)	Book et al. (1981)

Saraswati and Vijayaghavan 1967; Blinc et al. 1970; Gesi and Ozawa 1973; Strukov et al. 1986; Łukaszewicz et al. 1996a, b; Andriyevsky et al. 2001, 2002, 2005; Andriyevsky and Czapla 2004). Results of detailed vibrational spectroscopic investigation of 2Gly·HNO_3 are reported by Baran et al. (1994, 1995b). Němec et al. (1998) discovered a new 2Gly·H_2SeO_3 crystal in the system Glycine + H_2SeO_3 + H_2O in addition to previously known Gly·H_2SeO_3 crystal (see Natarajan et al. 1992, in Sect. 5.1.24). This was the first salts of $HSeO_3^-$ anion with dimeric cation. The crystal was characterized structurally and by vibrational spectra. No phase transition was found in the temperature range 90–380 K. The systems Glycine + HBF_4 + H_2O and Glycine + $HClO_4$ + H_2O were studied by Ghazaryan et al. (2010a). The crystals of 2Gly·HBF_4 and 2Gly·$HClO_4$ as well as respective Gly·HBF_4 and Gly·$HClO_4$ were obtained. Crystals with 1:1 composition were very hygroscopic and were characterized only by their IR and Raman spectra, while crystals with 2:1 composition were characterized also structurally. The crystals of 2Gly·HBF_4 and 2Gly·$HClO_4$ comprising (Gly...Gly$^+$) dimeric cation and the same space group are not isostructural because of different numbers of formula units. The crystal of 2Gly·HBF_4 is the first salt with BF_4^- anion. The crystal of 2Gly·$HClO_4$ was obtained and studied independently of the work of Ghazaryan et al. (2010a) also by Panicker et al. (2011) and Panicker and Sakuntala (2010, 2011). Two polymorphs of 2Gly·$HReO_4$ (a monoclinic, Rodrigues et al. 2009, and a triclinic one, Dekola et al. 2009) were obtained by slow evaporation of aqueous solution containing stoichiometric quantities of components. This is the first (and yet the only) case of obtaining a salt with dimeric cation containing ReO_4^- anion. Crystal structures of both polymorphs have been determined. Phase transitions were found in triclinic form in result of thermal (DSC) and dielectric measurements. Peaks on the DSC curve were observed at 257 K, 267 K and 376 K, 382 K. The first and yet the only salt with dimeric cation comprising fumarate anion is glycine glycinium hydrogen fumarate monohydrate (Natarajan et al. 2009b), which was obtained from aqueous solution containing equimolar quantities of components. The crystal was characterized also by Joseph and Raja (2012). It is worth noting that similar salt with hydrogen maleate anion is not formed with glycine glycinium cation (see Petrosyan et al. 2012a) although such salt with other amino acid is known (see Table 5.37 below). The crystal of 2Gly·picric acid was known since nineteenth century as "glycine picrate" because it was obtained from solution containing equimolar quantities of components. Levene and Van Slyke (1912) showed that the real composition of this compound is 2Gly·picric acid. Guha (1970) knew of the old paper and first determined the centrosymmetric space group ($P2_1/c$, $Z = 4$) and unit cell parameters of the compound. However, detailed structure was not published. Kai et al. (1994) aimed to obtain and determine the structure of glycine picrate obtained from solution with equimolar quantities of components. However, apparently not knowing the works of Levene and Van Slyke (1912) and Guha (1970) they found that the composition of the "glycine picrate" actually does not correspond to 1:1 compound, but is glycine glycinium picrate. Later "glycine picrate" was studied as well by two groups (Uma Devi et al. 2008b; Shakir et al. 2009a). Moreover, "remarkable second harmonic generation in centrosymmetric crystal" (Shakir et al. 2009a) and ferroelectricity in centrosymmetric "glycine picrate" (Shakir et al. 2009b) were found. Ghazaryan et al. (2011b) reinvestigated the structure and vibrational spectra. Glycine glycinium picrate was obtained at both stoichiometric and equimolar molar ratios of components. Comparison of these two crystals allowed explaining that second harmonic generation signals in "glycine picrate" obtained at equimolar ratio are caused by incorporated picric acid impurity. It is also worth of note that Shakir et al. (2009a) probably after Petrosyan (2010b) published an Erratum (Shakir et al. 2010), where in addition to the correction of the name of the compound they wrote that "...the ratio 1:1 should be read as 2:1." However, it should not been done, because in fact the ratio was 1:1, because crystals obtained at 2:1 ratio do not show any second harmonic generation signals (Ghazaryan et al. 2011b). Tylczynski and Wiesner (2010) showed that the conclusion of Shakir et al. (2009a, b) on presence of ferroelectricity in "glycine picrate" was erroneous.

Considering the salts of glycine one may say that the first known salts with 2:1 composition were the salts of glycine (2Gly·HCl, 2Gly·HNO_3). The existence of dimeric cations was established structurally first on the salts of glycine (2Gly·HBr, 2Gly·HCl). The first ferroelectrics with dimeric cations (triglycine sulfate) and with 2:1 composition (2Gly·HNO_3) also were the salts of glycine. Due to investigation of salts of glycine the following 9 anions in the salts with dimeric cations were found: Cl^-, Br^-, I^-, NO_3^-, $HSeO_3^-$, ReO_4^-, BF_4^-, picrate, and hydrogen fumarate.

5.2.1.2 Salts of Sarcosine with (A...AH) Type Dimeric Cations

Known salts of sarcosine with (A...AH) type dimeric cations are shown in Table 5.29. The first salt of sarcosine 2Sar·HBr was discovered by Bhattacharyya and Saha (1978b). The second salt was found by Chwaleba et al. (2006). This is the first and yet the only salt with methanesulfonate anion. The crystal has been studied structurally as well as by vibrational and high-resolution solid-state nuclear magnetic resonance spectroscopy (Chwaleba et al. 2006; Ilczyszyn et al. 2008b). The presence of short O-H...O hydrogen bond of dimeric cation is reflected in the infrared spectrum by very broad absorption band in the region 2,000–500 cm^{-1} centered at *ca.* 824 cm^{-1}. Later the number of salts of sarcosine with dimeric cations was increased significantly. Ghazaryan et al. (2012a) determined the structure of 2Sar·HCl and of 2Sar·HBr was redetermined. Vibrational spectra of 2Sar·HCl and 2Sar·HBr were studied and compared with spectra of Sar·HCl and Sar·HBr. The presence of dimeric cations in the structures of 2Sar·HCl and 2Sar·HBr is reflected in the infrared spectra. These crystals are isotypic and have polar symmetry (point symmetry mm2). Ghazaryan et al. (2013b) obtained two crystals Sar·HI and 2Sar·HI and studied them structurally and by vibrational spectroscopy. The crystal of Sar·HI is isotypic with respective bromide and chloride, while the crystal of 2Sar.HI in spite of the presence of dimeric cation ((Sar...Sar^+)·I^-) is not isostructural with respective bromide and chloride and has centrosymmetric structure at room temperature.

Fleck et al. (2012a) obtained crystals of Sar·HNO_3 and 2Sar·HNO_3 with dimeric cation. Structure determination of 2Sar·HNO_3 at room temperature revealed triclinic centrosymmetric structure. Structure determination at 150 K showed that phase transitions do not occur at least in this range of temperature and the centrosymmetric structure remains unchanged. Thermal expansion of the crystal of 2Sar·HNO_3 in the 200–290 K range as well as vibrational spectra of Sar·HNO_3 and 2Sar·HNO_3 were studied. The presence of dimeric cation in the structure of 2Sar·HNO_3 is reflected in the infrared spectrum.

The systems Sar + HBF_4 + H_2O and Sar + $HClO_4$ + H_2O were studied (Ghazaryan et al. 2012b). Both expected compounds Sar·HBF_4(Sar·$HClO_4$) and 2Sar·HBF_4(2Sar·$HClO_4$) were found in these systems. Crystals with 1:1 composition are very hygroscopic. Probably this circumstance was the reason why these systems were not studied before. The structures of 2Sar·HBF_4 and 2Sar·$HClO_4$, (which are not as hygroscopic as 1:1 compounds) were determined and turned out isostructural and have non-centrosymmetric orthorhombic symmetry. IR and Raman spectra of 2Sar·HBF_4 and 2Sar·$HClO_4$ are studied and compared with respective spectra of Sar·HBF_4 and Sar·$HClO_4$.

The crystal of Sar·CF_3COOH was obtained by evaporation from aqueous solution containing sarcosine and excess amount of CF_3COOH (see Rodrigues et al. 2000). It was interesting to know is it possible to obtain 2Sar·CF_3COOH from this system. We found that at the ratios 2:1 to 1:1 of sarcosine and CF_3COOH the known crystal Sar·CF_3COOH is formed, while in the interval 2.5:1 to 2:1 a new phase is formed. The IR spectra of Sar·CF_3COOH and that of the new phase are shown in Fig. 5.58. We believe that the new phase has 2Sar·CF_3COOH composition with dimeric cation. Unfortunately the quality of the single crystals was not good enough to determine its structure by XRD, because the final conclusion may be done on the basis of the results of XRD.

The first compound from the system sarcosine + L-tartaric acid + H_2O was sarcosinium hydrogen L-tartrate (I) obtained by Krishnakumar et al. (2001). As in previous case it was interesting to check, is it possible to obtain a salt with 2:1 composition in this system. Our studies showed that in addition to known compound (I) at least three phases exist in this system. Two of them (II and III) were obtained in form of single crystals and were characterized structurally and by vibrational spectroscopy.

From the solution with 2:1 molar ratio a new 2Sar · L-tartaric acid single crystal (phase (II)) is formed. At higher content of sarcosine compared to 2:1 molar ratio (2.5:1, 3:1, 3.2:1) one more new 2Sar · L-tartaric acid single crystal (phase (III)) is formed. Both phases have the same 2Sar · L-tartaric acid composition and the same orthorhombic symmetry (the same space group $P2_12_12_1$ and $Z = 4$). The phase (II) has not the structure (Sar $\cdots$ Sar^+) · L-$tartrate^-$ with dimeric cation (see Table 5.2), while the phase (III) has (see Table 5.29). The hydrogen L-tartrate anions form infinite chains due to O-H $\cdots$ O(2.510(2) Å) hydrogen bond.

The last salt in Table 5.29 is 2(Sar...Sar^+)·SiF_6^{2-}, which is the second salt of the 2(A...A^+)·SiF_6^{2-} type (the first salt was 2(Bet...Bet^+)·SiF_6^{2-}, see Table 5.30).

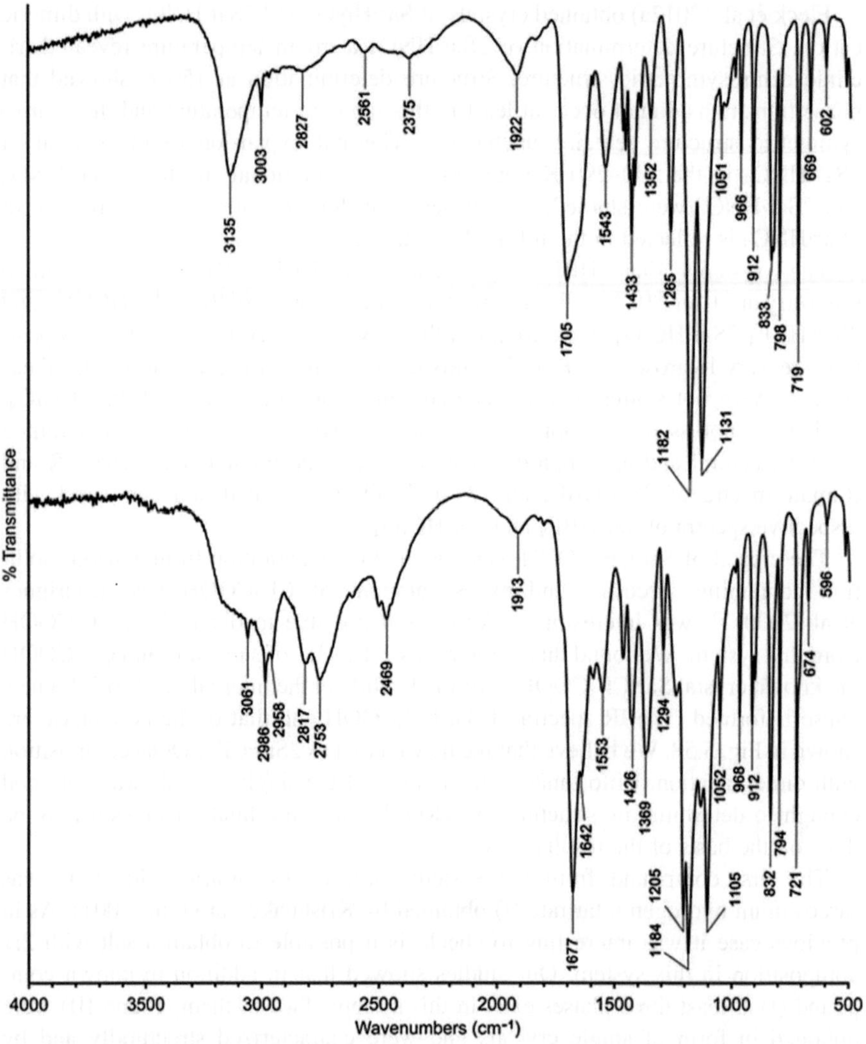

Fig. 5.58 IR spectra of Sar·CF_3COOH (*top*) as well as the new phase with the expected composition of 2 Sar·CF_3COOH (*bottom*)

It may be presented as (Sar…Sar$^+$)·0.5SiF_6^{2-}, because Sar and Sar$^+$ are in general position, while the hexafluorosilicate anion is in special position. Vibrational spectra of 2(Sar…Sar$^+$)·SiF_6^{2-} as well as thermal expansion parameters were studied too (Fleck et al. 2012b).

Due to investigation of salts of sarcosine the following three anions in the salts with dimeric cations were found: $CH_3SO_3^-$, CF_3COO^-, hydrogen L-tartrate.

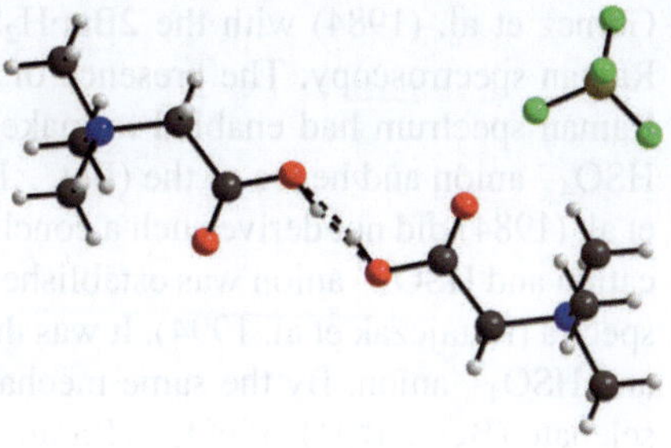

Fig. 5.59 Molecular structure of (Bet…Bet$^+$)·BF_4^-. The O2B-H2B…O1C hydrogen bond is indicated by a dashed line

5.2.1.3 Salts of Betaine with (A…AH) Type Dimeric Cations

Known salts of betaine with (A…AH) type dimeric cations are shown in Table 5.30. The first salt of betaine 2Bet·HCl·H_2O was obtained in 1990 by Chen and Mak. Nevertheless, now the number of known salts of betaine with dimeric cation is more than that of glycine and sarcosine. Respective bromide and iodide salts were not characterized structurally, but only by their infrared spectra (Godzisz et al. 2002).

Betaine betainium nitrate was obtained and investigated by Baran et al. (1995c). The structure was determined at room temperature as well as IR and Raman spectra of normal and deuterated samples were studied. In addition, for checking the existence of possible phase transitions IR spectra in wide temperature interval (13–298 K) were studied as well as by DSC method in the 100–470 K interval. First order phase transitions at 394 K and 439 K were observed as well as a weak probably second-order phase transition at 448 K was found. Melting with decomposition takes place at 460 K.

The salts with perchlorate and tetrafluoroborate anions as a rule are isostructural, even though there are some exceptions. The crystal of (Bet…Bet$^+$)·ClO_4^- was obtained and investigated by Godzisz et al. (2002) and the distance of O…O hydrogen bond in dimeric cation was determined. However, detailed structural data and symmetry of this crystal are not provided in this work. We found and identified by vibrational spectra at least two forms of 2Bet·$HClO_4$ and could determine the structure of one form shown in Table 5.30. This form is not isostructural with monoclinic centrosymmetric crystal (Bet…Bet$^+$)·BF_4^-, which was investigated by Ghazaryan et al. (2011a) and is shown in Fig. 5.59.

Very interesting results were obtained by several researchers in result of investigation of the systems betaine + H_2SO_4 + H_2O (Ratajczak et al. 1994) and betaine + H_2SeO_4 + H_2O (Baran et al. 1995d, e, 1997b). The system betaine + H_2SO_4 + H_2O was studied in the group of Prof. S. Haussühl (1989). Two salts (Bet·H_2SO_4·H_2O and 2Bet·H_2SO_4) were obtained and their crystallographic data were determined. However, the detailed structures were not determined. In case of Bet·H_2SO_4·H_2O one might suppose the Bet$^+$·HSO_4^-·H_2O structure, while in case of 2Bet·H_2SO_4 two mechanisms are possible: 2Bet$^+$·SO_4^{2-} and (Bet…Bet$^+$)·HSO_4^-. Based on the existing 2Gly$^+$·SO_4^{2-}. H_2O structure (see Table 5.1) one might suppose the same mechanism also in this case. It is worth to note that Prof. S. Haussühl provided

Gómez et al. (1984) with the 2Bet·H_2SO_4 crystals for investigation by means of Raman spectroscopy. The presence of characteristic line at *ca.* 1,040 cm^{-1} in the Raman spectrum had enabled to make a conclusion in respect of the presence of HSO_4^- anion and hence of the (Bet...Bet^+)·HSO_4^- mechanism. However, Gómez et al. (1984) did not derive such a conclusion. The presence of dimeric (Bet...Bet^+) cation and HSO_4^- anion was established by structure determination and vibrational spectra (Ratajczak et al. 1994). It was the first salt (yet the only) with dimeric cation and HSO_4^- anion. By the same mechanism is formed respective salt of hydrogen selenate (Bet...Bet^+)·$HSeO_4^-$ (Baran et al. 1997b). The structure and vibrational spectra of normal and deuterated samples are studied in detail. This is also the first and yet the only salt with $HSeO_4^-$ anion. Before publication of the results on the salt (Bet...Bet^+)·$HSeO_4^-$ Baran et al. (1995d) have discovered another salt from this system, namely, 4Bet·H_2SeO_4, which was the first salt with dimeric cation of the type 2(A...A^+)·Y^{2-} and the first (yet the only) salt with SeO_4^{2-} anion. In the dimeric (Bet...Bet^+) cation the very short O...O hydrogen bond is of O...H...O type, where oxygen atoms are crystallographically equivalent. The SeO_4^{2-} anion also is in special position. All oxygen atoms are equivalent with Se-O bond length equal to 1.432(9) Å. The crystal (and its deuterated analog) was studied also by IR and Raman spectroscopy. The presence of strong hydrogen bond is reflected in IR spectrum as a very broad absorption band with maximum at *ca.* 800 cm^{-1}. Investigation of the crystal by means of DSC (in the range 100–420 K), IR (in the range 11–300 K) and dielectric (in the range 100–300 K) revealed two phase transitions. One is the first order phase transition at *ca.* 135 K and at *ca.* 150 K (second order), which, however, have not the ferroelectric nature (Baran et al. 1995e).

Ilczyszyn et al. (2000) obtained and investigated the crystal of betaine betainium sulfamate. This is the first and yet the only salt with dimeric cation containing sulfamate anion. The authors determined its crystal and molecular structure. On the same crystal structure determination initially were performed at 250 K, then at 100 K and at last again at 250 K. Also the crystal was studied by methods of vibrational spectroscopy, including temperature dependencies in wide (12–300 K) region. Later these investigations were continued in another work (Ilczyszyn and Ilczyszyn 2003) with addition of ^{13}C NMR method. The crystal has monoclinic symmetry (space group $P2_1/c$, $Z = 4$). At this space group and $Z = 4$ one would have one independent anion and one dimeric cation if all atoms are in general position. Actually there is one independent anion in general position and two dimers each of which is in special positions. This is the first case of discovering of 2(A...A^+)·$2X^-$ type of salts (see also the Sect. 5.2.1.4). In the structure there are two different dimeric (Bet...H...Bet) cations with hydrogen bonds O1...H(10)...O1(i)(2.446 (2) Å) and O3...H(30)...O3(ii)(2.442(2) Å). These hydrogen bonds are reflected in IR spectrum as very broad absorption band with maximum at *ca.* 750 cm^{-1}. Expected phase transition at low temperature was not found neither by DSC method in 403–113 K nor by IR spectra. According to the structural data, part of the sulfamate anions at 250 K is disordered. At 100 K the structure is ordered due to inversion of the NH_2 groups.

The next salt is betaine betainium hydrogen oxalate (Rodrigues et al. 2001c). It crystallizes from aqueous solution containing stoichiometric ratio of components and has the same symmetry as in previous case. This is the first salt with dimeric cation containing hydrogen oxalate anion. Unlike the previous case here asymmetric unit contains one $HC_2O_4^-$ anion and dimeric (Bet...BetH) cation with short, but asymmetric hydrogen bond O-H...O(2.457(3) Å). The $HC_2O_4^-$ anions form infinite chain with hydrogen bond O-H...O(2.614(3) Å).

The last salt in Table 5.30 is $(Bet...BetH)_2SiF_6$ (Ghazaryan et al. 2013a). This is the first salt with dimeric cation containing SiF_6^{2-} anion and the second salt of $2(A...A^+)Y^{2-}$ type after $(Bet...BetH)_2SeO_4$. In spite of difference in the geometry of SeO_4^{2-} and SiF_6^{2-} anions, from which the first represents tetrahedron and the second is octahedron their crystal structures are very similar. They crystallize in the same space group and $Z = 8$ and unit cell parameters are close.

So, due to investigations of salts of betaine the first salt of $2(A...A^+)Y^{2-}$ type was discovered and also the following anions forming salts with dimeric cation: HSO_4^-, $HSeO_4^-$, $NH_2SO_3^-$, $HC_2O_4^-$, SeO_4^{2-} and SiF_6^{2-} were revealed.

5.2.1.4 Salts of β-Alanine with (A...AH) Type Dimeric Cations

Known salts of β-alanine with (A...AH) type dimeric cations are shown in Table 5.31.

The first salt of β-alanine with dimeric cation was β-alanine β-alaninium nitrate obtained by Sridhar et al. (2001c). The salt crystallizes in centrosymmetric structure. Dimeric cation represents two crystallographically equivalent β-alanine connected by symmetric O...H...O(2.467(2) Å) hydrogen bond. The authors consider that the crystal may change to the non-centrosymmetric space group *Cc*, triggering a structural phase transition at low temperature. The crystal was studied also by IR and Raman spectroscopy (Jinnah et al. 2004b). In the IR spectrum there is characteristic broad absorption band caused by asymmetric stretching vibration ν(OH) in the region 2,300–500 cm^{-1}. However, the authors assigned to ν(OH) a line at 2,965 cm^{-1} in the Raman spectrum. We do not consider this assignment correct.

The second salt of β-alanine with dimeric cation was β-alanine β-alaninium chloride obtained by Godzisz et al. (2003a), which also crystallizes in the same symmetry and Z. It is worth to note the closeness of unit cell parameters in spite of difference in the geometry of anions. As in previous case there is dimeric cation with symmetric hydrogen bond O...H...O(2.473(2) Å). IR and Raman spectra were analyzed. The presence of strong O...H...O hydrogen bond is reflected in the IR spectrum in form of broad absorption band with maximum at *ca.* 880 cm^{-1}. The crystal was studied also by ^{13}C solid-state NMR method. The authors also considered probable the possibility of phase transition with changing the symmetry. With this aim they determined the structure at 100 K and also studied by DSC method in the region 293–390 K. However, phase transition was not found in this temperature interval.

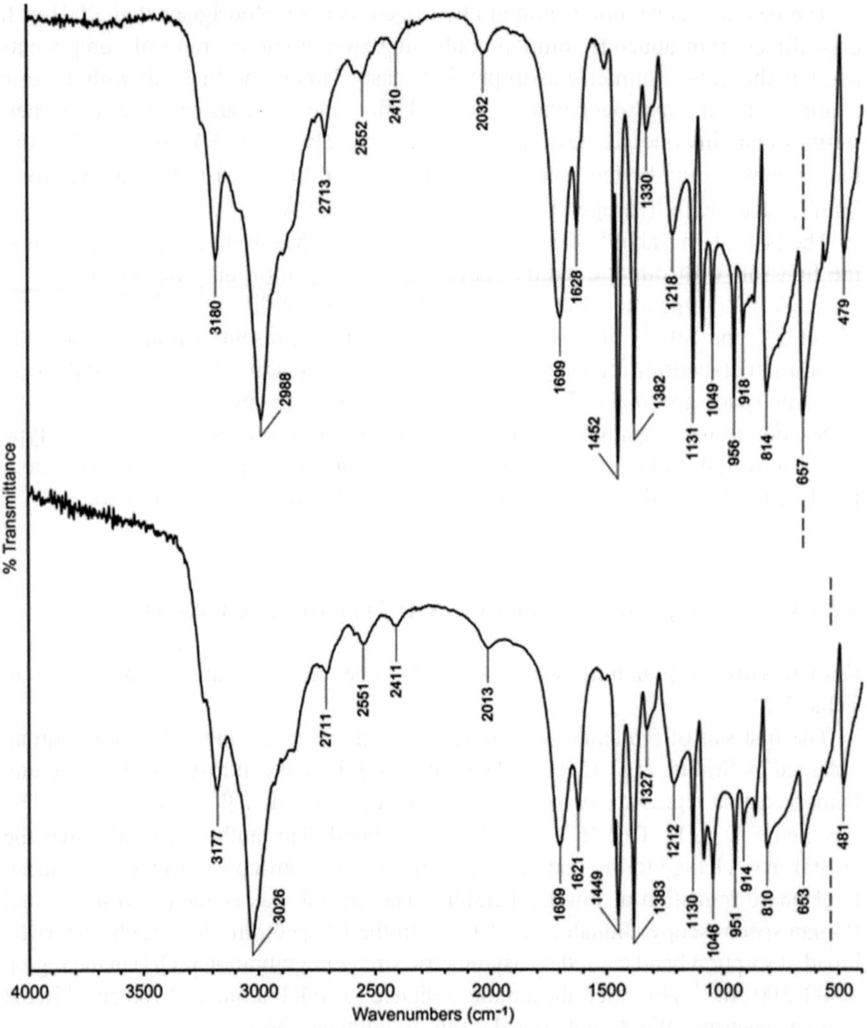

Fig. 5.60 IR spectra of 2β-Ala·HBr (*top*) and presumably 2β-Ala·HI (*bottom*). For explanation see text

Ilczyszyn et al. (2006) reported also the synthesis of crystal 2β-Ala·$HClO_4$ with O...O distance of hydrogen bond equal to 2.457 Å, however, detailed structural data were not reported.

Ghazaryan et al. (2011a) obtained crystals of 2β-Ala·HBr and 2β-Ala·HBF_4 and their structures were determined. From Table 5.31 it is clear that they have the same symmetry, close unit cell parameters and O...O distances in hydrogen bonds. Later the crystal of 2β-Ala·HI also was obtained in our group (this work). Although the structure was not determined yet, based on similarity of vibrational spectra (Fig. 5.60) with spectra of 2β-Ala·HBr one may assume that they are isostructural and 2β-Ala·HI also is characterized by presence of dimeric cation.

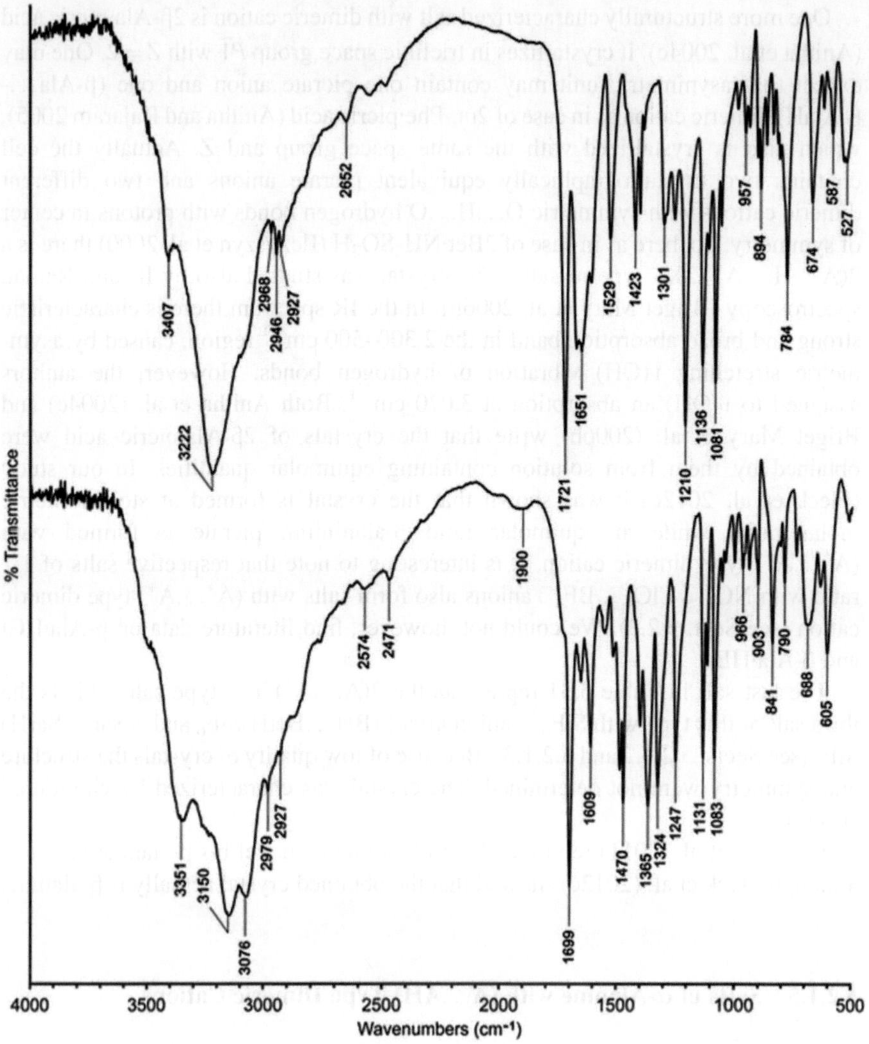

Fig. 5.61 IR spectra of (β-AlaH) L-tartrate (*top*) and (β-Ala…β-AlaH) L-tartrate (*bottom*)

Crystals (β-Ala…β-AlaH) L-tartrate and (β-AlaH) L-tartrate were obtained in our group from aqueous solution containing stoichiometric quantities of components by evaporation.

In Fig. 5.61 the IR spectra of (β-Ala…β-AlaH) L-tartrate and (β-AlaH) L-tartrate are shown. Characteristic broad absorption band is seen from the IR spectrum in the 1,800–900 cm^{-1} interval, which allows supposing the presence of (β-Ala…β-AlaH) dimeric cation in the structure of (β-Ala…β-AlaH) L-tartrate. However, structural confirmation is necessary and determination of symmetry, which should be non-centrosymmetric due to presence of optically active L-tartrate anion in the structure.

One more structurally characterized salt with dimeric cation is 2β-Ala·picric acid (Anitha et al. 2004c). It crystallizes in triclinic space group $P\overline{1}$ with $Z=2$. One may expect that asymmetric unit may contain one picrate anion and one (β-Ala. . .-β-AlaH) dimeric cation as in case of 2DL-Phe·picric acid (Anitha and Rajaram 2005), which also is crystallized with the same space group and Z. Actually the cell contains two crystallographically equivalent picrate anions and two different dimeric cations with symmetric O. . .H. . .O hydrogen bonds with protons in center of symmetry. So, here as in case of 2Bet·NH_2SO_3H (Ilczyszyn et al. 2000) there is a $2(A...H...A)^+ \cdot 2X^-$ type of salts. The crystal was studied also by IR and Raman spectroscopy (Briget Mary et al. 2006b). In the IR spectrum there is characteristic strong and broad absorption band in the 2,300–500 cm^{-1} region, caused by asymmetric stretching ν(OH) vibration of hydrogen bonds. However, the authors assigned to ν(OH) an absorption at 3,070 cm^{-1}. Both Anitha et al. (2004c) and Briget Mary et al. (2006b) write that the crystals of 2β-Ala·picric acid were obtained by them from solution containing equimolar quantities. In our study (Fleck et al. 2012c) it was shown that the crystal is formed at stoichiometric molar ratio, while at equimolar ratio β-alaninium picrate is formed with $(A^+...A^+)$ type dimeric cation. It is interesting to note that respective salts of 1:1 ratio with NO_3^-, ClO_4^-, BF_4^- anions also form salts with $(A^+...A^+)$ type dimeric cation (see Sect. 5.2.3). We could not, however, find literature data on β-Ala·HCl and β-Ala·HBr.

The last salt in Table 5.31 represents the $2(A...A^+) \cdot Y^{2-}$ type salt. This is the third salt of this type with SiF_6^{2-} anion after 2(Bet. . .BetH)SiF_6 and 2(Sar. . .SarH) SiF_6 (see Sects. 5.2.1.2 and 5.2.1.3). Because of low quality of crystals the structure and symmetry were not determined. The crystal was characterized by vibrational spectra.

Ponmani et al. (2011) reported about obtaining a "novel bis β-alanine picrate," although Fleck et al. (2012c) showed that the obtained crystal actually is β-alanine.

5.2.1.5 Salts of α-Alanine with (A. . .AH) Type Dimeric Cations

Known salts of α-alanine with (A. . .AH) type dimeric cations are shown in Table 5.32. As was mentioned before, the first attempts to obtain the salts of L-alanine with 2:1 composition was undertaken by Frost (see Sect. 5.2.1.1). However, the first such salt 2L-Ala·HNO_3 was obtained only in 2001 by Ramos Silva et al. (2001). It is interesting to note that before that interaction of nitric acid with L-alanine was studied and the salt (L-AlaH)NO_3 was obtained (see Sect. 5.1). Later the crystal of 2L-Ala·HNO_3 attracted attention as NLO crystal (Aravindan et al. 2007, 2008; Caroline and Vasudevan 2008; Caroline et al. 2011). Fleck and Petrosyan (2009) showed that Aravindan et al. (2007, 2008) actually obtained crystals of L-alanine and not 2L-Ala·HNO_3. Unfortunately for Gopalakrishnan and Ramasamy (2009) the arguments of Fleck and Petrosyan (2009) were not cogent. Unlike Aravindan et al. (2007, 2008); Caroline and Vasudevan (2008) and Caroline et al. (2011) obtained and characterized the crystals of 2L-Ala·HNO_3. Fleck and Petrosyan (2009) expressed their disagreement with assignments of some absorption bands, particularly, with assignment of ν(OH) of short O-H..O hydrogen bond. Hudson et al. (2009) performed an interesting investigation. With the aim to find a possible phase transition the structure of crystal was determined at 90 K, registered and analyzed the Raman spectra at room temperature and at 78 K as well as investigation was performed by inelastic neutron scattering at 25 K and DFT calculations. No phase transition was found, however, the authors note the presence of some anomalies connected with torsion vibrations of amino groups at 454 cm^{-1}, which need further investigation.

The next salt 2L-Ala·H_3PO_3·H_2O containing (L-Ala...L-AlaH) dimeric cation was obtained by Pankova and studied with coauthors (Lemanov et al. 2002; Smolin et al. 2003c; Yarmarkin et al. 2005). This is the first and yet the only salt with dimeric cation containing $H_2PO_3^-$ anion. Crystal structure was determined (Smolin et al. 2003c) and also piezoresponses and their temperature dependence (Lemanov et al. 2002) and pyroresponses and their temperature dependence (Yarmarkin et al. 2005) were studied. The temperature dependence of pyroresponses has not anomalies in the 140–340 K interval. This indicates that disappearing of piezoresponses above 220 K is not caused with phase transition and probably is caused by increasing of damping.

One may assume that Frost tried to obtain salts of alanine with 2:1 composition with HCl, HBr and HI, because he obtained respective salts of glycine. Apparently he could not obtain them. Now we consider results obtained in this direction during last time. Fischer (2006) could obtain the crystal of D-alanine D-alaninium bromide and determined its structure. However, the crystal was not obtained by direct interaction of D-alanine with HBr, but at attempting to obtain a salt of D-alanine with 2,3-dibromosuccinic acid. It is supposed that HBr was formed as result of an elimination reaction.

Tillmanns et al. (2006) and Fleck et al. (2008) made an attempt to obtain L-Ala·HI from solution containing excess of HI. However, instead of expected compound L-alanine L-alaninium triiodide was obtained. There are two independent formula units of (L-Ala...L-Ala^+)·I_3^- in the cell, where I_3^- has linear configuration $(I\text{-}I\text{-}I)^-$. This is the first and yet the only case of obtaining a salt with dimeric cation containing the I_3^- anion, the structure of which was determined. Thermal expansion parameters also were determined in the 298–383 K interval and thermal behavior in the 300–660 K interval. The onset of decomposition was recorded at *ca.* 122 °C and it melts at 155 °C.

Salts with composition 2L-Ala·HCl and 2L-Ala·HBr were obtained by direct interaction of L-alanine with HCl and HBr (Ghazaryan et al. 2011a). It was found that these salts are not isostructural with previously obtained compound 2D-Ala·HBr by Fischer (2006), because having the same space group they contain each two formula units in asymmetric unit. The O...O distance (2.414(5) Å) of one of dimeric cation in structure of 2L-Ala·HCl is equal to shortest previously known O...O distance found in structure of 2L-Pro·HNO_3 (2.414(3) Å) (see Sect. 5.2.1.7).

At the same time the O...O distance of one of dimeric cations in the structure of 2L-Ala·HBr is even shorter (2.381(12) Å), which is record short distance to date.

The last salts in the table are picrates. In the system L-Ala + picric acid + H_2O at least three phases were identified by IR spectra. Anhydrous sample 2L-Ala·picric acid was obtained at 2:1 ratio (Ghazaryan et al. 2011a). In some experiments initially crystals of L-alanine are formed and then as a second phase 2L-Ala·picric acid·H_2O is formed (Ghazaryan et al. 2012d), which was obtained also by recrystallization of samples obtained at 1.5:1 ratio. IR and Raman spectra of 2L-Ala·picric acid·H_2O were studied as well. The O...O distance in dimeric cation (2.553(2) Å) is longest among salts of α-alanine with dimeric cations.

Vela et al. (2010) reported the synthesis of a crystal "bis-L-alanine picrate." In the work of Ghazaryan et al. (2012d), based on results provided by Vela et al. (2010), it was pointed out that the obtained crystal actually is L-alanine, contaminated by picric acid. Another crystal called "bis-L-alanine triethanol amine" was reported (Vela et al. 2011), which actually is L-alanine as well (see Sect. 5.2.2).

So, investigation of salts of α-alanine with dimeric cation allowed revealing two new anions ($H_2PO_3^-$ and I_3^-) capable to form salts with dimeric cation, and also to reveal a crystal (2L-Ala·HBr) in which the O...O distance has record short value 2.381(12) Å.

5.2.1.6 Salts of Phenylalanine with (A...AH) Type Dimeric Cations

Known salts of phenylalanine with (A...AH) type dimeric cations are shown in Table 5.33. The first salt of phenylalanine with dimeric cation was (L-Phe...L-Phe^+) NO_3^- obtained by Srikrishnan et al. (1984), when they attempted to obtain L-Phe·HNO_3 from solution containing equimolar quantities of components. Detailed structural data were not provided, however. It was done later by Srinivasan et al. (2001e). Dimeric cation found in the crystal has unusual characteristics on which the authors of both works paid attention. At rather short O...O distance equal to 2.426(8) Å bond lengths of carboxyl groups of dimer are characteristic for rather isolated L-Phe^+ cations (bond lengths C11=O11(1.202(8) Å), C11-O12(1.313(6) Å) in one moiety are close to the respective values of the second moiety (C21-O21 (1.220(7) Å) and C21-O22(1.311(7) Å) (Srinivasan et al. 2001e). The position of proton in dimeric cation is not determined, however, the authors note that two residues may be connected by either a symmetric hydrogen bond or an asymmetric hydrogen bond with the H atom disordered over two positions. Three more papers were published on 2L-Phe·HNO_3: Crystals were obtained at stoichiometric 2:1 ratio using as solvent a mixture of methanol and water (Caroline and Vasudevan 2009; Amalanathan et al. 2011b) and water (Cyrac Peter et al. 2010a). Caroline and Vasudevan (2009) and Cyrac Peter et al. (2010a) grew crystals and characterized them by various experimental methods as possible NLO materials, while Amalanathan et al. (2011b) employed mainly DFT calculations. Caroline and Vasudevan (2009) called the obtained crystal "L-phenylalanine nitric acid" although it was identified by unit cell parameters with crystal reported by Srinivasan et al. (2001e). IR spectra of crystals registered by KBr pellet technique are provided in all three papers, however, some assignments, in our opinion, are not acceptable. In connection with uncertainty of the nature of the hydrogen bond based on XRD it were interesting conclusions of authors based on vibrational spectra. Caroline and Vasudevan (2009) and Cyrac Peter et al. (2010a) reported absorption bands at 3,444 cm^{-1} (Caroline and Vasudevan 2009) and 3,444.97 cm^{-1} (Cyrac Peter et al. 2010a). These bands were assigned to asymmetric stretching vibration of NH_3^+ group (Caroline and Vasudevan 2009) and stretching ν(OH) vibration of COOH group (Cyrac Peter et al. 2010a). This band is absent in the spectrum provided by Amalanathan et al. (2011b). Based on this these bands can be safely assigned to stretching ν(OH) vibration of water which was in KBr. Caroline and Vasudevan (2009) did not assign any band to stretching ν(OH) vibration of COOH group. The optimized model of the structure obtained by Amalanathan et al. (2011b), in our opinion, is rather far from reality. According to Amalanathan et al. (2011b), there are $CH(NH_2)COOH$ and $CH(NH_3^+)COOH$ groups where the distance between oxygen atoms of hydroxyl groups is equal to 2.611 Å. Based on this model bands at 3,031 and 3,510 cm^{-1} were assigned to O-H stretching mode, the last band is on the level of noises. Caroline and Vasudevan (2009) and Cyrac Peter et al. (2010a) determined thermal behavior of obtained crystals as well as NLO activity by powder method. The obtained results, however, differ significantly. According to Caroline and Vasudevan (2009) NLO activity is 0.26 compared to KDP, while according to Cyrac Peter et al. (2010a) it is 1.7 times higher than in KDP. According to Caroline and Vasudevan (2009) the crystal decomposes above 200 °C, while according to Cyrac Peter et al. (2010a) it decomposes above 168 °C.

It is interesting to note that simultaneously with Srinivasan et al. (2001e), Görbitz (2001) reported the existence of one more crystal in the L-Phe + $HNO_3 + H_2O$ system with 4L-Phe·HNO_3 composition, in the structure of which is dimeric (L-Phe...L-Phe$^+$) cation and additionally two L-Phe$^{\pm}$ zwitterions. The accuracy of structure determination, unfortunately, is not so high and determination of optimal conditions of obtaining this crystal and redetermination of structure is desirable.

Görbitz and Etter (1992) obtained the crystal of L-phenylalanine L-phenylalaninium formate and its crystal structure was determined at room temperature. This is the first and yet the only salt of amino acids with dimeric cation containing formate anion. Mebs et al. (2006) determined the structure at 25 K and also electron density distribution.

The salt L-phenylalanine L-phenylalaninium perchlorate obtained from aqueous solution containing stoichiometric quantities of components (Srinivasan and Rajaram 1997b) is the first salt of amino acids with dimeric cation containing perchlorate anion. Vibrational spectra (IR and Raman) were studied in (Rajkumar et al. 2002). The expected position of ν(OH) for O...O hydrogen bond with distance 2.440 Å is *ca.* 1,000 cm^{-1}. Rajkumar et al. (2002) assigned to ν(OH) of dimeric cation a weak peak at 2,581 cm^{-1}, which probably relates to overtone. The crystal of 2L-Phe·$HClO_4$ as possible NLO material was obtained and studied by Shankar et al. (2009) without quotation of previous works (Srinivasan and Rajaram 1997b; Rajkumar et al. 2002). Shankar et al. (2009) assigned the absorption band at 3,075 cm^{-1} to ν(OH). NLO activity estimated by powder method makes 0.5 compared to KDP. Cyrac Peter et al. (2010b) also obtained the crystal of 2L-Phe·$HClO_4$ and characterized it as possible NLO material. According to the data, the NLO activity of 2L-Phe·$HClO_4$ is 1.4 times higher than that of KDP. In both works decomposition of crystal was studied. According to Shankar et al. (2009), the crystal decomposes with endothermic effect, while according to Cyrac Peter et al. (2010b) it decomposes with exothermic effect. Recently, Elleuch et al. (2014) obtained the crystal of 2L-Phe·$HClO_4$, redeterminated its structure at room temperature, studied IR and Raman spectra and theoretically studied vibrational spectra and NLO properties. Strangely the authors do not discuss at all the hydrogen bond O-H. . .O between L-phenylalanine and L-phenylalaninium.

Ghazaryan et al. (2011a) obtained the crystal of 2L-Phe·HBF_4 and characterized it structurally and by vibrational spectroscopy. As expected the crystal is isostructural with respective perchlorate. Their spectra are very similar, which confirms their isostructurality. The IR spectra of 2L-Phe·$HClO_4$ and 2L-Phe·HBF_4 crystals obtained by us do not contain any absorption bands above 3,200 cm^{-1}, which indicates that respective bands in the spectra (Shankar et al. 2009; Cyrac Peter et al. 2010b) in this region are caused by water in KBr.

Ghazaryan et al. (2011a) reported a number of other salts of L-phenylalanine with dimeric cation. The salt 2L-Phe·H_2SeO_3 is the second salt with dimeric cation after 2Gly·H_2SeO_3 containing $HSeO_3^-$ anion as well as 2L-Phe·$H_2C_2O_4$ is the second salt after 2Bet·$H_2C_2O_4$ with $HC_2O_4^-$ anion. The quality of 2L-Phe·$H_2C_2O_4$ crystals did not allow to determine the structure with high accuracy. Nevertheless, a rough structure determination established that there are two types of phenylalanine phenylalaninium dimers in the structure.

In Fig. 5.62 the IR spectra of the 2L-Phe·HCl and 2L-Phe·HBr crystals are shown. The similarity of these spectra indicates on possible isostructurality of these crystals. Because of low quality of 2L-Phe·HCl crystals it was not possible to determine their structure. Determination of the structure of 2L-Phe·HBr showed that there are two crystallographically nonequivalent dimers (Fig. 5.63). One may suppose that the similar structure has also the crystal of 2L-Phe·HCl.

Alagar et al. (2005) obtained the crystal of L-phenylalanine L-phenylalaninium hydrogen malonate and determined its structure. This is the first and yet the only salt of amino acids with dimeric cation containing hydrogen malonate anion. Intramolecular hydrogen bond O-H. . .O in hydrogen malonate anion makes 2.413 (4) Å. This crystal was grown and characterized also by Prakash et al. (2011b) as possible NLO material. The crystal was obtained from aqueous solution with stoichiometric composition. The authors provided IR spectrum of the crystal. As in some other papers, Prakash et al. (2011b) assigned an absorption band at 3,425 cm^{-1} to ν(OH) stretching vibration of COOH group, which probably is caused by ν(OH) of water in KBr because for hydrogen bond of dimer with

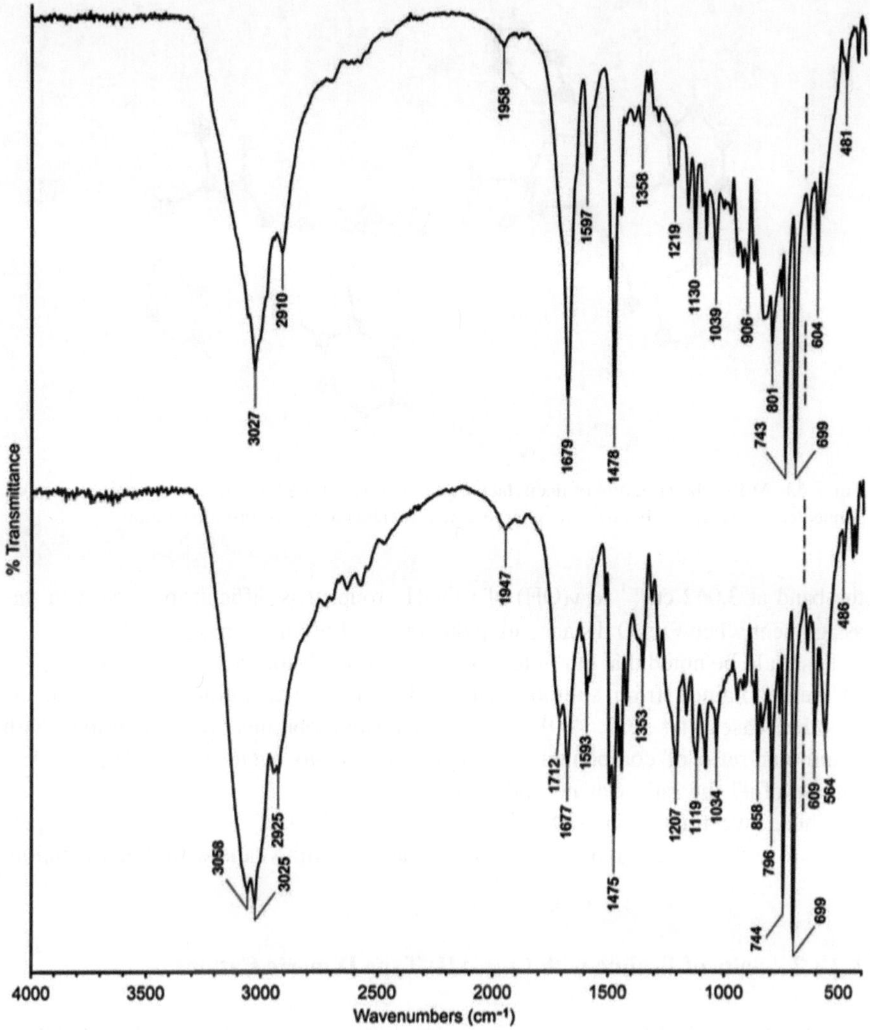

Fig. 5.62 IR spectra of 2L-Phe·HCl and 2L-Phe·HBr

O...O(2.429(4) Å) and hydrogen malonate anion with O...O(2.413(4) Å) absorption caused by ν(OH) should be expected below 2,000 cm^{-1}. NLO activity estimated by powder method makes 0.38 compared to KDP. The crystal decomposes above 180 °C.

The last salt in Table 5.33 is (DL-Phe...DL-PheH) picrate obtained by Anitha and Rajaram (2005) from aqueous solution containing equimolar quantities of DL-phenylalanine and picric acid. Triclinic cell contains two dimeric cations (L-Phe...D-PheH) and (D-Phe...L-PheH) and two picrate anions. Vibrational spectra of the crystal were studied by Briget Mary et al. (2006b). The authors indicated on the presence of absorption band caused by water absorbed by KBr, however, they assigned

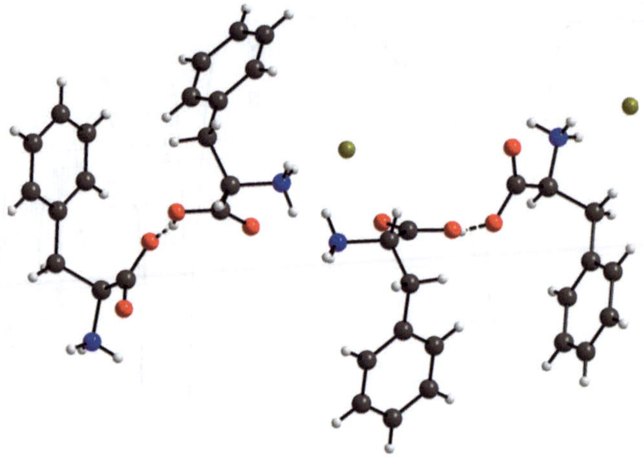

Fig. 5.63 Molecular structure of phenylalaninium phenylalanine bromide. Two dimeric cations – connected by hydrogen bonds – are charge-counterbalanced by two bromide anions

the band at 3,062 cm^{-1} to ν(OH) of COOH group. It is difficult to agree with this assignment when O. . .O distance in hydrogen bond of dimer makes 2.473(5) Å.

It should be noted that out of ten salts provided in Table 5.33 only three (## 3, 5, 8) were obtained from solution with stoichiometric quantities of components. In other cases (## 1, 2, 4, 9, 10) crystals were obtained from solutions with equimolar ratio of components, while in case of #6 even at ratio 1:2 and 1:3. In case of #7 the ratio was not indicated.

Thus, investigation of salts of phenylalanine allowed revealing formate, perchlorate and hydrogen malonate anions capable forming salts with dimeric cation.

5.2.1.7 Salts of Proline with (A. . .AH) Type Dimeric Cations

Known salts of proline with (A. . .AH) type dimeric cations are shown in Table 5.34. The first salt of proline (2DL-Pro·HCl) with dimeric cation was found by Swaminathan and Chacko (1981). There are four $(L\text{-Pro}...H...D\text{-Pro})^+Cl^-$ formula units or eight DL-Pro·0.5HCl formula units. The proton H^+ and anion Cl^- are in special positions. In dimeric cation proline moieties are connected with each other by symmetric O. . .H. . .O hydrogen bond. Later salts of proline with dimeric cation were found also for L-proline: 2L-Pro·HNO_3 (Pandiarajan et al. 2002b) and 2L-Pro·$HClO_4$ (Pandiarajan et al. 2002c). Before finding of 2L-Ala·HBr (see Sect. 5.2.1.5) the shortest length of hydrogen bond in salts of amino acids with dimeric cation remained the hydrogen bond in the structure of 2L-Pro·HNO_3 with length equal to 2.414(3) Å. Ghazaryan et al. (2011a) reported three more salts of L-proline with dimeric cation. The salt 2L-Pro·HBF_4 as expected proved to be isostructural with 2L-Pro·$HClO_4$. Ghazaryan et al. (2011a) indicated the value of $a = 16.4126(7)$ Å by mistake instead of $a = 7.7860(5)$ Å as indicated in Table 5.34. Two other salts are 2L-Pro·HCl and 2L-Pro·HBr, which are isostructural. The feature of these salts is that this is the first case among salts of amino acids with dimeric cation with symmetry higher than orthorhombic. All before known salts referred to triclinic, monoclinic or orthorhombic systems. Vibrational spectra of 2L-Pro·HNO_3 and 2L-Pro·$HClO_4$ crystals were studied by Pandiarajan et al. (2005a). IR spectra contain broad absorption bands characteristic for ν(OH) vibration of short O-H...O hydrogen bonds in the region 2,000–500 cm^{-1}. However, Pandiarajan et al. (2005a) assigned to ν(OH) the intensive Raman line at 3,005 cm^{-1} in the spectrum of 2L-Pro·HNO_3 and the shoulder at 3,016 cm^{-1} in the Raman spectrum of 2L-Pro·$HClO_4$. In addition we do not agree with assignment of a shoulder at 1,326 cm^{-1} in the spectrum of 2L-Pro·HNO_3 to asymmetric stretching vibration of nitrate anion. In our opinion very strong absorption band at 1,385 cm^{-1} relates to this vibration. The crystal of 2L-Pro·HNO_3 was grown and characterized by Selvaraju and Kirubavathi (2013) as possible NLO material. According to this work, the SHG efficiency estimated by powder method is 2.8 times higher than in KDP. IR and Raman spectra are provided and discussed. Strong absorption band at 1,384 cm^{-1} is assigned to NH_2^+ twist, which in our opinion relates to asymmetric stretching vibration of nitrate anion. The IR spectrum contains very strong and broad band, which in our opinion relates to ν(OH) vibration of O-H...O hydrogen bond in dimeric cation.

5.2.1.8 Salts of Leucine with (A...AH) Type Dimeric Cations

Known salts of leucine with (A...AH) type dimeric cations are shown in Table 5.35. The salt 2L-Leu·HCl was obtained by Rolski et al. (1965). Then Hadži and Marciszewski (1967) based on the IR spectrum concluded on the existence of dimeric L-leucine L-leucinium dimeric cation with strong hydrogen bond. At last Golič and Hamilton confirmed the presence of dimeric cation (L-Leu...L-LeuH) with short O...O distance 2.429(7) Å by crystal and molecular structure determination (Golič and Hamilton 1972). Ghazaryan et al. (2011a) obtained the crystal of 2L-Leu·HBr, which as expected is isostructural with respective chloride. Also crystals of 2L-Leu·ClO_4 and 2L-Leu·HBF_4 were obtained by Ghazaryan et al. (2011a). The difficult with their obtaining is connected with that they are possible to obtain by evaporation from solution containing equimolar ratio of components, while from stoichiometric ratio L-leucine is formed. The crystals of L-Leu·$HClO_4$ and L-Leu·HBF_4 were possible to obtain from solution containing respective components with 1:2 molar ratio.

Anitha et al. (2005a, b) obtained for the first time the salt L-leucine L-leucinium picrate from aqueous solution containing stoichiometric quantities of components. The salt crystallizes in triclinic system with $Z = 2$, that is, there are two independent dimeric cations and two picrate anions. Bhagavannarayana et al. (2011) obtained the crystal of L-leucine L-leucinium picrate and characterized as possible NLO material. The IR spectrum is shown, thermal behavior was determined and SHG efficiency was estimated by powder method, which was 1.5 times higher than in KDP powder. Amino acid salts of picrates usually have high and very high NLO properties due to the presence of picrate anion. So, NLO properties of this crystal based on this estimation should be consider as low NLO activity. The spectrum contains a band at 3,465 cm^{-1}, which is probably caused by water in KBr. The authors assigned this band to unclear Ct=O vibration. Probably this is misprint. On DSC curve there are endothermic peak at 186 °C and exothermic peak at 206 °C, which probably relate to melting and decomposition. The authors interpreted these peaks as "melting and complete melting." The crystal was studied also by Guidara et al. (2013). Experimental IR and Raman spectra as well as DFT calculations on vibrational spectra and NLO properties were provided and discussed. Unfortunately the optimized model of molecular structure is far from real structure. Particularly, the dimeric (L-Leu...L-LeuH) cation is absent in optimized model. As a result the calculated value of ν(OH) (3,565 cm^{-1}) is far from expected value for O-H...O hydrogen bonds in the structure of the L-leucine L-leucinium picrate crystal. Based on this calculation the authors assigned a narrow peak at 3,400 cm^{-1} to ν(OH), while for hydrogen bonds with O...O 2.454(5) Å and 2.471(1) Å the ν(OH) absorption band should be expected below 1,500 cm^{-1} according to well-known Novak's correlation.

5.2.1.9 Salts of Valine with (A...AH) Type Dimeric Cations

Known salts of valine with (A...AH) type dimeric cations are shown in Table 5.36. The first salt of valine with dimeric cation (L-Val...L-ValH)ClO_4·H_2O was obtained by Pandiarajan et al. (2001a) from aqueous solution containing stoichiometric quantities of components. The O...O distance of dimeric cation (2.562(4) Å) coincides with longest O...O distance found in the structure of 2Gly·HBr (see Sect. 5.2.1.1). The perchlorate anion is partly disordered. The water molecule forms OW1-HW2...O1A(2.806(4) Å) with one of oxygen atoms of carboxylate group, while the other hydrogen atom HW1 does not practically form hydrogen bond. There is a contacts OW1-HW1...O4(3.050(12) Å) with HW1...O4(2.45(7) Å) with one of disordered oxygen atom of anion. IR and Raman spectra of the crystal were studied (Pandiarajan et al. 2005b). The IR and Raman spectra provided by Pandiarajan et al. (2005b) differ significantly from the spectra of the sample of (L-Val...L-ValH)ClO_4·H_2O obtained in our group. In addition, Pandiarajan et al. (2005b) assigned to ν(OH) of O-H...O hydrogen bond with O...O(2.562 (4) Å) the band at 3,014 cm^{-1} and Raman line at 3,020 cm^{-1}, while this band should be expected near 2,000 cm^{-1}.

Ghazaryan et al. (2011a) obtained crystals of 2L-Val·HCl and 2L-Val·HBr. Vibrational spectra are shown in (Fig. 5.64). The quality of 2L-Val·HCl did not allow to determine its structure. The structure of 2L-Val·HBr was determined. Moitra et al. (2010) could obtain suitable crystals of 2L-Val·HCl for structure

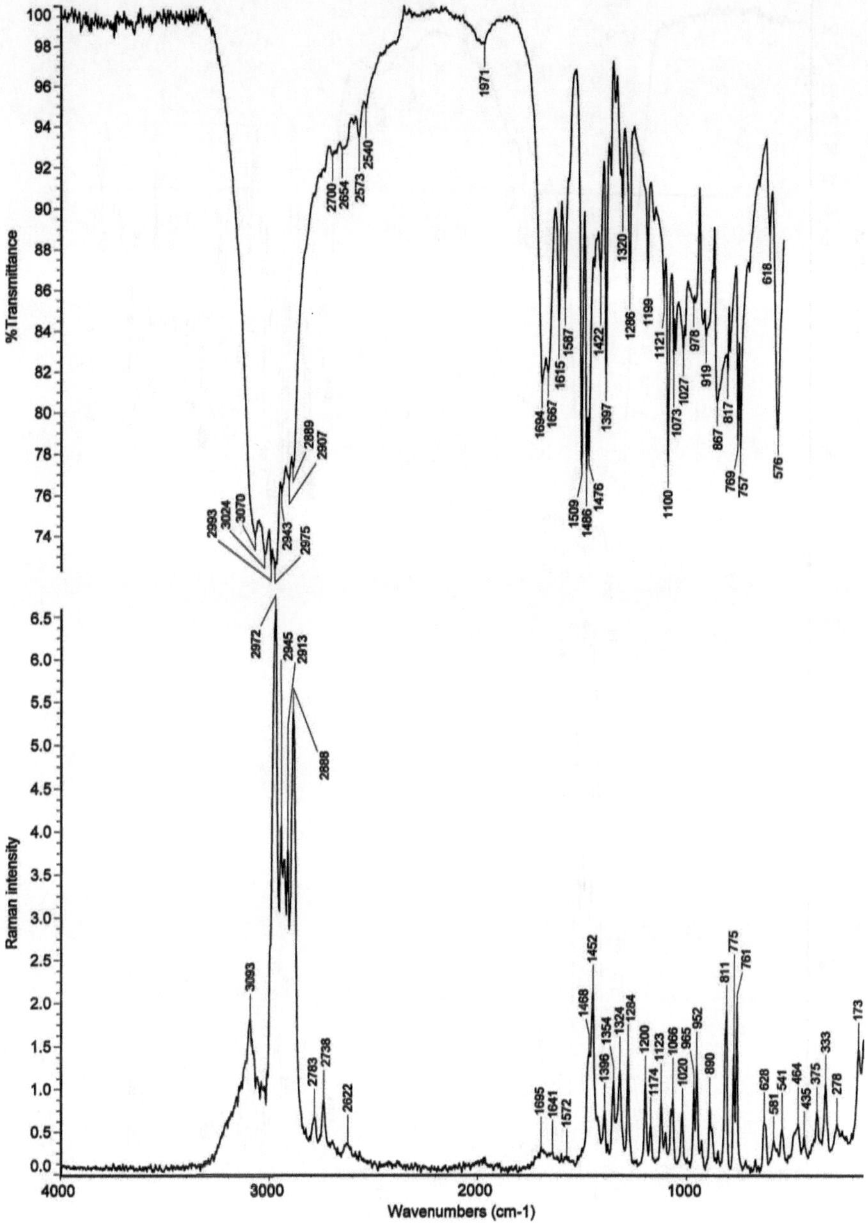

Fig. 5.64 IR and Raman spectra of 2L-Val·HCl *this page*) and 2L-Val·Hr *next page*)

determination The IR spectrum was registered by Kr pellet technique thermal behavior was determined the transmittance spectrum in UV Vis region and NLO activity $I^2 = 0\ 44 I^2_{KDP}$) Comparison of structures of 2L-Val·HCl Moitra et al 2010) and 2L-Val·Hr Ghazaryan et al 2011a) shows that they are

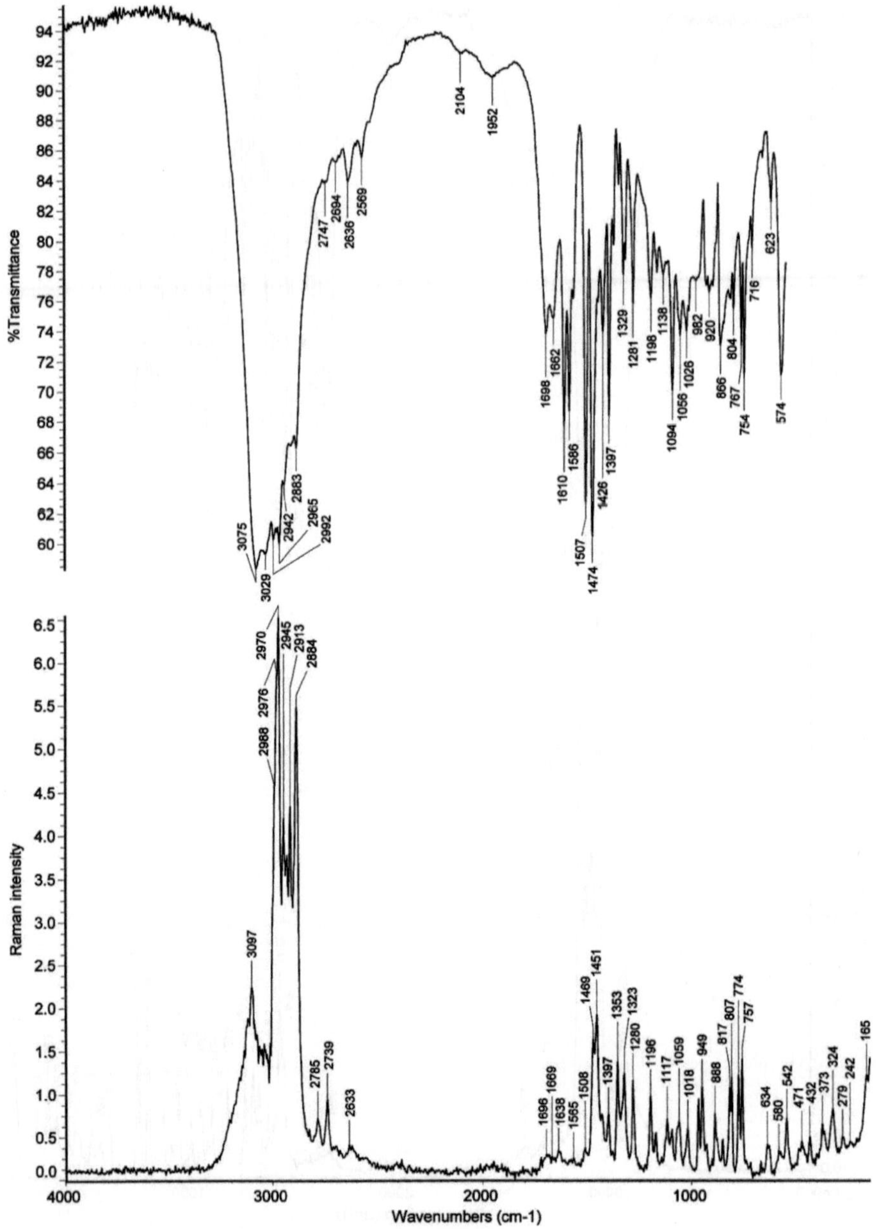

Fig. 5.64 (continued)

similar but not isostructural. It should be noted that the title of 2L-Val·HCl "L-valine L-valinium hydrochloride" used by Moitra et al. (2010) is not correct, should be either "di-L-valine hydrochloride" or "L-valine L-valinium chloride."

5.2.1.10 Salts of Methionine with (A. . .AH) Type Dimeric Cations

Known salts of methionine with (A. . .AH) type dimeric cations are shown in Table 5.37. The first salt of methionine with dimeric cation (L-Met. . .L-MetH) $ClO_4 \cdot H_2O$ was obtained by Sridhar et al. (2002e) from aqueous solution containing stoichiometric quantities of components. The structure of crystal was determined at 105 K, vibrational spectra (IR, Raman) were studied by Briget Mary et al. (2004). In our group we studied the system L-Met + HBF_4 + H_2O. The crystal of 2L-Met·HBF_4 was obtained from aqueous solution with stoichiometric quantities of components by evaporation at room temperature. The compound L-Met·HBF_4 proved to be very hygroscopic. It was possible to obtain from aqueous solution with equimolar ratio of components in a closed vessel in presence of silica gel as water absorber. IR spectra of these compounds are shown in Fig. 5.65. Absorption bands caused by ν(C=O) are at 1,759 and 1,717 cm^{-1} for L-Met·HBF_4 and 2L-Met·HBF_4 respectively. Very strong absorption bands at *ca.* 1,020 cm^{-1} are caused by asymmetric stretching vibration of BF_4^- anion. Absorption band in the spectrum of L-Met·HBF_4 is caused by adsorbed water due to hygroscopicity of crystal.

Natarajan et al. (2008c) found a new salt (L-Met. . .L-MetH) hydrogen maleate from aqueous solution containing stoichiometric quantities of components and determined its structure and by powder method estimated its NLO activity ($I^2 = 0.9I^2_{KDP}$). This is the first and yet the only case of salts of amino acids with dimeric cation containing hydrogen maleate anion. Natarajan et al. (2010c) provided results of further investigation of the crystal, i.e., IR and UV–Vis spectra, and also TG, DTA curves. The crystal decomposes near 200 °C. The authors write also about observation an endothermic peak on DTA curve at 122.27 °C caused by melting, which, however, is absent in provided figure. The last salt in the table is (DL-Met. . .DL-MetH) picrate obtained by Anitha et al. (2006b) from aqueous solution also containing stoichiometric quantities of components. The salt crystallizes in triclinic system and contains two formula units in the cell with (L-Met. . .D-MetH) and (D-Met. . .L-MetH) dimeric cations. Briget Mary et al. (2006a) studied vibrational spectra of (DL-Met. . .DL-MetH) picrate, as well as of (DL-Val. . .DL-ValH) picrate. Ghazaryan et al. (2012d) noted that some assignments in this paper are questionable.

So, investigation of salts of methionine with dimeric cation allowed revealing for the first time a new anion, namely, hydrogen maleate capable to form salt with dimeric cation.

5.2.1.11 Salts of Other Amino Acids with (A. . .AH) Type Dimeric Cations

Book et al. (1981) obtained the crystal of (L-Trp. . .L-TrpH)$HgCl_3$ and determined its crystal and molecular structure. In this work for the first time dimeric cation L-tryptophan L-tryptophanium was found in result of structure determination.

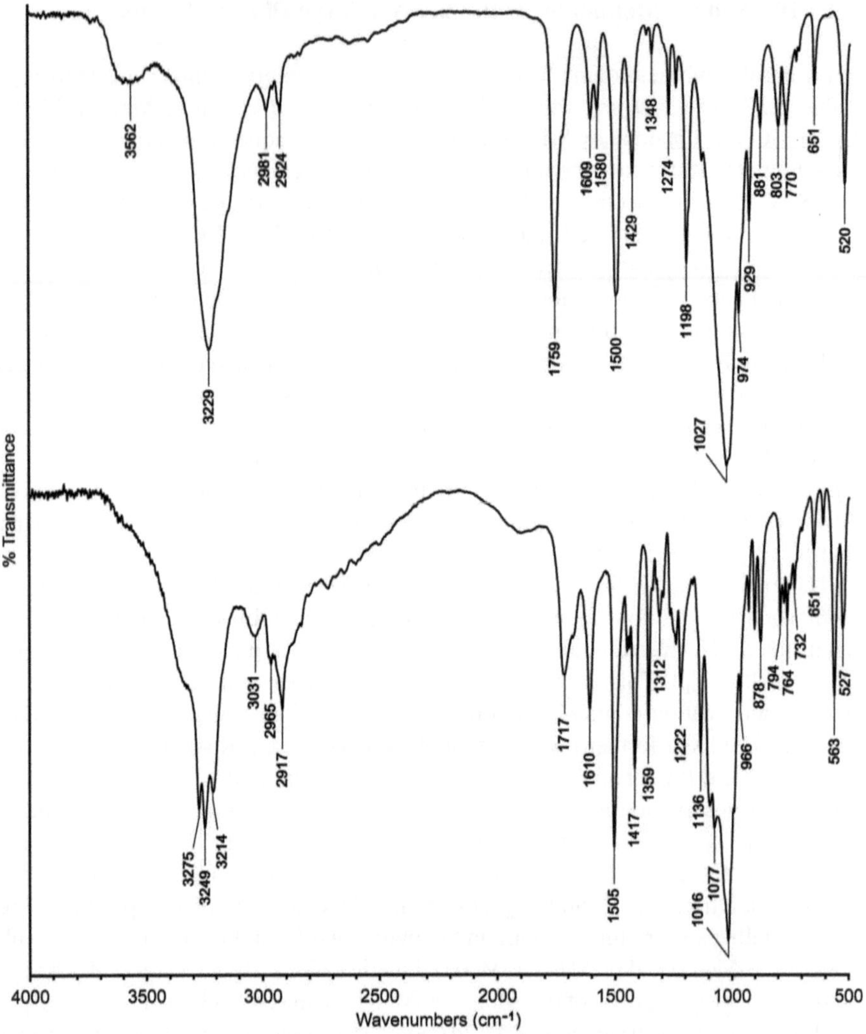

Fig. 5.65 IR spectra of L-Met·HBF_4 (*top*) and 2L-Met·HBF_4 (*bottom*)

In addition, a new anion $HgCl_3^-$ capable to form salt of amino acids with dimeric cation was found. It should be noted, however, that as far back as 1966 Takiagawa et al. (1966) obtained crystals of L-Trp·HCl and L-Trp·HBr and determined their structures. In addition to the crystal of L-Trp·HBr the authors obtained one more crystal with 2L-Trp·HBr·0.5H_2O composition and determined its symmetry and cell parameters. The quality of crystals did not allow, however, determining its crystal structure and establishing the presence of dimeric cation. IR spectrum can provide useful information on presence of dimeric cation with short O-H…O hydrogen bond (Ghazaryan et al. 2011a). With aim to obtain additional vibrational spectroscopic information on presence of dimeric cation in structure of

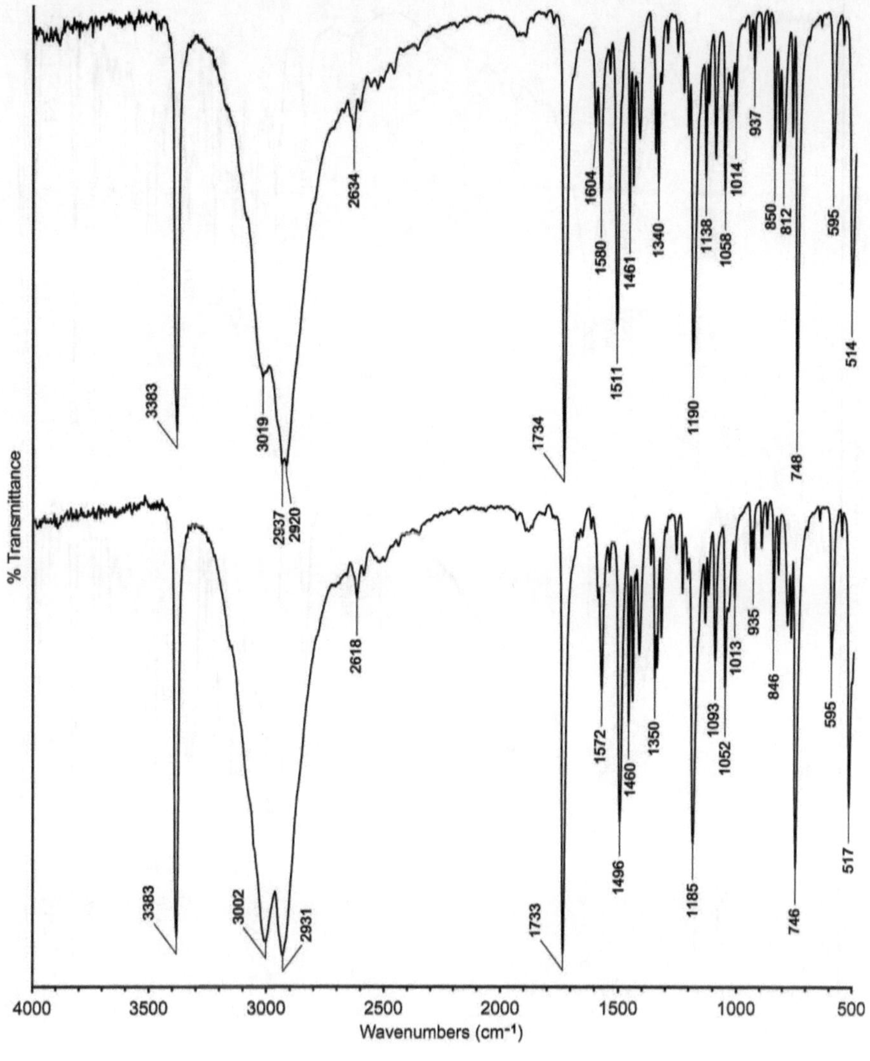

Fig. 5.66 IR spectra of L-Trp·HCl (*top*) and L-Trp·HBr (*bottom*)

2L-Trp·HBr·0.5H_2O and also to check the possibility of obtaining respective chloride we returned to investigation of L-Trp + HCl + H_2O and L-Trp + HBr + H_2O systems. We found and optimized the conditions of obtaining L-Trp·HCl, L-Trp·HBr and 2L-Trp·HBr·0.5H_2O. In addition we obtained two new salts: 2L-Trp·HCl and 2L-Trp·HBr. It was established that as in case of L-threonine salts with 1:1 composition are more convenient to obtain at 1:1.5 molar ratio of L-tryptophan and respective acid. The spectra of L-Trp·HCl and L-Trp·HBr are shown in Fig. 5.66. At equimolar ratio a mixture of salts with 2:1 and 1:1 compositions are formed. The crystals of 2L-Trp·HBr·0.5H_2O were obtained from solution with molar ratio

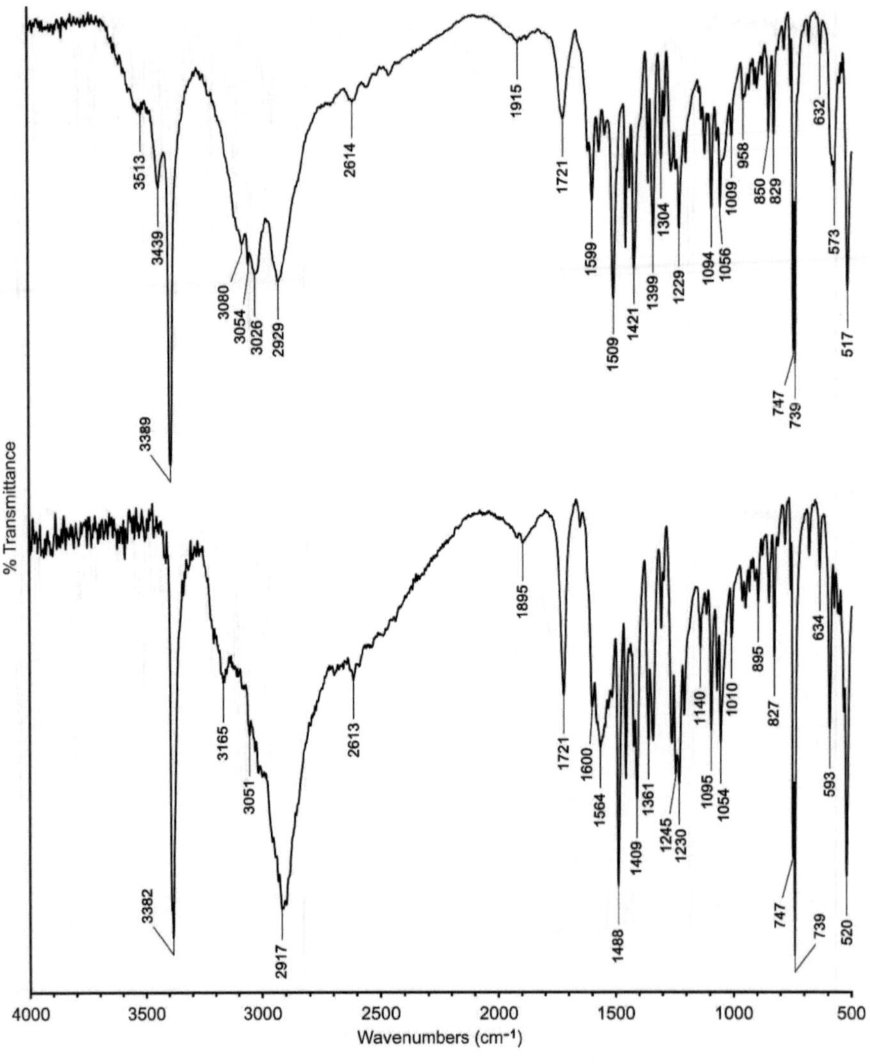

Fig. 5.67 IR spectra of 2L-Trp·HBr·0.5H_2O (*top*) and 2L-Trp·HBr (*bottom*)

1.7:1 at room or slightly lower temperature. At slightly higher temperature and at the same molecular ration anhydrous 2L-Trp·HBr is formed. Respective anhydrous chloride 2L-Trp·HCl was obtained at room temperature from solution containing L-tryptophan and HCl in molar ratio 1.5:1. The spectra of 2L-Trp·HBr·0.5H_2O and 2L-Trp·HBr are shown in Fig. 5.67 and in Fig. 5.68 IR and Raman spectra of 2L-Trp·HCl are shown. Absorption at 3,513 cm^{-1}, 3,439 cm^{-1} observed in the spectrum of 2L-Trp·HBr·0.5H_2O above characteristic strong band at 3,389 cm^{-1} relating to ν(NH) of indole group we assign to ν(OH) of crystal water. There is no absorption in the spectra of L-Trp·HCl, L-Trp·HBr and 2L-Trp·HCl, 2L-Trp·HBr in

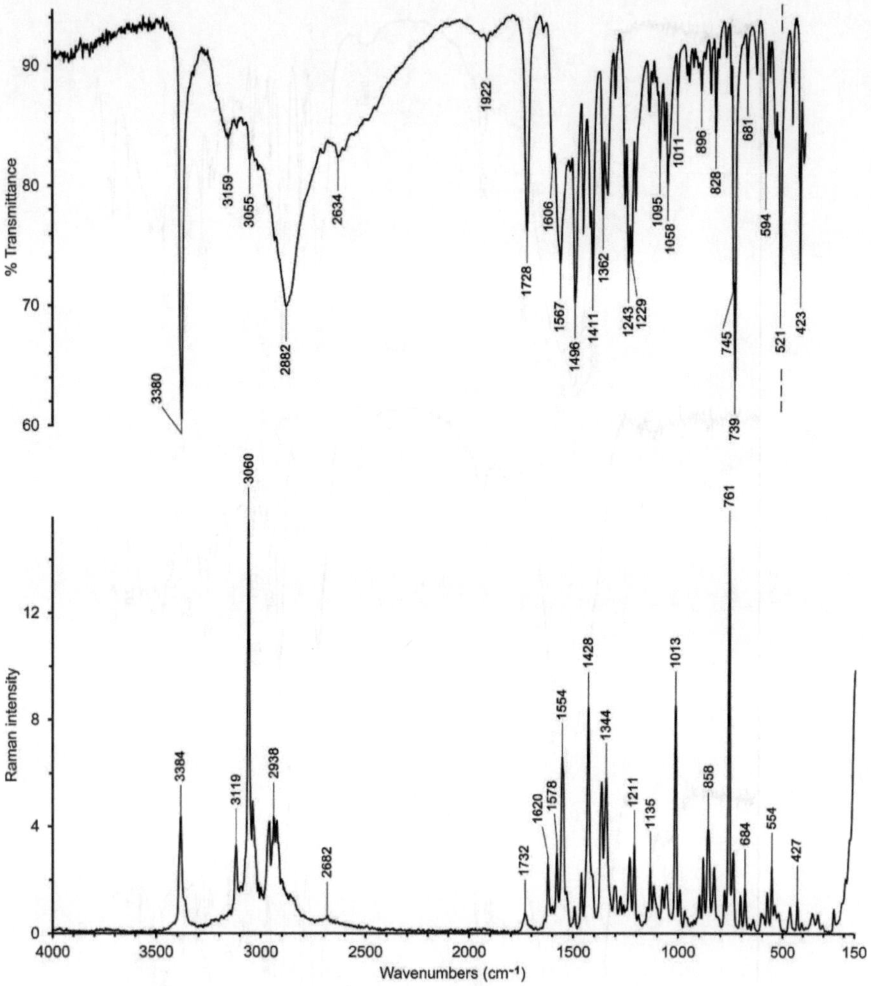

Fig. 5.68 IR and Raman spectra of 2L-Trp·HCl

this region. Absorption bands caused by ν(C=O) in the spectra of 2L-Trp·HCl (1,728 cm^{-1}), 2L-Trp·HBr·0.5H_2O (1,721 cm^{-1}) and 2L-Trp·HBr (1,721 cm^{-1}) are somewhat lower than in the spectra of L-Trp·HCl (1,734 cm^{-1}) and L-Trp·HBr (1,733 cm^{-1}). Comparison of the spectra of 2:1 and 1:1 compositions in the region 1,700–650 cm^{-1} shows that absorption peaks in the spectra of salts with 2:1 composition are superposed on broad band in this region caused by presence of O-H. . .O hydrogen bond with dimeric cation. So, based on IR spectra one may conclude on presence of dimeric cations in the structures of 2L-Trp·HCl and 2L-Trp·HBr·0.5H_2O and 2L-Trp·HBr.

Preliminary research performed in our group showed that L-cysteine also can form a number of salts with dimeric cations. In Fig. 5.69 as examples IR spectra of 2L-Cys·HCl (upper), 2L-Cys·HBr (middle) and 2L-Cys·HI·nH_2O are shown. The

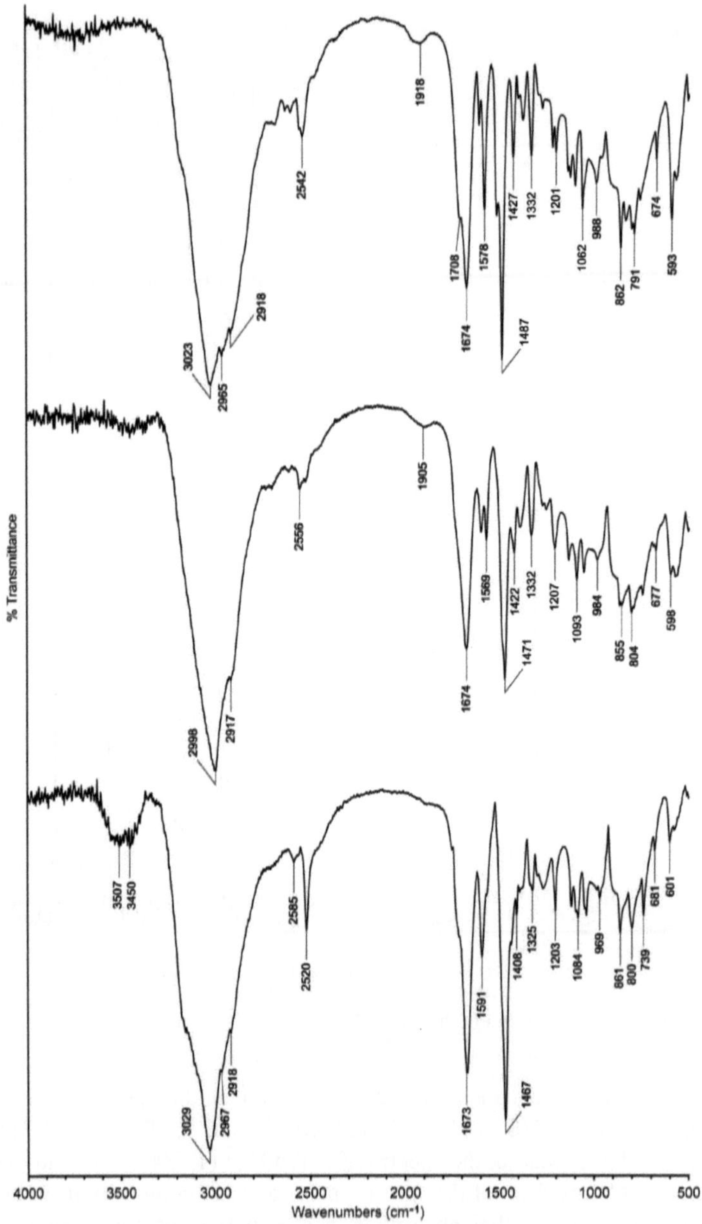

Fig. 5.69 IR spectrum of 2L-Cys·HCl (*top*), 2L-Cys·HBr (*middle*) and 2L-Cys·HI·nH_2O (*bottom*)

spectra of chloride and bromide are similar and probably they are isostructural. The iodide is hydrated. The presence of characteristic broad absorption below 1,500 cm^{-1} indicates on presence of dimeric (L-Cys...L-CysH) dimeric cations with short O-H...O hydrogen bonds. The presence of bands near 1,674 cm^{-1} also is

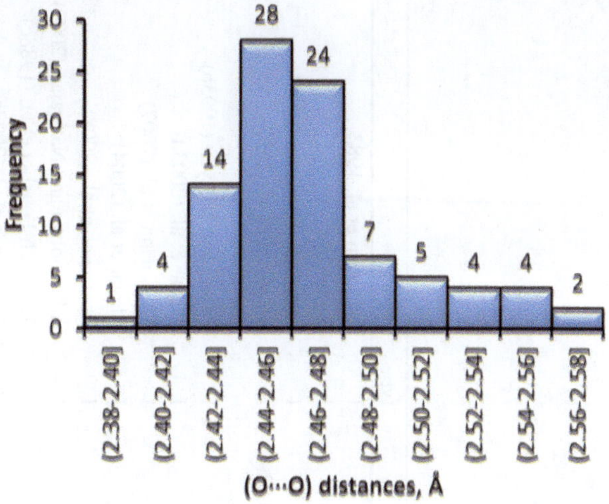

Fig. 5.70 Histogram of hydrogen bonds of salts with 2:1 composition and (A. . .A$^+$) type dimeric cation

indicative of presence of strong hydrogen bonds in dimers. Previously any salts of cysteine with dimeric cation were not known.

In Fig. 5.70 the histogram of hydrogen bonds in structures of salts of amino acids with 2:1 composition and (A. . .A$^+$) type dimeric cations are shown. The shortest O-H. . .O(2.381(12) Å) hydrogen bond was found in the structure of (L-Ala. . .L-AlaH)Br and the longest O-H. . .O(2.562 Å) hydrogen bond in the structures of (Gly. . .GlyH)Br (2.562(3) Å) and (L-Val. . .L-ValH)ClO_4·H_2O (2.562(4) Å). The majority of hydrogen bonds fall in the 2.42–2.48 Å interval (*ca.* 71 %) and with maximal frequency at 2.45 ± 0.01 Å.

5.2.1.12 Compounds with 2:1 Composition, but Without (A. . .AH) Dimeric Cation

In Sects. 5.2.1.1, 5.2.1.2, 5.2.1.3, 5.2.1.4, 5.2.1.5, 5.2.1.6, 5.2.1.7, 5.2.1.8, 5.2.1.9, 5.2.1.10, and 5.2.1.11 more than 80 salts of amino acids with composition 2A·HX (or 4A·H_2Y) containing (A. . .A$^+$) (or (A. . .AH)) type dimeric cations were considered. There are, however, a number of compounds of amino acids with acids of 2A·HX composition, which do not form (A. . .AH) type dimeric cation. These compounds collected from various tables are shown in Table 5.39. They can be divided into two groups. The majority of them represents molecular adducts without protonation of respective amino acid and hence without formation of A$^+$ cation capable to form (A. . .A$^+$) dimeric cation. The second group, which includes four compounds, represents salts. Here in result of proton transfer to amino acid formation of respective cation and anion takes place. However, the cation of amino acid does not form dimer with zwitterion, but forms hydrogen bond with anion or water molecule.

Table 5.39 Compounds with 2:1 composition, but without (A. . .AH) dimeric cation

#	Crystal	s.g., Z	*a*, *b*, *c* (Å), *α*, *β*, *γ* (°)	type	Refs
1	2Gly·Te(OH)$_6$·2H_2O[a]	$P2_1/n$, 2	7.117(4), 12.201(5), 7.936(4), $\beta = 110.46(4)$	I	Andersen et al. 1983
2	2Gly·Te(OH)$_6$	–	–	I	Botto (1987)
3	2Sar·Te(OH)$_6$	$P2_1$, 2	10.045(8), 13.080(9), 5.106(4), $\beta = 96.09(1)$	I	Averbuch-Pouchot (1988)
4	2Bet·Te(OH)$_6$	$P\bar{1}$, 1	7.164(5), 8.325(5), 10.035(6), $\alpha = 108.58(5)$, $\beta = 123.82(4)$, $\gamma = 96.02(5)$	I	Ilczyszyn et al. (1992)
5	2DL-Pro·succinic acid	$P2_1/c$, 2	5.254(1), 17.480(1), 10.230(1), $\beta = 119.60(6)$	I	Prasad and Vijayan (1993b)
6	2DL-Val·fumaric acid	$C2/c$, 4	24.417(4), 7.5713(10), 10.013(2), $\beta = 109.268(10)$	I	Alagar et al. (2003d)
7	2L-Val·fumaric acid	$C2$, 2	22.308(1), 7.2294(4), 5.3070(2), $\beta = 91.205(4)$	I	Klussmann et al. (2007)
8	2DL-Val·succinic acid	$C2/c$, 4	24.373(2), 7.5744(6), 10.0284(9), $\beta = 106.134(2)$	I	Alagar et al. (2004)
9	2Sar·pyromellitic acid	$P2_1/c$, 2	8.8894(3), 5.4118(2), 20.2205(7), $\beta = 104.388(2)$	I	Domingos et al. (2008)
10	2L-Pro·X·H_2O	$P2_1$, 2	9.8171(6), 10.4167(11), 10.1713(12),$\beta = 112.078(9)$	I	Athimoolam and Natarajan (2007)
11	2DL-Ala·H_3PO_4	$C2/c$, 4	14.442(7), 10.352(6), 9.062(4), $\beta = 119.08(4)$	I	Averbuch-Pouchot et al. (1988c)
12	(L-PheH)H_2PO_4·L-Phe	$P2_1$, 2	13.441(1), 4.8760(4), 15.470(2), $\beta = 97.12(1)$	II	Ravikumar et al. (2002a)
13	(L-SerH)H_2PO_4·L-Ser·H_2O	$P2_1$, 2	4.704(5), 10.512(5), 13.573(3), $\beta = 98.73(5)$	II	Smolin et al. (2003b)
14	(DL-ValH) picrate·DL-Val	$P2_1/n$, 4	5.3822(4), 23.221(2), 16.579(2), $\beta = 94.695(7)$	II	Anitha et al. (2004b)
15	(SarH) L-tartrate·Sar	$P2_12_12_1$, 4	9.581(2), 9.809(2), 15.811(3)	II	Fleck et al. (2013a)

[a]The authors considered that the composition of the compound is 2Gly·Te(OH)$_6$·H_2O (see the text), I-molecular adduct, II-salt, X = 4-aminobenzoic acid

Three adducts formed by glycine, betaine and sarcosine with telluric acid are known. The compound 2Gly·Te$(OH)_6$·2H_2O was obtained by Andersen et al. (1983), who determined its structure. Qui et al. (1984) determined its structure at 120 K, which allowed determining the positions of hydrogen atoms, and in another study (Qui et al. 1987) the structure of crystal was determined by neutron diffraction method. The unit cell contains two Te$(OH)_6$ molecules, four glycine molecules and four water molecules. At $Z=2$ the composition therefore corresponds to the 2Gly·Te$(OH)_6$·2H_2O formula. Strangely, the authors of all these three structural works erroneously considered that the compound is monohydrate. The molecular mass given by Andersen et al. (1983) and Qui et al. (1987) assumed to be equal $M_r=397.8$, which corresponds to monohydrate, Qui et al. (1984) assumed as $M_r=406$, while molecular mass corresponding to dihydrate is $M_r=415.87$. Telluric acid retaining approximately octahedral symmetry is in special position. Irma Botto (1987) studied thermal behavior and vibrational spectra. She also assumed that the crystal is monohydrate. Above 65 °C dehydration occurs, which allowed the author to obtain also anhydrous sample. In this paper, the IR and Raman spectra and also wavenumbers of peaks of anhydrous sample are provided as well. The Raman spectrum of "monohydrate" does not contain the lines caused by stretching and deformation vibrations of water molecule. We believe having in view easiness of dehydration that the Raman spectrum ascribed to "monohydrate" actually relates to anhydrous sample. It is noteworthy that the author does not provide the Raman spectrum of anhydrous sample, because probably the Raman spectra of "monohydrate" and anhydrous were the same.

Averbuch-Pouchot (1988) obtained the sarcosine analog and Ilczyszyn et al. (1992) obtained the betaine analog. Both crystals were obtained at room temperature by evaporation of aqueous solution containing stoichiometric quantities of components. However, unlike glycine analog they do not contain water of crystallization. In the structure of 2Sar·Te$(OH)_6$ telluric acid as well as sarcosine molecules are in general position, hence asymmetric unit contains one molecule of telluric acid retaining its octahedral structure and two crystallographically independent sarcosine molecules having zwitterionic structure. Telluric acid forms four O-H…O hydrogen bonds with oxygen atoms of carboxylate group of both sarcosine molecules, and also two O-H…O hydrogen bonds with neighboring molecules of telluric acid. So, in the structure there is a chain of telluric acids surrounded by sarcosine molecules.

The structure of 2Bet·Te$(OH)_6$ more reminds the structure of 2Gly·Te$(OH)_6$·2H_2O rather sarcosine analog. Here telluric acid also is in special position (the Te atom is in center of symmetry) and its neighborhood has four betaine molecules connected with hydrogen bonds. The crystal of 2Bet·Te$(OH)_6$ was studied in detail by infrared spectroscopy (Ilczyszyn et al. 1992).

By similar way are formed the compounds: 2L-Pro·succinic acid (Prasad and Vijayan 1993b), 2DL-Val·fumaric acid (Alagar et al. 2003d), 2L-Val·fumaric acid (Klussmann et al. 2007), 2DL-Val·succinic acid (Alagar et al. 2004) and also 2Sar·pyromellitic acid (Domingos et al. 2008). Dicarboxylic (succinic and fumaric) acids are in special positions (in center of symmetry in cases of ##5, 6, 8 and on twofold axis in case of #7) relative to which are symmetrically arranged respective amino acids.

The compound 2Sar·pyromellitic acid was obtained when Domingos et al. (2008) attempted to prepare a copper compound containing pyromellitic acid and sarcosine as ligands. Copper hydroxyfluoride (0.5 mmol) were added to a 20 ml warmed ethanolic solution containing 1.5 mmol of pyromellitic acid and 1.5 mmol of sarcosine. After a few weeks transparent, colorless crystals are formed. The structure of 2Sar·pyromellitic acid reminds the considered structures by that the center of pyromellitic acid (1,2,4,5-benzenetetracarboxylic acid) is in center of symmetry. Sarcosine molecules are arranged symmetrically by both sides of pyromellitic acid. Carboxyl groups form hydrogen bonds with the same oxygen atom of sarcosine.

In case of 2L-Pro·4-aminobenzoic acid monohydrate due to non-centrosymmetrical molecule and optically active proline molecule the compound crystallizes in non-centrosymmetrical structure.

One more compound with 2:1 composition, but without formation of dimeric cation is 2DL-Ala·H_3PO_4 (Averbuch-Pouchot et al. 1988c), which has an interesting structure. The P atom is on a twofold axis and the site of one H atom (the H2) is statistical. There are O-H...O(2.449(2) Å) hydrogen bond between phosphate group and oxygen atom of alanine, and hydrogen bond O-H...O(2.479(2) Å) between phosphate groups. The authors in result of structure determination interpreted the compound 2DL-Ala·H_3PO_4 as molecular adduct, however, because of presence of short hydrogen bond between alanine and phosphate group we do not rule out that in result of more reliable determination of the position of hydrogen atom (e.g. by neutron diffraction) it will appear that the compound is possible interpreting as a salt.

The crystal of L-phenylalanine L-phenylalaninium dihydrogenphosphate was obtained by Ravikumar et al. (2002a) by evaporation of aqueous solution containing stoichiometric quantities of components. Asymmetric unit contains one $H_2PO_4^-$ anion, one L-phenylalanine molecule in form of zwitterion and one L-phenylalaninium cation. However, L-Phe and L-PheH do not form dimeric cation. The L-PheH cation forms O-H...O(2.543(9) Å) hydrogen bond with $H_2PO_4^-$ anion, which in turn forms O-H...O(2.535(9) Å) hydrogen bond with zwitterionic L-Phe. IR and Raman spectra of 2L-Phe·H_3PO_4 are reported as well (Ravikumar et al. 2006). Sujatha et al. (2010) reported crystals of 2L-Phe·H_3PO_4 as possible NLO material. NLO activity was estimated by powder methods as 1.2 times higher than in KDP.

One more salt with 2:1 composition containing $H_2PO_4^-$ anion is 2L-Ser·H_3PO_4·H_2O (Smolin et al. 2003b). Asymmetric unit contains zwitterionic L-Ser, L-SerH cation, $H_2PO_4^-$ anion and one water molecule. In this case dimeric cation is not formed because the carboxyl group of L-SerH cation forms O-H...O(2.562 Å) hydrogen bond with water molecule. Dihydrogenphosphate anion forms rather strong O-H...O (2.481 Å) hydrogen bond with oxygen atom of carboxylate group of L-serine zwitterion, and also with nearest symmetrically equivalent anion O-H...O(2.570 Å). These important hydrogen bonds with distances 2.562 Å and 2.570 Å are absent in table of hydrogen bonds given by Smolin et al. (2003b). It would be interesting to obtain anhydrous 2L-Ser·H_3PO_4 crystal from suitable solvent.

There is a number of picrates of amino acids (both L and DL) with 2:1 composition with dimeric cation, however, the case of 2DL-Val·picric acid is an exception. The crystal of 2DL-Val·picric acid was obtained (Anitha et al. 2004b) by slow evaporation at room temperature of an aqueous solution containing equimolar quantities of components. Structure determination showed that valinium cation forms O-H. . .O(2.684(4) Å) hydrogen bond with deprotonated oxygen atom of picrate anion. Cationic valinium forms hydrogen bond with zwitterionic valine only N-H. . .O type hydrogen bonds.

The last salt sarcosine sarcosinium hydrogen L-tartrate is unique in Table 5.39 by that there are two forms of the same salt with the same symmetry, but with different formation mechanisms (Fleck et al. 2013a). The first form is formed from aqueous solution containing stoichiometric quantities of components. The asymmetric unit contains sarcosine in zwitterionic state, sarcosinium cation and hydrogen L-tartrate anion. The sarcosinium cation forms a strong O-H. . .O(2.4553(15) Å) hydrogen bond with carboxylate group of the hydrogen L-tartrate anion, while the carboxyl group of the hydrogen L-tartrate anion forms an O-H. . .O(2.5065(14) Å) hydrogen bond with zwitterion. The second form is formed at greater content of sarcosine compared to 2:1 molar ratio (2.5:1, 3:1, 3.2:1), which contains dimeric (Sar. . .SarH) cation and hydrogen L-tartrate anion considered in Sect. 5.2.1.2. In this work the vibrational spectra of both forms are studied.

5.2.1.13 Salts with (A. . .A^+) Cation, But Not with 2:1 Composition

In Sects. 5.2.1.1, 5.2.1.2, 5.2.1.3, 5.2.1.4, 5.2.1.5, 5.2.1.6, 5.2.1.7, 5.2.1.8, 5.2.1.9, 5.2.1.10, and 5.2.1.11 salts of amino acids with 2A·HX or 4A·H_2Y compositions containing dimeric (A. . .A^+) cation were considered. However, there exists considerable number of salts of amino acids, which not having 2A·HX or 4A·H_2Y compositions, nevertheless, contain dimeric (A. . .A^+) cation. In Table 5.40 such type salts are collected. Now short consideration of these salts will be performed.

In 1956 simultaneously with the first structure determination of 2Gly·HBr with dimeric cation new ferroelectrics triglycine sulfate (TGS) and triglycine selenate (TGSe) (Matthias et al. 1956) and also one more its analog triglycine fluoroberyllate (Pepinsky et al. 1957) were discovered. Structure determination of TGS showed (Hoshino et al. 1959) that the structure consists of sulfate anion (SO_4^{2-}), GlyH cation and dimeric (Gly. . .GlyH) cation with short hydrogen bonds. Later these crystals were investigated in detail, especially, TGS in view of its great technological importance. There is comprehensive bibliography on TGS, survey of which is out of our task. During long time the number of salts formed by A^+(A. . .A^+)Y^{2-} remains invariable. However, due to Němec (1998) and Němec and Mička (2001b) two more crystals: L-Val^+(L-Val. . .L-Val^+)SO_4^{2-} (TVS) and L-Val^+(L-Val. . .L-Val^+)SeO_4^{2-}

Table 5.40 Salts with (A...A$^+$) cation but not with 2:1 composition

#	Crystal	s.g., Z	*a*, *b*, *c* (Å), *α*, *β*, *γ* (°)	O...O, (Å)	Refs
1	GlyH(Gly...GlyH)SO_4 (TGS)	$P2_1$, 2	9.416(7), 12.646(1), 5.734(3), $\beta = 110.33(3)$	2.470(9)	Choudhury and Chitra (2008)
2	GlyH(Gly...GlyH)SeO_4 (TGSe)	$P2_1$, 2	9.544, 12.719, 5.841, $\beta = 110.46$		Olejnik et al. (1977)
3	GlyH(Gly...GlyH)BeF_4 ($TGBeF_4$)	$P2_1$, 2	9.385(1), 12.632(2), 5.699(1), $\beta = 110.31$	2.4759(4)	Waśkowska et al. (1978)
4	L-ValH(L-Val...L-ValH)SO_4	$P2_12_12_1$, 4	5.782(4), 17.271(4), 23.511(2)	2.5249(12)	Němec (1998)
5	L-ValH(L-Val...L-ValH)SeO_4	–	–	–	Němec and Mička (2001b)
6	GlyH(Gly...GlyH)SiF_6	$P2_1/n$, 4	5.6134(11), 12.900(3), 19.554(4), $\beta = 91.31(3)$	2.458(3)	Fleck et al. (2013b)
7	SarH(Sar...SarH)SiF_6·$2H_2O$	*Pnma*, 4	11.705(2), 14.709(3), 11.113(2)	2.466(2)	Fleck et al. (2012b)
8	L-AlaH(L-Ala...L-AlaH)SiF_6·H_2O	$P2_1$, 2	11.3321(5), 5.8734(3), 13.9471(6), $\beta = 90.806(1)$	2.553(2)	Ghazaryan et al. (2012c)
9	L-ProH(L-Pro...L-ProH)SiF_6·H_2O	$P2_12_12_1$, 4	5.7740(2), 10.0468(3), 38.3966(12)	2.420(4)	Fleck et al. (2013b)
10	(L-Phe...L-PheH)NO_3·2L-Phe	*C*2, 4	30.285(8), 5.3022(13), 22.317(6), $\beta = 102.263(4)$, at 150 K	2.443(9)	Görbitz (2001)
11	(Sar...SarH)(5-nitrosalicylic acid...5-nitrosalicylate)	$P\bar{1}$, 2	5.532(2), 9.760(4), 11.731(5) Å, $\alpha = 68.107(7)$, $\beta = 89.179(7)$, $\gamma = 77.830(7)$	2.492(3)	Jin et al. (2004)
12	(L-Trp...L-TrpH)picrate·picric acid	$P2_12_12_1$, 4	7.7745(7), 15.462(2), 31.088(3)	2.470(6)	Petrosyan et al. (2013b)
13	(GlyH)F	*Pbca*, 32	15.610(2), 14.001(2), 15.662(3)	2.446(2)	Fleck et al. (2010)
14	Na(Gly...GlyH)$_3$(SiF_6)$_2$	$P\bar{1}$, 1	5.5364(1), 10.9918(2), 11.8432(2), $\alpha = 107.631(1)$, $\beta = 102.397(1)$, $\gamma = 93.562(1)$	2.4495(12), 2.4309(16)	Moolya et al. (2007)
15	(Bet...BetH)(BetH)$_2$(Tf_2N)$_3$[a]	$P2_1/c$, 2	8.2731(6), 13.0086(10), 24.9865(18), $\beta = 95.181(3)$	2.639(4), 2.450(5)	Nockemann et al. (2006)

[a]Tf_2N = bis(trifluoromethylsulfonyl)imide

(TVSe) were found, which are formed by the same mechanism. The crystals of TVSe are formed from aqueous solution with molar ratio 1.5:1 (Němec and Mička 2001b). At equimolar proportion 2L-Val·H_2SeO_4·H_2O is formed (see Sect. 5.1.13). The structure was determined only for TVS, however, based on similarity of vibrational spectra is assumed that they are isostructural. Vibrational spectra of TVSe in wide temperature region were investigated, and also by DSC method, which allowed to find a phase transition near 138 K.

New crystalline salts of amino acids forming by the mechanism of TGS (A^+(A...A^+)Y^{2-}) were found during searching of salts of amino acids with H_2SiF_6 acid. Four salts (## 6,7,8,9) showed in Table 5.40 were found. Their crystal structures were determined and vibrational spectra were analyzed (Fleck et al. 2012b, 2013b; Ghazaryan et al. 2012c, 2014b). In these crystals also phase transitions are possible and the crystals deserve for further investigation. In SarH (Sar...SarH)SiF_6·$2H_2O$ a phase transition was found near 180 K caused by disappearing of disordering, in result of which the symmetry change from centrosymmetric *Pnma* to non-centrosymmetric $P2_12_12_1$.

In the system L-Phe + HNO_3 + H_2O in addition to the compound 2L-Phe·HNO_3 considered in Sect. 5.2.1.6 one more compound is formed with 4L-Phe·HNO_3 composition found by Görbitz (2001). The crystal was obtained from aqueous solution containing dilute nitric acid and L-phenylalanine by slow evaporation, however, molar ratio was not indicated. Asymmetric unit contains one nitrate ion, one phenylalaninium cation and additionally three phenylalanine molecules in zwitterionic forms. The L-phenylalaninium cation forms short O-H...O(2.443 (9) Å) hydrogen bond with one of zwitterions, thus forming dimeric cation (L-Phe...L-PheH). Although the structure was determined at 150 K, the accuracy of structure determination is not high ($R = 0.113$). It is possible that this is the result of non-optimal ratio of components in solution. For example C-O bond length of one molecule in zwitterionic form is equal to 1.298(10) Å), which is more characteristic for C-OH bond. In our opinion repeated investigation and more accurate structure determination of this crystal is desired.

The second salt in Table 5.40 is sarcosine 5-nitrosalicylic acid having the 1:1 composition (Jin et al. 2004), was obtained from aqueous solution at equimolar ratio of components, which, however, is not a simple salt "sarcosinium 5-nitrosalicylate" as one may expect, but has an interesting structure. At triclinic $P\bar{1}$ symmetry the cell contains two formula units. One proton from two molecules of 5-nitrosalicylic acid transferred to two molecules of sarcosine, which shared the proton between them, forming thus a dimeric cation (Sar(1/2H)...(1/2H)Sar) with O...O(2.492(3) Å). Remaining 5-nitrosalicylic acid and 5-nitrosalicylate also formed dimeric anion sharing the remaining proton between them with O...O (2.463(3) Å) hydrogen bond. In addition there is intramolecular hydrogen bond between hydroxyl group and carbonyl oxygen atom of carboxyl group with O...O (2.536(3) Å).

One more salt with 1:1, but containing dimeric cation is L-tryptophan picrate obtained and investigated by Ishida et al. (1993). According to the authors, the structure of the crystal consists of two pairs: (1) L-tryptophanium picrate and

(2) L-tryptophan in form of zwitterion and picric acid. Reinvestigation of "L-tryptophan picrate" (Petrosyan et al. 2013b) showed that there is in the structure a dimeric L-tryptophan L-tryptophanium cation, picrate anion and picric acid. In contrast to the previous case picrate anion and picric acid do not form a dimeric anion. The crystal "L-tryptophanium picrate" was obtained with using as solvents ethanol and acetone by Gowri et al. (2011). In the work of Petrosyan et al. (2013b), however, was shown that crystals obtained from ethanol are identical to that obtained from aqueous solution (Petrosyan et al. 2013b; Ishida et al. 1993), while the IR spectrum of samples obtained from acetone does not coincide neither with spectrum of crystals obtained from water and ethanol (Petrosyan et al. 2013b) nor with the spectrum provided by Gowri et al. (2011) for "L-tryptophanium picrate."

One more interesting salt with 1:1 composition, but containing dimeric cation and dimeric anion is Gly·HF (Fleck et al. 2010). The system Gly + HF + H_2O was studied by Frost (1942) with aim to obtain a compound 2Gly·HF after obtaining 2Gly·HCl, 2Gly·HBr and 2Gly·HI, however, the desired 2Gly·HF was not able to obtain. The synthesis of 2Gly·HF (Khandpekar 2003) and Gly·HF (Schmid 1968; Selvaraju et al. 2006a; Vijayan et al. 2009a) was reported. However, convincing evident, and moreover, structure determinations were not reported. We investigated this system (Fleck et al. 2010) with aim to obtain (Gly...GlyH)F, (GlyH)F and (GlyH)(F...HF). Experiments were carried out in polyethylene vessel by slow evaporation at room temperature. It was established that at 2:1 ratio γ-form of glycine is formed, while at ratios 1:1 and 1:2 Gly·HF is formed. The interval 2:1–1:1 was investigated in more detail, however, apart from γ-form of glycine and Gly·HF other phases were not obtained. Structure determination of Gly·HF showed that the compound crystallizes in centrosymmetrical space group *Pbca*, $Z = 32$, $Z' = 4$. It was found that three cations, namely, GlyH(A), GlyH(B), GlyH(C) form strong O-H...F type hydrogen bonds with F(1), F(2), F(3) fluoride anions, while the fourth GlyH(D) and F(4) have other interesting structure. Similarly a dimeric anion is formed with F(4): $(F\text{-}(1/2H)...(1/2H)...F)^-$. Unit cell parameters of the crystal obtained by Fleck et al. (2010) do not correspond with parameters provided by Selvaraju et al. (2006a) and Vijayan et al. (2009a). It is well known that HF reacts with glass. Therefore glass vessels are not used in experiments with HF. In the above works, however, it was not indicated what kind of vessels were used. We decided to check, is it possible to obtain other crystals with using glass vessels, particularly the crystal obtained by the above authors. The result, however, was neither Gly·HF nor the crystal with parameters given by Selvaraju et al. (2006a) and Vijayan et al. (2009a), but the previously known species obtained by Moolya et al. (2007). So, at least for us the nature of crystals obtained by Selvaraju et al. (2006a) and Vijayan et al. (2009a) is not clear. Moolya et al. (2007) used glass vessel for the synthesis of Gly·HF. In result of reaction of HF taken in excess with Na_2O and SiO_2 contained in glass wall and glycine in solution a new compound $Na(Gly...GlyH)_3(SiF_6)_2$ was formed, which contains dimeric cations. At space group $P\bar{1}$ and $Z = 1$ the sodium ion and one of dimeric cations are in special position (1(−)), while the second dimeric cation and SiF_6^{2-} are in general position. The O...O distance of O-H...O hydrogen bond of cation in general position is equal to 2.4495(12) Å), and the O...O distance of O-H...O hydrogen bond in special position is equal to 2.4309(16) Å). Thus, due to violation of the rules of experiments with HF a new salt was discovered, which is the first structurally characterized salt containing amino acid and hexafluorosilicate anion, is the first salt containing SiF_6^{2-} anion and dimeric cation, and also the first salt of hexafluorosilicate containing mixed cation containing amino acid.

The last compound in Table 5.40 has unusual structure. The crystal of (Bet...BetH)$(BetH)_2(Tf_2N)_3$ was obtained from the mixture of the ionic liquids betaine bis (trifluoromethylsulfonyl)imide and 1-butyl-3-methylimidazolium bis (trifluoromethylsulfonyl) imide. The melting point of (Bet...BetH)$(BetH)_2(Tf_2N)_3$ crystal is 70 °C. In the structure there is a tetramer, which contains two different hydrogen bonds. Asymmetric hydrogen bond O1-H23...O3(2.639(4) Å) and symmetric hydrogen bond O4-(1/2H24)...(1/2H24)-O4 with O...O distance equal to 2.450(5) Å). One of Tf_2N moieties is in general position, while the second is in special position and is disordered. The O...O(2.639(4) Å) distance is the longest, if one consider this pair also as dimer.

5.2.2 *Salts with (A⁺...A⁺) Type Dimeric Cations*

In this section we will consider the next type of dimeric cations, namely, $(A^+...A^+)$. The first three salts with dimeric cations of this type: (β-AlaH)$FeCl_4$, (β-AlaH)$_2CuCl_4$ and (β-AlaH)$CuBr_4$, and later also (β-AlaH)$_2SnI_4$, were investigated by Kirkwood and Willett (1979) and Willett et al. (1981, 1983, 2002). Then other similar salts of β-alanine were found (Table 5.41). The crystal of (β-AlaH)ClO_4 was studied also by vibrational spectroscopic methods (Pandiarajan et al. 2005c) and the crystal (β-AlaH)$HC_2O_4\cdot0.5H_2O$ (#10) was obtained and investigated by XRD, vibrational spectroscopy, ^{13}C NMR and DSC also by Godzisz et al. (2003b). Apart from these 11 salts of β-alanine there known also two salts of other than (β-alanine amino acid: (DL-AspH)$NO_3\cdot H_2O$ (#12) and (DL-ValH)ClO_4 (#13) in structures of which dimeric cation of $(A^+...A^+)$ type present as well. The most frequent type of $(A^+...A^+)$ cation is formed by type A shown in Fig. 5.71A. In this case (β-AlaH) cation has *gauche*-conformation and cations in dimer are connected with each other by center of symmetry. Eight dimers out of 11 shown in Table 5.41 belong to this type. Dimeric $(A^+...A^+)$ cation found in the structure of (β-AlaH) NO_3 (Němec et al. 1999b) is similar to type A, however, the cations in dimer are not connected by center of symmetry (type B Fig. 5.71B. In this case O21, O22, O11, O12 are not in the same plane unlike the type A and O21...O12 and O22...O11 distances are different. The next type (type C) was found in the structure of (β-AlaH)$_2$ (biphenyl-4,4′-disulfonate) (#5). Here the cations in dimer also are connected by center of symmetry, however, (β-AlaH) has *trans*-conformation (Fig. 5.71C). Theoretically one may assume by analogy with type B the possibility of existence of type C, but without center of symmetry, which, however, was not found yet. The fourth type was found in the structure of (β-AlaH)$_2SnI_4$, where the two amino ends of both *gauche*-oriented cations are facing in the same direction

Table 5.41 Salts with ($A^+ \ldots A^+$) type dimeric cations

#	Crystal	s.g., Z	a, b, c (Å), α, β, γ (°)	O…O, (Å)	type	Ref
1	(β-AlaH)$FeCl_4$	$P2_1/c$, 4	6.966(1), 11.994(2), 12.444(3), $\beta = 91.01(1)$	2.640(5)	A	Kirkwood and Willett (1979)
2	$(\beta\text{-AlaH})_2CuCl_4$	$I2/c$, 4	7.402(2), 7.734(1), 24.236, $\beta = 91.82(1)$	2.682(1)	A	Willett et al. (1981)
3	$(\beta\text{-AlaH})_2CuBr_4$	$I2/c$, 4	7.761(1), 8.027(1), 24.295(7), $\beta = 92.49(2)$	2.704(1)	A	Willett et al. (1983)
4	$(\beta\text{-AlaH})_2$ SnI_4	$P2_1/c$, 4	20.604(3), 13.1461(17), 8.4823(11), $\beta = 95.75(3)$	2.629(7), 2.646(7)	D	Willett et al. (2002)
5	$(\beta\text{-AlaH})_2$BPDS[a]	$P\bar{1}$, 1	5.4565(5), 7.3922(7), 13.4798(12), $\alpha = 99.197(2)$, $\beta = 90.039(2)$, $\gamma = 101.465(2)$	2.690(2)	C	Liao et al. (2001)
6	$(\beta\text{-AlaH})_2Re_2Cl_8$	$C2/c$, 4	21.0493(13), 7.7180(5), 13.2011(9), $\beta = 116.397(1)$	2.64(5)	A	Shtemenko et al. (2003)
7	(β-AlaH)NO_3	$Pca2_1$, 4	14.8135(11), 7.3338(5), 11.9121(12)	2.638(3), 2.643(3)	B	Němec et al. (1999b)
8	(β-AlaH)ClO_4	$P2_1/n$, 4	7.024(3), 7.556(4), 14.102(4), $\beta = 97.52(4)$	2.726(4)	A	Pandiarajan et al. (2001b)
9	(β-AlaH)BF_4	$P2_1/n$, 4	6.9778(14), 7.4946(15), 13.760(3), $\beta = 97.64(3)$	2.698(2)	A	This work
10	(β-AlaH)X[a]	$C2/c$, 8	22.373(3), 5.6925(7), 14.917(3), $\beta = 115.31(1)$	2.750(2)	A	Krishnakumar et al. (2002)
11	(β-AlaH) picrate	$P\bar{1}$, 2	4.5764(9), 12.060(2), 12.291(3), $\alpha = 104.72(3)$, $\beta = 97.69(3)$, $\gamma = 95.83(3)$	2.636(2)	A	Fleck et al. (2012c)
12	(DL-AspH) $NO_3 \cdot H_2O$	$P\bar{1}$, 2	5.680(1), 7.331(2), 10.812(1), $\alpha = 76.74(1)$, $\beta = 105.33(1)$, $\gamma = 96.41(1)$	2.657		Bahadur and Rajaram (1995b)
13	(DL-ValH)ClO_4	$P\bar{1}$, 8	10.3005(6), 13.0905(8), 14.6105(9), $\alpha = 81.288(2)$, $\beta = 88.447$ (2), $\gamma = 75.965(2)$	2.638(1)[b]		Sridhar et al. (2003)

[a]BPDS = (biphenyl-4,4′-disulfonate) $^{2-}$, X = $HC_2O_4^- \cdot 0.5H_2O$, PES = (S)-(−)-1-phenylethanesulfonate
[b]Out of four ValH independent cations only one forms ($A^+ \ldots A^+$) type dimeric cation

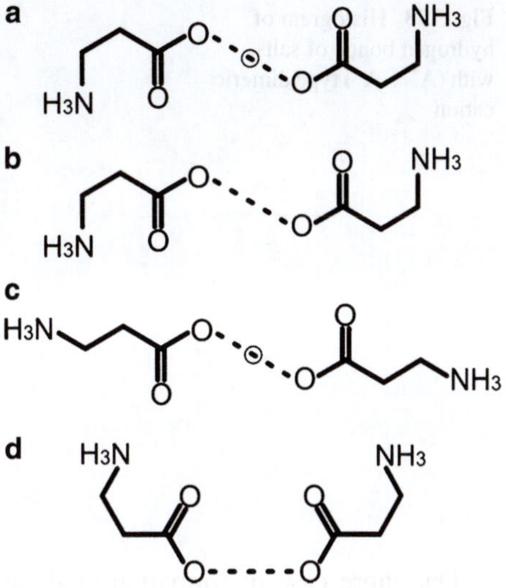

Fig. 5.71 Types of dimeric cations in β-alaninium salts. Two gauche-oriented molecules linked by an inversion center (*A*), two gauche-oriented molecules without inversion center (*B*), two trans-oriented molecules linked by an inversion center (*C*), and two gauche-oriented molecules facing in the same direction (*D*)

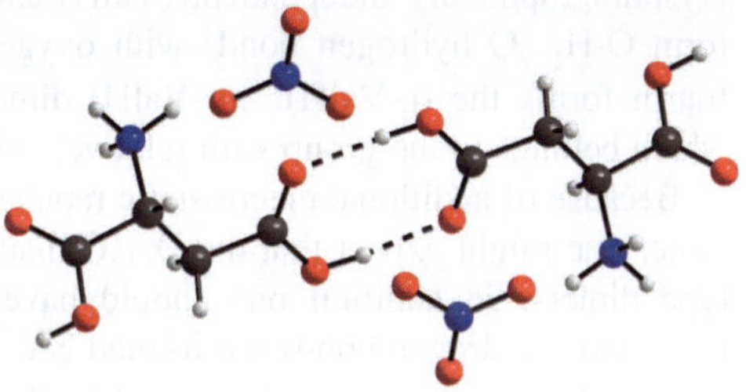

Fig. 5.72 Molecular structure of DL-aspartic acid nitrate monohydrate. The dimeric cation is formed via two symmetric hydrogen bonds

(type D, Fig. 5.71D). It is obvious that here also center of symmetry is absent and O3…O1 and O4…O2 distances are different.

The O…O distances of 11 salts of *β*-alanine in Table 5.41 may be divided into two groups. In one group (## 1,4,6,7,11) the O…O distances are relatively short (the mean value is 2.639 Å), while in second group (## 2,3,5,8,9,10) the O…O distances are longer (the mean value is 2.708 Å). It should be noted, however, that according to Godzisz et al. (2003b) the O…O distance in #10 is equal to 2.659(16) Å, which is intermediate between these two groups. In 1995 it was reported on obtaining of (DL-Asp)$NO_3 \cdot H_2O$ and its structure was determined (Bahadur and Rajaram 1995b). In centrosymmetrical triclinic cell there are two formula units. The L-AspH and D-AspH cations are connected with each other by centrosymmetrical hydrogen bond by means of terminal carboxyl groups and form ($A^+ \ldots A^+$) type dimer shown in Fig. 5.72. Having in view that the dimeric cation is formed by terminal carboxyl groups one may consider it as an analog of dimeric cations of *β*-alanine. This dimeric cation is similar to the type A cation of *β*-alanine. The O…O(2.657 Å) distance may be considered as intermediate. IR and Raman spectra of (DL-Asp)$NO_3 \cdot H_2O$ were studied by Rajkumar et al. (1998).

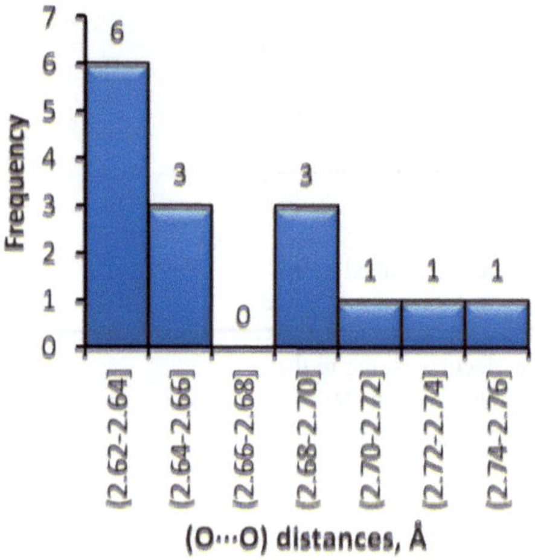

Fig. 5.73 Histogram of hydrogen bonds of salts with (A^+...A^+) type dimeric cation

One more case of formation of dimeric cation of (A^+...A^+) type, where the amino acid is not β-alanine is (DL-ValH)ClO_4 (Sridhar et al. 2003). There are four crystallographically independent (ValH) cations in the structure, three of which form O-H...O hydrogen bonds with oxygen atoms of ClO_4^- anions, while the fourth forms the (L-ValH)...(D-ValH) dimeric cation, with O...O(2.638(1) Å), which belongs to the group with relatively short O...O distances.

Because of additional electrostatic repulsion between cations in (A^+...A^+) type dimer one might expect that the O...O distances will be longer than in (A...A^+) type dimers. In addition one should have in view that in case of (A^+...A^+) type dimer hydrogen bonds are formed between neutral carboxyl groups, while in case of (A...A^+) type dimer between COOH and COO^- groups. Indeed, the longest O...O distance in structure of (Gly...GlyH)Br makes 2.562(3) Å (see Table 5.28), while the shortest O...O distance found in the structure of (β-AlaH)$_2SnI_4$ makes 2.629(7) Å (see Table 5.41). The only exception is the O...O(2.639(4) Å distance found in the structure of unusual tetramer of (Bet...BetH)(BetH)$_2$, in addition to short O...O(2.450(5) Å) distance (see Table 5.40).

In Fig. 5.73 the histogram of hydrogen bonds in crystals with (A^+...A^+) type dimeric cations are shown, which reflects the existence of two types, with relatively short and relatively long O...O distances.

5.2.3 *Salts of Amino Acids with (A^+...A^{2+}) Type Dimeric Cations*

The (A^+...A^{2+}) type dimeric cation on example of (L-Lys^+...L-Lys^{2+}) cation was discovered in the group of Rajaram (Srinivasan et al. 2001f, g) in the structures of mixed salts (L-Lys^+...L-Lys^{2+})2Cl^-.ClO_4^- (#1) and (L-Lys^+...L-Lys^{2+})2Cl^-.NO_3^- (#5) with different anions. The work of Briget Mary et al. (2005) deals with investigation of vibrational spectra of (L-Lys^+...L-Lys^{2+})2Cl^-.NO_3^-, however, the interpretation of the spectra is not correct, because as shown by Ghazaryan et al. (2010b) and Petrosyan et al. (2009b), the provided spectra do not belong to this crystal, but to L-Lys^{2+}·Cl^-·NO_3^-. The crystal of (L-Lys^+...L-Lys^{2+})2Cl^-.NO_3^- was grown and characterized as possible NLO material (Vasudevan et al. 2011b, c), and the crystal of (L-Lys^+...L-Lys^{2+})2Cl^-.ClO_4^- was grown and characterized by Vasudevan et al. (2012).

Petrosyan et al. (2009b) started a systematic investigation of mixed salts of amino acids with different anions. Particularly, investigations on obtaining of salts with (A^+...A^{2+}) type dimeric cation were continued. New analogs with (L-Lys^+...L-Lys^{2+}) were obtained shown in Table 5.42, particularly, the existence of salts with compositions (L-Lys^+...L-Lys^{2+})2BF_4^-.Cl^-, (L-Lys^+...L-Lys^{2+}) 2ClO_4^-.Cl^- and respective bromides was established. Also for the first time some mixed salts with dimeric (L-Orn^+...L-Orn^{2+}) cation were obtained. The crystal of (L-Orn^+...L-Orn^{2+})2Cl^-.ClO_4^- has the same symmetry and close cell parameters as in case of respective salt of L-lysine, however, the crystal of (L-Orn^+...L-Orn^{2+}) 2Cl^-.NO_3^- unlike the respective salt of L-lysine crystallizes in monoclinic system.

The dimeric (L-His^+...L-His^{2+}) cation was found in the structure of (L-His^+...L-His^{2+}) NO_3^-.SiF_6^{2-} mixed salt (Ghazaryan et al. 2012e), while in structure of (L-His^+...L-His^{2+})·L-His^{2+}·3NO_3^-·SO_4^{2-}·H_2O it presents in addition to the L-His^{2+} cation. Crystals with (A^+...A^{2+}) type dimeric cation are observed only in case of mixed salts. In more detailed they are considered in Sect. 5.3.

In Fig. 5.74 the histogram of O-H...O hydrogen bonds in structures of salts with (A...A^+) dimeric cations are shown. It includes salts with 2:1 and other compositions and also mixed salts with (A^+...A^{2+}) type dimeric cations of L-lysine, L-ornithine and L-histidine. The histogram is similar to that of salts with 2:1 composition with the exception of the presence of O-H...O(2.639(4) Å) hydrogen bond in the structure of (Bet...BetH)$(BetH)_2(Tf_2N)_3$ (see Sect. 5.2.1.13), where unusual tetramer exists.

5.3 Mixed Salts of Amino Acids with Different Anions and Cations

5.3.1 Introductory Notes

In Sect. 5.1 simple salts of amino acids consisting of protonated amino acid as cation and definite anion were considered, and in Sect. 5.2 salts of amino acids containing three types of dimeric cations: (1) (A...A^+), (2) (A^+...A^+) and (3) (A^+...A^{2+}) were considered. Salts of first two types also contain anions of one definite type. From chemical and crystallochemical and also from viewpoint of obtaining new materials salts of amino acids with different anions, and also with different cations are of obvious interest.

Vilminot and Cot (1973) obtained double salts (NH_4)(GlyH)SO_4 and K(GlyH) SO_4 in which glycine acts as one of cations. Crystals were obtained by evaporation

Table 5.42 Salts of amino acids with ($A^{+}...A^{2+}$) type dimeric cations

#	Crystal	s.g., Z	a, b, c (Å), α, β, γ (°)	O...O, (Å)	Refs
1	(L-Lys^{+}...L-Lys^{2+})·$2Cl^{-}$·ClO_4^{-}	$P2_12_12_1$, 4	5.0264(6), 20.822(2), 20.897(2)	2.453(7)	Srinivasan et al. (2001f)
2	(L-Lys^{+}...L-Lys^{2+})·$2Br^{-}$·ClO_4^{-}	–	–	–	This work
3	(L-Lys^{+}...L-Lys^{2+})·$2Cl^{-}$·BF_4^{-}	$P2_12_12_1$, 4	5.0228(2), 20.7017(6), 20.7853(6)	2.431(3)	Ghazaryan et al. (2010c)
4	(L-Lys^{+}...L-Lys^{2+})·$2Br^{-}$·BF_4^{-}	–	–	–	This work
5	(L-Lys^{+}...L-Lys^{2+})·$2Cl^{-}$·NO_3^{-}	$P2_12_12_1$, 4	4.9846(2), 20.604(2), 20.663(2)	2.425(8)	Srinivasan et al. (2001g)
6	(L-Lys^{+}...L-Lys^{2+})·$2Br^{-}$·NO_3^{-}	–	–	–	Petrosyan et al. (2009b)
7	(L-Lys^{+}...L-Lys^{2+})·$2BF_4^{-}$·Cl^{-}	$P2_1$, 2	11.0866(4), 5.2110(2), 19.9942(8), β = 97.163(2)	2.424(4)	Petrosyan et al. (2009b)
8	(L-Lys^{+}...L-Lys^{2+})·$2ClO_4^{-}$·Cl^{-}	$P2_1$, 2	11.2238(11), 5.2268(6), 20.104(2), β = 98.791(5)	2.466	Petrosyan et al. (2009b)
9	(L-Lys^{+}...L-Lys^{2+})·$2BF_4^{-}$·Br^{-}	–	–	–	This work
10	(L-Lys^{+}...L-Lys^{2+})·$2ClO_4^{-}$·Br^{-}	–	–	–	This work
11	(L-Orn^{+}...L-Orn^{2+})·$2Cl^{-}$·NO_3^{-}	$P2_1$, 2	11.126(4), 5.1564(17), 16.612(6), β = 100.808(14)	2.447(10)	This work
12	(L-Orn^{+}...L-Orn^{2+})·$2Br^{-}$·NO_3^{-}	–	–	–	Petrosyan et al. (2009b)
13	(L-Orn^{+}...L-Orn^{2+})·$2Cl^{-}$·BF_4^{-}	–	–	–	Petrosyan et al. (2009b)
14	(L-Orn^{+}...L-Orn^{2+})·$2Cl^{-}$·ClO_4^{-}	$P2_12_12_1$, 4	5.0875(6), 18.8122(8), 20.4059(9)	2.427(4)	This work
15	(L-Orn^{+}...L-Orn^{2+})·$2Br^{-}$·BF_4^{-}	–	–	–	This work
16	(L-Orn^{+}...L-Orn^{2+})·$2Br^{-}$·ClO_4^{-}	–	–	–	This work
17	(L-His^{+}...L-His^{2+})·NO_3^{-}·SiF_6^{2-}	$P2_1$, 2	5.9967(3), 19.914(1), 8.4646(5), β = 99.770(2)	2.507(2)	Ghazaryan et al. (2012e)
18	(L-His^{+}...L-His^{2+})·L-His^{2+}·X[a]	$P2_12_12_1$, 4	6.014(2), 10.536(3), 50.232(8)	2.441(9)	This work

[a]X = $3NO_3^{-}$·SO_4^{2-}·H_2O

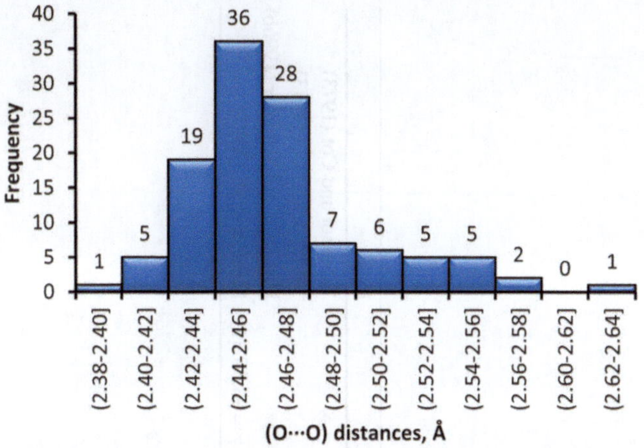

Fig. 5.74 Histogram of O-H...O hydrogen bonds of all salts with dimeric cations except (A^+...A^+) type cations

of aqueous solution containing equimolar quantities of ammonium sulfate (or potassium) and diglycine sulfate at ambient temperature. The structure of $(NH_4)(GlyH)SO_4$ was determined by XRD Vilminot et al. (1974) and neutron diffraction (Vilminot et al. (1976). Double salts $(NH_4)(GlyH)SO_4$ and K(GlyH) SO_4 crystallize in centrosymmetrical structure (space group $P2_1/c$). Later Haussühl and Schreuer (1996b) obtained a similar salt, but with betaine $(NH_4)(BetH)SO_4$ with $P\overline{1}$ symmetry, without quotation of previous publications (Vilminot and Cot 1973; Vilminot et al. 1974, 1976). Crystals were obtained by evaporation of aqueous solution containing equimolar quantities of betaine in form of monohydrate and ammonium hydrogen sulfate. Large crystals were grown by evaporation method at 32 °C. Crystal structure, elastic properties and thermal expansion coefficients were determined.

One more salt with different cations and interesting structure is Na $(Gly...GlyH)_3(SF_6)_2$ (Moolya et al. 2007). The authors decided to obtain Gly·HF. It is known that the HF acid react with glass, therefore for reactions with participation of HF glass vessels is not used, but vessels, e.g. from polyethylene. The authors fortunately did not know or ignored this rule, which allowed to discover a new salt with different cations, salt containing ($Gly...Gly^+$) dimeric cations and the first salt containing amino acid and SiF_6^{2-} anion the structure of which was determined. In Table 5.43 mixed salts with different cations containing amino acid cation are provided.

Now we consider salts of amino acids with different anions. Saari and Anderson patented chemical compounds or mixtures of compounds having cariostatic properties (Saari and Anderson 1977), among them mixed fluoride-phosphates of L-lysine, L-hydroxylysine, L-ornithine, L-arginine and L-histidine. However, the authors have not provided any data (IR spectra, powder X-ray diffraction patterns, unit cell parameters, not speaking already about determination the structures) confirming the structure of the compounds. Probably they represent stoichiometric mixtures.

Table 5.43 Double salts, which include amino acids as cation

#	Crystal	s.g., Z	*a*, *b*, *c* (Å), *α*, *β*, *γ* (°)	Refs
1	$(NH_4)(GlyH)SO_4$	$P2_1/c$, 4	8.262(7), 10.073(8), 8.632(8), $\beta = 92.66(6)$	Vilminot and Cot (1973)
2	$K(GlyH)SO_4$	$P2_1/c$, 4	5.272(6), 15.377(10), 8.635(8), $\beta = 94.47(6)$	Vilminot and Cot (1973)
3	$(NH_4)(BetH)SO_4$	$P\bar{1}$, 2	6.829(1), 7.239(2), 10.282(2), $\alpha = 88.78(2)$, $\beta = 88.18(2)$, $\gamma = 82.71(2)$	Haussühl and Schreuer (1996b)
4	$Na(Gly\ldots GlyH)_3(SiF_6)_2$	$P\bar{1}$, 1	5.5364(1), 10.9918(2), 11.8432(2), $\alpha = 107.631(1)$, $\beta = 102.397(1)$, $\gamma = 93.562(1)$	Moolya et al. (2007)

The first mixed salts (L-Lys^{+}...L-Lys^{2+})2Cl^{-}.ClO$_4^{-}$ and (L-Lys^{+}...L-Lys^{2+}) 2Cl^{-}.NO$_3^{-}$ with different anions were obtained by Srinivasan et al. (2001f, g) with determination of their crystal and molecular structures. Then one more salt with 2L-Orn^{2+}·Cl^{-}·NO$_3^{-}$·SO$_4^{2-}$ composition was obtained (Ramaswamy et al. 2004b). In the same period Shtemenko et al. (2003) synthesized a salt with 4Gly^{+}·2Cl^{-}·Re$_2$Cl$_8^{2-}$ composition in course of obtaining the salt of 2Gly^{+}·Re$_2$Cl$_8^{2-}$·H$_2$O crystal.

Our interest to the mixed salts of amino acids with different anions arose after finding of salts of L-arginine with doubly charged cation L-Arg^{2+}·2X^{-} (Petrosyan et al. 2000b). It was intended to check the possibility of existence of L-Arg^{2+}·X(1)$^{-}$·X(2)$^{-}$ type salts, where X(1)$^{-}$ and X(2)$^{-}$ are different anions. Initial experiments with various combinations of following acids: HCl, HBr, HNO$_3$, HIO$_3$, H$_3$PO$_4$ and HCOOH did not give positive results (Petrosyan 2004). However, the discovery of mixed salts (Srinivasan et al. 2001f, g; Ramaswamy et al. 2004b) strengthened the hope of possibility of their obtaining and allowed to look to this problem wider. Based on formation mechanisms of simple salts (Petrosyan 2004) and two formation mechanisms of mixed salts, 15 possible formation mechanisms of mixed salts with different anions were proposed (Petrosyan et al. 2009b). These mechanisms served as a basis for goal-seeking search and their systematic investigation. Below in Sect. 5.3.2 we will consider the results obtained in our group and also by other authors.

5.3.2 *Mixed Salts of Amino Acids with Different Anions*

5.3.2.1 Mixed Salts of Amino Acids with Singly Charged Cations

In Table 5.44 mixed salts with singly charged glycinium and L-lysinium cations are shown. Of these, only the structure of 4Gly^{+}·2Cl^{-}·Re$_2$Cl$_8^{2-}$ was determined (Shtemenko et al. 2003).

The quality of mixed salts of singly charged L-lysinium did not allow determining their structures. However, based on the IR spectra (Ghazaryan 2014) we assume that 2L-Lys^{+}·Cl^{-}·BF$_4^{-}$ and 2L-Lys^{+}·Cl^{-}·ClO$_4^{-}$ salts are formed. In Fig. 5.75 the IR spectrum of 2L-Lys^{+}·Cl^{-}·BF$_4^{-}$ together with spectra of two forms of anhydrous L-Lys^{+}·Cl^{-} and L-Lys^{+}·BF$_4^{-}$ are shown. In the IR spectrum of 2L-Lys^{+}·Cl^{-}·BF$_4^{-}$ there are characteristic bands of singly charged L-lysinium cation and also BF$_4^{-}$ anion. The spectrum of 2L-Lys^{+}·Cl^{-}·BF$_4^{-}$ is not the sum of L-Lys^{+}·Cl^{-} and L-Lys^{+}·BF$_4^{-}$ spectra, which confirms the formation of the 2L-Lys^{+}·Cl^{-}·BF$_4^{-}$ salt. The salt 2L-Lys^{+}·Cl^{-}·ClO$_4^{-}$ has similar spectrum (Ghazaryan 2014). The final conclusion on existence of these salts will be possible to make after structure determination.

Table 5.44 Mixed salts with different anions of singly charged glycinium and L-lysinium cations

#	Crystal	s.g., Z	*a*, *b*, *c* (Å), *α*, *β*, *γ* (°)	Refs
1	$4Gly^{+}\cdot 2Cl^{-}\cdot Re_2Cl_8^{2-}$	*P4/n*, 2	11.7356(4), 11.7356(4), 9.7051(5), at 223 K	Shtemenko et al. (2003)
2	2L-$Lys^{+}\cdot Cl^{-}\cdot BF_4^{-}$	–	–	Ghazaryan (2014)
3	2L-$Lys^{+}\cdot Cl^{-}\cdot ClO_4^{-}$	–	–	Ghazaryan (2014)

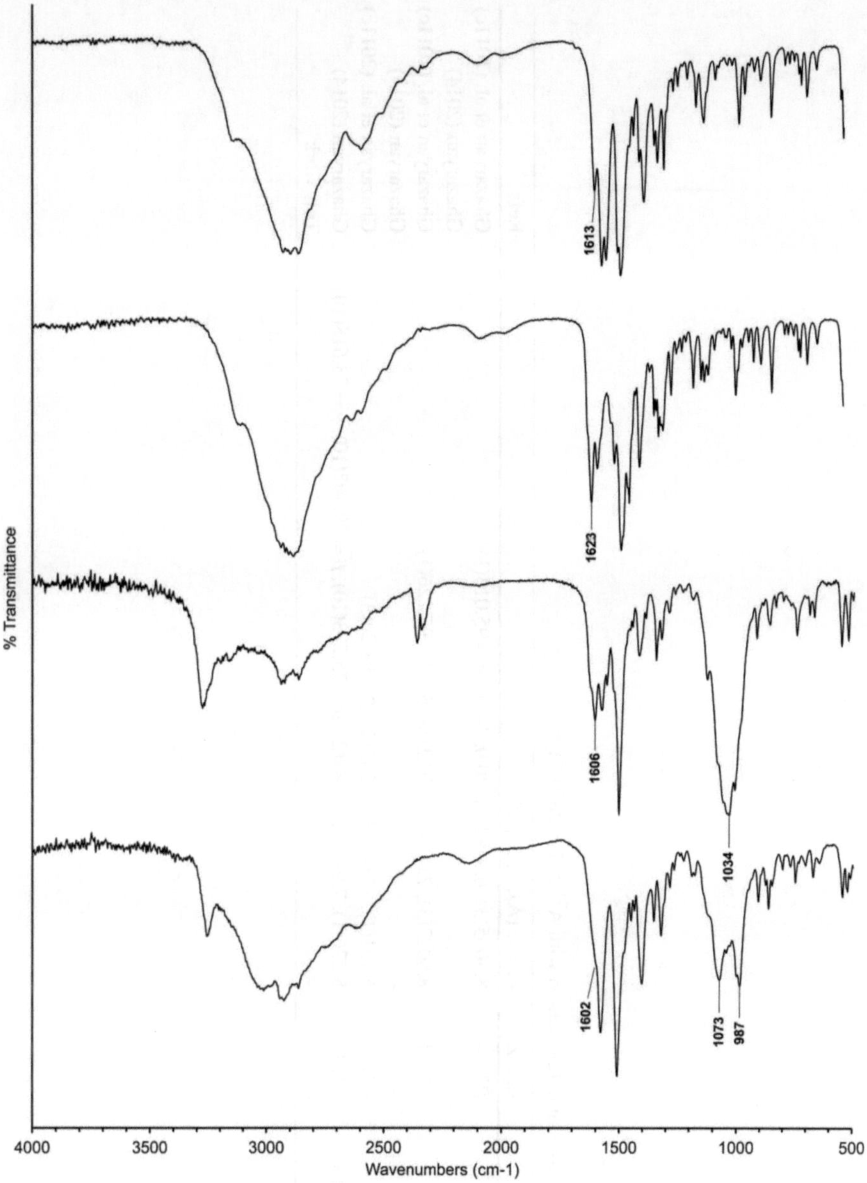

Fig. 5.75 IR spectra of 2L-Lys^{+}·Cl^{-}·BF_4^{-}, L-Lys^{+}·BF_4^{-} and two (β and α) forms of anhydrous L-Lys^{+}·Cl^{-} (from *bottom* to *top*)

5.3.2.2 Mixed Salts of A^{2+}·$X(1)^{-}X(2)^{-}$ Type with L-Lysinium(2+) Cation

We established the possibility of formation of A^{2+}·$X(1)^{-}$·$X(2)^{-}$ type salts with doubly charged L-lysinium(2+) cation. The obtained salts are collected in Table 5.45. These salts are formed from aqueous solution with stoichiometric

Table 5.45 Mixed salts of L-lysinium(2+) cation with $A^{2+}\cdot X(1)^{-}\cdot X(2)^{-}$ type

#	Crystal	s.g., Z	*a*, *b*, *c* (Å), *α*, *β*, *γ* (°)	Refs
1	L-$Lys^{2+}\cdot Cl^{-}\cdot BF_4^{-}$	$P2_1$, 4	8.8005(3), 6.9492(2), 20.0344(6), $\beta=95.013(1)$	Ghazaryan et al. (2011c)
2	L-$Lys^{2+}\cdot Br^{-}\cdot BF_4^{-}$	–	–	Ghazaryan (2014)
3	L-$Lys^{2+}\cdot Cl^{-}\cdot ClO_4^{-}$	$P2_1$, 4	8.8022(3), 7.0096(2), 20.2169(6), $\beta=95.226(1)$	Ghazaryan et al. (2011c)
4	L-$Lys^{2+}\cdot Br^{-}\cdot ClO_4^{-}$	–	–	Ghazaryan (2014)
5	L-$Lys^{2+}\cdot Cl^{-}\cdot NO_3^{-}$	$P2_1$, 2	5.161(1), 13.085(3), 8.543(2), $\beta=98.26(3)$	Ghazaryan et al. (2011c)
6	L-$Lys^{2+}\cdot Br^{-}NO_3^{-}\cdot H_2O$	$P1$, 1	5.276(1), 7.973(1), 8.387(2), $\alpha=75.773(10)$, $\beta=79.407(10)$, $\gamma=73.618(10)$	Ghazaryan (2014)
7	L-$Lys^{2+}\cdot BF_4^{-}\cdot ClO_4^{-}$	–	–	This work

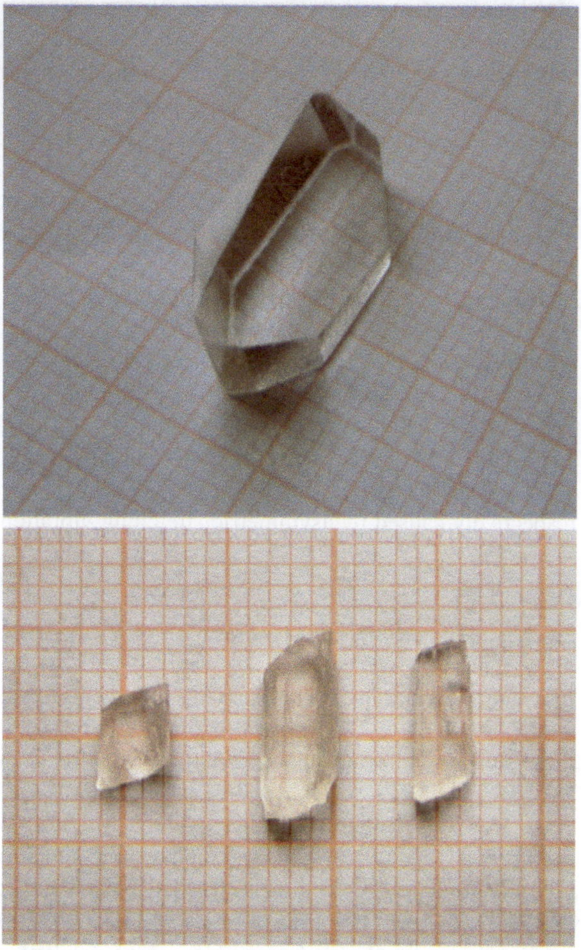

Fig. 5.76 Crystals of L-$Lys^{2+}{\cdot}Cl^{-}{\cdot}NO_3^{-}$ (*top*) and L-$Lys^{2+}{\cdot}Br^{-}{\cdot}NO_3^{-}{\cdot}H_2O$ (*bottom*)

ratio by slow evaporation. The structures and vibrational spectra of L-$Lys^{2+}{\cdot}Cl^{-}{\cdot}BF_4^{-}$, L-$Lys^{2+}{\cdot}Cl^{-}{\cdot}ClO_4^{-}$ and L-$Lys^{2+}{\cdot}Cl^{-}{\cdot}NO_3^{-}$ were determined and discussed by Ghazaryan et al. (2010c). The spectra of L-$Lys^{2+}{\cdot}Br^{-}{\cdot}BF_4^{-}$ and L-$Lys^{2+}{\cdot}Br^{-}{\cdot}ClO_4^{-}$ are similar to respective spectra of chlorides, which allow supposing that their structures also are similar. However, the respective bromide of L-$Lys^{2+}{\cdot}Cl^{-}{\cdot}NO_3^{-}$ crystallizes as monohydrate with triclinic symmetry: L-$Lys^{2+}{\cdot}Br^{-}{\cdot}NO_3^{-}{\cdot}H_2O$. Crystals of L-$Lys^{2+}{\cdot}Cl^{-}{\cdot}NO_3^{-}$ and L-$Lys^{2+}{\cdot}Br^{-}{\cdot}NO_3^{-}{\cdot}H_2O$ grown by evaporation at 25 °C are shown in Fig. 5.76. It is interesting to note that the structure of L-$Lys^{2+}{\cdot}Cl^{-}{\cdot}NO_3^{-}$ differs from that of L-$Lys^{2+}{\cdot}Cl^{-}{\cdot}BF_4^{-}$, L-$Lys^{2+}{\cdot}Cl^{-}{\cdot}ClO_4^{-}$ by that its asymmetric unit has one formula unit, while asymmetric units of last two crystals have each two formula units. Carboxyl groups of L-lysinium(2+) cations in all three structures form O-H...Cl hydrogen bonds with chloride anions. In the structure of L-Lys^{2+}·Br^-·NO_3^-·H_2O carboxyl group of cation forms hydrogen bond not with bromide or nitrate anions, but with water molecule: O1-H1...O1W(2.575(6) Å). Probably this is the reason of its thermal behavior. Heating of L-Lys^{2+}·Br^-·NO_3^-·H_2O above 50 °C leads to its melting without dehydration. The IR spectrum of sample obtained by cooling of liquid heated up to 70 °C coincides with initial spectrum of monohydrate. Our attempts to obtain respective mixed salts with fluoride were not successful. In each case the simple salt L-Lys·HF was formed.

However, we consider that the following salts are possible to obtain: L-Lys^{2+}·I^-·BF_4^-, L-Lys^{2+}·I^-·ClO_4^-, L-Lys^{2+}·I^-·NO_3^-, L-Lys^{2+}·Cl^-·Br^-, L-Lys^{2+}·Cl^-·I^-, L-Lys^{2+}·Br^-·I^-. We did not check them yet.

Formation of L-Lys^{2+}·BF_4^-·ClO_4^- was confirmed by its IR spectrum. The presence of cation is reflected by presence of ν(C=O) of COOH group at 1,736 cm^{-1}, and the presence of BF_4^- and ClO_4^- anions is reflected by presence of characteristic strong absorption bands with peaks at 1,015 and 1,050 cm^{-1} respectively caused by asymmetric stretching vibrations.

5.3.2.3 Mixed Salts of L-Lysine with (A^+...A^{2+}) Type Dimeric Cations

In result of further investigations we could increase the number of salts of L-lysine with (L-Lys^+...L-Lys^{2+}) dimeric cation in addition to two known such salts. All these salt are collected in Table 5.46. It was found that in addition to salts with $2Cl^-.BF_4^-$ and $2Cl^-.ClO_4^-$ anions, salts with $2BF_4^-.Cl^-$ and $2ClO_4^-.Cl^-$ also can exist. However, unlike the first two salts, which crystallize in the orthorhombic system, the second two salts crystallize in monoclinic system. In Fig. 5.77 the vibrational spectra of salts with $2Cl^-.BF_4^-$ and $2BF_4^-.Cl^-$ anions are shown. The O...O distances in dimeric cations are close: 2.431(3) Å and 2.424(4) Å respectively. However, the difference of anions (and respectively the difference of N-H...F and N-H...Cl hydrogen bonds) leads to a significant difference of spectra in 2,800–3,300 cm^{-1} region. The characteristic feature of the presence of dimeric cation with strong hydrogen bond is the presence of the band near 1,680 cm^{-1}. As one may expect the intensity of the line of $\nu_s(BF_4)$ (near 772 cm^{-1}) nearly two times higher in case of $2BF_4^-.Cl^-$ anions compared with case of $2Cl^-.BF_4^-$ anions.

We consider that other salts with dimeric (L-Lys^+...L-Lys^{2+}) cations are possible to obtain with following anions: $2I^-$·ClO_4^-, Cl^-·Br^-·ClO_4^-, Cl^-·I^-·ClO_4^-, Br^-·I^-·ClO_4^-, Cl^-·Br^-·BF_4^-, Cl^-·I^-·BF_4^-, Br^-·I^-·BF_4^-, $2I^-$·NO_3^-,Cl^-·Br^-·NO_3^-, Cl^-·I^-·NO_3^-, Br^-·I^-·NO_3^-, $2BF_4^-$·I^-, $2ClO_4^-$·I^-, Cl^-·BF_4^-·ClO_4^-, Br^-·BF_4^-·ClO_4^-, I^-·BF_4^-·ClO_4^-, Cl^-·NO_3^-·ClO_4^-, Br^-·NO_3^-·ClO_4^-, I^-·NO_3^-·ClO_4^-, Cl^-·NO_3^-·BF_4^-, Br^-·NO_3^-·BF_4^-, I^-·NO_3^-·BF_4^-, NO_3^-·BF_4^-·ClO_4^-.

Table 5.46 Mixed salts of L-lysine with ($A^{+}...A^{2+}$) type dimeric cations

#	Crystal	s.g., Z	a, b, c (Å), α, β, γ (°)	Refs
1	(L-Lys^{+}...L-Lys^{2+})·$2Cl^{-}$·ClO_4^{-}	$P2_12_12_1$, 4	5.0264(6), 20.822(2), 20.897(2)	Srinivasan et al. (2001f)
2	(L-Lys^{+}...L-Lys^{2+})·$2Br^{-}$·ClO_4^{-}	$P2_12_12_1$, 4	5.1084(10), 20.634(4), 21.039(4)	This work
3	(L-Lys^{+}...L-Lys^{2+})·$2Cl^{-}$·BF_4^{-}	$P2_12_12_1$, 4	5.0228(2), 20.7017(6), 20.7853(6)	Ghazaryan et al. (2010b); Petrosyan et al. (2009b)
4	(L-Lys^{+}...L-Lys^{2+})·$2Br^{-}$·BF_4^{-}	$P2_12_12_1$, 4	5.1612(5), 20.843(2), 21.055(2)	This work
5	(L-Lys^{+}...L-Lys^{2+})·$2I^{-}$·BF_4^{-}	–	–	This work
6	(L-Lys^{+}...L-Lys^{2+})·$2Cl^{-}$·NO_3^{-}	$P2_12_12_1$, 4	4.9846(2), 20.604(2), 20.663(2)	Srinivasan et al. (2001g)
7	(L-Lys^{+}...L-Lys^{2+})·$2Br^{-}$·NO_3^{-}	–	–	Petrosyan et al. (2009b)
8	(L-Lys^{+}...L-Lys^{2+})·$2BF_4^{-}$·Cl^{-}	$P2_1$, 2	11.0866(4), 5.2110(2), 19.9942(8), $\beta = 97.163(2)$	This work; Petrosyan et al. (2009b)
9	(L-Lys^{+}...L-Lys^{2+})·$2ClO_4^{-}$·Cl^{-}	$P2_1$, 2	11.2238(11), 5.2268(6), 20.104(2), $\beta = 98.791(5)$	This work; Petrosyan et al. (2009b)
10	(L-Lys^{+}...L-Lys^{2+})·$2BF_4^{-}$·Br^{-}	–	–	This work
11	(L-Lys^{+}...L-Lys^{2+})·$2ClO_4^{-}$·Br^{-}	–	–	This work

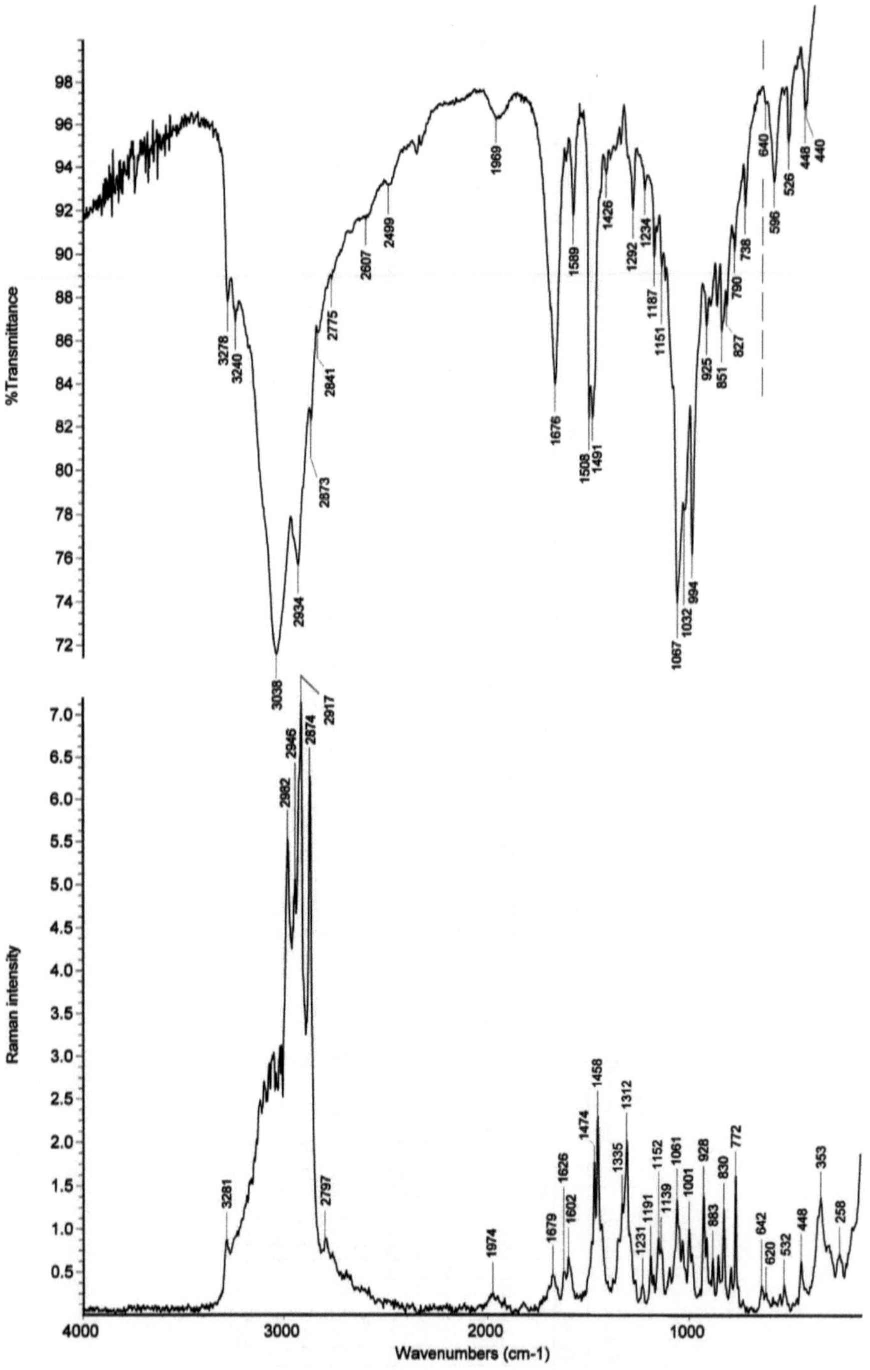

Fig. 5.77 IR spectra of (L-Lys$^+$…L-Lys^{2+}) $2Cl^-{\cdot}BF_4^-$ and (L-Lys$^+$…L-Lys^{2+}) $2BF_4^-{\cdot}Cl^-$

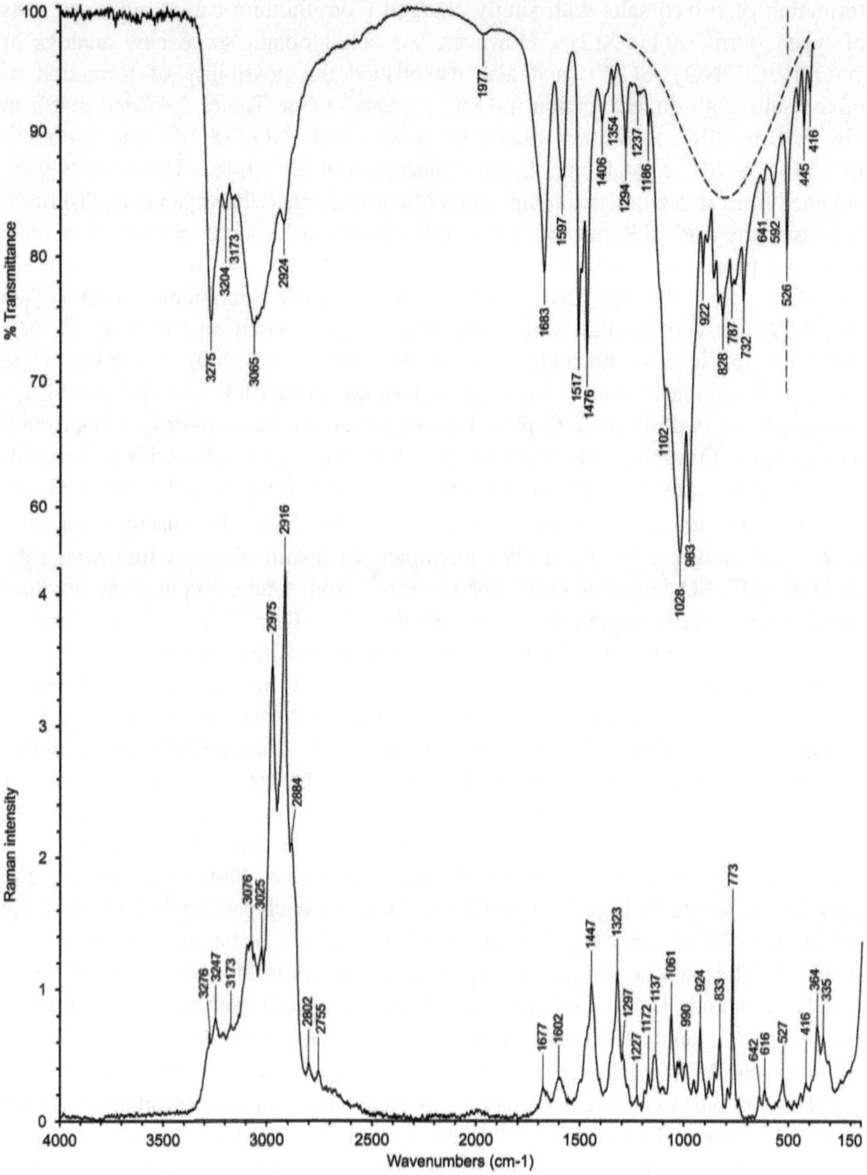

Fig. 5.77 (continued)

5.3.2.4 Mixed Salts of L-Ornithinium(2+) Cation

Because of some similarity of lysine and ornithine it was interesting to investigate the possibility of formation of similar salts also with ornithine. We could not observe the formation of mixed salts with singly charged L-ornithinium cation, and also salts of type L-Orn^{2+}·$X(1)^-$·$X(2)^-$. However, we could obtain some new analogs of 2L-Orn^{2+}·Cl^-·NO_3^-·SO_4^{2-}, and also established the possibility of formation of mixed salts with dimeric cation (L-Orn^+...L-Orn^{2+}) (see Tables 5.47 and 5.48). In Fig. 5.78 the IR and Raman spectra of 2L-Orn^{2+}·Cl^-·NO_3^-·SO_4^{2-} and 2L-Orn^{2+}·Br^-·NO_3^-·SO_4^{2-} crystals are shown (Ghazaryan et al. 2011c). The crystals were obtained from stoichiometric composition of components (Ghazaryan et al. 2011c) in contrast to the work of Ramaswamy et al. (2004b) where the components were taken in equimolar proportion. Ghazaryan et al. (2011c) also obtained isostructural 2L-Orn^{2+}·2-Cl^-·SO_4^{2-} and 2L-Orn^{2+}·2Br^-·SO_4^{2-} crystals. Their vibrational spectra (see Fig. 5.79) are very similar, which one might expect based on similarity of their structures. In Fig. 5.80 molecular structure of 2L-Orn^{2+}·2Cl^-·SO_4^{2-} is shown. Here also L-ornithinium(2+) cation is in general position, while $Cl(1)^-$, $Cl(2)^-$ and SO_4^{2-} anions are on twofold axis. Carboxyl group of cation forms hydrogen bond with sulfate anion. Oxygen atoms O12 and O12a form hydrogen bonds with two crystallographically equivalent cations O1-H1...O12 and O1-H1...O12a with O...O distances equal to 2.561(1) Å. In case of 2L-Orn^{2+}·2Br^-·SO_4^{2-} this distance is slightly longer and makes 2.583(6) Å. We attempted to obtain also the following salts 2L-Orn^{2+}·2F^-·SO_4^{2-} and 2L-Orn^{2+}·2NO_3^-·SO_4^{2-} from solutions containing stoichiometric quantities of components. In first case the salt 2L-Orn^{2+}·SO_4^{2-}·HF was formed, which may be obtained also from stoichiometric composition. Strictly speaking it is not a mixed salt, because in the structure the neutral HF molecule is present instead of fluoride ion. This salt by its structure and spectra displays great similarity with 2L-Orn^{2+}·SO_4^{2-}·H_2O (Fleck et al. 2014; see Table 5.20). It will be interesting to obtain and compare the salt 2L-Orn^{2+}·2I^-·SO_4^{2-}. In second case obtained salt is proved to be 3L-Orn^{2+}·2NO_3^-·2SO_4^{2-}, which, however, was not able to obtain from solution with stoichiometric composition. Hexagonal symmetry of 3L-Orn^{2+}·2NO_3^-·2SO_4^{2-} was not observed previously in crystals of salts of ornithine. Molecular structure and packing are shown in Figs. 5.81 and 5.82. There are each two crystallographycally independent NO_3^- and SO_4^{2-} anions, which are on axis 6_3 having thus point symmetry 3. The sulfate oxygen atom O8 is an acceptor of two hydrogen bonds (O1-H...O8 and N1-H...O8) whereas O7 does not participate in any, resulting in one very short and three moderately long S-O distances (S2-O7(1.424(4) Å) and S2-O8(1.480(2) Å)). This also is true for O5 and O6. The atom O6 is an acceptor for N-H...O type two bonds. In this case the discrepancy is not so prominent (S1-O5 (1.458(3) Å) and S1-O6(1.484(2) Å)). The structure is found to be highly polar (see Fig. 5.82). All the sulfate tetrahedra "point" into one direction, namely along [001]. Vibrational spectra were discussed in (Fleck et al. 2014).

We consider that some other similar salts may be obtained: 2L-Orn^{2+}·2I^-·SO_4^{2-}, 2L-Orn^{2+}·Cl^-·Br^-·SO_4^{2-}, 2L-Orn^{2+}·Cl^-·I^-·SO_4^{2-}, 2L-Orn^{2+}·Br^-·I^-·SO_4^{2-} and 2L-Orn^{2+}·I^-·NO_3^-·SO_4^{2-}.

Table 5.47 Mixed salts of L-ornithinium(2+) cation

#	Crystal	s.g., Z	*a*, *b*, *c* (Å), *α*, *β*, *γ* (°)	Refs
1	$2\text{L-Orn}^{2+}\cdot Cl^{-}\cdot NO_3^{-}\cdot SO_4^{2-}$	*C*2, 2	13.872(5), 9.578(6), 9.178(1), $\beta = 125.350(14)$	Ramaswamy et al. (2004b)
2	$2\text{L-Orn}^{2+}\cdot Br^{-}\cdot NO_3^{-}\cdot SO_4^{2-}$	–	–	Ghazaryan et al. (2011c)
3	$2\text{L-Orn}^{2+}\cdot 2Cl^{-}\cdot SO_4^{2-}$	*C*2, 2	13.9500(6), 9.2300(4), 9.3169(7), $\beta = 127.260(1)$	Ghazaryan et al. (2011c)
4	$2\text{L-Orn}^{2+}\cdot 2Br^{-}\cdot SO_4^{2-}$	*C*2, 2	14.129(2), 9.389(1), 9.437(3), $\beta = 127.628(5)$	Ghazaryan et al. (2011c)
5	$3\text{L-Orn}^{2+}\cdot 2NO_3^{-}\cdot 2SO_4^{2-}$	$P6_3$, 2	10.0952(8), 10.0952(8), 17.791(2), $\gamma = 120$	Fleck et al. (2014)

Table 5.48 Mixed salts of L-ornithine with ($A^{+}...A^{2+}$) type dimeric cations

#	Crystal	s.g., Z	a, b, c (Å), α, β, γ (°)	Refs
1	(L-Orn^{+}...L-Orn^{2+})·$2Cl^{-}$·NO_3^{-}	$P2_1$, 2	11.126(4), 5.1564(17), 16.612(6), $\beta = 100.808(14)$	Petrosyan et al. (2009b); this work
2	(L-Orn^{+}...L-Orn^{2+})·$2Cl^{-}$·BF_4^{-}	–	–	Petrosyan et al. (2009b)
3	(L-Orn^{+}...L-Orn^{2+})·$2Cl^{-}$·ClO_4^{-}	$P2_12_12_1$, 4	5.0875(6), 18.8122(8), 20.4059(9)	Petrosyan et al. (2009b); this work
4	(L-Orn^{+}...L-Orn^{2+})·$2Br^{-}$·NO_3^{-}	–	–	Petrosyan et al. (2009b)
5	(L-Orn^{+}...L-Orn^{2+})·$2Br^{-}$·BF_4^{-}	–	–	This work
6	(L-Orn^{+}...L-Orn^{2+})·$2Br^{-}$·ClO_4^{-}	–	–	This work

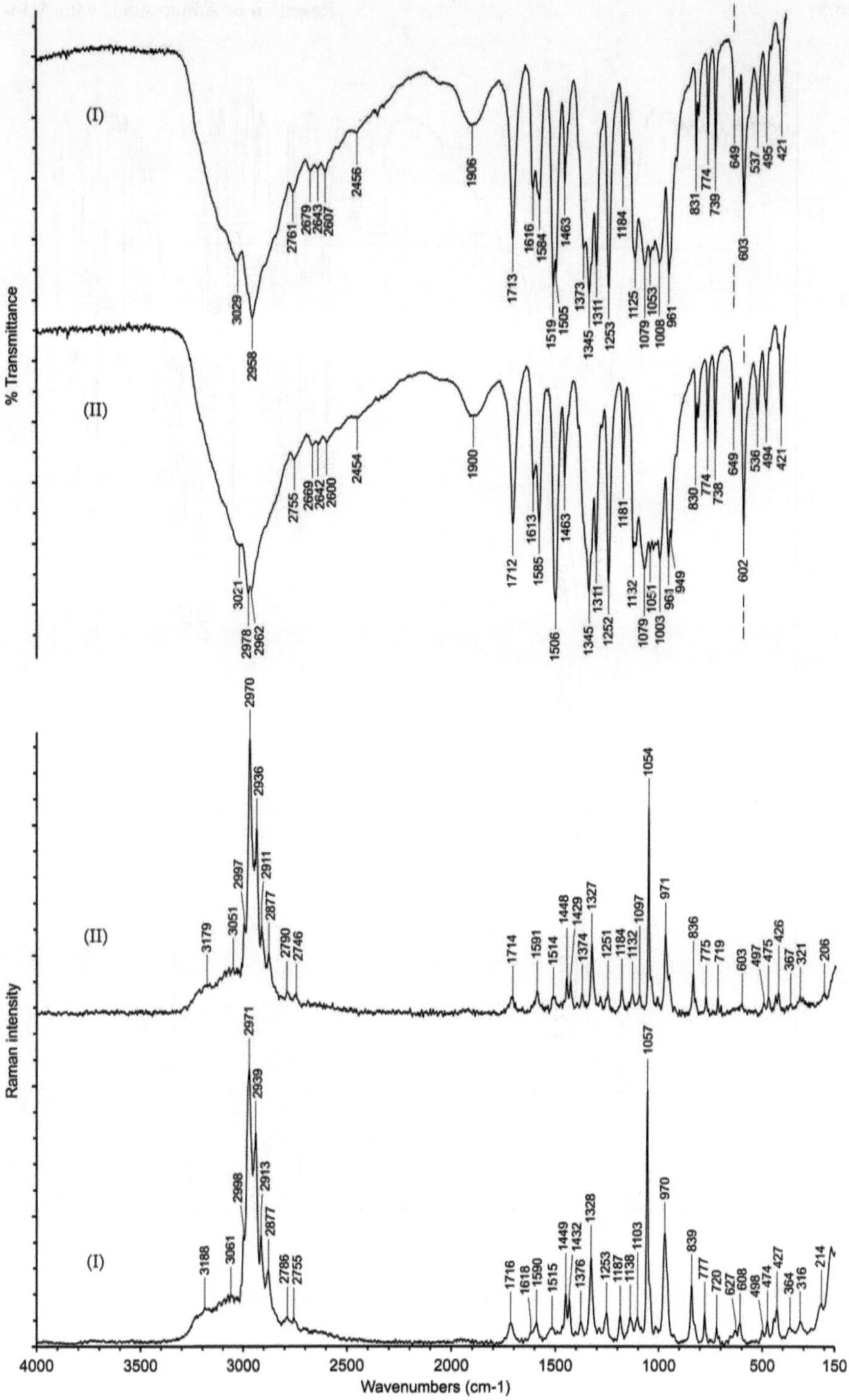

Fig. 5.78 IR and Raman spectra of 2L-Orn^{2+}·Cl^{-}·NO$_3$$^{-}$·SO$_4$$^{2-}$ (I) and 2L-Orn^{2+}·Br^{-}·NO$_3$$^{-}$·SO$_4$$^{2-}$ (II)

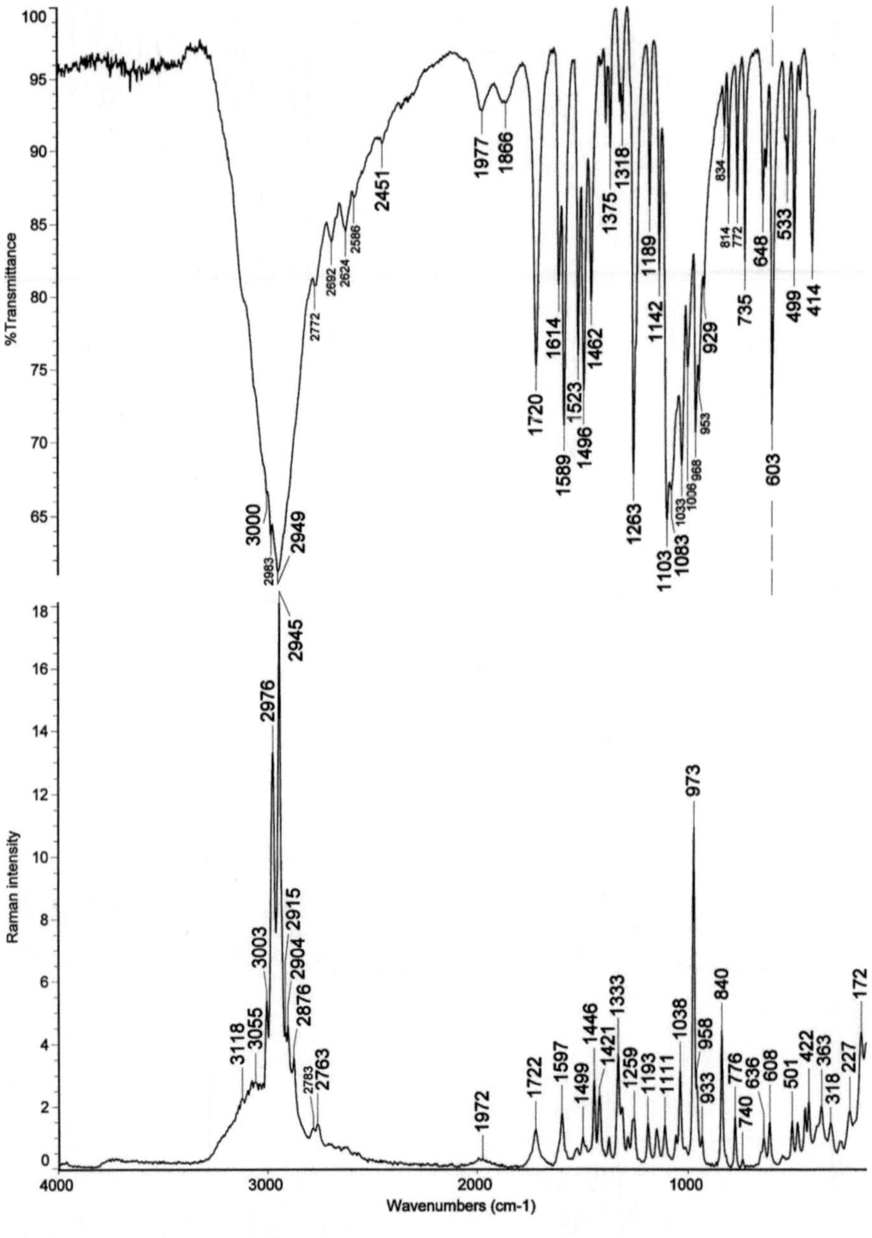

Fig. 5.79 IR spectra of $2\text{L-Orn}^{2+}{\cdot}2\text{Cl}^{-}{\cdot}\text{SO}_4^{2-}$ and $2\text{L-Orn}^{2+}{\cdot}2\text{Br}^{-}{\cdot}\text{SO}_4^{2-}$

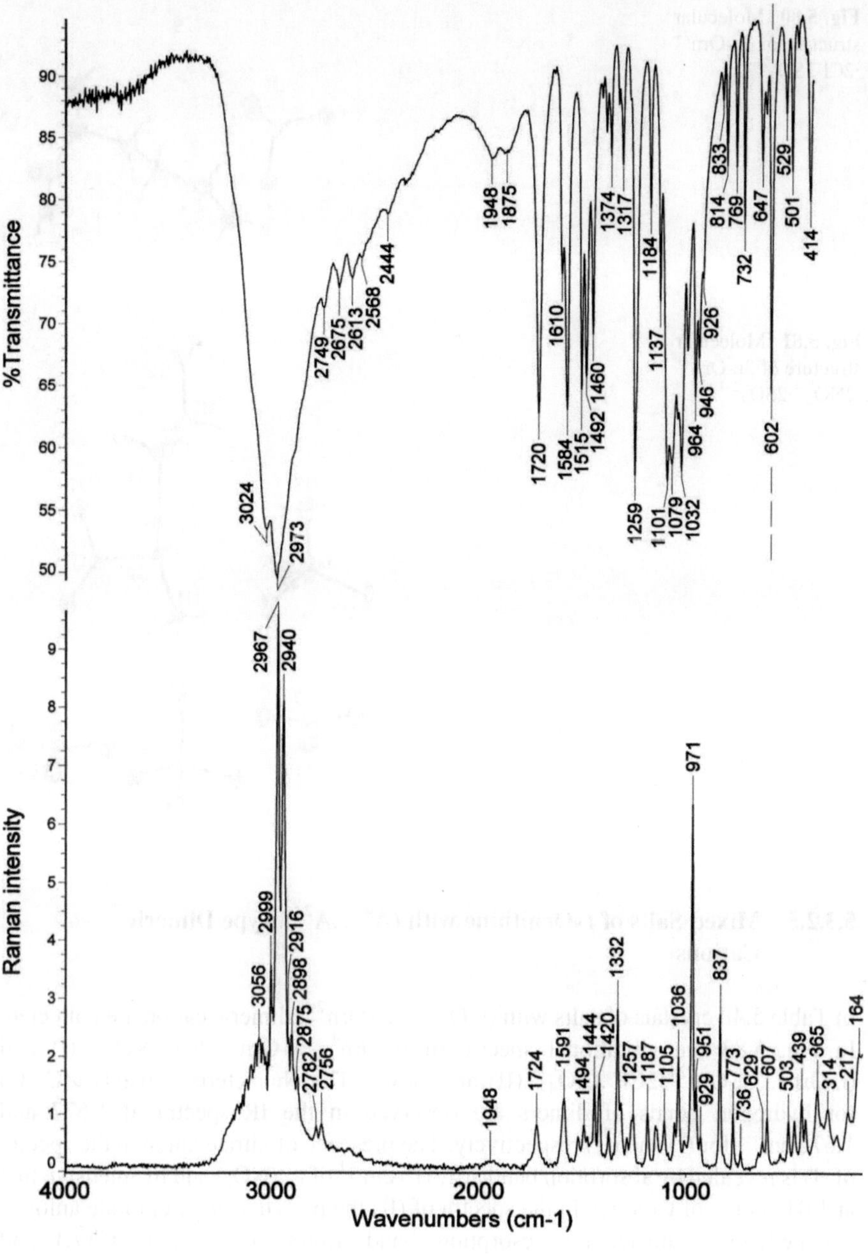

Fig. 5.79 (continued)

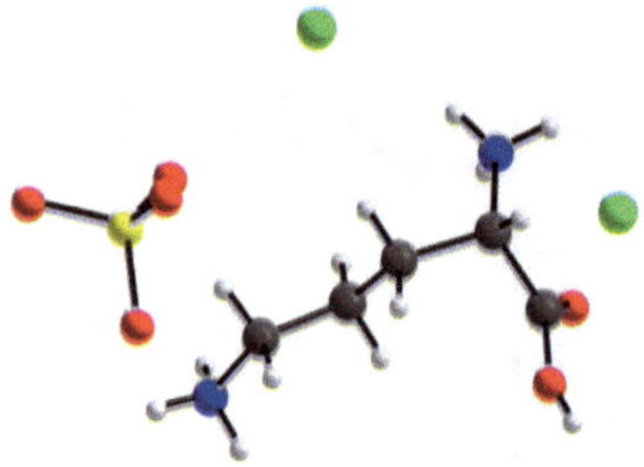

Fig. 5.80 Molecular structure of 2L-Orn^{2+} $\cdot 2Cl^{-} \cdot SO_4^{2-}$

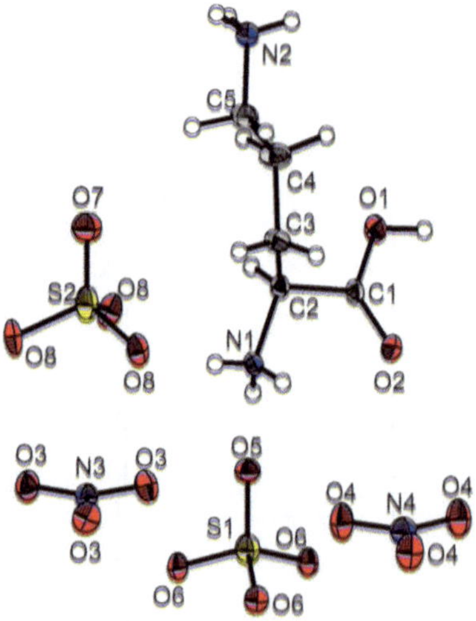

Fig. 5.81 Molecular structure of 3L-Orn^{2+} $\cdot 2NO_3^{-} \cdot 2SO_4^{2-}$

5.3.2.5 Mixed Salts of L-Ornithine with (A^{+}...A^{2+}) Type Dimeric Cations

In Table 5.48 crystals of salts with (L-Orn^{+}...L-Orn^{2+}) dimeric cation are collected. In Fig. 5.83 the vibrational spectra of (L-Orn^{+}...L-Orn^{2+})$2Cl^{-}.NO_3^{-}$ (I) and (L-Orn^{+}...L-Orn^{2+})$2Cl^{-}.ClO_4^{-}$ (II) are shown. The characteristic bands ν(C=O) for hydrogen bonds of dimers are observed in the IR spectra at 1,677 and 1,676 cm^{-1} for (I) and (II) respectively. The presence of nitrate anion in the spectra of (I) is revealed by absorption band at 1,318 cm^{-1} of $\nu_{as}(NO_3)$ and by intensive line at 1,042 cm^{-1} of $\nu_s(NO_3)$. In the spectra of (II) the presence of perchlorate anion is revealed by characteristic absorption band 1,068 cm^{-1} of $\nu_{as}(ClO_4)$ and 627, 618 cm^{-1}(δ_{as}) in the IR spectrum and very intensive line at 934 cm^{-1} ($\nu_s(ClO_4)$) and 633, 621 cm^{-1} (δ_{as}) and 472, 458 cm^{-1} (δ_s) in the Raman spectrum. The characteristic absorption of ν(OH) in dimers (I) and (II) apparently are revealed in the region *ca.* 800 cm^{-1}.

Conclusions on existence of other salts in Table 5.48 were made based on their vibrational spectra.

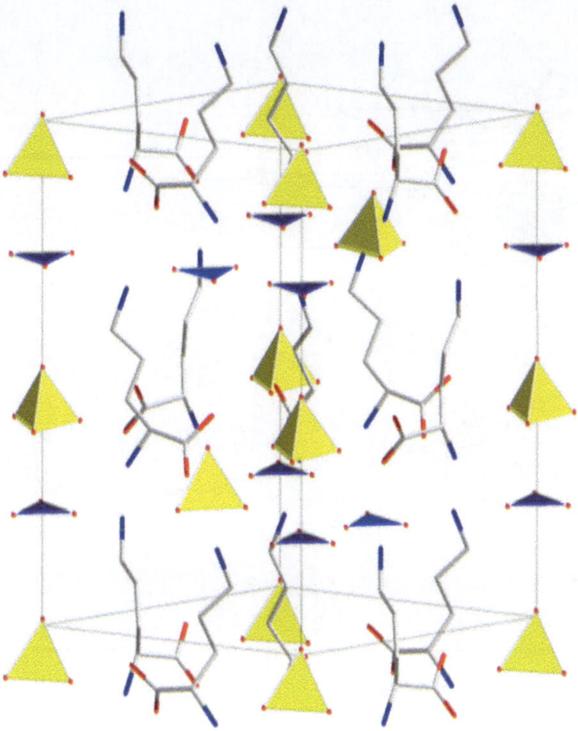

Fig. 5.82 Packing diagram of $3\text{L-Orn}^{2+}\cdot 2NO_3^-\cdot 2SO_4^{2-}$

We consider that other salts with ($\text{L-Orn}^+\ldots\text{L-Orn}^{2+}$) dimeric cation and the following anions are possible: $2I^-\cdot NO_3^-$, $2I^-\cdot BF_4^-$, $2I^-\cdot ClO_4^-$, $Cl^-\cdot Br^-\cdot NO_3^-$, $Cl^-\cdot I^-\cdot NO_3^-$, $Br^-\cdot I^-\cdot NO_3^-$, $Cl^-\cdot Br^-\cdot BF_4^-$, $Cl^-\cdot I^-\cdot BF_4^-$, $Br^-\cdot I^-\cdot BF_4^-$, $Cl^-\cdot Br^-\cdot ClO_4^-$, $Cl^-\cdot I^-\cdot ClO_4^-$, $Cl^-\cdot I^-\cdot ClO_4^-$, $Br^-\cdot I^-\cdot ClO_4^-$, $Cl^-\cdot NO_3^-\cdot BF_4^-$, $Br^-\cdot NO_3^-\cdot BF_4^-$, $I^-\cdot NO_3^-\cdot BF_4^-$, $Cl^-\cdot NO_3^-\cdot ClO_4^-$, $Br^-\cdot NO_3^-\cdot ClO_4^-$, $I^-.NO_3^-.$ ClO_4^-.

5.3.2.6 Mixed Salts of L-Argininium(2+) Cation

In Table 5.49 mixed salts of L-arginine are collected. They are formed by three mechanisms: (1) $A^{2+}\cdot X(1)^-\cdot X(2)^-$, (2) $2A^{2+}\cdot 2X(1)^-\cdot X(2)^-\cdot X(3)^-$ and (3) $2A^{2+}\cdot 3X(1)^-\cdot X(2)^-$.

In system L-Arg + HF + HNO_3 + H_2O two mixed salts were found: $\text{L-Arg}^{2+}\cdot F^-\cdot NO_3^-$ and $2\text{L-Arg}^{2+}\cdot(F^-\ldots HF)\cdot 3NO_3^-$. In triclinic cell of $\text{L-Arg}^{2+}\cdot F^-\cdot NO_3^-$ (Fig. 5.84) there are two formula units. Carboxyl groups of both cations form hydrogen bonds with fluoride ions, which additionally form each two N-H...F type hydrogen bonds. Nitrate anions form only N-H...O type hydrogen bonds. Hydrogen bonds O-H...F formed by $F(1)^-$ and $F(2)^-$ ions are very short: 2.399

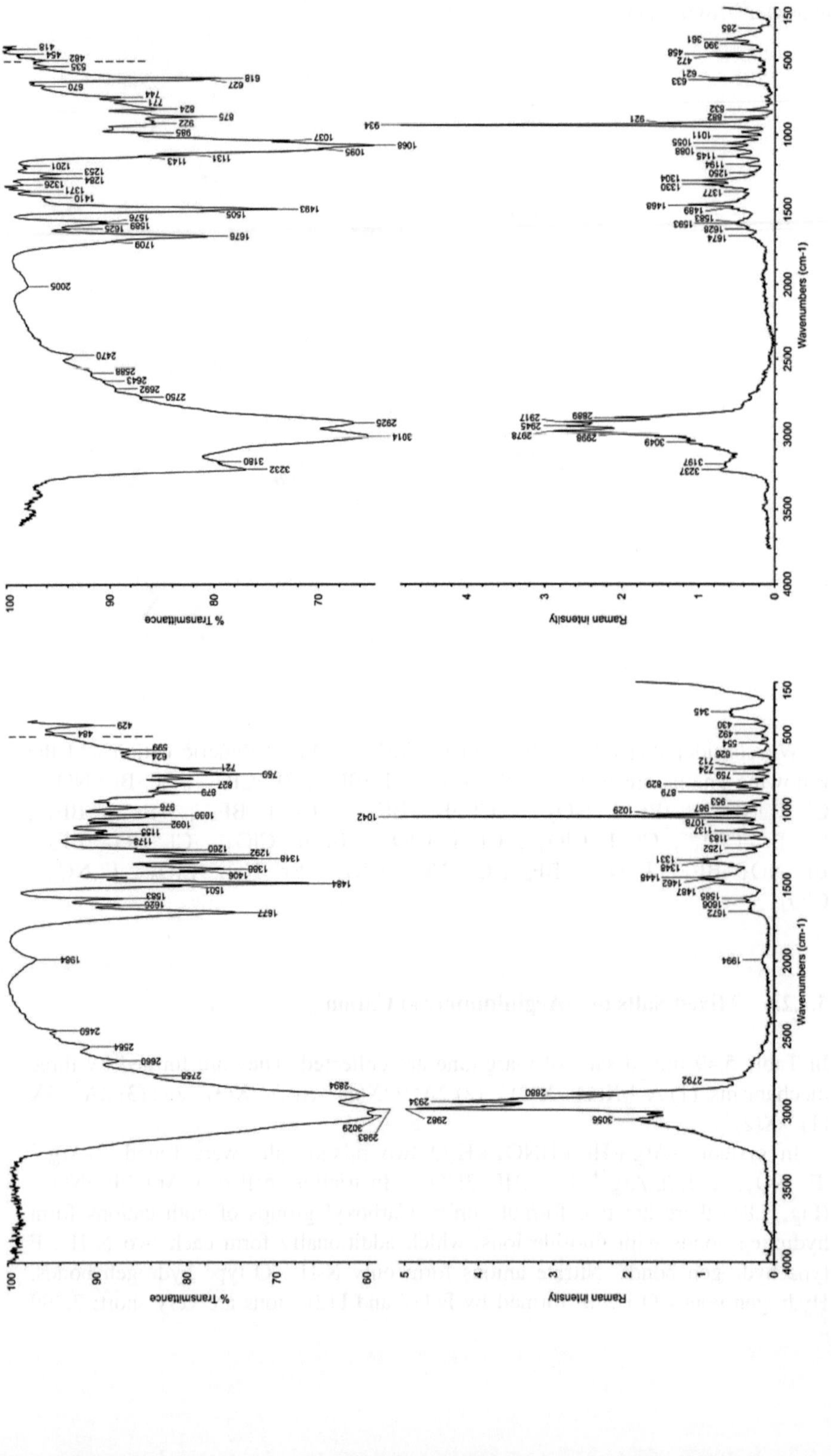

Fig. 5.83 IR and Raman spectra of (L-Orn^{+}…L-Orn^{2+})2Cl^{-}.NO$_3^{-}$ (I) and (L-Orn^{+}…L-Orn^{2+})2Cl^{-}·ClO$_4^{-}$ (II)

Table 5.49 Mixed salts of L-argininium(2+) cation

#	Crystal	s.g., Z	a, b, c (Å), α, β, γ (°)	Refs
1	L-$Arg^{2+}\cdot F^{-}\cdot NO_3^{-}$	$P1$, 2	5.0130(3), 10.6161(7), 11.0352(7), $\alpha = 87.471(3)$, $\beta = 86.038(3)$, $\gamma = 78.157(6)$	Ghazaryan et al. (2013c)
2	L-$Arg^{2+}\cdot NO_3^{-}\cdot BF_4^{-}$ (β)	–	–	Petrosyan et al. (2009b)
3	L-$Arg^{2+}\cdot NO_3^{-}\cdot BF_4^{-}$ (β′)	$C2$, 4	27.230(7), 6.6626(13), 7.7260(15), $\beta = 106.27(3)$	This work
4	L-$Arg^{2+}\cdot NO_3^{-}\cdot BF_4^{-}$ (γ)	–	–	Petrosyan et al. (2009b)
5	L-$Arg^{2+}\cdot NO_3^{-}\cdot ClO_4^{-}$ (α)			Petrosyan et al. (2009b)
6	L-$Arg^{2+}\cdot NO_3^{-}\cdot ClO_4^{-}$ (β)	–	–	Petrosyan et al. (2009b)
7	L-$Arg^{2+}\cdot NO_3^{-}\cdot ClO_4^{-}$ (β′)	–	–	This work
8	L-$Arg^{2+}\cdot NO_3^{-}\cdot ClO_4^{-}$ (γ)	–	–	Petrosyan et al. (2009b)
9	L-$Arg^{2+}\cdot NO_3^{-}\cdot ClO_4^{-}$ (δ)	$P2_1$, 4	11.7898(4), 7.9058(3), 14.9714(5), $\beta = 90.948(1)$	Ghazaryan et al. (2013c)
10	2L-$Arg^{2+}\cdot 2Cl^{-}\cdot NO_3^{-}\cdot ClO_4^{-}$	–	–	Petrosyan et al. (2009b)
11	2L-$Arg^{2+}\cdot 3Cl^{-}\cdot ClO_4^{-}$	$C222_1$, 8	9.520(2), 14.547(3), 18.135(4)	This work
12	2L-$Arg^{2+}\cdot(F^{-}\ldots HF)\cdot 3NO_3^{-}$	$P2_12_12_1$, 4	7.3960(4), 12.8672(7), 26.990(2)	Ghazaryan et al. (2013c)

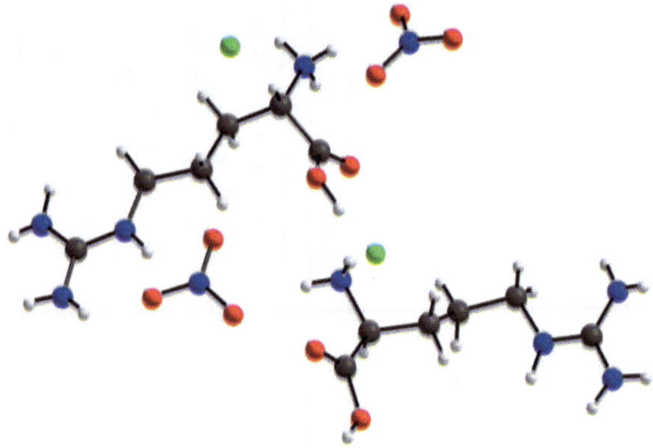

Fig. 5.84 Molecular structure of L-Arg^{2+}·F^-·NO_3^-

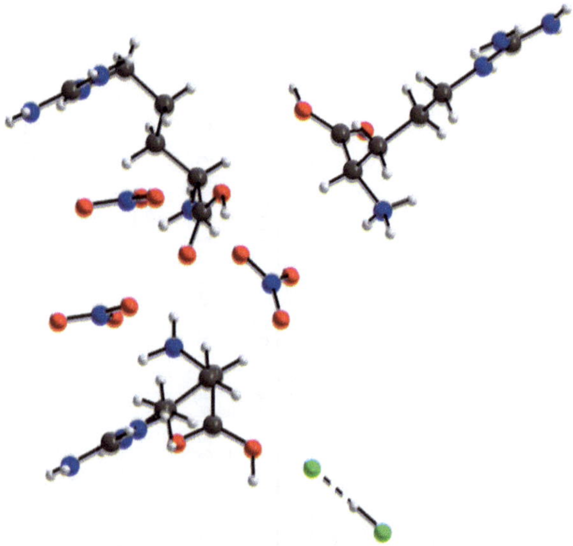

Fig. 5.85 Molecular structure of 2L-Arg^{2+}·(F^-...HF)·$3NO_3^-$

(3) Å and 2.371(3) Å respectively. Of these, hydrogen bond with distance 2.371 (3) Å is shortest among all before known of O-H...F type hydrogen bonds. The shortest before known bonds were: 2.382(1) Å (Ramos Silva et al. 2004) or 2.3839 (13) Å (Petrosyan et al. 2005c) in the structure of L-His^{2+}·$2F^-$ and 2.3888(13) Å (Szafran et al. 2004) in the structure of *N*-methylpiperidine betaine hydrofluoride, which, however, was determined at 100 K and at room temperature may be somewhat longer. Orthorhombic cell of 2L-Arg^{2+}·(F^-...HF)·$3NO_3^-$ (Fig. 5.85) contains one independent formula unit. The fluoride ion $F(2)^-$ forms relatively weak hydrogen bond with HF molecule with F1...F2 distance equal to 2.682(2) Å.

One of L-Arg^{2+} cations forms O-H...O hydrogen bond with one of nitrate anions with O...O distance equal to 2.629(2) Å. The other cation forms O-H...F hydrogen bond with second fluoride ion $F(2)^-$ with record short O...F distance equal to 2.341 (2) Å. The fluoride $F(2)^-$ ion forms one more N-H...F bond. The distance O1A...F2(2.341(2) Å) is the shortest not only among O-H...F type bonds, but it is also among F-H...O type hydrogen bonds. Previously it was considered that F-H...O type hydrogen bonds have shorter O...F distances than O-H...F type hydrogen bonds. Indeed, the shortest distance O...F in hydrogen bond F-H...O the structure of $CsH_2PO_3{\cdot}HF$ (Troyanov et al. 2004) equal to 2.356 (3) Å is significantly shorter than 2.371(3) Å. Thus, the distance O...F(2.341(2) Å) found in the structure of 2L-$Arg^{2+}{\cdot}(F^-...HF){\cdot}3NO_3^-$ crystal is the shortest among known O-H...F and F-H...O types hydrogen bonds.

In the systems L-Arg + HNO_3 + HBF_4($HClO_4$) + H_2O the salts L-$Arg^{2+}{\cdot}NO_3^-{\cdot}BF_4^-$ and L-$Arg^{2+}{\cdot}NO_3^-{\cdot}ClO_4^-$ are formed, which display propensity toward polymorphism. From aqueous solution containing L-arginine, HNO_3 and $HClO_4$ in molar ratios 1:1:1 the salt L-$Arg^{2+}{\cdot}NO_3^-{\cdot}ClO_4^-$ was obtained, which was labeled later as α-form, because it was found that at attempts to obtain salts with compositions 2L-$Arg^{2+}{\cdot}3NO_3^-{\cdot}ClO_4^-$ and 2L-$Arg^{2+}{\cdot}NO_3^-{\cdot}3ClO_4^-$ salts with the same composition (1:1:1) but other forms are formed. Recrystallization of obtained forms, and also using obtained forms as seeds allowed to identify by vibrational spectra a number of forms of the L-$Arg^{2+}{\cdot}NO_3^-{\cdot}ClO_4^-$ salt. Their vibrational spectra were studied by Ghazaryan (2014). Further works need to obtain these forms as single crystals and determining their structures. In case of L-$Arg^{2+}{\cdot}NO_3^-{\cdot}BF_4^-$ three forms were obtained and were labeled as β, β′ and γ based on similarity of their spectra with respective forms of L-$Arg^{2+}{\cdot}NO_3^-{\cdot}ClO_4^-$ salt. It was possible to determine the structure of β′-form of L-$Arg^{2+}{\cdot}NO_3^-{\cdot}BF_4^-$ and δ-form of L-$Arg^{2+}{\cdot}NO_3^-{\cdot}ClO_4^-$. Asymmetric unit of β′-form L-$Arg^{2+}{\cdot}NO_3^-{\cdot}BF_4^-$ contains one formula unit. Carboxyl group of L-argininium(2+) cation forms O-H...F hydrogen bond with BF_4^- anion, while NO_3^- anion forms only N-H...O type hydrogen bonds. Fluorine atoms of BF_4^- anion are disordered. We assume that the β′-form L-$Arg^{2+}{\cdot}NO_3^-{\cdot}ClO_4^-$ has the same structure. Asymmetric unit of δ-form of L-$Arg^{2+}{\cdot}NO_3^-{\cdot}ClO_4^-$ contains two formula units. Carboxyl groups of both L-argininium(2+) cations form O-H...O hydrogen bonds with nitrate anions. Structural details and discussion of vibrational spectra are provided by Ghazaryan et al. (2013c).

The crystal of 2L-$Arg^{2+}{\cdot}3Cl^-{\cdot}ClO_4^-$ is formed from aqueous solution containing stoichiometric quantities of components. Asymmetric unit contains one L-argininium (2+) cation in general position, two chloride ions and perchlorate anion. One of chloride ions is in general position, while the second in special position, on twofold axis as well as perchlorate anion. Oxygen atoms of perchlorate anion are disordered. Carboxyl group of cation forms hydrogen bond with chloride ion, which is in general position. We attempted to obtain also a salt L-$Arg^{2+}{\cdot}Cl^-{\cdot}ClO_4^-$ from stoichiometric composition, however, a viscous solution is formed in result of evaporation. After long staying (more than a year) the formed crystals were the known crystals with 2L-$Arg^{2+}{\cdot}3Cl^-{\cdot}ClO_4^-$ composition.

We consider that other similar salts of L-argininium(2+) cation also are possible: 2L-Arg^{2+}·$2Br^-$·NO_3^-·ClO_4^-, 2L-Arg^{2+}·$2I^-$·NO_3^-·ClO_4^-, 2L-Arg^{2+}·$2Cl^-$·NO_3^-·BF_4^-, 2L-Arg^{2+}·$2Br^-$·NO_3^-·BF_4^-, 2L-Arg^{2+}·$2I^-$·NO_3^-·BF_4^-, 2L-Arg^{2+}·$3Br^-$·ClO_4^-, 2L-Arg^{2+}·$3I^-$·ClO_4^-, 2L-Arg^{2+}·$3Cl^-$·BF_4^-, 2L-Arg^{2+}·$3Br^-$·BF_4^-, 2L-Arg^{2+}·$3I^-$·BF_4^-.

5.3.2.7 Mixed Salts of L-Histidinium(2+) and (L-Histidinium (+)...L-Histidinium(2+)) Cations

In Table 5.50 mixed salts of L-histidine are collected. Mixed salts of L-histidine are formed by larger number of mechanisms: (1) A^{2+}·$X(1)^-$·$X(2)^-$, (2) $2A^{2+}$·$2X^-$·Y^{2-}, (3) (A^+...A^{2+})·X^-·Y^{2-}, (4) $2A^{2+}$·H_3O^+·X^-·$2Y^{2-}$, (5) (A^+...A^{2+})·A^{2+}·$3X^-$·Y^{2-}, (6) $2A^{2+}$·$Y(1)^{2-}$·$Y(2)^{2-}$.

It is possible to obtain by evaporation bulk crystals of L-His^{2+}·F^-·NO_3^- from solution containing L-histidine, HNO_3 and HF in molar ratios 1:1:1. However, these crystals in contrast to respective crystals of L-arginine (L-Arg^{2+}·F^-·NO_3^-) are not stable and being out of solution over time become white. One may suppose that this is caused by release of fluorine in form of HF from surface layers similar to that observed in case of Gly·HF. The IR spectrum of fully white sample of L-His^{2+}·F^-·NO_3^- showed that it transformed into L-His^+·NO_3^- (Ghazaryan 2014). Crystals with higher quality are formed at somewhat excess of HF in solution.

The crystals of L-His^{2+}·Cl^-·HSO_4^- are formed from solution containing L-histidine, HCl and H_2SO_4 in molar ratios 1:1:1. Respective crystals with bromide ion L-His^{2+}·Br^-·HSO_4^- are formed from solution containing L-histidine, HBr and H_2SO_4 in molar ratios 1:1:1, however, contaminated with L-His^{2+}·SO_4^{2-}. Vibrational spectra of L-His^{2+}·Cl^-·HSO_4^- and L-His^{2+}·Br^-·HSO_4^- are very similar, which indicate that they are isostructural. In the structure of L-His^{2+}·Cl^-·HSO_4^- carboxyl group of cation forms O11-H1...O2(2.583(1) Å) with HSO_4^- anion, which in turn forms hydrogen bond O4-H4...Cl1(2.9754(9) Å) with chloride ion. One may hope that respective salt with iodide ion L-His^{2+}·I^-·HSO_4^- is also possible to obtain.

From solutions containing L-histidine, HNO_3, HBF_4 as well as L-histidine, HNO_3, $HClO_4$ in molar ratios 1:1:1 mixed salts L-His^{2+}·NO_3^-·BF_4^- and L-His^{2+}·NO_3^-·ClO_4^- are formed (Petrosyan et al. 2010b). Based on isostructurality of many salts of perchlorates and tetrafluoroborates it was possible to expect that these crystals will be also isostructural. This would seem to indicate the same symmetry, the same number of formula units and similar values of unit cell parameters. However, closer inspection revealed an interesting distinction in the structures of these crystals and their physical properties. The volume of ClO_4^- anion is larger than that of BF_4^- anion (mean value of Cl-O makes 1.429 Å, while B-F is 1.389 Å). Therefore, one might expect that the volume of unit cell of L-His^{2+}·NO_3^-·ClO_4^- will be larger than that of L-His^{2+}·NO_3^-·BF_4^-. However, in contrast to expectation unit volume of L-His^{2+}·NO_3^-·ClO_4^- (1191.29(8) $Å^3$) is lesser than that of L-His^{2+}·NO_3^-·BF_4^- (1226.9(5) $Å^3$). This distinction occurs due to different conformations

Table 5.50 Mixed salts of L-histidinium(2+) and (L-histidinium(+)...L-histidinium(2+)) cations

#	Crystal	s.g., Z	a, b, c (Å), α, β, γ (°)	Refs
1	L-His^{2+}·F^-·NO_3^-	–	–	Petrosyan et al. (2009b)
2	L-His^{2+}·Cl^-·HSO_4^-	$P2_12_12_1$, 4	5.3758(2), 10.9154(4), 19.2452(8)	Ghazaryan et al. (2013c)
3	L-His^{2+}·Br^-·HSO_4^-	–	–	This work
4	L-His^{2+}·NO_3^-·BF_4^-	$P2_12_12_1$, 4	7.430(2), 9.670(2), 17.072(3)	Petrosyan et al. (2010b)
5	L-His^{2+}·NO_3^-·ClO_4^-	$P2_12_12_1$, 4	6.7829(3), 10.2466(3), 17.1404(6)	Petrosyan et al. (2010b)
6	2L-His^{2+}·2Cl^-·SiF_6^{2-}	$C2$, 2	27.037(2), 6.9849(5), 5.3630(4), $\beta = 91.928(4)$	Petrosyan et al. (2014)
7	2L-His^{2+}·2Br^-·SiF_6^{2-}	$C2$, 2	27.063(6), 7.1013(17), 5.5076(11), $\beta = 92.141(11)$	Petrosyan et al. 2014)
8	2L-His^{2+}·2I^-·SiF_6^{2-}	–	–	This work
9	2L-His^{2+}·2BF_4^-·SiF_6^{2-}·2H_2O	$P2_12_12$, 2	10.0250(6), 14.8608(6), 8.7280(4)	Ghazaryan et al. (2013a); Petrosyan et al. (2014)
10	2L-His^{2+}·2ClO_4^-·SiF_6^{2-}·2H_2O	$P2_12_12$, 2	10.1549(8), 14.1549(8), 8.7536(7)	Petrosyan et al. (2014)
11	(L-His^+...L-His^{2+})·NO_3^-·SiF_6^{2-}	$P2_1$, 2	5.9967(3), 19.914(1), 8.4646(5), $\beta = 99.770(2)$	Ghazaryan et al. (2012e)
12	2L-His^{2+}·H_3O^+·NO_3^-·2SO_4^{2-}	$C2$, 2	16.646(3), 7.2959(15), 9.6221(19), $\beta = 99.64(3)$	This work
13	(L-His^+...L-His^{2+})·L-His^{2+}·X^a	$P2_12_12_1$, 4	6.014(2), 10.536(3), 50.232(8)	This work
14	2L-His^{2+}·SO_4^{2-}·SiF_6^{2-}·3H_2O	$P1$, 1	8.5005(17), 8.7896(18), 9.7286(19), $\alpha = 71.49(3)$, $\beta = 86.17(3)$, $\gamma = 62.60(3)$	Petrosyan et al. (2014)

[a] $X = 3NO_3^-·SO_4^{2-}·H_2O$

of cations and packing. The carbon backbone of the L-histidinium(2+) cation in L-His^{2+}·NO_3^-·ClO_4^- has *gauche*-conformation (torsion angle C1C2C3C4 is 55.8 (2)°) in contrast to the nearly completely extended *trans*-conformation in L-His^{2+} ·NO_3^-·BF_4^- (respective torsion angle is 171.3(1)°). This results in different dispositions of anions, hydrogen bonds, especially in high-frequency region, and also in physical properties. Observation of powder SHG revealed strong signals in L-His^{2+} ·NO_3^-·BF_4^- and absence of noticeable signals in L-His^{2+}·NO_3^-·ClO_4^-. Having in view the polymorphism of respective salts of L-arginine one may assume the existence of polymorphic form of L-His^{2+}·NO_3^-·BF_4^- isostructural with obtained crystal L-His^{2+}·NO_3^-·ClO_4^- and vice versa the existence of yet unobserved form of L-His^{2+} ·NO_3^-·ClO_4^-, which will be isostructural to L-His^{2+}·NO_3^-·BF_4^- with respective physical properties.

The salt 2L-His^{2+}·2Cl$^-$·SiF_6^{2-} is formed by the mechanism $2A^{2+}$·$2X^-$·Y^{2-}, and its bromide analog (Petrosyan et al. 2014) (respective iodide analog was obtained recently (this work) and identified by IR spectroscopy) as well as 2L-His^{2+}·2BF_4^-·SiF_6^{2-}·2H_2O (Ghazaryan et al. 2013a) and its perchlorate analog (Petrosyan et al. 2014). The asymmetric unit of 2L-His^{2+}·2Cl$^-$·SiF_6^{2-}contains one L-His^{2+} cation, and one chloride ion in general position, and also one SiF_6^{2-} anion in special position. The twofold axis passes through the F1-Si1-F2 atoms. Its composition may be representing as L-His^{2+}·Cl$^-$·0.5SiF_6^{2-} or 2L-His^{2+}·2Cl$^-$·SiF_6^{2-}. Similarly asymmetric unit of 2L-His^{2+}·2BF_4^-·SiF_6^{2-}·2H_2O contains one L-His^{2+} cation, one tetrafluoroborate anion, which is disordered, and one water molecule in general position, and also one SiF_6^{2-} anion in special position. The twofold axis is perpendicular to the F3-Si1-F3 direction. Similarly is formed its perchlorate analog. We attempted to obtain also crystals of salts formed by $2A^{2+}$·X(1)$^-$·X(2)$^-$·Y^{2-} mechanism from solutions containing L-histidine, HBF_4, $HClO_4$, H_2SiF_6 in 2:1:1:1 molar ratios as well as L-histidine, HCl, HBr, H_2SiF_6 in the same molar ratios. One might expect that in case of formation of these salts they will not be isostructural to crystals of salts formed by $2A^{2+}$·$2X^-$·Y^{2-} mechanism, because in the structure of these salts there are not two different positions for X^- anions as in case, for example, (A^+. . .A^{2+}) ·2X(1)$^-$·X(2)$^-$ for L-lysine, L-ornithine or in case of $2A^{2+}$·$2X^-$·Y^{2-}, where A^{2+} is L-Orn^{2+} and X^-, Y^{2-} are Cl^- (Br^-) and SO_4^{2-} (see Tables 5.46, 5.47, and 5.48). In spite of that vibrational spectra of crystals obtained from solutions containing HBF_4 and $HClO_4$ indicate the presence in the structure of crystal both BF_4^- and ClO_4^- anions, these spectra are very similar to the spectra of 2L-His^{2+}·2BF_4^-·SiF_6^{2-}·2H_2O and 2L-His^{2+}·2ClO_4^-·SiF_6^{2-}·2H_2O crystals. X-ray diffraction analysis showed that crystals containing both BF_4^- and ClO_4^- anions have the same structure. This indicates that in the structure the BF_4^- and ClO_4^- anions occupy the same positions, that is, the crystal represents a solid solution, 2L-His^{2+}·$(BF_4^-)_{(2-2x)}(ClO_4^-)_{2x}$·$SiF_6^{2-}$·2$H_2O$ with the same space group $P2_12_12$, $Z=2$, and cell parameters $a=10.088(3)$ Å, $b=14.874(4)$ Å, $c=8.755(2)$ Å similar to parameters of ## 9, 10 in Table 5.50. The same conclusion may be to do for 2L-His^{2+}· $Cl^-_{(2-2x)}Br^-_{2x}$·SiF_6^{2-}·2H_2O crystal.

A crystal with interesting structure (L-His$^+$. . .L-His^{2+})·NO_3^-·SiF_6^{2-} was found in the L-His + HNO_3 + H_2SiF_6 + H_2O system. It is formed from solution with

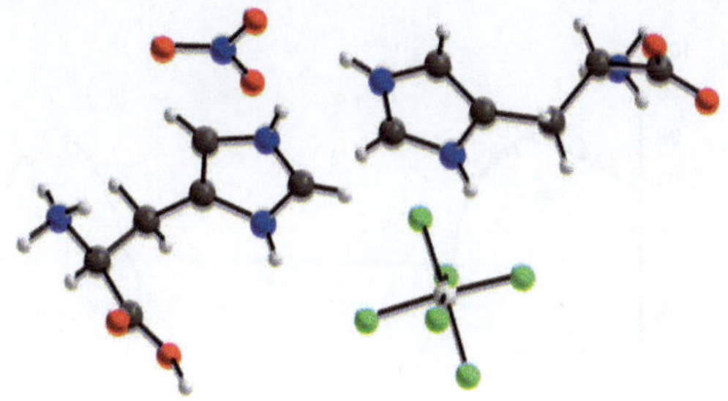

Fig. 5.86 Molecular structure of (L-His^{+}...L-His^{2+})·NO_3^-·SiF_6^{2-}

stoichiometric ratios of components. In Fig. 5.86 the molecular structure of the crystal is shown. In the structure in addition to NO_3^- and SiF_6^{2-} anions there is dimeric (L-His^{+}...L-His^{2+}) cation with O1A-H1A...O1B(2.507(2) Å) hydrogen bond. This is the first case of discovering of (A^{+}...A^{2+}) type dimeric cation for histidine. Vibrational spectra of the crystal are provided and discussed by Ghazaryan et al. (2012e). The presence of mentioned O-H...O hydrogen bond of dimeric cation in the structure of crystal is reflected by presence of broad absorption band of ν(OH) centered at *ca.* 1,350 cm^{-1}.

In the system L-His + HNO_3 + H_2SO_4 + H_2O two mixed salts were found: 2L-His^{2+}·H_3O^+·NO_3^-·2SO_4^{2-} and (L-His^{+}...L-His^{2+})·L-His^{2+}·3NO_3^-·SO_4^{2-}·H_2O. Their vibrational spectra are shown in Figs. 5.87 and 5.88. The first salt is noticeable by that it probably is the first salt of amino acids in structure of which hydroxonium ion is present. In second salt there is a dimeric (L-His^{+}...L-His^{2+}) cation in which the hydrogen bond O-H...O(2.441(9) Å) is shorter than in case of (L-His^{+}...L-His^{2+})·NO_3^-·SiF_6^{2-} (2.507(2) Å).

The last salt in Table 5.50 is 2L-His^{2+}·SO_4^{2-}·SiF_6^{2-}·3H_2O, in the structure of which there are two doubly charged anions. Its vibrational spectra are shown in Fig. 5.89. Carboxyl groups of L-histidinium(2+) cations form rather strong O-H...O hydrogen bonds with two of three water molecules with O...O distances 2.491 Å and 2.501 Å. Water molecules as proton donors form hydrogen bonds with SO_4^{2-} and SiF_6^{2-} anions.

Simple and mixed salts of amino acids containing hexafluorosilicate (SiF_6^{2-}) anion may find application in dentistry as preparations having anticaries activity (Petrosyan et al. 2012b; Brsikyan et al. 2012).

5.3.3 Concluding Remarks

5.3.3.1 Other Reports on Mixed Salts of Amino Acids

There are several publications on mixed salts of amino acids apart from mentioned above. In this section we will consider them.

Rajagopal et al. (2003f) reported on obtaining the "L-histidinium hemihydrochloride tartrate tartaric acid dihydrate" and reported its structure. According to the

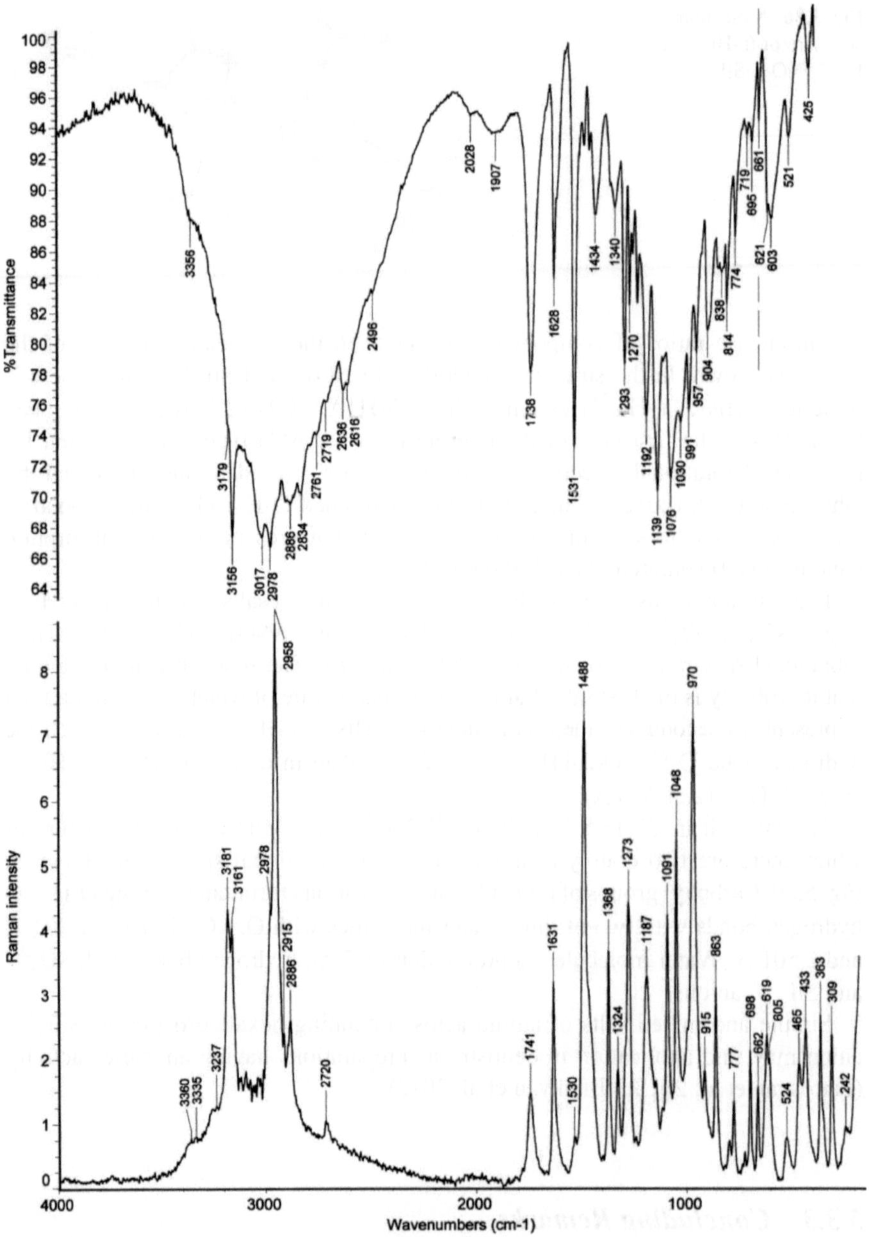

Fig. 5.87 Vibrational spectra of $2L\text{-His}^{2+}\cdot H_3O^+\cdot NO_3^-\cdot 2SO_4^{2-}$

authors they obtained crystals from solution containing L-His·HCl and tartaric acid (without specifying what form of tartaric acid was used) in molar ratio 1:2 and came to conclusion that it represents L-His$^+$·tartrate$^-$·0.5HCl·tartaric acid·$2H_2O$. At the

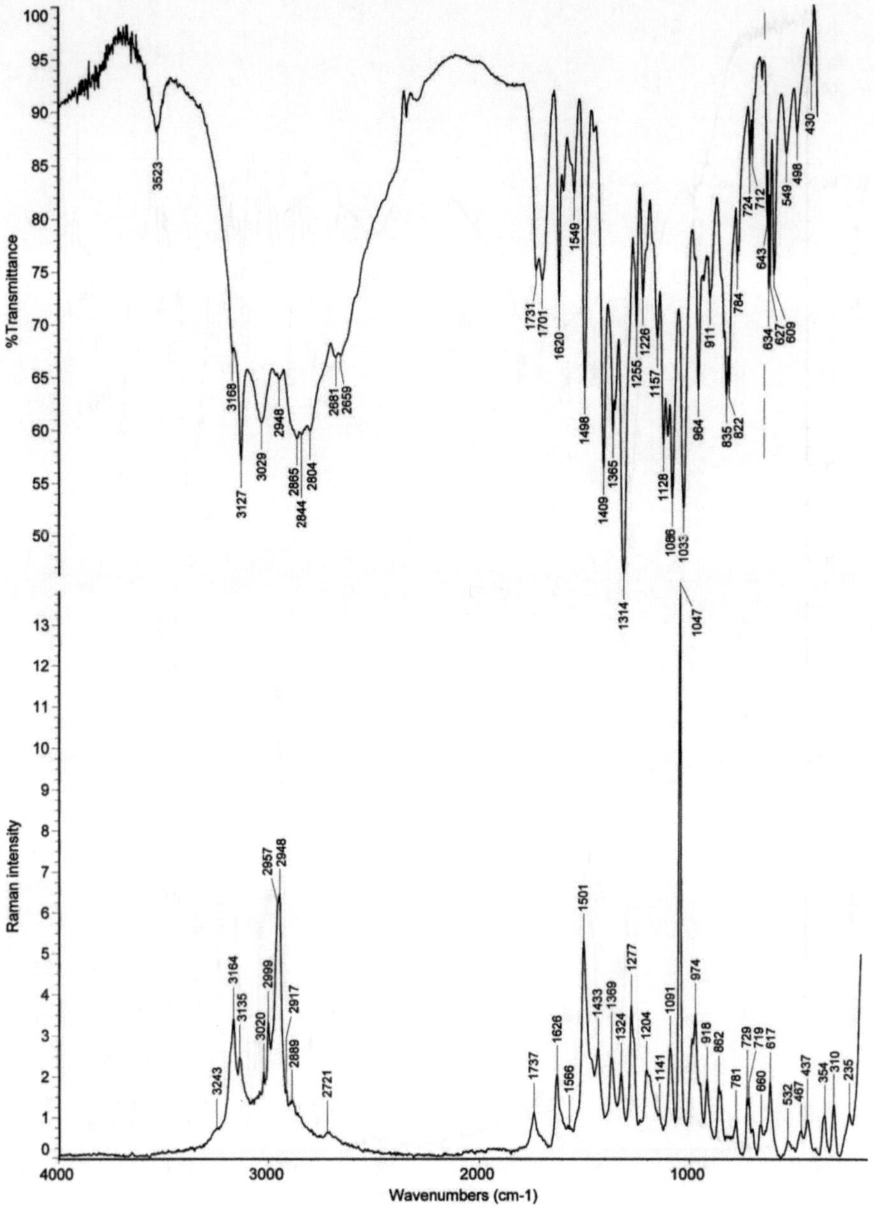

Fig. 5.88 Vibrational spectra of (L-His^{+}…L-His^{2+})·L-His^{2+}·3NO_3^{-}·SO_4^{2-}·H_2O

same time as if HCl is in the special position, on twofold axis of space group *C*2. Measured density makes $\rho_m = 1.65(2)$ g/cm^3 and calculated value is $\rho_c = 1.644$ g/cm^3. Analysis of data provided in the paper show that in spite of very low value of $R = 0.0383$ the structural model is questionable. The biggest question is the assumption of presence of "$H^{+}Cl^{-}$". It is difficult to imagine the presence of

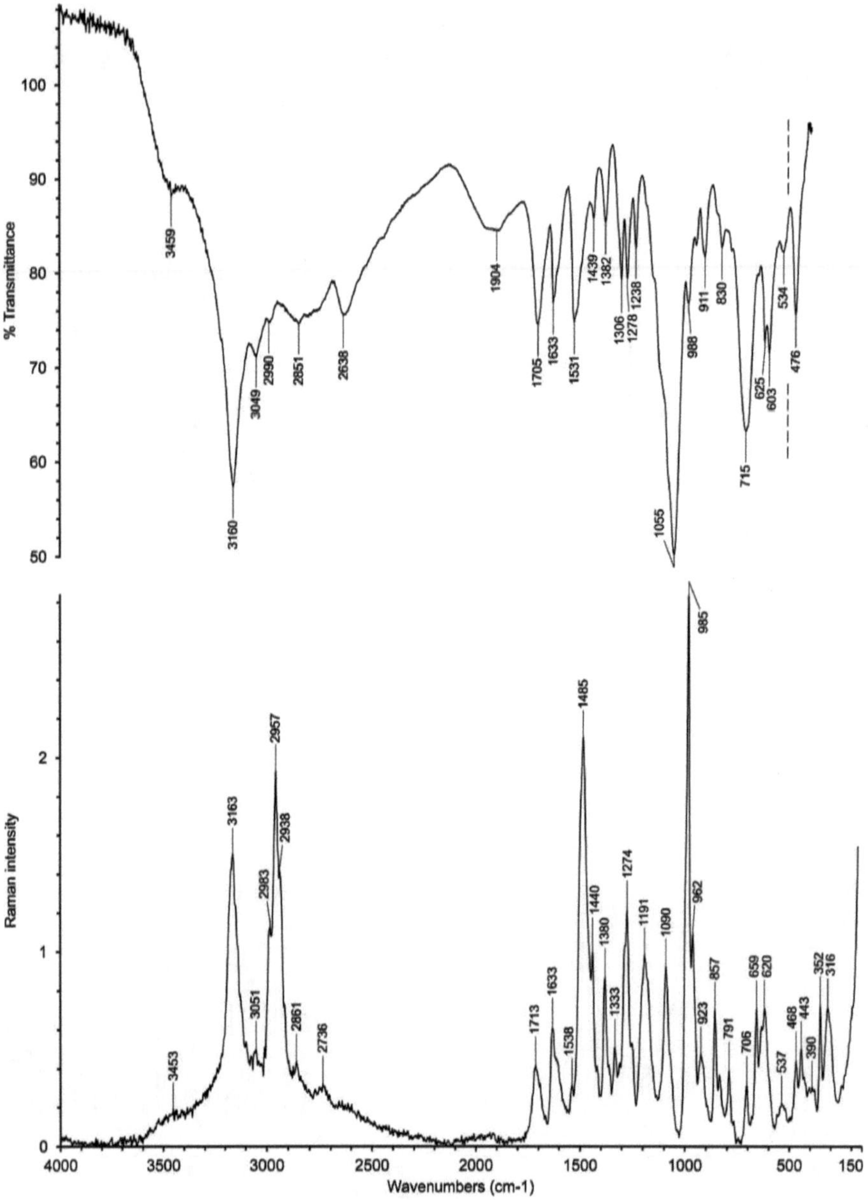

Fig. 5.89 Vibrational spectra of 2L-$His^{2+}\cdot SO_4^{2-}\cdot SiF_6^{2-}\cdot 3H_2O$

"HCl" in the presence of the anion of weaker tartaric acid. If we assume the existence of "H^+Cl^-", then it is unclear where H^+ is. It does not manifest itself, and in fact it has not even its place in the structure, which violates the most important principle of electroneutrality. The presence of neutral molecule of tartaric acid is also questionable. The presence of H8 hydrogen atom at O8 is not confirmed by bond lengths C10-O7(1.237 Å) and C10-O8(1.264 Å), which are close to bond lengths of COO^- group of hydrogen tartrate anion with C14-O13 (1.227 Å) and C14-O14(1.268 Å). From table of hydrogen bonds for O-H8…O6 it is evident (H…O(2.58 Å)) that there is not hydrogen bond and it is more probable that there is not H8 atom at O8. Therefore the hydrogen bond O8-H8…Cl1 should be interpreted in a different way. All these forced us to analyze the structure in more detail based on CIF (CSD code: OJEPIC). Next, we will use the notation adopted in CIF. It is evident that hydrogen atom H4 at O10 atom (they are correspond to H8 and O8 in notation of the authors) does not form any hydrogen bond. Therefore it is very probable that H4 atom is absent at O10 atom and hence the neutral tartaric acid molecule actually is hydrogen tartrate anion. So, there are two hydrogen tartrate anions, each of which forms usual infinite head-to-tail chain with hydrogen bonds O11-H5…O16(2.538 Å) and O5-H1…O10(2.493 Å). Supposed Cl1 is surrounded by eight oxygen atoms with distances: Cl1…O10(2.437 Å) (two bonds), Cl1…O1 (2.549 Å) (two bonds), Cl1…O8(2.439 Å) (two bonds), Cl1…O2(2.404 Å) (two bonds). Such dispositions and mentioned Cl1…O distances indicate, that Cl1 cannot be chloride ion and is rather metallic cation. Having in view octahedral coordination and mentioned distances it is most probable that Cl1 is actually Ca^{2+}. The calculated density of crystal when $0.5Ca^{2+}$ is instead of 0.5HCl makes 1.651 g/cm^3. So, in our opinion the composition of the crystal corresponds to L-His$^+$·0.5 Ca^{2+} ·2tartrate$^-$·$2H_2O$ with $Z=4$ or 2L-His$^+$. Ca^{2+}·4tartrate$^-$·$4H_2O$ with $Z=2$.

So, in our opinion this crystal needs further study and may become a mixed salt of amino acid with different cations.

Anandan and Jayavel (2011) obtained the crystal "tartaric acid mixed L-histidine hydrochloride monohydrate" from aqueous solution containing equimolar quantities of L-His·HCl·H_2O and tartaric acid (without specifying what form of tartaric acid was used). Unit cell parameters, IR spectrum and thermogram of grown crystal are provided. The obtained crystal actually is L-His·HCl·H_2O as it is clear from its characteristics. Interpretation and the IR spectrum and justification of incorporation of tartaric acid into the structure of crystal are not acceptable. The crystal of L-His·HCl·H_2O does not contain COOH group in contrast to statement of the authors and hence assignments of some vibrations to C=O of COOH group are erroneous.

Siddheswaran et al. (2006) reported the synthesis of a crystal "L-arginine hydrochlorofluoride monohydrate (LAHClF)," which was according to the authors obtained from aqueous solution containing equimolar quantities of L-Arg$^+$·Cl^-·H_2O and L-Arg$^+$·F^-·H_2O. Actually there is not known crystal with L-Arg$^+$·F^-·H_2O composition. It is known the L-Arg$^+$·F^- crystal (see Sect. 5.1.22), which is not isostructural with L-Arg$^+$·Cl^-·H_2O. From provided unit cell parameters and IR spectrum it is clear that the obtained crystal actually is L-Arg$^+$·Cl^-·H_2O. Justification based on the IR spectrum that the crystal contains both F^- and Cl^- ions is fully erroneous. The authors assigned some bands in IR spectrum to C-Cl and C-F, which is erroneous. Even if the ions are incorporated into the crystal they cannot be revealed in the IR spectrum because they do not form covalent C-Cl and C-F bonds. The authors note that crystal water starts to release at 102 °C. Nevertheless the authors state in their conclusion that the sample is stable up to 224.7 °C.

Jeyanthi et al. (2010) reported the crystal "L-lysine glycine dihydrochloride (L-LGHCl)." According to the authors it was obtained from aqueous solution containing 1 mol glycine, 0.5 mol HCl and 1 mol L-Lys·HCl. Why the obtained crystal should have the composition "L-Lys·Gly·2HCl" is not clear. The authors provided unit cell parameters of crystal "L-LGHCl" ($a = 5.871$ Å, $b = 13.315$ Å, $c = 7.489$ Å, $\beta = 97.76°$). It is clear from these parameters that the obtained crystal actually is L-Lys·HCl·2H_2O (compare with unit cell parameters shown in Table 5.21).

Shankar et al. (2008) reported on "tetra L-lysine alanine mono hydrochloride dihydrate (TLAMHCl)." Actually the obtained crystal is L-Lys·HCl·2H_2O. Critical consideration of this paper was made by Petrosyan and Ghazaryan (2009).

Two reports on the synthesis of the crystals of "2Gly·HCl·HF" (Khandpekar and Pati 2010a) and "α-glycine sulpho-nitrate (GLSN)" (Khandpekar and Pati 2010b) were published, although it turned out that both were erroneous (as was shown by Petrosyan et al. 2011b).

In all these papers except Rajagopal et al. (2003f) the authors proposed new mixed salts without confirmation by structure determination, in some cases (see above) it turned out that no new species have been synthesized at all. This sad fact underlines the importance of careful study and structure determination whenever crystals are available.

5.3.3.2 Amino Acids, Anions and Formation Mechanisms of Mixed Salts with Different Anions

Mixed salts of amino acids with different cations and anions are less studied (especially with different cations) field of salts of amino acids. We hope that our work will stimulate further investigations in this direction. In addition to the first three mixed salts with different anions of L-lysine ((L-Lys^+...L-Lys^{2+})2Cl^-.ClO_4^-, Srinivasan et al. 2001f) and (L-Lys^+...L-Lys^{2+})2Cl^-.NO_3^-, Srinivasan et al. 2001g) and L-ornithine (2L-Orn^{2+}·Cl^-·NO_3^-·SO_4^{2-}, Ramaswamy et al. 2004b) discovered by the group of Rajaram we found that L-arginine and L-histidine also capable to form mixed salts. In spite of that only one mixed salt with different anions of glycine (4Gly^+·2Cl^-·$Re_2Cl_8^{2-}$) is known (Shtemenko et al. 2003), nevertheless this work is important because shows that mixed salts with different anions are not limited by above mentioned four amino acids capable to form doubly charged cation. These four salts are formed by three different formation mechanisms (1) (A^+...A^{2+})·2$X(1)^-$·$X(2)^-$, (2) 2A^{2+}·$X(1)^-$·$X(2)^-$·Y^{2-}, (3) 4A^+·2X^-· Y^{2-}). Now ten more mechanisms are revealed: (4) 2A^+·$X(1)^-$·$X(2)^-$, (5) A^{2+} ·$X(1)^-$·$X(2)^-$, (6) 2A^{2+}·2$X(1)^-$·$X(2)^-$· $X(3)^-$, (7) 2A^{2+}·3$X(1)^-$·$X(2)^-$, (8) 2A^{2+}·2X^-·Y^{2-}, (9) 2A^{2+}·$Y(1)^{2-}$·$Y(2)^{2-}$, (10) 2A^{2+}·$(H_{2n+1}O_n)^+$·X^-·2Y^{2-}, (11) 3A^{2+}·2X^-·2Y^{2-}, (12) (A^+...A^{2+})·X^-·Y^{2-}, (13) (A^+...A^{2+})·A^{2+}·3X^-·Y^{2-}. Of these 13 mechanisms 11 are confirmed structurally. Based on these mechanisms other possible mechanisms may be proposed (Table 5.51). In Table 5.52 new mixed salts formed by 12 of 13 mentioned mechanisms are shown obtained in addition to above mentioned four known salts. In addition to Cl^-, ClO_4^-, NO_3^-, SO_4^{2-} and $Re_2Cl_8^{2-}$ known anions some new anions were revealed: F^-, $(F...HF)^-$, Br^-, I^-, BF_4^-, HSO_4^-, SiF_6^{2-}. Of course it is possible to obtain additional mixed salts. Some examples of possible new salts are shown in Table 5.53.

Table 5.51 Some possible formation mechanisms of mixed salts of amino acids with different anions

#	Mechanism	#	Mechanism
1	$2A^{+}{\cdot}X(1)^{-}{\cdot}X(2)^{-}$	25	$3A^{2+}{\cdot}2Y(1)^{2-}{\cdot}Y(2)^{2-}$
2	$3A^{+}{\cdot}X(1)^{-}{\cdot}X(2)^{-}{\cdot}X(3)^{-}$	26	$(A^{+}\ldots A^{2+}){\cdot}X(1)^{-}{\cdot}X(2)^{-}{\cdot}X(3)^{-}$
3	$3A^{+}{\cdot}2X(1)^{-}{\cdot}X(2)^{-}$	27	$(A^{+}\ldots A^{2+}){\cdot}X(1)^{-}{\cdot}X(2)^{-}$
4	$3A^{+}{\cdot}X^{-}{\cdot}Y^{2-}$	28	$(A^{+}\ldots A^{2+}){\cdot}X^{-}{\cdot}Y^{2-}$
5	$4A^{+}{\cdot}X(1)^{-}{\cdot}X(2)^{-}{\cdot}X(3)^{-}{\cdot}X(4)^{-}$	29	$(A^{+}\ldots A^{2+}){\cdot}(H_{2n+1}O_n)^{+}{\cdot}X(1)^{-}{\cdot}X(2)^{-}{\cdot}Y^{2-}$
6	$4A^{+}{\cdot}2X(1)^{-}{\cdot}X(2)^{-}{\cdot}X(3)^{-}$	30	$(A^{+}\ldots A^{2+}){\cdot}(H_{2n+1}O_n)^{+}{\cdot}2X^{-}{\cdot}Y^{2-}$
7	$4A^{+}{\cdot}X(1)^{-}{\cdot}X(2)^{-}{\cdot}Y^{2-}$	31	$(A^{+}\ldots A^{2+}){\cdot}(H_{2n+1}O_n)^{+}{\cdot}Y(1)^{2-}{\cdot}Y(2)^{2-}$
8	$4A^{+}{\cdot}2X^{-}{\cdot}Y(2)^{2-}$	32	$(A^{+}\ldots A^{2+}){\cdot}A^{+}{\cdot}X(1)^{-}{\cdot}X(2)^{-}{\cdot}X(3)^{-}{\cdot}X(4)^{-}$
9	$4A^{+}{\cdot}Y(1)^{2-}{\cdot}Y(2)^{2-}$	33	$(A^{+}\ldots A^{2+}){\cdot}A^{+}{\cdot}2X(1)^{-}{\cdot}X(2)^{-}{\cdot}X(3)^{-}$
10	$A^{2+}{\cdot}X(1)^{-}{\cdot}X(2)^{-}$	34	$(A^{+}\ldots A^{2+}){\cdot}A^{+}{\cdot}3X(1)^{-}{\cdot}X(2)^{-}$
11	$2A^{2+}{\cdot}X(1)^{-}{\cdot}X(2)^{-}{\cdot}X(3)^{-}{\cdot}X(4)^{-}$	35	$(A^{+}\ldots A^{2+}){\cdot}A^{+}{\cdot}2X(1)^{-}{\cdot}2X(2)^{-}$
12	$2A^{2+}{\cdot}2X(1)^{-}{\cdot}X(2)^{-}{\cdot}X(3)^{-}$	36	$(A^{+}\ldots A^{2+}){\cdot}A^{+}{\cdot}X(1)^{-}{\cdot}X(2)^{-}{\cdot}Y^{2-}$
13	$2A^{2+}{\cdot}3X(1)^{-}{\cdot}X(2)^{-}$	37	$(A^{+}\ldots A^{2+}){\cdot}A^{+}{\cdot}2X^{-}{\cdot}Y^{2-}$
14	$2A^{2+}{\cdot}X(1)^{-}{\cdot}X(2)^{-}{\cdot}Y^{2-}$	38	$(A^{+}\ldots A^{2+}){\cdot}A^{+}{\cdot}Y(1)^{2-}{\cdot}Y(2)^{2-}$
15	$2A^{2+}{\cdot}2X^{-}{\cdot}Y^{2-}$	39	$(A^{+}\ldots A^{2+}){\cdot}A^{2+}{\cdot}X(1)^{-}{\cdot}X(2)^{-}{\cdot}X(3)^{-}{\cdot}X(4)^{-}{\cdot}X(5)^{-}$
16	$2A^{2+}{\cdot}X^{-}{\cdot}HY^{-}{\cdot}Y^{2-}$	40	$(A^{+}\ldots A^{2+}){\cdot}A^{2+}{\cdot}2X(1)^{-}{\cdot}X(2)^{-}{\cdot}X(3)^{-}{\cdot}X(4)^{-}$
17	$2A^{2+}{\cdot}Y(1)^{2-}{\cdot}Y(2)^{2-}$	41	$(A^{+}\ldots A^{2+}){\cdot}A^{2+}{\cdot}2X(1)^{-}{\cdot}2X(2)^{-}{\cdot}X(3)^{-}$
18	$2A^{2+}{\cdot}(H_{2n+1}O_n)^{+}{\cdot}X^{-}{\cdot}Y(1)^{2-}{\cdot}Y(2)^{2-}$	42	$(A^{+}\ldots A^{2+}){\cdot}A^{2+}{\cdot}2X(1)^{-}{\cdot}3X(2)^{-}$
19	$2A^{2+}{\cdot}(H_{2n+1}O_n)^{+}{\cdot}X^{-}{\cdot}2Y^{2-}$	43	$(A^{+}\ldots A^{2+}){\cdot}A^{2+}{\cdot}3X(1)^{-}{\cdot}X(2)^{-}{\cdot}X(3)^{-}$
20	$3A^{2+}{\cdot}X(1)^{-}{\cdot}X(2)^{-}{\cdot}Y(1)^{2-}{\cdot}Y(2)^{2-}$	44	$(A^{+}\ldots A^{2+}){\cdot}A^{2+}{\cdot}X(1)^{-}{\cdot}X(2)^{-}{\cdot}X(3)^{-}{\cdot}Y^{2-}$
21	$3A^{2+}{\cdot}X(1)^{-}{\cdot}X(2)^{-}{\cdot}2Y^{2-}$	45	$(A^{+}\ldots A^{2+}){\cdot}A^{2+}{\cdot}X(1)^{-}{\cdot}X(2)^{-}{\cdot}X(3)^{-}{\cdot}Y^{2-}$
22	$3A^{2+}{\cdot}2X^{-}{\cdot}Y(1)^{2-}{\cdot}Y(2)^{2-}$	46	$(A^{+}\ldots A^{2+}){\cdot}A^{2+}{\cdot}3X^{-}{\cdot}Y^{2-}$
23	$3A^{2+}{\cdot}2X^{-}{\cdot}2Y^{2-}$	47	$(A^{+}\ldots A^{2+}){\cdot}A^{2+}{\cdot}X^{-}{\cdot}Y(1)^{2-}{\cdot}Y(2)^{2-}$
24	$3A^{2+}{\cdot}Y(1)^{2-}{\cdot}Y(2)^{2-}{\cdot}Y(2)^{2-}$		

Table 5.52 New mixed salts of amino acids with different anions

#	Mechanism	#	Mechanism
DA1.1	$\mathbf{2A^{+}{\cdot}X(1)^{-}{\cdot}X(2)^{-}}$	28	$2\text{L-Orn}^{2+}{\cdot}2Cl^{-}{\cdot}SO_4^{2-}$
1	$2\text{L-Lys}^{+}{\cdot}Cl^{-}{\cdot}BF_4^{-}$	29	$2\text{L-Orn}^{2+}{\cdot}2Br^{-}{\cdot}SO_4^{2-}$
2	$2\text{L-Lys}^{+}{\cdot}Cl^{-}{\cdot}ClO_4^{-}$	30	$2\text{L-His}^{2+}{\cdot}2Cl^{-}{\cdot}SiF_6^{2-}$
DA2.1	$\mathbf{A^{2+}{\cdot}X(1)^{-}{\cdot}X(2)^{-}}$	31	$2\text{L-His}^{2+}{\cdot}2Br^{-}{\cdot}SiF_6^{2-}$
3	$\text{L-His}^{2+}{\cdot}Cl^{-}{\cdot}HSO_4^{-}$	32	$2\text{L-His}^{2+}{\cdot}2I^{-}{\cdot}SiF_6^{2-}$
4	$\text{L-His}^{2+}{\cdot}Br^{-}{\cdot}HSO_4^{-}$	33	$2\text{L-His}^{2+}{\cdot}2BF_4^{-}{\cdot}SiF_6^{2-}{\cdot}2H_2O$
5	$\text{L-His}^{2+}{\cdot}NO_3^{-}{\cdot}F^{-}$	34	$2\text{L-His}^{2+}{\cdot}2ClO_4^{-}{\cdot}SiF_6^{2-}{\cdot}2H_2O$
6	$\text{L-His}^{2+}{\cdot}NO_3^{-}{\cdot}BF_4^{-}$	**DA2.8**	$\mathbf{2A^{2+}{\cdot}Y(1)^{2-}{\cdot}Y(2)^{2-}}$
7	$\text{L-His}^{2+}{\cdot}NO_3^{-}{\cdot}ClO_4^{-}$	35	$2\text{L-His}^{2+}{\cdot}SO_4^{2-}{\cdot}SiF_6^{2-}{\cdot}3H_2O$
8	$\text{L-Arg}^{2+}{\cdot}NO_3^{-}{\cdot}F^{-}$	**DA2.10**	$\mathbf{2A^{2+}{\cdot}(H_{2n+1}O_n)^{+}{\cdot}X^{-}{\cdot}2Y^{2-}}$
9	$\text{L-Arg}^{2+}{\cdot}NO_3^{-}{\cdot}BF_4^{-}$ (β)	36	$2\text{L-His}^{2+}{\cdot}(H_3O)^{+}{\cdot}NO_3^{-}{\cdot}2SO_4^{2-}$
10	$\text{L-Arg}^{2+}{\cdot}NO_3^{-}{\cdot}BF_4^{-}$ (β′)	**DA2.14**	$\mathbf{3A^{2+}{\cdot}2X^{-}{\cdot}2Y^{2-}}$
11	$\text{L-Arg}^{2+}{\cdot}NO_3^{-}{\cdot}BF_4^{-}$ (γ)	37	$3\text{L-Orn}^{2+}{\cdot}2NO_3^{-}{\cdot}2SO_4^{2-}$
12	$\text{L-Arg}^{2+}{\cdot}NO_3^{-}{\cdot}ClO_4^{-}$ (α)	**DA3.2**	$\mathbf{(A^{+}\cdots A^{2+}){\cdot}2X(1)^{-}{\cdot}X(2)^{-}}$
13	$\text{L-Arg}^{2+}{\cdot}NO_3^{-}{\cdot}ClO_4^{-}$ (β)	38	$(\text{L-Lys}^{+}\cdots\text{L-Lys}^{2+}){\cdot}2Cl^{-}{\cdot}BF_4^{-}$
14	$\text{L-Arg}^{2+}{\cdot}NO_3^{-}{\cdot}ClO_4^{-}$ (β′)	39	$(\text{L-Lys}^{+}\cdots\text{L-Lys}^{2+}){\cdot}2Br^{-}{\cdot}BF_4^{-}$
15	$\text{L-Arg}^{2+}{\cdot}NO_3^{-}{\cdot}ClO_4^{-}$ (γ)	40	$(\text{L-Lys}^{+}\cdots\text{L-Lys}^{2+}){\cdot}2I^{-}{\cdot}BF_4^{-}$
16	$\text{L-Arg}^{2+}{\cdot}NO_3^{-}{\cdot}ClO_4^{-}$ (δ)	41	$(\text{L-Lys}^{+}\cdots\text{L-Lys}^{2+}){\cdot}2Br^{-}{\cdot}ClO_4^{-}$

17	L-Lys^{2+}·Cl$^-$·NO$_3^-$	42	(L-Lys$^+$···L-Lys^{2+})·2Br$^-$·NO$_3^-$
18	L-Lys^{2+}·Cl$^-$·BF$_4^-$	43	(L-Lys$^+$···L-Lys^{2+})·2BF$_4^-$·Cl$^-$
19	L-Lys^{2+}·Cl$^-$·ClO$_4^-$	44	(L-Lys$^+$···L-Lys^{2+})·2BF$_4^-$·Br$^-$
20	L-Lys^{2+}·Br$^-$·NO$_3^-$	45	(L-Lys$^+$···L-Lys^{2+})·2ClO$_4^-$·Cl$^-$
21	L-Lys^{2+}·Br$^-$·BF$_4^-$	46	(L-Lys$^+$···L-Lys^{2+})·2ClO$_4^-$·Br$^-$
22	L-Lys^{2+}·Br$^-$·ClO$_4^-$	47	(L-Orn$^+$···L-Orn^{2+})·2Cl$^-$·NO$_3^-$
23	L-Lys^{2+}·NO$_3^-$·ClO$_4^-$	48	(L-Orn$^+$···L-Orn^{2+})·2Cl$^-$·BF$_4^-$
DA2.3	**2A^{2+}·2X(1)$^-$·X(2)$^-$·X(3)$^-$**	49	(L-Orn$^+$···L-Orn^{2+})·2Cl$^-$·ClO$_4^-$
24	2L-Arg^{2+}·2Cl$^-$·NO$_3^-$·ClO$_4^-$	50	(L-Orn$^+$···L-Orn^{2+})·2Br$^-$·NO$_3^-$
DA2.4	**2A^{2+}·3X(1)$^-$·X(2)$^-$**	51	(L-Orn$^+$···L-Orn^{2+})·2Br$^-$·BF$_4^-$
25	2L-Arg^{2+}·3NO$_3^-$·(F$^-$···HF)	52	(L-Orn$^+$···L-Orn^{2+})·2Br$^-$·ClO$_4^-$
26	2L-Arg^{2+}·3Cl$^-$·ClO$_4^-$	**DA3.3**	**(A$^+$···A^{2+})·X$^-$·Y^{2-}**
DA2.5	**2A^{2+}·X(1)$^-$·X(2)$^-$·Y^{2-}**	53	(L-His$^+$···L-His^{2+})·NO$_3^-$·SiF$_6^{2-}$
27	2L-Orn^{2+}·Br$^-$·NO$_3^-$·SO$_4^{2-}$	**DA5.8**	**(A$^+$···A^{2+})·A^{2+}·3X$^-$·Y^{2-}**
DA2.6	**2A^{2+}·2X$^-$·Y^{2-}**	54	(L-His$^+$···L-His^{2+})·L-His^{2+}·3NO$_3^-$·SO$_4^{2-}$·H$_2$O

Table 5.53 New possible mixed salts of L-Lys, L-Orn, L-Arg and L-His

1	L-$Lys^{2+}\cdot I^-\cdot NO_3^-$	42	(L-Orn^+…L-Orn^{2+})$\cdot Cl^-\cdot I^-\cdot ClO_4^-$
2	L-$Lys^{2+}\cdot I^-\cdot BF_4^-$	43	(L-Orn^+…L-Orn^{2+})$\cdot Br^-\cdot I^-\cdot ClO_4^-$
3	L-$Lys^{2+}\cdot I^-\cdot ClO_4^-$	44	(L-Orn^+…L-Orn^{2+})$\cdot 2I^-\cdot ClO_4^-$
4	L-$Lys^{2+}\cdot BF_4^-\cdot ClO_4^-$	45	(L-Orn^+…L-Orn^{2+})$\cdot Cl^-\cdot NO_3^-\cdot BF_4^-$
5	L-$Lys^{2+}\cdot Cl^-\cdot Br^-$	46	(L-Orn^+…L-Orn^{2+})$\cdot Br^-\cdot NO_3^-\cdot BF_4^-$
6	L-$Lys^{2+}\cdot Cl^-\cdot I^-$	47	(L-Orn^+…L-Orn^{2+})$\cdot I^-\cdot NO_3^-\cdot BF_4^-$
7	L-$Lys^{2+}\cdot Br^-\cdot I^-$	48	(L-Orn^+…L-Orn^{2+})$\cdot Cl^-\cdot NO_3^-\cdot ClO_4^-$
8	(L-Lys^+…L-Lys^{2+})$\cdot Cl^-\cdot Br^-\cdot ClO_4^-$	49	(L-Orn^+…L-Orn^{2+})$\cdot Br^-\cdot NO_3^-\cdot ClO_4^-$
9	(L-Lys^+…L-Lys^{2+})$\cdot Cl^-\cdot I^-\cdot ClO_4^-$	50	(L-Orn^+…L-Orn^{2+})$\cdot I^-\cdot NO_3^-\cdot ClO_4^-$
10	(L-Lys^+…L-Lys^{2+})$\cdot Br^-\cdot I^-\cdot ClO_4^-$	51	2L-$Orn^{2+}\cdot Cl^-\cdot Br^-\cdot SO_4^{2-}$
11	(L-Lys^+…L-Lys^{2+})$\cdot 2I^-\cdot ClO_4^-$	52	2L-$Orn^{2+}\cdot Cl^-\cdot I^-\cdot SO_4^{2-}$
12	(L-Lys^+…L-Lys^{2+})$\cdot Cl^-\cdot Br^-\cdot BF_4^-$	53	2L-$Orn^{2+}\cdot 2I^-\cdot SO_4^{2-}$
13	(L-Lys^+…L-Lys^{2+})$\cdot Cl^-\cdot I^-.BF_4^-$	54	2L-$Orn^{2+}\cdot I^-\cdot NO_3^-\cdot SO_4^{2-}$
14	(L-Lys^+…L-Lys^{2+})$\cdot Br^-\cdot I^-\cdot BF_4^-$	55	L-$His^{2+}\cdot I^-\cdot HSO_4^-$
15	(L-Lys^+…L-Lys^{2+})$\cdot 2I^-\cdot BF_4^-$	56	2L-$Arg^{2+}\cdot 2Br^-\cdot NO_3^-\cdot ClO_4^-$
16	(L-Lys^+…L-Lys^{2+})$\cdot Cl^-\cdot Br^-\cdot NO_3^-$	57	2L-$Arg^{2+}\cdot 2I^-\cdot NO_3^-\cdot ClO_4^-$
17	(L-Lys^+…L-Lys^{2+})$\cdot Cl^-\cdot I^-\cdot NO_3^-$	58	2L-$Arg^{2+}\cdot 2Cl^-\cdot NO_3^-\cdot BF_4^-$
18	(L-Lys^+…L-Lys^{2+})$\cdot Br^-\cdot I^-\cdot NO_3^-$	59	2L-$Arg^{2+}\cdot 2Br^-\cdot NO_3^-\cdot BF_4^-$
19	(L-Lys^+…L-Lys^{2+})$\cdot 2I^-\cdot NO_3^-$	60	2L-$Arg^{2+}\cdot 2I^-\cdot NO_3^-\cdot BF_4^-$
20	(L-Lys^+…L-Lys^{2+})$\cdot Cl^-\cdot BF_4^-\cdot ClO_4^-$	61	2L-$Arg^{2+}\cdot 3Br^-\cdot ClO_4^-$
21	(L-Lys^+…L-Lys^{2+})$\cdot Br^-\cdot BF_4^-\cdot ClO_4^-$	62	2L-$Arg^{2+}\cdot 3I^-\cdot ClO_4^-$

22	$(\text{L-Lys}^{+}\ldots\text{L-Lys}^{2+})\cdot I^{-}\cdot BF_4^{-}\cdot ClO_4^{-}$	63	$2\text{L-Arg}^{2+}\cdot 3Cl^{-}\cdot BF_4^{-}$
23	$(\text{L-Lys}^{+}\ldots\text{L-Lys}^{2+})\cdot Cl^{-}\cdot NO_3^{-}\cdot ClO_4^{-}$	64	$2\text{L-Arg}^{2+}\cdot 3Br^{-}\cdot BF_4^{-}$
24	$(\text{L-Lys}^{+}\ldots\text{L-Lys}^{2+})\cdot Br^{-}\cdot NO_3^{-}\cdot ClO_4^{-}$	65	$2\text{L-Arg}^{2+}\cdot 3I^{-}\cdot BF_4^{-}$
25	$(\text{L-Lys}^{+}\ldots\text{L-Lys}^{2+})\cdot I^{-}\cdot NO_3^{-}\cdot ClO_4^{-}$	66	$2\text{L-Orn}^{2+}\cdot Cl^{-}\cdot NO_3^{-}\cdot SeO_4^{2-}$
26	$(\text{L-Lys}^{+}\ldots\text{L-Lys}^{2+})\cdot Cl^{-}\cdot NO_3^{-}\cdot BF_4^{-}$	67	$2\text{L-Orn}^{2+}\cdot Br^{-}\cdot NO_3^{-}\cdot SeO_4^{2-}$
27	$(\text{L-Lys}^{+}\ldots\text{L-Lys}^{2+})\cdot Br^{-}\cdot NO_3^{-}\cdot BF_4^{-}$	68	$2\text{L-Orn}^{2+}\cdot I^{-}\cdot NO_3^{-}\cdot SeO_4^{2-}$
28	$(\text{L-Lys}^{+}\ldots\text{L-Lys}^{2+})\cdot I^{-}\cdot NO_3^{-}\cdot BF_4^{-}$	69	$2\text{L-Orn}^{2+}\cdot 2Cl^{-}\cdot SeO_4^{2-}$
29	$(\text{L-Lys}^{+}\ldots\text{L-Lys}^{2+})\cdot NO_3^{-}\cdot BF_4^{-}\cdot ClO_4^{-}$	70	$2\text{L-Orn}^{2+}\cdot 2Br^{-}\cdot SeO_4^{2-}$
30	$(\text{L-Lys}^{+}\ldots\text{L-Lys}^{2+})\cdot 2BF_4^{-}\cdot I^{-}$	71	$2\text{L-Orn}^{2+}\cdot 2I^{-}\cdot SeO_4^{2-}$
31	$(\text{L-Lys}^{+}\ldots\text{L-Lys}^{2+})\cdot 2ClO_4^{-}\cdot I^{-}$	72	$2\text{L-Orn}^{2+}\cdot Cl^{-}\cdot Br^{-}\cdot SeO_4^{2-}$
32	$(\text{L-Orn}^{+}\ldots\text{L-Orn}^{2+})\cdot 2I^{-}\cdot NO_3^{-}$	73	$2\text{L-Orn}^{2+}\cdot Cl^{-}\cdot I^{-}\cdot SeO_4^{2-}$
33	$(\text{L-Orn}^{+}\ldots\text{L-Orn}^{2+})\cdot Cl^{-}\cdot Br^{-}\cdot NO_3^{-}$	74	$2\text{L-Orn}^{2+}\cdot Br^{-}\cdot I^{-}\cdot SeO_4^{2-}$
34	$(\text{L-Orn}^{+}\ldots\text{L-Orn}^{2+})\cdot Cl^{-}\cdot I^{-}\cdot NO_3^{-}$	75	$3\text{L-Orn}^{2+}\cdot 2NO_3^{-}\cdot 2SeO_4^{2-}$
35	$(\text{L-Orn}^{+}\ldots\text{L-Orn}^{2+})\cdot Br^{-}\cdot I^{-}\cdot NO_3^{-}$	76	$\text{L-His}^{2+}\cdot I^{-}\cdot HSeO_4^{-}$
36	$(\text{L-Orn}^{+}\ldots\text{L-Orn}^{2+})\cdot Cl^{-}\cdot Br^{-}\cdot BF_4^{-}$	77	$\text{L-His}^{2+}\cdot Br^{-}\cdot HSeO_4^{-}$
37	$(\text{L-Orn}^{+}\ldots\text{L-Orn}^{2+})\cdot Cl^{-}\cdot I^{-}\cdot BF_4^{-}$	78	$\text{L-His}^{2+}\cdot Cl^{-}\cdot HSeO_4^{-}$
38	$(\text{L-Orn}^{+}\ldots\text{L-Orn}^{2+})\cdot Br^{-}\cdot I^{-}\cdot BF_4^{-}$	79	$2\text{L-His}^{2+}\cdot H_3O^{+}\cdot NO_3^{-}\cdot 2SeO_4^{2-}$
39	$(\text{L-Orn}^{+}\ldots\text{L-Orn}^{2+})\cdot 2I^{-}\cdot BF_4^{-}$	80	$(\text{L-His}^{+}\ldots\text{L-His}^{2+})\cdot\text{L-His}^{2+}\cdot 3NO_3^{-}\cdot SeO_4^{2-}\cdot H_2O$
40	$(\text{L-Orn}^{+}\ldots\text{L-Orn}^{2+})\cdot Cl^{-}\cdot Br^{-}\cdot ClO_4^{-}$	81	$2\text{L-His}^{2+}\cdot SeO_4^{2-}\cdot SiF_6^{2-}\cdot 3H_2O$
41	$2\text{L-Orn}^{2+}\cdot Br^{-}\cdot I^{-}\cdot SO_4^{2-}$	–	–

Acknowledgments We sincerely thank our colleague Dr. Ghazaryan V.V. for fruitful collaboration and for his help in preparation of this chapter. Unpublished results included in this book were obtained with his participation.

References

Aakeröy CB, Bahra GS, Brown CR, Hitchcock PB, Patel Y, Seddon KR (1995) L-Proline 2,5-dihydroxybenzoic acid (1/1): a zwitterion co-crystal. Acta Chem Scand 49:762–767

Aarthy A, Anitha K, Athimoolam S, Bahadur SA, Rajaram RK (2005) *L*-Asparaginium nitrate. Acta Crystallogr E61:o2042–o2044

Adams MJ, Hodgkin DC, Raeburn UA (1970) Crystal structure of a complex of mercury (II) chloride and histidine hydrochloride. J Chem Soc A 2632–2635

Adhikari S, Kar T (2012) Experimental and theoretical studies on physicochemical properties of L-leucine nitrate-a probable nonlinear optical material. J Cryst Growth 356:4–9

Adhikari S, Kar T (2013) Experimental and theoretical characterization of semiorganic nonlinear optical material *L*-leucine hydrobromide. Mater Res Bull 48:1612–1617

Adhikari S, Seth SK, Kar T (2013) Structural studies and physicochemical properties of l-valine hydrochloride monohydrate. CrystEngComm 15:7372–7379

Aggarwal MD, Choi J, Wang WS, Bhat K, Lal RB, Shields AD, Penn BG, Frazier DO (1999) Solution growth of a novel nonlinear optical material: L-histidine tetrafluoroborate. J Cryst Growth 204:179–182

Aggarwal MD, Stephens J, Barta AK, Lal RB (2003) Bulk crystal growth and characterization of semiorganic nonlinear optical materials. J Optoelectron Adv Mater 5(3):555–562

Ahmed AB, Feki H, Abid Y, Boughzala H, Mlayah A (2008) Structural, vibrational and theoretical studies of *L*-histidine bromide. J Mol Struct 888:180–186

Ahmed AB, Feki H, Abid Y, Boughzala H, Mlayah A (2009) Crystal structure, vibrational spectra and theoretical studies of *L*-histidinium dihydrogen phosphate-phosphoric acid. J Mol Struct 920:1–7

Ahmed AB, Feki H, Abid Y, Boughzala H, Minot C (2010) Crystal studies, vibrational spectra and non-linear optical properties of *L*-histidine chloride monohydrate. Spectrochim Acta A 75:293–298

Ahmed AB, Elleuch N, Feki H, Abid Y, Minot C (2011) Vibrational spectra and non linear optical properties of *L*-histidine oxalate: DFT studies. Spectrochim Acta A 79:554–561

AIP (1993) Aleksei Aleksandrovich Shternberg (1911-1993). Crystallogr Rep 38:585–586

Akhmanov SA, Kovrigin AI, Kolosov VA, Piskarskas AS, Fadeev VV, Khokhlov RV (1966) Tunable optical parametric oscillator on a KDP crystal. JETP Lett 3:372–378 (In Russian)

Akkurt M, Özütrk S, Ramajothi J, Büyükgüngör O, Dhanuskodi S (2004) *L*-Histidinium tetrafluorosuccinate. Acta Crystallogr E60:o481–o483

Alagar M, Krishnakumar RV, Nandhini MS, Natarajan S (2001a) *L*-Alanine maleate. Acta Crystallogr E57:o855–o857

Alagar M, Krishnakumar RV, Natarajan S (2001b) *L*-Phenylalaninium maleate. Acta Crystallogr E57:o968–o970

Alagar M, Krishnakumar RV, Mostad A, Natarajan S (2001c) *DL*-Valinium maleate at 150K. Acta Crystallogr E57:o1102–o1104

Alagar M, Krishnakumar RV, Mostad A, Natarajan S (2002a) Bis(*DL*-serinium) oxalate dihydrate. Acta Crystallogr E58:o114–o116

Alagar M, Krishnakumar RV, Mostad A, Natarajan S (2002b) *DL*-Methioninium maleate. Acta Crystallogr E58:o396–o398

Alagar M, Nandhini MS, Krishnakumar RV, Mostad A, Natarajan S (2003a) *DL*-Phenylalaninium maleate at 123K. Acta Crystallogr E59:o209–o211

Alagar M, Krishnakumar RV, Nandhini MS, Cameron TS, Natarajan S (2003b) Bis(*DL*-aspartic acid) oxalate. Acta Crystallogr E59:o108–o110

Alagar M, Krishnakumar RV, Rajagopal K, Nandhini MS, Natarajan S (2003c) *L*-Phenylalanine fumaric acid. Acta Crystallogr E59:o952–o954

Alagar M, Krishnakumar RV, Nandhini MS, Natarajan S (2003d) *DL*-Valine-fumaric acid (2/1). Acta Crystallogr E59:o857–o859

Alagar M, Nandhini MS, Krishnakumar RV, Ravikumar K, Natarajan S (2004) *DL*-Valine-succinic acid (2/1). Acta Crystallogr E60:o1009–o1011

Alagar M, Krishnakumar RV, Parimala Devi P, Natarajan S (2005) *L*-Phenylalanine *L*-phenylalaninium malonate. Acta Crystallogr E61(4):o992–o994

Albers J (1988) Betaine compounds – a new family with ferroelectric and incommensurate phase. Ferroelectrics 78(1):3–10

Albers J, Klöpperpieper A, Rother HJ, Haussühl S (1988) Ferroelectricity in betaine phosphite. Ferroelectrics 81(1):27–30

Aleksandrov KS, Haussühl S (1975) Elastic and thermoelastic properties of *L*-glutamic acid hydrochloride. Z Kristallogr 142:328–331

Al-Karaghouli AR, Koetzle TF (1975) Neutron diffraction study of *L*-phenylalanine hydrochloride. Acta Crystallogr B31:2461–2465

Al-Karaghouli AR, Cole FE, Lehmann MS, Miskell CF, Verbist JJ, Koetzle TF (1975) Precision neutron diffraction structure determination of protein and nucleic acid components. XVII. Molecular and crystal structure of the amino acid glycine hydrochloride. J Chem Phys 63:1360–1366

Almeida JMA, Miranda MAR, Remédios CMR, Melo FE, Freire PTC, Sasaki JM, Cardoso LP, dos Santos AO, Kycia S (2003) Piezoelectric coefficients of *L*-arginine hydrochloride monohydrate obtained by X-ray multiple diffraction using synchrotron radiation. J Appl Crystallogr 36:1348–1351

Almeida JMA, Miranda MAR, Avanci LH, de Menezes AS, Cardoso LP, Sasaki JM (2006) Piezoelectric coefficients d_{14}, d_{16}, d_{34} and d_{36} of an *L*-arginine hydrochloride monohydrate crystal by X-ray three-beam diffraction. J Synchrotron Radiat 13:435–439

Amalanathan M, Hubert Joe I, Prabhu SS (2010) Charge transfer interaction and terahertz studies of a nonlinear optical *L*-glutamine picrate: a DFT study. J Phys Chem A114:13055–13064

Amalanathan M, Hubert Joe I, Rastogi VK (2011a) Molecular structure, vibrational spectra and nonlinear optical properties of *L*-valine hydrobromide: DFT study. J Mol Struct 985:48–56

Amalanathan M, Hubert Joe I, Rastogi VK (2011b) Density functional theory studies on molecular structure and vibrational spectra of NLO crystal *L*-phenylalanine phenylalaninium nitrate for THz application. J Mol Struct 1006:513–526

Ambujam K, Rajarajan K, Selvakumar S, Vetha Potheher I, Joseph GP, Sagayaraj P (2006) Growth and characterization of a novel NLO crystal bis-glycine hydrogen chloride (BGHC). J Cryst Growth 286(2):440–444

Ambujam K, Rajarajan K, Selvakumar S, Madhavan J, Mohamed G, Sagayaraj P (2007) Growth and characterization of gel grown single crystals of bis-glycine hydrogen chloride (BGHC). Opt Mater 29(6):657–662

Anandan P, Jayavel R (2011) Crystal growth and characterization of semiorganic single crystals of *L*-histidine family for NLO applications. J Cryst Growth 322:69–73

Anandan P, Saravanan T, Vasudevan S, Mohan Kumar R, Jayavel R (2010a) Crystal growth and characterization of *L*-tyrosine bromide (LTB) nonlinear optical single crystals. J Cryst Growth 312:837–841

Anandan P, Parthipan G, Saravanan T, Mohan Kumar R, Bhagavannarayana G, Jayavel R (2010b) Crystal growth, structural and optical characterization of a semi-organic single crystal for frequency conversion applications. Physica B 405:4951–4956

Anandan P, Saravanan T, Parthipan G, Mohan Kumar R, Bhagavannarayana G, Ravi G, Jayavel R (2011) Crystal growth, structural and thermal studies of amino acids admixture *L*-arginine phosphate monohydrate single crystals. Solid State Sci 13:915–922

Anandan P, Vetrivel S, Jayavel R, Vedhi C, Ravi G, Bhagavannarayana G (2012a) Crystal growth, structural and photoluminescence studies of *L*-tyrosine hydrobromide semi organic single crystal. J Phys Chem Solids 73:1296–1301

Anandan P, Jayavel R, Saravanan T, Parthipan G, Vedhi C, Mohan Kumar R (2012b) Crystal growth and characterization of *L*-hydrochloride monohydrate semiorganic optical single crystals. Opt Mater 34:1225–1230

Anandan P, Parthipan G, Pazhanivel K, Ravi G, Jayavel R (2014) Growth and characterization of potassium halides mixed *L*-arginine phosphate monohydrate semi organic nonlinear optical single crystals. Optik 125:8–10

Anandha Babu G, Mohanapriya SK, Ramasamy P, Chandramohan A (2011) Studies on the synthesis, growth, crystal structure, and physical properties of a novel nonlinear optical: glycine 3,5-dihydroxybenzoic acid. J Cryst Growth 318:1021–1025

Anbuchezhiyan M (2010) Growth and characterization of some nonlinear optical crystals of amino acid complexes. PhD thesis, SRM University, Kattankulathur

Anbuchezhiyan M, Ponnusamy S, Muthamizhchelvan C (2009a) Synthesis and characterization of a new organic nonlinear optical crystal: *L*-Phenylalaninium maleate. Spectrochim Acta A 74:917–923

Anbuchezhiyan M, Ponnusamy S, Muthamizhchelvan C (2009b) Synthesis, characterization and nonlinear optical studies of L-leucinium oxalate: a single crystal. Optoelectron Adv Mater-Rapid Commun 3(11):1161–1167

Andersen L, Lindqvist O, Moret J (1983) Diglycine-telluric acid monohydrate, $Te(OH)_6.2NH_2CH_2COOH.H_2O$. Acta Crystallogr C39:57–58

Ando O, Ashida T, Sasada Y, Kakudo M (1967) The crystal structure of *L*-valine hydrochloride. Acta Crystallogr 23:172–173

Andrade LCR, Costa MRR, Paixão JA, Agostinho Moreira J, Almeida A, Chaves MR, Klopperpieper A (1999) Crystal structure of glycinium arsenate. Z Kristallogr NCS 214:535–536

Andrade LCR, Costa MRR, Paixão JA, Santos ML, Agostinho Moreira J, Chaves MR, Almeida A (2002) Crystal structure of trimethylglycine 2-hydroxy-1,2,3-propanetricarboxylic acid (1:1) adduct, $C_6O_7H_8.C_5NO_2H_{11}$. Z Kristallogr NCS 217:77–78

Andreas LB, Mehta AK, Mehta MA (2007) Determination of global structure from distance and orientation constraints in biological solids using solid-state NMR spectroscopy. J Am Chem Soc 129:15233–15239

Andreev RB, Vetrov KV, Voitsekhovskii VN, Volosov VD, Nikiforuk IV, Nikolaeva BP, Yakobson VE (1990) Growth of *d*-LAP crystals and investigation of its basic nonlinear optical properties. Bull AS USSR Ser Phys 54:2491–2493 (In Russian)

Andreeva NV, Gavrilova LA, Ennan AA (1983) Hexafluorosilicates of amino acids. Zhurn Neorg Khimii 28(7):1720–1724 (In Russian)

Andriyevsky B, Czapla Z (2004) Refractive and dilative ferroelectric anomalies of DGN crystals. Ferroelectrics 302:39–41

Andriyevsky B, Czapla Z, Kardash V, Dacko S, Dumka Y, Kurlyak V (2001) Optical and dilatative properties of diglycine nitrate crystals. Ukr J Phys Opt 2(4):211–216

Andriyevsky B, Czapla Z, Dumka Y, Dacko S, Kardash V (2002) Dilatative and refractive properties of diglycine nitrate crystals in the range of phase transition. Mater Sci Eng B95(1):14–18

Andriyevsky B, Ciepluch-Trojanek W, Romanyuk M, Patryn A, Jaskolski M (2005) Band structure and optical properties of diglycine nitrate crystal. Physica B364(1–4):78–84

Angeli Mary PA, Dhanuskodi S (2001) EPR and optical absorption studies of VO^{2+} doped *L*-arginine phosphate monohydrate single crystals-part I. Spectrochim Acta A 57:2345–2353

Angelova O, Velikova V, Kolev T, Radomirska V (1996) Crystalline complexes involving amino acids. I. L-arginine hydrogen squarate. Acta Crystallogr C52(12):3252–3256

Anitha K, Rajaram RK (2005) *DL*-Phenylalanine *DL*-phenylalaninium picrate. Acta Crystallogr E61:o589–o591

Anitha K, Sridhar B, Rajaram RK (2004a) *L*-Valinium picrate. Acta Crystallogr E60:o1530–o1532

Anitha K, Annavenus S, Sridhar B, Rajaram RK (2004b) *DL*-Valine *DL*-valinium picrate. Acta Crystallogr E60:o1722–o1724

Anitha K, Sridhar B, Rajaram RK (2004c) β-Alanine β-alaninium picrate. Acta Crystallogr E60: o1630–o1632

Anitha K, Athimoolam S, Rajaram RK (2005a) *L*-Asparaginium picrate. Acta Crystallogr E61: o1463–o1465
Anitha K, Athimoolam S, Rajaram RK (2005b) L-leucine L-leucinium picrate. Acta Crystallogr E61(6):o1604–o1606
Anitha K, Athimoolam S, Natarajan S (2006a) *L*-Prolinium picrate and 2-methylpyridinium picrate. Acta Crystallogr C62:o567–o570
Anitha K, Athimoolam S, Rajaram RK (2006b) DL-methionine DL-methioninium picrate. Acta Crystallogr E62(1):o8–o10
Antipin MY, Lyssenko KA, Suponitsky KY, Karapetyan HA, Petrosyan AM (2006) Infrared and Raman spectra, structure and electron density distribution of *L*-arginine dioxalate. J Mol Struct 792–793:194–200
Aoki K, Nagano K, Iitaka Y (1971) The crystal structure of *L*-arginine phosphate monohydrate. Acta Crystallogr B27:11–23
Aravindan A, Srinivasan P, Vijayan N, Gopalakrishnan R, Ramasamy P (2007) Investigations on the growth, optical behaviour and factor group of an NLO crystal: *L*-alanine alaninium nitrate. Cryst Res Technol 42(11):1097–1103
Aravindan A, Srinivasan P, Vijayan N, Gopalakrishnan R, Ramasamy P (2008) A comparative study on the growth and characterization of nonlinear optical amino acid crystal: *L*-Alanine (LA) and *L*-alanine alaninium nitrate (LAAN). Spectrochim Acta A71(2):297–304
Arjunan S, Mohan Kumar R, Mohan R, Jayavel R (2008a) Nucleation kinetics and growth aspects of organic nonlinear optical *L*-arginine trifluoroacetate crystals. Cryst Res Technol 43:417–422
Arjunan S, Mohan Kumar R, Mohan R, Jayavel R (2008b) Growth and dielectric, mechanical, thermal and etching studies of an organic nonlinear optical *L*-arginine trifluoroacetate (LATF) single crystal. Mater Res Bull 43:2018–2025
Arjunan S, Bhaskaran A, Mohan Kumar R, Mohan R, Jayavel R (2010) Effect of rare-earth dopants on the growth and structural, optical, electrical and mechanical properties of *L*-arginine phosphate single crystals. J Alloy Compd 506:784–787
Arjunan S, Bhaskaran A, Mohan Kumar R, Mohan R, Jayavel R (2012a) Effect of iodic acid dopant on the growth and structural, optical, and electrical properties of L-arginine phosphate single crystals. Mater Manuf Process 27:49–52
Arjunan V, Mariusz Marchewka K, Kalaivani M (2012b) Synthesis, vibrational and quantum chemical investigations of hydrogen bonded complex betaine dihydrogen selenite. Spectrochim Acta A 96:744–758
Arkhipov SG, Zakharov BA, Boldyreva EV (2013) Semi-maleate of *L*- and *DL*-serinium: the first example of chiral and racemic serinium salts with the same composition and stoichiometry. Acta Crystallogr C69:517–521
Arun KJ, Jayalekshmi S (2008a) Growth and characterization of glycinium oxalate crystals for nonlinear optical applications. Optoelectron Adv Mater-Rapid Commun 2(11):701–706
Arun KJ, Jayalekshmi S (2008b) Studies on the microstructure and microhardness of L-alaninium oxalate single crystals for optical devices. Optoelectron Adv Mater-Rapid Commun 2 (12):802–805
Arun KJ, Jayalekshmi S (2009) Growth and characterization of nonlinear optical single crystals of L-alaninium oxalate. J Miner Mater Charact Eng 8(8):635–646
Arun KJ, Jayalekshmi S (2011) Investigation on the nonlinear optical properties of glycinium oxalate single crystals. Optoelectron Lett 7(2):136–138
Aruna S, Anuradha A, Thomas PC, Mohamed MG, Rajasekar SA, Vimalan M, Mani G, Sagayaraj P (2007a) Growth, optical and thermal studies of *L*-arginine perchlorate-A promising non-linear optical single crystal. Indian J Pure Appl Phys 45:524–528
Aruna S, Bhagavannarayana G, Palanisamy M, Thomas PC, Varghese B, Sagayaraj P (2007b) Growth, morphological, mechanical and dielectric studies of semi organic NLO single crystal: *L*-argininium perchlorate. J Cryst Growth 300:403–408

Aruna S, Bhagavannarayana G, Sagayaraj P (2007c) Investigation on the physicochemical properties of nonlinear optical (NLO) single crystal: *L*-histidinium dinitrate. J Cryst Growth 304:184–190

Aruna S, Vimalan M, Thomas PC, Thamizharasan K, Ambujam K, Madhavan J, Sagayaraj P (2007d) Growth and characterization of semi organic nonlinear optical LHPCL crystals. Cryst Res Technol 42:180–185

Arunmozhi G, Jayavel R, Subramanian C (1997) Experimental determination of metastable zone width, induction period and interfacial energy of LAP family crystals. J Cryst Growth 178:387–392

Arunmozhi G, De M, Gomes E (2004) Metastability and crystal growth kinetics on *L*-arginine phosphate. Cryst Growth Technol 39:34–39

Asaji T, Seliger J, Žagar V, Sekiguchi M, Watanabe J, Gotoh K, Ishida H, Vrtnik S, Dolinšek J (2007) Phase transition and temperature dependent electronic state of an organic ferroelectric, phenazine-chloranilic acid (1:1). J Phys: Condens Matter 19:226203

Athimoolam S, Natarajan S (2007) Hydrogen-bonding features in the 1:2 adduct of 4-aminobenzoic acid and *L*-proline. Acta Crystallogr C63:o283–o286

Atwood JL, Dalgarno SJ, Hardie MJ, Raston CL (2005) Selective single crystal complexation of *L*- or *D*-leucine by p-sulfonatocalix[6]arene. Chem Commun 337–339

Avdienko KI et al (1980) Lithium iodate. Growth of crystals, their properties and application. (Edited by Bogdanov S.V.). Novosibirsk, Nauka (In Russian)

Averbuch-Pouchot MT (1988) Crystal structure of a new telluric acid adduct: $Te(OH)_6.2(CH_3NHCH_2COOH)$. Z Kristallogr 183:285–291

Averbuch-Pouchot MT (1993a) Structures of glycinium phosphite and glycylglycinium phosphite. Acta Crystallogr C49:815–818

Averbuch-Pouchot MT (1993b) Crystal structure of N-methylglycinium mono-hydrogen-phosphite, $(CH_3NH_2CH_2COOH)(HPO_3H)$. Z Kristallogr 207:149–150

Averbuch-Pouchot MT (1993c) Crystal structure of 3-ammonium propionic acid monohydrogen-phosphite, $(H_3NC_2H_4COOH)(HPO_3H)$. Z Kristallogr 208(2):257–258

Averbuch-Pouchot MT (1993d) Crystal structure of *L*-histidinium phosphite and a structure reinvestigation of the monoclinic form of *L*-histidine. Z Kristallogr 207:111–120

Averbuch-Pouchot MT, Durif A, Guitel JC (1988a) Structure of glycine monophosphate and glycine *cyclo*-triphosphate. Acta Crystallogr C44:99–102

Averbuch-Pouchot MT, Durif A, Guitel JC (1988b) Structure of glycine *cyclo*-tetraphosphate. Acta Crystallogr C44:888–890

Averbuch-Pouchot MT, Durif A, Guitel JC (1988c) Structures of β-alanine, DL-alanine and sarcosine monophosphates. Acta Crystallogr C44:1968–1972

Averbuch-Pouchot MT, Durif A, Guitel JC (1988d) Structure of *L*-histidinium dihydrogenmonophosphate monohydrate. Acta Crystallogr C44:890–892

Ayyar RR (1968) Crystal structure of *L*(+)-cysteine hydrochloride monohydrate. Z Kristallogr 126:227–239

Ayyar RR, Chandrasekharan R (1967) Crystal structure of *L*-tryptophan hydrobromide. Curr Sci 36(6):139–143

Ayyar RR, Srinivasan R (1965) Crystal structure of *L*(+)-cysteine hydrochloride monohydrate. Curr Sci 34(15):449–450

Bahadur SA (1992) Crystal structure analyses of aminoacid inorganic acid complexes. PhD Thesis, Madurai Kamaraj University, Madurai

Bahadur SA, Athimoolam S (2007) *DL*-Methioninium dihydrohenphosphate. Acta Crystallogr E63:o1952–o1954

Bahadur SA, Athimoolam S (2009) Chain C(5), C(8), $C_2^1(4)$ and $C_1^2(4)$ motifs in DL-histidinium perchlorate. Anal Sci X-Ray Struct Anal Online 25:15–16

Bahadur SA, Rajaram RK (1995a) Crystal structure of *DL*-alanine nitrate. Z Kristallogr 210:279–281

Bahadur SA, Rajaram RK (1995b) Crystal and molecular structure of *DL*-aspartic acid nitrate monohydrate. Z Kristallogr 210:276–278

Bahadur SA, Rajaram RK, Nathaji M (1991) Structure of diprotonated *DL*-histidinium dinitrate. Acta Crystallogr C47:1420–1423

Bahadur SA, Rajaram RK, Nethaji M, Natarajan S (1993) Crystal structure of *L*-leucine nitrate. Z Kristallogr 203:93–100

Bakke O, Mostad A (1980) The structure and conformation of tryptophan in the crystal of the pure racemic compound and the hydrogen oxalate. Acta Chem Scand B34:559–570

Balakrishnan T, Ramamurthi K (2009) Crystal growth, structural, optical, mechanical and thermal properties of a new nonlinear optical single crystal: *L*-ornithine monohydrochloride. Spectrochim Acta A 72:269–273

Balakrishnan T, Ramamurthi K, Thamotharan S (2013) Glycine-phthalic acid (1/1). Acta Crystallogr E69:o57

Balashova EV, Lemanov VV (2003) Dielectric and acoustic properties of some betaine and glycine compounds. Ferroelectrics 285:179–205

Balashova EV, Lemanov VV, Albers J, Klöpperpieper A (1998) Acoustic properties of betaine phosphite in the vicinity of phase transitions. Phys Solid State 40:995–1000

Balashova EV, Lemanov VV, Pankova GA (2001) Acoustic and dielectric anomalies in the temperature range of a ferroelectric phase transition in glycine phosphate crystals. Phys Solid State 43:1328–1335

Balasubramanian D, Jayavel R, Murugakoothan P (2009) Studies on the growth aspects of organic L-alanine maleate: a promising nonlinear optical crystal. Nature 1(3):216–221

Balasubramanian D, Murrugakoothan P, Jayavel R (2010) Synthesis, growth and characterization of organic nonlinear optical bis-glycine maleate (BGM) single crystals. J Cryst Growth 312:1855–1859

Baran J, Śledź M, Jakubas R, Bator G (1977) Ferroelectric phase transition in deuterated glycinium phosphate crystals. Phys Rev B55:169–172

Baran J, Barnes AJ, Ratajczak H (1994) Polarized Raman spectra of diglycine nitrate single crystal in the lattice-vibration region. J Mol Struct 325:71–75

Baran J, Drozd M, Lis T, Sledz M, Barnes AJ, Ratajczak H (1995a) Crystal structure and vibrational spectra of betaine hydrogen selenate monohydrate. J Mol Struct 372:29–40

Baran J, Barnes AJ, Ratajczak H (1995b) Polarized infrared and Raman spectra of diglycine nitrate single crystal. Spectrochim Acta A51(2):197–214

Baran J, Drozd M, Głowiak T, Śledź M, Ratajczak H (1995c) Crystal structure, phase transitions and vibrational spectra of bis(betaine) nitrate. J Mol Struct 372:131–144

Baran J, Drozd M, Lis T, Ratajczak H (1995d) and vibrational spectra of the tetra(betaine) selenate crystal. J Mol Struct 372(2–3):151–159

Baran J, Drozd M, Lis T, Ratajczak H (1995e) Phase transitions in the tetra(betaine) selenate crystal. J Mol Struct 372(2–3):145–150

Baran J, Bator G, Jakubas R, Śledź M (1996) Dielectric dispersion and vibrational studies of a new ferroelectric, glycinium phosphate crystal. J Phys Condens Matter 8:10647–10658

Baran J, Czapla Z, Drozd MK, Ilczyszyn MM, Marchewka M, Ratajczak H (1997a) Polarised FTIR and Raman spectra of betaine phosphite single crystal I. Paraelectric phase. J Mol Struct 403:17–37

Baran J, Barnes AJ, Marchewka MK, Pietrashko A, Ratajczak H (1997b) Structure and vibrational spectra of the bis(betaine)-selenic acid molecular crystal. J Mol Struct 416(1–3):33–42

Baran J, Barnes AJ, Engelen B, Panthöfer M, Pietraszko A, Ratajczak H, Sledz M (2000) Structure and polarized IR and Raman spectra of the solid complex betaine-trichloroacetic acid. J Mol Struct 550:21–41

Baran J, Łukaszewicz K, Pietraszko A, Śledź M (2002) Structural investigations of the ferroelectric glycinium hydrogenphosphite (GPI) and its deuterated analogue (DGPI) I. X-ray diffraction studies of the crystal structure of paraelectric and ferroelectric phases. J Mol Struct 611:155–168

Baran J, Trzebiatowska M, Ratajczak H (2004) Polarised IR and Raman spectra of monoglycine dihydrogenphosphate single crystal. J Mol Struct 708:127–144

Baran J, Drozd MA, Ratajczak H (2010) Polarised IR and Raman spectra of monoglycine nitrate single crystal. J Mol Struct 976:226–242

Baran J, Barnes AJ, Ratajczak H (2012) The polarized IR and Raman spectra of the diglycine hydrochloride crystal. J Mol Struct 1009:55–68

Baraniraj T, Philominathan P (2010) Growth and characterization of NLO based *L*-arginine maleate dihydrate single crystal. Spectrochim Acta A 75:74–76

Barker CE, Eimerl D, Velsko SP (1991) Temperature-insensitive phase matching for second-harmonic generation in deuterated *l*-arginine phosphate. J Opt Soc Am B8:2481–2492

Bass M, Franken PA, Hill AE, Peters CW, Weinreich G (1962a) Optical mixing. Phys Rev Lett 8(1):18

Bass M, Franken PA, Ward JF, Weinreich G (1962b) Optical rectification. Phys Rev Lett 9(11):446–448

Beja AM, Paixão JA, Silva MR, da Veiga LA (2000) Crystal structure of betainium perchlorate hydrate, $[(CH_3)_3NCH_2COOH]ClO_4.H_2O$. Z Kristallogr NCS 215:581–582

Belhouas R, Bouacida S, Boudaren C, Daran JC, Roisnel T (2012) (±)-Bis(1-carboxy-2-phenyl-ethanaminium) hexafluorosilicate(VI). Acta Crystallogr E68:o1791–o1792

Bena Jothy V, Vijayakumar T, Sajan D, Jayakumar VS, Hubert Joe I (2006) Vibrational spectra and *ab initio* computations of sarcosinium oxalate monohydrate. Laser Phys 16:1242–1252

Benali-Cherif N, Benguedouar L, Cherouana A, Bendjeddou L, Merazig H (2002) *L*-Histidinium dinitrate. Acta Crystallogr E58:o822–o824

Bendheif L, Benali-Cherif N, Benguedouar L, Bouchouit K, Merazig H (2003) *L*-Valinium hydrogenphosphite. Acta Crystallogr E59:o141–o142

Bennett I, Davidson AGH, Harding MM, Morrelle I (1970) The crystal structure of *DL*-histidine hydrochloride dihydrate. Acta Crystallogr B26:1722–1729

Berrah F, Benali-Cherif N, Lamraoui H (2005) DL-Leucinium nitrate. Acta Crystallogr E61: o1517–o1519

Bhaduri D, Saha NN (1979) Crystal and molecular structure of DL-lysine hydrochloride. J Cryst Mol Struct 9:311–316

Bhagavannarayana G, Kumar S, Shakir M, Kushawaha SK, Maurya KK, Malhotra R, Rao KR (2010) Unidirectional growth of *L*-cysteine hydrochloride monohydrate: first time observation as nonlinear optical material and its characterization. J Appl Crystallogr 43:710–715

Bhagavannarayana G, Riscob B, Shakir M (2011) Growth and characterization of *L*-leucine *L*-leucinium picrate single crystal: a new nonlinear optical material. Mater Chem Phys 126 (1–2):20–23

Bhar GC, Rudra AM, Datta PK, Roy UN, Wadhawan VK, Sasaki T (1995) A comparative study of laser second harmonic generation in some crystals. Pramana: J Phys 44:45–53

Bhat H (1994) Growth and characterization of some novel crystals for nonlinear optical applications. Bull Mater Sci 17:1233–1249

Bhat TN, Vijayan M (1976) X-ray studies on crystalline complexes involving amino acids. I. Crystal structure of L-lysine L-aspartate. Acta Crystallogr B32:891–895

Bhat TN, Vijayan M (1977) X-ray studies of crystalline complexes involving amino acids. II. The crystal structure of L-arginine L-glutamate. Acta Crystallogr B33(6):1754–1759

Bhat TN, Vijayan M (1978) X-ray studies of crystalline complexes involving amino acids. III. The structure of the twinned pseudosymmetric crystals of a complex between histidine and aspartic acid. Acta Crystallogr B34:2556–2565

Bhatt H, Murli C, Garg N, Deo MN, Chitra R, Choudhury RR, Sharma SM (2012) High pressure phase transformations in bis(glycinium)oxalate-An infrared absorption study. Chem Phys Lett 532:57–62

Bhattacharyya SC, Saha NN (1978a) Crystal and molecular structure of sarcosine hydrochloride. J Cryst Mol Struct 8(3):105–113

Bhattacharyya SC, Saha NN (1978b) Crystal and molecular structure of disarcosine hydrobromide. J Cryst Mol Struct 8(5):209–215

Blessing RH (1986) Hydrogen bonding and thermal vibrations in crystalline phosphate salts on histidine and imidazole. Acta Crystallogr B42:613–621, Erratum: B43:407 (19873)

Blessing RH, McGandy EL (1972) ACA Abstr Papers (Winter), p 68 (Quoted from CSD, code: HISTPA)

Blinc R, Jamšek-Vilfan M, Lahajnar G, Hajduković G (1970) NMR study of the ferroelectric transitions in diglycine nitrate and trissarcosine calcium chloride. J Chem Phys 52 (12):6407–6411

Book L, Carty AJ, Chieh C (1981) Five-coordinated halogenomercurate(II) complexes: crystal structures of bis(L-tryptophanium) trichloromercurate(II) and bis(D, L-homocysteine thiolactonium) tetrabromomercurate(II). Can J Chem 59:138–143

Boopathi K, Rajesh P, Ramasamy P (2012) Investigation on growth, structural, optical, thermal, dielectric and mechanical properties of organic *L*-prolinium trichloroacetate single crystals. Mater Res Bull 47:2299–2305

Botto IL (1987) Thermal and vibrational behaviour of $Te(OH)_6.2NH_2CH_2COOH.H_2O$. J Less-Common Met 128:47–55

Bouchouit K, Sofiani Z, Benali-Cherif N, Bendheif L, Migalska-Zalas A, Kityk IV, Sahraoui B (2005) Third order nonlinear optical properties of hybrid mono crystals with n-conjugated systems. Transparent Optical Networks, 2005. Proc 2005 7th Intern Conf 2:367–371

Braga D, Chelazzi L, Ciabatti I, Grepioni F (2013) From 3D channelled frameworks to 2D layered structures in molecular salts of L-serine and DL-serine with oxalic acid. New J Chem 37:97–104

Briget Mary M, Umadevi M, Pandiarajan S, Ramakrishnan V (2004) Vibrational spectral studies of L-methionine L-methioninium perchlorate monohydrate. Spectrochim Acta A 60 (11):2643–2651

Briget Mary M, Umadevi M, Ramakrishnan V (2005) Vibrational spectral analysis of *L*-lysine *L*-lysinium dichloride nitrate. Spectrochim Acta A 61:3124–3130

Briget Mary M, Sasirekha V, Ramakrishnan V (2006a) Vibrational spectral analysis of *DL*-valine *DL*-valinium and *DL*-methionine *DL*-methioninium picrates. Spectrochim Acta A 65:955–963

Briget Mary M, Sasirekha V, Ramakrishnan V (2006b) Spectral investigations of amino acid picrates. Spectrochim Acta A 65(2):414–420

Brsikyan NA, Andriasyan LH, Badalyan GR, Harutyunyan AV, Petrosyan AM, Ghazaryan VV (2012) Comparative morphology of dentinal tubules occlusion at the use of different desensitizing agents in experiment. New Armenian Med J 6(4):52–55

Buerger MJ, Barney E, Hahn T (1956) The crystal structure of diglycine hydrobromide. Z Kristallogr 108:130–144

Bugayong RR, Sequeira A, Chidambaram R (1972) A neutron diffraction study of the structure of *L*-lysine monohydrochloride dihydrate. Acta Crystallogr B28:3214–3219

Bye E, Mostad A, Romming C (1973) Crystal structure of *DL*-tryptophan formate. Acta Chem Scand 27:471–484

Cano FH, Martinez-Carrera S (1974) The crystal structure of diglycine sulphate monohydrate. Acta Crystallogr B30:2729–2732

Capasso S, Mattia CA, Mazzarella L, Zagari A (1983) *L*-Lysine sulphate, $C_6H_{16}N_2O_2^{2+}.SO_4^{2-}$: a novel conformation of the *L*-lysine side chain. Acta Crystallogr C39:281–283

Caroline ML, Vasudevan S (2008) Growth and characterization of an organic nonlinear optical material: *L*-alanine alaninium nitrate. Mater Lett 62(15):2245–2248

Caroline ML, Vasudevan S (2009) Growth and characterization of *L*-phenylalanine nitric acid, a new organic nonlinear optical material. Mater Lett 63:41–44

Caroline ML, Prakash M, Geetha D, Vasudevan S (2011) Growth, structural, vibrational, optical, laser and dielectric aspects of *L*-alanine alaninium nitrate single crystal. Spectrochim Acta A 79(5):1936–1940

Carvalho JF, Hernandes AC, Nunes FD, de Moraes LBOA, Misoguti L, Zilio ASC (1997) LAP single crystal growth free of microorganisms by an accurately controlled solvent evaporation technique. J Cryst Growth 173:487–491

Chacko KK, Chandrasekharan R, Chandrasekharan G, Kalyanaraman AR, Mallikarjunan M, Ayyar RR, Sabesan MN, Srikrishnan T, Srinivasan R, Subramanian E, Rao ST, Venkatesan K (1966) Crystallographic data for some amino acids, dipeptides and related compounds. Curr Sci 35(21):5229–5533

Chandra NR, Prabu MM, Venkatraman J, Suresh CG, Vijayan M (1998) X-ray studies of crystalline complexes involving amino acids. XXXIII. Crystal structures of L- and DL-arginine complexed with oxalic acid and a comparative study of amino acid-oxalic acid complexes. Acta Crystallogr B54(3):257–263

Chandrasekaran J, Ilayabarathi P, Maadeswaran P, Mohamedkutty P (2011a) Spectroscopic, thermal and optical properties of glycine oxalate (GOA) – an organic optical crystal. Rasāyan J Chem 4(2):387–392

Chandrasekaran J, Ilayabarathi P, Maaadeswaran P (2011b) Crystal growth, structure and characterizations of an organic optical material-L-alanine oxalate (LAO). Rasāyan J Chem 4 (2):425–430

Chaney MO, Seely O, Steinrauf LK (1971) The crystal structure of *L*-leucine hydriodide. Acta Crystallogr B27:544–548

Chapman RP, Bryce DL (2007) A high-field solid-state $^{35/37}$Cl NMR and quantum chemical investigation of the chlorine quadrupolar and chemical shift tensors in amino acids hydrochlorides. Phys Chem Chem Phys 9:6219–6230

Charoen-In U, Ramasamy P, Manyum P (2010) Comparative study on L-alaninium maleate single crystal grown by Sankaranarayanan-Ramasamy (SR) method and conventional slow evaporation solution technique. J. Crystal Growth 312:2369–2375

Charoen-In U, Ramasamy P, Manyum P (2011) Unidirectional growth of organic nonlinear optical *L*-arginine maleate dihydrate single crystal by Sankaranarayanan-Ramasamy (SR) method and its characterization. J Cryst Growth 318:745–750

Chemla DS, Zyss J (eds) (1987) Nonlinear optical properties of organic molecules and crystals, vol 1. Academic Press, Orlando

Chen XM, Mak TCW (1990) Crystal structure of bis(betaine) hydrochloride monohydrate. J Mol Struct 240:69–75

Chen T, Zheng L, Xu D (1988) Vibrational spectra of single crystal LAP and DLAP. Acta Opt Sin 8:884–891 (In Chinese)

Chen H, Xue L, Che YX, Zheng JM (2006) Synthesis and crystal structure of glycine·3,5-dihydroxybenzoic acid. Jiegou Huaxue (Chin J Struct Chem) 25(2):229–231

Chen T, Sun Z, Song C, Ge Y, Luo J, Lin W, Hong M (2012) Bulk crystal growth and optical and thermal properties of the nonlinear optical crystal *L*-histidinium-4-nitriphenolate 4-nitrophenol (LHPP). Cryst Growth Des 12:2673–2678

Cheng Z, Cheng YF, Guo LP (1995) A neutron diffraction study for the crystal structure of the deuterium (hydrogen) *L*-arginine phosphate monohydrate. Jiegou Huaxue (Chin J Struct Chem) 14:29–32

Cheng Z, Cheng YF, Guo LP, Xu D (1997) A neutron diffraction study of hydrogen bonding in the deuterium (hydrogen) *L*-arginine phosphate monohydrate. Z Kristallogr 212:221–225

Cherouana A, Benali-Cherif N, Bendjeddou L, Merazig H (2002) Diglycinium sulfate. Acta Crystallogr E58:o1351–o1353

Chiba A, Ueki T, Ashida T, Sasada Y, Kakudo M (1967) The crystal structure of L-ornithine hydrochloride. Acta Crystallogr 22:863–870

Chibata I, Sumi A, Ito H, Ohtsuki O, Izutsu N (1983) Crystalline salt of basic *L*-amino acid with *L*-malic acid and process for the preparation thereof. US Patent #4420432

Chimpri AS, Gryl M, DosSantos LHR, Krawczuk A, Macchi P (2013) Correlation between accurate electron density and linear optical properties in amino acid derivatives: L-histidinium hydrogen oxalate. Cryst Growth Des 13:2995–3010

Chitra R, Choudhury RR (2007) Investigation of hydrogen-bond network in bis(glycinium) oxalate using single-crystal neutron diffraction and spectroscopic studies. Acta Crystallogr B63:497–504

Chitra R, Thiruvenkatam V, Choudhury RR, Hosur MV, Guru Row TN (2006) Bis(glycinium) oxalate: evidence of strong hydrogen bonding. Acta Crystallogr C62:o274–o277

Choudhury RR, Chitra R (2008) Single crystal diffraction study of triglycine sulphate revisited. Pramana-J Phys 71(5):911–915

Choudhury RR, Roussel P, Capet F, Chitra R (2009) Charge-density analysis of hydrogen-bonded complexes of glycine by the maximum entropy method. J Mol Struct 938:229–237

Chruszcz M, Yasukawa SH, Ferrara JD, Minor W (2004) L-methioninium chloride and L-selenomethioninium chloride at 103K. Acta Crystallogr C60:o868–o871

Chwaleba D, Ciunik Z, Ilczyszyn MM, Ilczyszyn M (2006) Sarcosine-methanesulfonic acid(2:1) crystal: x-ray structure, vibrational properties and nature of hydrogen bonds. J Mol Struct 791 (1–3):61–69

Chwaleba D, Ilczyszyn MM, Ilczyszyn M, Ciunik Z (2007) Glycine-methanesulfonic acid (1:1) and glycine-*p*-toluenesulfonic acid (1:1) crystals: comparison of structures, hydrogen bonds, and vibrations. J Mol Struct 831:119–134

Clastre J (1964) Structures cristallines du bromhydrate de bétaïne et du chrorhydrate de bétaïne. C R Acad Sci Paris 259:3267–3269

Colaneri MJ, Peisach J (1992) An electron spin-echo envelope modulation study of Cu(II)-doped single crystals of *L*-histidine hydrochloride monohydrate. J Am Chem Soc 114:5335–5341

Cyrac Peter A, Vimalan M, Sagayaraj P, Rajesh Kumar T, Madhavan J (2010a) Linear and nonlinear optical properties of NLO active *L*-Phenylalanine *L*-Phenylalaninium nitrate single crystals. Int J Chem Tech Res 2(3):1445–1453

Cyrac Peter A, Vimalan M, Sagayaraj P, Madhavan J (2010b) Thermal, optical, mechanical and electrical properties of a novel NLO active *L*-phenylalanine *L*-phenylalaninium perchlorate single crystals. Phys B Condens Matter 405(1):65–71

Dacko S, Czapla Z, Baran J, Drozd M (1996) Ferroelectricity in Gly.H_3PO_3. Phys Lett A 223:217–220

Dammak T, Fourati N, Abid Y, Boughzala H, Mlayah A, Minot C (2007) Structural, vibrational and ab initio studies of *L*-histidine oxalate. Spectrochim Acta A 66:1097–1101

Dauter Z, Hempel A, Jedrzejczak H (1977) Crystallographic data of hexaiodotellurates (IV) of some amino acids. Z Kristallogr 146:320–321

Davidovich RL, Logvinova VB, Zemnukhova LA, Udovenko AA, Kondratyuk IP (1991) Complex compounds of antimonium (III) fluoride and oxofluoride with glycine. Koord Khim (Russ) (Coord Chem) 17:1342–1348

Davidovich RL, Logvinova VB, Kaidalova TA, Gerasimenko AV (2007) Synthesis, and study of hybrid organic-inorganic glycinium fluorozirconates. Russ J Inorg Chem 52:742–748

Davydov BL, Derkacheva LD, Dunina VV, Zhabotinskii ME, Zolin BF, Koreneva LG, Samokhina MA (1970) Connection of charge transfer with second harmonic generation of OCG. JETP Lett 12:24–26 (In Russian)

Davydov BL, Derkacheva LD, Dunina VV, Zhabotinskii ME, Zolin BF, Koreneva LG, Samokhina MA (1971) Charge transfer and second harmonic generation of laser radiation in molecular crystals. Opt Spectrosc 30(3):503–507 (In Russian)

Davydov BL, Katovshchikov SG, Nefedov VA (1977) New nonlinear organic materials for second harmonic generation of neodymium laser radiation. Kvantovaya Elektron 4(1):214–219 (In Russian)

Dawson B (1977) *DL*-1-Amino-1,2-dicarboxyethane (aspartic acid) hydrochloride. Acta Crystallogr B33:882–884

Dawson B, Mathieson AML (1951) The crystal structures of some α-amino acids. A preliminary X-ray examination. Acta Crystallogr 4:475–477

de Matos Gomes E, Nogueira E, Fernandes I, Belsley M, Paixão JA, Matos Beja A, Ramos Silva M, Martín-Gil J, Martín-Gil F, Mano JF (2001) Synthesis, structure, thermal and non-linear optical properties of *L*-argininium hydrogen selenite. Acta Crystallogr B57:828–832 de Matos Gomes E, Rodrigues VH, Costa MMR, Belsley MS, Cardoso PJM, Gonçalves CF, Proença F (2006) Unusual supramolecular assembly and nonlinear optical properties of *L*-histidinium hydrogen malate. J Solid State Chem 179:2521–2528

de Menezes AS, dos Santos AO, Almeida JMA, Sasaki JM, Cardoso LP (2007) Piezoelectric coefficients of *L*-histidine hydrochloride monohydrate obtained by synchrotron x-ray Renninger scanning. J Phys Condens Matter 19:106218 (9 p)

De Sousa GP, Freire PTC, Lima JA Jr, Mendes Filho J, Melo FEA (2011) High-pressure Raman spectra of *L*-histidine hydrochloride monohydrate crystal. Vib Spectrosc 57:102–107

Debrus S, Marchewka MK, Baran J, Drozd M, Czopnik R, Pietraszko A, Ratajczak H (2005) *L*-lysine-*L*-tartaric acid: new molecular complex with nonlinear optical properties. Structure, vibrational spectra and phase transitions. J Solid State Chem 178:2880–2896

Deepthy A, Bhat HL (2000) Ultrasonic investigation of elastic properties and phase transition in ferroelectric glycine phosphate $NH_3CH_2COOH_3PO_3$ single crystals. Phys Rev B 62:8752–8758

Deepthy A, Bhat HL (2001) Growth and characterization of ferroelectric glycine phosphate single crystals. J Cryst Growth 226:287–293

Dega-Szafran Z, Komasa A, Grundwald-Wyspiańska M, Szafran M, Buczak G, Katrusiak A (1997) Hydrogen bonding and proton localization in complexes of carboxybetaines with phenols and carboxylic acids. J Mol Struct 404:13–23

Dekola T, Isakov D, De Matos Gomes E, Nunes Rodrigues VH, Costa MMR (2009) Polarized hydrogen bonds and dielectric properties of glycine glycinium perrhenate. Eur Phys J B68:519–522

Delfino M, Dougherty JP, Zwicker WK, Choy MM (1976) Solution growth and characterization of *L*(+) glutamic acid hydrochloride single crystals. J Cryst Growth 36:267–272

Delfino M, Loiacono GM, Fitzpatrick BJ, Smith WA (1977) Deuterium isotope effect in l(+)-glutamic acid hydrochloride. J Solid State Chem 20:391–395

Delfino M, Loiacono GM, Nicolosi JA (1978) Halide effect in *L*(+) glutamic acid halogen acid salts. J Solid State Chem 23:289–296

DeLoach LD (1994) Microscopic single-crystal refractometry as a function of wavelength. J Opt Soc Am B 11:1186–1196

Dhanalakshmi B, Ponnusamy S, Muthamizhchelvan C (2010) Growth and characterization of a solution grown, new organic crystal: *L*-histidine-4-nitrophenolate 4-nitrophenol (LHPP). J Cryst Growth 313:30–36

Dhanaraj G, Srinivasan MR, Bhat HL (1991a) Vibrational spectroscopic study of *L*-arginine phosphate monohydrate (LAP), a new organic non-linear crystal. J Raman Spectrosc 22:177–181

Dhanaraj G, Shripathi T, Bhat HL (1991b) Growth and defect characterization of *L*-arginine phosphate monohydrate. J Cryst Growth 113:456–464

Dhanaraj G, Srinivasan MR, Bhat HL, Jayanna HS, Subramanyam SV (1992) Thermal and electrical properties of the novel organic nonlinear crystal *L*-arginine phosphate monohydrate. J Appl Phys 72:3464–3467

Dhanuskodi S, Angeli Mary PA (2002) EPR and optical absorption studies of Cu^{2+} doped *L*-arginine phosphate monohydrate single crystals-part II. Spectrochim Acta A 58:1473–1481

Dhanuskodi S, Ramajothi J (2004) Crystal growth, thermal and optical studies of *L*-histidine tetrafluoroborate: a semiorganic NLO material. Cryst Res Technol 39(7):592–597

Dhanuskodi S, Vasantha K (2004) Structural, thermal and optical characterizations of a NLO material: L-alaninium oxalate. Cryst Res Technol 39(3):259–265

Dhanuskodi S, Angeli Mary PA, Vasantha K (2003) Spectroscopic and optical studies on pure and doped single crystals of sulphate-mixed *L*-arginine phosphate monohydrate-a nonlinear optical crystal. Spectrochim Acta A 59:927–935

Dhanuskodi S, Angeli Mary PA, Sambasiva Rao P (2005) Single crystal ESR and optical absorption studies of Cu^{2+} ion in *L*-arginine sulphophosphate monohydrate-a nonlinear optical crystal. Spectrochim Acta A 61:721–725

Di K (2010) (2*S**)-2-Ammonio-3-(1*H*-indole-3-yl)-propionate pyridine-2,4-dicarboxylic acid ethanol solvate. Acta Crystallogr E66:o1125–o1126

Di Blasio B, Pavone V, Pedone C (1977a) Glycine hydrochloride $C_2H_5NO_2$.HCl, *L*-alanine hydrochloride $C_3H_7NO_2$.HCl. Cryst Struct Commun 6:745–748

Di Blasio B, Pavone V, Pedone C (1977b) *L*-methionine hydrochloride, $C_5H_{10}NO_2S$.HCl. Cryst Struct Commun 6:845–848

Di Blasio B, Napolitano G, Pedone C (1977c) *DL*-Valine hydrochloride. Acta Crystallogr B33:542–545

Dittrich B, Munshi P, Spackman MA (2007) Redetermination, invariom-model and multipole refinement of *L*-ornithine hydrochloride. Acta Crystallogr B63:505–509

Dmitriev VG, Gurzadyan GG, Nikogosyan DN (1999) Handbook of nonlinear optical crystals. Springer, New York

Domingos SR, Silva MR, Martins ND, Beja AM, Paixão JA (2008) Pyromellitic acid-sarcosine (1/2). Acta Crystallogr E64:o826

Domingos SR, Silva PSP, Buma WJ, Garcia MH, Lopes NC, Paixão JA, Silva MR, Woutersen S (2012) Amplification of the linear and nonlinear optical response of a chiral molecular crystal. J Chem Phys 136:134501

Donohue J, Caron A (1964) Refinement of the crystal structure of histidine hydrochloride monohydrate. Acta Crystallogr 17:1178–1180

Donohue J, Lavine LR, Rollett JS (1956) The crystal structure of histidine hydrochloride monohydrate. Acta Crystallogr 9:655–662

Dörffel M, Narz T, Haussühl S (1989) Struktur und Phasenübergänge des monoklinen Betainphosphits, $(CH_3)_3NCH_2COO.H_3PO_3$. Z Kristallogr 186:71–73

Dow J, Jensen LH, Mazumdar SK, Srinivasan R, Ramachandran GN (1970) Refinement of the structure of arginine hydrochloride monohydrate. Acta Crystallogr B26(11):1662–1671

Drebushchak TN, Bizyaev SV, Boldyreva EV (2008) Bis(*DL*-cysteinium) oxalate. Acta Crystallogr C64:o313–o315

Drozd M, Marchewka MK (2006) The structure, vibrational spectra and nonlinear optical properties of the *L*-lysine x tartaric acid complex-theoretical studies. Spectrochim Acta A 64:6–23

Dubský JV, Maitner J (1936) Addition compounds of neutral salts with glycine. Práace Mor Přirodověd Spol V Brnè 6(3); Chem Obzor 10, Abstract Sect 45 (1935). Quoted from: Chemical Abstracts 30, 7478

Easwaran KRK (1966) NMR study of ferroelectric diglycine nitrate. J Phys Soc Jpn 21(1):61–64

Ebdrup S, Pettersson I, Rasmussen BH, Deussen HJ, Jensen AF, Mortensen SB, Fleckner J, Pridal L, Nygaard L, Sauerberg P (2003) Synthesis and biological and structural characterization of the dual-acting peroxisome proliferator-activated receptor α/γ agonist ragaglitazar. J Med Chem 46:1306–1317

Edwin B, Rastogi VK, Hubert Joe I (2010) Structural conformation and vibrational studies of nonlinear optical material *L*-prolinium picrate. AIP Conf Proc 1267:1196–1197

Edwin B, Amalanathan M, Hubert Joe I (2012) Vibrational spectra and natural bond orbital analysis of organic crystal *L*-prolinium picrate. Spectrochim Acta A 96:10–17

Egorov VM, Lemanov VV (2006) Differential scanning calorimetry study of the phase transition in diserine sulfate monohydrate crystals. Phys Solid State 48(10):1954–1958

Eimerl D, Velsko S, Davis L, Wang F, Loiacono G, Kennedy G (1989a) Deuterated L-arginine phosphate a new efficient nonlinear crystal. IEEE J Quantum Electron 25:179–193

Eimerl D, Marion J, Graham EK, McKinstry HA, Haussühl S (1989b) Elastic constants and thermal fracture of $ArGaSe_2$ and *d*-LAP. IEEE J Quantum Electron 27:142–145

Eimerl D, Velsko S, Davis L, Wang F (1990) Progress in nonlinear optical materials for high power lasers. Prog Cryst Growth Ch 20:59–113

Elleuch N, Ahmed AB, Feki H, Abid Y, Minot C (2014) Vibrational spectra, optical properties. NBO and HOMO-LUMO analysis of L-Phenylalanine L-Phenylalalinium Perchlorate: DFT calculations. Spectrochim Acta A 121:129–138

Ennan AA, Gavrilova LA, Chebotarev AN, Andreeva NV (1981) Glycine complexes of fluorocontaining acids. Zhurn Neorg Khimii 26(2):374–378 (In Russian)
Ennan AA, Chebotarev AN, Gavrilova LA, Andreeva NV (1983) Hydrofluorides and hexafluorosilicates of monoaminocarboxylic acids. Zhurn Neorg Khimii 28(5):1117–1124 (In Russian)
Espinosa E, Wyncke B, Brehat F, Veintemillas S, Molins E, Lecomte C (1994) Far-infrared reflectivity spectra of *L*-arginine phosphate monohydrate (LAP). Infrared Phys Technol 35:625–632
Espinosa E, Veintemillas S, Miravitlles C, Molins E (1995) Hydrogen bonding in crystalline L-HisH + ·H2PO4-·H2O. Z Kristallogr 210:195–198
Espinosa E, Lecomte C, Molins E, Veintemillas S, Molins E, Cousson A, Paulus W (1996) Electron density study of a new non-linear optical material: *L*-arginine phosphate monohydrate (LAP). Comparison between X-X and X-(X + N) refinements. Acta Crystallogr B52:519–534
Espinosa E, Wyncke B, Bréhat F, Gerbaux X, Veintemillas S, Molins E (1997) Infrared vibrational spectra of L-histidinium dihydrogen orthophosphate orthophosphoric acid (LHP). Infrared Phys Technol 38:449–458
Esthaku Peter M, Ramasamy P (2012) Growth and characterization of an organic nonlinear optical crystal: glycinium trichloroacetate. Mater Chem Phys 137:258–263
Evlanova NF, Moldozhanova GT, Pashina ZS, Rashkovich LN, Shekunov BY (1990) Study of crystallization of *L*-arginine chloride monohydrate. Kristallografiya 35:1527–1533
Faamau JW, Tiekink ERT (1993) New data for diglycine hydrochloride. Z Kristallogr 2004(2):277–282
Faria JLB, Almeida FM, Pilla O, Rossi F, Sasaki JM, Melo FEA, Mendes Filho J, Freire PTC (2004) Raman spectra of *L*-histidine hydrochloride monohydrate crystal. J Raman Spectrosc 35:242–248
Fehst I, Paasch M, Hutton SL, Braune M, Böhmer R, Loidl A, Dörffel R, Narz T, Haussühl S, McIntyre GJ (1993) Paraelectric and ferroelectric phases of betaine phosphite: structural, thermodynamic, and dielectric properties. Ferroelectrics 138:1–10
Fischer A (2006) D-Alanine alaninium bromide. Acta Crystallogr E62(12):o5786–o5788
Fischer MS, Templeton DH, Zalkin A (1970) Solid state structure and chemistry of the choline halides and their analogues. Redetermination of the betaine hydrochloride structure, $[(CH_3)_3NCH_2COOH]^+Cl^-$. Acta Crystallogr B26:1392–1397
Flaig R, Koritsanszky T, Dittrich B, Wagner A, Luger P (2002) Intra- and intermolecular topological properties of amino acids: a comparative study of experimental and theoretical results. J Am Chem Soc 124:3407–3417
Fleck M (n.d.) $(GlyH)NO_3$. CCDC #928883
Fleck M, Petrosyan AM (2009) Comments on papers reporting IR-spectra and other data of alleged *L*-Alanine Alaninium Nitrate and L-Alanine Sodium Nitrate crystals. Cryst Res Technol 44 (7):769–772
Fleck M, Petrosyan AM (2010) Difficulties in the growth and characterization of non-linear optical materials: a case study of salts of amino acids. J Cryst Growth 312(15):2284–2290
Fleck S, Weiss A (1984) Hydrogen bonds in hydrohalides of amino acids. A ^{79,81}Br and ^{127}I NQR study. Ber Bunsenges Phys Chem 88:956–963
Fleck M, Lengauer C, Bohatý L, Tillmanns E (2008) Synthesis, crystal structures and thermal behaviour of novel *L*-alanine halogenide compounds. Acta Chim Slov 55:880–888
Fleck M, Ghazaryan VV, Petrosyan AM (2010) Glycine hydrogen fluoride: remarkable hydrogen bonding in the dimeric glycine glycinium cation. J Mol Struct 984:83–88
Fleck M, Ghazaryan VV, Petrosyan AM (2012a) Crystal structure at 296 and 150K, vibrational spectra and thermal behaviour of sarcosine sarcosinium nitrate. Z Kristallogr 227(12):819–824
Fleck M, Ghazaryan VV, Petrosyan AM (2012b) Hexafluorosilicates of sarcosine. Solid State Sci 14:952–963
Fleck M, Ghazaryan VV, Petrosyan AM (2012c) β-Alaninium picrate: a new salt with di--β-alaninium dimeric cation. J Mol Struct 1019:91–96

Fleck M, Ghazaryan VV, Petrosyan AM (2013a) Two forms of sarcosine sarcosinium hydrogen *L*-tartrate. J Mol Struct 1045:95–103

Fleck M, Ghazaryan VV, Petrosyan AM (2013b) Amino acid hexafluorosilicates – an overview. Z Kristallogr 228(5):240–249

Fleck M, Ghazaryan VV, Bezhanova LS, Atanesyan AK, Petrosyan AM (2013c) New crystals of *L*-histidine maleates. J Mol Struct 1035:407–415

Fleck M, Ghazaryan VV, Petrosyan AM (2014) Mixed salts of amino acids: *L*-ornithinium(2+) sulfate hydrogen fluoride and tri-*L*-ornithinium(2+) dinitrate disulfate. JUET Res J Sci Tech 1:11–25

Fonseca TL, Sabino JR, Castro MA, Georg HC (2010) A theoretical investigation of electric properties of *L*-arginine phosphate monohydrate including environment polarization effects. J Chem Phys 133:144103

Franken PA, Hill AE, Peters CW, Weinreich G (1961) Generation of optical harmonics. Phys Rev Lett 7(4):118–119

Frey MN, Koetzle TF, Lehmann MS, Hamilton WC (1973) Precision neutron diffraction structure determination of protein and nucleic acid components. X. A comparison between the crystal and molecular structures of L-tyrosine and L-tyrosine hydrochloride. J Chem Phys 58 (6):2547–2556

Frost WS (1942) Bis(amino acid) derivatives. I. Diglycine halogen acid addition products. J Am Chem Soc 64(1942):1286–1287

Fu TY, Scheffer JR, Trotter J (1997) 4-[2,4,6,-tris(1-methylethyl)benzoyl]benzoic acid (*S*)-(-)-Proline. Acta Crystallogr C53:1259–1263

Fuchs BA, Syn CK, Velsko SP (1989) Diamond turning of *L*-arginine phosphate, a new organic nonlinear crystal. Appl Optics 28:4465–4472

Fuess H, Hohlwein D, Mason SA (1977) Neutron diffraction study of *L*-histidine hydrochloride monohydrate. Acta Crystallogr B33:654–659

Fujii I (2009) Crystal structure of *L*-(*S*)-tryptophan *D*-(*R*)-mandelate 1.5H2O. X-Ray Struct Anal Online 25:35–36

Fujii I, Hirayama N (2002) Chiral space formed by (+)-(1S)-1,1′-binaphthalene-2,2′-diyl phosphate: recognition of aliphatic *L*-α-amino acids. Helv Chim Acta 85:2946–2960

Fujii I, Baba H, Takahashi Y (2005a) Crystal structures of *L*-(R)-cysteine-mandelic acid diastereomers. Anal Sci 21:x175–x176

Fujii I, Watadani T, Nunomura S, Takahashi Y (2005b) Crystal structure of (*R*)-2-Phenoxypropionic acid-(*S*)-valine. Anal Sci 21:x41–x42

Furberg S, Solbakk J (1973) Crystal structure of arginine diethyl phosphate. Acta Chem Scand 27(4):1226–1232

Ganesh RB, Kannan V, Sathyalakshmi R, Ramasamy P (2007) The growth of *L*-glutamic acid hydrochloride crystals by Sankaranarayanan-Ramasamy (SR) method. Mater Lett 61:706–708

Garfinkel D, Edsall JT (1958) Raman spectra of amino acids and related compounds. VIII. Raman and infrared spectra of imidazole, 4-methylimidazole and histidine. J Am Chem Soc 80:3807–3812

Garret CGB, Robinson FNH (1966) Miller's phenomenological rule for computing nonlinear susceptibilities. IEEE J Quantum Electron 2:328–329

Gartland GL, Freeman GR, Bugg CE (1974) Crystal structures of tryptamine picrate and D, L-tryptophan picrate-methanol, two indole donor-acceptor complexes. Acta Crystallogr B30:1841–1849

Geng YL, Xu D, Sun DL, Zhang GH, Du W, Liu HY, Wang XQ (2004a) Growth morphology of {100} faces of *L*-Arginine phosphate monohydrate single crystals investigated by atomic force microscopy. Cryst Res Technol 39:712–717

Geng YL, Xu D, Sun DL, Wang XQ, Du W, Liu HY, Zhang GH (2004b) AFM study of the surface morphology of the {100} cleavage planes of *L*-arginine phosphate monohydrate single crystals. Cryst Res Technol 39:759–762

Geng YL, Xu D, Wang YL, Du W, Liu HY, Zhang GH, Wang XQ, Sun DL (2005a) Atomic force microscopy study on crystal growth of Cu^{2+}-doped *L*-arginine phosphate monohydrate crystals. J Cryst Growth 273:624–628

Geng YL, Xu D, Sun DL, Du W, Liu HY, Zhang GH, Wang XQ (2005b) Atomic force microscopy studies on growth mechanisms of LAP crystals grown in solution containing excessive amount of *L*-arginine. Mater Chem Phys 90:53–56

Geng YL, Xu D, Wang XQ, Yu GW, Zhang GH, Zhang HB (2005c) AFM study of surface morphology of {100} cleavage planes of *L*-arginine acetate crystals. J. Crystal Growth 282:208–213

Gerasimenko AV, Pushilin MA, Davidovich RL (2008) Disordering of the $[NbOF_5]^{2-}$ complex anions in bis(glycinium) pentafluoridooxidoniobate(V) dihydrate. Acta Crystallogr C64: m358–m361

Gerasimenko AV, Davidovich RL, Logvinova VB (2011) Crystal structure of layered zirconium pentafluorides of methylammonium, glycine, and β-alanine. J Struct Chem 52:524–530

Gesi K, Ozawa K (1973) Effect of hydrostatic pressure on the ferroelectric phase transition in diglycine nitrate. Jpn J Appl Phys 12(6):951–952

Ghazaryan VV (2011) Crystal growth and characterization of *L*-threoninium chloride and *L*-threoninium bromide, new nonlinear optical crystals. Proc SPIE 7998:79980G

Ghazaryan VV (2014) Preparation and investigation of mixed salts of amino acids with different anions. PhD thesis, Yerevan (In Armenian)

Ghazaryan VV, Fleck M, Petrosyan AM (2010a) Crystal structures and vibrational spectra of novel compounds with glycine glycinium cations. J Mol Struct 977(1–3):117–129

Ghazaryan VV, Fleck M, Petrosyan AM (2010b) Mixed salts of amino acids: on the vibrational spectra of mixed salts containing a L-lysinium(+)...L-lysinium(2+) dimeric cation. J Mol Struct 982(1–3):145–151

Ghazaryan VV, Fleck M, Petrosyan AM (2010c) Mixed salts of amino acids: L-lysinium(2+) chloride nitrate, L-lysinium(2+) chloride tetrafluoroborate and L-lysinium(2+) chloride perchlorate. J Mol Struct 984(1–3):268–275

Ghazaryan VV, Fleck M, Petrosyan AM (2011a) New salts of amino acids with dimeric cations. Proc SPIE 7998:79980F

Ghazaryan VV, Fleck M, Petrosyan AM (2011b) Glycine glycinium picrate – reinvestigation of the structure and vibrational spectra. Spectrochim Acta A 78(1):128–132

Ghazaryan VV, Fleck M, Petrosyan AM (2011c) Mixed salts of amino acids: new analogs of the di-*L*-ornithinium(2+) chloride nitrate sulfate crystal. J Crystallization Phys Chem 2(1):7–16

Ghazaryan VV, Fleck M, Petrosyan AM (2012a) Sarcosine sarcosinium chloride and sarcosine sarcosinium bromide. J Mol Struct 1020:160–166

Ghazaryan VV, Fleck M, Petrosyan AM (2012b) Sarcosine sarcosinium tetrafluoroborate and sarcosine sarcosinium perchlorate: synthesis, structure and vibrational spectra. J Mol Struct 1021:130–137

Ghazaryan VV, Fleck M, Petrosyan AM (2012c) Hexafluorosilicates of alanine. Z Kristallogr 227(9):646–655

Ghazaryan VV, Fleck M, Petrosyan AM (2012d) Structure and vibrational spectra of *L*-alanine *L*-alaninium picrate monohydrate. J Mol Struct 1015:51–55

Ghazaryan VV, Fleck M, Petrosyan AM (2012e) Mixed salts of amino acids: L-histidinium(+)... L-histidinium(2+) nitrate-hexafluorosilicate. J Mol Struct 1026:140–144

Ghazaryan VV, Fleck M, Petrosyan AM (2013a) Salts of amino acids with hexafluorosilicate anion, J. Crystal Growth 362:162–166

Ghazaryan VV, Fleck M, Petrosyan AM (2013b) Iodides of sarcosine. J Mol Struct 1032:35–40

Ghazaryan VV, Fleck M, Petrosyan AM (2013c) Mixed salts of amino acids with different anions. J. Crystal Growth 362:182–188

Ghazaryan VV, Fleck M, Petrosyan AM (2014a) Glycine glycinium bromide. CCDC #929566

Ghazaryan VV, Fleck M, Petrosyan AM (2014b) New chemical analogs of triglycine sulfate. J Cryst Growth. http://dx.doi.org/10.1016/j.jcrysgro.2013.11.054

Ghazaryan VV, Suponitsky KY, Lyssenko KA, Atanesyan AK, Petrosyan AM (in prep)

Giordmaine JA (1962) Mixing of light beams in crystals. Phys Rev Lett 8(1):19–20

Giordmaine JA, Miller RC (1965) Tunable coherent parametric oscillation in $LiNbO_3$ at optical frequencies. Phys Rev Lett 14(24):973–976

Globelny P, Mukherjee A, Dessiraju GR (2011) Drug-drug co-crystals: temperature-dependent proton mobility in the molecular complex of isoniazid with 4-aminosalicylic acid. CrystEngComm 13:4358–4364

Głowiak T, Szemik AW (1986) Crystal structure and spectroscopic properties of glycinium monophenylphosphate. J Cryst Spectrosc Res 16(1):79–89

Głowiak T, Wnęk I (1985) Crystal and molecular structure of glycinium diphenylphosphate. J Cryst Spectrosc Res 15(2):157–171

Gnanasekaran P, Madhavan J (2010) Synthesis, structural, FT-IR and non-linear optical studies of pure and lanthanum doped L-arginine acetate crystals. Asian J Chem 22:109–144

Godzisz D, Ilczyszyn MM, Ilczyszyn M (2002) Classification and nature of hydrogen bonds to betaine. X-ray, ^{13}C CP MAS and IR description of low barrier hydrogen bonds. J Mol Struct 606(1–3):123–137

Godzisz D, Ilczyszyn M, Ciunik Z (2003a) β-Alanine-hydrochloride (2:1) crystal: structure, ^{13}C NMR and vibrational properties, protonation character. Spectrochim Acta A 59(2):235–244

Godzisz D, Ilczyszyn M, Ilczyszyn MM (2003b) β-Alanine-oxalic acid (1:1) hemihydrates crystal: structure, ^{13}C NMR and vibrational properties, protonation character. Spectrochim Acta A 59:681–693

Gokul Raj S, Ramesh Kumar G, Mohan R, Pandi S, Jayavel R (2005) Structural, optical and dielectric studies on solution-grown semi-organic *L*-histidine tetrafluoroborate single crystal. Mater Chem Phys 90(1):144–147

Gokul Raj S, Ramesh Kumar G, Raghavalu T, Mohan R, Jayavel R (2006a) *L*-histidinium tetrafluoroborate. Acta Crystallogr E62(3):o1178–o1180

Gokul Raj S, Ramesh Kumar G, Mohan R, Jayavel R (2006b) *L*-histidinium trifluoroacetate. Acta Crystallogr E62:o5–o7

Gokul Raj S, Ramesh Kumar G, Mohan R, Jayavel R, Varghese B (2006c) *L*-histidinium trichloroacetate. Acta Crystallogr E62:o1704–o1706

Gokul Raj S, Ramesh Kumar G, Raghavalu T, Kumar P, Mohan R, Jayavel R (2006d) Structural, spectral, linear and nonlinear optical properties of new nonlinear optical *L*-histidinium trichloroacetate crystals. Spectrochim Acta A 65:1021–1024

Gokul Raj S, Ramesh Kumar G, Mohan R, Varghese B, Jayavel R (2006e) Crystal structure of single crystals of nonlinear optical *L*-histidinium trichloroacetate. J Mol Struct 825:158–164

Gokul Raj S, Ramesh Kumar G, Mohan R, Jayavel R, Varghese B (2007) Crystal structure and vibrational analysis of novel nonlinear optical *L*-histidine tetrafluoroborate (L-HFB) single crystals. Phys Status Solidi B 244(2):558–568

Golič L, Hamilton WC (1972) The structure of di-*L*-Leucine hydrochloride: an X-ray diffraction study. Acta Crystallogr B28(4):1265–1271

Gómez E, Claus R, Pierce RM, Ushioda S (1984) The Raman-active long wave phonon modes of betaine sulfate. Z Kristallogr 168:67–74

Gonsago CA, Albert HM, Karthikeyan J, Sagayaraj P, Pragasam AJA (2012a) Crystal structure, optical and thermal studies of a new organic nonlinear optical material: *L*-histidinium maleate 1.5-hydrate. Mater Res Bull 47:1648–1652

Gonsago CA, Pandi S, Albert HM, Pragasam AJA (2012b) *L*-histidinium maleate crystals for nonlinear frequency conversion. Int J Appl Phys Math 2:54–57

Gopalakrishnan R, Ramasamy P (2009) Reply to “Comments on papers reporting IR-spectra and other data of L-alanine alaninium nitrate and L-alanine sodium nitrate crystals” by M. Fleck and A.M. Petrosyan. Cryst Res Technol 44(7):773–775

Görbitz CH (2001) *L*-Phenylalanine nitrate (4/1). Acta Crystallogr E57(2):o192–o194

Görbitz CH, Etter MC (1992) Structure of *L*-phenylalanine *L*-phenylalaninium formate. Acta Crystallogr C48(7):1317–1320

Görbitz CH, Husdal J (1998) Cocrystallizing agents for amino acids. II. The crystal structures of *L*-histidine.4,5-imidazoledicarboxylic acid (1:1) and *L*-lysine.4,5-imidazoledicarboxylic acid (1:1). Acta Chem Scand 52:218–226

Gowri S, Uma Devi T, Sajan D, Bheeter SR, Lawrence N (2011) Spectral, thermal and optical properties of *L*-tryptophanium picrate: a nonlinear optical single crystal. Spectrochim Acta A 81:257–260

Grandin A, Borel MM, Raveau B (1985) Intercalation of primary diamines and amino acids in the layer structure oxide $HTiNbO_5$. J Solid State Chem 60(3):366–375

Guenifa F, Bendjeddou L, Cherouana A, Dahaoui S, Lecomte C (2009) *DL*-Asparaginium perchlorate. Acta Crystallogr E65:o2264–o2265

Guenifa F, Bendjeddou L, Cherouana A, Dahaoui S, Lecomte C (2012) *DL*-Tyrosinium chloride dihydrate. Acta Crystallogr E68:o3227–o3228

Guha S (1970) A preliminary report on the structure of glycocyamine hemihydrate, diglycine monopicrate and 4-(N-phenylpiperizino)-6-methoxy quinaldine. Indian J Phys 44:267–269

Guha S, Mazumdar SK, Saha NN (1969) The crystal and molecular structure of *L*-ornithine hydrochloride. Z Kristallogr 129:84–100

Guidara S, Feki H, Abid Y (2013) Vibrational spectral studies and non-linear optical properties of *L*-leucine *L*-leucinium picrate: a density functional theory approach. Spectrochim Acta A 115:437–444

Gulam Mohamed M, Vimalan M, Jesudurai JGM, Madhavan J, Sagayaraj P (2007) Growth and characterization of pure and doped nonlinear optical *l*-arginine acetate single crystals. Cryst Res Technol 42:948–954

Günay N, Pir H, Avci D, Atalay Y (2013) NLO and NBO analysis of sarcosine-maleic acid by using HF and B3LYP calculations. J Chem 2013:712130, 16 p

Hadži D, Marciszewski H (1967) Hydrogen bonding in dileucine hydrochloride (with reference to triglycine sulphate). Chem Commun (1):2–3

Hahn T, Buerger MJ (1957) The crystal structure of diglycine hydrochloride, $2(C_2H_5O_2N).HCl$. Z Kristallogr 108:419–453

Haja Hameed AS, Rohani S (2007) Nucleation studies and surface SHG analysis of *L*-arginine phosphate monohydrate (LAP) family crystals. Mater Lett 61:5142–5144

Haja Hameed AS, Ravi G, Hossain MDM, Ramasamy P (1999) Growth and characterization of *L*-arginine phosphate family crystals. J. Crystal Growth 204:333–340

Haja Hameed AS, Ravi G, Ramasamy P (2001) Inhibition of microbial growth, study of solution stability, growth and characterization of potassium fluoride mixed *L*-arginine phosphate single crystals. J Cryst Growth 229:547–552

Haja Hameed AS, Ravi G, Ilangovan R, Nixon Azariah A, Ramasamy P (2002a) Growth and characterization of deuterated analog of *L*-arginine phosphate. J Cryst Growth 237–239:890–893

Haja Hameed AS, Ravi G, Mahalingam T, Ramasamy P (2002b) Growth of stubbier habit LAP2 single crystals and their characterisation studies. Mater Sci Eng B95:61–66

Haja Hameed AS, Anandan P, Jayavel R, Ramasamy P, Ravi G (2003a) Synthesis, growth and characterization of nonlinear optical material: *L*-arginine fluoride. J Cryst Growth 249:316–320

Haja Hameed AS, Ravi G, Jayavel R, Ramasamy P (2003b) Nucleation kinetics, growth and characterization of *d*LAP, dLAP:KF and dLAP:NaN_3 crystals. J Cryst Growth 250:126–133

Haja Hameed AS, Karthikeyan C, Ravi G, Rohani R (2011) Characterization studies on the additives mixed *L*-arginine phosphate monohydrate (LAP) crystals. Physica B 406:1363–1367

Harbison G, Herzfeld J, Griffin RG (1981) Nitrogen-15 chemical shift tensors in *L*-histidine hydrochloride monohydrate. J Am Chem Soc 103:4752–4754

Hasebe K, Masuno K, Nanitani S, Asahi T (1998) Structural phase transitions of betaine phosphite studied by X-ray scattering. Ferroelectrics 217(1):75–81

Hasebe K, Tsuchiya S, Kawamura Y, Asahi T (2001) X-ray study of phase transition in betaine phosphite. J Phys Soc Jap 70:3300–3305

Hasegawa K, Ishikawa K, Kawaoka R, Sano C, Iitani K, Komatsu H, Nagashima N (1998) Complexes between *L*-Leucine and its precipitants. Acta Crystallogr C54:637–641

Haussühl S (1988) Ferroelastic phase transition and elastic properties of orthorhombic betaine hydrogen maleinate, (CH3)3NCH2COO. (CH2)2(COOH)2. Solid State Commun 68:963–966

Haussühl S (1989) Elastic and thermoelastic properties of twelve adducts of betaine, $(CH_3)_3NCCOO$, with H_2O, HCl, HBr, HI, HNO_3, H_2SO_4, H_3PO_3, H_3PO_4, 1,4-toluene sulfonic acid, $MnCl_2$ and KBr. Z Kristallogr 188:311–320

Haussühl E, Haussühl S (1995) Elastic properties of sulfamic acid and sulfamates of Li, Na, K, Rb, Cs, Tl, NH_4, $C(NH_2)_3$ and $(CH_3)_3NCH_2COOH$. Z Kristallogr 210:269–275

Haussühl S, Schreuer J (1996a) Crystal structure, thermal expansion and elastic properties of triclinic betaine hydrogen dihydrogen triiodate, $((CH_3)_3NCH_2COO)H_2(IO_3)_3$. Z Kristallogr 211:903–907

Haussühl S, Schreuer J (1996b) Crystal structure, thermal expansion and elastic properties of triclinic betaine hydrogen ammonium sulfate $((CH_3)_3NCH_2COOH)NH_4SO_4$. Z Kristallogr 211 (2):79–83

Haussühl E, Schreuer J (2001) Crystal structure and elastic properties of betaine fumarate and of betaine maleate, two isomers of ((CH3)3NCH2COOH.OOC(CH)2COOH. Z Kristallogr 216:616–622

Haussühl S, Wang J (1989) Elastic and thermoelastic properties of monoclinic betaine hydrogen oxalate hydrate (CH3)3NCH2COO. (COOH)2.H2O. Z Kristallogr 187:249–251

Haussühl S, Chrosch J, Gnanam F, Fiorentini E, Recker K, Wallrafen F (1990) Crystal growth and physical properties of monoclinic *L*-arginine hydrochloride monohydrate, $C_6H_{14}O_2N_4HCl.H_2O$, and *L*-arginine hydrobromide monohydrate, $C_6H_{14}O_2N_4HBr.H_2O$. Cryst Res Technol 25:617–623

Haussühl S, Karapetyan HA, Petrosyan AM (2003) Crystal structure and properties of *L*-arginine diphosphate. Z Kristallogr 218:501–506

Haussühl S, Karapetyan HA, Sukiasyan RP, Petrosyan AM (2006) Crystal structure and properties of orthorhombic *L*-arginine formate. Cryst Growth Design 6(9):2041–2046

Hempel A, Dauter Z, Szwabski S (1977a) Preliminary crystal data on hexabromoselenates(IV) of some amino acids. Z Kristallogr 146:318–319

Hempel A, Dauter Z, Pastuszak R (1977b) Crystal data on amino acids salts containing anions of the type MeX_6^{2-}. Z Kristallogr 146(4–6):317

Henschel D, Moers O, Wijaya K, Wirth A, Blaschette A, Jones PG (2002) Polysulfonylamines, CLIII [1]. Weak hydrogen bonding with activated methyl donors: crystal structures of cholinium, betainium and dimethyl[2-(dimethylamino)ethyl]ammonium-dimesylamide. Z Naturforsch B: Chem Sci 57:534–546

Herbstein FH, Kapon M (1979) The crystal structure of trimesic acid, its hydrates and complexes. V. *L*- (or *DL*-) histidinium trimesate-1/3 acetone. Acta Crystallogr B35:1614–1619

Herbstein FH, Kapon M, Maor I, Reisner GM (1981) The structure of trimesic acid, its hydrates and complexes. VI. Glycine-trimesic acid monohydrate, $H_3N^+CH_2COO^-.C_9H_6O_6.H_2O$. Acta Crystallogr B37:136–140

Herreros-Cedrés J, Hernández-Rodríguez C, González-Díaz B, Guerrero-Lemus R (2003) Temperature-dependent optical anisotropy of L-arginine phosphate single crystal. Appl Phys B 77:607–612

Hiramatsu H, Miki N, Takeuchi H (2010) Effect of conformation and hydrogen bonding on the C=C and NCN stretching Raman bands of N-deuterated histidinium. J Raman Spectrosc 41:1708–1713

Hoshino S, Okaya Y, Pepinsky R (1959) Crystal structure of the ferroelectric phase of $(glycine)_3.H_2SO_4$. Phys Rev 115(2):323–330

Hoshino S, Sato S, Toyoda K (1963) Crystal structure and phase transition of di-glycine nitrate. Jap J Appl Phys 2(8):519–520

Hu ZQ, Xu DJ, Xu YZ, Wu JY, Chiang MY (2002) (*R*)-Mandelic acid (*S*)-alanine hemihydrate. Acta Crystallogr C58:o612–o614

Hu ZQ, Xu DJ, Xu YZ (2004a) (*S*)-Alanine-(*S*)-mandelic acid (1/1). Acta Crystallogr E60:o269–o271

Hu ZQ, Xu DJ, Xu YZ, Wu JY, Chiang MY (2004b) Optical separation of racemic phenylalanine, and structure of complex consisting of *R*-phenylalanine and *R*-mandelic acid. Chin J Struc Chem 23(1):38–40

Hubert Joe I (2010) FT-Raman and IR spectral analysis and DFT computations of NLO crystal *L*-valine hydrobromide. AIP Conf Proc 1267(1):1194–1195

Hubert Joe I, Philip D, Aruldhas G, Botto IL (1991) Raman and IR spectra of β-alanine and sarcosine monophosphates. J Raman Spectrosc 22:423–425

Hübschle CB, Dittrich B, Luger P (2002) *L*-tryptophan formic acid solvate at 183K. Acta Crystallogr C58:o540–o542

Hudson MR, Allis DG, Ouellette W, Hudson BS (2009) Inelastic neutron scattering and Raman spectroscopic investigation of L-alanine alaninium nitrate, homologue of a ferroelectric material. Phys Chem Chem Phys 11(41):9474–9483

Hussani SS, Dhumane NR, Dongre VG, Shirsat MD (2009) Growth and characterization of an NLO material-crystal of triglycine acetate. Mater Sci-Poland 27(2):365–372

Ilayabarathi P, Chandrasekaran J, Maadeswaran P (2013) Synthesis, growth and characterization of *L*-tyrosine hydrochloride a semi-organic nonlinear optical crystal. Optik 124:1125–1127

Ilczyszyn MM, Ilczyszyn M (2003) Raman, infrared and ^{13}C NMR studies on betaine-sulfamic acid (2:1) crystal and its hydrogen bonds. J Raman Spectrosc 34:693–704

Ilczyszyn MM, Lis T, Baran J, Ratajczak H (1992) Structure and IR spectra of the solid complex of bis(betaine)-telluric acid. J Mol Struct 265:293–310

Ilczyszyn MM, Barnes AJ, Pietraszko A, Ratajczak H (1995a) Structure and vibrational spectra of the solid complex of betaine-sulphuric acid monohydrate. J Mol Struct 354:109–118

Ilczyszyn MM, Lis T, Ratajczak H (1995b) Structure and polarized infrared and Raman spectra of the complex of betaine and maleic acid (1:1). J Mol Struct 372:9–27

Ilczyszyn MM, Wierzejewsk M, Ciunik Z (2000) Crystal structure and vibrational spectra of bis (betaine) sulfamate. Phys Chem Chem Phys 2(15):3503–3510

Ilczyszyn M, Godzisz D, Ilczyszyn MM (2002) Structural and vibrational properties of betainium perchlorate monohydrate crystal and character of its hydrogen bonds. J Mol Struct 611:103–118

Ilczyszyn M, Godzisz D, Ilczyszyn MM (2003) Sarcosine-maleic acid (1:1) crystal: structure, ^{13}C NMR and vibrational properties, protonation character. Spectrochim Acta A 59:1815–1828

Ilczyszyn M, Godzisz D, Ilczyszyn MM (2006) ^{13}C chemical shift tensors of hydrogen bonded amino acids: relation between experimental and calculated results. Chem Phys 323 (2–3):231–242

Ilczyszyn M, Chwaleba C, Ciunik Z, Ilczyszyn MM (2008a) Structural role of hydrogen bond networks in amino acid–acid systems. (II) The network with weakly polarizable OHO hydrogen bonds in sarcosine–p-toluenesulfonic acid (1:1) crystal. Chem Phys 352:57–64

Ilczyszyn M, Chwaleba C, Mierzwicki K, Ilczyszyn MM (2008b) Structural role of hydrogen bond networks in aminoacid-acid systems. (I) The network with highly polarizable OHO hydrogen bonds in sarcosine-methanesulfonic acid (2:1) crystal. Chem Phys 351(1–3):99–105, Erratum: Chem Phys 355:196 (2009)

Ilczyszyn MM, Ilczyszyn M, Jesariew D, Baran J, Piecha A (2013) Non-centrosymmetric betaine-selenious crystal. Vibrational, X-ray, calorimetric, and dielectric studies. Vib Spectrosc 66:50–62

Ilyukhin AB, Davidovich RL (1999) Effect of a cation on stereochemical activity of lone electron pair in structures of ethylenediaminetetraacetatoantimonates(III) Cat[Sb(Edta)].xH_2O [Cat^+=Cs, NMe_4, ½{H_2En}, NH4, Tl, K, 1/2Mn, 1/2Cd, $NH_3CH_2CH_2COOH$]. Crystallogr Rep 44(2):204–213

Isakov D, de Matos Gomes E, Belsley MS, Rodrigues VH, Costa MMR (2012) Synthesis, structure and non-linear optical properties of *L*-argininium perrhenate crystal. CrystEngComm 14:3767–3771

Ishida T, Nagata H, In Y, Doi M, Inoue M, Extine MW, Wakahara A (1993) X-Ray crystal structure of *L*-tryptophan-picric acid charge-transfer complex and comparison with *DL*-tryptophan-picric acid complex. Chem Pharm Bull 41:433–438

Ittyachan R, Sagayaraj P (2002) Growth and characterisation of *L*-arginine diphosphate crystal. J Cryst Growth 243:356–360

Ittyachan R, Sagayaraj P (2003a) Growth and characterization of a new NLO *L*-histidine bromide crystal. J Cryst Growth 249:557–560

Ittyachan R, Sagayaraj P (2003b) Growth and characterization of a new nonlinear optical *L*-histidine diphosphate single crystal. J Cryst Growth 249:553–556

Ittyachan R, Sagayaraj P, Kthandapani B (2003) *L*-Argininium bis(dihydrogen phosphate). Acta Crystallogr E59:o886–o888

Ittyachan R, Thomas PC, Anand DP, Palanichamy M, Sagayaraj P (2005a) Growth and characterization of semiorganic non-linear optical LHB single crystal. Mater Chem Phys 93:272–276

Ittyachan R, Raja SXJ, Rajasekar SA, Sagayaraj P (2005b) Crystallization and characterization of L-histidinium perchlorate single crystal. Mater Chem Phys 90:10–15

Janarthanan S, Kumar TK, Pandi S, Anand DP (2009) Growth and spectroscopic studies of L-argininium formate NLO single crystals. Indian J Pure Appl Phys 47:332–336

Janczak J, Perptuo GJ (2007) *L*-Leucinium perchlorate. Acta Crystallogr C63:o117–o119

Jarmelo S, Reva I, Rozenberg M, Ramos Silva M, Beja AM, Fausto R (2008) Crystal and molecular structure of *DL*-serine hydrochloride studied by X-ray diffraction, low-temperature Fourier transform infrared spectroscopy and DFT(B3LYP) calculations. J Phys Chem B112:8032–8041

Jeyakumari AP, Dhanuskodi S, Manivannan S (2006) Phase matchable semiorganic NLO material for frequency doubling: *L*-Arginine tetrafluoroborate. Spectrochim Acta A 63:91–95

Jeyanthi CE, Rajarajan K, Siddheswaran R (2010) Investigation on growth and properties of a nonlinear optical crystal, *L*-lysine glycine dihydrochloride (*L*-LGHCl). Int J Mater Sci 5 (3):431–439

Jiang M, Xu D, Tan ZK (1983) A new phase-matchable nonlinear optic *L*-arginine phosphate monohydrate (LAP). Program and abstracts of VII international conference on crystal growth, Stuttgart, 12–16 Sept 1983, p 267

Jin ZM, Pan YJ, Hu ML, Shen L, Li MC (2003) The salt of l-prolinium picrate. Cryst Res Technol 38(11):1009–1012

Jin Y, Che YX, Wei RM, Zheng JM (2004) Preparation and crystal structure of sarcosine 5-nitrosalicylic acid organic adduct. Jiegou Huaxue (Chin J Struct Chem) 23(11):1292–1294

Jinnah MMA, Umadevi M, Ravikumar B, Ramakrishnan V (2004a) Infrared and laser Raman studies of bis(*L*-threoninium) sulphate monohydrate. Spectrochim Acta A 69:2977–2983

Jinnah MMA, Umadevi M, Ramakrishnan V (2004b) Vibrational spectral studies of (β-alanine) β-alaninium nitrate. J Raman Spectrosc 35:956–960

Jiyang W, Gnanam FD, Haussühl S (1986) Crystal growth and physical properties of monoclinic betaine fluoroborate monohydrate, $(CH_3)_3NCH_2COO\ HBF_4.H_2O$. Z Kristallogr 175:155–158

Joema SE, Ramalingam S, Perumal S, Selvarajan P (2011a) Synthesis, growth and studies of pure and ammonium sulfate-doped L-histidine bromide (LHB) single crystals. World J Sci Technol 1(6):33–39

Joema SE, Perumal S, Ramalingam S, Sevarajan P (2011b) Studies on structural, optical, mechanical and thermal properties of undoped and urea-doped L-histidine bromide (LHB) single crystals. Recent Res Sci Technol 3(3):63–68

Johnson MN, Feeder N (2004a) (*R*)-histidinium (2*R*,3*R*)-tartrate. Acta Crystallogr E60:o1476–o1477

Johnson MN, Feeder N (2004b) *D*-histidinium (2*S*,3*S*)-tartrate. Acta Crystallogr E60:o1374–o1375

Johnson MN, Feeder N (2004c) *DL*-histidine *DL*-tartrate. Acta Crystallogr E60:o1273–o1274

Joseph AA, Raja CR (2012) Studies on growth and spectral characterization of diglycine fumarate monohydrate single crystals. Spectrochim Acta A 97:1120–1124

Joseph AA, Ebenezar IJD, Raja CR (2011) Spectroscopic studies on the influence of 10 mol% of glycine on nonlinear optical crystal *L*-valinium picrate. Spectrochim Acta A 82:410–413

Joseph AA, Ebenezar IJD, Raja CR (2012) Crystal growth, spectral and NMR studies of nonlinear optical crystal: *L*-valinium picrate. Optik 123:1436–1439

Kai T, Goto M, Furuhata K, Takayanagi H (1994) Crystal structure of glycine picrate. Anal Sci 10(2):359–360

Kalaiselvi D, Mohan Kumar R, Jayavel R (2008a) Growth and characterization of nonlinear optical *L*-arginine maleate dihydrate single crystals. Mater Lett 62:755–758

Kalaiselvi D, Mohan Kumar R, Jayavel R (2008b) Crystal growth, thermal and optical studies of semiorganic nonlinear optical material: *L*-lysine hydrochloride dihydrate. Mater Res Bull 43:1829–1835

Kalaiselvi D, Mohan Kumar R, Jayavel R (2008c) Single crystal growth and properties of semiorganic nonlinear optical *L*-arginine hydrochloride monohydrate crystals. Cryst Res Technol 43:851–856

Kalaiselvi D, Mohan Kumar R, Jayavel R (2008d) Growth, optical and thermal studies of nonlinear optical *L*-arginine perchlorate single crystal. Cryst Res Technol 43:645–650

Kalyanaraman AR, Srinivasan R (1971) The crystal structure of *DL*-ornithine hydrobromide. Acta Crystallogr B27:1420–1427

Kannan V, Bairava Ganesh R, Ramasamy P (2006) Growth of new nonlinear optical crystals of hydrochlorides of *L*-histidine from solution. Cryst Growth Design 6:1876–1880

Kar T (2012) Amino acid-precursors for synthesizing nonlinear optical materials. Prog Cryst Growth Charact Mater 58:74–83

Karapetyan HA (2008) *L*-Argininium ethyl sulfate. Acta Crystallogr E64:o1982

Karapetyan HA, Petrosyan HA, Petrosyan AM (2003) Nonlinear optical crystalline salts of *L*-histidine. In: Proceedings of conference on "Laser Physics-2003", Ashtarak, 14–17 Oct 2003, pp 152–155

Karle IL, Raganathan D, Haridas V (1996) A persistent preference for layer motifs in self-assemblies of squarates and hydrogen squarates by hydrogen bonding [X-H…O; X=N, O, or C]: a crystallographic study of five organic salts. J Am Chem Soc 118:7128–7133

Kavitha N, Arivanandhan M, Ramamoorthy K, Ragavendran K, Sankaranarayanan K (2004) Microbial inhibition, growth of Li^+-doped LAP single crystals and their characterization. Opt Mater 26:275–280

Kavun VY, Didenko NA, Makarenko NV, Slobodyuk AB, Merkulov EB, Uvarov NF, Zemnukhova LA (2012) Thermal properties and ion mobility in complex antimony(III) fluorides with α-amino acids. Russ J Inorg Chem 57(9):1262–1266

Kavuru P, Aboarayes D, Arora KK, Clarke HD, Kennedy A, Marshall L, Ong TT, Perman J, Pujari T, Wojtas Ł, Zaworotko MJ (2010) Hierarchy of supramolecular synthons: persistent hydrogen bonds between carboxylates and weakly acidic hydroxyl moieties in cocrystals of zwitterions. Cryst Growth Design 10:3568–3584

Khandpekar MM (2003) Thermal evaporation of diglycine hydrogen fluoride crystals. Indian J Pure Appl Phys 41:704–706

Khandpekar MM, Pati SP (2010a) Growth, characterization, non-linear optical properties and dislocation studies in new GCF crystals. J Cryst Growth 312:1150–1153

Khandpekar MM, Pati SP (2010b) Growth and characterisation of new non linear optical material α-glycine sulpho-nitrate (GLSN) with stable dielectric and light dependent properties. Optics Commun 283:2700–2704

Khanna RK, Horak M, Lippincott ER (1966) Infrared studied on some ferroelectric compounds of glycine. Spectrochim Acta 22:1801–1811

Kim YK (2000) Physical, chemical and optical properties of aqueous *L*-arginine phosphate (LAP) solution. J Mater Sci 35:873–880

Kim YK, Lal RB (1998) Physical and optical properties of aqueous *L*-arginine phosphate (LAP) solution for single-crystal growth. J Mater Sci Lett 17:1363–1365

Kimoto H, Saigo K, Ohashi Y, Hasegawa M (1989) Molecular recognition in the formation of conglomerate crystal. 2. The role of arenesulfonic acid in the conglomerate crystals of amino acid salts. Bull Chem Soc Jpn 62:2189–2195

Kinsel GR, Zhao Q, Narayanasamy J, Yassin F, Dias HVR, Niesner B, Prater K, St. Marie C, Ly L, Marynick DS (2004) Arginine/2,5-dihydroxybenzoic acid clusters: an experimental and computational study of the gas-phase and solid-state systems. J Phys Chem A108:3153–3161

Kirfel A, Wallrafen F (1985) The crystal structure of *L*(+)-glutamic acid hydroiodide, *L*-$C_5H_9NO_4$. HI. Z Kristallogr 172:121–128

Kirkwood B, Willett RD (1979) β-Alaninium tetrachloroferrate(III). Acta Crystallogr B35:1883–1884

Kirubavathi K, Selvaraju K (2008) Growth and characterization of glycine hydrobromide single crystal for nonlinear optical applications. Mod Phys Lett B 22(21):2035–2042

Kirubavathi K, Selvaraju K, Vijayan N, Kumararaman S (2008a) Synthesis growth and characterization of *L*-valinium picrate a new nonlinear optical crystal. Spectrochim Acta A 71:288–291

Kirubavathi K, Selvaraju K, Valluvan R, Vijayan N, Kumararaman S (2008b) Synthesis, growth, structural, spectroscopic and optical studies of a new semiorganic nonlinear optical crystal: *L*-valine hydrochloride. Spectrochim Acta A 69:1283–1286

Kistenmacher TJ, Sorrell T (1974) Further conformational aspects of the amino-acid histidine: crystal and molecular structure of histidine dihydrochloride. J Cryst Mol Struct 4:419–432

Kistenmacher TJ, Hunt DJ, Marsh RE (1972) The crystal and molecular structure of *L*-N-acetylhistidine monohydrate: an application of direct methods to space group *P*1. Acta Crystallogr B28:3352–3361

Kitamura M (1993) Crystallization behavior and transformation kinetics of *L*-histidine polymorphs. J Chem Eng Jpn 26:303–307

Kitaoka Y, Yokotani A, Sasaki T (1989) Third harmonic generation of pulsed Nd:YAG laser by an organic optical nonlinear crystal *L*-arginine phosphate monohydrate. Tech Rep Osaka Univ 39 (1966):173–179

Kleinman DA (1962) Nonlinear dielectric polarization in optical media. Phys Rev 126 (6):1977–1979

Klussmann M, Izumi T, White AJP, Armstrong A, Blackmond DG (2007) Emergence of solution-phase homochirality via crystal engineering of amino acids. J Am Chem Soc 129:7657–7660

Koetzle TF, Lehmann MS, Verbist JJ, Hamilton WC (1972) Precision neutron structure determination of protein and nucleic acid components. VII. The crystal and molecular structure of the amino acid L-lysine monohydrochloride dihydrate. Acta Crystallogr B28:3207–3214

Koetzle TF, Golic L, Lehmann MS, Verbist JJ, Hamilton WC (1974) Precision neutron diffraction structure determination of protein and nucleic acid components. XV. Crystal and molecular structure of the amino acid L-valine hydrochloride. J Chem Phys 60:4690–4696

Kolev T, Stahl R, Preut H, Bleckmann P, Radomirska V (1998a) Crystal structure of *L*-(+)-serinium hydrogensquarate, $C_7H_9NO_7$. Z Kristallogr NCS 213:169–170

Kolev T, Stahl R, Preut H, Koniczek L, Bleckmann P, Radomirska V (1998b) Crystal structure of bis((L)-(-)-asparaginium hydrogensquarate) monohydrate, (C8H10N2O7)2.H2O. Z Kristallogr NCS 213:167–168

Kolev T, Koleva BB, Spiteller M (2007) Spectroscopic, theoretical and structural characterization of *L*-threonyl-*L*-serine and *L*-serine. Amino Acids 33:719–725

Kolev T, Mayer-Figge H, Seidel WR, Sheldrick WS, Spiteller M, Koleva BB (2009) New structural motifs and properties of squaric acid anions in presence of the *L*-lysinium counterion. J Mol Struct 919:246–254

Koreneva LG, Zolin BF, Davydov BL (1975) Molecular crystals in nonlinear optics. Nauka, Moscow

Koreneva LG, Zolin BF, Davydov BL (1985) Nonlinear optics of molecular crystals. Nauka, Moscow

Koteeswari P, Mani P, Suresh S (2012) Optical and dielectric studies on *L*-valinium picrate single crystal. J Cryst Proc Technol 2:117–120

Kotler Z, Hierle R, Josse D, Zyss J, Masse R (1992) Quadratic nonlinear optical properties of new transparent and highly efficient organic-inorganic crystal: 2-amino-5-nitropyridinium dihydrogen phosphate (2A5NPDP). J Opt Soc Am B9:534–547

Kozlova OG (1967) Rost kristallov (Crystal growth). Moscow State Univ, Moscow, In Russian

Kozlova OG (1972) Rost i morfologia kristallov (Growth and morphology of crystals), 2nd edn. Moscow State University, Moscow, In Russian

Kozlova OG (1980) Rost i morfologia kristallov (Growth and morphology of crystals), 3rd edn. Moscow State University, Moscow, In Russian

Krause JA, Baures PW, Eggleston D (1991) Histidyl conformations and short N-H. . .N hydrogen bonds: structure of D, L-histidyl-L, D-histidine pentahydrate. Acta Crystallogr B47:506–511

Kripal R, Pandey S (2011) Single crystal ESR and optical absorption study of Cr^{3+} doped *L*-histidine hydrochloride monohydrate. J Phys Chem Solids 72:67–72

Krishnakumar V, Nagalakshmi R (2006) Polarised Raman and infrared spectral analysis of L-alanine oxalate ($C_5H_9NO_6$)-a non-linear optical single crystal. Spectrochim Acta A 64:736–743

Krishnakumar RV, Natarajan S (1996) Crystal structure of an adduct of sarcosine with sucrose. Carbohydr Res 287:117–122

Krishnakumar RV, Nandhini MS, Natarajan S (1998) Sarcosinium oxalate monohydrate. Acta Crystallogr C54, IUC9800063

Krishnakumar RV, Nandhini MS, Natarajan S (2001) Sarcosinium tartrate. Acta Crystallogr C57:165–166

Krishnakumar RV, Nandhini MS, Natarajan S (2002) β-Alaninium oxalate hemihydrate. Acta Crystallogr E58:o117–o119

Krishnakumar V, Nagalakshmi R, Manohar S, Kocsis L (2008) Probes on *L*-lysine monohydrochloride dihydrate: a semiorganic nonlinear optical crystal. Spectrochim Acta A 71:471–479

Krishnakumar V, Nagalakshmi R, Manohar S, Ozga K, Claudet B, Piasecki M, Kityk IV, Kocsis L, Pisarek J (2010) Elastooptical spectra of novel *L*-lysine monohydrochloride dehydrate single crystals. Int J Mod Phys B 24:629–645

Krummeck H, Knorr K, Haussühl S (1994) The ferroelastic phase transition of betaine borate and maleinate. J Phys Condens Matter 6:L111–L114

Kulik M, Pazio A, Wozniak K (2013) Bis(*L*-serinium) oxalate dihydrate: polymorph II. Acta Crystallogr E69:o1667–o1668

Kumar RP, Athimoolam S, Bahadur SA, Rajaram RK (2005) *L*-histidinium sulphate. Acta Crystallogr E61:o2419–o2421

Kumar PP, Manivannan V, Tamilselvan S, Senthil S, Raj VA, Sagayaraj P, Madhavan J (2008) Growth and characterization of a pure and doped nonlinear optical *L*-histidine acetate single crystals. Opt Commun 281:2989–2995

Kumara Swamy KC, Kumaraswamy S, Kommana P (2001) Very strong C-H. . .O, N-H. . .O, and O-H. . .O hydrogen bonds involving a cyclic phosphate. J Am Chem Soc 123:12642–12649

Kurtz SK, Perry TT (1968) A powder technique for the evaluation of nonlinear optical materials. J Appl Phys 39(8):3798–3813

Kuzminov YS (1987) Electrooptical and nonlinear optical crystal of lithium niobate. Nauka, Moscow (In Russian)

Kyriakos M, Karlheinz D (1995) Method for the preparation of salts of *L*-ornithine. US Patent #5405761

Lamberts K, Englert U (2012) *DL*-Alaninium iodide. Acta Crystallogr E68:o1846

Langhoff CA (1990) Theoretical studies of the nonlinear optical properties of *L*-arginine phosphate. Mater Res Soc Symp Proc 173:677–682

Lazar AN, Danylyuk O, Suwinska K, Coleman AW (2006) The solid-state structure of calix[4] arene dihydroxyphosphonic acid-*L*-lysine complex. J Mol Struct 825:20–25

Lehmann MS, Koetzle TF, Hamilton WC (1972) Precision neutron diffraction structure determination of protein and nucleic acid components. IV. The crystal and molecular structure of the amino acid L-histidine. Int J Pept Protein Res 4:229–239

Lemanov VV, Popov SN, Pankova GA (2002) Piezoelectric properties of crystals of some protein aminoacids and their related compounds. Phys Solid State 44(10):1929–1935

Lemanov VV, Popov SN, Pankova GA (2011) Phase transitions in sarcosine phosphate single crystals. Phys Solid State 53:1274–1276

Lemanov VV, Yarmarkin VK, Egorov VM, Pankova GA, Zaitseva NV, Markova LA (2012) Phase transitions in crystals of protein amino acids with trapped drops of water solutions. Phys Solid State 54(2):346–349

Levene PA, Van Slyke DD (1912) The composition and properties of glycocoll picrate and the separation of glycocoll from alanine. J Biol Chem 12:285–294

Levintow L, Greenstein J (1951) Preparation of *D*-ornithine, *D*-citrulline, and *D*-arginine. J Biol Chem 188(2):643–646

L'Haridon P, Lang J, Pastuszak R, Dobrowolski J (1978) Structure cristalline du composé $PtCl_6$-*L*-lysine. Acta Crystallogr B34:2436–2439

Li ZS, Chai JS (2007) Glycinium 3-nitrophtalate. Acta Crystallogr E63:o2857–o2859

Li A, Xu C, Li A, Ming N (2000) Study of coloration, microbe inhibition during the growth of L-arginine phosphate monohydrate single crystals. J Cryst Growth 220:291–295

Li MC, Lin CS, Li YQ, He L, Jin ZM (2006) Redetermination of *L*-valinium chloride monohydrate. Acta Crystallogr E62:o2235–o2236

Li J, Liang ZP, Tai XS (2009) Crystal structure of (*S*)-2-amino-3-(1*H*-indole-3-yl) propanoic acidate-acitic acid-water (1:1:1), $C_{11}H_{12}N_2O_2.CH_3COOH.H_2O$. Z Kristallogr NCS 224:153–154

Liang P (2008) A cocrystal of pyridine-2,4-dicarboxylic acid and serine. Acta Crystallogr E64:o43

Liao CZ, Feng XL, Yao JH, Cai J (2001) Bis(β-alaninium) biphenyl-4,4'-disulfonate. Acta Crystallogr C57:1215–1216

Lima RJC, Feire PTC, Sasaki JM, Melo FEA, Mendes Filho J (2002) Temperature-dependent Raman study of *L*-arginine hydrochloride monohydrate single crystal. J Raman Spectrosc 33:625–630

Liu WJ, Ferrari C, Zha M, Zanotti L, Jiang SS (2000) X-ray topographic characterization of *L*-arginine phosphate monohydrate crystals. Cryst Res Technol 35:1215–1219

Liu XJ, Wang ZY, Xu D, Wang XQ, Song YY, Yu WT, Guo WF (2007a) Investigation on the micro-crystallization of *L*-arginine trifluoroacetate (LATF) crystals. J Alloy Compd 441:323–326

Liu XJ, Wang ZY, Yu GW, Sun ZH, Zhang GH, Wang XQ, Zhu LY, Yu G, Xu D (2007b) Growth morphology of {101} surface of *L*-arginine trifluoroacetate crystals investigated by AFM. J Phys Chem Solids 68:608–610

Liu XJ, Wang ZY, Duan A, Yu G, Zhang GH, Yu GW, Wang XQ, Sun ZH, Xu D (2007c) AFM investigation of the {101} morphology of *L*-arginine trifluoroacetate (LATF) crystals. Solid State Sci 9:527–530

Liu XJ, Wang ZY, Zhang G, Yu G, Wang X, Duan A, Fan J, Xu D (2007d) Atomic force microscopy studies on {101} surfaces of *L*-arginine trifluoroacetate single crystals. J Phys Chem C 111:14165–14169

Liu XJ, Wang ZY, Zhang G, Wang X, Duan A, Sun Z, Zhu L, Xu D (2007e) Crystal growth of high quality nonlinear optical crystals of *L*-arginine trifluoroacetate. J Cryst Growth 308:130–132

Liu XJ, Wang ZY, Zhang G, Yu G, Wang X, Xu D (2008a) Nucleation growth mechanism and defects of nonlinear optical crystals of *L*-Arg.CF_3COOH. Mater Lett 62:1986–1988

Liu XJ, Wang ZY, Duan A, Zhang G, Wang X, Sun Z, Zhu L, Yu G, Sun G, Xu D (2008b) Measurement of *L*-arginine trifluoroacetate crystal nucleation kinetics. J Cryst Growth 310:2590–2592

Liu XJ, Wang ZY, Wang XQ, Zhang GH, Xu SX, Duan A, Zhang SJ, Sun ZH, Xu D (2008c) Morphology and physical properties of *L*-arginine trifluoroacetate crystals. Cryst Growth Design 8:2270–2274

Liu XJ, Wang Z, Zhang G, Yu G, Duan A, Wang X, Xu D (2009) Distinct growth phenomenon observed on *L*-Arg.CF_3COOH crystals. Curr Appl Phys 9:22–25

Liu XJ, Xu D, Wei XQ, Ren MJ, Zhang GH (2010a) Growth of Cu^{2+} and Mg^{2+} doped nonlinear optical LATF crystals and their characterization. Mater Sci Eng B 166:203–208

Liu XJ, Wang PJ, Zhang Z, Xu D, Zhang GH, Wang XQ (2010b) Imaging of surface morphologies of *L*-arginine trifluoroacetate crystals. Curr Appl Phys 10:715–717

Liu XJ, Xu D, Ren MJ, Zhang GH, Wei XQ, Wang J (2010c) An examination of the growth kinetics of *L*-arginine trifluoroacetate (LATF) crystals from induction period and atomic force microscopy investigations. Cryst Growth Design 10:3442–3447

Liu XJ, Wang P, Xu D, Wei X (2011) Observation of the kinetic roughening of *L*-arginine trifluoroacetate (LATF) crystals. Cryst Growth Design 11:791–795

Liu XJ, Yan S, Ren M (2013) Theoretical investigations of optical properties of *L*-arginine trifluoroacetate crystal. Mater Chem Phys 143:286–291

Losev EA, Zakharov BA, Drebushchak TN, Boldyreva EV (2011) Glycinium semi-malonate and a glutaric acid-glycine cocrystal: new structures with O-H...O hydrogen bonds. Acta Crystallogr C67:o297–o300

Lucia Rose ASJ, Selvarajan P, Perumal S (2010) Synthesis, growth and spectroscopic studies of *L*-alanine hydrogen chloride (LAHC) crystals. Recent Res Sci Technol 2(3):76–79

Lucia Rose ASJ, Selvarajan P, Perumal S (2011a) Studies on growth and characterization of an NLO crystal: *L*-alanine hydrogen chloride (LAHC). Mater Chem Phys 130:950–955

Lucia Rose ASJ, Selvarajan P, Perumal S (2011b) Growth, structural, spectral, mechanical and dielectric characterization of RbCl-doped *L*-alanine hydrogen chloride monohydrate single crystals. Physica B 406:412–417

Łukaszewicz K, Pietraszko A, Stepien-Damm J (1996a) Redetermination of the crystal structure of paraelectric diglycine nitrate. Part I. Crystal structure at 220K and 293K. Pol J Chem 70 (11):1414–1418

Łukaszewicz K, Pietraszko A, Stepien-Damm J (1996b) Redetermination of the crystal structure of paraelectric diglycine nitrate. Part II. Short-range order at 220K and 293K. Pol J Chem 70 (12):1550–1553

MacDonald JC, Yigit M, Mychajlonka K (2005) Two concomitant polymorphs of a supramolecular model of the Asp...His...Ser catalytic triad. Cryst Growth Design 5(6):22248–22255

Machado FLA, Sousa LLL, Cunha RO, Cabral FAO, Rodrigues AR, Carvalho JF, Santana RC (2010) Specific heat measurements in pure and in (Cu, Mn, Fe, Ni)-doped single-crystals of *L*-arginine phosphate monohydrate. J Phys Chem Solids 71:862–866

Machida M, Ikeda H, Kakiuchi T, Ishibashi T, Hasebe K (2003) NMR study of phase transitions in betaine phosphite $(CH_3)_3NCH_2COOH.H_2PO_3$. J Phys Soc Jap 72:2250–2255

Madden JJ, McGandy EL, Seeman NC (1972a) The crystal structure of the orthorhombic form of *L*-(+)-histidine. Acta Crystallogr B28:2377–2382

Madden JJ, McGandy EL, Seeman NC, Harding MM, Hoy A (1972b) The crystal structure of the monoclinic form of *L*-histidine. Acta Crystallogr B28:2382–2389

Madhavan J, Aruna S, Prabha K, Packium Julius J, Joseph GP, Selvakumar S, Sagayaraj P (2006a) Growth and characterization of a novel NLO crystal *L*-histidine hydrofluoride dihydrate (LHHF). J Cryst Growth 293:409–414

Madhavan J, Aruna S, Ambujam K, Pragasam JA, Ravikumar SM, Gulam Mohamed M, Sagayaraj P (2006b) Crystallization and characterization of nonlinear optical *L*-histidinium dihydrogen orthophosphate orthophosphoric acid single crystal. Cryst Res Technol 41:997–1001

Madhavan J, Aruna S, Thomas PC, Vimalan M, Rajasekar SA, Sagayaraj P (2007a) Growth and characterization of *L*-histidine hydrochloride monohydrate single crystals. Cryst Res Technol 42:59–64

Madhavan J, Aruna S, Anuradha A, Remanand D, Potheher IV, Thamizharasan K, Sagayaraj P (2007b) Growth and characterization of a new nonlinear optical *L*-histidine acetate single crystals. Opt Mater 29:1211–1216

Magome E, Machida M, Tamura Y, Komukae M (2007) Isotope effect of glycinium phosphate $NH_3CH_2COOH.H_2PO_3$. J Phys Soc Jap 76:124601

Maiman TH (1960) Stimulated optical radiation in ruby. Nature 187(4736):493–494

Majerz I, Malarski Z, Sobczyk L (1997) Proton transfer and correlations between the C-O, O-H, N-H and O. . .N bond lengths in amine phenolates. Chem Phys Lett 274:361–364

Maker PD, Terhune RW, Nisenoff M, Savage CM (1962) Effects of dispersion and focusing on the production of optical harmonics. Phys Rev Lett 8(1):21–22

Mallik T, Kar T (2005a) Optical, thermal and structural characterization of an NLO crystal, *L*-arginine perchlorate. J. Crystal Growth 274:251–255

Mallik T, Kar T (2005b) Synthesis, crystal structure and solubility of $C_6H_{14}N_4O_2$, $C_4H_4O_4$, $2H_2O$. Sci Tech Adv Mater 6:508–512

Mallik T, Kar T (2005c) Synthesis, growth and characterization of a new nonlinear optical crystal: l-arginine maleate dihydrate. Cryst Res Technol 40:778–781

Mallik T, Kar T (2007) Crystallization and characterization of nonlinear optical material *L*-arginine formomaleate. Mater Lett 61:3826–3828

Manimaran D, John CJ, Rastogi VK, Hubert Joe I (2013) Growth and vibrational spectral investigation of nonlinear optical *L*-Argininium perchlorate-DFT study. Spectrochim Acta A 109:173–178

Marchewka MK, Drozd M (2013) L-leucine perchlorate: new molecular complex with nonlinear optical properties. Vibrational, calorimetric and theoretical studies. Cent Eur J Chem 11 (8):1264–1277

Marchewka MK, Debrus S, Ratajczak H (2003a) Vibrational spectra and second harmonic generation in molecular complexes of L-lysine with L-tartaric, D, L-malic, acetic, arsenous, and fumaric acids. Cryst Growth Design 3(4):587–592

Marchewka MK, Debrus S, Pirtraszko A, Barnes AJ, Ratajczak H (2003b) Crystal structure, vibrational spectra and nonlinear optical properties of *L*-histidinium-*L*-tartrate hemihydrates. J Mol Struct 656(1–3):265–273

Marcy HO, Warren LF, Webb MS, Ebbers CA, Velsko SP, Kennedy G, Catella GC (1992) Second-harmonic generation in zinc tris(thiourea) sulfate. Appl Optics 31:5051–5061

Marcy HO, Rosker MJ, Warren LF, Cunningham PH, Thomas CA, Deloach LA, Velsko SP, Ebbers CA, Liao JH, Kanatzidis MG (1995) *L*-Histidine tetrafluoroborate: a solution-grown semiorganic crystal for nonlinear frequency conversion. Opt Lett 20:252–254

Marder SR, Perry JW, Schaefer WP (1989) Synthesis of organic salts with large second-order optical nonlinearities. Science 245:626–628

Martin Britto Dhas SA, Natarajan S (2007a) Growth and characterization of a new organic NLO material: glycine nitrate. Optics Commun 278:434–438

Martin Britto Dhas SA, Natarajan S (2007b) Growth and characterization of *L*-prolinium tartrate – a new organic NLO material. Cryst Res Technol 42:471–476

Martin Britto Dhas SA, Natarajan S (2008a) Growth and characterization of two new NLO materials from the amino acid family: *L*-Histidine nitrate and *L*-Cysteine tartrate monohydrate. Optics Commun 281:457–462

Martin Britto Dhas SA, Natarajan S (2008b) Growth and characterization of a new potential second harmonic generation material from the amino acid family: *L*-Valinium picrate. Cryst Res Technol 43(8):869–873

Martin Britto Dhas SA, Suresh M, Raji P, Ramachandran K, Natarajan S (2007) Photoacoustic studies on two new organic materials: *L*-threonine and *L*-prolinium tartrate. Cryst Res Technol 42:190–194

Martin Britto Dhas SA, Bhagavannarayana G, Natarajan S (2008a) Growth and characterization of a new potential NLO material from the amino acid family-*L*-prolinium picrate. J Cryst Growth 310:3535–3539

Martin Britto Dhas SA, Suresh J, Bhagavannarayana G, Natarajan S (2008b) Growth and characterization of a new organic non-linear optical (NLO) material: *L*-histidinium trifluoroacetate. Open Crystallogr J 1:46–50

Mata I, Espinosa E, Molins E, Veintemillas S, Maniukiewicz W, Lecomte C, Cousson A, Paulus W (2006) Contributions to the application of the transferability principle and the multipolar modeling of H atoms: electron-density study of L-histidinium dihydrogen orthophosphate orthophosphoric acid. I. Acta Crystallogr A62:365–378

Mathivanan V, Raghavalu T, Kovendhan M, Gokul Raj S, Ramesh Kumar G, Mohan R, Kumar KS (2007) Synthesis, growth and characterization of *L*-histidinium acetate dihydrate single crystals. Cryst Res Technol 42:895–898

Mathivanan V, Raghavalu T, Kovendhan M, Kumar KS, Gokul Raj S, Ramesh Kumar G, Mohan R (2008) Synthesis and characterization of a new nonlinear optical single crystal: *L*-lysinium trifluoroacetate. Cryst Res Technol 43:248–252

Matthias BT, Miller CE, Remeika JP (1956) Ferroelectricity of glycine sulfate. Phys Rev 104 (3):849–850

Mazumdar SK, Srinivasan R (1964) X-ray analysis of *L*-arginine hydrohalides. Curr Sci 33:573–575

Mazumdar SK, Srinivasan R (1966) The crystal structure of *L*-arginine monohydrobromide monohydrate. Z Kristallogr 123(Suppl):186–205

Mazumdar SK, Venkatesan K, Mez HC, Donohue J (1969) The crystal structure of *L*-arginine hydrochloride. Z Kristallogr 130:328–339

Mazumder A, Kar T, Gupta SPS (1995) Infrared spectroscopy and thermal studies of as-grown *L*-arginine phosphate monohydrate crystals. Jpn J Appl Phys 34(Part 1):5717–5720

Mebs S, Messerschmidt M, Luger P (2006) Experimental charge density of an *L*-phenylalanine formic acid complex with a short hydrogen bond determined at 25K. Z Kristallogr 221 (9):656–664

Meena M, Mahadevan CK (2008) Growth and dielectric properties of *L*-arginine acetate and *L*-arginine oxalate single crystals. Mater Lett 62:3742–3744

Meena M, Mahadevan CK (2010) Effect of added impurities on the electrical properties of *L*-arginine acetate single crystals. Archiv Appl Sci Res 2:185–199

Meera K, Muralidharan R, Dhanasekaran R, Manyum P, Ramasamy P (2004) Growth of nonlinear optical material: *L*-arginine hydrochloride and its characterisation. J Cryst Growth 263:510–516

Miller RC (1964) Optical second harmonic generation in piezoelectric crystals. Appl Phys Lett 5(1):17–19

Minemoto H, Ozaki Y, Sonoda N, Sasaki T (1993) Intracavity second harmonic generation using a deuterated organic ionic crystal. Appl Phys Lett 63:3565–3567

Minkov VS, Boldyreva EV (2008) *L*-Cysteinium semioxalate. Acta Crystallogr C64:o344–o348

Minkov VS, Boldyreva EV (2009) *DL*-Cysteinium semioxalate. Acta Crystallogr C65:o245–o247

Minkov VS, Boldyreva EV (2011) *L*-Cysteinium semioxalate: a new monoclinic polymorph or a hydrate? Acta Crystallogr C67:o139–o142

Minkov VS, Boldyreva EV (2012) The effect of partial methylation of the glycine amino group on crystal structure in N, N-dimethylglycine and its hemihydrates. Acta Crystallogr C68:o283–o287

Minkov VS, Boldyreva EV, Drebushchak TN, Görbitz CH (2012) Stabilizing structures of cysteine-containing crystal with respect to variations of temperature and pressure by immobilizing amino acid side chains. CrystEngComm 14:5943–5954

Mishra AK, Muril C, Garg N, Chitra R, Sharma SM (2010) Pressure-induced structural transformations in bis(glycinium)oxalate. J Phys Chem B114:17084–17091

Mitsui Y, Tsuboi M, Iitaka Y (1969) The crystal structure of *DL*-proline hydrochloride. Acta Crystallogr B25:2182–2192

Mohan Kumar R, Gopalakrishnan N, Jayavel R, Ramasamy P (1999) Investigation on the nucleation kinetics of *L*-arginine phosphate single crystals. Cryst Res Technol 34:1265–1268

Mohandas T, Inbaseelan CRT, Saravanan S (2013) Glycine-*D*-tartaric acid (1/1). Acta Crystallogr E69:o236

Mohandoss R, Dhanuskodi S, Jayalakshmy MS, Philip J, Bhagavannarayana G (2012) Structural, electrical, thermal and optical properties of the nonlinear optical crystal *L*-arginine fluoride. Cryst Res Technol 47:620–629

Moitra S, Kar T (2008) Second harmonic generation of a new nonlinear optical material *L*-valine hydrobromide. J Cryst Growth 310:4539–4543

Moitra S, Seth SK, Kar TG (2010) Synthesis, crystal structure, characterization and DFT studies of L-valine L-valinium hydrochloride. J Cryst Growth 312(12–13):1977–1982

Molčanov K, Kojić-Prodić B (2010) Salts and co-crystals of chloranilic acid with organic bases: is it possible to predict a salt formation? CrystEngComm 12:925–939

Monaco SB, Davis LE, Velsko SP, Wang FT, Eimerl D, Zalkin A (1987) Synthesis and characterization of chemical analogs of L-arginine phosphate. J Cryst Growth 85(1–2):252–255

Moolya BN, Dharmaprakash SM (2005) X-ray diffraction, vibrational spectra and thermal studies of nonlinear optical *L*-tyrosine hydrochloride single crystals. Indian J Phys 79(12):1423–1425

Moolya BN, Darmaprakash SM (2006) Nonlinear optical diglycine hydrochloride: synthesis, crystal growth and structural characteristics. J Cryst Growth 293(1):86–92

Moolya BN, Dharmaprakash SM (2006) Synthesis, growth and characterization of nonlinear optical crystal: *L*-tyrosine hydrobromide. J Cryst Growth 290:498–503

Moolya BN, Dharmapraksh SM (2007) Growth and characterization of nonlinear optical diglycinehydrobromide single crystals. Mater Lett 61(17):3559–3562

Moolya BN, Rai C, Dharmaprakash SM, Harrison WTA (2007) Sodium tris(glycinium)bis (hexafluorosilicate) glycine trisolvate. Acta Crystallogr C63:m312–m314

Moovendaran K, Srinivasan BR, Sundar JK, Martin Britto Dhas SA, Natarajan S (2012) Structural, vibrational and thermal studies of a new nonlinear optical material: *L*-Asparagine–*L*-tartaric acid. Spectrochim Acta A 92:388–391

Moovendaran K, Martin Britto Dhas SA, Natarajan S (2013) Growth and characterization of *L*-histidinium 2-nitrobenzoate single crystals: a new NLO material. Optik 124:3117–3119

Mostad A, Natarajan S (1985) Crystal and molecular structure of *DL*-methionine nitrate. Z Kristallogr 172:175–182

Mostad A, Natarajan S (1995) Crystal and molecular structure of *L*-(*S*)-histidine dihidrochloride (monoclinic form). Z Kristallogr 210:114–117

Mostad A, Natarajan S (1996) Crystal structures of an adduct of sarcosine with sulfuric acid (at 140K). Cryst Res Technol 31(3):295–300

Mostad A, Nistol KA, Romming C, Natarajan S (1995) Crystal and molecular structure of *L*-(+)-histidine acetate dihydrate (at 123 K). Z Kristallogr 210:352–354

Mukerji S, Kar T (1998a) Thermal and spectroscopic studies of as-grown *L*-arginine hydrochloride monohydrate crystals. Mater Chem Phys 57:72–76

Mukerji S, Kar T (1998b) Structural, thermal and spectroscopic investigation of nonlinear optical crystal *L*-arginine hydrobromide monohydrate. Mater Res Bull 33:619–626

Mukerji S, Kar T (1999a) Surface micromorphology of different crystallographic faces of *L*-arginine hydrochloride monohydrate etched in organic solvents. Jpn J Appl Phys 38 (Part 1, No. 2A):832–837

Mukerji S, Kar T (1999b) Etch pit study of different crystallographic faces of *L*-arginine hydrobromide monohydrate (LAHBr) in alcohols. J Cryst Growth 200:543–549

Mukerji S, Kar T (1999c) Vicker's microhardness studies of *L*-arginine hydrobromide monohydrate crystals (LAHBr). Cryst Res Technol 34:1323–1328

Mukerji S, Kar T (2000a) Microhardness study of the nonlinear optical *L*-arginine hydrochloride monohydrate. Metal Mater Trans 31A:3087–3090

Mukerji S, Kar T (2000b) Knoop microhardness anisotropy and Young's modulus of *L*-arginine hydrochloride monohydrate and *L*-arginine hydrobromide monohydrate. Mater Res Bull 35:711–717

Mukhopadhyay BP, Ghosh S, Banerjee A (1995) Role of water molecules in protein-nucleic acid interactions: visualization of a model highly hydrated complex structure of Inosine 5′-monophosphate and *L*-Serine ($2C_{10}H_{13}N_4O_8P.C_3H_7NO_3.12H_2O$) at atomic resolution. J Chem Crystallogr 25(8):477–485

Muralidharan R, Rkumar M, Jayavel R, Ramasamy P (2003) Growth and characterization of *L*-arginine acetate single crystals: a new NLO material. J Cryst Growth 259:321–325

Muralidharan S, Nagapandiselvi P, Srinivasan T, Gopalakrishnan R, Velmurugan D (2013) L-Histidinium p-toluenesulfonate. Acta Crystallogr E69(5):o804

Mythili P, Kanagasekaran T, Gopalakrishnan R (2008) Growth and characterization of glycinium oxalate (GOX) single crystals. Mater Lett 62:2185–2188

Nabakhtiani GN, Furmanova NG, Chachkhiani LG (1992) X-ray structural study of *L*-arginine hydrobromide monohydrate $C_6H_{14}O_2N_4HBr.H_2O$. Kristallografia 37:1185–1192

Nagashima N, Sano C, Kawakita T, Iitaka Y (1992) The crystal structure of di-*L*-phenylalanine sulphate monohydrate. Anal Sci 8:723–725

Nagashima N, Kishimoto S, Tanabe T, Iitaka Y (1993) Crystal structure of *L*-histidine and α-ketoisocaproic acid complex ethanol solvate. Anal Sci 9:743–745

Nagata H, In Y, Tomoo K, Doi M, Ishida T, Wakahara A (1995) Structural feature and molecular interaction of basic amino acid-picric acid complexes by X-ray crystal analyses. Chem Pharm Bull 43(11):1836–1843

Nalbandyan AG, Sukiasyan RP, Novikov EV, Van Stryland EW, Petrosyan AM (2000) Investigation of growth conditions of lysine trihydrogen iodate crystals. In: Proceedings of conference on Laser Physics-99, Ashtarak, 18–22 Oct 1999, pp 128–132

Nandhini MS, Krishnakumar RV, Natarajan S (2001a) Glycinium oxalate. Acta Crystallogr C57:115–116

Nandhini MS, Krishnakumar RV, Natarajan S (2001b) *L*-Alanine oxalate. Acta Crystallogr E57: o633–o635

Nandhini MS, Krishnakumar RV, Natarajan S (2001c) *DL*-Alaninium oxalate. Acta Crystallogr E57:o666–o668

Nandhini MS, Krishnakumar RV, Malathi R, Rajan SS, Natarajan S (2001d) *DL*-Threoninium oxalate. Acta Crystallogr E57:o769–o771

Nandhini MS, Krishnakumar RV, Natarajan S (2001e) *L*-Prolinium tartrate. Acta Crystallogr C57:423–424

Narayanan P, Venkataraman S (1975) Crystal and molecular structure of monoglycine nitrate. J Cryst Mol Struct 5:15–26

Natarajan S, Zangrando E (1992) Crystal structure of bisglycine hydrobromide – a reinvestigation. Proc Indian Acad Sci Chem Sci 104(4):483–487

Natarajan S, Muthukrishnan C, Bahadur SA, Rajaram RK, Rajan SS (1992) Reinvestigation of the crystal structure of diglycine hydrochloride. Z Kristallogr 198:265–270

Natarajan S, Martin Britto SA, Ramachandran E (2006) Growth, thermal, spectroscopic, and optical studies of *L*-alaninium maleate, a new organic nonlinear optical material. Cryst Growth Design 6:137–140

Natarajan S, Shanmugam G, Martin Britto Dhas SA (2008a) Growth and characterization of a new semi organic NLO material: *L*-tyrosine hydrochloride. Cryst Res Technol 43(5):561–564

Natarajan S, Chitra GP, Martin Britto Dhas SA, Athimoolam S (2008b) Growth, structural, thermal and optical studies on *L*-glutamic acid hydrobromide – a new semiorganic NLO material. Cryst Res Technol 43:713–719

Natarajan S, Devi NR, Martin Britto Dhas SA, Athimoolam S (2008c) Crystal growth and structure of L-methionine L-methioninium hydrogen maleate-a new NLO material. Sci Technol Adv Mater 9:025012 (5 p)

Natarajan V, Arivanandhan M, Sankaranarayanan K, Ramasamy P (2009a) Growth aspects and characteristic properties of pure and Li-doped *L*-arginine acetate (LAA) single crystals: a promising nonlinear optical material. J Cryst Growth 311:572–575

Natarajan S, Kalyanasundar A, Suresh J, Martin Britto Dhas SA, Lakshman PLN (2009b) Glycinium hydrogen fumarate glycine solvate monohydrate. Acta Crystallogr E65:o462

Natarajan S, Umamaheswaran M, Sundar JK, Suresh J, Martin Britto Dhas SA (2010a) Structural, spectroscopic and nonlinear optical studies on a new efficient organic donor-acceptor crystal for second harmonic generation: *L*-Threoninium picrate. Spectrochim Acta A 77:160–163

Natarajan S, Hema V, Sundar JK, Suresh J, Lakshman PLN (2010b) *L*-Asparagine-*L*-tartaric acid (1/1). Acta Crystallogr E66:o2239

Natarajan S, Devi NR, Martin Britto Dhas SA, Athimoolam S (2010c) Growth, thermal and optical studies of a new organic NLO material: L-methionine L-methioninium hydrogen maleate. Optoelectron Adv Mat-Rapid Commun 4(4):516–519

Natarajan S, Moovendaran K, Sundar JK, Bhagavannarayana G, Martin Britto Dhas SA (2011) Influence of picric acid on the SHG efficiency of *L*-prolinium tartrate crystals. J Miner Mater Charact Eng 10:913–921

Natarajan S, Moovendaran K, Sundar JK, Ravikumar K (2012) Crystal structure of *L*-histidinium 2-nitrobenzoate. J Amino Acids 2012:463183

Neelam R, Vijayan N, Thukral K, Maurya KK, Haranath D, Bhagavannarayana G, Verma S, Wahab MA (2013) Crystalline perfection, optical and third harmonic generation analyses of non-linear optical single crystal of *L*-lysine acetate. Spectrochim Acta A 105:192–199

Němec I (1998) Preparation and study of addition compounds of aminoacids with inorganic oxyacids. PhD thesis, Charles University in Prague, Faculty of Science

Němec I, Mička Z (1999) FTIR and FT Raman study of *L*-leucine addition compound with nitric acid. J Mol Struct 482–483:23–28

Němec I, Mička Z (2001a) FTIR and FT Raman study of *L*-leucine-selenic acid addition compound. J Mol Struct 563–564:295–299

Němec I, Mička Z (2001b) Tri-*L*-valine selenate-study of vibrational spectra and structural phase transition. J Mol Struct 563–564:289–294

Němec I, Cisarová I, Mička Z (1998) Study of the glycine-selenious acid addition compounds: crystal structure of glycine hydrogen selenite and vibrational spectra and DSC measurement of diglycine hydrogen selenious acid crystals. J Solid State Chem 140(1):71–82

Němec I, Cisarová I, Mička Z (1999a) The crystal structure, vibrational spectra and DSC measurement of mono-L-alaninium nitrate. J Mol Struct 476:243–253

Němec I, Gyepes R, Mička Z (1999b) The crystal structure, vibrational spectra and DSC measurements of mono-β-alaninium nitrate. J Mol Struct 476:203–213

Němec I, Cisarová I, Mička Z (2001) The crystal structure and vibrational spectra of mono-*L*-valinium nitrate: DSC, FTIR, and X-ray diffractional study of low-temperature phase transition. J Solid State Chem 158:1–13

Němec I, Gyepes R, Mička Z, Troj\u00e1nek F (2002) Novel materials for second harmonic generation of *L*-valine and selenic acid. Mater Res Soc Proc 725:9.2.1–P9.2.6

Nepomnyashaya VN, Shternberg AA, Gavrilova IV (1961) Laboratornaya metodikavyrashivania krupnykh ogranennykh kristallov i orientirovannykh blokov sulfata litia (Laboratory technique for growing of large faceted crystals and oriented blocks of lithiumsulphate). In: Rost kristallov, III:209–295 (In Russian)

Nockemann P, Thijs B, Pittois S, Thoen J, Glorieux C, Van Hecke K, Van Meervelt L, Kirchner B, Binnemans K (2006) Task-specific ionic liquid for solubilizing metal oxides. J Phys Chem B110:20978–20992

Novak A (1979) Vibrational spectroscopy of hydrogen bonded systems. In: Infrared and Raman spectroscopy of biological molecules. Proceedings of the NATO Advanced Study Institute, Athens, 22–31 Aug 1978. D. Reidel Publishing Company, Dordrecht, pp 279–303

Oda K, Koyama H (1972) A refinement of the crystal structure of histidine hydrochloride monohydrate. Acta Crystallogr B28:639–642

Ojala WH, Sudbeck EA, Lu LK, Richardson TI, Lovrien RE, Gleason WB (1996) Complexes of lysine, histidine, and arginine with sulfonated azo dyes: model systems for understanding the biomolecular recognition of glycosaminoglycans by proteins. J Am Chem Soc 118:2131–2142

Okamura K, Aoe K, Hiramatsu H, Nishimura N, Sato T, Hashimoto K (1997) Crystal structure of diastereomeric 1:1 complex of (*R*)- and (*S*)-phenylalanine (*S*)-mandelic acid. Anal Sci 13:315–317

Olejnik S, Łukaszewicz K, Lis T (1975) Diglycine selenate. Acta Crystallogr B31:1785–1787

Olejnik S, Łukaszewicz K, Wratislava AU (1977) Mat F As. 43 CSD code: TGLYSE10

Ondráček J, Walzelová M, Mička Z, Novotný J (1992) Structure of glycine-selenious acid (1/1). Acta Crystallogr C48:391–392

Owens C, Bhat K, Wang WS, Tan A, Aggarwal MD, Penn BG, Fraizier DO (2001) Bulk growth of high quality nonlinear optical crystals of L-arginine tetrafluoroborate (L-AFB). J Cryst Growth 225:465–469

Ozga K, Krishnakumar V, Kityk IV, Jasik-Ślęzak J (2008) *L*-lysine monohydrochloride dihydrate as novel elasto- and electrooptical materials. Mater Lett 62:4597–4600

Packiam Julius J, Rajasekar SA, Pragasam JAA, Joseph V, Sagayaraj P (2004a) Growth, microhardness and optical characterization of L-arginine hydrofluoride (LAHF) single crystals. Mater Sci Eng B107:259–263

Packiam Julius J, Rajasekar SA, Rajasekar SA, Selva Kumar S, Stephen A, Sagayaraj P (2004b) Studies on the growth and characterization of *L*-argininium formate single crystals. J Cryst Growth 267:619–623

Padmaja L, Vijayakumar T, Hubert Joe I, Nair CPR, Jayakumar VS (2006) Vibrational spectral studies and the nonlinear optical properties of a novel NLO material *L*-prolinium tartrate. J Raman Spectrosc 37:1427–1441

Paixão JA, Matos Beja A, Ramos Silva M, Alte da Veiga L, Martín-Gil J, Martín-Gil F, de Matos Gomes E (1997) Crystal structure of betaine dihydrogen selenite, $C_5H_{13}NO_5Se$. Z Kristallogr NCS 212:51–52

Paixão JA, Ramos Silva M, Matos Beja A, Eusébio E (2006) Crystal structure and properties of *L*-tryptophanium hydrogen selenite. Polyhedron 25:2021–2025

Pal T, Kar T (2002) Single crystal growth and characterization of the nonlinear optical crystal *L*-arginine hydrofluoride. J Cryst Growth 234:267–271

Pal T, Kar T (2003) Vickers microhardness studies of *L*-arginine halide mixed crystals. Mater Sci Eng A354:331–336

Pal T, Kar T (2005a) Studies on growth defects and mechanical properties of nonlinear optical crystal: *L*-arginine hydrofluoride. J Cryst Growth 276:247–252

Pal T, Kar T (2005b) Studies of microhardness anisotropy and Young's modulus of nonlinear optical crystal *L*-arginine hydrochlorobromo monohydrate. Mater Lett 59:1400–1404

Pal T, Kar T (2005c) Optical, mechanical and thermal studies of nonlinear optical crystal *L*-arginine acetate. Mater Chem Phys 91:343–347

Pal T, Kar T (2006a) Studies on surface micromorphology and growth mechanism of nonlinear optical crystal: L-arginine hydrochlorobromo monohydrate. J Cryst Growth 289:202–206

Pal T, Kar T (2006b) Studies on mechanical properties of an organic nonlinear optical crystal. Mater Sci Eng A 437:235–239

Pal T, Kar T, Zhou GY, Wang XQ, Wang D, Chen XF, Yang ZH (2002) Growth and characterization of nonlinear optical material, LAHClBr-a new member of *L*-arginine halide family. J Cryst Growth 235:523–528

Pal T, Kar T, Bocelli G, Rigi L (2003) Synthesis, growth, and characterization of *L*-arginine acetate crystal: a potential NLO material. Cryst Growth Design 3:13–16

Pal T, Kar T, Bocelli G, Rigi L (2004) Morphology, crystal structure, and thermal and spectral studies of semiorganic nonlinear optical crystal LAHClBr. Cryst Growth Design 4:743–747

Pan F, Wong MS, Bosshard C, Günter P (1996) Crystal growth and characterization of the organic salt 4-N, N-dimethylamino-4'-N'-methyl-stilbazolium tosylate (DAST). Adv Mater 8:592–595

Pandiarajan S, Sridhar B, Rajaram RK (2001a) L-Valine L-valinium perchlorate monohydrate. Acta Crystallogr E57(5):o466–o468

Pandiarajan S, Sridhar S, Rajaram RK (2001b) β-Alaninium perchlorate. Acta Crystallogr E57: o1130–o1132

Pandiarajan S, Sridhar B, Rajaram RK (2002a) L-Methioninium nitrate. Acta Crystallogr E58: o882–o884

Pandiarajan S, Sridhar B, Rajaram RK (2002b) Bis(L-proline) hydrogen nitrate. Acta Crystallogr E58(8):o862–o864

Pandiarajan S, Sridhar B, Rajaram RK (2002c) Bis(L-proline) hydrogen(1+) perchlorate. Acta Crystallogr E58:o74–o76

Pandiarajan S, Umadevi M, Briget Mary M, Rajaram RK, Ramakrishnan V (2004) Infrared and Raman spectroscopic studies of L-methioninium nitrate. J Raman Spectrosc 35:907–913

Pandiarajan S, Umadevi M, Sasirekha V, Rajaram RK, Ramakrishnan V (2005a) FT-IR and FT-Raman spectral studies of bis(L-proline) hydrogen nitrate and bis(L-proline) hydrogen perchlorate. J Raman Spectrosc 36:950–961

Pandiarajan S, Umadevi M, Rajaram RK, Ramakrishnan V (2005b) Infrared and Raman spectroscopic studies of L-valine L-valinium perchlorate monohydrate. Spectrochim Acta A 62:630–636

Pandiarajan S, Rajaram RK, Ramakrishnan V (2005c) Vibrational spectral studies of β-alaninium perchlorate. J Raman Spectrosc 36:785–790

Panicker L, Sakuntala T (2010) DSC and Raman studies of diglycine-perchlorate (DGPCl) doped TGS. AIP Conf Proc 1313:334–336

Panicker L, Sakuntala T (2011) Raman spectroscopic and DSC studies of diglycine-perchlorate (DGPCl). Vib Spectrosc 57:270–274

Panicker L, Mathur P, Mobin SM (2011) Crystal structure and phase transition of diglycine perchlorate. J Chem Crystallogr 41:147–154

Parthasarathy R (1966) The structure of *L*-valine hydrochloride. Acta Crystallogr 21:422–426

Pepinsky R, Okaya Y, Jona F (1957) Ferroelectricity and structure of tri-glycine fluoroberyllate and isomorphs. Bull Am Phys Soc 2(4):220

Pepinsky R, Vedam K, Hoshino S, Okaya Y (1958) Ferroelectricity in di-glycine nitrate $(NH_2CH_2COOH)_2.HNO_3$. Phys Rev 111(2):430–432

Pereira Silva PS, Domingos SR, Ramos Silva M, Paixão JA, Matos Beja A (2008) N, N, V-Triphenylguanidinium 5-nitro-2,4-dioxo-1,2,3,4-tetrahydropyrimidin-1-ide. Acta Crystallogr E64:o1082–o1083

Periasamy M, Venkatraman L, Thomas KRJ (1997) New methods of resolution and enrichment of enantiomeric excesses of 1,1'-bi-2-naphthol. J Org Chem 62:4302–4306

Petrosyan AM (2004) Formation mechanisms of nonlinear optical crystalline salts of L-arginine and L-histidine. In: Proceedings of conference on "Laser Physics-2003", Ashtarak, 14–17 Oct 2003, pp 148–151

Petrosyan AM (2006a) Vibrational spectra of *L*-arginine tetrafluoroborate and *L*-arginine perchlorate. Vib Spectrosc 41:97–100

Petrosyan AM (2006b) L-arginine ethylsulfate. In: Proceedings of conference on Laser Physics-2005, Ashtarak, 11–14 Oct 2005, pp 123–126

Petrosyan HA (2007a) Nonlinear optical properties of L-histidine salts. In: Proceedings of conference on Laser Physics-2007, Ashtarak, 9–12 Oct 2007, pp 60–63

Petrosyan AM (2007b) Vibrational spectra of L-histidine perchlorate and L-histidine tetrafluoroborate. Vib Spectrosc 43:284–289

Petrosyan AM (2008a) Comment on the paper by Sankar R, R. Muralidharan, C.M. Rahgavan, Mohan Kumar R, Jayavel R, "Synthesis, growth, and characterization of nonlinear optical material *L*-arginine iodate crystal", Materials Letters 62(1), 133–136 (2008). Mater Lett 62:3305–3306

Petrosyan AM (2008b) Comment on the paper "Growth and characterization of a new non-linear optical *L*-histidinium dihydrogen phosphate single crystal", V. Rajendran et al., Materials Letters 61 (2007) 3477–3479. Mater Lett 62:3782

Petrosyan HA (2009) Preparation and investigation of nonlinear optical crystals on the basis of *L*-histidine. Dissertation, Yerevan, 139 p (In Russian)

Petrosyan AM (2010a) Salts of *L*-histidine as nonlinear optical materials: a review. J Cryst Phys Chem 1(1):33–56

Petrosyan AM (2010b) Comment on "Growth and characterization of glycine picrate single crystal" by Uma Devi T et al., Spectrochim. Acta A71 (2008) 340–343. Spectrochim Acta A 75:1176

Petrosyan AM, Ghazaryan VV (2009) Vibrational spectra of *L*-lysine monohydrochloride dihydrate and its two anhydrous forms. J Mol Struct 917(1):56–62

Petrosyan AM, Petrosyan HA (2002) New salts of L-histidine. In: Proceedings of conference on "Laser Physics-2002", Yerevan-Ashtarak, 15–18 Oct 2002, pp 83–88

Petrosyan AM, Shishkin VA (1996) Correlation between structural, infrared and nuclear quadrupole resonance data of iodates. Z Naturforsch 51a:667–671

Petrosyan AM, Sukiasyan RP (2008) Vibrational spectra of *L*-arginine nitrates. J Mol Struct 874:51–56

Petrosyan AM, Sukiasyan RP, Nalbandyan AG, Kochikyan RV, Aghajanyan AE (1995) Search for new nonlinear optical crystals. In: Proceedings of conference on Laser Physics-95, Ashtarak, 17–22 Oct 1995, pp 15–22 (In Russian)

Petrosyan AM, Terzyan SS, Burbelo VM, Sukiasyan RP (1998) Investigation of some new nonlinear optical crystals by means of NQR, IR and X-ray diffraction methods. Z Naturforsch 53a:528–536

Petrosyan AM, Sukiasyan RP, Terzyan SS, Burbelo VM (1999) Interaction of lysine with iodic acid. Acta Crystallogr B55:221–225

Petrosyan AM, Burbelo VM, Tamazyan RA, Karapetyan HA, Sukiasyan RP (2000a) NQR in alanine and lysine iodates. Z Naturforsch 55(1/2):199–206

Petrosyan AM, Sukiasyan RP, Karapetyan HA, Terzyan SS, Feigelson RS (2000b) Growth and investigation of new nonlinear optical crystals of LAP family. J Cryst Growth 213 (1/2):103–111

Petrosyan AM, Feigelson RS, Van Stryland EW, Sukiasyan RP, Karapetyan HA (2002) New class of nonlinear optical crystals among arginine salts. Proc SPIE 4751:217–222

Petrosyan AM, Karapetyan HA, Bush AA, Sukiasyan RP (2004a) Crystal structure and characterization of *L*-arginine dichloride monohydrate and *L*-arginine dibromide monohydrate. Mater Chem Phys 84(1):79–86

Petrosyan AM, Alchangyan SV, Petrosyan HA (2004b) Growth of L-histidine perchlorate crystals. Abstracts of 14-th international conference on Crystal Growth (ICCG-14), 9–13 Aug 2004, Grenoble, p 537

Petrosyan AM, Karapetyan HA, Sukiasyan RP, Aghajanyan AE, Morgunov VG, Kravchenko EA, Bush AA (2005a) Crystal structure and characterization of L-arginine chlorate and L-arginine bromate. J Mol Struct 752:144–152

Petrosyan AM, Sukiasyan RP, Karapetyan HA, Antipin MA, Apreyan RA (2005b) *L*-arginine oxalates. J Cryst Growth 275(1/2):e1927–e1933

Petrosyan HA, Karapetyan HA, Antipin MY, Petrosyan AM (2005c) Nonlinear optical crystals of *L*-histidine salts. J Cryst Growth 275:e1919–e1925

Petrosyan HA, Karapetyan HA, Petrosyan AM (2005d) *L*-histidine trifluoroacetate. Report presented at the conference on Laser Physics–2005, Ashtarak, 11–14 Oct 2005

Petrosyan HA, Karapetyan HA, Petrosyan AM (2006) L-histidine nitrates. J Mol Struct 794:160–167

Petrosyan HA, Karapetyan HA, Atanesyan AK, Petrosyan AM (2007) Nonlinear optical crystals of *L*-histidine oxalates. In: Proceedings of conference on "Laser Physics-2007", Ashtarak, 9–12 Oct 2007, pp 56–59

Petrosyan HA, Apreyan RA, Hovhannesyan AA, Atanesyan AK, Petrosyan AM (2009a) Nonlinear optical activity and spectroscopy of *L*-nitrohistidine monohydrate. J Contemp Phys (Armenian Acad Sci) 44(1):43–49

Petrosyan HA, Karapetyan HA, Ghazaryan VV (2009b) New approach for searching nonlinear optical materials among salts of amino acids. In: Proceedings of conference on "Laser Physics-2008", Ashtarak, 14–17 Oct 2008, pp 63–66

Petrosyan HA, Karapetyan HA, Atanesyan AK, Petrosyan AM (2010a) *L*-histidine sulfates. J Mol Struct 963(2–3):168–174

Petrosyan AM, Fleck M, Ghazaryan VV (2010b) Mixed salts of amino acids: syntheses, crystal structure and vibrational spectra of *L*-histidinium(2+) nitrate-perchlorate and *L*-histidinium (2+) nitrate-tetrafluoroborate. Z Kristallogr 225(9):388–395

Petrosyan AM, Ghazaryan VV, Fleck M (2011a) On the infrared spectrum of *L*-lysinium(2+) sulfate. Spectrochim Acta A 79:2020–2022

Petrosyan AM, Ghazaryan VV, Fleck M (2011b) Comments on the papers recently published by M.M. Khandpekar et al. Optics Commun 284:4295–4296

Petrosyan AM, Ghazaryan VV, Fleck M (2012a) On the existence of "bis-glycine maleate". J Cryst Growth 359:129–131

Petrosyan AM, Ghazaryan VV, Fleck M, Harutyunyan AV, Andriasyan LH, Brsikyan NA (2012b) Hexafluorosilicates of amino acids having anti-caries activity. Armenian Patent #2695

Petrosyan AM, Ghazaryan VV, Fleck M (2013a) On the existence of "*L*-threonine formate", "*L*-alanine lithium chloride" and "bis *L*-alanine lithium chloride" crystals. Spectrochim Acta A 105:623–625

Petrosyan AM, Fleck M, Ghazaryan VV (2013b) Reinvestigation of *L*-tryptophan picrate: establishment of the existence of the *L*-tryptophan *L*-tryptophanium dimeric cation. Spectrochim Acta A 104:486–491

Petrosyan AM, Fleck M, Ghazaryan VV (2014) New mixed salts of *L*-histidinium(2+) comprising hexafluorosilicate anion. J Cryst Growth. doi:http://dx.doi.org/10.1016/j.jcrysgro.2013.10.060

Petrov TG, Treivus EB, Kasatkin AP (1967) Vyrashivanie kristallov iz rastvorov (Growing crystals from solutions). Nedra, Leningrad (In Russian)

Petrov TG, Treivus EB, Kasatkin AP (1969) Growing crystals from solutions. Plenum Publishing Corporation, New York

Petrov TG, Treivus EB, Punin YO, Kasatkin AP (1983) Vyrashivanie kristallov izrastvorov (Growing crystals from solutions), 2nd edn. Nedra, Leningrad (In Russian)

Philip D, Aruldhas G (1995) Infrared, polarized Raman, and SERS spectra of betaine hydrogen oxalate monohydrate. J Solid State Chem 114:129–137

Piret P, Meunier-Piret J, Verbist J, Van Meerssche M (1972) Structure crystalline de derives d'acides amines VI. Le composé HI.2glycine. Bull. Soc. Chim. Belges 81:539–546

Ponmani SLM, Selvarajan P, Balasundari N, Jencylin D (2011) Growth and characterization of a novel bis β-alanine picrate (BBAP) single crystal. Int J Curr Res 2(1):086–091

Portalone G (2010) Supramolecular association in proton-transfer adducts containing benzamidinium cations. I. Four molecular salts with uracil derivatives. Acta Crystallogr C66:o295–o301

Prabu MM, Nagendra HG, Suresh S, Vijayan M (1996) X -ray studies on crystalline complexes involving amino acids and peptides XXXI. Effect of chirality on ionization state, stoichiometry and aggregation in the complexes of oxalic acid with *L*- and *DL*-histidine. J Biomol Struct Dyn 14:387–392

Pragasam AJA, Selvakumar S, Thamizharasan K, Prem Anand D, Sagayaraj P (2005a) Growth and optical characterization of Cu- and Mg- substituted *L*-arginine di phosphate single crystals. J Cryst Growth 280:271–278

Pragasam AJA, Selvakumar S, Madhavan J, Prem Anand D, Sagayaraj P (2005b) Effect of metallic substitution on the optical, mechanical and photoconducting properties of *L*-arginium diphosphate single crystals. Indian J Pure Appl Phys 43:463–468

Pragasam AJA, Madhavan J, Gulam Mohamed M, Selvakumar S, Ambujam K, Sagayaraj P (2006) Growth and characterization of amino acid (glycine and valine) substituted diphosphate single crystals. Opt Mater 29:173–179

Prakash M, Greetha D, Caroline ML (2011a) Crystal growth and characterization of *L*-phenylalaninium trichloroacetate – A new organic nonlinear optical material. Physica B 406:2621–2625

Prakash M, Greetha D, Caroline ML, Ramesh PS (2011b) Crystal growth, structural, optical, dielectric and thermal studies of an amino acid based organic NLO material: *L*-Phenylalanine *L*-phenylalaninium malonate. Spectrochim Acta A 83(1):461–466

Prasad GS, Vijayan M (1990) X-ray studies of crystalline complexes involving amino acids. XIX. Effects of change in chirality in the succinic acid with *DL* and *L*-arginine. Int J Pept Protein Res 35(4):357–364

Prasad GS, Vijayan M (1991) X-ray studies on crystalline complexes involving amino acids and peptides. XXIII. Variability in ionization state, conformation and molecular aggregation in the complexes of succinic acid with *DL*- and *L*-lysine. Acta Crystallogr B47:927–935

Prasad GS, Vijayan M (1993a) X -ray studies on crystalline complexes involving amino acids and peptides XXIV. Ionization states and novel aggregation patterns in the complexes of succinic acid with DL- and L-histidine. Biopolymers 33(2):283–292

Prasad GS, Vijayan M (1993b) X-ray studies on crystalline complexes involving amino acids and peptides. XXV. Structures of DL-proline hemisuccinic acid and glycyl-L-histidinium semisuccinate monohydrate and a comparative study of amino-acid and peptide complexes of succinic acid. Acta Crystallogr B49:348–356

Pratap JV, Vijayan M, Ravishankar R (2000) X-ray studies on crystalline complexes involving amino acids and peptides. XXXV. Invariance and variability in amino acid aggregation in the complexes of maleic acid with L-histidine and L-lysine. Acta Crystallogr B56:690–696

Preema CT, Kumar LB, Anurandha A, Aruna S, Joseph GP, Sagayaraj P (2006) Growth and characterization of nonlinear optical single crystals of *L*-arginine diiodate. J Cryst Growth 290:560–564

Preema CT, Aruna S, Nadhavan J, Ittyachan R, Pragasam JA, Jesudurai JGM, Prabha K, Sagayaraj P (2007) Growth and thermal studies of nonlinear optical *L*-argininium diiodate. L-argininium dinitrate and L-argininium hydrochloride bromide single crystals. Indian J Pure Appl Phys 45:591–595

Preethy Menon C, Philip J, Deepthy A, Bhat HL (2001) Thermal properties of glycine phosphate across ferroelectric phase transition: a photopyroelectric study. Mater Res Bull 36:2407–2414

Puccetti G, Perigaud A, Badan J, Ledoux I, Zyss J (1993) 5-Nitrouracil: a transparent and efficient nonlinear organic crystal. J Opt Soc Am 10:733–744

Pushilin MA, Gerasimenko AV, Davidovich RL (2007) Bis(*DL*-valinium) pentafluoridooxidoniobate(V). Acta Crystallogr E63:m2086

Qui DT, Vicat J, Durif A (1984) Structure of telluric acid-glycine (1:2) monohydrate, $Te(OH)_6.2C_2H_5NO_2.H_2O$, at 120K and determination of hydrogen positions. Acta Crystallogr C40(1):181–184

Qui DT, Lambert-Andron B, Boucherle JX (1987) Neutron refinement of telluric acid-glycine (1/2) monohydrate. Acta Crystallogr C43(5):907–909

Raghavalu T, Kumar KS, Kumar GR, Raj SG, Mohan R (2007) *L*-Lysinium trichloroacetate. Acta Crystallogr E63:o1706–o1707

Ragowska P, Cyrański MK, Sporzyński A, Ciesieelski A (2006) Evidence for strong heterodimeric interactions of phenylboronic acids with amino acids. Tetrahedron Lett 47:1389–1393

Raj KR, Murugakoothan P (2011) Studies on the effect of Ba^{2+} on growth, structural, morphology, optical and mechanical properties of *L*-valinium picrate. J Miner Mater Charact Eng 10 (10):973–987

Raj KR, Murugakoothan P (2013) Effect of Co^{2+} on the growth and physical properties of potential nonlinear optical *L*-asparaginium picrate crystal. Optik 124:2696–2700

Raj KR, Bhagavannarayana G, Murugakoothan P (2013) Studies on the effect of nickel on growth, structural, optical, electrical, thermal and mechanical properties of *L*-valinium picrate. Optik 124:493–500

Rajagopal K, Krishnakumar RV, Mostad A, Natarajan S (2001a) Glycinium maleate. Acta Crystallogr E57:o751–o753

Rajagopal K, Krishnakumar RV, Natarajan S (2001b) β-Alaninium maleate. Acta Crystallogr E57: o922–o924

Rajagopal K, Nandhini MS, Krishnakumar RV, Mostad A, Natarajan S (2002a) Sarcosinium maleate at 123K. Acta Crystallogr E58:o478–o480
Rajagopal K, Nandhini MS, Krishnakumar RV, Natarajan S (2002b) L-Alanine tartrate. Acta Crystallogr E58:o1306–o1308
Rajagopal K, Krishnakumar RV, Nandhini MS, Mostad A, Natarajan S (2002c) *DL*-Valinium trichloroacetate at 123K. Acta Crystallogr E58:o279–o281
Rajagopal K, Krishnakumar RV, Nandhini MS, Mostad A, Natarajan S (2003a) β-Alaninium trichloroacetate at 105K. Acta Crystallogr E59:o206–o208
Rajagopal K, Krishnakumar RV, Nandhini MS, Cameron TS, Natarajan S (2003b) L-Phenylalaninium trichloracetate. Acta Crystallogr E59:o1084–o1086
Rajagopal K, Krishnakumar RV, Mostad A, Natarajan S (2003c) *DL*-Methioninium trichloroacetate at 123K. Acta Crystallogr E59:o31–o33
Rajagopal K, Krishnakumar RV, Nandhini MS, Malathi R, Rajan SS, Natarajan S (2003d) L-Leucinium oxalate. Acta Crystallogr E59:o878–o880
Rajagopal K, Krishnakumar RV, Mostad A, Natarajan S (2003e) *L*-Prolinium trichloroacetate at 105K. Acta Crystallogr E59:o277–o279
Rajagopal K, Krishnakumar RV, Nandhini MS, Natarajan S (2003f) *L*-histidinium hemihydrochloride tartrate tartaric acid dihydrate. Acta Crystallogr E59:o955–o958
Rajagopal K, Franklin RD, Krishnakumar RV, Ravikumar K, Natarajan S (2004a) *DL*-Threonine trichloroacetate. Acta Crystallogr E60:o1355–o1357
Rajagopal K, Ramachandran E, Mostad A, Natarajan S (2004b) *DL*-Threoninium maleate at 150K. Acta Crystallogr E60:o386–o388
Rajagopal K, Krishnakumar RV, Natarajan S (2005) *DL*-Phenylalaninium trichloroacetate. Acta Crystallogr E61:o203–o205
Rajan Babu D, Jayaraman D, Mohan Kumar R, Ravi G, Jayavel R (2003) Growth aspects of semiorganic nonlinear optical *L*-arginine tetrafluoroborate single crystals. J Cryst Growth 250:157–161
Rajendran KV, Jayaraman D, Jayavel R, MohanKumar R, Ramasamy P (2001) Growth and characterization of non-linear optical L-hystidine tetrafluoroborate (L-HFB) single crystals. J Cryst Growth 224(1–2):122–127
Rajendran KV, Jayaraman D, Jayavel R, Ramasamy P (2003a) Effect of pH on the growth and characterization of L-HFB single crystal. J Cryst Growth 254(3–4):461–468
Rajendran KV, Rajasekaran R, Jayaraman D, Jayavel R, Ramasamy P (2003b) Experimental determination of metastable zone width, induction period, interfacial energy and growth of non-linear optical L-HFB single crystals. Mater Chem Phys 81:50–55
Rajendran KV, Jayaraman D, Jayavel R, Ramasamy P (2003c) Growth and characterization of nonlinear optical crystal: *L*-histidinium bromide. J Cryst Growth 255:361–368
Rajendran V, Shamala D, Loganayaki M, Ramasamy P (2007) Growth and characterization of a new non-linear optical L-histidinium dihydrogen phosphate single crystal. Mater Lett 61:3477–3479
Rajkumar BJM, Ramakrishnan V (2000) Infrared and Raman spectra of *L*-valine nitrate and *L*-leucine nitrate. J Raman Spectrosc 31:1107–1112
Rajkumar BJM, Ramakrishnan V (2001) Vibrational spectroscopic study of *DL*-methionine dihydrogen phosphate. Spectrochim Acta A 57:247–254
Rajkumar BJM, Ramakrishnan V (2002) Infrared and laser Raman studies of *L*-phenylalanine *L*-phenylalaninium perchlorate and bis(*DL*-phenylalaninium) sulphate monohydrate. Spectrochim Acta A 58(9):1923–1934
Rajkumar BJM, Ramakrishnan V, Rajaram RK (1998) Infrared and Raman spectra of DL-aspartic acid nitrate monohydrate. Spectrochim Acta A 54:1527–1532
Rajkumar BJM, Ramakrishnan V, Bahadur SA (1999) Infrared and Raman spectra of *DL*-histdininium dinitrate and *L*-histidinium sulphamate. J Raman Spectrosc 30:589–594
Ramajothi J, Dhanuskodi S (2003) Optical and microhardness studies of semiorganic nonlinear optical material: *L*-histidine tetrafluoroborate. Cryst Res Technol 38:986–991

Ramajothi J, Dhanuskodi S (2006) Crystal growth, thermal and optical studies on phase matchable new organic NLO material for blue-green laser generation. J Cryst Growth 289:217–223

Ramajothi J, Dhanuskodi S (2007) Crystal growth, thermal and optical studies on a semiorganic nonlinear optical material for blue-green laser generation. Spectrochim Acta A 68:1213–1219

Ramajothi J, Dhanuskodi S, Akkurt M (2008) Crystal growth of phase matchable new organic nonlinear material for UV laser generation. Spectrochim Acta A 69:1271–1276

Ramanathan CR, Periasamy M (1998) Resolution of C_2-symmetric 9,10-dihydro-9,10, ethanoanthracene-11,12-dicarboxylic acid and 2,3-diphenylsuccinic acid using (*S*)-proline. Tetrahedron Asymmetry 9:2651–2656

Ramaswamy S, Sridhar B, Ramakrishnan V, Rajaram RK (2001) *L*-argininium dinitrate. Acta Crystallogr E57:o872–o874

Ramaswamy S, Sridhar B, Ramakrishnan V, Rajaram RK (2002) *L*-ornithine nitrate. Acta Crystallogr E58:o646–o648

Ramaswamy S, Umadevi M, Rajaram RK, Ramakrishnan V (2003a) Infrared and Raman spectral studies of *L*-ornithine nitrate. J Raman Spectrosc 34:806–812

Ramaswamy S, Rajaram RK, Ramakrishnan V (2003b) Vibrational spectroscopic studies of *L*-argininium dinitrate. J Raman Spectrosc 34:50–56

Ramaswamy S, Sridhar B, Ramakrishnan V, Rajaram RK (2004a) Bis(*DL*-methioninium) sulfate. Acta Crystallogr E60:o1691–o1693

Ramaswamy S, Sridhar B, Ramakrishnan V, Rajaram RK (2004b) Bis(*L*-ornithinium) chloride nitrate sulfate. Acta Crystallogr E60:o768–o770

Ramaswamy S, Sridhar B, Ramakrishnan V, Rajaram RK (2005) Vibrational spectra of bis (L-ornithinium) chloride nitrate sulfate. J Raman Spectrosc 36(1):12–17

Ramesh Babu R, Vijayan N, Gopalakrishnan R, Ramasamy P (2006a) Growth and characterization of *L*-lysine monohydrochloride dihydrate (*L*-LMHCl) single crystal. Cryst Res Technol 41:405–410

Ramesh Babu R, Sethuraman K, Gopalakrishnan R, Ramasamy P (2006b) Growth of *L*-lysine monohydrochloride dihydrate bulk single crystal by Sankaranarayanan-Ramasamy (SR) method. J Cryst Growth 297:356–360

Ramesh Babu R, Sethuraman K, Vijayan N, Bhagavannarayana G, Gopalakrishnan R, Ramasamy P (2006c) Etching and dielectric studies on *L*-lysine monohydrochloride dihydrate single crystal. Cryst Res Technol 41:906–910

Ramesh Kumar G, Gokul Raj S, Mohan R, Jayavel R (2006) Structural, thermal, linear and nonlinear optical studies on new NLO crystal: *DL*-Threonium trichloroacetate. J Rare Earths 24:249–252

Ramos Silva M, Matos Beja A, Paixão JA, Alte da Veiga L, Martin-Gil J (1999a) Crystal structure of betainium hexafluorotantalum(IV) dihydrate, $(C_5H_{12}NO_2)_2TaF_6.2H_2O$. Z Kristallogr NCS 214(4):477–478

Ramos Silva M, Matos Beja A, Paixão JA, Alte da Veiga L, Martin-Gil J (1999b) L-argininum tris (acetato-O, O′)dioxouranium(III)-acetic acid-water (1/1/1). Acta Crystallogr C55:2039–2041

Ramos Silva M, Matos Beja A, Paixão JA, Alte da Veiga L, Martin-Gil J (1999c) Crystal structure of *L*-arginine diphenylacetate $C_6H_{15}N_4O_2^+ \cdot C_{14}H_{11}O_2^-$. Z Kristallogr NCS 214(3):326–328

Ramos Silva M, Paixão JA, Matos Beja A, Alte da Veiga L (2000a) Crystal structure of the nonlinear optical compound *L*-arginine fluoride. J Chem Cryst 30(6):411–414

Ramos Silva M, Paixão JA, Matos Beja A, Alte da Veiga L (2000b) Very short F-H$\cdots$F hydrogen bond in *L*-argininium fluoride hydrogen fluoride. Acta Crystallogr B56(1):104–106

Ramos Silva M, Paixão JA, Matos Beja A, Alte da Veiga L (2000c) Crystal structure of L-argininium hexafluorovanadate trihydrate, $(C_6H_{15}N_4O_2)_3VF_6.3H_2O$. Z Kristallogr NCS 215:178–180

Ramos Silva M, Paixão JA, Matos Beja A, Alte da Veiga L (2001) Strong hydrogen-bonded amino acid dimers in *L*-alanine alaninium nitrate. Acta Crystallogr C57(7):838–840

Ramos Silva M, Paixão JA, Matos Beja A (2003) *L*-argininium bis (trifluoroacetate). Acta Crystallogr E59:o1912–o1914

Ramos Silva M, Matos Beja A, Paixão JA (2004) Very short O-H. . .hydrogen bond in L-histidinium difluoride. J Fluor Chem 125:695–699

Ramos Silva M, Paixão JA, Matos Beja A (2005a) Crystal structure of L-tryptophanium phosphite, $(C_{11}H_{13}N_2O_2)[H_2PO_3]$. Z Kristallogr NCS 220:487–488

Ramos Silva M, Matos Beja A, Paixão JA (2005b) Tris(L-argininium) hexafluorochromate(III) trihydrate. Acta Crystallogr E61:m2459–m2461

Rani JS, Mahadevan CK, Antony Arockiaraj M, Rajasekar S, Vimalan M (2013) Growth and characterization of semiorganic nonlinear optical LHHCl-LHB mixed crystals. Archiv Appl Sci Res 5(2):213–221

Rao ST (1969) Crystal structure of *L*-valine hydrochloride monohydrate. Z Kristallogr 128:339–351

Rao SM, Batra AK, Cao C, Lal RB (1990) Etch pit study of different crystallographic faces of L-arginine phosphate (LAP). J Cryst Growth 106:481–482

Rashkovich LN, Shekunov BY (1991) Study of the growth mechanism of *L*-arginine chloride monohydrate (LACh) crystals. J Cryst Growth 112:183–191

Ratajczak H, Pietrashko A, Baran J, Barnes AJ, Tarnavski YU (1994) Structure and polarized IR and Raman spectra of the solid complex of bis(betaine)-sulphuric acid. J Mol Struct 327 (2–3):297–312

Ratajczak H, Barycki J, Pietraszko A, Baran J, Debrus S, May M, Venturini J (2000) Preparation and structural study of a novel nonlinear molecular material: the L-histidinium dihydrogen-arsenate orthoarsenic acid crystal. J Mol Struct 526:269–278

Rathyalakshmi R, Bhagavannarayana G, Ramasamy P (2009) Growth and characterization of a new NLO material: *L*-glutamic acid hydro bromide [*L*-GluHBr]. Mater Res Bull 44:1097–1101

Ravi G, Srinivasan K, Anbukumar S, Ramasamy P (1994) Growth and characterization of sulphate mixed L-arginine phosphate and ammonium dihydrogen phosphate/potassium dihydrogen phosphate mixed crystals. J Cryst Growth 137:598–604

Ravikumar B, Sridhar B, Rajaram RK (2001a) *DL*-Phenylalaninium dihydrogen phosphate. Acta Crystallogr E57:o1078–o1080

Ravikumar B, Sridhar B, Rajaram RK (2001b) *D*-Serinium *D*-serine nitrate. Acta Crystallogr E57 (8):o682–o684

Ravikumar B, Sridhar B, Rajaram RK (2002a) *L*-Phenylalanine *L*-phenylalaninium dihydrogen-phosphate. Acta Crystallogr E58:o123–o125

Ravikumar B, Sridhar B, Rajaram RK (2002b) *DL*-Threonine dihydrogen phosphate. Acta Crystallogr E58:o1185–o1187

Ravikumar B, Sridhar B, Rajaram RK (2002c) *DL*-Valinium dihydrogen phosphate. Acta Crystallogr E58:o879–o881

Ravikumar B, Sridhar B, Rajaram RK (2004) *L*-Ornithinium sulfate monohydrate. Acta Crystallogr E60:o2093–o2095

Ravikumar B, Rajaram RK, Ramakrishnan V (2006) Raman and IR spectral studies of *L*-phenyl-alanine *L*-phenylalaninium dihydrogenphosphate. J Raman Spectrosc 37:597–605

Ravishankar R, Nagasuma C, Vijayan M (1998) X-ray studies of crystalline complexes involving amino acids. XXXIV. Novel mode of aggregation, interaction patterns and chiral effects in the maleic acid complexes of *DL*- and *L*-arginine. J Biomol Struct Dyn 15(6):1093–1100

Rayar SL, Selvarajan P (2010) Structural, mechanical, FTIR, SHG and thermal studies of L-HTFA single crystals grown by solution method. Recent Res Sci Technol 2(10):77–81

Recker R, Wallrafen F, Haussühl S, Krumbe W (1986) Crystal growth and physical properties of *L*-glutamic acid halides *L*-$C_5H_9NO_4$.HX (X=Cl, Br, I). Z Kristallogr 177:125–132

Remédios CMR, Paraguassu W, Lima JA Jr, Freire PTC, Mendes-Filho J, Melo FEA, de Menezes AS, dos Santos AO, Cardoso LP, Miranda MAR (2008) Effect of Ni(II) doping on the structure of *L*-histidine hydrochloride monohydrate crystals. J Phys Condens Matter 20:275209 (6 p)

Renuka N, Vijayan N, Brijesh Rathi, RameshBabu R, Nagarajan K (2012) Synthesis, growth and optical properties of semi organic non linear optical single crystal: *L*-Arginine acetate. Optik 123:189–192

Rieckhoff KE, Peticolas WL (1965) Optical second-harmonic generation in crystalline amino acids. Science 147(3658):610–611

Riscob B, Kushwaha SK, Shakir M, Nagarajan K, Maurya KK, Haranath D, Roy SDD, Bhagavannarayana G (2011a) Crystalline perfection, optical and dielectric studies on L-histidine nitrate: a nonlinear optical material. Physica B 406:4440–4446

Riscob B, Shakir M, Sudar JK, Natarajan S, Wahab MA, Bhagavannarayana G (2011b) Synthesis, growth, crystal structure and characterization of a new organic material: glycine glutaric acid. Spectrochim Acta A 78:543–548

Riscob B, Shakir M, Vijayan N, Maurya KK, Wahab MA, Bhagavannarayana G (2012) Unidirectional crystal growth and crystalline perfection of L-arginine phosphate monohydrate. J Appl Crystallogr 45:679–685

Robert R, Justin Raj C, Krishnan S, Uthrakumar R, Dinakaran S, Jerome Das S (2010) Spectral, optical and mechanical studies on *L*-histidine hydrochloride (LHC) single crystals grown by unidirectional growth technique. Physica B 405:3248–3252

Robertson G, Dunn MH (1993) Excimer pumped deuterated *L*-arginine phosphate optical parametric oscillator. Appl Phys Lett 62:3405–3407

Robson W, Selim ASM (1952) The direct isolation of arginine and lysine picrates from gelatin hydrolysates. Biochem J 52:318–319

Rodrigues VH, Paixão JA, Costa MMRR, Matos Beja A (2000) Sarcosinium trifluoroacetate. Acta Crystallogr C56:1053–1055

Rodrigues VH, Paixão JA, Costa MMRR, Matos Beja A (2001a) Conformation of cationic N, N-dimethylglycine in dimethylglycinium trifluoroacetate. Acta Crystallogr C57:417–420

Rodrigues VH, Paixão JA, Costa MMRR, Matos Beja A (2001b) Betainium trifluoroacetate. Acta Crystallogr C57:761–763

Rodrigues VH, Paixão JA, Costa MMRR, Matos Beja A (2001c) Betaine betainium hydrogen oxalate. Acta Crystallogr C57(2):213–215

Rodrigues VH, Paixão JA, Costa MMRR, Matos Beja A (2002) Glycinium trifluoroacetate. Acta Crystallogr C58:o658–o660

Rodrigues VH, Matos Beja A, Paixão JA, Costa MMRR (2006a) Glycinium trichloroacetate. Acta Crystallogr C62:o71–o72

Rodrigues VH, Costa MMRR, De Matos Gomes E, Nogueira E, Belslsey MS (2006b) *L*-Phenylalanine-4-nitrophenol (1/1). Acta Crystallogr C62:o699–o701

Rodrigues VH, Costa MMRR, Dekola T, De Matos Gomes E (2009) Bis[glycinium(0.5+)] perrhenate. Acta Crystallogr E65(1):m19

Rodrigues VH, Costa MMRR, Belsley M, De Matos Gomes E (2012) *L*-tryptophan 4-nitrophenol trisolvate. Acta Crystallogr E68:o920

Rolski S, Zdunska A, Popko A (1965) *L*-Leucine hydrochlorides. Acta Polon Pharm 20:141–145, CA:62:1737d

Román P, Gutiérrez-Jorrilla JM, Luque A, Vegas A (1987) The MoO3-histidine-HClO4 system. Synthesis, spectroscopic, and structural study of *L*-histidinium perchlorate. J Crystallogr Spectrosc Res 17:585–595

Roy S, Singh DD, Vijayan M (2005) X -ray studies on crystalline complexes involving amino acids and peptides. XLII. Adipic acid complexes of L- and DL- arginine and supramolecular association in arginine-dicarboxylic acid complexes. Acta Crystallogr B61(1):89–95

Ruby A, Raj SAC (2012) Synthesis, growth, spectroscopic, optical and thermal studies of glycinium maleate single crystals. Adv Appl Sci Res 3:1677–1685

Ruby Nirmala L, Prakash JTH (2013) Investigation on the influence of foreign metal ions in crystal growth and characterization of *L*-Alaninium Maleate (LAM) single crystals. Spectrochim Acta A 115:778–782

Rudert R, Schulz B, Reck G, Vollhardt D, Kriwanek J (2000) N-n-Alkyl N, N-dimethylammonioacetic acid bromides: the first complete series of crystal and molecular structure determinations of an amphiphilic compound with alkyl chain lengths n = 1,. . ., 16. Acta Crystallogr B56:124–131

Saari AL, Anderson RH (1977) Amino acid derivatives. US Patent 406438

Saenger W, Wagner KG (1972) An X-ray study of the hydrogen bonding in the crystalline *L*-arginine phosphate monohydrate complex. Acta Crystallogr B28:2237–2244

Sajan D, Binoy J, Pradeep B, Krishna KV, Kartha VB, Hubert Joe I, Jayakumar VS (2004) NIR-FT Raman and infrared spectra and ab initio computations of glycinium oxalate. Spectrochim Acta A 60:173–180

Sajan D, Joseph L, Vijayan N, Karabacak M (2011) Natural bond orbital analysis, electronic structure, non-linear properties and vibrational spectral analysis of L-histidinium bromide monohydrate: a density functional theory. Spectrochim Acta A 81:85–98

Salunke DM, Vijayan M (1981) Specific interactions involving guanidyl group observed in crystal structures. Int J Pept Protein Res 18:348–351

Salunke DM, Vijayan M (1982) X-ray studies of crystalline complexes involving amino acids and peptides. VII. *L*-arginine *L*-aspartate. Acta Crystallogr B38(4):1328–1330

Salunke DM, Vijayan M (1983) X-ray studies on crystalline complexes involving amino acids and peptides. IX. Crystal structure of L-ornithine L-aspartate hemihydrate. Int J Pept Protein Res 22:154–160

Salunke DM, Vijayan M (1984) Crystal structure of the amino acid-vitamin complex lysine pantothenate. Biochim Biophys Acta (BBA)-Gen Subjects 798(2):175–179, Preliminary report has been made in: Acta Crystallogr A 37:C65 (1981)

Salvestrini JP, Zaccaro J, Ibanez A, Fontana MD (1998) Investigation of electrooptic modulation from organic-inorganic crystals. J Appl Phys B67:761–763

Sangeetha K, Ramesh Babu R, Ramamurthi K, Jai Prakash, Khan SA (2011a) Spectral studies on Ag8+ ions irradiated LAHCl.H_2O and LAHBr.H_2O single crystals. Spectrochim Acta A 79:884–888

Sangeetha K, Ramesh Babu R, Bhagavannarayana G, Ramamurthi K (2011b) Structural, spectral, optical and dielectric properties of copper and glycine doped LAHCl single crystals. Spectrochim Acta A 79:1017–1023

Sangeetha K, Ramesh Babu R, Bhagavannarayana G, Ramamurthi K (2011c) Unidirectional growth and characterization of *L*-arginine monohydrochloride monohydrate single crystals. Mater Chem Phys 130:487–492

Sangwal K, Sanz F (1997) On the formation of dislocation hallow cores on cleaved {100} faces of *L*-arginine phosphate monohydrate single crystals. Cryst Res Technol 32:261–270

Sangwal K, Veintemillas-Verdaguer S, Torrent-Burgués J, Rodriguez-Clemente R (1994) Decoration of growth and dissolution steps on the surface of *L*-arginine phosphate monohydrate crystals. J Cryst Growth 140:447–450

Sangwal K, Veintemillas-Verdaguer S, Torrent-Burgués J (1995a) Growth habit and surface morphology of *L*-arginine phosphate monohydrate single crystals. J Cryst Growth 155:135–143

Sangwal K, Veintemillas-Verdaguer S, Torrent-Burgués J (1995b) On the formation of dislocation etch pits on *L*-arginine monohydrate single crystals. J Cryst Growth 154:364–369

Sangwal K, Veintemillas-Verdaguer S, Torrent-Burgués J (1996) Study of growth dislocations in *L*-arginine phosphate monohydrate single crystals by chemical etching. J Mater Sci 31:6299–6304

Sangwal K, Torrent-Burgués J, Sanz F, Gorostiza P (1997a) Atomic force microscopic study of step bunching and macrostep formation during the growth of *L*-arginine phosphate monohydrate single crystals. J Cryst Growth 172:209–218

Sangwal K, Servat J, Sanz F, Torrent-Burgués J (1997b) AFM study of etching of cleaved {100} faces of *L*-arginine phosphate monohydrate single crystals I Dislocation etch pits and step bunching. J Cryst Growth 180:263–273

Sangwal K, Torrent-Burgués J, Sanz F, Servat J (1997c) AFM study of etching of cleaved {100} faces of *L*-arginine phosphate monohydrate single crystals II. Two-dimensional nucleation formation of spiral elevations and decoration layers. J Cryst Growth 180:274–279

Sangwal K, Torrent-Burgués J, Sanz F, Servat J (1997d) Observations of cleavage steps, slip traces and dislocation hallow core on cleaved {100} faces of *L*-arginine phosphate monohydrate single crystals by atomic force microscopy. Surf Sci 374:387–396

Sankar R, Muralidharan R, Rahgavan CM, Mohan Kumar R, Jayavel R (2008a) Synthesis, growth, and characterization of nonlinear optical material *L*-arginine iodate crystal. Mater Lett 62:133–136

Sankar R, Muralidharan R, Raghavan CM, Jayavel R (2008b) Structural, thermal, mechanical and optical properties of *L*-arginine diiodate crystal: a new nonlinear optical material. Mater Chem Phys 107:51–56

Sankar D, Menon VR, Sagayaraj P, Madhavan J (2010) Thermal, mechanical, electrical, linear and nonlinear optical properties of *L*-arginine dihydrofluoride single crystal. Physica B 405:192–197

Sankaranarayanan K, Ramasamy P (2005) Unidirectional seeded single crystal growth from solution of benzophenone. J Cryst Growth 280:467–473

Santana RC, Carvalho JF, Amaral SR, Vencato I, Pelegrini F, Terrile MC, Hernandes AC, Calvo R (2002) Electron spin resonance of Cu^{2+} impurities in *L*-arginine phosphate monohydrate single crystals. J Phys Chem Solids 63:1857–1862

Santarsiero BD, Marsh RE (1983) N, N-dimethylglycine hydrochloride. J Crystallogr Spectrosc Res 13:245–251

Saraswathi NT, Vijayan M (2001) X -ray studies on crystalline complexes involving amino acids and peptides. XXXVIII. Crystal structures of the complexes of L-arginine and L-histidine with glutaric acid and a comparative study of amino-glutaric acid complexes. Acta Crystallogr B57:842–849

Saraswathi NT, Vijayan M (2002a) X -ray studies on crystalline complexes involving amino acids and peptides. XL. Conformational variability, recurring and new features of aggregation and effect of chirality in the malonic acid complexes of *DL*- and *L*-arginine. Acta Crystallogr B58:1051–1056

Saraswathi NT, Vijayan M (2002b) X -ray studies on crystalline complexes involving amino acids and peptides XXXIX. Crystal structures of malonic acid complexes of *DL*- and *L*-histidine. Preservation of aggregation pattern on reversal of chirality. Acta Crystallogr B58(4):734–739

Saraswathi NT, Manoj M, Vijayan M (2001) X-ray studies on crystalline complexes involving amino acids and peptides. XXXVII. Novel aggregation patterns and effect of chirality in the complexes of DL- and L-lysine with glutaric acid. Acta Crystallogr B57:366–371

Saraswathi NT, Roy S, Vijayan M (2003) X-ray studies on crystalline complexes involving amino acids and peptides. XLI. Commonalities in aggregation and conformation revealed by the crystal structures of the pimelic acid complexes of L-arginine and DL-lysine. Acta Crystallogr B59:641–646

Saraswati V, Vijayaghavan R (1967) Proton and deuteron magnetic resonances in glycine compounds. J Phys Soc Jpn 23(3):590–597

Sasaki T, Yokotani A, Fujioka K, Yamanaka T, Nakai S (1989) Synthesis and second harmonic generation of deuterated L-arginine phosphate monohydrate crystal. Tech Rep Osaka Univ 39(1950):25–32

Sathyalakshmi R, Kannan V, Bairava Ganesh R, Ramasamy P (2007) Synthesis, growth and characterization of single crystals of pure and thiourea doped *L*-glutamic acid hydrochloride. Cryst Res Technol 42:78–83

Sato Y (1966) Optical study of X-ray irradiation effect on the phase transition in triglycine sulfate and diglycine nitrate. J Chem Phys 45(1):275–282

Sato S (1968) Crystal structure and phase transition of $(glycine)_2HNO_3$. I. Crystal structure determination of ferro- and paraelectric phases. J Phys Soc Jpn 25(1):185–201

Schaak G (1990) Experimental results on phase transitions in betaine compounds. Ferroelectrics 104(1):147–158

Schaffrin R, Trotter J (1970) Crystal structure of *DL*-ornithine hydrobromide. J Chem Soc (A) 25–27

Scheins S, Dittrich B, Messerschmidt M, Paulmann C, Luger P (2004) Atomic volumes and charges in a system with a strong hydrogen bond: *L*-tryptophan formic acid. Acta Crystallogr B60:184–190

Schildkamp W, Spilker J (1984) Structural and antiferroelectric phase transitions in betaine phosphate, $(CH_3)_3NCH_2COO.H_3PO_4$. Z Kristallogr 168:159–171

Schildkamp W, Schäfer G, Spilker J (1984) The crystal structure of ferroelastic betaine arsenate, $(CH_3)_3NCH_2COO.H_3AsO_4$, compared with antiferrodistortive betaine phosphate, $(CH_3)_3NCH_2COO.H_3PO_4$. Z Kristallogr 168:187–195

Schmid H (1968) Addition compounds of amino acids and hydrofluoric acid or soluble fluorides, and method of preparing the same. US Patent # 3,413,326

Seely O Jr (1966a) Dissertation. CSD code: LTYRHI. University of Illinois, Urbana

Seely O Jr (1966b) The crystal structure analysis of *L*-arginine dihydriodide and *L*-tyrosine hydriodide. PhD thesis, University of Illinois, Urbana, 183 p

Selkti M, Coleman AW, Nicolis I, Douteau-Guevel N, Villain F, Tomas A, de Rango C (2000) The first example of a substrate spanning the calix[4]arene bilayer: the solid state complex of *p*-sulfonatocalix[4]arene with *L*-lysine. Chem Commun 161–162

Selvaraj M, Thamotharan S, Roy S, Vijayan M (2007) X-ray studies on crystalline complexes involving amino acids and peptides. XLIV. Invariant features of supramolecular association and chiral effects in the complexes of arginine and lysine with tartaric acid. Acta Crystallogr B63:459–468

Selvaraju K, Kirubavathi K (2013) Synthesis, growth and characterization of a nonlinear optical crystal: Bis *L*-proline hydrogen nitrate. Spectrochim Acta A 115:537–543

Selvaraju K, Valluvan R, Kumararaman S (2006a) New nonlinear optical material: glycine hydrofluoride. Mater Lett 60:2848–2850

Selvaraju K, Valluvan R, Kumararaman S (2006b) Experimental determination of metastable zonewidth, induction period, interfacial energy and growth of non-linear optical *L*-glutamic acid hydrochloride single crystals. Mater Lett 60:1565–1569

Selvaraju K, Valluvan R, Kirubavathi K, Kumararaman S (2007) Investigation on the nucleation kinetics of *L*-arginine acetate single crystals. Mater Lett 61:3041–3044

Selvaraju K, Kirubavathi K, Vijayan N, Kumararaman S (2009) Synthesis, growth, structural, spectroscopic, optical, thermal and mechanical studies of a semi-organic nonlinear optical crystal: *L*-glutamic acid hydrochloride. Mod Phys Lett B23:861–869

Senthil A, Ramesh Babu R, Balamurugan N, Ramasamy P (2009a) Unidirectional growth of largest *L*-LMHCl dihydrate crystal by SR method. J Cryst Growth 311:544–547

Senthil S, Pari S, Joseph GP, Sagayaraj P, Madhavan J (2009b) Thermal, mechanical, electrical, linear and nonlinear optical properties of a nonlinear optical L-ornithine monohydrochloride single crystal. Physica B 404:2336–2339

Senthilkumar S, Briget Mary M, Ramakrishnan V (2007) Infrared and Raman spectroscopic studies of *L*-valinium picrate. J Raman Spectrosc 38:288–294

Sequeira A, Rajagopal H, Chidambaram R (1972) A neutron diffraction study of the structure of *L*-glutamic acid.HCl. Acta Crystallogr B28:2514–2519

Sethuram M, Rajasehakaran MV, Dhandapani M, Amirthaganesan G, NizamMohideen M (2013) *L*-histidinium dipicrate dihydrate. Acta Crystallogr E69:o957–o958

Sethuram M, Dhandapani M, Sethu Raman M, Amirthaganesan G, Senthilkumar K (2014) Structure and spectral properties of *L*-histidinium dipicrate dihydrate. Spectrochim Acta A 118:102–111

Shamala N, Venkatesan K (1972) *L*-glutamine hydrochloride, $C_5H_{11}ClN_2O_3$. Cryst Struct Commun 1:227–229

Shan Y, Huang SD (1999) Crystal structure of *L*-cysteine *L*-tartrate monohydrate, $C_7H_{15}NO_9S$. Z Kristallogr NCS 214:41–42

Shankar VS, Sankar R, Siddheswaran R, Jayavel R, Murugakoothan P (2008) Growth and characterization of tetra *L*-lysine alanine mono hydrochloride dihydrate (TLAMHCl), a new semiorganic nonlinear optical single crystal. Mater Chem Phys 109:119–124

Shankar VS, Siddheswaran R, Sankar R, Jayavel R, Murugakoothan P (2009) Growth and characterization of new semiorganic nonlinear optical single crystal *L*-Phenylalanine *L*-Phenylalaninium perchlorate (LPPAPC). Mater Lett 63(3–4):363–365

Shanmugavadivu R, Ravi G, Nixon Azariah A, Haja Hameed AS, Thenappan T (2004a) Synthesis, growth and characterization of new mixed analogs of LAP family crystals. Mater Sci Eng B 113:269–273

Shanmugavadivu R, Ravi G, Jayavel R, Mohankumar R, Nixon Azariah A (2004b) Growth and characterization of L-arginine fluro phosphate-a new additive for fluent growth microbes. J Cryst Growth 271:252–256

Sharma A, Thamotharan S, Roy S, Vijayan M (2006) X-ray studies on crystalline complexes involving amino acids and peptides. XLIII. Adipic acid complexes of L- and DL-lysine. Acta Crystallogr C62:o148–o152

Shikanai F, Komukae M, Czapla Z, Osaka T (2002) Crystal structure of $NH_3CH_2COOHH_2PO_3$ in the ferroelectric phase. J Phys Soc Jap 71:498–503

Shiraiwa T, Mitazaki H, Watanabe T, Kurokawa H (1997) Optical resolution by preferential crystallization of *DL*-methionine hydrochloride. Chirality 9(1):48–51

Shkir M, Abbas H (2014) Physico chemical properties of *L*-asparagine *L*-tartaric acid single crystals: a new nonlinear optical material. Spectrochim Acta A 118:172–176

Shakir M, Kushwaha SK, Maurya KK, Manju Arora, Bhagavannarayana G (2009a) Growth and characterization of glycine picrate – remarkable second-harmonic generation in centrosymmetric crystal. J Cryst Growth 311:3871–3875

Shakir M, Singh BK, Kumar B, Bhagavannarayana G (2009b) Ferroelectricity in glycine picrate: an astonishing observation in a centrosymmetric crystal. Appl Phys Lett 95:252902

Shakir M, Kushwaha SK, Maurya KK, Manju Arora, Bhagavannarayana G (2010) Erratum to "Growth and characterization of glycine picrate – remarkable secondharmonic generation in centrosymmetric crystal" by Mohd. Shakir, S.K. Kushwaha, K.K. Maurya, Manju Arora, G. Bhagavannarayana [J Cryst Growth 311:3871–3875 (2009)]. J Cryst Growth 312:1663

Shtemenko AV, Kozhura OV, Pasenko AA, Domasevitch KV (2003) New octachlorodirhenate (III) salts: solid state manifestation for a certain conformational flexibility of the $[Re_2Cl_8]^{2-}$ ion. Polyhedron 22:1547–1552

Shternberg AA (1961) Kristally v prirode i tekhnike (Crystals in nature and engineering). Uchpedgiz, Moscow (In Russian)

Siddheswaran R, Rankar R, Rathnakumari M, Murugakoothan P, Jayavel R, Sureshkumar P (2006) Growth and characterization of a new semi-organic non-linear optical crystal *L*-arginine hydrochlorofluorido monohydrate (LAHClF). Surf Rev Lett 13:803–808

Siddheswaran R, Shankar VS, Jayavel R, Murugakoothan P (2009) Studies on growth and properties of *L*-tyrosine hydrochloride (LTHCL) crystal. Int J Mater Sci 4(3):205–213

Silvestrova IM, Nabakhtiani GN, Kozin VB, Kuznetsov VA, Pisarevskii YV (1992) Elastic, piezoelectric and dielectric properties of LAP crystals. Kristallografiya 37:1535–1541 (In Russian). (English translation: Sov Phys-Crystallogr 37:831–834 (1992)

Singh N, Kumar B (2011) Structural, optical and dielectric studies of glycinium trifluoroacetate single crystal. Physica B 406:2152–2157

Singh S, Lal B (2008) Crystalline fiber growth of dye-doped *L*-arginine phosphate by the laser-heated pedestal growth technique. J Cryst Growth 310:2039–2042

Singh N, Singh BK, Gupta MK, Kumar B (2012) Optical and dielectric studies of solution grown glycinium maleate single crystal. Optics Commun 285:659–664

Sinha N, Sahas, Singh BK, Kumar K, Singh N, Gupta MK, Budakoti GC, Kumar B (2009) Solution growth and comparative characterization of L-HFB single crystals. Cryst Res Technol 44:167–172

Sivabarathy M, Natarajan S, Ramakrishnan SK, Ramachandran K (2004) Phase transition in L-alaninium oxalate by photoacoustics. Bull Mater Sci 27(5):403–407

Sivanesan T, Natarajan V, Pandi S (2010) Growth and characterization of L-glycine formate single crystal. Asian J Chem 22(6):4229–4233

Śledź M, Baran J (2004) Structural investigations of the ferroelectric glycinium hydrogenphosphite (GPI) crystal and its deuterated analogue (DGPI). II: Polarized vibrational studies at room temperature. J Mol Struct 706:15–48

Slimane NM, Cherouana A, Bendjeddou L, Dahaoui S, Lecomte C (2009) *DL*-Asparaginium nitrate. Acta Crystallogr E65:o2180–o2181

Smolin YI, Lapshin AE, Drozdova IA (2002) Crystal structure of diserinesulfate monohydrate. Phys Solid State 44(10):1972–1975

Smolin YI, Lapshin AE, Pankova GA (2003a) Crystal structure of *L*-alanine phosphate. Crystallogr Rep 48(2):283–285

Smolin YI, Lapshin AE, Pankova GA (2003b) Crystal structure of di-(L-serine) phosphate monohydrate $[C_3O_3NH_7]_2H_3PO_4H_2O$. Phys Solid State 45(10):1893–1897

Smolin YI, Lapshin AE, Pankova GA (2003c) Crystal structure of di-(L-alanine) monophosphite monohydrate $[C_3O_2NH_7]_2.H_3PO_3.H_2O$. J Struct Chem 44(3):511–515

Smolin YI, Lapshin AE, Pankova GA, Osipova YN (2004) Crystal structure of *DL*-alanine sulfate. J Struct Chem 45(2):327–330

Smolin YI, Lapshin AE, Pankova GA (2005) Crystal structure of *L*-serine phosphate. Crystallogr Rep 50(1):58–60

Sobiestianskas R, Grigas J, Czapla Z, Dacko S (1993) Critical dielectric relaxation in ferroelectric betaine phosphite in the microwave region. Phys Status Solidi (a) 136:223–228

Soman J, Vijayan M (1988) X-ray studies on crystalline complexes involving amino acids and peptides. XVI. Structure of L-ornithine D-aspartate monohydrate. Acta Crystallogr C44:1794–1797

Soman J, Vijayan M (1989) X-ray studies of crystalline complexes involving amino acids. XVIII. Crystal structures of a new form *L*-arginine *D*-glutamate and a comparative study of amino acid crystal structure containing molecules of the same and mixed chirality. J Biosci 14(2):111–125

Soman J, Suresh CG, Vijayan M (1988) X-ray studies on crystalline complexes involving amino acids and peptides. XV. Crystal structures of L-lysine D-glutamate and L-lysine D-aspartate monohydrate and the effect of chirality on molecular aggregation. Int J Pept Protein Res 32:352–360

Soman J, Rao T, Radhakrishnan R, Vijayan M (1989) X-ray studies of crystalline complexes involving amino acids. Part XX. Crystal structure of DL-arginine acetate monohydrate and DL-lysine acetate and comparison with the corresponding L-amino acid complexes. J Biomol Struct Dyn 7(2):269–277

Soman J, Vijayan M, Ramakrishnan B, Guru Row TN (1990) X-ray studies of crystalline complexes involving amino acids and peptides. XVII. Chirality and molecular aggregation: the crystal structures of DL-arginine DL-glutamate monohydrate and DL-arginine DL-aspartate. Biopolymers 29:533–542

Song YJ, Rienstra CM, NcDermott AE (2001) N-H bond stretching in histidine complexes: a solid-state NMR study. Magn Reson Chem 39:S30–S36

Sridhar B, Srinivasan N, Rajaram RK (2001a) Bis(*L*-threoninium) sulfate monohydrate. Acta Crystallogr E57:o581–o583

Sridhar B, Srinivasan N, Rajaram RK (2001b) Tri(*L*-isoleucinium) sulfate bisulfate. Acta Crystallogr E57:o558–o560

Sridhar B, Srinivasan N, Rajaram RK (2001c) Bis(β-alanine) hydrogen nitrate. Acta Crystallogr E57(11):o1004–o1006

Sridhar B, Srinivasan N, Rajaram RK (2002a) *DL*-Phenylalaninium nitrate. Acta Crystallogr E58: o1103–o1105

Sridhar B, Srinivasan N, Rajaram RK (2002b) Bis(*L*-tyrosinium) sulfate monohydrate. Acta Crystallogr E58:o211–o214

Sridhar B, Srinivasan N, Rajaram RK (2002c) Bis(*L*-glutamic acid) sulfate hemihydrate. Acta Crystallogr E58:o272–o276

Sridhar B, Srinivasan N, Dalhus B, Rajaram RK (2002d) A triclinic polymorph of *L*-arginine chloride. Acta Crystallogr E58:o747–o749

Sridhar B, Srinivasan N, Rajaram RK (2002e) *L*-Aspartic acid nitrate- *L*-aspartic acid (1/1). Acta Crystallogr E58(12):o1372–o1374

Sridhar B, Srinivasan N, Dalhus B, Rajaram RK (2002f) L-Methionine L-methioninium perchlorate monohydrate. Acta Crystallogr E58(7):o779–o781

Sridhar B, Srinivasan N, Dalhus B, Rajaram RK (2003) *DL*-Valinium perchlorate. Acta Crystallogr E59:o28–o30

Srikrishnan T, Narasinga Rao S, Parthasarathy R (1984) Structure and conformation of nitrates amino acids: crystal and molecular structure of *L*-histidine nitrate, *L*-phenylalanine nitrate, *DL*-methionine nitrate. Acta Crystallogr A40:c92

Srimahaprom W, Flood AE (2013) Crystal growth and optical resolution of *DL*-methionine hydrochloride by preferential crystallization from aqueous solution. J Cryst Growth 362:88–92

Srinivasan R (1956) Unit cell and space group of some amino acids. Acta Crystallogr 9:1039–1040

Srinivasan R (1959a) The crystal structure of *L*-tyrosine hydrohalides. Part I. L-tyrosine hydrobromide. Proc Ind Acad Sci 49:340–373

Srinivasan R (1959b) The crystal structure of *L*-tyrosine hydrohalides. Part II. L-tyrosine hydrochloride. Proc Ind Acad Sci 50:19–50

Srinivasan N, Rajaram RK (1997a) Crystal structure of *L*-arginine perchlorate. Z Kristallogr NCS 212(4):311–312

Srinivasan N, Rajaram RK (1997b) *L*-phenylalanine *L*-phenylalaninium perchlorate. Acta Crystallogr C53(11):1711–1713

Srinivasan MR, Bhat HL, Narayanan PS (1983) Studies on diglicine sulfate. Ferroelectrics 46(1):191–199

Srinivasan N, Rajaram RK, Raj DDJ (1997) Crystal structure of bis(*L*-valine nitrate), $C_5H_{12}N_2O_5$. Z Kristallogr NCS 212:313–314

Srinivasan N, Sridhar B, Rajaram RK (2001a) Bis(*DL*-phenylalaninium) sulfate monohydrate. Acta Crystallogr E57:o772–o774

Srinivasan N, Sridhar B, Rajaram RK (2001b) Bis(*DL*-methioninium) sulfate. Acta Crystallogr E57:o746–o748

Srinivasan N, Sridhar B, Rajaram RK (2001c) Bis(*DL*-aspartic acid) sulfate. Acta Crystallogr E57: o679–o681

Srinivasan N, Sridhar B, Rajaram RK (2001d) Bis (*L*-histidinium sulfate) trihydrate. Acta Crystallogr E57:o607–o609

Srinivasan N, Sridhar B, Rajaram RK (2001e) *L*-Phenylalanine-nitric acid (2/1). Acta Crystallogr E57(10):o916–o918

Srinivasan N, Sridhar B, Rajaram RK (2001f) Hydrogen bis[*L*-lysinium(2+)] dichloride perchlorate. Acta Crystallogr E57(9):o875–o877

Srinivasan N, Sridhar B, Rajaram RK (2001g) *L*-Lysine *L*-Lysinium dichloride nitrate. Acta Crystallogr E57(9):o888–o890

Srinivasan N, Sridhar B, Rajaram RK (2002) *DL*-Valinium nitrate. Acta Crystallogr E58:o95–o97

Srinivasan P, Kanagasekaran T, Gopalakrishnan R, Bhagavannarayana G, Ramasamy P (2006) Studies on the growth and characterization of L-asparaginium picrate (LASP) – a novel nonlinear optical crystal. Cryst Growth Design 6:1663–1670

Srinivasan P, Kanagasekaran T, Gopalakrishnan R (2008a) A highly efficient nonlinear optical donor-acceptor single crystal: *L*-valinium picrate. Cryst Growth Design 8(7):2340–2345

Srinivasan P, Kanagasekaran T, Gopalakrishnan R (2008b) Factor group analysis and hyperpolarizability studies of nonlinear optical L-asparaginium picrate (LASP) crystals. Spectrochim Acta A 71:592–596

Srinivasan P, Vidyalakshmi Y, Gopalakrishnan R (2008c) Studies on the synthesis, growth, crystal structure, and nonlinear optical properties of a novel nonlinear optical crystal: *L*-argininium-4-nitro phenolate monohydrate (LARP). Cryst Growth Design 8(7):2329–2334

Steiner T, Majerz I, Wilson CC (2001) First O-H-N hydrogen bond with a centered proton obtained by thermally induced proton migration. Angew Chem Int Ed 40:2651–2654

Stewart K (2009) Redetermination of *L*-tryptophan hydrobromide. Acta Crystallogr E65:o1291–o1292

Strohmeier M, Stueber D, Grant DM (2003) Accurate ^{13}C and ^{15}N chemical shift and ^{14}N quadrupolar coupling constant calculations in amino acid crystals: Zwitterionic, hydrogen-bonded systems. J Phys Chem A107:7629–7642

Strukov BA, Taraskin SA, Savilova ZV (1986) Thermodynamical properties of diglycine nitrate (DGN) crystals. Phys Status Solidi (a) 95:447–451

Su JR, Xu DJ (2005) *R*)-Methionininium-(*R*)-mandelate-(*R*)-mandelic acid (1/1/2. Acta Crystallogr E61:o1933–o1935

Su JR, Zhu ZG, Zhu CF, Xu D, Yuan DR (2000a) Ultrasonic studies on a low temperature attenuation peak in *L*-arginine phosphate monohydrate single crystal. J Alloy Compd 310:139–143

Su JR, Zhu CF, Zhu ZG, Xu D, Yuan DR (2000b) Experimental evidence for the unusual damping parameter in L-arginine phosphate monohydrate single crystal. Mater Res Bull 35:977–983

Subramanian E (1967) The crystal structure of *L*-Leucine hydrobromide. Acta Crystallogr 22:910–917

Sudbeck EA, Gleason WB, Etter MC (1995) Crystal structures and hydrogen bond analysis of three arenesulfonate salts of L-alanine, glycine, and L-serine. J Chem Cryst 25:555–567

Sudhakar V, Vijayan M (1980) X-ray studies of crystalline complexes involving amino acids. IV. The structure of L-arginine L-ascorbate. Acta Crystallogr B36(1):120–125

Sudhakar V, Bhat TN, Vijayan M (1980) X-ray studies on crystalline complexes involving amino acids. V. The structure of L-serine-L-ascorbic acid. Acta Crystallogr B36:125–128

Sujatha T, Cyrac Peter A, Vimalan M, Merline Shyla J, Madhavan J (2010) Thermal, mechanical, optical and conductivity studies of a novel NLO active *L*-phenylalanine *L*-phenylalaninium dihydrogenphosphate single crystal. Physica B 405:3365–3370

Sukiasyan RP (2005) New salts of *L*-lysine. In: Proceedings of conference on Laser Physics-2004, Ashtarak, 12–15 Oct 2004, pp 105–108

Sukiasyan RP, Karapetyan HA, Martirosyan GG, Petrosyan AM (2007) *L*-arginine diperchlorate. A new salt with doubly charged cation. In: Proceedings of conference on "Laser Physics-2006", Ashtarak, 10–13 Oct 2006, pp 113–116

Sukiasyan RP, Karapetyan HA, Petrosyan AM (2008) *L*-Lysine dioxalate. J Mol Struct 888:230–237

Sukiasyan RP, Karapetyan HA, Atanesyan AK, Apreyan RA (2014) Crystal structure, thermal and spectral studies and growth of nonlinear optical *L*-lysinium fluoride crystal. Spectrochim Acta A 123:206–210

Sun ZH, Xu D, Wang XQ, Liu XJ, Yu G, Zhang GH, Zhu LY, Fan HL (2007a) Growth and characterization of the nonlinear optical crystal: L-arginine trifluoroacetate. Cryst Res Technol 42:812–816

Sun ZH, Yu W-T, Fan J-D, Xu D, Wang XQ (2007b) *L*-Argininium(+) maleate(-) dihydrate. Acta Crystallogr E63:o2805–o2807

Sun ZH, Fan JD, Zhang GH, Wang XQ, Xu D (2008a) L-Lysinium trifluoroacetate. Acta Crystallogr E64:o393–o394

Sun ZH, Zhang GH, Wang XQ, Cheng XF, Liu XJ, Zhu LY, Fan HL, Yu G, Xu D (2008b) Growth and characterization of the nonlinear optical single crystal: *L*-lysine acetate. J Cryst Growth 310:2842–2847

Sun ZH, Zhang L, Xu D, Wang XQ, Liu XJ, Zhang GH (2008c) Theoretical calculation and vibrational spectral analysis of L-arginine trifluoroacetate. Spectrochim Acta A 71:663–668

Sun ZH, Yu WT, Cheng XF, Wang XQ, Zhang GH, Yu G, Fan HL, Xu D (2008d) Synthesis, crystal structure and vibrational spectroscopy of a nonlinear optical crystal: L-arginine maleate dihydrate. Opt Mater 30:1001–1006

Sun ZH, Xu D, Wang XQ, Zhang GH, Yu G, Zhu LY, Fan HL (2009a) Growth and characterization of the nonlinear optical single crystal: L-lysinium trifluoroacetate. Mater Res Bull 44:925–930

Sun ZH, Zhang GH, Wang XQ, Yu G, Zhu LY, Fan HL, Xu D (2009b) Nucleation kinetics, microcrystallization and surface morphology studies of the nonlinear optical single-crystal *L*-lysinium trifluoroacetate. J Cryst Growth 311:3455–3460 (Erratum: J Cryst Growth 311:4152 (2009))

Sun ZH, Zhang G, Wang X, Gao Z, Cheng X, Zhang S, Xu D (2009c) Growth, morphology, thermal, spectral linear, and nonlinear optical properties of *L*-arginine bis(trifluoroacetate) crystal. Cryst Growth Design 9:3251–3259

Sun GH, Sun XT, Sun ZH, Wang XQ, Liu XJ, Zhang GH, Xu D (2009d) Growth and characterization of a nonlinear optical crystal: L-histidine trifluoroacetate. J Cryst Growth 311:3904–3910

Sun GH, Zhang GH, Wang XQ, Sun ZH, Xu D (2010) Nucleation kinetics and studies of deuterated *L*-arginine trifluoroacetate crystal. Mater Chem Phys 122:524–527

Sun GH, Zhang GH, Sun ZH, Wang XQ, Xu D (2011a) Growth, crystal structure and characterization of a nonlinear optical crystal: deuterated *L*-arginine trifluoroacetate. Mater Chem Phys 127:265–270

Sun ZH, Sun WM, Chen CT, Zhang GH, Wang XQ, Xu D (2011b) *L*-Arginine trifluoroacetate salt bridges in its solid state compound: the low-temperature three dimensional structural determination of *L*-arginine bis(trifluoroacetate) crystal and its vibrational spectral analysis. Spectrochim Acta A 83:39–45

Sun GH, Zhang GH, Wang XQ, Xu D (2011c) Nucleation kinetics, micro-crystallization and etching studies of L-histidine trifluoroacetate crystal. J Cryst Growth 316:132–136

Suponitsky KY, Sukiasyan RP, Petrosyan AM (in prep)

Surekha R, Thilagavathy SR, Sagayaraj P, Ambujam K (2014a) Synthesis, optical, dielectric, thermal and mechanical properties of a nonlinear optical amino acid crystal: Bis-glycine hydrobromide. Optik 125:934–938

Surekha R, Sagayaraj P, Ambujam K (2014b) Third order nonlinear optical, luminescence and electrical properties of bis glycine hydrobromide single crystals. Opt Mater 36:945–949

Suresh S (2012) Optical and dielectric properties of *L*-histidinium trifluoroacetate NLO single crystal. Res J Chem Sci 2(2):83–86

Suresh J, Natarajan S (2006) DL-Valinium trifluoroacetate. Acta Crystallogr E62:o3331–o3333

Suresh CG, Vijayan M (1983a) X-ray studies on crystalline complexes involving amino acids and peptides. X. Head-to-tail sequences in the crystal structure of L-lysine acetate. Int J Pept Protein Res 22:617–621

Suresh CG, Vijayan M (1983b) X-ray studies of crystalline complexes involving amino acids. VIII. Head to tail arrangement and a specific interaction in the crystal structure of L-arginine acetate. Int J Pept Protein Res 21(3):223–226

Suresh CG, Vijayan M (1987) X-ray studies on crystalline complexes involving amino acids and peptides. Part XIV: closed conformation and head-to-tail arrangement in a new crystal form of *L*-histidine *L*-aspartate monohydrate. J Biosci 12(1):13–21

Suresh S, Vijayan M (1995a) X-ray studies on crystalline complexes involving amino acids and peptides. XXVIII. Recurrence of characteristic aggregation and interaction patterns in the crystal structures of DL- and L-lysine formate. Acta Crystallogr B51:353–358

Suresh S, Vijayan M (1995b) Variability in ionization state, stoichiometry and aggregation in histidine complexes with formic acid (X-ray studies on crystalline complexes involving amino acids and peptides. Part XXIX). J Biosci 20(2):225–234

Suresh S, Vijayan M (1996) X-ray studies on crystalline complexes involving amino acids and peptides XXX. Structural invariance and optical resolution through interactions with an achiral molecule in the histidine complexes of glycolic acid. Acta Crystallogr B52(5):876–881

Suresh CG, Ramaswamy J, Vijayan M (1986) X-ray studies of crystalline complexes involving amino acids and peptides. XIII. Effect of chirality on molecular aggregation: the crystal structures of *L*-arginine *D*-aspartate and *L*-arginine *D*-glutamate trihydrate. Acta Crystallogr B42:473–478

Suresh S, Padmanabhan S, Vijayan M (1994a) X-ray studies of crystalline complexes involving amino acids. XXVII. Effect of chirality, specific interactions and characteristic aggregation patterns in the structures of arginine and its complexes with formic acid. J Biomol Struct Dyn 11(6):1425–1435

Suresh S, Prasad GS, Vijayan M (1994b) X -ray studies on crystalline complexes involving amino acids and peptides XXVI. Crystal structures of two forms of L-histidine acetate and a comparative study of the amino acid complexes of acetic acid. Int J Pept Protein Res 43 (2):139–145

Suresh J, Krishnakumar RV, Natarajan S (2005) *L*-Phenylalanine-benzoic acid (1/1). Acta Crystallogr E61:o3625–o3627

Suresh J, Krishnakumar RV, Natarajan S (2006) *L*-Argininium trifluoroacetate. Acta Crystallogr E62:o3127–o3129

Suresh S, Ramanand A, Jayaraman D, Mani P (2010) Mechanical and dielectric studies of *L*-prolinium tartrate (LPT) NLO single crystals. Optoelectron Adv Mater-Rapid Commun 4:1743–1746

Suzuki A, Matsuoka Y (1991) Organic nonlinear optical material, Japan Kokai Tokkyo Koho JP 03 03,834 [91 33,834] (Cl.G02 F1/35) 14 Feb 1991, Appl 89/168 872, 4 p (In Japanese)

Swaminathan S, Chacko KK (1981) *DL*-proline hemihydrochloride, $C_5H_9NO_2.1/2HCl$. Cryst Struct Commun 10:469–472

Swaminathan P, Srinivasan R (1975) Studies in molecular structure, symmetry, and conformation X: crystal and molecular structure of *DL*-allothreonine hydrobromide. J Cryst Mol Struct 5:203–214

Syamala D, Rajendran V, Natarajan RK, Moorthy Babu S (2007) Effect of sodium fluoroborate (NaBF4) doping on the NLO properties of *L*-histidine single crystals. Cryst Growth Design 7:1695–1698

Szafran M, Dega-Szafran Z, Thaimattam R, Jaskólski M (2004) Crystal and molecular structure of N-methylpiperidine betaine hydrofluoride. J Mol Struct 706:49–55

Taira Z, Watson WH (1977) The structure of a 1:1 mixed crystal of *L*-glutamic acid and *L*-pyroglutamic acid, $C_5H_9NO_4.C_5H_7NO_3.H_2O$, and a refinement of the structure of pyroglutamic acid $C_5H_7NO_3$. Acta Crystallogr B33:3823–3827

Takahashi Y, Fujii I (2004) Crystal structure of (*R*)-2-Phenoxypropionic acid-(*S*)-alanine. Anal Sci 20:x77–x78

Takayanagi Y, Iizumi K, Miyazawa M, Yamaya N, Tamura H (1981) Anhydrous *L*-lysine monohydrochloride in α-crystalline form and preparation thereof. US Patent 4256917

Takiagawa T, Ashida T, Sasada Y, Kakudo M (1966) The crystal structures of *L*-tryptophan hydrochloride and hydrobromide. Bull Chem Soc Jpn 39:2369–2378

Tamilselvan S, Cyrac Peter A, Mahadevan CK, Madhavan J, Vimalan M (2011) Electrical properties of *L*-prolinium tartrate (LPT) – a new organic nonlinear optical material. Archiv Appl Sci Res 3:180–185

Tamilselvan S, Vimalan M, Vetha Potheher I, Jeyasekaran R, Yogam F, Madhavan J (2014) Generation of 532 nm laser radiation and phase matching properties of organic nonlinear optical material. Optik 125:164–169

Tani K, Naganawa A, Ishida A et al (2002) A practical synthesis and biological evaluation of 9-halogenated PGF analogues. Bioorg Med Chem 10:1883–1894

Terzyan SS, Karapetyan HA, Sukiasyan RB, Petrosyan AM (2004) *L*-arginine nitrates. J Mol Struct 687(1–3):111–117

Thomas PC, Thomas J, Packiam Julius J, Madhavan J, Selvakumar S, Sagayaraj P (2005a) Growth and characterization of *L*-argininium dinitrate. J Cryst Growth 277:303–307

Thomas R, Gopalan RS, Kulkarni GU, Rao CNR (2005b) Hydrogen bonding patterns in the coccrystals of 5-nitrouracil with several donor and acceptor molecules. Beilstein J Org Chem 1:15

Thomas PC, Aruna S, Anuradha A, Packiam Julius J, Pragasam AJA, Sagayaraj P (2006) Growth and characterization of semiorganic NLO crystals of LAHClBr. Cryst Res Technol 41:1231–1235

Thomas R, Pal S, Datta A, Marchewka MK, Ratajczak H, Pati SK, Kulkarni GU (2008) Charge density analysis of two proton transfer complexes: understanding hydrogen bonding and determination of in-crystal dipole moments. J Chem Sci 6:613–620

Thulasidhass R, Mohana Rao JK (1979) The crystal structure of glycine orthophosphate. Curr Sci 48:626–627

Tillmanns E, Fleck M, Bohatý L (2006) New compounds of *L*-Alanine with inorganic materials. Acta Crystallogr A62(Suppl):s109

Timofeeva TV, Kuhn GH, Nesterov VV, Nesterov VN, Frazier DO, Penn BG, Antipin MYU (2003) Cocrystal of 1,1-dicyano-2-(4-hydroxy-3-methoxyphenyl)-ethene with L-proline and induced conformational polymorphism of 1,1-dicyano-2-(4-hydroxy-3-methoxyphenyl)-ethene. Cryst Growth Design 3(3):383–391

Trommel J, Bijvoet JM (1954) Crystal structure and absolute configuration of the hydrochloride and hydrobromide of *D*(-)-isoleucine. Acta Crystallogr 7:703–709

Trotter J (1962) Crystallographic data for di(phenylethynyl) mercury, potassium 2,4,6,2′,4′,6′-hexanitrodiphenylamide, and dl-α-alanine hydrochloride. Can J Chem 40:1218–1220

Troyanov SI, Morozov IV, Snigireva EM, Ling CD, Kemnitz E (2004) Neutron diffraction study of the HF adducts containing a hydrogen bond F-H…O. J Solid State Chem 177:3224–3228

Tumanov NA, Boldyreva EV, Shikina NE (2010) Two new structures in the glycine-oxalic acid system. Acta Crystallogr C66:o279–o283

Tylczynski Z, Wiesner M (2010) Comment on "Ferroelectricity in glycine picrate: an astonishing observation in a centrosymmetric crystal" [Appl Phys Lett 95:252902 (2009)]. Appl Phys Lett 96:126101

Udovenko AA, Zemnukhova LA, Kovaleva EV, Davidovich RL (2005) Crystal structure of monohydrate and anhydrous β-alaninium tetrafluoroantimonates(III). Russ J Coord Chem 31(4):225–229

Udovenko AA, Makarenko NV, Davidovich RL, Zemnukhova LA, Kovaleva EV (2010a) Crystal structure of *L*-leucinium tetrafluoroantimonate(III). J Struct Chem 51(5):904–908

Udovenko AA, Davidovich RL, Zemnukhova LA, Kovaleva EV, Makarenko NV (2010b) Crystal structure of *DL*-serinium tetrafluoroantimonate (III). J Struct Chem 51(3):540–544

Uma Devi T, Lowrence N, Ramesh Babu R, Ramamurthi K (2008a) Growth and characterization of *L*-prolinium single crystal: a promising NLO crystal. J Cryst Growth 310:116–123

Uma Devi T, Lawrence N, Ramesh Babu R, Ramamurthi K (2008b) Growth and characterization of glycine picrate single crystal. Spectrochim Acta A 71:340–343

Uma Devi T, Lawrence N, Ramesh Babu R, Ramamurthi K, Bhagavannarayana G (2009) Studies on *L*-valinium picrate single crystal: a promising NLO crystal. J Miner Mater Charact Eng 8(4):393–403

Unruh HG, Heim J, Hermann T (1995) The ferroelastic transition of betaine fumarate. Z Phys B 97:133–138

Varughese KI, Srinivasan R (1976a) Studies in molecular structure, symmetry and conformation – Part XIV. Crystal and molecular structure of *L*-isoleucine hydrochloride monohydrate form II. Pramana 6(4):189–195

Varughese KI, Srinivasan R (1976b) Studies in molecular structure, symmetry and conformation. XIII. Crystal and molecular structure of D-alloisoleucine hydrochloride monohydrate. Acta Crystallogr B32:994–998

Vasantha K, Dhanuskodi S (2004a) The structural, thermal and optical characterization of a new nonlinear optical material. J Cryst Growth 263:466–472

Vasantha K, Dhanuskodi S (2004b) Single crystal growth and characterization of phase-matchable *L*-arginine maleate: a potential nonlinear optical material. J Cryst Growth 269:333–341

Vasudevan V, Ramesh Babu R, Ramamurthi K (2011a) Refractive index, birefringence, third-order non-linearity and piezoelectric resonance studies of *L*-lysine monohydrochloride dihydrate single crystals. Physica B 406:4100–4104

Vasudevan V, Ramesh Babu R, Ramamurthi K (2011b) Unidirectional growth of *L*-lysine *L*-lysinium dichloride nitrate (*L*-LLDN) single crystals by the SR method. Physica B 406:936–940

Vasudevan V, Ramesh Babu R, Reicher Nelcy A, Bhagavannarayana G, Ramamurthi K (2011c) Synthesis, growth, optical, mechanical and electrical properties of *L*-lysine *L*-lysinium dichloride nitrate (*L*-LLDN) single crystal. Bull Mater Sci 34(3):469–475

Vasudevan V, Ramesh Babu R, Ramamurthi K (2012) Synthesis, growth and characterization of *L*-lysinium(+)…*L*-Lysinium(2+) dichloride perchlorate (LLDP) single crystals by Sankaranarayanan-Ramasamy method. Spectrochim Acta A 99:259–265

Vasudevan P, Sankar S, Jayaraman D (2013a) Synthesis and characterization of L-methioninium nitrate single crystal for nonlinear optical applications. Int J Chem Tech Res 5(4):2018–2026

Vasudevan P, Gokul Raj S, Sankar S (2013b) Synthesis and characterization of nonlinear optical *L*-arginine semi-oxalate single crystal. Spectrochim Acta A 106:210–215

Vasudevan P, Sankar S, Jayaraman D (2013c) Synthesis, optical and electrical studies of nonlinear optical crystal: *L*-arginine semi-oxalate. Bull Korean Chem Soc 34:128–132

Vasudevan P, Sankar S, Gokul Raj S (2013d) Studies on second harmonic generation efficiency of organic material *L*-arginine maleate dihydrate. Optik 124:4155–4158

Vayá I, Jiménez MC, Miranda MA (2005) Stereodifferentiation in the fluorescence of naproxen-arginine salts in the solid state. Tetrahedron Asymmetry 16:2167–2171

Vela T, Selvarajan P, Freeda TH (2010) Growth and characterization of bis-*L*-alanine picrate (BLAP) single crystals. J Exp Sci 1(8):29–32

Vela T, Selvarajan P, Freeda TH (2011) Synthesis, growth and material characterization of bis *L*-alanine triethanol amine (BLATEA) single crystals grown by slow evaporation technique. J Miner Mater Charact Eng 10(10):959–972

Velsko S, Eimerl D (1986) New frequency conversion materials for high power lasers. Proc SPIE 622:171–174

Velsko S, Eimerl D (1988) Nonlinear materials for frequency conversion. Proc SPIE 895:152–155

Venkatararamanan V, Dhanaraj G, Bhat HL (1994) Growth mechanism, dislocation etching and mechanical properties of *L*-arginine phosphate. J Cryst Growth 140:336–342

Venkatraman J, Prabu M, Vijayan M (1997) X-ray studies on crystalline complexes involving amino acids and peptides. XXXII. Effect of chirality on ionisation state, stoichiometry and aggregation in the complexes of oxalic acid with *DL*- and *L*-lysine. J Pept Res 50:77–87

Vijayakumar T, Hubert Joe I, Reghunadhan Nair CP, Jayakumar VS (2009) Vibrational spectral studies on charge transfer and ionic hydrogen-bonding interactions of nonlinear optical material *L*-arginine nitrate hemihydrates. J Raman Spectrosc 40:18–30

Vijayan N, Bhagavannarayana G, Sharma SN, Das S (2009a) Synthesis, growth and structural perfection of nonlinear optical material of glycine hydrofluoride (GHF). J Mater Sci 44:3457–3461

Vijayan N, Bhagavannarayana G, Nagarajan K, Upadhyaya V (2009b) Synthesis and growth of nearly perfect single crystal of *L*-histidine bromide (LHB) and its structural, optical and electrical characterizations. Mater Chem Phys 115:656–659

Vijayan N, Bhagavannarayana G, Maurya KK, Sharma SN, Gopalakrishnan R, Jayabharathi J, Ramasamy P (2012) Synthesis, growth of single crystals and their characterization: *L*-alaninium maleate (LAM). Optik 123:604–608

Vilminot S, Cot L (1973) Sur deux nouveaux sulfates de glycine et d'ion monovalent. C R Acad Sci Paris Ser C 277:1355–1358

Vilminot S, Philippot E, Cot L (1974) Structure du sulfate d'ammonium et de glycinium $NH_4NH_3CH_2COOHSO_4$. Acta Crystallogr B30:2602–2606

Vilminot S, Philippot E, Lehmann M (1976) Affinement de la structure de $NH_4NH_3CH_2COOHSO_4$ par diffraction neutronique. Acta Crystallogr B32:1817–1822

Vimalan M, Flora XH, Tamilselvan S, Jeyasekaran R, Sagayaraj P, Mahadevan CK (2010a) Optical, thermal, mechanical and electrical properties of a new NLO material: Mono-L-alaninium nitrate (MAN). Arch Phys Res 1(3):44–53

Vimalan M, Rajesh Kumar T, Tamilselvan S, Sagayaraj P, Mahadevan CK (2010b) Growth and properties of novel organic nonlinear optical crystal: l-alaninium tartrate (LAT). Physica B 405:3907–3913

Vivek P, Murugakoothan P (2013) Linear and nonlinear optical properties of a new organic NLO *L*-asparaginium *L*-tartrate(AST) single crystal. Optik 124:3510–3513

Wang CC, Racette CW (1965) Measurement of parametric gain accompanying optical difference frequency generation. Appl Phys Lett 6(8):169–171 [Errata 6(10):213]

Wang R, Wang H, Zheng J, Che Y (1994) Crystal structure of *L*-arginine fluoroborate. Jiegou Huaxue (Chin J Struct Chem) 13(2):159–161

Wang LN, Wang XQ, Zhang GH, Liu XT, Sun ZH, Sun GH, Wang L, Yu WT, Xu D (2011) Single crystal growth, crystal structure and characterization of a novel crystal: *L*-arginine 4-nitrophenolate 4-nitrophenol dihydrate (LAPP). J Cryst Growth 327:133–139

Wang L, Zhang G, Liu X, Wang L, Wang X, Zhu L, Xu D (2014) A novel *L*-arginine salt nonlinear optical crystal: *L*-arginine *p*-nitrobenzoate monohydrate (LANB). J Mol Struct 1058:155–162

Warren LF (1990) New developments in "semiorganic" nonlinear optical crystals. Intern SAMPE Electron Conf 1990 (Electronic Materials-Our Future) 4:388–397

Waśkowska A, Olejnik S, Łukaszewicz K, Ciechanowicz-Rutkowska M (1978) The crystal structure of triglycine fluoberylate in ferroelectric and paraelectric phase. Ferroelectrics 22 (1):855–861

Wei Y, McDermott AE (1999) Effect of hydrogen bonding on ^{1}H chemical shifts. In: Modeling NMR chemical shifts. Gaining insights into structure and environment. ACS Symp Ser 732:177–193

Wilkinson HS, Harrison WTA (2005) A 1:1 adduct of *DL*-threonine and arsenic acid. Acta Crystallogr C61:m254–m255

Willett RD, Jardine FH, Rouse I, Wong RJ, Landee CP, Numata M (1981) Crystal structure, magnetic susceptibility, and EPR study of bis-(β-alaninium) tetrachlorocuprate(II): spin-diffusion effects in a two-dimensional square planar ferromagnet with anisotropic and anti-symmetric exchange. Phys Rev B24:5372–5381

Willett RD, Wong RJ, Numata M (1983) Magnetic susceptibility and EPR study of bis-(-β-alaninium) tetrabromocuprate(II). Inorg Chem 22:3189–3194

Willett RD, Maxcy KR, Twamley B (2002) Polytypism in columnar group 14 halide salts: structures of $(Et_2NH_2)_3Pb_3X_9.nH_2O$ (X=Cl, Br) and $(\beta\text{-alaninium})_2SnI_4$. Inorg Chem 41:7024–7030

Wright DA, Marsh RE (1962) The structure of l-lysine monohydrochloride dihydrate. Acta Crystallogr 15:54–64 (Correction: 16:431)

Wu K, Liu C, Mang C (2007) Theoretical studies on vibrational spectra and nonlinear optical property of *L*-arginine phosphate monohydrate crystal. Opt Mater 29:1129–1137

Xian L, Lu G (2007) Huaxue Yanjiu Yu Yingyong (Chin) (Chem Res Appln) 19:474 (CSD code: ARGHCL11)

Xu D, Jiang MH, Tan ZK (1983a) A new phase matchable nonlinear optic crystal *L*-arginine phosphate monohydrate. Acta Chim Sin 41(6):570–573 (In Chinese)

Xu D, Jiang MH, Tan ZK (1983b) A new phase matchable nonlinear optic crystal- *L*-arginine phosphate monohydrate (LAP). Acta Chim Sin 1(2):230–233 (English version)

Xu D, Wang XD, Yu WT, Xu SX, Zhang GH (2003) Crystal structural and characterization of a novel organic nonlinear optical crystal: *L*-arginine trifluoroacetate. J Cryst Growth 253:481–487, Erratum: J Cryst Growth 257:432–433

Yamada K, Sato A, Shimizu T, Yamazaki T, Yokoyama S (2008) *L*-Alanine hydrochloride monohydrate. Acta Crystallogr E64:o806

Yan B, Wang HD, Chen ZD (2001) A novel supramolecular compound with three-dimensional hydrogen-bonded network $[Hbet]_3[Fe(CN)_6].4H_2O$: synthesis, characterization and crystal structure. J Mol Struct 569:65–70

Yang W, Jin T, Zhang S (1996) Crystal structure of L-tyrosine perchlorate. Acta Sci Nat Univ Pek 32(4):413–414

Yarmarkin VK, Shul'man SG, Pankova GA, Lemanov VV (2005) Pyroelectric properties of some compounds based on protein aminoacids. Phys Solid State 47(11):2135–2137

Yarmarkin VK, Shul'man SG, Lemanov VV (2006) Phase transition in diserine sulfate monohydrate crystals. Phys Solid State 48(8):1568–1571

Yasuda N, Kaneda A, Czapla Z (1997) Effects of hydrostatic pressure on the paraelectric-ferroelectric phase transition in deuterated glycinium phosphate crystals. J Phys Condens Matter 9:L447–L450

Yip WH, Ru-Ji W, Mak CWT (1990) Structure of a new form of betaine hydrochloride. Acta Crystallogr C46:717–719

Yogam F, Potheher IV, Cyrac Peter A, Tamilselvan S, Vimalan M, Sagayaraj P (2011) Growth and characterization of organic nonlinear optical single crystal of L-asparaginium picrate (LASP). Archiv Appl Sci Res 3(1):267–276

Yogam F, Potheher IV, Vimalan M, Jeyasekaran R, Kumar TR, Sagayaraj P (2012) Growth and physicochemical properties of *L*-phenylalaninium maleate: a novel nonlinear optical crystal. Spectrochim Acta A 95:369–373

Yokotani A, Sasaki T, Yoshida K, Nakai S (1989) Extremely high damage threshold of a new nonlinear crystal *L*-arginine phosphate and its deuterium compound. Appl Phys Lett 55:2692–2693

Yokotani A, Sasaki T, Yoshida K, Nakai S, Yamanaka C (1990) Growth and characterization of deuterated *L*-arginine phosphate monohydrate, a new nonlinear crystal, for efficient harmonic generation of fusion experiment lasers. J Cryst Growth 99:815–819

Yoshida H, Nakatsuka M, Fujita H, Sasaki T, Yoshida K (1997) High-energy operation of a stimulated Brillouin scattering mirror in an *L*-Arginine phosphate monohydrate crystal. Appl Optics 36:7783–7787

Yoshioka R, Okamura K, Yamada S, Aoe K, Da-te T (1998) The role of water-solvation in the optical resolution of *DL*-leucine with (S)-(-)-1-phenylethanesulfonic acid. Characterization and X-ray crystal structures of their diastereometric salts. Bull Chem Soc Jpn 71:1109–1116

Zakharov BA, Boldyreva EV (2011) DL-Alaninium semi-oxalate monohydrate. Acta Crystallogr C67:o47–o51

Zakharov BA, Boldyreva E (2013) A high-pressure single-crystal to single-crystal phase transition in DL-alaninium semi-oxalate monohydrate with switching-over hydrogen bonds. Acta Crystallogr B69:271–280

Zakharov BA, Kolesov BA, Boldyreva EV (2011) Monitoring selected hydrogen bonds in crystal hydrates of amino acid salts: combining variable-temperature single-crystal X-ray diffraction and polarized Raman spectroscopy. Phys Chem Chem Phys 13:13106–13116

Zakharov BA, Losev EA, Kolesov BA, Drebushchak VA, Boldyreva EV (2012) Low-temperature phase transition in glycine-glutaric acid co-crystals studied by single-crystal X-ray diffraction, Raman spectroscopy and differential scanning calorimetry. Acta Crystallogr B68:287–296

Zakharov BA, Losev EA, Boldyreva EV (2013) Polymorphism of "glycine-glutaric acid" co-crystals: the same phase at low temperatures and high pressures. CrystEngComm 15:1693–1697

Zalkin A, Eimerl D, Velsko SP (1989) Structure of (+)-*L*-arginine diarsenate. Acta Crystallogr C45 (5):812–813

Zeghouan O, Bendjeddou L, Cherouana A, Dahaoui S, Lecomte C (2012) *L*-Leucinium fluoride monohydrate. Acta Crystallogr E68:o2959–o2960

Zemnukhova LA, Davidovich RL (1998) ^{121,123}Sb NQR study of antimony (III) fluoride complexes. Russ Chem Bull 47:2164–2168

Zernike F, Berman PR (1965) Generation of far infrared as a difference frequency. Phys Rev Lett 15(26):999–1001

Zernike F, Mitwinter JE (1970) Applied nonlinear optics. Wiley, New York

Zhang H, Wang X, Chang X, Xiao W, Zhang K, Teo BK (1997) An assessment of ionic and hydrogen bonding in the crystal structure of (*DL*-serine.$H^+)_2.SO_4^{2-}.H_2O$. Acta Crystallogr C53:1286–1288

Zhang YJ, Shu Z, Xu W, Chen G, Li ZG (2008a) *L*-Glutamic acid hydrochloride at 153K. Acta Crystallogr E64:o446

Zhang Y, Li H, Xi B, Che Y, Zheng J (2008b) Growth and characterization of *L*-histidine nitrate single crystal, a promising semiorganic NLO material. Mater Chem Phys 108:192–195

Zhao X, Harbison GS (2006) Deuterium quadrupolar tensors of *L*-histidine hydrochloride monohydrate-d_7. J Phys Chem B110:25059–25065

Zheng JM, Che YX, Wang RJ, Wang HG (1994a) Preparation and crystal structure of adduct dihydrate of *L*-arginine and maleic acid. Gaodeng Xuexiao Xuaxue Xuebao (Chem J Chin Univ) 15:557–560

Zheng JM, Che YX, Wang RJ, Wang HG (1994b) Crystal structure of 1:1 adduct of glycine and *m*-nitrobenzoic acid. Acta Phys -Chim Sin 10(1):64–68

Zobetz E, Preisinger A (1989) Kristallstructur und Phasenumwandlung von Betain-Borat, $(CH_3)_3NCH_2COO.B(OH)_3$. Monatsh Chem 120:291–298

Yamamoto A, Sasaki J, Yoshida K, Sakai S, Yamamoto K (1999) Growth and characterization of [illegible] Zn [illegible] a new nonlinear crystal for [illegible] generation [illegible] J Cryst Growth [illegible]

[illegible]

Zakharov [illegible] (2010) A high-pressure [illegible] [illegible]

[illegible]

Zalkin A, [illegible] (1963) [illegible] Acta Cryst [illegible]

Zechner [illegible]

[illegible] (1998) [illegible] NMR study of [illegible]

[illegible]

[illegible] B (1970) [illegible] New York

Zhang H, Wang X, [illegible] crystal structure of [illegible] Cryst Technol [illegible]

Zhang YJ, [illegible]

Zhang [illegible] single crystal [illegible]

Zhao X, [illegible] study of [illegible] Phys Chem [illegible]

Zheng [illegible] Xuebao [illegible]

Zhou M, [illegible] Mater Sci [illegible]

Zhu [illegible] Monatsh Chem [illegible]

Index

M. Fleck and A.M. Petrosyan, *Salts of Amino Acids: Crystallization, Structure and Properties*, DOI 10.1007/978-3-319-06299-0,

Springer Nature Customer Service Center GmbH
Heidelberg, Germany

MIX
Papier aus verantwortungsvollen Quellen
Paper from responsible sources
FSC® C105338

If you have any concerns about our products,
you can contact us on
ProductSafety@springernature.com

In case Publisher is established outside the EU,
the EU authorized representative is:
Springer Nature Customer Service Center GmbH
Europaplatz 3, 69115 Heidelberg, Germany

Printed by Libri Plureos GmbH
in Hamburg, Germany

